# Inhaltsverzeichnis.
## Contents. — Table des matières.

# FORTSCHRITTE DER CHEMIE ORGANISCHER NATURSTOFFE

# PROGRESS IN THE CHEMISTRY OF ORGANIC NATURAL PRODUCTS

# PROGRÈS DANS LA CHIMIE DES SUBSTANCES ORGANIQUES NATURELLES

HERAUSGEGEBEN VON     EDITED BY     RÉDIGÉ PAR

## L. ZECHMEISTER
CALIFORNIA INSTITUTE OF TECHNOLOGY, PASADENA

### ACHTZEHNTER BAND
EIGHTEENTH VOLUME     DIX-HUITIÈME VOLUME

VERFASSER     AUTHORS     AUTEURS

P. W. BRIAN · H. BROCKMANN · J. F. GROVE · M. HEIDELBERGER
A. KJÆR · J. MACMILLAN · M. PAILER · J. ROCHE · N. VAN THOAI
O. VÖLKER · J. W. WILLIAMS · L. ZECHMEISTER

MIT 65 ABBILDUNGEN     WITH 65 FIGURES     AVEC 65 ILLUSTRATIONS

WIEN · SPRINGER - VERLAG · 1960

ISBN-13: 978-3-7091-7161-5    e-ISBN-13: 978-3-7091-7159-2
DOI: 10.1007/978-3-7091-7159-2

## Selected Subjects in Sedimentation Analysis, with Some Applications to Biochemistry. By J. W. Williams, Department of Chemistry, The University of Wisconsin, Madison, Wisconsin.................... 434

## Structure and Immunological Specificity of Polysaccharides.

By MICHAEL HEIDELBERGER, College of Physicians and Surgeons, Columbia University, New York, and Institute of Microbiology, Rutgers, the State University, New Brunswick, New Jersey ..................

# Die Actinomycine.

## Von H. Brockmann, Göttingen.

### Mit 1 Abbildung.

### Inhaltsübersicht.

# I. Einleitung.

Die Actinomycine sind gelbrote, sehr toxische Antibiotica, die von verschiedenen *Streptomyces*-Arten produziert werden (vgl. S. 9). Ihr Molekül gliedert sich in einen aus Aminosäuren und N-Methylaminosäuren aufgebauten Peptidteil sowie einen Farbstoffteil, der im folgenden kurz Chromophor genannt wird. Die Actinomycine sind demnach Vertreter einer neuartigen Naturstoffklasse, die man in Analogie zu den Chromoproteiden als *Chromopeptide* bezeichnen kann (*31*). Die bisher näher untersuchten Actinomycine haben alle den gleichen Chromophor und unterscheiden sich nur in der Struktur ihres Peptidteils.

Zur Actinomycin-Synthese befähigte *Streptomyces*-Stämme bilden stets mehrere Actinomycine, deren Menge und Peptidstruktur von der Natur des Stammes und in gewissen Grenzen auch von der Zusammensetzung des Kulturmediums abhängt. Da diese Actinomycingemische einheitlich kristallisieren und durch fraktionierte Kristallisation schwer oder meistens gar nicht zu trennen sind, hat man sie zunächst für einheitliche Stoffe gehalten.

Der Name *Actinomycin* stammt von WAKSMAN und WOODRUFF (*116, 117*). Sie benannten so ein kristallisiertes, rotes, stark giftiges Antibioticum, das sie 1940 aus der Kulturlösung von *Actinomyces antibioticus* isolierten. Um es von einer farblosen Begleitsubstanz (zunächst Actinomycin B genannt) zu unterscheiden, kennzeichneten sie es zusätzlich durch den Buchstaben A. Der Name Actinomycin B für die farblose Begleitsubstanz wurde später aufgegeben.

Actinomycin A* (*111*), wie sich später herausstellte ein Gemisch aus mehreren Actinomycinen (*114*), wurde von WAKSMAN und TISHLER (*115*) eingehender untersucht. Es reagierte schwach basisch, war optisch aktiv ([α]$_D^{25}$: — 320° ± 5°, in Äthanol) und enthielt C-Methyl- und N-Methylgruppen. Mol.-Gew. und Analysenwerte paßten auf $C_{41}H_{56}N_6O_{11}$. Reduktionsversuche ließen auf Vorliegen einer oder mehrerer chinoider Gruppen schließen.

Bei verschiedenen grampositiven Bakterien-Arten zeigte „Actinomycin A"** hohe antibiotische Wirksamkeit. So hemmte es die Ver-

---

\* WAKSMANS Actinomycin A ist das erste Antibioticum, das in kristallisierter Form isoliert werden konnte.

\*\* Um anzudeuten, daß ein Actinomycin*gemisch* vorliegt, ist der Name hier und im folgenden in Anführungszeichen gesetzt.

*Literaturverzeichnis: SS. 48—54.*

mehrung von *Sarcina lutea* und *Bac. subtilis* bis zur Verdünnung $1 : 10^8$. Relativ unempfindlich dagegen waren gramnegative Bakterien. Auch fungistatische Wirkung wurde beobachtet (*95*). „Actinomycin A" medizinisch als Antibioticum zu verwenden, verbot sich wegen seiner großen Giftigkeit (*96*). Schon 10 $\gamma$ reichten aus, um innerhalb 24 bis 48 Stunden eine 20 g schwere Maus zu töten (*115*).

Da Actinomycin bildende *Streptomyces*-Arten weit verbreitet und Actinomycine leicht zu isolieren sind, ist verständlich, daß man verhältnismäßig frühzeitig auf sie aufmerksam geworden ist und nach WAKSMAN verschiedene Autoren die Isolierung roter, dem ‚Actinomycin A" ähnlicher Antibiotica beschrieben haben [WELSCH, 1946 (*118*); UMEZAWA, 1947 (*107*); KOCHOLATY, 1948 (*88*); TRUSSELL und RICHARDSON, 1948 (*105*); LEHR und BERGER, 1949 (*90*); SARLET, 1949 (*98*)]. Eingehender untersucht wurden sie zunächst nicht. Später, als die auf S. 4 beschriebenen Trennungsverfahren zur Verfügung standen, hat man einige dieser Präparate als Actinomycingemische charakterisiert (vgl. S. 9).

Erst 1949, neun Jahre nach der Isolierung des „Actinomycins A", wurden Arbeiten veröffentlicht, die einen ersten Einblick in die Konstitution der Actinomycine brachten. TODD und DALGLIESH (*68*) sowie TODD, DALGLIESH, JOHNSON und VINING (*67*) stellten fest, daß ein von LEHR und BERGER (*90*) isoliertes Antibioticum „X 45" dem Actinomycin A sehr ähnlich war und beide beim Kochen mit konz. Salzsäure *L*-Threonin, *D*-Valin, *L*-Prolin, *L*-N-Methylvalin und Sarkosin lieferten. Und gleichzeitig und unabhängig davon fanden BROCKMANN und GRUBHOFER (*30, 31*) sowie BROCKMANN, GRUBHOFER, KASS und KALBE (*33*), daß ein von ihnen aus *Streptomyces chrysomallus*-Kulturen (*91*) gewonnenes, dem „Actinomycin A" ähnliches, kristallisiertes, rotes Antibioticum durch Säurehydrolyse zu *L*-Threonin, *D*-Valin, *D-allo*-Isoleucin, *L*-Prolin, *L*-N-Methylvalin und Sarkosin abgebaut wurde. Damit war gezeigt, daß die roten Antibiotica vom Typ des Actinomycins A Chromopeptide sind. Da die Identität von Actinomycin A mit Antibioticum „X 45" fraglich war und daher TODD und Mitarb. ihr Antibioticum provisorisch durch den Buchstaben B kennzeichneten, haben BROCKMANN und GRUBHOFER (*30, 31*) ihr dem Actinomycin B ähnliches, aber nicht mit ihm identisches rotes Antibioticum Actinomycin C genannt.

Kurz darauf wurden „Actinomycin C"** und „Actinomycin B"** als Gemische verschiedener Actinomycine erkannt und Verfahren zur analytischen und präparativen Trennung solcher Gemische ausgearbeitet [BROCKMANN und PFENNIG (*56, 57*); BROCKMANN und GRÖNE (*23—25*)]. Nachdem damit einheitliche Actinomycine zugänglich geworden waren (*7*), gelang es, die Konstitution des Actinomycin-Chromophors aufzuklären [BROCKMANN und GRÖNE (*26, 27*); BROCKMANN und MUXFELDT (*45—47, 50*)] und bei den wichtigsten Actinomycinen die Aminosäure-Sequenz ihres Peptidteils zu ermitteln [BROCKMANN, BOHNSACK, FRANCK, GRÖNE, MUXFELDT und SÜLING (9); BROCKMANN und BOLDT (12, 13); BROCKMANN und PETRAS (55); BROCKMANN und MANEGOLD (35, 38)].

Da die Actinomycine stark giftig sind, ist verständlich, daß man irgendeine therapeutische Verwendung zunächst für ausgeschlossen hielt. Das 1949 erschienene Standardwerk „Antibiotics" (3) faßt die Ergebnisse der pharmakologischen Prüfung von „Actinomycin A" in dem Satz zusammen:

"In view of the great toxicity of this substance it is unlikely to have any therapeutic application."

Daß die Actinomycine dennoch medizinisch von Interesse sind, zeigte 1952 als erster HACKMANN (76—82), der im Tierversuch bei verschiedenen Tumor-Arten eine beachtliche cytostatische Wirkung von „Actinomycin C" fand. Auf Anregung HACKMANNS hat dann SCHULTE (101 bis 103) „Actinomycin C" klinisch geprüft und dabei eine günstige Wirkung bei Lymphogranulomatose (HODGKINscher Krankheit) festgestellt, die den Anlaß gab, „Actinomycin C" (unter dem Warenzeichen Sanamycin ⓡ der Farbenfabriken Bayer A. G., Leverkusen) in den Handel zu bringen (8).

Ausführliche Literaturangaben über die cytostatische Wirksamkeit der Actinomycine im Tierversuch und über ihre klinische Anwendung finden sich bei FARBER (69) und bei ZEPF und BERG (121).

Der folgende Bericht faßt zusammen, was bisher über die Chemie der Actinomycine bekanntgeworden ist.

## II. Gewinnung und Eigenschaften der Actinomycine.

### 1. Trennung von Actinomycingemischen.

Durch fraktionierte Kristallisation lassen sich einheitliche Actinomycine nur in Ausnahmefällen gewinnen; nur dann nämlich, wenn in dem aus Kulturlösung oder Mycel isolierten Actinomycingemisch e i n e Komponente der Menge nach so überwiegt, daß die anderen als geringfügige Beimengungen gelten können und als solche in der Mutterlauge zurückbleiben. Bei der Mehrzahl der Actinomycin bildenden Stämme jedoch ist das Mengenverhältnis der Actinomycine nicht so extrem zugunsten einer Komponente verschoben. Solche Gemische verhalten sich bei fraktionierter Kristallisation wie einheitliche Substanzen.

Daß ein Actinomycinpräparat, das nach den klassischen Reinheitskriterien als einheitlich gelten darf, dennoch ein Gemisch sein kann, zeigte sich zuerst bei der fraktionierten Gegenstromverteilung von „Actinomycin C" [BROCKMANN, GRUBHOFER, KASS und KALBE (31, 33)]. Um dieses Verfahren auf die in lipophilen Solvenzien gut, in Wasser dagegen sehr schwer löslichen Actinomycine anwenden zu können, benötigte man ein Lösungsmittel-System, dessen wäßrige Phase ausreichende Mengen Actinomycin aufnimmt. Das zuerst verwendete Phasenpaar Äther/5,6%ige Salzsäure (*31, 33*) genügte zwar dieser Anforderung, war jedoch nur für kurzdauernde Verteilungen brauchbar, weil die Actinomycine gegen verdünnte Säure nicht beständig sind. Es wurde daher durch das System Methyl-*n*-butyläther + *n*-Dibutyläther (7 : 3)/30%ige Harnstofflösung ersetzt, in dem man „Actinomycin C" durch eine über 180 Stufen geführte Verteilung zum erstenmal präparativ in die Actinomycine $C_1$, $C_2$ und $C_3$ auftrennen konnte [BROCKMANN und PFENNIG (*56, 57*)]. Die Indexzahlen geben die Reihenfolge der Verteilungsmaxima wieder.

Eine präparativ und analytisch wirklich brauchbare Trennungsmethode für Actinomycingemische wurde die Gegenstromverteilung aber erst, als man in den Natriumsalzen aromatischer Sulfonsäuren (*p*-Xylolsulfonsäure, *β*-Naphthalinsulfonsäure, Anthrachinon-*β*-sulfonsäure oder Naphthionsäure) Lösungsvermittler für die wäßrige Phase fand, die dem Harnstoff weit überlegen sind. Für präparative Trennungen in der Gegenstrom-Verteilungsapparatur bewährte sich Methyl-*n*-butyläther oder Mischungen dieses Äthers mit *n*-Dibutyläther in Kombination mit einer 5%igen Lösung von Natrium-*p*-xylolsulfonat [BROCKMANN und PFENNIG (*57*)] oder besser noch 1,25—2%igen Lösung von Natrium-*β*-naphthalinsulfonat [BROCKMANN und GRÖNE (*23, 25*)]. Vergrößerung der Natrium-*β*-naphthalinsulfonat-Konzentration sowie Zugabe von Butanol zur mobilen Phase erhöhen das Aufnahmevermögen des Systems und erlauben, das Lösungsvermögen jeder Phase so zu bemessen, daß für jedes Actinomycingemisch der zur Trennung am besten geeignete Verteilungskoeffizient erreicht wird.

Auch zur Verteilungschromatographie auf Papier oder an der Cellulose-Säule eignen sich derartige Lösungsmittel-Systeme, doch muß dann der Verteilungskoeffizient mobile Phase/stationäre Phase kleiner sein als beim Verteilen in der Apparatur; oder anders ausgedrückt: Die stationäre Phase muß mehr Actinomycin aufnehmen können als die mobile. Dieser Forderung genügt z. B. das Phasenpaar *n*-Dibutyläther/10%ige Lösung von Natriumnaphthalin-1,6-disulfonat, in dem zum erstenmal ein Actinomycingemisch („Actinomycin C") im Ring-Papierchromatogramm sowie an der Cellulose-Säule in seine Komponenten zerlegt werden konnte (*23, 25*).

Einen zur Verteilungschromatographie von Actinomycinen besonders geeigneten Lösungsvermittler fand SCHMIDT-KASTNER (*100*) im Natrium-*m*-kresotinat, das in 5—10%iger wäßriger, mit *m*-Kresotinsäure auf etwa pH 5 eingestellter Lösung mit *n*-Dibutyläther-Butanol, *n*-Dibutyläther-Butylacetat oder ähnlichen lipophilen Gemischen kombiniert werden kann. Für die präparative Trennung von Actinomycingemischen ist die Verteilungschromatographie an Cellulose-Säulen der Gegenstromverteilung überlegen, denn sie braucht weniger Lösungsmittel, ist bequemer und eignet sich besser zur Erfassung von Komponenten, die nur in kleiner Menge vorliegen.

Die von Brockmann und Gröne ausgearbeiteten Trennungsmethoden haben die Grundlage für die weitere Bearbeitung der Actinomycine geliefert, denn nunmehr war es möglich, a) einheitliche Actinomycine für Versuche zur Konstitutionsermittlung zu gewinnen, b) die von *Streptomyces*-Arten gebildeten Actinomycingemische zu analysieren, c) zu untersuchen, wie viele Actinomycine es gibt.

Einige Lösungsmittel-Systeme, die sich in Untersuchungen von Brockmann und Mitarbeitern zur Verteilungschromatographie von Actinomycingemischen bewährt haben, sind in *Tabelle 1* zusammengestellt.

Tabelle 1. Lösungsmittel-Systeme zur Verteilungschromatographie von Actinomycingemischen.

| Nr. | Mobile Phase | Stationäre Phase | Literatur |
|---|---|---|---|
| 1 | $n$-Dibutyläther | Na-1,6-disulfonat (10%) | (23, 25) |
| 2 | $n$-Dibutyläther + Butanol (3 : 2) | Na-$m$-kresotinat (10%) | (25) |
| 3 | $n$-Dibutyläther + Butylacetat (1 : 3) | Na-$m$-kresotinat (10%) | (25) |
| 4 | $n$-Butylacetat | Na-$m$-kresotinat (10%) | |
| 5* | $n$-Butanol + $n$-Dibutyläther + $n$-Dipropyläther (2 : 1 : 7) | Na-$m$-kresotinat (7%) | (100) |
| 6** | Methyl-$n$-butyläther | Na-$m$-kresotinat (5%) | (53) |
| 7** | Butylacetat + $n$-Dibutyläther (3 : 1) | Na-$m$-kresotinat (5%) | (36) |

\* Geeignet zur Trennung von Actinomycin $X_{0\delta}$ und $C_1$ und von F-Actinomycinen.

\*\* Geeignet zur Trennung von Actinomycin $X_0$-Fraktionen.

Daß sich diese Systeme in der mobilen Phase durch Verwendung anderer lipophiler Solvenzien und in der stationären durch andere Lösungsvermittler ändern lassen, war zu erwarten und ist später von verschiedenen Autoren gezeigt worden [Vining und Waksman (109); Goss und Katz (74); Roussos und Vining (97); Zepf und Berg (121); Gregory, Vining und Waksman (75); Vining, Gregory und Waksman (108)].

## 2. Von *Streptomyces*-Arten produzierte Actinomycingemische.

### Die Actinomycingemische C, X und I.

Mit ihren Trennungsmethoden haben Brockmann und Gröne (24, 25) die Actinomycine von einundzwanzig *Streptomyces*-Stämmen untersucht, die aus 2140 Actinomyceten-Stämmen als Actinomycinbildner ausgemustert worden waren. Elf dieser Stämme, zur Species *Str. chrysomallus* gehörend, produzierten Actinomycingemische, die in ihrer Zusammensetzung dem „Actinomycin C" (30, 31) ähnlich waren. Wie dieses enthielten sie als Hauptkomponenten die Actinomycine $C_2$ und $C_3$, in geringerer

Menge Actinomycin $C_1$ und außerdem zwei Komponenten ($C_0$ und $C_{0a}$) mit kleineren $R_F$-Werten als $C_1$.

Acht Stämme bildeten ein in Kristallform und Löslichkeit von „Actinomycin C" verschiedenes Actinomycingemisch. Es wurde provisorisch mit dem Buchstaben X belegt (34), da über seine Beziehung zum „Actinomycin A" oder „Actinomycin B" zunächst nichts bekannt war. Gegenstromverteilung (34, 25) trennte in die Hauptkomponente $X_2$ und zwei als Actinomycin $X_0$ und $X_1$ bezeichnete Nebenkomponenten. Im Papierchromatogramm und an der Cellulose-Säule zeigten sich neben $X_0$, $X_1$ und $X_2$ weitere Nebenfraktionen (vgl. *Tabelle 2*).

Tabelle 2. Zusammensetzung von Actinomycingemischen verschiedener *Streptomyces*-Stämme (25).

(Die mit * bezeichneten Präparate wurden kristallisiert.)

| Actinomycingemisch I | | Actinomycingemisch X | | | Actinomycingemisch C | | |
|---|---|---|---|---|---|---|---|
| | (%) | | (%) | (%) | | (%) | (%) |
| $I_{0a}$ | 0,5 | $X_{0a}$⎫ $X_0$⎬ | 5,1 | 19,6 | $C_{0a}$⎫ $C_0$*⎬ | — | 8,9 |
| $I_0$* | 7,0 | | | | | — | |
| $I_1$* ($C_1$) | 86,5 | $X_1$* ($C_1$) | 5,3 | 21,9 | $C_1$* | 10,3 | 13,0 |
| $I_2$ ($C_2$) | 5,5 | $X_{1a}$* | — | 3,7 | $C_2$* | 40,7 | 37,2 |
| $I_3$ | 0,5 | $X_2$* | 88,6 | 54,8 | $C_3$* | 49,0 | 40,8 |
| | | $X_3$* | 0,5 | — | | | |
| | | $X_4$ | 0,5 | — | | | |

Zwei von den einundzwanzig Stämmen endlich lieferten ein provisorisch mit dem Buchstaben I belegtes Actinomycingemisch, das aus einer Hauptkomponente $I_1$ und mehreren in geringer ($I_0$, $I_2$) oder sehr geringer Menge ($I_{0a}$, $I_3$) anwesenden Nebenkomponenten bestand. Tabelle 2 bringt Beispiele für die prozentuale Zusammensetzung der mit I, X oder C bezeichneten Actinomycingemische.

Die Frage, ob die Produktion mehrerer Actinomycine dadurch zu erklären ist, daß die Stämme Populationen von Varianten und Mutanten sind, von denen jede Variante oder Mutante nur ein einziges Actinomycin aufbaut, ist von BROCKMANN und PFENNIG (57) sowie von PFENNIG (94) untersucht worden und kann nach deren Befunden verneint werden. Zwar waren die Stämme tatsächlich Populationen, doch bildeten auch Einsporkulturen ihrer Varianten Actinomycingemische.

Bei einem papierchromatographischen Vergleich der I-, X- und C-Komponenten zeigten a) $C_{0a}$ und $I_{0a}$, b) $C_1$ und $I_1$, c) $C_2$ und $I_2$ in zwei verschiedenen Lösungsmittel-Systemen den gleichen $R_F$-Wert und konnten nach den damaligen Erfahrungen als identisch angesehen werden (vgl. S. 11). Das gleiche fand man später für $X_1$ und $C_1$ [BROCKMANN und DÖRING (15)]. Damit wurden die Symbole $I_1$, $X_1$ und $I_2$ überflüssig und konnten durch $C_1$ bzw. $C_2$ (vgl. Tabelle 2) ersetzt werden.

Da die Actinomycine $C_1$ und $C_2$ zur Zeit dieses Vergleichs bereits kristallisiert vorlagen, schien es zweckmäßig, die Symbole $C_1$ und $C_2$ beizubehalten.

Verschieden in ihren $R_F$-Werten waren die Fraktionen $C_{0a}$, $C_0$, $C_1$, $C_2$, $C_3$, $X_{0a}$, $X_0$, $X_{1a}$, $X_2$, $X_3$, $X_4$ und $I_0$, von denen die Actinomycine $C_0$, $C_1$, $C_2$, $C_3$, $X_{1a}$, $X_2$, $X_3$ und $I_0$ zur Kristallisation gebracht und $C_1$, $C_2$, $C_3$, $X_{1a}$ und $X_2$ durch Aminosäureanalyse (S. 16) charakterisiert wurden.

Ersetzt man in Tabelle 2 die überflüssig gewordenen Symbole $I_1$ sowie $X_1$ durch $C_1$ und ferner $I_2$ durch $C_2$, so lassen sich die Actinomycingemische I, X und C folgendermaßen charakterisieren:

*Actinomycingemisch I.* Enthält überwiegend Actinomycin $C_1$ ($I_1$) und kein Actinomycin $X_2$ oder nur Spuren davon. „Neben-actinomycine" sind in geringer Menge vorhanden: Actinomycin $C_2$ ($I_2$) und eine als $I_0$ bezeichnete Actinomycinfraktion.

*Actinomycingemisch X.* Enthält als Hauptkomponente Actinomycin $X_2$, in geringer Menge Actinomycin $C_1$ ($X_1$) und eine als $X_0$ bezeichnete Fraktion (die unter anderen das unten beschriebene Actinomycin $X_{0\beta}$ enthält).

*Actinomycingemisch C.* Enthält neben den Hauptkomponenten Actinomycin $C_2$ und $C_3$ kleinere Mengen Actinomycin $C_1$ und $C_0$.

Zu der gleichen Einteilung nativer Actinomycingemische kamen später CORBAZ, ETTLINGER, KELLER-SCHIERLEIN und ZÄHNER (66) bei der Untersuchung von neunundzwanzig Actinomycin bildenden *Streptomyces*-Stämmen, die bei der Durchmusterung einiger tausend Actinomyceten-Stämme angefallen waren.

### Actinomycingemisch Z.

Einen vierten Typ von Actinomycingemischen haben kürzlich BOSSI, HÜTTER, KELLER-SCHIERLEIN, NEIPP und ZÄHNER (6) aus einem *Str. fradiae*-Stamm isoliert. Durch Papierchromatographie gelang es ihnen, dieses kristallisierte, antibiotisch hochwirksame Actinomycingemisch Z in die Komponenten $Z_0$, $Z_1$, $Z_2$, $Z_3$, $Z_4$ und $Z_5$ zu zerlegen (Indexzahlen in der Reihenfolge der $R_F$-Werte wie bei den anderen Actinomycinen), von denen $Z_1$ und $Z_5$ kristallisierten.

Da die Z-Actinomycine in ihrem Absorptionsspektrum mit den anderen bisher bekannten Actinomycinen übereinstimmen, haben sie wahrscheinlich den gleichen Chromophor. Bemerkenswerte Unterschiede gegenüber den anderen Actinomycinen fanden sich jedoch im Peptidteil ihres Moleküls, denn allen fehlt das Prolin und alle enthalten — außer Threonin, Sarkosin, Valin und N-Methylvalin — das bisher in Actinomycinen nicht gefundene N-Methylalanin. Quantitative Aminosäure-Analysen sind bisher nicht veröffentlicht worden.

## Actinomycin bildende Streptomyces-Arten.

Die Fähigkeit, Actinomycine aufzubauen, wurde bei *Str. antibioticus, Str. griseus, Str. chrysomallus, Str. fradiae, Str. michiganensis* und *Str. parvulus* beobachtet. Zur Systematik der Actinomycin bildenden *Streptomyces*-Stämme, die hier nicht erörtert werden kann, haben neuerdings CORBAZ, ETTLINGER, KELLER-SCHIERLEIN und ZÄHNER (66) sowie FROMMER (71) Beiträge geliefert; vgl. ferner: WAKSMAN und GREGORY (113); WAKSMAN, GEIGER und REYNOLDS (112).

## Identifizierung früher beschriebener Actinomycingemische.

Papierchromatographische Untersuchungen haben gezeigt, daß „Actinomycin B" (S. 4) ein Gemisch gewesen ist, das man zur X-Gruppe rechnen kann (25). Die zu verschiedenen Zeiten gewonnenen und von verschiedenen Seiten untersuchten „Actinomycin A"-Präparate (S. 3) haben außer dem als Hauptkomponente vorliegenden Actinomycin $C_1$ wechselnde Mengen von Actinomycin $X_2$ sowie Nebenfraktionen mit kleinerem $R_F$-Wert als Actinomycin $C_1$ enthalten. Infolgedessen hat man solche Präparate teils als Gemische des Typs I (25), teils als solche des Typs X bzw. B* angesehen (109). Ring-Papierchromatogramme von „Actinomycin A"-Präparaten haben WAKSMAN, KATZ und VINING veröffentlicht (114).

Wie GOSS, KATZ und WAKSMAN (74), GOSS und KATZ (73) sowie KATZ, PIENTA und SIVAK (86) zeigen konnten, hing die Zusammensetzung der „Actinomycin A"-Präparate nicht nur von der Art der Nährlösung, sondern auch von der Bebrütungs-dauer der Stämme ab. Nach kurzer Bebrütung fand man Gemische mit relativ viel Actinomycin $X_2$ und Komponenten mit kleinerem $R_F$-Wert als Actinomycin $C_1$, nach längerer dagegen Gemische vom Typ I.

Zwei andere, früher beschriebene Actinomycinpräparate, das Actinomycin J (107) und das Actinomycin S-67 (98, 118), haben KELLER-SCHIERLEIN und Mitarb. (6) als Actinomycingemisch X bzw. C charakterisiert. Nach den gleichen Autoren kann Actinomycin M als Actinomycingemisch X angesehen werden.

Daß die Zusammensetzung nativer Actinomycingemische von der Zusammensetzung der Kulturlösung abhängt, zeigte sich zuerst beim Actinomycingemisch C, bei dem das Verhältnis Actinomycin $C_2$ : Actinomycin $C_3$ mit Natriumnitrat als Stickstoffquelle größer war als mit Glykokoll (57, 94). Ähnliches fand man bei Stämmen, die Actinomycingemisch X erzeugten. Hier war, wenn die Nährlösung den Stickstoff als Glykokoll enthielt, erheblich mehr $X_0$-Fraktion und Actinomycin $C_1$ ($= X_1$) im Gemisch als bei Stickstoffzufuhr in Form von Natriumnitrat (34). In keinem Fall jedoch ließ sich das Mengenverhältnis der Komponenten durch Abänderung der Kulturbedingungen umkehren. Das gelang erst WAKSMAN und Mitarb. (73, 74, 75) sowie unabhängig

---

* Als Actinomycinkomplex B bezeichnen ROUSSOS und VINING (97) Gemische, in denen das Verhältnis Actinomycin $X_2$ : Actinomycin $C_1$ etwa 2 : 1 ist. Vgl. dazu S. 11.

davon MARTIN und PAMPUS (93), die bei Actinomycingemisch X produzierenden Stämmen eingehend untersucht haben, wie die Zusammensetzung des Actinomycingemisches von den Kulturbedingungen abhing.

Ob man durch geeignete Kulturbedingungen auch die Ausbeute an solchen Actinomycinen erhöhen kann, die in den Gemischen I, X und C nur in sehr geringer Menge vorliegen, hat zuerst FROMMER (72) bei den Actinomycinen $X_3$ und $X_4$ untersucht. Wie er fand, lagen beide erst dann in greifbarer Menge vor, als bereits die Autolyse des Mycels eingesetzt hatte. Actinomycin $X_3$ konnte bei diesen Versuchen zum erstenmal in kristallisierter Form isoliert werden.

Schließlich ist im Zusammenhang mit den vorstehenden Befunden noch zu erwähnen, daß FROMMER (70) aus degenerierten, zur Synthese von Actinomycingemisch I nicht mehr fähigen *Streptomyces*-Kulturen durch Auslese Stämme gewinnen konnte, die alle wieder Actinomycingemisch I bildeten, d. h. die Degeneration hatte keine Varianten oder Mutanten entstehen lassen, die Actinomycingemisch X oder C produzieren.

### *Actinomycingemische der dirigierten Biosynthese.*

Nachdem die Actinomycine als Chromopeptide erkannt waren, lag die Frage nahe, ob man Actinomycin bildende Stämme durch Verabreichung von Aminosäuren dazu bringen kann, diese in den Peptidteil ihrer Actinomycine einzubauen. Eine solche „Dirigierung" der Biosynthese gelang zum erstenmal SCHMIDT-KASTNER (99, 100). Er hat eine größere Zahl von Aminosäuren und Aminosäure-derivaten, u. a. Valin, Norvalin, Leucin, Isoleucin, Sarkosin, Threonin, Methionin, der Kulturlösung von *Streptomyces*-Stämmen zugesetzt, die aus den üblichen Nährstoffen (Glycerin als Kohlenstoff- und Nitrat als Stickstoff-Lieferant) die Actinomycingemische I, X und C aufbauten, und gefunden, daß nun neben den normalerweise entstehenden neue Actinomycine auftraten.

Näher untersucht wurden bisher die nach Zugabe von *DL*-Isoleucin entstehenden E-Actinomycine und die nach Sarkosin-Zusatz gebildeten F-Actinomycine.

Die kristallisiert isolierten Actinomycine $E_1$ und $E_2$ unterscheiden sich von den anderen bisher bekannten Actinomycinen durch ihren Gehalt an N-Methyl-isoleucin, das entweder in einer (Actinomycin $E_1$) oder in beiden Peptidketten (Actinomycin $E_2$) das N-Methylvalin vertritt. Die Actinomycine $F_1$, $F_2$, $F_3$ und $F_4$ enthalten drei bzw. vier Mole Sarkosin und demgegenüber nur ein Mol bzw. kein Prolin (vgl. *Tabelle 3*, S. 16). Das der Kulturlösung zugesetzte Sarkosin kann somit beim Aufbau der Peptidkette das Prolin ganz oder zum Teil ersetzen.

Ein bevorzugter Einbau von Valin wurde erreicht, als man diese Aminosäure zur Nährlösung von Stämmen gab, die Actinomycingemisch C bilden. Hauptkomponente war nun mit einem Anteil von etwa 80% das

Actinomycin $C_1$ (2 Mol Valin enthaltend), während die Synthese von Actinomycin $C_2$ (1 Mol Valin, 1 Mol *allo*-Isoleucin enthaltend) und Actinomycin $C_3$ (2 Mol *allo*-Isoleucin enthaltend) in den Hintergrund trat.

Bemerkenswert ist, daß sich die F-Actinomycine nur durch eine Kombination von Adsorptionschromatographie und Verteilungschromatographie trennen lassen. Im Ringchromatogramm hatten die Actinomycine $F_1$ und $F_3$ den gleichen $R_F$-Wert wie Actinomycin $F_5$ und $C_2$. Bei der Suche nach neuen Actinomycinen wird man sich daher in Zukunft nicht mehr wie bisher allein auf die Aussage des Ring-Papierchromatogrammes verlassen dürfen (vgl. S. 7).

### 3. Nomenklatur der Actinomycine.

Die Nomenklatur der Actinomycine ist zur Zeit unübersichtlich und erfordert daher eine etwas ausführlichere Erörterung. Diese Unübersichtlichkeit liegt zur Hauptsache daran, daß a) bei den Actinomycinen im Gegensatz zu anderen Naturstoffklassen neben der Nomenklatur der reinen Actinomycine auch eine solche nativer Actinomycingemische existiert und b) einige Autoren für bereits bekannte und durch ihren Aminosäuregehalt charakterisierte Actinomycine neue Bezeichnungen eingeführt haben.

*Nomenklatur nativer Actinomycingemische.* Seit sich die anfangs für einheitlich gehaltenen Actinomycinpräparate A, B und C als Gemische mehrerer Actinomycine entpuppt haben, besteht für den Chemiker streng genommen kein Anlaß mehr, solche Gemische durch besondere Buchstaben zu kennzeichnen. Daß man es dennoch getan hat, läßt sich rechtfertigen; denn Actinomycingemische, wie beispielsweise die vom Typ Z, E und F, weisen in der Zusammensetzung ihres Peptidteils Eigentümlichkeiten auf, durch die sie sich voneinander und von den mit I, X und C gekennzeichneten Gemischen unterscheiden. Und auch der Einteilung in die letztgenannten drei Gruppen kann eine gewisse Berechtigung nicht abgesprochen werden.

Bezweifeln kann man dagegen die Notwendigkeit, Actinomycingemische nach dem Mengenverhältnis Actinomycin $X_2$ : Actinomycin $C_1$ weiter zu unterteilen.

Während nach Tabelle 2 (S. 7) Gemische, in denen das Verhältnis Actinomycin $X_2$ : Actinomycin $C_1$ 17 : 1 bis 2 : 1 beträgt, zum Typ X gerechnet werden, bezeichnen Roussos und Vining (97) Gemische, in denen das Mengenverhältnis Actinomycin $X_2$ : Actinomycin $C_1$ etwa 2 : 1 ist, als B-Typ, solche, in denen es etwa 1 : 3 ist, als Typ A und solche, die praktisch aus Actinomycin $C_1$ bestehen und andere Actinomycine nur in Spuren enthalten, nach dem Vorschlag von Waksman (114) als D-Typ. Warum ein Präparat, das neben Actinomycin $C_1$ nur Spuren anderer Actinomycine enthält, eine andere Bezeichnung als $C_1$ erhalten soll, ist nicht einzusehen. Wollte man dieses Vorgehen verallgemeinern, so müßte man für alle Naturstoffe, die durch Spuren verwandter Verbindungen verunreinigt sind, besondere Bezeichnungen einführen.

Die einzelnen Komponenten ihrer Actinomycingemische unterscheiden Roussos und Vining (97) dadurch, daß sie an den Buchstaben des Gemisches der Reihenfolge der $R_F$-Werte entsprechend römische Indexzahlen hängen. Eine Komponente, die dem $R_F$-Wert und dem Aminosäuregehalt nach mit Actinomycin $C_1$ identisch ist, nennen sie $A_{IV}$ und $B_{IV}$, eine offensichtlich mit Actinomycin $X_2$ identische Komponente Actinomycin $A_V$ oder $B_V$ und eine, wohl mit Actinomycin $X_{0\beta}$ (52, 53) identische Actinomycin $A_I$ bzw. $B_I$. Allerdings geben sie selber nichts über die Identität bzw. Nicht-identität ihrer Präparate mit dem damals bereits bekannten und durch Aminosäuregehalt charakterisierten Actinomycinen $C_1$, $C_2$, $C_3$, $X_{0\beta}$ und $X_2$ an.

Zusammenfassend läßt sich sagen, daß eine Nomenklatur und damit eine Systematik von Actinomycingemischen nur begrenzten Wert hat und daher auf wenige Typen beschränkt bleiben sollte. Nur so lassen sich fruchtlose Diskussionen darüber vermeiden, ob ein natives Gemisch in diese oder jene Gruppe einzuordnen ist. Um Verwechslungen auszuschließen, sollte man, wie neuerdings von Waksman (114) vorgeschlagen, Actinomycingemische als solche bezeichnen und beispielsweise statt Actinomycin X nur noch die Bezeichnung Actinomycingemisch X oder Actinomycinkomplex X verwenden.

*Nomenklatur reiner Actinomycine.* Reine Actinomycine werden von den meisten Autoren (6, 66, 100, 120, 121) nach Brockmann und Mitarb. so benannt, daß man an den Buchstaben des Gemisches, aus dem sie zuerst isoliert worden sind, eine arabische Indexzahl hängt, die ihrer Lage im Chromatogramm entspricht. Je kleiner die Indexzahl, desto kleiner ist der $R_F$-Wert. Eine erste Erweiterung dieser Nomenklatur wurde notwendig, als man aus Actinomycingemischen in geringer Menge Komponenten isolierte, deren Zonen zwischen denen schon bekannter Actinomycine lagen. Solche „Neben-actinomycine" hat man durch zusätzliche, kleine, lateinische Buchstaben gekennzeichnet (25, 121). Sie werden dem Symbol desjenigen Actinomycins angehängt, das den nächst kleineren $R_F$-Wert hat. $X_{1a}$ ist somit die Bezeichnung für ein Actinomycin, das in kleiner Menge aus einem Actinomycingemisch X isoliert wurde [Brockmann und Manegold (36)] und dessen Zone zwischen der $X_1$- und $X_2$-Zone des Chromatogrammes liegt. Und Actinomycin $C_{2a}$ [Brockmann und Franck (22)] ist ein aus Actinomycingemisch C abgetrenntes „Neben-actinomycin", dessen Zone man zwischen der $C_2$- und $C_3$-Zone findet [vgl. dazu Zepf und Berg (121)].

Zu einer zweiten Erweiterung der Nomenklatur wurde man gezwungen, als es gelang, die $X_0$-Fraktion von Actinomycingemischen X in mehrere Komponenten zu zerlegen. Hier hat man die Unterfraktionen in der Reihenfolge ihrer $R_F$-Werte durch zusätzliche kleine griechische Buchstaben unterschieden. Actinomycin $X_{0\gamma}$ (36) hat somit einen größeren $R_F$-Wert als Actinomycin $X_{0\beta}$.

Über die Frage, ob man antibiotisch wirksame, amorphe Fraktionen, die bei der Trennung von Actinomycingemischen in sehr geringer Menge anfallen und zunächst nur durch ihre $R_F$-Werte charakterisiert sind, benennen soll, sind die Meinungen geteilt. WAKSMAN, KATZ und VINING (*114*) lehnen es ab, BROCKMANN, PAMPUS und MANEGOLD (*53*) halten es für richtig.

Nach der vorstehenden Nomenklatur sind die Actinomycine der Tabelle 3 (S. 16) von Anfang an eindeutig gekennzeichnet worden und das gleiche gilt für die Komponenten des Actinomycingemisches Z.

Zu Doppelbezeichnungen kam es dagegen in folgenden Fällen. Bei der Analyse von Actinomycingemischen vom Typ I wurden, wie schon erwähnt, zwei Komponenten zunächst als $I_1$ und $I_2$ bezeichnet, die sich im Laufe der gleichen Unteras Actinomycin $C_1$ bzw. $C_2$ identifizieren ließen (*25*). Ferner hielten BROCKMANN und GRÖNE (*25*) ein als Actinomycin $X_1$ bezeichnetes Präparat für ein Isomeres des Actinomycins $C_1$, während später BROCKMANN und DÖRING (*15*) zeigen konnten, daß beide identisch sind und die Bezeichnung Actinomycin $X_1$ demnach zu streichen ist.

Neuerdings haben WAKSMAN, KATZ und VINING (*114*) angeregt, die eben erläuterte Nomenklatur der Actinomycine aufzugeben und die reinen Actinomycine nur noch durch römische Ziffern voneinander zu unterscheiden. BROCKMANN, PAMPUS und MANEGOLD (*53*) befürchten, daß dadurch die Nomenklatur der Actinomycine unübersichtlicher wird und halten die Einführung einer neuen Nomenklatur nur dann für sinnvoll, wenn sie die Aminosäuresequenz des Peptidteils zum Ausdruck bringt. Wie das in einfacher Weise möglich ist, wird auf S. 46 erörtert.

## 4. Eigenschaften und Aminosäuregehalt reiner Actinomycine.

*Physikalische Eigenschaften.* Die reinen Actinomycine sind gelbrote, gut kristallisierende Verbindungen. Ihre Kristalle können Wasser und Lösungsmittel enthalten, was bei Analysen und physikalischen Messungen zu berücksichtigen ist. In den meisten organischen Solvenzien sind sie recht gut, in Wasser dagegen sehr schwer löslich. Sie schmelzen unscharf unter Zersetzung ($\sim 250°$), so daß Charakterisierung und Identifizierung durch Schmelzpunkt und Misch-Schmelzpunkt entfällt.

Bemerkenswert ist die Größe der spezifischen Drehung. Sie liegt bei den meisten Actinomycinen in Methanol oder Chloroform zwischen — 300° und — 350°. Das bei allen Actinomycinen gleiche Absorptionsspektrum hat in Cyclohexan im sichtbaren Gebiet zwei Maxima bei 446 m$\mu$ und 424 m$\mu$ und im UV-Gebiet eins bei 241 m$\mu$ (*Abb. 1*, S. 14). Auch im IR-Spektrum sind keine charakteristischen Unterschiede bei den einzelnen Actinomycinen zu erkennen.

Während sich die Actinomycine in der spezifischen Drehung sowie im Schmelzpunkt und Spektrum wenig oder gar nicht voneinander unterscheiden, sind andererseits selbst bei Vertretern, die in ihrer Struktur so ähnlich sind wie die Actinomycine $C_1$, $C_2$ und $C_3$, die Löslichkeitsunterschiede groß genug, um eine glatte Trennung durch fraktionierte Gegenstromverteilung oder Verteilungschromatographie zu gestatten.

*Chemische Eigenschaften.* Die Actinomycine sind sehr schwache Basen, die bei potentiometrischer Titration in Eisessig mit Perchlorsäure 1 Äquivalent Säure verbrauchen. Aus Lösungsmitteln, die mit Wasser begrenzt mischbar sind, lassen sie sich mit Salzsäure, deren Konzentration über 10% liegt, unter Farbumschlag von Gelb nach Weinrot ausschütteln. Die gleiche Halochromie tritt auf, wenn über Actinomycin-Zonen eines

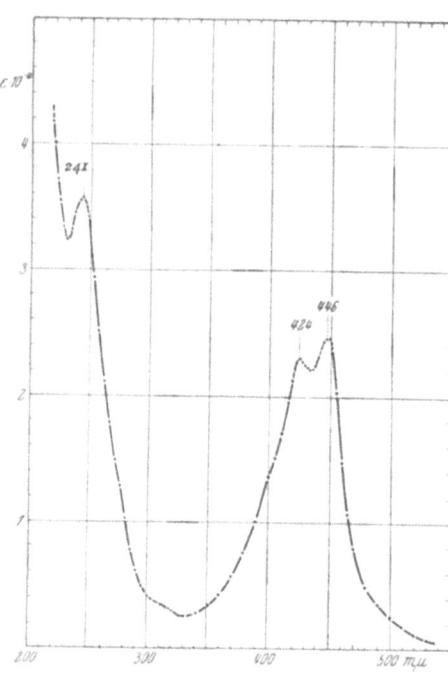

Papierchromatogrammes Chlorwasserstoff geblasen wird — eine Reaktion, mit der man die Actinomycine von Desamino-actinomycinen und anderen gelben Verbindungen unterscheiden kann. Gegen Säure und mehr noch gegen Alkali sind die Actinomycine empfindlich; 10%ige Salzsäure verwandelt sie bei 40° innerhalb 15 Stunden in Desamino-actinomycine, und in 0,1 *n*-methanolischem Alkali ist bei 20° bereits nach 4 Stunden kein Ausgangsmaterial mehr vorhanden.

*Molekulargewicht.* Kryoskopische Bestimmungen in Phenol, Cyclohexan und in Kohlenstofftetrabromid gaben Werte zwischen 768 und 1000 (*115*) und waren demnach unbrauchbar. Bewährt haben sich dagegen nach BROCKMANN und Mitarb. chemische Methoden zur

Abb. 1  Absorptionskurve von Actinomycin $C_3$ in Cyclohexan.

Molekulargewichtsbestimmung, die sich auf die chinoide Struktur des Actinomycin-Chromophors gründen. Dank dieser Struktur werden die Actinomycine bei katalytischer Hydrierung schon unter milden Bedingungen leicht zu hellgelben Leukoverbindungen reduziert, die an der Luft quantitativ wieder in Actinomycine übergehen — ein Verhalten, das ermöglicht, das Mol.-Gew. recht genau durch katalytische Hydrierung zu bestimmen. Unter der plausiblen Annahme, daß nur e i n e chinoide Gruppe vorliegt, kam man an Hand des Wasserstoffverbrauches zu Werten um 1200 (*33, 84*). Eine Fehlerquelle dieses Verfahrens liegt darin, daß unter Umständen andere Gruppen in geringem Umfang mithydriert und daher zu niedrige Mol.-Gew.-Zahlen gefunden werden. Diese Möglichkeit ist so gut wie ausgeschlossen, wenn das chinoide

System durch Redox-Titration erfaßt wird. Wie BROCKMANN und VOHWINKEL (60) zeigen konnten, läßt sich so mit Titan-III-chlorid in 50%igem Eisessig oder mit Chrom-II-chlorid in 70%igem Methanol das Mol.-Gew. der Actinomycine bei Einwaagen von 5—15 mg auf ± 2,5% und bei größeren Einwaagen sogar auf ± 1% genau bestimmen.

Auch die potentiometrische Titration in Eisessig mit Perchlorsäure, bei der die Actinomycine, wie schon erwähnt, als schwache, einsäurige Basen reagieren, kann zur Mol.-Gew.-Bestimmung dienen [BROCKMANN und MEYER (41)] und liefert Werte, die befriedigend mit denen der Redox-Titration übereinstimmen.

*Aminosäuregehalt.* Bei der quantitativen Aminosäure-Analyse der Actinomycine waren zwei Fehlerquellen zu beseitigen, die durch die eigentümliche Struktur der Actinomycine bedingt sind und daher bei Proteinen und normalen Polypeptiden keine Rolle spielen. Die eine betrifft das Threonin, die andere die Methylaminosäuren. Threonin, gegen heißes, wäßriges Alkali empfindlich und auch gegen konz. Salzsäure nicht ganz beständig, wird bei Totalhydrolyse der Actinomycine durch die oxydierende Wirkung des Chromophors in erheblichem Umfang zerstört. Dieser Verlust läßt sich auf zweierlei Weise vermeiden: a) Durch Hydrolyse mit Jodwasserstoffsäure anstelle von Salzsäure, wobei das Threonin zu α-Aminobuttersäure reduziert und als solche bestimmt wird [BROCKMANN, GRÖNE und TIMM (29)], und b) durch katalytische Hydrierung des betreffenden Actinomycins zur Dihydro-verbindung und deren Hydrolyse mit konz. Salzsäure unter Wasserstoff [BROCKMANN und MANEGOLD (37)].

Die zweite Fehlerquelle lag bei der Umsetzung der Methylaminosäuren mit Ninhydrin. Denn unter Bedingungen, wie sie zur Bestimmung der Aminosäuren üblich sind, war die Farbstoffausbeute nur etwa 40%. Durch Abänderung der Reaktionsbedingungen (höhere Konzentration des Reagens, höherer Methylglykolgehalt der Reagenzlösung, längeres Erhitzen, Einhalten enger pH-Grenzen) gelang es, die Farbstoffausbeute auf 95% zu erhöhen (37).

Trennung und Bestimmung der Aminosäuren führte man zunächst in Anlehnung an BOISSONNAS (5) aus [BROCKMANN, BOHNSACK und GRÖNE (10)]. Auch Ionenaustauscher-Säulen wurden benutzt (37), doch gelang es hier nicht, die im HJ-Hydrolysat vorliegenden Aminosäuren in einem Gang zu trennen. Beim Eluieren mit wäßrigen Pufferlösungen blieben Sarkosin und N-Methylvalin in einer Fraktion zusammen und beim Eluieren mit propanolischer Pufferlösung Valin und α-Aminobuttersäure (aus Threonin).

Bei den meisten Analysen hydrolysierte man die Dihydroverbindung der Actinomycine mit konz. Salzsäure unter Wasserstoff, trennte die Aminosäuren papierchromatographisch und bestimmte sie durch photometrische Ausmessung ihrer blauen Ninhydrinflecken (37). Auch Überführung der Amino- und N-Methyl-aminosäuren in gelbe 2,4-Dinitrophenylderivate mit anschließender papier-chromatographischer Trennung und Kolorimetrie der Derivate nach Koch und Weidel (87) hat sich bewährt (119).

Tabelle 3. Aminosäuregehalt von Actinomycinen in Molen*.

| Name und Literatur | Thre | Val | a-Ileu | Pro | Hypro | a-Hypro | Opro | Sar | Meval | Me-ileu | Literatur** |
|---|---|---|---|---|---|---|---|---|---|---|---|
| Actinomycin $C_1$ (24, 25) . | 2 | 2 | | 2 | | | | 2 | 2*** | | (10) |
| Actinomycin $C_2$ (56,57,25) | 2 | 1 | 1 | 2 | | | | 2 | 2*** | | (10) |
| Actinomycin $C_{2a}$ (22) ... | 2 | 1 | 1 | 2 | | | | 2 | 2*** | | (37) |
| Actinomycin $C_3$ (56,57,25) | 2 | | 2 | 2 | | | | 2 | 2*** | | (10) |
| Actinomycin $X_{0\beta}$ (52, 53) | 2 | 2 | | 1 | 1 | | | 2 | 2*** | | (53) |
| Actinomycin $X_{0\gamma}$ (36) ... | 2 | 2 | | 1 | | | | 3 | 2*** | | (37) |
| Actinomycin $X_{0\delta}$ (38) ... | 2 | 2 | | 1 | | 1 | | 2 | 2*** | | (37) |
| Actinomycin $X_{1a}$ (36) ... | 2 | 2 | | | | | 1 | 3 | 2 | | (37) |
| Actinomycin $X_2$ (34, 25) . | 2 | 2 | | 1 | | | 1 | 2 | 2*** | | (10, 35) |
| Actinomycin $E_1$ (99) .... | 2 | | 2 | 2 | | | | 2 | 1*** | 1*** | (39) |
| Actinomycin $E_2$ (99) .... | 2 | | 2 | 2 | | | | 2 | | 2*** | (39) |
| Actinomycin $F_1$ (99) .... | 2 | 1 | 1 | | | | | 4 | 2*** | | (39) |
| Actinomycin $F_2$ (99) .... | 2 | 1 | 1 | 1 | | | | 3 | 2*** | | (39) |
| Actinomycin $F_3$ (99) .... | 2 | | 2 | | | | | 4 | 2*** | | (39) |
| Actinomycin $F_4$ (99) .... | 2 | | 2 | 1 | | | | 3 | 2*** | | (39) |

  * Die angegebenen Werte sind aus Aminosäureanalyse und Mol.-Gew. be-rechnet und auf ganze Zahlen abgerundet.
  ** Literatur über Aminosäureanalysen.   *** C-terminal.

| | |
|---|---|
| a-Ileu = *allo*-Isoleucin. | Val = Valin. |
| Pro = Prolin. | Meval = N-Methylvalin. |
| Hypro = γ-Hydroxyprolin. | Sar = Sarkosin. |
| a-Hypro = γ-*allo*-Hydroxyprolin. | Me-ileu = N-Methyl-isoleucin. |
| Opro = γ-Oxoprolin. | Thre = Threonin. |

Die Ergebnisse der bisherigen Analysen sind in *Tabelle 3* zusammen-gestellt. Soweit bekannt, sind die C-terminalen Methylaminosäuren (vgl. S. 33) gekennzeichnet. Gemeinsam ist allen 15 Actinomycinen der Tabelle 3 ein Gehalt an a) insgesamt 10 Aminosäuren, b) 2 Mol Threonin, c) mindestens 2 Mol Sarkosin.

Bemerkenswert ist, daß die einzelnen Amino- bzw. N-Methylamino-säuren, wenn überhaupt, nur in engen Grenzen variiert werden, nämlich entweder durch Substitution mit einer Methylgruppe (Valin → *allo*-Isoleucin; N-Methylvalin → N-Methyl-isoleucin) oder durch γ-Hydroxy-lierung bzw. Ketonisierung von Prolin (γ-Hydroxyprolin, γ-Ketoprolin).

Die Konfiguration der Aminosäuren wurde bei den Actinomycinen $C_1$, $C_2$ und $C_3$ ermittelt. Bei ihnen liegt Threonin, Prolin und N-Methylvalin in der *L*-Konfiguration, Valin und *allo*-Isoleucin in der *D*-Konfiguration vor. Das gleiche gilt für die Actinomycine $X_{0\beta}$, $X_{0\delta}$ und $X_2$, die durch Reduktion mit Actinomycin $C_1$ sterisch verknüpft werden konnten. Bei den anderen Actinomycinen ist über die Konfiguration ihrer Aminosäuren noch nichts bekannt, doch ist anzunehmen, daß die Verhältnisse analog sind, wie bei den vorhergenannten Actinomycinen.

## III. Die Konstitution der Actinomycine.

Um die Konstitution der Actinomycine aufzuklären, mußten ihrer Chromopeptidstruktur entsprechend zwei dem Thema nach ganz verschiedene Aufgaben in Angriff genommen werden: a) Die Strukturermittlung einer farbigen Verbindung, nämlich des Chromophors, und b) die Aufklärung des Peptidteils. Von diesen Aufgaben ist für eine Reihe von Actinomycinen die erste ganz und die zweite bis auf die Stellung zweier Lactongruppen gelöst.

### 1. Die Konstitution des Chromophors.

#### *Despeptido-actinomycin.*

Für die Aufklärung des Actinomycin-Chromophors schien es zweckmäßig, ihn vorher vom Peptidteil abzulösen. Alkalihydroxyd kam dafür nicht in Frage, denn es zerstört den Chromophor schon unter milden Bedingungen; und auch Kochen mit Säure schien zunächst ungeeignet, weil der Chromophor in braune, undefinierbare Produkte überging. Günstiger verlief der Abbau mit Bariumhydroxyd, denn dabei entstand aus ,,Actinomycin C" eine kristallisierte, rote Verbindung, die keine Amino- oder N-Methyl-aminosäuren des Peptidteils mehr enthielt und daher Despeptido-actinomycin genannt wurde [BROCKMANN und GRUBHOFER (*31, 32*)]. Das gleiche Abbauprodukt haben später JOHNSON, TODD und VINING (*84*) beim Bariumhydroxyd-Abbau von ,,Actinomycin B" gefaßt und als Actinomycinol B bezeichnet.

Daß der Chromophor durch wäßriges Alkalihydroxyd schnell zerstört wird, beim Kochen mit Bariumhydroxyd dagegen, wenn auch in geringer Ausbeute, eine rote Verbindung liefert, liegt offenbar an den verschiedenartigen Kationen. Sind Natrium- oder Kalium-Ionen vorhanden, so verläuft die Reaktion im homogenen System. Mit Barium-Ionen dagegen fallen schon die ersten Abbauprodukte als unlösliche Bariumsalze aus und können, wenn überhaupt, nur noch im heterogenen System weiter reagieren.

Despeptido-actinomycin ist, wie sein Absorptionsspektrum (*31*) von Anfang an vermuten ließ, ein Umwandlungsprodukt des Chromophors. Seine Untersuchung hat daher zu dessen Konstitutionsaufklärung nur wenig beigetragen. Aber da es immerhin dem Chromophor entstammt,

hat es sich in anderer Hinsicht nützlich erwiesen; zur Beantwortung der
Frage nämlich, ob die verschiedenen Actinomycine alle den gleichen
Chromophor haben. Denn Actinomycine mit gleichem Chromophor
sollten das gleiche Despeptido-actinomycin liefern, auch wenn dieses
ein Umlagerungsprodukt ist; Actinomycine mit verschiedenem Chromophor
dagegen nicht.

Wie BROCKMANN und VOHWINKEL (59, 61) zeigen konnten, entsteht
aus den Actinomycinen $C_1$, $C_2$, $C_3$ und $X_2$ beim Bariumhydroxyd-Abbau
das gleiche Despeptido-actinomycin und dasselbe fand man später beim
Actinomycin $X_{0\beta}$ [BROCKMANN, PAMPUS und MANEGOLD (53)]. Die
fünf Actinomycine unterscheiden sich demnach nur in der Struktur
des Peptidteils. In gleicher Weise wird sich auch bei den anderen
Actinomycinen klären lassen, ob sie denselben Chromophor enthalten.
Da somit dem Despeptido-actinomycin eine gewisse Bedeutung für die
Actinomycin-Chemie nicht abzusprechen ist, und es überdies einer bisher
kaum untersuchten Verbindungsklasse angehört, soll seine Konstitutionsaufklärung und Synthese kurz beschrieben werden.

*Konstitution des Despeptido-actinomycins (43, 48).* Despeptido-actinomycin, $C_{15}H_{11}NO_5$ (*14, 42*), enthält zwei C-Methylgruppen und reagiert
sowohl basisch (Bildung eines Perchlorates) wie sauer. Acetylierung und
Methylierung mit und ohne gleichzeitige Reduktion zeigten das Vorliegen
von zwei Hydroxy- und zwei Chinon-Carbonylgruppen. Da Despeptido-
actinomycin — obgleich ihm Carboxygruppen fehlen — stärker sauer
ist als Benzoesäure, muß eine der beiden Hydroxygruppen im chinoiden
Ring stehen. Übereinstimmend damit kondensiert es mit o-Phenylen-
diamin leicht zu einem Phenazinderivat.

Despeptido-actinomycin ist gegen Alkali sehr beständig, wird durch
Methanol-HCl nicht methyliert und gibt mit Hypojodit 1 Mol Jodoform,
Reaktionen, aus denen hervorging, daß e i n e der beiden C-Methylgruppen
im chinoiden Ring steht. Zum gleichen Ergebnis führten vergleichende
Redoxpotentialmessungen von Despeptido-actinomycin und Modell-
verbindungen (*110*). Damit ergaben sich für Despeptido-actinomycin
und sein o-Phenylendiamin-Kondensationsprodukt die Teilformeln (I)
bzw. (II).

(I.)                                        (II.)

Wichtige Auskunft über das Grundgerüst des Despeptido-actinomycins erhielt man durch spektroskopische Messungen. Denn sie zeigten, daß die Absorptionskurven von a) Acridon und Dihydro-despeptido-actinomycin-pentaacetat sowie b) von 1,3,4-Trimethoxy-acridon und Dihydro-despeptido-actinomycin-tetramethyläther auffallend ähnlich sind. Daraus ließ sich schließen, daß Dihydro-despeptido-actinomycin-pentaacetat und Dihydro-despeptido-actinomycin-tetramethyläther Derivate des Acridons (III) und dementsprechend nach (IV) bzw. (IVa) zu formulieren sind.

Im Einklang mit diesen Formeln stand die Beobachtung, daß energische, acetylierende Reduktion das Despeptido-actinomycin in ein Dihydro-desoxy-despeptido-actinomycin-tetraacetat verwandelt, das seinem Spektrum nach ein Acridinderivat ist. Denn Acridon geht unter den gleichen Bedingungen in Acridin über. Für das Dihydro-desoxy-despeptido-actinomycin-tetraacetat ergab sich damit die Teilformel (V).

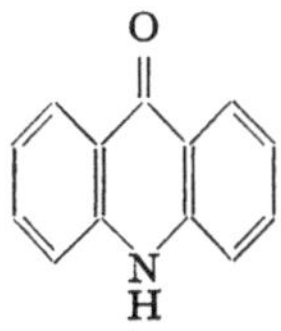

(III.) Acridon.

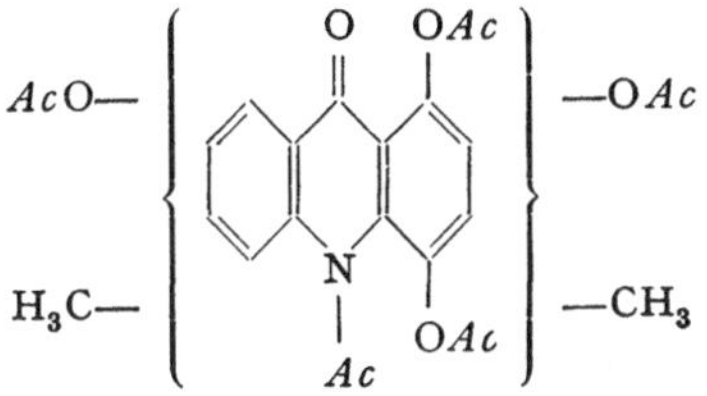

(IV.)
(IVa.) $OAc = OCH$ ; $NAc = NH$.

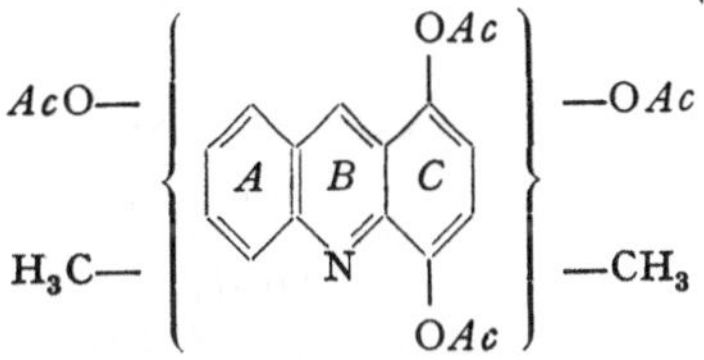

(V.)

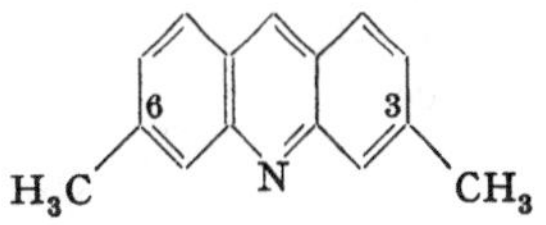

(VI.) 3,6-Dimethyl-acridin.

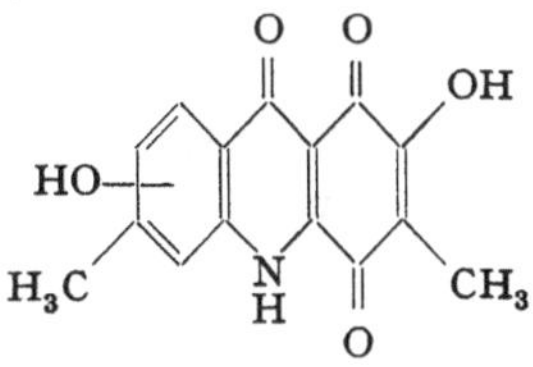

(VII.)

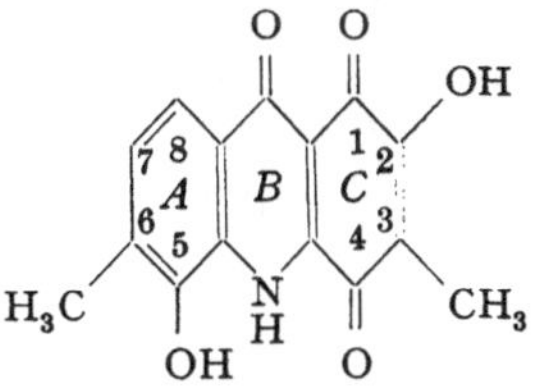

(VIII.) Despeptido-actinomycin.

Bestätigt und erweitert wurden diese Befunde durch die Zinkstaubdestillation des Despeptido-actinomycins, bei der 3,6-Dimethyl-acridin (VI) entstand. Dadurch war die Stellung der beiden C-Methylgruppen bewiesen, und aus (I) wurde somit Teilformel (VII).

Da Despeptido-actinomycin mit Blei-II-acetat eine tiefblaue Lösung gibt und diese Reaktion nur bei 5- oder 7-Hydroxy-acridon-chinonen-(1,4) eintritt, kamen in (VII) nur $C_{(5)}$ und $C_{(7)}$ als Träger der Hydroxygruppe in Betracht. Die Entscheidung brachte eine systematische Untersuchung über die Absorptionsinkremente von Methyl- und Methoxygruppen bei Methyl- und Methoxy-acridonen, die eindeutig zugunsten von $C_{(5)}$ entschied [BROCKMANN, MUXFELDT und HAESE (51)]. Damit war für Despeptido-actinomycin die Konstitution (VIII) bewiesen.

Nachdem bekannt war, daß Despeptido-actinomycin ein Dihydroxy-dimethyl-acridon-chinon-(1,4) ist, haben ANGYAL und Mitarb. (1, 2) sowie HANGER, HOWELL und JOHNSON (83) die Stellung der Hydroxy- und Methylgruppe an $C_{(5)}$ bzw. $C_{(6)}$ dadurch bewiesen, daß sie „Actinomycin B" (S. 3) mit alkalischem Wasserstoffperoxyd zu 7-Methyl-benzoxazolon-4-carbonsäure abbauten (XLII, S. 30). Ferner kamen sie — durch einen nicht näher erläuterten spektroskopischen Vergleich von Despeptido-actinomycin (Actinomycinol B) mit 3-Hydroxy-2-methoxy-10-methylacridon-chinon-(1,4) — zu dem Schluß, daß in Ring $C$ die Hydroxygruppe an $C_{(2)}$ und die Methylgruppe an $C_{(3)}$ steht.

*Synthese des Despeptido-actinomycins.* Die Despeptido-actinomycin-Formel (VIII) wurde von BROCKMANN und MUXFELDT (44, 49) durch

(XIV.) Despeptido-actinomycin.

*Formelübersicht 1.* Synthese von Despeptido-actinomycin.

Synthese bestätigt *(Formelübersicht 1)*. Ausgangsverbindungen waren das Kaliumsalz der 2-Chlor-3-methoxy-4-methylbenzoesäure (IX) und das 2,3-Dimethoxy-6-nitro-5-amino-toluol (X), die nach Jourdan-Ullmann zur 6'-Nitro-6,3',4'-trimethoxy-5,5'-dimethyl-diphenylamincarbonsäure-(2) (XI) kondensiert wurden. Verbindung (XI) ließ sich durch Polyphosphorsäure in 4-Nitro-1,2,5-trimethoxy-3,6-dimethyl-acridon (XII) verwandeln, dessen Reduktion zum 4-Amino-1,2,5-trimethoxy-3,6-dimethyl-acridon (XIII) führte. Entmethylierung von (XIII) und anschließende Luftoxydation in alkalischer Lösung lieferte 2,5-Dihydroxy-3,6-dimethyl-acridonchinon-(1,4) (XIV), das mit Despeptido-actinomycin identisch ist.

Eine zweite Despeptido-actinomycin-Synthese, ausgehend von 3-Methoxy-4-methylanthranilsäure und 2-Methoxy-3-methylbenzochinon, wurde später von HANGER, HOWELL und JOHNSON *(83)* beschrieben.

## *Farbige Abbauprodukte der Säurehydrolyse.*

Nachdem sicher war, daß der Chromophor weder durch Alkali- noch durch Bariumhydroxyd unverändert vom Peptidteil abzulösen ist, und auch fermentative Abbauversuche erfolglos verlaufen waren, wurde erneut versucht, eine Abspaltung durch Säure zu erreichen. Da drastische Säureeinwirkung die Amino- und Methylaminosäuren zwar quantitativ abspaltet, den Chromophor aber zum großen Teil in braune, amorphe Produkte verwandelt, haben BROCKMANN und Mitarb. die Säurehydrolyse der Actinomycine systematisch untersucht, um zu sehen, ob man den Peptidteil auch unter milderen Bedingungen abspalten kann. Dabei ergab sich folgendes.

Erwärmt man ein Actinomycin mit 10%iger Salzsäure 4 Stunden auf 60°, so wird aus dem Chromophor eine Aminogruppe als Ammoniak abgespalten und durch eine Hydroxygruppe ersetzt, während der Peptidteil intakt bleibt. In fast quantitativer Ausbeute entsteht dabei ein kristallisiertes Desamino-actinomycin (vgl. S. 47) [BROCKMANN und FRANCK *(16, 17)*]. Damit war gezeigt: Durch längere Hydrolyse eines Actinomycins mit 10%iger Salzsäure wird man niemals den Actinomycin-Chromophor selbst, sondern bestenfalls nur den Chromophor des entsprechenden Desamino-actinomycins erhalten, eine Einschränkung, die für die Konstitutionsaufklärung des Chromophors jedoch bedeutungslos war. Denn da sich die Actinomycine von den Desamino-actinomycinen nur in einem Substituenten des Chromophors ($NH_2$ statt OH) unterscheiden, war die Konstitutionsaufklärung des Desamino-actinomycin-Chromophors gleichbedeutend mit der des Actinomycin-Chromophors.

Läßt man 10%ige Salzsäure 120 Stunden bei 80° auf ein Actinomycin einwirken, so werden aus dem zunächst auftretenden Desamino-actinomycin die Amino- und Methylaminosäuren nach und nach abgespalten und es entstehen (neben anderen) schließlich rote Abbauprodukte, die keine Amino- oder Methylaminosäuren mehr enthalten, im Spektrum aber noch den Desamino-actinomycinen ähnlich sind. Aus einem solchen Hydrolysat von ` Actinomycingemisch C* wurden zwei rote Abbauprodukte, das *Actinocinin* und das *Desamino-actinocinyl-threonin*, isoliert [BROCKMANN und GRÖNE (*26*, *27*)] und bald darauf zwei weitere, die im folgenden als „gelbes" und „rotes" Abbauprodukt bezeichnet sind [BROCKMANN und MUXFELDT (*45*)]. Die Untersuchung dieser vier Abbauprodukte hat, wie nun gezeigt werden soll, zur Konstitutionsaufklärung des Actinomycin-Chromophors geführt [BROCKMANN und MUXFELDT (*46*, *47*, *50*)].

Das nur in geringer Menge entstandene gelbe Abbauprodukt war 2,5-Dihydroxy-toluchinon (XV). Das rote, mit der Bruttoformel $C_{14}H_{11}NO_3$, enthielt zwei C-Methylgruppen und eine schwach saure, acetylierbare Hydroxygruppe. Mit Zinn-II-chlorid gab es ein blaues Semichinon und mit *o*-Phenylendiamin kondensierte es zu einem braunroten Phenazinderivat, das sich nicht acetylieren ließ. Da das rote Abbauprodukt seinem Oxydationspotential nach kein *o*-Chinon sein konnte, kam als Reaktionspartner für das *o*-Phenylendiamin nur eine Hydroxy-chinon-Gruppierung in Betracht.

Die sehr geringe Basizität des Abbauproduktes ließ darauf schließen, daß sein Stickstoffatom einem heterocyclischen Ring angehört.

Diese Befunde und die Annahme, daß das 2,5-Dihydroxy-toluchinon (XV) im Laufe der Säurehydrolyse aus dem roten Abbauprodukt entsteht, erlaubten, diesem die Teilformel (XVI) zuzuschreiben. Ihr entsprechend mußte der chinoide Ring *C* so mit dem Rest des Moleküls verknüpft sein, daß diese Verknüpfung bei Säurehydrolyse unter Bildung von (XV) gelöst werden kann.

Ersetzt man in (XVI) die beiden Methylgruppen durch Wasserstoff, so erhält man (XVIa), die Teilformel der Stammverbindung des roten Abbauproduktes. Auch für diese Stammverbindung waren die drei für (XVI) charakteristischen Umsetzungen mit a) Zinn-II-chlorid, b) *o*-Phenylendiamin und c) konz. Salzsäure zu erwarten, denn C-Methylgruppen haben auf diese Reaktionen keinen Einfluß. Damit aber lief die Strukturaufklärung von (XVIa) auf die Beantwortung folgender Frage hinaus: Welche heterocyclische Verbindung $C_{12}H_7NO_3$ gibt a) mit Zinn-II-chlorid ein blaues Semichinon, b) mit *o*-Phenylendiamin ein Phenazinderivat, und c) mit Säure 2,5-Dihydroxy-benzochinon (statt 2,5-Dihydroxytoluchinon wie beim roten Abbauprodukt)? Eine Durchmusterung

---

* Da Actinomycin $C_1$, $C_2$ und $C_3$, die Komponenten dieses Gemisches, den gleichen Chromophor enthalten, war zur Gewinnung von Desamino-actinomycin-Chromophor oder dessen Abbauprodukten eine vorherige Trennung der drei Komponenten nicht erforderlich.

heterocyclische Verbindungen hat gezeigt, daß es nur e i n e gibt, die alle drei Reaktionen eingeht; diese ist das schon lange bekannte 3-Hydroxy-phenoxazon-(2) (XVIIa). Sie war demnach als Stammverbindung des roten Abbauproduktes anzusehen und damit wurde dessen Teil-formel (XVI) zur Formel (XVII).

Welche von den acht nach Formel (XVII) möglichen Stellungen die beiden C-Methylgruppen des roten Abbauproduktes einnehmen, haben BROCKMANN und MUXFELDT (*46, 50*) — da für Abbauversuche zu wenig Material zur Verfügung stand — durch Synthese entschieden *(Formelübersicht 2)*. Als sie nämlich 2,5-Dihydroxy-toluchinon (XV) nacheinander mit den vier isomeren o-Amino-methylphenolen konden-sierten, erhielten sie aus (XV) und 3-Amino-2-hydroxy-toluol (XVIII) ein kristallisiertes 3-Hydroxy-dimethyl-phenoxazon, das mit dem roten Abbauprodukt identisch war.

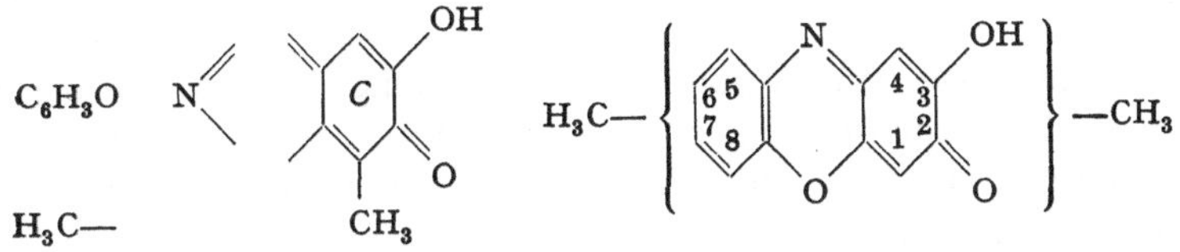

(XVI.)
(XVIa.) $CH_3 = H.$

(XVII.)
(XVIIa.) $CH_3 = H.$
3-Hydroxy-phenoxazon-(2).

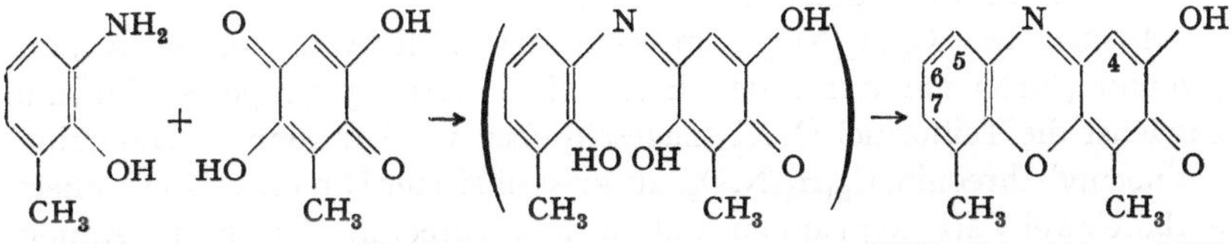

(XVIII.) 3-Amino-2-hydroxy-toluol.   (XV.) 2,5-Dihydroxy-toluchinon.

(XIX.) „Rotes Abbauprodukt".

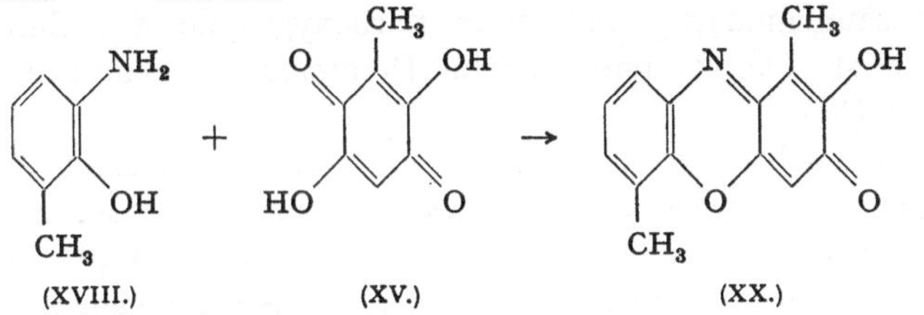

(XVIII.)   (XV.)   (XX.)

*Formelübersicht 2.* Synthese des „roten Abbauproduktes".

Bei dieser Kondensation könnten an sich zwei Verbindungen, nämlich (XIX) und (XX), entstehen, je nachdem, welche Chinoncarbonyl-gruppe von (XV) mit der Aminogruppe von (XVIII) kondensiert. Ver-suche, die hier nicht zu erörtern sind, haben gezeigt, daß erwartungsgemäß die nur von e i n e m o-Substituenten flankierte Carbonylgruppe in (XV) bevorzugt reagiert und das Kondensationsprodukt und damit auch das rote Abbauprodukt die Konstitution (XIX) hat.

Das rote Abbauprodukt (XIX), sein Monoacetat und sein blaues, semichinoides Zinn-II-chlorid-Reduktionsprodukt sind in ihrem Absorptionsspektrum den Desamino-actinomycinen bzw. deren Acetaten und grünen Reduktionsprodukten sehr ähnlich. Und ebenso wie die Desamino-actinomycine wird (XIX) von wäßrigem Alkali mit braunroter Farbe aufgenommen und dann schnell verändert. Da das rote Abbauprodukt (XIX) somit noch alle charakteristischen Eigenschaften des Desamino-actinomycin-Chromophors zeigt, andererseits aber nicht weiter abgebaut werden kann, ohne diese Eigenschaften zu verlieren, konnte als sicher gelten, daß man mit ihm das Kernstück des Desamino-actinomycin-Chromophors in Händen hatte. Als nächstes mußte nun geklärt werden, a) wie dieses Kernstück in den Desamino-actinomycinen mit dem Peptidteil verbunden ist, und b) ob sich bei der Säurehydrolyse außer diesem Peptidteil noch andere Gruppen vom Kernstück ablösen. Antwort auf diese Fragen gab die Untersuchung des Actinocinins und Desamino-actinocinyl-threonins, der beiden anderen roten Abbauprodukte der Säurehydrolyse. Beide sind spektroskopisch dem roten Abbauprodukt (XIX) und damit auch den Desamino-actinomycinen ähnlich, woraus zu schließen war, daß sie Derivate von (XIX) oder, anders ausgedrückt, Zwischenprodukte auf dem Wege vom Desamino-actinomycin zum roten Abbauprodukt (XIX) sind.

*Actinocinin*, $C_{15}H_{11}NO_5$, unterscheidet sich vom roten Abbauprodukt (XIX) nur durch den Besitz einer Carboxygruppe, so daß ihm zunächst die Teilformel (XXI) zugeschrieben werden konnte. Desamino-actinocinyl-threonin, $C_{20}H_{18}N_2O_9$, als kristallisierter Dimethylester isoliert, enthält zwei Carboxygruppen und ein Mol Threonin. Da dessen Aminogruppe im Desamino-actinocinyl-threonin nicht frei ist, war anzunehmen, daß sie säureamidartig mit einer Carboxygruppe des Chromophors verknüpft ist. Damit ergab sich für Desamino-actinocinyl-threonin die Teilformel (XXII).

Zunächst soll erörtert werden, wie beim Actinocinin (XXI) die Stellung der Carboxygruppe bewiesen wurde, für welche die Formel (XXI) vier Möglichkeiten offen ließ (an $C_{(4)}$, $C_{(5)}$, $C_{(6)}$ und $C_{(7)}$). Auch hier mußte die Synthese entscheiden, denn für Abbauversuche stand zu wenig Material zur Verfügung. Die Aussicht, dabei auf Anhieb das mit Actinocinin identische der vier Isomeren (XXI) zu gewinnen, wurde durch folgende Überlegungen vergrößert. Wenn Actinocinin (XXI) und Desamino-actinocinyl-threonin (XXII) Zwischenprodukte beim Abbau von Desamino-actinomycin zum roten Chromophor-Kernstück (XIX) sind, müssen aus (XXII) während der Säurehydrolyse die beiden Carboxygruppen (die freie und die mit Threonin verbundene) abgespalten werden,

(XXI.

HOOC—CH—NH—CO—
|
CHOH
|
CH₃

(XXII.)
Desamino-actinocinyl-threonin.

(XXIII.) Methoxy-actinocinin-methylester
(CO-Bande: 5,80 μ).

(XXIV.) Dimethoxy-dihydro-actinocinin-methylester
(CO-Bande: 5,89 μ).

(XXV.) 2-Amino-3-hydroxy-
4-methylbenzoesäure.

(XV.) 2,5-Dihydroxy-
toluchinon.

(XXVI.)
Actinocinin.

(XXVII.)

und zwar eine von ihnen leichter, denn sonst würde kein Actinocinin entstehen. Da Chinon-carbonsäuren relativ leicht zu decarboxylieren sind, war kaum zu bezweifeln, daß im Desamino-actinocinyl-threonin die eine der beiden Carboxygruppen an $C_{(4)}$ steht und diese beim Übergang in Actinocinin abgespalten wird.

Damit standen als Träger der Carboxygruppe des Actinocinins noch $C_{(5)}$, $C_{(6)}$ und $C_{(7)}$ zur Diskussion. Die Auswahl unter ihnen wurde durch die Beobachtung erleichtert, daß Methoxy-actinocinin-methylester eine Estercarbonyl-Bande bei 5,80 μ hat, während die Estercarbonyl-Bande des Dimethoxy-dihydro-actinocinin-methylesters bei 5,89 μ liegt. Eine Erklärung für diese Differenz bot die Annahme, daß die Carboxygruppe des Actinocinins an $C_{(5)}$ steht und der Methoxy-actinocinin-methylester sowie der Dimethoxy-dihydro-actinocinin-methylester dementsprechend nach (XXIII) bzw. (XXIV) zu formulieren sind. Denn dann würde die längerwellige Lage der Carbonylbande von (XXIV) durch Chelierung bedingt sein. Dieser Annahme gemäß wurde als erste durch Kondensation von 2-Amino-3-hydroxy-4-methylbenzoesäure (XXV) mit 2,5-Dihydroxy-toluchinon-(1,4) (XV) die 3-Hydroxy-1,8-dimethyl-phenoxazon-(2)-carbonsäure-(5) (XXVI) aufgebaut, die tatsächlich mit Actinocinin identisch ist.

Daß das Kondensationsprodukt nicht die Konstitution (XXVII) hat, zeigte sein Verhalten gegen siedende Salzsäure, denn dabei wurde es zu (XIX) abgebaut. Abgesehen davon war die Entstehung von (XXVII) auch deshalb unwahrscheinlich, weil bei der Kondensation von (XV) mit (XVIII) praktisch nur die von e i n e m o-Substituenten flankierte Carbonylgruppe von (XV) reagierte und nicht einzusehen war, warum sie sich bei der Kondensation mit (XXV) anders verhalten sollte.

COOH
|
CH₃—CHOH—CH
|
NH
|
C=O    COOH

(XXVIII.) Desamino-actinocinyl-L-threonin.

Mit dem Beweis, daß Actinocinin die Konstitution (XXVI) hat, war gleichzeitig auch bewiesen, daß die ausgeklammerten Substituenten der Desamino-actinocinyl-threonin-Formel (XXII) an $C_{(4)}$ und $C_{(5)}$ stehen. Wie der Vergleich mit einem synthetischen Präparat später zeigte, ist die $C_{(5)}$-Carboxygruppe Träger des Threoninrestes, sodaß Desamino-actinocinyl-threonin die Konstitution (XXVIII) hat [Brockmann und Mecke (40)].

Bestätigt und ergänzt wurden diese Befunde später dadurch, daß Brockmann und Petras (55) aus dem HBr-Hydrolysat von Actinomycin-gemisch C ein Abbauprodukt $C_{24}H_{25}N_3O_{11}$ isolierten und durch Synthese als Desamino-actinocinyl-bis-[L-threonin] (XXXIII) identifizierten.

### Konstitution des Actinomycin-Chromophors.

Von den beiden oben gestellten Fragen: a) wie das Kernstück (XIX) des Desamino-actinomycin-Chromophors mit dem Peptidteil verbunden ist und b) ob am Desamino-actinomycin-Chromophor außer dem Peptidteil noch andere Gruppen hängen, wird die erste durch die Konstitutionsformel von Desamino-actinocinyl-L-threonin (XXVIII) und Desamino-actinocinyl-bis-[L-threonin] (XXXIII) dahingehend beantwortet, daß der Chromophor über zwei Carboxygruppen an $C_{(4)}$ und $C_{(5)}$ mit dem Peptidteil verbunden ist.

Zur zweiten Frage ließ sich folgendes sagen. Die Abbauprodukte (XIX), (XXVI), (XXVIII) und (XXXIII) sind, wie erwähnt, in ihren Absorptionsspektren sehr ähnlich und das gleiche gilt für ihre blauen bzw. grünen semichinoiden Reduktionsprodukte. Wäre in den Desamino-actinomycinen

COOCH₃ / NH₂ ... (chemical structures)

(XXIX.)      (XXIX.)      (XXX.) Actinocin-dimethylester.      (XXXI.)
(XXIXa.) COOCH₃ = COOH.      (XXXa.) Actinocin. COOCH₃ = COOH.
(XXIXb.) COOCH₃ = CO · NH · CH₂ · COOCH₃.      (XXXb.) COOCH₃ = CO · NH · CH₂ · COOCH₃.

(XXXII.) Actinocinyl-bis-[L-threonin]-dimethylester.

(XXXIII.) Desamino-actinocinyl-bis-[L-threonin].

(XXXIV.) Desamino-actinomycine.
(XXXIVa.) Peptid = OH. Desamino-actinocin.      (XXXV.)

der Chromophor über $C_{(6)}$ oder (und) $C_{(7)}$ mit Resten verknüpft, die in den Abbauprodukten (XXVIII) und (XXXIII) nicht mehr vorhanden sind, so müßten diese Reste so beschaffen sein, daß sie im Absorptionsspektrum nicht in Erscheinung treten und bei Säurehydrolyse durch Wasserstoff ersetzt werden. Da es keine Substituenten gibt, die beide Bedingungen erfüllen, konnte als gesichert gelten, daß die 3-Hydroxy-1,8-dimethyl-phenoxazon-(2)-dicarbonsäure-(4,5) (XXXIVa) der Chromophor der Desamino-actinomycine ist und diese daher nach (XXXIV) zu formulieren sind. Und da die Desamino-actinomycine aus den Actinomycinen dadurch entstehen, daß im Chromophor eine Aminogruppe hydrolytisch durch eine Hydroxygruppe ersetzt wird, ergab sich aus (XXXIV) für die Actinomycine die Teilformel (XXXV). Ihr entsprechend ist der Chromophor der Actinomycine die 3-Amino-1,8-dimethyl-phenoxazon-(2)-dicarbonsäure-(4,5) (XXXa), die man als Actinocin bezeichnet hat (50). Dementsprechend wurde der Chromophor der Desamino-actinomycine Desamino-actinocin genannt (50).

## Synthese des Actinomycin-Chromophors.

Die Ergebnisse der Abbauversuche wurden durch die Synthese des Actinocin-dimethylesters bestätigt [BROCKMANN und MUXFELDT (47, 50)]. Sie gelang durch oxydative Kondensation von 2-Amino-3-hydroxy-4-methyl-benzoesäure-methylester (XXIX). Daß das dunkelrote, kristallisierte Kondensationsprodukt die Konstitution (XXX) hat und nicht das isomere Phenoxazim-derivat (XXXI) ist, zeigte sich beim Kochen mit Salzsäure. Denn dabei wurde das Kondensationsprodukt zu Actinocinin (XXVI, S. 25) abgebaut, was bei (XXXI) nicht möglich wäre. Wie dieser Abbau zum Actinocinin zeigt, wird den obigen Überlegungen (S. 25) entsprechend die im chinoiden Ring stehende Carboxygruppe am leichtesten als Kohlendioxyd abgespalten.

In analoger Weise wie (XXIX) kann man auch die Säure (XXIXa) oxydativ kondensieren und so zum Actinocin (XXXa) selbst kommen. Es zeichnet sich dadurch aus, daß es in den meisten organischen Solvenzien unlöslich ist.

Die Absorptionskurve des Actinocin-dimethylesters (XXX) ist derjenigen der Actinomycine zwar ähnlich, doch liegt das Hauptmaximum um 12 m$\mu$ kürzerwellig. Dieser Unterschied kommt daher, daß die Carboxygruppen in (XXX) verestert, in den Actinomycinen dagegen säureamidartig mit dem Peptidteil verbunden sind. Denn der durch oxydative Kondensation von N-[2-Amino-3-hydroxy-4-methylbenzoyl]-glycinmethylester (XXIXb) bereitete kristallisierte, gelbrote Actinocinyl-bis-[glycinmethylester] (XXXb) stimmt im Spektrum bis auf eine geringe Differenz in der Extinktion der Maxima mit den Actinomycinen überein. Völlige Übereinstimmung seiner molaren Absorptionskurve mit den

Kurven der Actinomycine zeigte der Actinocinyl-bis-[L-threoninmethyl-ester] (XXXII), der analog zu (XXXb) durch oxydative Kondensation von N-[2-Amino-3-hydroxy-4-methylbenzoyl]-L-threoninmethylester dargestellt wurde (*104*).

Da 10%ige Salzsäure die Actinomycine schnell in Desamino-actinomycine verwandelt, hielt man es zunächst für unmöglich, durch Säurehydrolyse zu einem Abbauprodukt der Actinomycine zu kommen, das noch den unveränderten Chromophor enthält. Später hat sich herausgestellt, daß dies doch gelingt, und zwar dann, wenn mit konz. Salzsäure abgebaut wird. Dabei bleibt, offenbar weil hier das Kation (XXXVIa) ↔ (XXXVIb) vorliegt, die Chromophor-Aminogruppe intakt, und aus dem Hydrolysat ließ sich Actinocinyl-bis-[L-threonin] als kristallisierter Dimethylester (XXXII, S. 27) isolieren [BROCKMANN und BOLDT (*12*)]. Dieser Befund schließt die Beweiskette für die Konstitution des Actinomycin-Chromophors. Ferner zeigt er ebenso wie der Abbau zu Desamino-actinocinyl-bis-[L-threonin] (XXXIII) (S. 27) eine neue Möglichkeit, noch nicht untersuchte Actinomycine daraufhin zu prüfen, ob sie den gleichen Chromophor enthalten wie die Actinomycine $C_1$, $C_2$, $C_3$, $X_{0\beta}$ und $X_2$.

### Reaktionen des Actinomycin-Chromophors.

Mit der Konstitutionsaufklärung des Actinomycin-Chromophors sind verschiedene, vorher nicht zu deutende Reaktionen der Actinomycine verständlich geworden (*20*). So kommt in der schwachen Basizität und der leichten Abspaltung der Chromophor-aminogruppe ihr Amidcharakter zum Ausdruck, der neben sterischer Hinderung und Wasserstoffbrückenbildung mit dafür verantwortlich zu machen ist, daß die Aminogruppe mit den üblichen Methoden nicht zu acetylieren ist.

Die schwach basischen, in Lösung gelbroten Actinomycine bilden mit konz. Mineralsäuren Salze, deren Lösungen weinrot sind. Diese für 3-Amino-phenoxazone-(2) charakteristische Halochromie beruht auf Bildung des resonanz-stabilisierten Kations (XXXVIa) ↔ (XXXVIb). Wäre die Aminogruppe des Chromophors der Protonen-Acceptor, so dürfte keine Halochromie auftreten.

Actinomycine lösen sich im wässerigen Alkali (dem z. B. Alkohol als Lösungsvermittler zugesetzt ist) mit gelbbrauner Farbe, die schnell verblaßt. Beim Ansäuern wird die Lösung wieder gelbbraun, und man kann Actinomycin aus ihr zurückgewinnen, sofern das Alkali nicht zu lange eingewirkt hat. Diese Ausbleichreaktion läßt sich spektroskopisch verfolgen, weil eine anfangs vorhandene Absorptionsbande bei 444 m$\mu$ allmählich abgebaut wird, während eine neue bei 343 m$\mu$ erscheint [ANGYAL und Mitarb. (*1*)]. Die Reaktion ist zweiter Ordnung, woraus sich schließen läßt, daß die Entfärbung nicht allein auf Ionisierung,

(XXXVIa.)                          (XXXVIb.)

(XXXVII.) $+ OH^{\ominus}$ / $+ H^{\oplus}$ (XXXVIII.) $+ OH^{\ominus}$ / $+ H^{\oplus}$

(XXXIX.) $+ H_2O$          (XL.) Peptid A.   (XLI.) Peptid B.   (XLII.) 7-Methyl-benzoxazolon-carbonsäure-(4).

(XLIII.)                          (XLIV.)

(XLV.)

sondern auch auf Anlagerung einer OH-Gruppe beruht. Die dabei zunächst entstehende Verbindung (XXXVIII) geht schnell in das Chinonimin-Derivat (XXXIX) über, das relativ kurzwellig absorbiert, weil sein Molekül wegen der sperrigen Substituenten am Stickstoff und in o-Stellung zur Iminogruppe nicht planar ist (1).

Bei der Oxydation von Actinomycinen mit alkalischem Wasserstoff-peroxyd (2) wurden zwei als Peptid A (XL) und B (XLI) bezeichnete Abbauprodukte gefaßt, von denen A durch Hydrolyse zu 7-Methyl-benzoxazolon-carbonsäure-(4) (XLII) abgebaut werden konnte [BULLOCK und JOHNSON (62)]. Aus Formel (XXXIX) folgt, daß im Abbauprodukt (XLII) der benzoide Ring des Actinomycin-Chromophors vorliegt.

Ohne Analogie ist bisher die merkwürdige Umlagerung des Actino-mycin-Chromophors in Despeptido-actinomycin. Sie beginnt zweifellos mit einer Öffnung des heterocyclischen Ringes unter Bildung von (XLIII), das sich wahrscheinlich mit (XLIV) ins Gleichgewicht setzt. In (XLIV) kondensiert dann offenbar die Carboxygruppe des benzoiden Ringes mit dem chinoiden Ring unter Abspaltung der Peptidreste und der anderen Carboxygruppe. Dabei entsteht zunächst (XLV) und daraus durch Verseifung der Iminogruppe Despeptido-actinomycin [BROCKMANN und MUXFELDT (47)]. Die Chromophor-Aminogruppe muß bei der Umlagerung eine Rolle spielen, denn die Desamino-actinomycine bilden keine Despeptido-actinomycine. Die Möglichkeit, daß der Ringschluß im Sinne einer Dieckmann-Cyclisierung verläuft, ist von HANGER, HOWELL und JOHNSON (83) diskutiert worden.

Zum Schluß dieses Abschnittes soll darauf hingewiesen werden, daß das 2-Amino-phenoxazon-(3)-Ringsystem Hauptbestandteil der als Insektenfarbstoffe weit verbreiteten Ommochrome ist, die BUTENANDT und Mitarb. (64) in ausgedehnten Untersuchungen aufgeklärt haben. Zu erwähnen ist ferner, daß Cinnabarin (Polystictin), ein aus *Coriolus sanguineus* und *Trametes cinnabaria* (Pilzarten, die auf faulendem Holz wachsen) isolierter roter Farbstoff, kürzlich als 4-Carboxy-3-amino-5-hydroxymethyl-phenoxazon-(2) identifiziert wurde [CAVILL, CLEZY, TETAZ und WERNER (65)].

## 2. Die Konstitution des Peptidteils.

Die Actinomycine $C_1$, $C_2$ und $C_3$ haben keine Carboxy-, Hydroxy-, Methoxy- oder Äthoxy-Gruppen und keine aliphatischen Amino-, Imino- oder Methylamino-Gruppen [BROCKMANN und FRANCK (17)], ein Er-gebnis, das folgende Aussagen über die Struktur des Peptidteils erlaubte: a) Beide Carboxygruppen des Chromophors (Actinocins) (XXXa, S. 27) sind mit Peptidketten verknüpft, b) die C-terminale Aminosäure jeder Peptidkette ist mit der Hydroxygruppe eines Threoninrestes verestert. Damit ergaben sich für die drei Actinomycine — unter der Annahme, daß die beiden Threoninreste nicht in derselben Kette stehen — je nachdem, ob das endständige Glied der Peptidkette mit dem Threonin der gleichen Kette oder dem der anderen verestert ist, die Grundformeln (XLVI) und (XLVII).

Mit (XLVI) und (XLVII) stand die Bilanz der Actinomycin-Abbauprodukte in bestem Einklang [BROCKMANN und FRANCK (21)]. Am Beispiel des Actinomycins $C_3$ zeigt dies *Tabelle 4*. Sie enthält die Summenformel a) der Amino- und Methylaminosäuren, b) des Ammoniaks, das bei Säurehydrolyse aus dem Chromophor abgespalten wird und c) des Desamino-actinomycin-Chromophors (XXXIVa, S. 27) (Desamino-actinocin), der in Form von Desamino-actinocinyl-threonin (XXVIII, S. 26) isoliert wurde (S. 22).

Tabelle 4. Bilanz der Actinomycin $C_3$-Abbauprodukte (21).

2 Mol Threonin .................. $C_8 H_{18} N_2 O_6$
2 Mol *allo*-Isoleucin.............. $C_{12}H_{26} N_2 O_4$
2 Mol Prolin.................... $C_{10}H_{18} N_2 O_4$
2 Mol Sarkosin.................. $C_6 H_{14} N_2 O_4$
2 Mol N-Methylvalin ............ $C_{12}H_{26} N_2 O_4$
1 Mol Ammoniak................. $H_3$ N
1 Mol Desamino-actinocin ........ $C_{16}H_{11} N O_7$
$$C_{64}H_{116}N_{12}O_{29}$$
— 13 $H_2O$     $H_{26}$      $O_{13}$
Actinomycin $C_3$...... $C_{64}H_{90} N_{12}O_{16}$ (Mol.-Gew. 1284)

*Literaturverzeichnis: SS. 48—54.*

Ließe sich die hydrolytische Spaltung des Actinomycins $C_3$ umkehren und aus den Abbauprodukten der Tabelle 4 unter Wasserabspaltung wieder Actinomycin $C_3$ aufbauen, so würden bei der Verknüpfung der Amino- und Methylaminosäuren zu zwei Peptidketten 8 Mol $H_2O$, bei der Kondensation der beiden Peptidketten mit dem Chromophor 2 Mol $H_2O$, bei der Lactonisierung von zwei Carboxygruppen nochmals 2 Mol $H_2O$ und bei der Umsetzung des Ammoniaks mit der Hydroxygruppe des Desamino-actinocins (XXXIVa) zum Actinocin (XXXa, S. 27) 1 Mol $H_2O$, insgesamt also 13 Mol $H_2O$ abgespalten. Subtrahiert man von der Gesamtsummenformel $C_{64}H_{116}N_{12}O_{29}$ aller Abbauprodukte 13 $H_2O$, so erhält man $C_{64}H_{90}N_{12}O_{16}$ als Bruttoformel des Actinomycins $C_3$. Auf diese Formel passen nicht nur die Analysen- und Mol.-Gew.-Zahlen von Actinomycin $C_3$, sondern entsprechend umgerechnet auch die von Desamino-actinomycin $C_3$, woraus zu schließen war, daß man in den Abbauprodukten der Tabelle 4 alle Teile des Actinomycin $C_3$-Moleküls in Händen hatte.

Die Grundformel (XLVI) bzw. (XLVII, S. 32) durch Ermittlung der *Aminosäure-Sequenz* zu einer Strukturformel zu entwickeln, gelang zuerst BROCKMANN und Mitarb. beim Actinomycin $C_3$. Da die hierbei benutzten Abbauverfahren auch bei den anderen Actinomycinen zu verwenden sind, genügt es, sie am Beispiel des Actinomycins $C_3$ zu erläutern.

## Actinomycin $C_3$.

Der Beweis, daß Actinomycin $C_3$ zwei Lactongruppen enthält, ließ sich, da Estergruppen nicht nachzuweisen waren, durch milde Alkalihydrolyse erbringen. Dabei wurden 2 Äquivalente Natriumhydroxyd verbraucht, und es entstand, ohne daß Aminosäuren oder Ammoniak abgespalten wurde, eine als Actinomycin $C_3$-Säure bezeichnete Verbindung, der man die Formel (XLVIII, S. 34) zuschreiben konnte [BROCKMANN und FRANCK (*19*)].

Erhitzen der Actinomycin $C_3$-Säure mit Acetanhydrid-Pyridin (*106*) (Dakin-West-Reaktion) gab ein Reaktionsprodukt, in dessen Totalhydrolysat N-Methylvalin fehlte, während Threonin, *allo*-Isoleucin, Prolin und Sarkosin nachzuweisen waren. Damit war gezeigt, daß *L*-N-Methylvalin in beiden Peptidketten der Actinomycin $C_3$-Säure C-terminal, und seine Carboxygruppe demnach im Actinomycin $C_3$ als Lacton-Gruppe vorliegt [BROCKMANN und FRANCK (*19*)]. Zum gleichen Ergebnis kam man durch Reduktion des Actinomycins $C_3$ mit Lithiumborhydrid, denn im Totalhydrolysat des Reduktionsproduktes fand sich statt N-Methylvalin das N-Methyl-valinol [BROCKMANN und DÖRING (*15*)].

Aufschluß über die dem N-Methylvalin benachbarte Aminosäure brachte die Hydrazinolyse des Actinomycins $C_3$, bei der 1,1—1,2 Mol N-Methylvalyl-sarkosinanhydrid gefaßt wurde [BROCKMANN, BOHNSACK und SÜLING (*11*)]. Da dieses keine Gruppen hat, mit denen es im

Actinomycinmolekül verankert sein könnte, mußte es während der Hydrazinolyse aus Sarkosyl-N-methylvalin oder einem Vorprodukt davon entstanden sein. Das aber bedeutete: N-Methylvalin ist im Peptidteil mit Sarkosin verknüpft, und zwar in beiden Peptidketten, denn nur dann kann N-Methylvalyl-sarkosinanhydrid in Ausbeuten von mehr als 1 Mol entstehen. Damit ergab sich für Actinomycin C$_3$ die Partial-Sequenzformel (IL), in der die Stellung des einen Threoninrestes durch die Desamino-actinocinyl-threonin-Formel (XXVIII, S. 26) bewiesen war, und in der offen gelassen ist, ob das N-Methylvalin mit der Hydroxygruppe des Threonins nach (XLVI) oder (XLVII, S. 32) verestert ist.

Da in Salzsäure-Hydrolysaten von Actinomycin C$_3$ *allo*-Isoleucyl-prolyl-sarkosin nachgewiesen wurde (*4*) und bei der Hydrazinolyse neben N-Methylvalyl-sarkosinanhydrid auch Prolyl-sarkosinanhydrid entstand (*11*), mußte Sarkosin mindestens in e i n e r Peptidkette über seine Methylaminogruppe mit Prolin und dieses über seine Iminogruppe mit *allo*-Isoleucin verbunden sein. Diese Befunde sowie die Annahme, daß a) die Lactonringe so angeordnet sind wie in (XLVI) und b) Actinomycin C$_3$ in vivo durch oxydative Kondensation aus zwei Molekülen eines 2-Amino-3-hydroxy-4-methylbenzoyl-peptidlactons (LII) oder einer Vorstufe davon entsteht und daher beide Peptidketten die gleiche Aminosäure-Sequenz haben, führten zum Vorschlag der Actinomycin C$_3$-Formel (LI) [BROCKMANN und Mitarb. (*9*)].

Um die so abgeleitete Aminosäure-Sequenz eindeutig zu beweisen, mußte gesichert werden, daß a) b e i d e Threoninreste am Chromophor stehen, b) b e i d e *allo*-Isoleucinreste mit Threonin verknüpft sind, und c) an b e i d e n *allo*-Isoleucinresten Prolin hängt.

Daß beide Carboxygruppen des Actinomycin-Chromophors Träger von *L*-Threoninresten sind, zeigte die auf S. 26 erwähnte Isolierung von Desamino-actinocinyl-bis-[*L*-threonin] (XXXIII, S. 27) [BROCKMANN

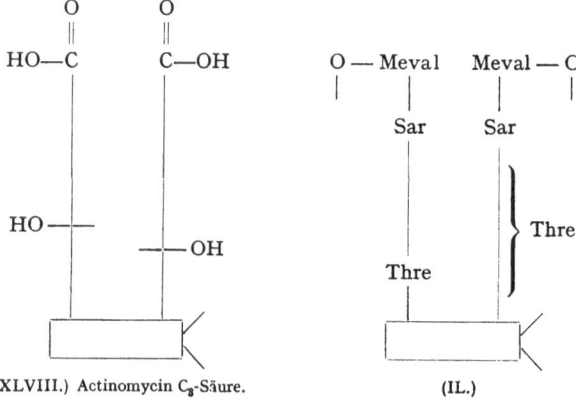

(XLVIII.) Actinomycin C$_3$-Säure.    (IL.)

*Literaturverzeichnis: SS. 48—54.*

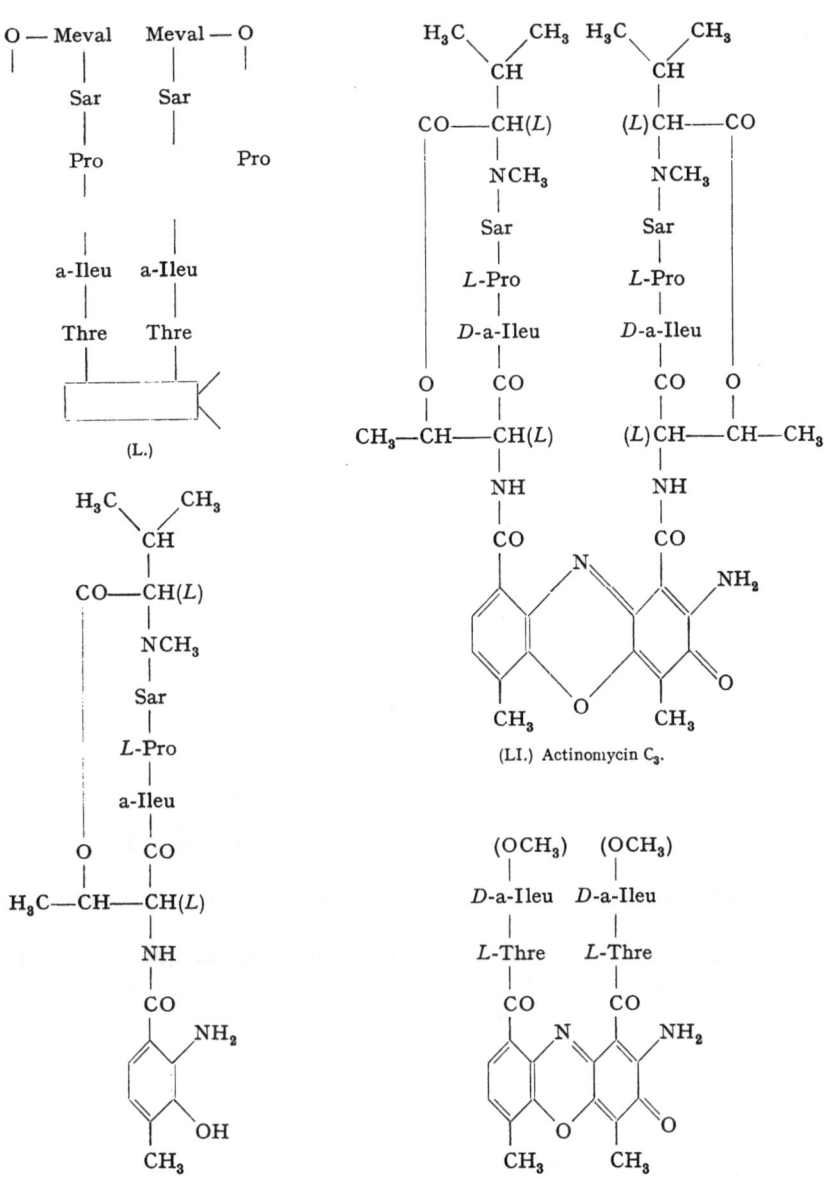

(LI.) Actinomycin C₃.

(LII.) 2-Amino-3-hydroxy-
4-methylbenzoyl-peptidlacton.

(LIII.) Actinocinyl-bis-[L-threonyl-D-*allo*-isoleucin-methylester].

und PETRAS (55)] und Actinocinyl-bis-[L-threonin-methylester] (XXXII)
[BROCKMANN und BOLDT (12)] aus Actinomycin C₃-Hydrolysaten. Und
daß beide Threoninreste mit *allo*-Isoleucin verbunden sind, ergab sich
aus der Untersuchung eines Abbauproduktes, das aus einem Salzsäure-

Hydrolysat von Actinomycin $C_3$ als Dimethylester isoliert und durch Vergleich mit einem synthetischen Präparat als Actinocinyl-bis-[L-threonyl-D-*allo*-isoleucin-methylester] (LIII) identifiziert wurde (*12*).

*Formelübersicht 3.*
Synthese des Actinocinyl-bis-[L-threonyl-D-*allo*-isoleucin-methylesters].

Die Synthese von (LIII) [Brockmann und Mecke (*40*)] ging aus vom L-Threonyl-D-*allo*-isoleucin-methylester (LV), den man mit 2-Nitro-3-hydroxy-4-methylbenzoyl-chlorid (LIV) zum N-[2-Nitro-3-hydroxy-4-methylbenzoyl]-L-threonyl-D-*allo*-isoleucin-methylester (LVI) umsetzte *(Formelübersicht 3)*. Katalytische Hydrierung von (LVI) führte zur Aminoverbindung (LVII), deren oxydative Selbstkondensation das gesuchte Actinocinyl-peptid (LVIII) lieferte.

Wie Formel (L) (S. 35) zeigt, war jetzt nur noch fraglich, ob der zweite Prolinrest zwischen Sarkosin und D-*allo*-Isoleucin in der rechten Kette steht (gleiche Zusammensetzung der Peptidketten) oder zwischen D-*allo*-Isoleucin und L-Prolin in der linken Kette. Die Entscheidung brachte eine kombinierte Alkali- und Säurehydrolyse, bei der es gelang, eine Peptidkette (wahrscheinlich die am chinoiden Ring) selektiv vom Chromophor abzuspalten [Brockmann und Sunderkötter (*58*)]. Das dabei anfallende gelbrote Abbauprodukt enthielt je ein Mol Threonin, *D-allo*-Isoleucin, Prolin, Sarkosin und N-Methylvalin — ein Beweis, daß beide Peptidketten die gleichen Aminosäuren enthalten und der zweite Prolinrest demnach in der rechten Kette von Formel (L) steht. Damit war die Aminosäure-Sequenz (LIX) der Actinomycin C$_3$-Formel (LI) einwandfrei bewiesen.

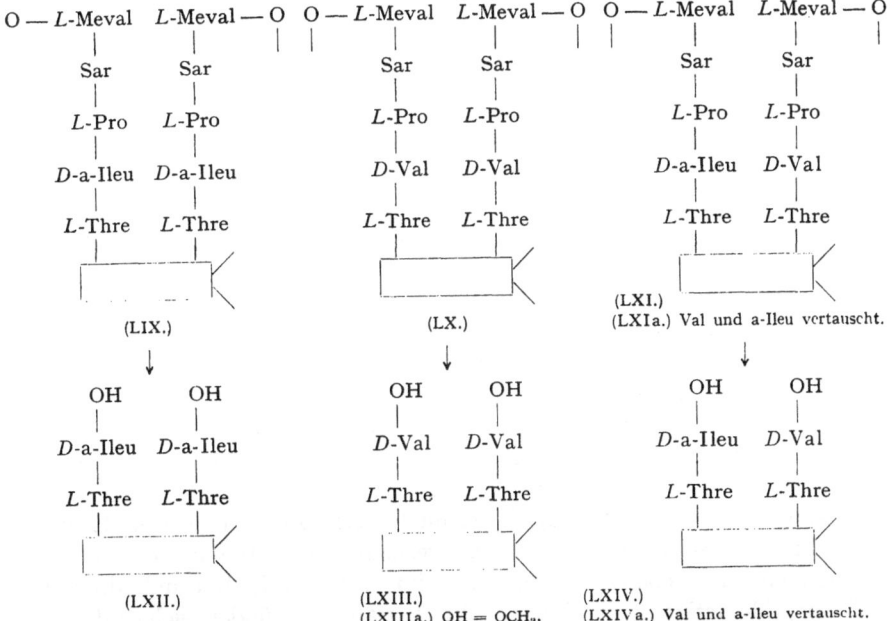

Noch nicht bewiesen ist dagegen die Lage der Lactonbrücken, die auch so sein könnte wie in Formel (XLVII) (S. 32). In diesem Fall bestünde das Actinomycin C$_3$ aus einem großen Peptidring, der aus zwei gleich strukturierten, über zwei Lactongruppen verbundenen Hälften aufgebaut ist und als „Brücke" den Chromophor enthält.

Zwischen beiden Formeln ließe sich entscheiden, wenn man den Chromophor oxydativ in zwei Teile aufspalten könnte, ohne dabei die Lactonbindung zu lösen. Gilt Formel (XLVI) bzw. (LI), so müßten bei einer solchen Spaltung zwei Peptidlacton-Derivate entstehen. Auch wenn sich nur eines von ihnen fassen und durch Aminosäure-Analyse charakterisieren ließe, wäre damit bewiesen, daß die Lactongruppen so angeordnet sind wie in (LI). Im anderen Fall (Formel XLVII) wäre bei Spaltung des Chromophors ein Abbauprodukt zu erwarten, das zwei Lactongruppen und zwei Mol Threonin, *D-allo*-Isoleucin, Prolin, Sarkosin und N-Methylvalin enthält.

Eine andere Möglichkeit, zwischen (XLVI) und (XLVII) zu entscheiden, wäre gegeben, wenn man 2 Moleküle (LII) oxydativ zu (LI) kondensieren und zeigen könnte, daß (LI) mit Actinomycin $C_3$ identisch ist.

### Actinomycin $C_1$ $(I_1, X_1, D)$.

Nachdem die Actinomycin $C_3$-Formel (LI) (S. 35) bekannt war, lag die Annahme nahe, daß Actinomycin $C_1$, das D-Valin statt D-allo-Isoleucin enthält (vgl. Tabelle 3, S. 16), nach (LX) zu formulieren ist. Daß dies zutrifft, konnte auf gleichem Wege bewiesen werden wie beim Actinomycin $C_3$. Abbau der Actinomycin $C_1$-Säure nach Dakin-West zeigte, daß die beiden N-Methylvalinreste C-terminal sind (4). Hydrazinabbau gab 1,1—1,2 Mol L-N-Methylvalyl-sarkosinanhydrid und 0,7 Mol L-Prolyl-sarkosinanhydrid [BROCKMANN, BOHNSACK und SÜLING (11)]. Hydrolyse mit konz. Salzsäure schließlich lieferte ein 2 Mol Threonin und 2 Mol Valin enthaltendes Abbauprodukt, dessen Dimethylester durch Vergleich mit einem synthetischen Präparat [BROCKMANN, BOLDT und PETRAS (13)] als Actinocinyl-bis-[L-threonin-D-valinmethylester] (LXIIIa) identifiziert wurde. Unter der Annahme, daß in jeder Peptidkette wie beim Actinomycin $C_3$ nur ein Prolinrest vorkommt, ergab sich aus diesen Befunden für Actinomycin $C_1$ die Aminosäure-Sequenz (LX, S. 37).

Die gleiche Sequenz (LX) haben BULLOCK und JOHNSON (63), nachdem die Actinomycin $C_3$-Formel (LI, S. 35) bekannt war, für ein mit dem Buchstaben D [VINING und WAKSMAN (109); MANAKER, GREGORY, VINING und WAKSMAN (92)] bezeichnetes Actinomycin abgeleitet, das in den $R_F$-Werten und im Aminosäuregehalt mit Actinomycin $C_1$ übereinstimmt und ebenso wie dieses beim Abbau mit Bariumhydroxyd Despeptido-actinomycin liefert. Ihr Verfahren gründet sich auf die Beobachtung, daß Actinomycine (LXV) durch alkalisches Wasserstoffperoxyd unter Verseifung der Lactongruppen in zwei Bruchstücke,

(LXV.) Actinomycine.  (LXVI.) Peptid A.  (LXVII.) Peptid B

Peptid A (LXVI) und Peptid B (LXVII), gespalten werden, von denen jedes eine Peptidkette des Actinomycins enthält. Aus A und B entstanden bei Totalhydrolyse: Threonin, Valin, Prolin, Sarkosin und N-Methylvalin — ein Beweis, daß die beiden Peptidketten des Actinomycins qualitativ und quantitativ die gleiche Zusammensetzung haben.

Daß in Peptid A und Peptid B N-Methylvalin C-terminal und über seine Methylaminogruppe mit Sarkosin verbunden ist (LXVI und LXVII), zeigte: a) die Dakin-West-Reaktion und b) die Pyrolyse, bei der aus beiden Peptiden N-Methylvalyl-sarkosinanhydrid entstand. Die Sequenz von Threonin, Valin und Prolin (LXVI und LXVII) wurde durch Partialhydrolyse ermittelt. In einem $n$-Salzsäure-Hydrolysat (2 Stunden, 100°) von Peptid A waren papierchromatographisch reichliche Mengen N-Methylvalin, Sarkosin und Prolin nachzuweisen, jedoch nur Spuren von Valin und kein Threonin. Längere Hydrolyse setzte mehr Valin frei und daneben auch Spuren von Threonin. Der Beweis für die Sequenz $R_1$-Thre-Val-Pro in (LXVI) gründet sich also darauf, daß unter den angewandten Bedingungen $R_1$ [Rest der 7-Methylbenzoxazolon-carbonsäure-(4), (XLII, S. 30)] nicht vom Threonin gelöst wird und die Bindung zwischen Threonin und Valin fester ist als zwischen Valin und Prolin. Für Peptid B, dessen Aminogruppe mit Oxalsäure verknüpft ist, wurde in analoger Weise die Sequenz (LXVII) abgeleitet.

Nach BULLOCK und JOHNSON (63) hat ihr Actinomycin D den gleichen Chromophor und die gleiche Aminosäure-Sequenz wie das bereits 1953 isolierte (57) Actinomycin $C_1$, ein Beweis, daß die beiden auch in den $R_F$-Werten übereinstimmenden Actinomycine (vgl. S. 12) identisch sind.

In der Konstitutionsformel, die BULLOCK und JOHNSON für ihr Actinomycin vorschlagen, liegen die Lactongruppierungen so wie in der Actinomycin $C_3$-Formel (LI). Befunde, die eine der Formel (XLVII) (S. 32) entsprechende Anordnung der Lactongruppen ausschließen, fehlen ebenso wie beim Actinomycin $C_3$.

### Actinomycin $C_2$.

Da Actinomycin $C_2$ 1 Mol Valin und 1 Mol *allo*-Isoleucin enthält und dadurch eine Mittelstellung zwischen Actinomycin $C_1$ (2 Mol Valin) und $C_3$ (2 Mol *allo*-Isoleucin) einnimmt, lag es nahe, bei ihm die Sequenz (LXI) oder (LXI a) (S. 37) zu vermuten. Daß diese Vermutung richtig war, wurde in gleicher Weise bewiesen wie beim Actinomycin $C_1$ und $C_3$; d. h. a) durch Reduktion des C-terminalen N-Methylvalins zu N-Methyl-valinol [BROCKMANN und DÖRING (15)], b) durch Hydrazinolyse, bei der mehr als 1 Mol N-Methylvalyl-sarkosinanhydrid entstand [BROCKMANN, BOHNSACK und SÜLING (11)], c) durch Abbau mit konz. Salzsäure zu einem Actinocinyl-peptid, das entweder nach (LXIV) oder (LXIVa) zu formulieren ist [BROCKMANN, BOLDT und PETRAS (13)], und d) durch

Isolierung eines bei kombinierter Alkali- und Säurehydrolyse entstehenden Abbauproduktes, das nur noch eine Peptidkette des Actinomycins $C_2$ und in dieser 1 Mol Prolin enthält [Brockmann und Döring (*15*)].

Zwischen den Actinomycin $C_2$-Formeln (LXI) und (LXIa) (S. 37) konnte noch nicht entschieden werden. An dieser Alternative wird sichtbar, wo die vorstehend beschriebenen Methoden zur Aufklärung des Peptidteils ihre Grenze erreichen. Bei allen Actinomycinen mit gleichen Peptidketten [*iso*-Actinomycine (*50*)] ermöglichen sie eine vollständige Sequenzanalyse. Bei ungleichen Peptidketten dagegen [*aniso*-Actinomycine (*50*)] lassen sie offen, welche Kette am benzoiden und welche am chinoiden Ring des Chromophors hängt. Beantwortet würde diese Frage durch eine Spaltung des Chromophors, bei der man mindestens eine der beiden Peptidketten als Abbauprodukt fassen und in ihrer Sequenz aufklären kann. Eine solche Spaltung erlaubt die oben erwähnte Oxydation mit alkalischem Wasserstoffperoxyd nach Johnson und Mitarb. (*2, 62*), mit der sich zweifellos entscheiden läßt, ob dem Actinomycin $C_2$ Formel (LXI) oder (LXIa) zukommt.

Zwei Isomere, die sich wie (LXI) und (LXIa) nur durch die Stellung der Peptidgruppen unterscheiden, sind bei allen *aniso*-Actinomycinen möglich. Bei allen steht man daher vor der Frage, ob die Zelle beide Isomeren aufbaut und ob sich die beiden trennen lassen. Abbauversuche, bei denen nur eine Peptidkette abgelöst wurde (*58, 15*), sprechen dafür, daß Actinomycin $C_2$ einheitlich und nicht ein Gemisch aus (LXI) und (LXIa) ist.

Ob das in geringer Ausbeute aus Actinomycingemisch C isolierte Actinomycin $C_{2a}$ (vgl. Tabelle 3, S. 16) das den Formeln (LXI) bzw. (LXIa) entsprechende Stellungsisomere von Actinomycin $C_2$ ist, oder ob es sich von diesem in der Aminosäure-Sequenz unterscheidet, muß noch geklärt werden.

### Actinomycin $X_2$.

Im Actinomycin $X_2$ wurden zunächst nur n e u n Aminosäuren gefunden [Brockmann, Bohnsack und Gröne (*10*)]. Wären sie wie im Actinomycin $C_3$ (LI, S. 35) in Gestalt von zwei Peptidlactonringen mit dem Chromophor (XXXa, S. 27) verknüpft, so hätte Actinomycin $X_2$ das Mol.-Gew. 1192. Durch Redoxtitration (*60*) wurde jedoch $1307 \pm 35$ gefunden — eine Diskrepanz, die vermuten ließ, daß noch nicht alle Bauelemente des Peptidteils in Form von Abbauprodukten gefaßt waren. Die Bestätigung dafür brachte die katalytische Hydrierung, bei der Actinomycin $X_2$ in Actinomycin $C_1$ überging [Brockmann und Mane-gold (*38*)]. Damit war gezeigt: Der Peptidteil des Actinomycins $X_2$ enthält ein Prolin-Derivat, das a) bei katalytischer Hydrierung in Prolin übergeht und b) in einer der beiden Peptidketten an gleicher Stelle steht wie das Prolin des Actinomycins $C_1$.

$$HOHC\text{---}CH_2$$
$$H_2C \qquad CH\text{---}COOH$$
$$N$$
$$H$$

(LXVIII.) Hydroxyprolin.

$$O=C\text{---}CH_2$$
$$H_2C \qquad CH\text{---}COOH$$
$$N$$
$$H$$

(LXIX.) $\gamma$-Oxoprolin.

```
O — L-Meval    L-Meval — O
    |              |
   Sar            Sar
    |              |
   Pro           Oxopro
    |              |
  D-Val          D-Val
    |              |
  L-Thre         L-Thre
    |              |
```

(LXX.)
(LXXa.) Oxopro = Hypro.
(LXXb.) Oxopro = allo-Hypro.
Hypro = Hydroxyprolin.
allo-Hypro = allo-Hydroxyprolin.
Oxopro = $\gamma$-Oxoprolin.

```
O — L-Meval    L-Meval — O
    |              |
   Sar            Sar
    |              |
  Oxopro          Pro
    |              |
  D-Val          D-Val
    |              |
  L-Thre         L-Thre
    |              |
```

(LXXI.)
(LXXIa.) Oxopro = Hypro.
(LXXIb.) Oxopro = allo-Hypro.

Um welches Prolinderivat es sich handelte, zeigte die Reduktion von Actinomycin $X_2$ mit Aluminium-isopropylat, bei der das 1 Mol Hydroxyprolin enthaltende Actinomycin $X_{0\beta}$ (Tabelle 3, S. 16) entstand. Denn eine Verbindung, aus der bei energischer Reduktion Prolin und bei milder Hydroxyprolin wird, konnte nur das erst in jüngster Zeit von KUHN und OSSWALD (*89*) synthetisierte $\gamma$-Oxoprolin (LXIX) sein. Da (LXIX) sehr leicht oxydierbar ist, wird es bei der Totalhydrolyse von Actinomycin $X_2$ mit konz. Salzsäure durch den oxydierend wirkenden Chromophor zerstört; verständlich daher, daß es zunächst übersehen wurde. Dagegen bleibt (LXIX) erhalten und kann im Papierchromatogramm nachgewiesen werden, wenn zur Totalhydrolyse von Actinomycin $X_2$ Jodwasserstoffsäure verwendet wird (*37*).

Durch seine Reduktion zum Actinomycin $C_1$ (LX, S. 37) ist die Aminosäure-Sequenz des Actinomycins $X_2$ so weit aufgeklärt, daß nur noch die Formeln (LXX) und (LXXI) zur Diskussion stehen. Um zwischen ihnen zu entscheiden, muß, wie beim Actinomycin $C_2$ erläutert, durch Spaltung des Chromophors mindestens eine Peptidkette des Actinomycins $X_2$ in Form eines Abbauproduktes gefaßt und in ihrer Sequenz aufgeklärt werden.

### Actinomycin $X_{0\beta}$.

Da Actinomycin $X_{0\beta}$ durch Reduktion aus Actinomycin $X_2$ entsteht, ist seine Aminosäure-Sequenz ebensoweit aufgeklärt wie die von $X_2$,

d. h. es ist nur noch zwischen den Formeln (LXXa) und (LXXIa) zu
entscheiden. Dank der engen Beziehung zwischen Actinomycin $X_2$
und $X_{0\beta}$ ist mit der vollständigen Aufklärung des einen auch die des
anderen erreicht.

### Actinomycin $X_{0\delta}$.

Hydriert man Actinomycin $X_2$ unter milden Bedingungen (Pd-
Methanol), so verwandelt es sich in ein Actinomycin, das im Gegensatz
zu $X_{0\beta}$ 1 Mol *allo*-Hydroxyprolin enthält und mit dem Symbol $X_{0\delta}$
gekennzeichnet wurde [Brockmann und Manegold (38)]. Für seine
Sequenz stehen die Formeln (LXXb) und (LXXIb) (S. 41) zur Diskussion.
Bemerkenswert ist, daß die Actinomycine $X_{0\beta}$ und $X_{0\delta}$, die sich nur
durch die Konfiguration eines C-Atoms (das im Hydroxyprolin die
OH-Gruppe trägt) unterscheiden, im Verteilungschromatogramm trennbar
und in ihrer antibiotischen Wirksamkeit deutlich verschieden sind.
Bei *B. subtilis* war die antibiotisch wirksame Grenzkonzentration von
Actinomycin $X_{0\delta}$ etwa fünfmal kleiner als bei $X_{0\beta}$. Bemerkenswert ist
ferner, daß bei der Reduktion mit Aluminium-isopropylat die Stereo-
spezifität der Reaktion davon abhängt, ob das $\gamma$-Oxoprolin im Peptid-
verband steht oder frei ist. Denn im Verband wird praktisch ausschließlich Hydroxyprolin gebildet, während aus Carbäthoxy-oxo-prolinäthylester
doppelt soviel *allo*-Hydroxyprolin wie Hydroxyprolin entsteht (89).
Bei der katalytischen Hydrierung gibt es diesen Unterschied nicht.
Hier wird $\gamma$-Oxoprolin, einerlei, ob frei (als Carbäthoxy-oxo-prolinäthylester) oder im Peptid gebunden, ausschließlich bzw. ganz überwiegend
zu *allo*-Hydroxyprolin reduziert.

### 3. Struktur-Variationen des Peptidteils.

Bisher sind achtzehn kristallisierte Actinomycine bekannt — die
in Tabelle 3 (S. 16) angeführten, Actinomycin $X_3$ (72) sowie die Actinomycine $Z_1$ und $Z_5$ — eine Zahl, die sich sicher noch erhöhen
wird und die Frage nahelegt, innerhalb welcher Grenzen die Zelle die
Struktur des Peptidteils variieren kann. Um diese Frage zu erörtern,
sollen die Aminosäure-Sequenzformeln (LIX), (LX), (LXI) (S. 37) und
(LXX) (S. 41) dadurch vereinfacht werden, daß der Chromophor (LXXII)
durch (LXXIII) symbolisiert und die Formeln um 90° im Uhrzeigersinn
gedreht werden. Für die Actinomycine $C_1$, $C_2$, $C_3$, $X_{0\beta}$, $X_{0\delta}$ und $X_2$ ergeben sich dann (LXXIV), (LXXV), (LXXVI), (LXXVII), (LXXIX)
und (LXXX) (*Tabelle 5*, S. 43). Bei (LXXV), (LXXVII), (LXXVIII),
(LXXIX) und (LXXX) deutet die S-förmige Umrahmung an, daß durch
Vertauschen der beiden Reste zwei Stellungsisomere möglich sind.

Ein Vergleich dieser Formeln zeigt, daß man die Actinomycine $C_2$,
$C_3$, $X_{0\beta}$, $X_{0\delta}$ und $X_2$ als Varianten einer „Grundstruktur" auffassen kann,

die durch das Actinomycin $C_1$ (LXXIV) repräsentiert wird. Und zwar besteht die Variation der Grundstruktur in: a) einer C-Methylierung eines oder beider Valinreste zu *allo*-Isoleucin und b) in der Oxydation eines Prolinrestes zu Hydroxyprolin oder $\gamma$-Oxoprolin.

Bei den übrigen Actinomycinen ist die Aminosäure-Sequenz noch nicht bewiesen, man kennt bis jetzt lediglich die C-terminalen Aminosäuren bei den E- und F-Actinomycinen. Läßt man die plausible Annahme gelten, daß auch diese Actinomycine Varianten der Struktur (LXXIV) sind oder, anders ausgedrückt, daß bei ihrem Aufbau die größtmögliche Strukturähnlichkeit mit (LXXIV) angestrebt wird, so kommt man für sie zu den Formeln (LXXVIII) und (LXXXI) bis (LXXXVII), die folgende

Tabelle 5. Aminosäure-Sequenzen verschiedener Actinomycine.

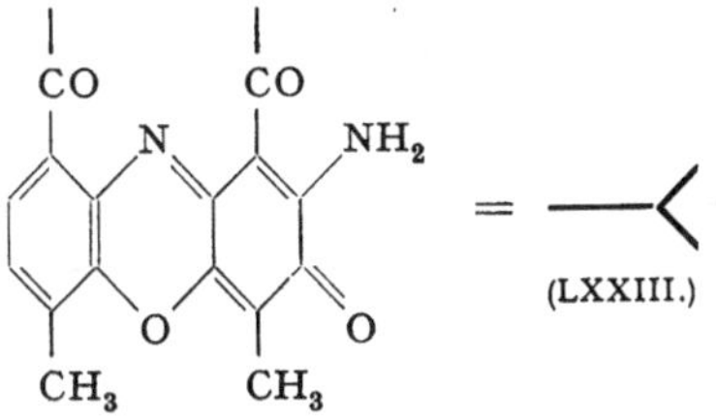

(LXXII.) (LXXIII.)

(LXXIV.) Actinomycin $C_1$.

(LXXV.) Actinomycin $C_2$.

(LXXVI.) Actinomycin $C_3$.

(LXXVII.) Actinomycin $X_{0\,\beta}$.

(LXXVIII.) Actinomycin $X_{0\,\gamma}$.

(LXXIX.) Actinomycin $X_{0\,\delta}$.

(Fortsetzung auf S. 44)

*(Fortsetzung der Tabelle 5.)*

$$\begin{array}{l} \text{— Thre—Val—}\boxed{\text{Oxopro}}\text{—Sar—Meval — O} \\ \text{— Thre—Val———}\boxed{\text{Pro}}\text{—Sar—Meval — O} \end{array}$$ (LXXX.) Actinomycin $X_2$.

$$\begin{array}{l} \text{— Thre—Val—}\boxed{\text{Oxopro}}\text{—Sar—Meval — O} \\ \text{— Thre—Val———}\boxed{\text{Sar}}\text{—Sar—Meval — O} \end{array}$$ (LXXXI.) Actinomycin $X_{1a}$.

$$\begin{array}{l} \text{— Thre—a-Ileu—Pro—Sar—}\boxed{\text{Meval}}\text{ — O} \\ \text{— Thre—a-Ileu—Pro—Sar—}\boxed{\text{Me-ileu}}\text{ — O} \end{array}$$ (LXXXII.) Actinomycin $E_1$.

$$\begin{array}{l} \text{— Thre—a-Ileu—Pro—Sar—Me-ileu — O} \\ \text{— Thre—a-Ileu—Pro—Sar—Me-ileu — O} \end{array}$$ (LXXXIII.) Actinomycin $E_2$.

$$\begin{array}{l} \text{— Thre—}\boxed{\text{Val}}\text{——Sar—Sar—Meval — O} \\ \text{— Thre—}\boxed{\text{a-Ileu}}\text{—Sar—Sar—Meval — O} \end{array}$$ (LXXXIV.) Actinomycin $\mathbf{F_1}$.

$$\begin{array}{l} \text{— Thre—}\boxed{\text{Val}}\text{——}\boxed{\text{Pro}}\text{—Sar—Meval — O} \\ \text{— Thre—}\boxed{\text{a-Ileu}}\text{—}\boxed{\text{Sar}}\text{—Sar—Meval — O} \end{array}$$ (LXXXV.) Actinomycin $\mathbf{F_2}$.

$$\begin{array}{l} \text{— Thre—a-Ileu—Sar—Sar—Meval — O} \\ \text{— Thre—a-Ileu—Sar—Sar—Meval — O} \end{array}$$ (LXXXVI.) Actinomycin $\mathbf{F_3}$.

$$\begin{array}{l} \text{— Thre—a-Ileu—}\boxed{\text{Pro}}\text{—Sar—Meval — O} \\ \text{— Thre—a-Ileu—}\boxed{\text{Sar}}\text{—Sar—Meval — O} \end{array}$$ (LXXXVII.) Actinomycin $\mathbf{F_4}$.

$R$—Thre—a-Ileu—Pro—Sar—Meval — O   (LXXXVIII.)

$R$—Thre—Val—Pro—Sar—Meval — O    (LXXXIX.)

$R$—Thre—a-Ileu—Hypro—Sar—Meval — O  (XC.)

$R$—Thre—Val—Hypro—Sar—Meval — O   (XCI.)

$R$—Thre—a-Ileu—Oxopro—Sar—Meval — O (XCII.)

$R$—Thre—Val—Oxopro—Sar—Meval — O  (XCIII.)

$$R = \begin{array}{l} \text{CO} \\ \text{NH}_2 \\ \text{OH} \\ \text{CH}_3 \end{array}$$

*Literaturverzeichnis: SS. 48—54.*

Zusammenhänge erkennen lassen: Im Actinomycin $X_{0\gamma}$ (LXXVIII) ist die Grundstruktur (LXXIV) abgewandelt durch Austausch eines Prolinrestes gegen Sarkosin; im Actinomycin $X_{1a}$ (LXXXI) dadurch, daß außer einem solchen Austausch auch noch das Prolin zu $\gamma$-Oxoprolin geworden ist.

Bei den E-Actinomycinen, die sich bilden, wenn der Kulturlösung $D,L$-Isoleucin zugesetzt wird, ist die Grundstruktur (LXXIV, S. 43) an zwei Stellen variiert, das Valin ist durch *allo*-Isoleucin oder Isoleucin* ersetzt und das N-Methylvalin ganz oder zur Hälfte durch N-Methylisoleucin — eine Variation der Grundstruktur (LXXIV), die man auch als drei- bzw. vierfache C-Methylierung auffassen kann (Umwandlung von 2 Mol Valin in Isoleucin und von 1 oder 2 Mol N-Methylvalin in N-Methyl-isoleucin).

Bei den F-Actinomycinen, die entstehen, wenn man den Actinomycinbildnern Sarkosin anbietet, wird dieses an Stelle von einem oder zwei Prolinresten eingebaut. Außerdem ist die Grundstruktur (LXXIV) durch C-Methylierung von einem oder zwei Valinresten so variiert wie beim Actinomycin $C_2$ bzw. $C_3$, so daß man auch sagen könnte: Die Actinomycine $F_1$ und $F_2$ sind Derivate von Actinomycin $C_2$, die aus diesem durch Austausch von einem oder zwei Prolinresten gegen Sarkosin hervorgehen. Die Actinomycine $F_3$ und $F_4$ sind in analoger Weise entstandene Derivate von Actinomycin $C_3$.

Wenn die Actinomycine in vivo durch oxydative Kondensation aus zwei Molekülen eines N-[2-Amino-3-hydroxy-4-methylbenzoyl]-peptides vom Typ (LII) (S. 35) entstehen, dann ist dank dieser Verdoppelungsreaktion, auch wenn die Vorprodukte vom Typ (LII) nur innerhalb enger Grenzen variiert werden, die Synthese zahlreicher Actinomycine möglich. Nimmt man z. B. an, das Vorprodukt (LXXXIX) (S. 44) des Actinomycins $C_1$ würde folgendermaßen abgeändert: a) durch Austausch von Valin gegen *allo*-Isoleucin, b) durch Oxydation von Prolin zu $\gamma$-Hydroxyprolin und c) durch Oxydation von Prolin zu $\gamma$-Oxoprolin, so stünden zur Actinomycin-Synthese die sechs Vorprodukte (LXXXVIII) bis (XCIII) zur Verfügung. Nützte die Zelle bei der oxydativen Kondensation nach dem Schema (LII) → (LI) (S. 35) alle Kombinationsmöglichkeiten aus, so könnten sich aus diesen sechs Vorprodukten 36 verschiedene Actinomycine bilden — eine Überlegung, die nicht nur fragen läßt, wieweit und mit welchen Mitteln die Zelle hier eine Auswahl trifft, sondern auch Zweifel aufkommen läßt, ob die bisher angewandten Trennungsverfahren ausreichen, um alle von der Zelle produzierten Actinomycine einzeln zu isolieren.

---

* Ob die E-Actinomycine Isoleucin oder *allo*-Isoleucin oder beide enthalten, ist noch nicht geklärt.

Die an den 15 Actinomycinen der Tabelle 3 (S. 16) zu erkennenden Variationen der Grundstruktur sind nicht die einzigen. Wie bereits erwähnt, enthalten die Z-Actinomycine neben N-Methylvalin auch noch N-Methyl-alanin, und es ist durchaus möglich, daß sich noch andere Actinomycin-Typen finden lassen. Das gilt ganz besonders für die dirigierte Biosynthese [vgl. dazu (*85*)], die ein eingehenderes Studium verdient.

Zum Schluß dieses Abschnittes mag darauf hingewiesen werden, daß sich aus den Formeln (LXXIV) bis (LXXXVII) eine Nomenklatur der Actinomycine entwickeln ließe, die in eindeutiger Weise die Sequenz des Peptidteils zum Ausdruck bringt. Und zwar dadurch, daß man bei den *iso*-Actinomycinen nur die Sequenz einer Kette angibt und bei den *aniso*-Actinomycinen nur die Reste, die in beiden Ketten verschieden sind. Actinomycin $C_1$ wäre dann als Actinomycin-[Thre-Val-Pro-Sar-Meval] zu bezeichnen, Actinomycin $C_2$ als Actinomycin-[Thre-$_{\text{a-Ileu}}^{\text{Val}}$-Pro-Sar-Meval], wenn das Valin im Sinne der Sequenzformel (LXXV) (S. 43) in der am benzoiden Ring hängenden Peptidkette steht, und als Actinomycin-[Thre-$_{\text{Val}}^{\text{a-Ileu}}$-Pro-Sar-Meval], wenn es der am chinoiden Ring hängenden Peptidkette angehört. Dort, wo diese Entscheidung noch aussteht, könnte man die in ihrer Stellung fraglichen beiden Reste in Klammern setzen. Actinomycin $C_2$ hätte dementsprechend zur Zeit noch die Bezeichnung Actinomycin-[Thre-[$_{\text{a-Ileu}}^{\text{Val}}$]-Pro-Sar-Meval].

## IV. Derivate der Actinomycine.

Derivate der Actinomycine sind in zweierlei Hinsicht von Interesse. Einmal, weil ihr biologisches Verhalten erkennen läßt, welche Teile des Actinomycin-Moleküls für die antibiotische, cytostatische und toxische Wirkung verantwortlich sind; und zum anderen, weil sich unter ihnen möglicherweise welche finden, bei denen das Verhältnis *cytostatische Wirksamkeit*: *Toxizität* günstiger ist als bei den Actinomycinen.

Man kann die bisher bekannten Actinomycin-Derivate in zwei Gruppen einteilen, je nachdem, ob sie durch eine Veränderung des Peptidteils oder des Chromophors entstehen.

### 1. Derivate, die durch Veränderung des Peptidteils entstehen.

Durch 0,1 *n*-Alkalihydroxyd werden bereits unter milden Bedingungen die beiden Lactonringe der Actinomycine geöffnet. Gleichzeitig verändert sich dabei das Threonin im Verband einer oder auch beider Peptidketten — wahrscheinlich unter Bildung von Aminocrotonsäure — und es entsteht neben Actinomycinsäure (*15*) ein Gemisch von sauren Abbauprodukten, die alle antibiotisch unwirksam sind. Da ihre Untersuchung noch nicht abgeschlossen ist, können sie hier unberücksichtigt bleiben.

Derivate, bei denen der Peptidteil ohne Öffnung der Lactonringe verändert ist, haben sich bisher nur von den Actinomycinen $X_{0\beta}$ und $X_{0\delta}$ gewinnen lassen, die Hydroxyprolin (LXVIII, S. 41) bzw. *allo*-Hydroxyprolin enthalten — und zwar dadurch, daß die Hydroxygruppe dieser beiden Aminosäuren mit Essigsäure oder Palmitinsäure verestert wurde [BROCKMANN, PAMPUS und MANEGOLD (53)]. Wie zu erwarten, steigt die Löslichkeit in unpolaren Solvenzien in der Reihenfolge Actinomycin < Acetat < Palmitat. Die antibiotische Wirksamkeit des Actinomycin $X_{0\beta}$-Acetates war bei *Bac. subtilis* um eine Zehnerpotenz geringer als beim Actinomycin $X_{0\beta}$.

### 2. Derivate, die durch Veränderung des Chromophors entstehen.

Umwandlungen des Actinomycin-Chromophors sind dadurch möglich, daß dessen Aminogruppe wie die eines vinylogen Säureamids reagiert *(Formelübersicht 4)*. Bei Säureeinwirkung wird sie unter Abspaltung von Ammoniak durch eine Hydroxygruppe ersetzt (XCIV) → (XCV). Die dabei entstehenden Desamino-actinomycine (XCV) (*17*) unterscheiden sich von den Actinomycinen (XCIV) u. a. dadurch, daß sie: a) antibiotisch

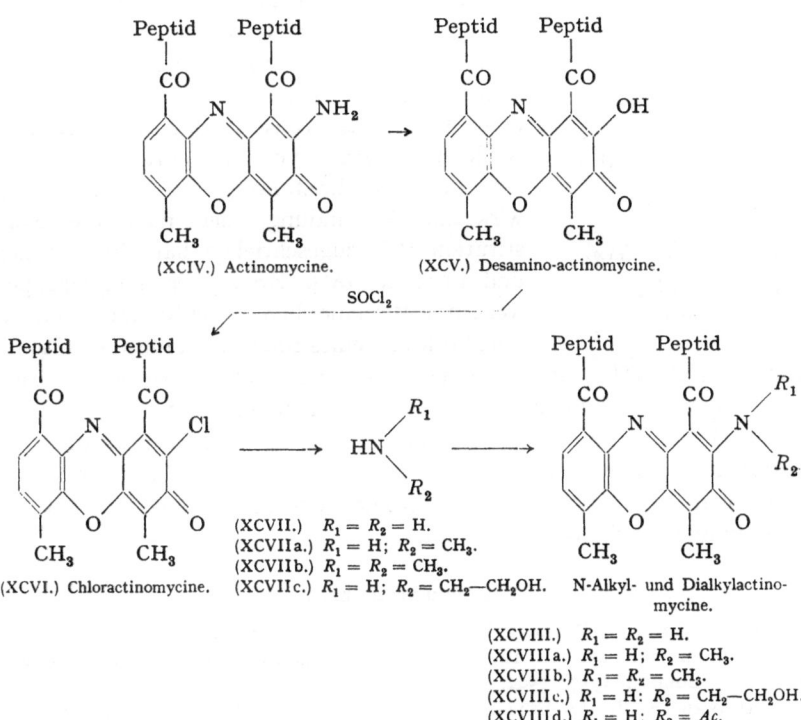

*Formelübersicht 4.* Actinomycin-Derivate.

unwirksam sind, b) ein anderes Absorptionsspektrum haben, c) mit konz. Salzsäure keine Rotfärbung geben und d) mit Zinn-II-chlorid ein tiefgrünes Semichinon bilden.

Beim Umsetzen mit Thionylchlorid tauschen die Desamino-actino-mycine (XCV) ihre Chromophor-Hydroxygruppe gegen Chlor aus und werden dabei zu roten, antibiotisch unwirksamen Chloractino-mycinen (XCVI) [BROCKMANN, GRÖNE und PAMPUS (28)]. Deren Chlor-atom ist so reaktionsfähig, daß man sie mit Ammoniak (XCVII) leicht in Actinomycine (XCVIII) zurückverwandeln kann. Verwendet man statt Ammoniak primäre (XCVIIa) oder sekundäre (XCVIIb) Amine, so entstehen N-Alkyl- bzw. N-Dialkyl-actinomycine (z. B. XCVIIIa; XCVIIIb) (Formelübersicht 4). Näher untersucht wurden bisher das N-Methyl-actinomycin $C_2$ (XCVIIIa) und das N-$\beta$-Hydroxyäthyl-actinomycin $C_3$ (XCVIIIc) [BROCKMANN, PAMPUS und MECKE (54)]. Das N-Methylderivat (XCVIIIc) hemmte das Wachstum von *Bac. subtilis* bis zur gleichen Grenzkonzentration wie Actinomycin $C_2$. Das N-$\beta$-Hydroxyäthyl-Derivat (XCVIIIc) dagegen war etwa zehnmal weniger wirksam als Actinomycin $C_3$.

Durch reduzierende Acetylierung mit Acetanhydrid und Zinkstaub (unter Zusatz von Pyridin oder Perchlorsäure) wurden aus Actinomycin-gemisch B (84) und Actinomycingemisch C (33) hellgelbe, kristallisierte Verbindungen erhalten, deren Acetylgehalt verschieden groß war. Hydrie-rende Acetylierung von Actinomycin $C_3$ lieferte eine hellgelbe, kristallisierte, antibiotisch un-wirksame Verbindung, der man die Kon-stitution (IC) zugeschrieben hat [BROCKMANN und FRANCK (18)]. Zwei ihrer Acetylgruppen werden in Methanol bereits bei Raumtemperatur unter Bildung eines roten, kristallisierten Mono-acetates (XCVIIId) abgespalten. Dieses läßt sich im Gegensatz zum Actinomycin $C_3$ nicht mit Perchlorsäure titrieren, zeigt mit Salzsäure keine Halochromie und wird durch kurze Salzsäureeinwirkung zu Actinomycin $C_3$ verseift. Durch direkte Acetylierung von Actinomycin $C_3$ ist es nicht zugänglich, weil die Chromophor-Aminogruppe sich wie die eines vinylogen Säureamides verhält.

### Literaturverzeichnis.

*1.* ANGYAL, S. J., E. BULLOCK, W. G. HANGER, W. C. HOWELL and A. W. JOHNSON: Actinomycin. Part III. The Reaction of Actinomycin with Alkali. J. Chem. Soc. (London) **1957**, 1592.
*2.* ANGYAL, S. J., E. BULLOCK, W. G. HANGER and A. W. JOHNSON: The Chromophor of Actinomycin. Chem. and Ind. **1955**, 1295.
*3.* Antibiotics. London: Oxford Univ. Press **1949**, p. 384.

*4.* BOHNSACK, G.: Zur Kenntnis des Peptidteils der Actinomycine. Dissert., Univ. Göttingen, 1955.

*5.* BOISSONNAS, R. A.: Dosage colorimétrique des acides aminés séparés par chromatographie sur papier. Helv. Chim. Acta **33**, 1975 (1950).

*6.* BOSSI, R., R. HÜTTER, W. KELLER-SCHIERLEIN, L. NEIPP und H. ZÄHNER: Stoffwechselprodukte von Actinomyceten, 14. Mitt. Actinomycin Z. Helv. Chim. Acta **41**, 1645 (1958).

*7.* BROCKMANN, H.: Chemie und Biologie der Actinomycine. Angew. Chem. **66**, 1 (1954).

*8.* BROCKMANN, H., A. BOHNE und H. FRIEDRICH: Zur Entstehungsgeschichte des H. B. F. 386, Actinomycin C Bayer. Dtsch. med. Wochschr. **79**, 437 (1954).

*9.* BROCKMANN, H., G. BOHNSACK, B. FRANCK, H. GRÖNE, H. MUXFELDT und C. H. SÜLING: Zur Konstitution der Actinomycine. Angew. Chem. **68**, 70 (1956).

*10.* BROCKMANN, H., G. BOHNSACK und H. GRÖNE: Der Aminosäuregehalt der Actinomycine. Naturwiss. **40**, 223 (1953).

*11.* BROCKMANN, H., G. BOHNSACK und C. H. SÜLING: Hydrazin-Spaltung von Actinomycinen. Angew. Chem. **68**, 66 (1956).

*12.* BROCKMANN, H. und P. BOLDT: Die Aminosäure-Sequenz von Actinomycin $C_3$. Naturwiss. **46**, 262 (1959).

*13.* BROCKMANN, H., P. BOLDT und H.-S. PETRAS: Die Aminosäure-Sequenz von Actinomycin $C_1$ und Actinomycin $C_2$. Naturwiss. **47**, 62 (1960).

*14.* BROCKMANN, H. und G. BUDDE: Zur Kenntnis des Despeptido-actinomycins. Naturwiss. **40**, 529 (1953).

*15.* BROCKMANN, H. und G. DÖRING: Zur Konstitution der Actinomycine. Chem. Ber. (im Druck).

*16.* BROCKMANN, H. und B. FRANCK: Abbau der Actinomycine zu Desamino-actinomycinen. Naturwiss. **41**, 451 (1954).

*17.* — — Abbau der Actinomycine zu Desamino-actinomycinen; XIV. Mitt. über Actinomycine; Antibiotica aus Actinomyceten, XXVI. Mitt. Chem. Ber. **87**, 1767 (1954).

*18.* — — Hydrierende Acetylierung von Actinomycinen. Angew. Chem. **68**, 68 (1956).

*19.* — — Aufspaltung der Actinomycine zu Actinomycinsäuren. Angew. Chem. **68**, 68 (1956).

*20.* — — Zur Acetylbestimmung bei Actinomycinen. Naturwiss. **42**, 180 (1955).

*21.* — — Bilanz der Actinomycin $C_3$-Abbauprodukte. Angew. Chem. **68**, 70 (1956).

*22.* — — Actinomycin $C_{2a}$, ein Isomeres des Actinomycins $C_2$. Naturwiss. **47**, 15 (1960).

*23.* BROCKMANN, H. und H. GRÖNE: Papierchromatographische Trennung der Actinomycine. Naturwiss. **40**, 222 (1953).

*24.* — — Reine Actinomycine. Naturwiss. **41**, 65 (1954).

*25.* — — Darstellung und Charakterisierung reiner Actinomycine; XII. Mitt. über Actinomycine; Antibiotica aus Actinomyceten XXIII. Mitt. Chem. Ber. **87**, 1036 (1954).

*26.* — — Neue farbige Abbauprodukte der Actinomycine. Angew. Chem. **68**, 66 (1956).

*27.* — — Abbau von Actinomycin C zu Actinocinin und Desamino-actinocinyl-threonin; Actinomycine, XVIII.; Antibiotica aus Actinomyceten, XXXIX. Chem. Ber. **91**, 773 (1958).

*28.* Brockmann, H., H. Gröne und G. Pampus: Chloractinomycine. Actinomycine, XX; Antibiotica aus Actinomyceten, XLI. Chem. Ber. **91**, 1916 (1958).

*29.* Brockmann, H., H. Gröne und J. Timm: Über den Threoningehalt der Actinomycine. Naturwiss. **42**, 125 (1955).

*30.* Brockmann, H. und N. Grubhofer: Actinomycin C. Naturwiss. **36**, 376 (1949).

*31.* — — Zur Kenntnis des Actinomycins C. Naturwiss. **37**, 494 (1950).

*32.* — — Abbau der. Actinomycine zu Despeptido-actinomycinen; VIII. Mitt. über Actinomycine; Antibiotica aus Actinomyceten, XVIII. Mitt. Chem. Ber. **86**, 1407 (1953).

*33.* Brockmann, H., N. Grubhofer, W. Kass und H. Kalbe: Über das Actinomycin C; Antibiotica aus Actinomyceten, V. Mitt. Chem. Ber. **84**, 260 (1951).

*34.* Brockmann, H., H. Linge und H. Gröne: Zur Kenntnis des Actinomycins X. Naturwiss. **40**, 224 (1953).

*35.* Brockmann, H. und J. H. Manegold: Überführung von Actinomycin $X_2$ in Actinomycin $X_1$. Naturwiss. **45**, 310 (1958).

*36.* — — Zur Kenntnis der Actinomycine. Chem. Ber. (im Druck).

*37.* — — Aminosäure-Analysen von Actinomycinen. Chem. Ber. (im Druck).

*38.* — — Reduktion von Actinomycin $X_2$ zu Actinomycin $C_1$, Actinomycin $X_{0\beta}$ und Actinomycin $X_{0\delta}$. Chem. Ber. (im Druck).

*39.* Brockmann, H., J. H. Manegold und G. Schmidt-Kastner: Der Aminosäuregehalt der E-Actinomycine und F-Actinomycine. Chem. Ber. (im Druck).

*40.* Brockmann, H. und R. Mecke: Unveröffentlicht.

*41.* Brockmann, H. und E. Meyer: Äquivalent- und Molekulargewichts-Bestimmungen durch potentiometrische Mikrotitration in nichtwäßrigen Lösungsmitteln. Chem. Ber. **86**, 1514 (1953).

*42.* Brockmann, H. und H. Muxfeldt: Zur Konstitution des Despeptido-actinomycins. Naturwiss. **41**, 500 (1954).

*43.* — — Die Konstitution des Despeptido-actinomycins. Angew. Chem. **67**, 617 (1955).

*44.* — — Die Synthese des Despeptido-actinomycins. Angew. Chem. **67**, 618 (1955).

*45.* — — Abbau des Actinomycins C zu 2,5-Dioxy-toluchinon und 3-Oxy-1,8-dimethyl-phenoxazon-(2). Angew. Chem. **68**, 67 (1956).

*46.* — — Konstitution und Synthese des Actinocinins. Angew. Chem. **68**, 67 (1956).

*47.* — — Konstitution und Synthese des Actinomycin-Chromophors. Angew. Chem. **68**, 69 (1956).

*48.* — — Die Konstitution des Despeptido-actinomycins; Actinomycine XVI. Mitt.; Antibiotica aus Actinomyceten, XXXV. Mitt. Chem. Ber. **89**, 1379 (1956).

*49.* — — Die Synthese des Despeptido-actinomycins; Actinomycine XVII. Mitt.; Antibiotica aus Actinomyceten, XXXVI. Mitt. Chem. Ber. **89**, 1397 (1956).

*50.* — — Konstitution und Synthese des Actinomycin-Chromophors; Actinomycine, XIX; Antibiotica aus Actinomyceten, XL. Chem. Ber. **91**, 1242 (1958).

*51.* Brockmann, H., H. Muxfeldt und G. Haese: Absorptionsinkremente von Methyl- und Methoxygruppen bei Methyl- und Methoxy-acridonen. Chem. Ber. **89**, 2174 (1956).

*52.* Brockmann, H. und G. Pampus: Actinomycin $X_{0\beta}$. Angew. Chem. **67**, 519 (1955).

53. BROCKMANN, H., G. PAMPUS und J. H. MANEGOLD: Actinomycin $X_{0\beta}$; zur Systematik und Nomenklatur der Actinomycine; Actinomycine, XXI; Antibiotica aus Actinomyceten, XLIII. Chem. Ber. 92, 1294 (1959).

54. BROCKMANN, H., G. PAMPUS und R. MECKE: N-Alkyl-actinomycine; Actinomycine, XXII; Antibiotica aus Actinomyceten, XLIV. Chem. Ber. 92, 3082 (1959).

55. BROCKMANN, H. und H.-S. PETRAS: Hydrolytischer Abbau von Actinomycinen. Naturwiss. 46, 400 (1959).

56. BROCKMANN, H. und N. PFENNIG: Auftrennung von Actinomycin C durch Gegenstromverteilung. Naturwiss. 39, 429 (1952).

57. — — Die Gewinnung reiner Actinomycine durch Gegenstromverteilung (V. Mitt. über Actinomycine). Z. physiol. Chem. (Hoppe-Seyler) 292, 77 (1953).

58. BROCKMANN, H. und W. SUNDERKÖTTER: Unveröffentlicht.

59. BROCKMANN, H. und K. VOHWINKEL: Zur Kenntnis der Farbstoffkomponente der Actinomycine. Naturwiss. 41, 257 (1954).

60. — — Molekulargewichtsbestimmung der Actinomycine und ihrer Abbauprodukte durch Redoxtitration. Angew. Chem. 67, 618 (1955).

61. — — Abbau der Actinomycine $I_1$ ($C_1$), $C_2$, $C_3$ und $X_2$ zu Despeptido-actinomycin; Actinomycine, XV. Mitt.; Antibiotica aus Actinomyceten, XXXIV. Mitt. Chem. Ber. 89, 1373 (1956).

62. BULLOCK, E. and A. W. JOHNSON: Actinomycin. Part IV. An Oxidative Degradation of Actinomycin B. J. Chem. Soc. (London) 1957, 1602.

63. — — Actinomycin. Part V. The Structure of Actinomycin D. J. Chem. Soc. (London) 1957, 3280.

64. BUTENANDT, A.: Über Ommochrome, eine Klasse natürlicher Phenoxazon-Farbstoffe. Angew. Chem. 69, 16 (1957).

65. CAVILL, G. W. K., P. S. CLEZY, J. R. TETAZ and R. L. WERNER: The Chemistry of Mould Metabolites-III; Structure of Cinnabarin (Polystictin). Tetrahedron 5, 275 (1959).

66. CORBAZ, R., L. ETTLINGER, W. KELLER-SCHIERLEIN und H. ZÄHNER: Zur Systematik der Actinomyceten. 2. Über Actinomycin bildende Streptomyceten. Arch. Mikrobiol. 26, 192 (1957).

67. DALGLIESH, C. E., A. W. JOHNSON, A. R. TODD and L. C. VINING: Actinomycin. Part I. Amino-acid-Content. J. Chem. Soc. (London) 1950, 2946.

68. DALGLIESH, C. E. and A. R. TODD: Actinomycin. Nature (London) 164, 830 (1949).

69. FARBER, S.: Clinical and Biological Studies with Actinomycins. Ciba Foundation Symposium on Amino Acids and Peptides with Antimetabolic Activity, London, 1958. p. 138.

70. FROMMER, W.: Untersuchungen an Actinomycin I-bildenden Streptomyceten und deren Actinomycinen. Arch. Mikrobiol. 31, 319 (1958).

71. — Zur Systematik der Actinomycin bildenden Streptomyceten. Arch. Mikrobiol. 32, 187 (1958/59).

72. — Über die Bildung des Actinomycin X; Versuche zur Gewinnung von Actinomycin $X_3$ und $X_4$. Arch. Mikrobiol. 34, 1 (1959).

73. GOSS, W. A. and E. KATZ: Actinomycin Formation by Streptomyces Cultures. Applied Microbiol. 5, 95 (1957).

74. GOSS, W. A., E. KATZ and S. A. WAKSMAN: Changes in the Composition of an Actinomycin Complex during Growth of a Streptomyces Culture. Proc. Nat. Acad. Sci. (USA) 42, 10 (1956).

75. GREGORY, F. J., L. C. VINING and S. A. WAKSMAN: Actinomycin. IV. Classification of the Actinomycins by Paper Chromatography. Antibiotics and Chemotherapy 5, 409 (1955).

76. HACKMANN, CHR.: Experimentelle Untersuchungen über die Wirkung von Actinomycin C (HBF 386) bei bösartigen Geschwülsten. Z. Krebsforsch. 58, 607 (1952).

77. — HBF 386 (Actinomycin C), ein cytostatisch wirksamer Naturstoff. Strahlentherapie 90, 296 (1953).

78. — Actinomycin C (Sanamycin): Biologische Eigenschaften und therapeutische Verwendung. Clinica Terapeut. (Roma) 7, 406 (1954).

79. — Zur Frage der medikamentösen Krebsbehandlung. Med. Klinik 49, 1539 (1954).

80. — Untersuchungen über den Einfluß des Sanamycins (Actinomycin C) auf tierische Organe: Milz, Thymus, Lymphknoten, Nebennieren, Keimdrüsen. Z. Krebsforsch. 60, 250 (1954).

81. — Stoffwechselprodukte aus Mikroorganismen (Antibiotika) als antineoplastische Mittel. Dtsch. med. Wochschr. 80, 812 (1955).

82. HACKMANN, CHR. und G. SCHMIDT-KASTNER: Über die cytostatische Wirkung verschiedener neuer biosynthetischer Actinomycine bei experimentellen Tumoren. Z. Krebsforsch. 61, 607 (1957).

83. HANGER, W. G., W. C. HOWELL and A. W. JOHNSON: Actinomycin. Part VI. The Structure and Synthesis of Actinomycinol. J. Chem. Soc. (London) 1958, 496.

84. JOHNSON, A. W., A. R. TODD and L. C. VINING: Actinomycin. Part II. Studies on the Chromophoric Grouping. J. Chem. Soc. (London) 1952, 2672.

85. KATZ, E. and W. A. GOSS: Influence of Amino-Acids on Actinomycin Biosynthesis. Nature (London) 182, 1668 (1958).

86. KATZ, E., P. PIENTA and A. SIVAK: The Role of Nutrition in the Synthesis of Actinomycin. Applied Microbiol. 6, 236 (1958).

87. KOCH, G. und W. WEIDEL: Über die Receptorsubstanz für den Phagen T 5. IV. Mitt.: Eine einfache Methode zur quantitativen Bestimmung von Aminosäuren und ihre Anwendung für Vergleiche zwischen Receptorsubstanz und mutativem Abwandlungsprodukt. Z. physiol. Chem. (Hoppe-Seyler) 303, 213 (1956).

88. KOCHOLATY, W., R. JUNOWICZ-KOCHOLATY and A. KELNER: Actinomycin A Produced by a Soil *Actinomyces* Different from *A. antibioticus*. Arch. Biochemistry 17, 191 (1948).

89. KUHN, R. und G. OSSWALD: Neue Synthese von $\beta$-Pyrrolidonen; Darstellung von D,L-$\gamma$-Oxo-prolin, D,L-*allo*-Hydroxy-prolin und 4-Äthoxy-pyrrol-carbonsäure-(2). Chem. Ber. 89, 1423 (1956).

90. LEHR, H. and J. BERGER: The Isolation of a Crystalline Actinomycin-like Antibiotic. Arch. Biochemistry 23, 503 (1949).

91. LINDENBEIN, W.: Über einige chemisch interessante Actinomycetenstämme und ihre Klassifizierung. Arch. Mikrobiol. 17, 361 (1952).

92. MANAKER, R. A., F. J. GREGORY, L. C. VINING and S. A. WAKSMAN: Actinomycin. III. The Production and Properties of a New Actinomycin. Antibiotics Annual 1954/55, 853.

93. MARTIN, H. H. und G. PAMPUS: Untersuchung über die Bildung des Actinomycins X. Über die Beeinflussung des Komponentenverhältnisses beim Actinomycin X durch Variation der mikrobiologischen Kulturbedingungen. Arch. Mikrobiol. 25, 90 (1956).

*94.* Pfennig, N.: Untersuchungen an Actinomycin-bildenden Strahlenpilz-stämmen und deren Actinomycinen. Arch. Mikrobiol. **18**, 327 (1953).

*95.* Reilly, H. C., A. Schatz and S. A. Waksman: Antifungal Properties of Antibiotic Substances. J. Bacteriol. **49**, 585 (1945).

*96.* Robinson, H. J. and S. A. Waksman: Studies on the Toxicity of Actino-mycin. J. Pharmacol. exp. Therapeut. **74**, 25 (1942).

*97.* Roussos, G. G. and L. C. Vining: Isolation and Properties of Pure Actino-mycins. J. Chem. Soc. (London) **1956**, 2469.

*98.* Sarlet, H.: Note sur l'actinomycine produite par *Streptomyces S 67*. Enzymologia **14**, 49 (1950).

*99.* Schmidt-Kastner, G.: Actinomycin E und Actinomycin F, zwei neue bio-synthetische Actinomycingemische. Naturwiss. **43**, 131 (1956).

*100.* — Über neue biosynthetische Actinomycine. Medizin und Chemie, Abh. med.-chem. Forschungsst. Farbenfabr. Bayer AG. **5**, 463 (1956).

*101.* Schulte, G.: Erfahrungen mit neuen cytostatischen Mitteln bei Hämoblastosen und Carcinomen und die Abgrenzung ihrer Wirkungen gegen Röntgentherapie. Z. Krebsforsch. **58**, 500 (1952).

*102.* — Weitere Erfahrungen mit Sanamycin bei der Behandlung der Lympho-granulomatose. Strahlentherapie **94**, 491 (1954).

*103.* Schulte, G. und H. Lings: Erfahrungen mit neuen cytostatischen Mitteln bei Leukosen und Lymphogranulomatosen und die Abgrenzung ihrer Wirkungen gegen Röntgentherapie. Strahlentherapie **90**, 301 (1953).

*104.* Troemel, G.: Beitrag zum Abbau der Actinomycine. Diplomarbeit, Univ. Göttingen, 1959.

*105.* Trussell, P. C. and E. M. Richardson: Actinomycin from a new *Streptomyces*. Canad. J. Res. **C 26**, 27 (1948).

*106.* Turner, R. A. and G. Schmerzler: A New Method for Identifying C-terminal Residues in Peptides. J. Amer. Chem. Soc. **76**, 949 (1954).

*107.* Umezawa, H., S. Hayano, T. Takeuchi and Y. Mizuhara: Studies on the Antibacterial Substance from a Strain of Actinomyces. I. A Crystalline Antibacterial Substance of a Strain of Actinomyces. J. Penicillin (Japan) **1**, 129 (1947).

*108.* Vining, L. C., F. J. Gregory and S. A. Waksman: Actinomycin. V. Chromato-graphic Separation of the Actinomycin Complexes. Antibiotics and Chemo-therapy **5**, 417 (1955).

*109.* Vining, L. C. and S. A. Waksman: Paper Chromatographic Identification of the Actinomycins. Science (Washington) **120**, 389 (1954).

*110.* Vohwinkel, K.: Untersuchung über das Chinonsystem der Actinomycine und ihrer farbigen Abbauprodukte. Diplomarbeit, Univ. Göttingen, 1954.

*111.* Waksman, S. A.: Actinomycin. I. Historical—Nature and Cytostatic Action. Antibiotics and Chemotherapy **4**, 502 (1954).

*112.* Waksman, S. A., W. B. Geiger and D. M. Reynolds: Strain Specifity and Production of Antibiotic Substances. VII. Production of Actinomycin by Different Actinomycetes. Proc. Nat. Acad. Sci. (USA) **32**, 117 (1946).

*113.* Waksman, S. A. and F. J. Gregory: Actinomycin. II. Classification of Organisms Producing Different Forms of Actinomycin. Antibiotics and Chemotherapy **4**, 1050 (1954).

*114.* Waksman, S. A., E. Katz and L. C. Vining: Nomenclature of Actinomycins. Proc. Nat. Acad. Sci. (USA) **44**, 602 (1958).

*115.* Waksman, S. A. and M. Tishler: The Chemical Nature of Actinomycin, an Antimicrobial Substance Produced by *Actinomyces antibioticus*. J. Biol. Chem. **142**, 519 (1942).

*116.* WAKSMAN, S. A. and H. B. WOODRUFF: Bacteriostatic and Bactericidal Substances Produced by a Soil *Actinomyces*. Proc. Soc. exp. Biol. Med. **45,** 609 (1940).

*117.* — — *Actinomyces antibioticus* a new Soil Organism Antagonistic to Pathogenic and Non-pathogenic Bacteria. J. Bacteriol. **42,** 231 (1941).

*118.* WELSCH, M.: Production d'actinomycine ou d'une substance voisine par un *Streptomyces* distinct de *S. antibioticus* WAKSMAN et WOODRUFF. Bull. soc. chim. biol. (Paris) **28,** 557 (1946).

*119.* WIMMER, E.: Die quantitative Bestimmung des Aminosäuregehaltes von Actinomycinen nach der DNP-Methode. Diplomarbeit, Univ. Göttingen, 1959.

*120.* ZEPF, K.: Nachweis neuer Spurenkomponenten des Actinomycin-C-Komplexes. Experientia **14,** 207 (1958).

*121.* ZEPF, K. und H. BERG: Zur Trennung von Actinomycingemischen durch Verteilungsverfahren. Pharmazie **14,** 396 (1959).

*(Eingelaufen am 7. Dezember 1959.)*

# Natürlich vorkommende Nitroverbindungen.

Von **M. Pailer**, Wien.

## Einleitung.

Natürlich vorkommende Nitroverbindungen sind noch nicht sehr lange bekannt, und es war geradezu ein Kuriosum, als man im Jahre 1949 feststellen konnte, daß das Chloramphenicol, ein Stoffwechselprodukt eines *Streptomyces*-Stammes, eine Nitrogruppe enthält. Bald darauf gelang die Charakterisierung der beiden aus pflanzlichem Material isolierten Glucoside Hiptagin und Karakin als Derivate der β-Nitropropionsäure, und in der Folgezeit wurde dieser Nitrokörper noch in weiteren Pflanzen nachgewiesen. Schließlich gelang dann die Konstitutionsermittlung von zwei Aristolochiasäuren, den Inhaltsstoffen verschiedener Aristolochiaarten und ihre Charakterisierung als Nitrophenanthren-derivate.

Die Zahl der bisher aufgefundenen Nitroverbindungen ist demnach noch nicht groß, ihre Mannigfaltigkeit in bezug auf Konstitution und Vorkommen läßt aber darauf schließen, daß man noch weitere Vertreter auffinden wird. Die Frage nach der Biogenese und der Rolle dieser

Stoffe in der Pflanze wurde natürlich schon öfter gestellt, konnte aber bisher noch nicht in befriedigender Weise beantwortet werden.

# I. Chloramphenicol.

Für einen kurzen Bericht in diesen *Fortschritten* vgl. VAN TAMELEN (*71*).

## 1. Isolierung und Konstitutionsermittlung.

Im Jahre 1947 gelang BURKHOLDER aus Erdproben, die in der Gegend von Caracas in Venezuela gesammelt worden waren, die Isolierung eines *Streptomyces*-Stammes, der dann, in Kulturen gezüchtet, ein stark antibiotisch wirkendes Substrat lieferte. Aus diesem konnten EHRLICH und Mitarbeiter (*19*) die Wirksubstanz in kristalliner Form isolieren, die sie Chloromycetin nannten. Das Antibiotikum wird heute sehr viel verwendet und allgemein als Chloramphenicol bezeichnet, während Chloromycetin ein Warenzeichen der Firma Parke, Davis & Comp. wurde, in deren Laboratorien die ersten Untersuchungen über diesen interessanten Stoff durchgeführt wurden. Mit der näheren Charakterisierung des *Streptomyces*-Stammes haben sich CARTER und Mitarbeiter (*12*) beschäftigt und ihn wegen seiner Herkunft *Streptomyces venezuelae* bezeichnet.

Chloramphenicol ist eine relativ stabile, neutrale Substanz vom Schmp. 150°, löslich in vielen organischen Lösungsmitteln, wenig löslich in Wasser, und im Hochvakuum ohne Zersetzung sublimierbar. Die Verbindung ist optisch aktiv [$(\alpha)_D^{25}$ + 19° (in Äthanol); — 25,5° (in Essigester)]. Als Bruttoformel wurde $C_{11}H_{12}O_5N_2Cl_2$ ermittelt, wobei das Chlor nicht ionogen gebunden ist.

Die Konstitutionsaufklärung des Chloramphenicols wurde von REBSTOCK und Mitarbeitern (*64*) durchgeführt. Das Studium des UV-Absorptionsspektrums im Vergleich mit anderen aromatischen Verbindungen machte das Vorhandensein einer Nitrogruppe sehr wahrscheinlich. Dieses Ergebnis wurde zuerst mit Skepsis aufgenommen, weil bis zu diesem Zeitpunkt noch von keinem Naturstoff mit einer Nitrogruppe berichtet worden war. Die UV-Absorption zeigte weiter, daß das Chlor wahrscheinlich nicht am Benzolring haftet und daß eine Alkylgruppe in *p*-Stellung zur Nitrogruppe vorhanden sein dürfte.

Die chemische Untersuchung konnte diese Ergebnisse bestätigen und führte dann zur völligen Aufklärung der Konstitution des Chloramphenicols. CO-Gruppen waren nicht nachweisbar. Hydrierung mit Palladiumoxyd als Katalysator führte unter Aufnahme von fünf Molen Wasserstoff zur Bildung einer Base, und in der Lösung ließ sich nun außerdem ionogenes Chlor feststellen. Die Base zeigte eine dem *p*-Toluidin sehr ähnliche UV-Absorption.

*Literaturverzeichnis: SS. 78—82.*

Die Natur des einen Stickstoffatoms konnte durch folgende Unter-
suchungen ermittelt werden. Im Chloramphenicol ließ sich nach VAN
SLYKE keine Aminogruppe nachweisen, sie war aber nach längerem
Stehen des Antibiotikums mit 0,1-$n$ NaOH eindeutig feststellbar.

Die bereits erwähnte Nitrogruppe war durch einen positiven Test
mit $Fe(OH)_2$ sowie durch andere Reaktionen nachzuweisen. So gab
Chloramphenicol nach Reduktion mit Zinn und Salzsäure, Diazotierung
und Kupplung mit $\beta$-Naphthol einen roten Farbstoff. Das durch Zinn-
Salzsäure reduzierte Produkt zeigte dieselbe, dem $p$-Toluidin sehr ähnliche
UV-Absorptionskurve wie das katalytisch hydrierte.

Chloramphenicol, mit Essigsäureanhydrid in Pyridin behandelt,
lieferte ein Diacetylderivat, dessen Hydrolyse nach der Methode von
KUNZ und HUDSON (36) zeigte, daß ein Di-O-acetylprodukt vorlag. Aus
der Hydrolyselösung konnte Chloramphenicol wiedergewonnen werden.

Saure und alkalische Hydrolyse von Chloramphenicol gab als Spalt-
stücke eine flüchtige organische Säure und eine kristalline, optisch aktive
Base der Formel $C_9H_{12}O_6N_2$. Die saure Komponente erwies sich als
Dichloressigsäure, und von der Base wurden zur näheren Charakterisierung
vorerst das N-$p$-Nitrobenzoyl-, N-Acetyl- und N,O,O-Triacetylderivat
hergestellt. Behandlung der Base mit Dichloressigsäuremethylester
lieferte wieder ein Dichloracetamid, das mit dem natürlichen Chloramphenicol identisch war. Während nun Chloramphenicol unter den
Bedingungen, wie sie für die Bestimmung vizinaler OH-Gruppen üblich
sind, durch Perjodsäure nicht angegriffen wurde, ließ sich die Base
nach der Hydrolyse durch zwei Äquivalente des Reagens in je ein Mol
$NH_3$, HCHO und $p$-Nitrobenzaldehyd spalten. Außerdem war Ameisensäure nachweisbar. Dadurch war das $C_{(2)}$ der aliphatischen Kette als
Träger der Aminogruppe ermittelt und die Überlegungen, welche im
Zusammenhang mit dem Spektrum angestellt wurden, weitgehend
bestätigt.

Die sterischen Verhältnisse ließen sich durch den zum Ephedrin
analogen Aufbau der Spaltbase klären. Da sich nämlich die Basen durch
ihre Eigenschaften, vor allem die Drehung (Vergleich verschiedener
Derivate), der Pseudo-Ephedrinreihe zuordnen ließen, ergab sich als
Struktur für das Chloramphenicol die Formel eines $D$-(—)-$threo$-1-$p$-
Nitrophenyl-2-dichloramido-1,3-propandiols (I).

(I.) Chloramphenicol.

## 2. Synthesen.

Die auf dem Abbauweg ermittelte Konstitution des Chloramphenicols ließ sich durch mehrere Synthesen bestätigen.

Nach Untersuchungen von CONTROULIS, REBSTOCK und CROOKS (*15*) wurde das 2-Amino-1-phenyl-1,3-propandiol (III) durch Kondensation von Benzaldehyd mit $\beta$-Nitroäthanol und Reduktion des Kondensationsproduktes (II) mit Wasserstoff und Pd als amorphe Masse, einem Gemisch der den beiden Stereoisomeren entsprechenden Racemate, erhalten. Die Trennung erfolgte durch die verschiedene Löslichkeit in $CHCl_3$, und zwar wurde eine kristalline (A) und eine amorphe Fraktion (B) erhalten.

$$\langle C_6H_5 \rangle - CHO + \underset{\underset{NO_2}{|}}{CH_2} - CH_2OH \rightarrow \langle C_6H_5 \rangle - CHOH - \underset{\underset{NO_2}{|}}{CH} - CH_2OH \xrightarrow{H_2(Pd)}$$

(II.)

$$\rightarrow \langle C_6H_5 \rangle - CHOH - \underset{\underset{NH_2}{|}}{CH} - CH_2OH$$

(III.) 2-Amino-1-phenyl-1,3-propandiol.

Die kristallisierte Fraktion A wurde nun durch geeignete Nitrierung und Dichloracetylierung in die dem Chloramphenicol entsprechende Verbindung übergeführt. Das UV-Spektrum entsprach dann vollkommen dem des natürlichen Antibiotikums, beim mikrobiologischen Test erwies sich die Verbindung aber als vollkommen wirkungslos. Das Racemat A hatte also *erythro*-Konfiguration.

Die Fraktion B lieferte ein kristallines Triacetat, das sich mit $HNO_3$ in *p*-Stellung nitrieren ließ (IV) ($Ac = CH_3 \cdot CO$).

$$O_2N - \langle C_6H_4 \rangle - \underset{\underset{OAc}{|}}{CH} - \underset{\overset{NH \cdot Ac}{|}}{CH} - CH_2O \cdot Ac \xrightarrow{H^+} O_2N - \langle C_6H_4 \rangle - \underset{\underset{OH}{|}}{CH} - \underset{\overset{NH_2}{|}}{CH} - CH_2OH \rightarrow$$

(IV.)                                                         (V.)

$$\xrightarrow{Cl_2CHCOOCH_3} O_2N - \langle C_6H_4 \rangle - \underset{\underset{OH}{|}}{CH} - \underset{\overset{NH-CO-CHCl_2}{|}}{CH} - CH_2OH$$

(VI.)

Nach saurer Hydrolyse wurde aus der so erhaltenen Base (V) das N-Dichloracetamid (VI) hergestellt, das den Schmp. 150,5—151,5° hatte und bei der Testung mit *Shigella paradysenteriae* 50% der Wirkung des natürlichen Chloramphenicols aufwies. Die Trennung in die optischen Antipoden erfolgte über das Salz der Base (V) mit *d*-Camphersulfosäure.

*Literaturverzeichnis: SS. 78—82.*

Die (+)- und die (—)-Base wurde jede für sich in das Dichloracetamid übergeführt. Das Derivat der linksdrehenden Base stimmte in allen physikalischen und chemischen Eigenschaften mit dem natürlichen Chloramphenicol überein.

Eine weitere Synthese, bereits mit dem Ziel einer technischen Herstellung durchgeführt, haben LONG und TROUTMAN (*38*) beschrieben. Sie setzten α-Benzamido-acetophenon mit Formaldehyd zur Verbindung (VII) um, die dann katalytisch reduziert wurde (VIII). Die weiteren Schritte folgten dann, wie oben beschrieben wurde. Die Trennung der *erythro-* und der *threo*-Form war durch Umkristallisieren aus Essigester möglich, da die *threo*-Verbindung, das Ausgangsmaterial für die Chloramphenicol-Synthese, schwerer löslich ist. Sie wurde diacetyliert, nitriert, verseift und dichloracetyliert, wodurch schließlich das Chloramphenicol erhalten wurde.

$$\text{C}_6\text{H}_5\text{—CO—CH}_2 \xrightarrow[\text{NaHCO}_3]{\text{HCHO}} \text{C}_6\text{H}_5\text{—CO—CH—CH}_2\text{OH} \rightarrow$$
$$\underset{\text{NH—CO—C}_6\text{H}_5}{|} \qquad \underset{\text{NH—CO—C}_6\text{H}_5}{|}$$

(VII.)

$$\rightarrow \text{C}_6\text{H}_5\text{—CH—CH—CH}_2\text{OH}$$
$$\underset{\text{OH}}{|} \; \underset{\text{NH—CO—C}_6\text{H}_5}{|}$$

(VIII.)

Die Synthese ließ sich dann noch einfacher gestalten (*39*), da es möglich war, direkt vom *p*-Nitroacetophenon auszugehen. Das α-Bromderivat dieser Verbindung (IX) wurde mit Hexamethylentetramin

$$\text{O}_2\text{N—C}_6\text{H}_4\text{—CO—CH}_2\text{Br} \xrightarrow{(\text{CH}_2)_6\text{N}_4}$$

(IX.)

$$\rightarrow \text{O}_2\text{N—C}_6\text{H}_4\text{—CO—CH}_2[(\text{CH}_2)_6\text{N}_4]^+\text{Br}^- \xrightarrow[\text{HCl}]{\text{C}_2\text{H}_5\text{OH}}$$

(X.)

$$\rightarrow \text{O}_2\text{N—C}_6\text{H}_4\text{—CO—CH}_2\text{—NH}_2 \xrightarrow{(\text{CH}_3\text{CO})_2\text{O}} \text{O}_2\text{N—C}_6\text{H}_4\text{—CO—CH}_2 \xrightarrow[\text{NaHCO}_3]{\text{HCHO}}$$
$$\underset{\text{NH—CO—CH}_3}{|}$$

(XI.) (XII.)

$$\rightarrow \text{O}_2\text{N—C}_6\text{H}_4\text{—CO—CH—CH}_2\text{OH} \xrightarrow[\text{Prop.}]{\text{Al-iso-}} \text{O}_2\text{N—C}_6\text{H}_4\text{—CH—CH—CH}_2\text{OH}$$
$$\underset{\text{NH—CO—CH}_3}{|} \qquad \underset{\text{OH}}{|} \overset{\text{NH—CO—CH}_3}{|}$$

(XIII.) (XIV.)

*Formelübersicht 1.* Synthese des Chloramphenicols (*39*).

umgesetzt (X) und das Produkt zum entsprechenden Amin (XI) verseift. Das Acetylderivat von (XI) gab mit Formaldehyd die Verbindung (XIII), welche bei der Reduktion mit Al-iso-Propylat (XIV) lieferte (*Formelübersicht 1, S. 59*).

Bei diesen Synthesen entsteht nach der Reduktion mit Al-iso-Propylat hauptsächlich das *DL-threo*-1-*p*-Nitrophenyl-2-acetamido-1,3-propandiol, während die Menge an *erythro*-Verbindung gering ist. Der Syntheseweg der Long-Troutmanschen Chloramphenicol-Darstellung konnte nach einer Mitteilung von Weigner (*78*) weiter verkürzt werden, indem man gleich anstatt des Acetylderivates (XII) die entsprechende N-Dichloracetylverbindung verwendete.

Das ganze Gebiet des Chloramphenicols wurde wegen der praktischen Bedeutung des Antibiotikums intensiv synthetisch bearbeitet und führte zu einer sehr großen Zahl von Publikationen und Patenten, die als Ziel einerseits die möglichst rationelle Darstellung dieses Pharmazeutikums und andererseits die Umgehung bereits geschützter Verfahren hatten. Ein lückenloser Bericht über alle diese synthetischen Arbeiten würde der Zielsetzung dieses Artikels nicht entsprechen und muß daher einer anders orientierten Zusammenfassung vorbehalten werden. Nachfolgend werden daher neben den bereits beschriebenen Synthesen, welche die durch Abbau ermittelte Konstitution des Chloramphenicols bestätigen, einige andere Darstellungsmöglichkeiten nur kurz skizziert.

So ging man bei einer Reihe weiterer Synthesen vom Phenylserin aus. Alberti und Mitarbeiter (*1*) stellten dieses aus Benzaldehyd und Glykokoll her, wobei hauptsächlich die *threo*-Form anfiel. Der Phenylserinester wurde nun mit LiAlH$_4$ reduziert und das so erhaltene *DL-threo*-Serinol acetyliert und nitriert. Verseifung und nachfolgende Dichloracetylierung führten zum Racemat des Chloramphenicols.

Ebenfalls von einem Phenylserinderivat ausgehend, beschrieben Kollonitsch und Mitarbeiter (*33*) eine weitere Chloramphenicol-Synthese. Aus den Diastereomeren des β-Phenylserin-O-methyläthers (XV) mit dem niederen Schmelzpunkt wurde die *threo*-Form des β-Phenylserinolmethyläthers (XVI) auf zwei Wegen hergestellt. Einerseits wurde der Äthylester mit LiAlH$_4$ reduziert und andererseits wurde die Phthalylverbindung von (XV) nach Rosenmund zum 2-Phenyl-2-methoxy-1-phthaliminopropionaldehyd, dann weiter mit Al-isopropylat zum Alkohol reduziert und mit Hydrazin die Phthalsäure wieder abgespalten. Das O,N-Diacetat von (XVI) wurde nitriert und entacetyliert. Die entstandene Base (XVII) gab nach Entmethylierung das *threo*-1-(*p*-Nitrophenyl)-2-amino-1,3-dioxypropan. Verbindung (XVII) wurde mit Weinsäure oder Dibenzoylweinsäure in die optischen Antipoden gespalten. Das linksdrehende Isomere von (XVII) konnte zu einer Verbindung entmethyliert werden, die mit der aus dem Chloramphenicol erhaltenen

*Literaturverzeichnis: SS. 78—82.*

Base identisch war und mit Pentachloraceton in guter Ausbeute den Naturstoff lieferte.

$$C_6H_5 \cdot CH(OCH_3) \cdot CH(NH_2) \cdot COOH \qquad C_6H_5 \cdot CH(OCH_3) \cdot CH(NH_2) \cdot CH_2OH$$
$$\text{(XV.)} \qquad\qquad\qquad\qquad \text{(XVI.)}$$

$$O_2N \cdot C_6H_5 \cdot CH(OCH_3) \cdot CH(NH_2) \cdot CH_2OH$$
$$\text{(XVII.)}$$

Für eine weitere Synthese von KOLLONITSCH und Mitarbeitern (33) diente als Ausgangsmaterial der gewöhnliche Zimtalkoholmethyläther (*trans*-Isomere). Er wurde in methylalkoholischer Lösung in Gegenwart von gelbem Bleioxyd mit Brom zur Reaktion gebracht. So bildete sich das 1-Phenyl-2-brom-1,3-dimethoxypropan (XVIII), das durch Ammonolyse β-Phenylserinol-dimethyläther (XIX) lieferte. Das N-Acetat von (XIX) war identisch mit der aus *threo*-β-Phenylserinol-N-acetat durch Methylierung hergestellten Verbindung. Auf diese Weise war seine Konfiguration bewiesen. Das N-Acetat von (XIX) wurde nitriert, entacetyliert und entmethyliert; so wurde *threo*-β-(*p*-Nitrophenyl)-serinol erhalten.

$$C_6H_5 \cdot CH(OCH_3) \cdot CH(CH_2O \cdot CH_3) \cdot Br \qquad C_6H_5 \cdot CH(OCH_3) \cdot CH(CH_2 \cdot OCH_3) \cdot NH_2$$
$$\text{(XVIII.)} \qquad\qquad\qquad\qquad \text{(XIX.)}$$

Eine vom Phenylserin ausgehende Synthese, die aber keine neuen Gesichtspunkte aufweist, wurde von BENDAS und BERGMANN (2) beschrieben.

Zimtalkohol, und zwar das Acetat (XX), diente auch FODOR und Mitarbeitern (22) als Ausgangsmaterial:

$$C_6H_5 \cdot CH = CH \cdot CH_2OAc \xrightarrow{HNO_2} DL\text{-}erythro\text{-}C_6H_5 \cdot CH(NO) \cdot CH(NO_2) \cdot CH_2OAc \rightarrow$$
$$\xrightarrow[H_2SO_4]{(CH_3CO)_2O} DL\text{-}threo\text{-}C_6H_5 \cdot CH(OAc) \cdot CH(NO_2) \cdot CH_2OAc \xrightarrow{HCl,\ H_2O}$$
$$\rightarrow DL\text{-}threo\text{-}C_6H_5 \cdot CH(OH) \cdot CH(NO_2) \cdot CH_2OH \rightarrow \text{Nitrierung des Acetylderivats usw.}$$

Auch TAGUCHI und TOMOEDA (70) verwenden für ihre Synthesen Zimtalkohol. Dieser wird zuerst zu *DL*-(XXI) bromiert:

$$DL\text{-}C_6H_5\text{—}CHBr \cdot CHBr \cdot CH_2OH + C_6H_5CN \xrightarrow[\text{Äther}]{HCl}$$
$$\text{(XXI.)}$$

$$\rightarrow DL\text{-}C_6H_5 \cdot CHBr\text{—}CHBr \cdot CH_2OC(:NH) \cdot C_6H_5 \cdot HCl \rightarrow \text{freie Base von (XXII)} \rightarrow$$
$$\text{(XXII.)}$$

$$\xrightarrow[\text{in trockenem Toluol}]{\text{Kochen}} DL\text{-}erythro\text{-}C_6H_5 \cdot CH(Br) \cdot CH(NH \cdot CH_2 \cdot C_6H_5)CH_2Br \rightarrow$$
$$\text{(XXIII.)}$$

$$\rightarrow DL\text{-}threo\text{-}C_6H_5\text{—}CH(OCH_2 \cdot C_6H_5)CH(NH_2) \cdot CH_2Br$$
$$\text{(XXIV.)}$$

Die *threo*-Konfiguration wurde durch katalytische Reduktion von (XXIV) zu *DL*-Norephedrin und auch zum *DL-threo*-$C_6H_5$—CH(OH) · CH(CH$_2$OH)NH · CH$_2$ · C$_6$H$_5$ bestätigt.

Durch selektive Reduktion von 2-substituierten 5-(*p*-Nitrophenyl)-4-carbäthoxy-oxazolinen (XXV) mit LiAlH$_4$ konnten die entsprechenden 4-Oxymethyloxazoline erhalten werden, die durch Hydrolyse mit verdünnter Ameisensäure N-Acyl-$\beta$-(*p*-nitrophenyl)-serinole lieferten. Daraus wurde durch weitere Hydrolyse und Dichloracetylierung rac. Chloramphenicol erhalten (*20*). Die Synthese konnte weiter vereinfacht werden, indem man vom 2-Dichlormethyl-5-(*p*-nitrophenyl)-4-carbäthoxy-oxazolin ausging.

FUNKE und KORNMANN (*26*) stellten für ihre Chloramphenicol-Synthesen aus Phenyl-1-brom-2-propandiol-(1,3) vor der Ammonolyse das Acetal (XXVI) her, weil bei der direkten Umsetzung mit Ammoniak der Ammonstickstoff nicht an $C_{(2)}$, sondern an $C_{(1)}$ eintritt. Das erhaltene Aminoacetal wurde mit HCl gespalten und lieferte so die Base der *threo*-Reihe. Die weitere Synthese des Chloramphenicols erfolgte in üblicher Weise.

$$(p)\ O_2N{-}C_6H_4{-}\underset{\displaystyle\underset{\displaystyle\underset{R}{|}}{C}}{\overset{\displaystyle O}{\underset{\displaystyle|}{C}}H}{-}\underset{N}{\overset{|}{C}}H{-}COOR$$

(XXV.)

$$C_6H_5{-}\underset{\displaystyle\underset{C_6H_5}{|}}{\underset{\displaystyle O}{CH}}{-}\underset{\displaystyle CH{-}O}{\underset{|}{CH}}{-}CH_2 \quad \overset{Br}{|}$$

(XXVI.)

Schließlich sei noch eine Chloramphenicol-Synthese von IKUMA (*32*) kurz skizziert:

$$DL\text{-}threo\text{-}C_6H_5{-}CHCl \cdot CH(NH \cdot Ac) \cdot CH_2OH \xrightarrow{Ac_2O}$$

$$\rightarrow DL\text{-}threo\text{-}C_6H_5 \cdot CHCl \cdot CH(NH \cdot Ac)CH_2O \cdot Ac \rightarrow$$

$$\xrightarrow[\text{HNO}_3]{\text{H}_2\text{SO}_4} DL\text{-}threo\text{-}p\text{-}O_2N \cdot C_6H_4{-}CHCl \cdot CH(NH \cdot Ac) \cdot CH_2OAc \xrightarrow[\text{Kochen}]{\text{2 Std., 5\% HCl}}$$

$$\rightarrow DL\text{-}threo\text{-}p\text{-}O_2N \cdot C_6H_4 \cdot CHCl \cdot CH(NH \cdot Ac) \cdot CH_2OH \xrightarrow[40°]{\text{NaOH(C}_2\text{H}_5\text{OH)}}$$

$$\rightarrow DL\text{-2-Methyl-4-hydroxymethyl-5-}p\text{-nitrophenyl-}trans\text{-2-oxazolin} \rightarrow$$

$$\xrightarrow[50°]{\text{22\% HCl (alk.)}} DL\text{-}threo\text{-}p\text{-}O_2N{-}C_6H_4 \cdot CH(OAc) \cdot CH(NH_2 \cdot HCl)CH_2OH \rightarrow$$

$$\rightarrow DL\text{-}threo\text{-}p\text{-}O_2N \cdot C_6H_4 \cdot CH(OH) \cdot CH(NH_2) \cdot CH_2OH \rightarrow$$

$$\xrightarrow[\text{mit Weinsäure}]{\text{Trennung}} D\text{-Form} \xrightarrow{\text{CHCl}_2\text{CO}_2\text{CH}_3} (I).$$

*Literaturverzeichnis: SS. 78—82.*

### 3. Herstellung des Chloramphenicols auf biologischem Wege.

Chloramphenicol wird auch auf biologischem Weg, und zwar wie die meisten anderen Antibiotika, in Submerskulturen gewonnen, worüber wieder eine Reihe von Arbeiten, vor allem Patente, in der einschlägigen Literatur zu finden sind. Als Beispiel sei nachfolgend ein Verfahren der Firma Parke, Davis & Co., das von OLIVE (49) veröffentlicht wurde, kurz beschrieben.

Aus einer Schräg-agarkultur des Pilzes wird eine Sporensuspension in Wasser gewonnen, die in einem sterilen Gefäß weiter mit sterilem Seifenwasser verdünnt wird. Die homogene Suspension dient zur Beimpfung eines Vorpropagier-gefäßes von 50 Gallonen Inhalt. Hier wird 24 Stunden vergoren, dann in ein 500-Gls.-Gefäß übergeführt und schließlich in einen 5000-Gls.-Fermenter gedrückt, wo noch 76 Stunden vergoren wird. Die Gärtemperatur beträgt 28—30°, der pH-Wert etwa 7. Laufend wird belüftet und gerührt. Die Filtration erfolgt dann über eine Rotationsfilterpresse unter Verwendung eines Filterhilfsmaterials und Entfärbungsmittels. Die Rohlösung wird nun in einem Gegenstromextraktor mit Amylacetat extrahiert, dieses mit Schwefelsäure, Bicarbonatlösung und Wasser gewaschen und dann im Vakuum unter 35° weiter eingeengt. In der Kälte über Nacht kristallisiert das Chloramphenicol aus, das abgesaugt und aus Wasser umkristallisiert wird.

### 4. Bestimmung des Chloramphenicols.

Wegen der großen Bedeutung des Chloramphenicols wurde eine Reihe von biologischen und chemischen Methoden zur qualitativen und quantitativen Bestimmung des Chloramphenicols im Serum, Plasma, pharmazeutischen Präparaten usw. ausgearbeitet.

Viele der bisher beschriebenen Bestimmungsmethoden sind kolorimetrische, die vor allem darauf basieren, daß die Nitrogruppe reduziert, das Amin diazotiert und zu einem Farbstoff gekuppelt wird. So bestimmen GLAZKO, WOLF und DILL (28) sowie SANTI und SEREMBE (67) Chloramphenicol im Plasma und Serum, indem sie die Nitrogruppe mit $SnCl_2$ reduzieren, dann diazotieren und mit Bratton-Marschall-Reagens ($\alpha$-Naphthyläthylendiamino-chlorhydrat) kuppeln. Es wird ein roter Farbstoff erhalten, der leicht kolorimetriert werden kann. Diese Methode ist auch für Mikrobestimmungen ausgearbeitet worden (3). Ganz ähnliche kolorimetrische Bestimmungsmöglichkeiten sind in einer Anzahl von weiteren Arbeiten veröffentlicht worden (23, 27, 46, 48, 58).

Auf etwas andere Art läßt sich, ebenfalls kolorimetrisch, das Chloramphenicol nach FREEMAN (24) (Reaktion der Nitrogruppen des Chloramphenicols mit einer Lösung von Tetraäthylammoniumhydroxyd in Dimethylformamid-Aceton) und TRUHAUT (73) (Farbreaktion mit Alkali und Pyridin) bestimmen.

Farbreaktionen der Spaltprodukte des Chloramphenicols beschreiben NOGUEIRA und Mitarbeiter (47). Danach geben die Produkte nach der alkalischen Hydrolyse Blaufärbung mit Phenolen und gelb- bis dunkelgrüne Färbung mit Nitroprussiat. Saures Hydrolysat wird blau mit Thymol, blau-rot mit Guajakol, gelb-grün bis lichtblau mit heißem $NH_4$-Metavanadat, rosa bis dunkelgrün mit Na-Selenit und blau mit Na-Phosphomolybdat.

Ebenfalls auf der Spaltung des Chloramphenicols und Bestimmung von Bruchstücken bzw. von Elementen (Cl, N) basieren weitere Bestimmungsmöglichkeiten.

Döll (*17*) kocht Chloramphenicol mit 40%iger NaOH und kolorimetriert die farbige Na-Verbindung des entstandenen *p*-Nitrophenols. Puga (*62*) verseift, oxydiert die Spaltbase mit $NaJO_4$ zum $p\text{-}O_2N \cdot C_6H_4 \cdot CHO$, der mit 2,4-Dinitrophenylhydrazin eine bestimmbare orange Färbung gibt. Nach Pesez (*59*) wird die bei der Spaltung entstehende Dichloressigsäure dadurch bestimmt, daß sie zur Glyoxylsäure verseift und diese zum Hydrazon umgesetzt wird. Dieses zeigt im alkalischen Milieu eine bestimmbare Blauviolettfärbung. Farbreaktionen des Chloramphenicols mit $Cu^{++}$, $Co^{++}$, $Ni^{++}$ werden zur Bestimmung, vor allem neben Sulfonamiden, empfohlen (*65*).

Chloramphenicol kann man auch indirekt azotometrisch nach Van Slyke quantitativ bestimmen. Dazu reduziert man entweder die Nitro- zur Aminogruppe und setzt sie mit $HNO_2$ um (*44*), oder man spaltet das Chloramphenicol, hydrolysiert und bestimmt die Spaltbase in gleicher Weise (*27*). Bei indirekten titrimetrischen Methoden ermittelt man den Cl-Gehalt nach der Verseifung durch Titration nach Volhard (*65*) oder ebenfalls argentometrisch nach Reduktion mit Zn und $H_2SO_4$ (*75*). Eine weitere titrimetrische Methode wird von Valseth und Wickstrøm beschrieben (*76*); Chloramphenicol wird nach vorheriger Hydrolyse mit Jodat oxydiert und mit Arsenit-Jod zurücktitriert.

Schließlich ist noch die Verwendung der Papierchromatographie (*68*) und der Polarographie (*30*) für die Chloramphenicol-Bestimmung zu erwähnen.

Die biologische Bestimmung erfolgt im allgemeinen nach der Methode der Verdünnungsreihe oder des Plattentestes, wobei als Testorganismen *E. coli*, *B. subtilis*, *B. cereus*, *B. dysenteriae* (*41*), Salmonellaarten, *Staph. aureus* (*41*) u. dgl. verwendet werden.

# II. β-Nitropropionsäure.

Im Jahre 1920 isolierte Gorter (*29*) aus der Wurzelrinde eines in Java vorkommenden Baumes, *Hiptage Madablota* Gaertn. (Familie Malpighiaceae), ein neues Glucosid, das Hiptagin. Dieses lieferte bei der Behandlung mit verdünnter Salzsäure neben Glucose, Ammoniak und $CO_2$ eine farblose, kristalline Substanz vom Schmp. 68°. Die Verbindung hatte die Bruttoformel $C_3H_5O_4N$, war eine starke Säure, konnte aber nicht näher charakterisiert werden. 1934 konnte Carrie (*9*) das Glucosid Karakin, das aus den Beeren des neuseeländischen Baumes *Corynocarpus laevigata* Forst. (Familie Anacardiaceae) isoliert wurde, in gleicher Weise spalten wie Gorter das Hiptagin, und er erhielt hiebei ebenfalls neben Glucose, $CO_2$ und $NH_3$ eine kristallisierte Säure vom Schmp. 68°, die sich mit der Hiptaginsäure identisch erwies. Mit dem Karakin hat sich 1943 auch Carter (*10*) beschäftigt, ohne allerdings die Konstitution der Verbindung zu ermitteln. 1949 gelang es dann Carter und McChesney (*11*), die Hiptaginsäure als β-Nitropropionsäure zu identifizieren. Vorher schon hatte man die Leguminose *Indigofera endecaphylla* Jacq. („trailing indigo") aus dem Orient nach Hawai und Lateinamerika eingeführt und festgestellt, daß sich nach Fütterung mit dieser Pflanze bei den Rindern schwere Vergiftungserscheinungen zeigten. Morris, Pagán und Warmke (*42*) (1954) konnten das giftige Prinzip isolieren und als $\beta$-Nitropropionsäure charakterisieren. Die Autoren gaben der Vermutung Ausdruck, daß die $\beta$-Nitropropionsäure möglicherweise eine wichtige Rolle im Nitratstoffwechsel spielen könnte. Diese Annahme scheint dadurch bekräftigt, daß inzwischen PAILER und NOWOTNY (53) in den Wurzeln einer weiteren Pflanze, des wohlriechenden Veilchens (*Viola odorata*, Fam. Violaceae) ebenfalls $\beta$-Nitropropionsäure nachweisen konnten.

Zur Isolierung der $\beta$-Nitropropionsäure wurde der methanolische Extrakt der Droge eingedampft, der Rückstand mit 2 n-HCl aufgenommen und die saure Lösung mit Äther ausgeschüttelt. Mit Hilfe einer Austauschersäule ließ sich aus der ätherischen Lösung die $\beta$-Nitropropionsäure abtrennen und charakterisieren. Sie wurde u. a. auch katalytisch zum $\beta$-Alanin hydriert.

Sehr interessant ist die Tatsache, daß $\beta$-Nitropropionsäure auch ein Stoffwechselprodukt von Pilzen ist. BUSH und GOTH (5) isolierten und untersuchten Flavicin, eine dem Penicillin ähnliche Substanz aus Kulturen von *Aspergillus flavus*. Bei der Reinigung dieses Antibiotikums fanden dann BUSH, GOTH und DICKISON (6) noch farblose Kristalle vom Schmp. 67,5—68,5° und dem Äquivalentgewicht 124, welche keine antibiotische Wirkung hatten, und die in einer Menge von 900 mg pro 90 l Kulturlösung anfielen. BUSH, TOUSTER und BROCKMAN (7) wiesen nach, daß diese Substanz ebenfalls $\beta$-Nitropropionsäure war. Die Autoren machten sich auch Gedanken über die Herkunft der $\beta$-Nitropropionsäure und erklärten, daß nicht zu sagen sei, ob die Verbindung durch enzymatischen Abbau einer Substanz aus dem Nährmedium, dem „Corn steep liquor", oder durch Aufbau aus einfachen Bausteinen entstanden sei.

$\beta$-Nitropropionsäure fanden auch NAKAMURA und SHIMODA (45) in Kulturen von *Aspergillus oryzae* neben dem neuen Antibiotikum „Oryzacidin". Das Nährmedium bestand aus einer Lösung von Zucker, Pepton und Mineralsalzen; Stickstoff war nur im Pepton vorhanden.

RAISTRICK und STÖSSL (63) berichten, daß *Penicillium atrovenetum* G. SMITH beachtliche Mengen $\beta$-Nitropropionsäure produziert, wenn man Raulin-Thom-Lösung als Kulturmedium verwendet. Diese Lösung besteht aus Mineralsalzen, Zucker, Weinsäure und Ammonsalzen. $\beta$-Nitropropionsäure wird auch gebildet, wenn man *P. atrovenetum* in Czapek-Dox-Lösung züchtet, in welcher Natriumnitrat die einzige Stickstoffquelle ist. Paradoxerweise ist dann die Ausbeute an $\beta$-Nitropropionsäure nur 10% der Menge, welche aus der Raulin-Thom-Lösung isoliert werden konnte.

RAISTRICK und STÖSSL (63) haben die Bildung der $\beta$-Nitropropionsäure in verschiedenen Entwicklungsstadien von *P. atrovenetum* G. SMITH studiert und gefunden, daß die Produktion von $\beta$-Nitropropionsäure mit der Entwicklung des Pilzes zu- und bei Beendigung des Wachstums wieder abnimmt. Außerdem wurden 65% des Ammoniakstickstoffes, der von dem Pilz umgesetzt und welcher nicht für den Aufbau des Mycels verwendet wurde, in der β-Nitropropionsäure wiedergefunden.

# III. Aristolochiasäuren.

## 1. Über ältere Arbeiten.

Die Familie Aristolochiaceae umfaßt ungefähr 180 Vertreter, die auf der Welt weit verbreitet sind und von denen etwa zwanzig in Europa, und zwar im Mittelmeergebiet, heimisch sind. Viele dieser Aristolochia-arten spielen seit alters her in der Volksmedizin eine wichtige Rolle und wurden vor allem als Wundheilmittel und, wie der Name sagt, als wehen-fördernde Droge empfohlen.

Griechen und Römer schätzten in der Geburtshilfe die *A. rotunda* und die *A. clematitis* L., und auch in Indien verwendete man bereits seit Jahrtausenden die *A. indica* L. Das hohe Ansehen, welches die Aristolochia bereits bei den griechischen und römischen Ärzten des Altertums genoß, kommt am besten in folgendem Zitat zum Ausdruck: „Zu den edelsten Gewächsen gehört auch die Aristolochia." (C. PLINIUS Sec., Naturalis hist. lib. XXV.)

Die medizinische Schule von Salerno erwähnt die Aristolochia im 11. Jahrhundert, und das ganze Mittelalter hindurch waren besonders die Wurzeln von *A. rotunda* und *A. cava* L. vielfach in Anwendung. Eine Reihe unangenehmer Nebenwirkungen, wie Erbrechen, Durchfall, hämorrhagische Nephritis, Leber-verfettung u. dgl., ließen die Drogen wieder in Vergessenheit geraten. Erst in neuerer Zeit wird die Aristolochia wieder als Heilmittel verwendet. Heute sind die Wurzeln von *A. reticulata* NUTTAL und *A. serpentaria* als *Radix serpentariae* in den Vereinigten Staaten von Amerika und jene von *A. indica* in Indien als *Aristolochia* offizinell. In neuerer Zeit beschäftigt man sich auch wieder mit der *A. clematitis* L., der Osterluzei, die besonders in den östlichen europäischen Ländern in der Volks-medizin verwendet wird.

Neben dieser medizinischen Anwendung der Aristolochia wird diese besonders in Brasilien, Indien und Mexiko schon seit langem gegen Schlangenbiß verwendet.

Es war naheliegend, daß man sich verhältnismäßig früh mit der experimentellen Untersuchung der Wirksamkeit dieser interessanten Droge beschäftigte. So konnte bereits MURRAY (*43*) (1793) feststellen, daß Schlangen vor der „betäubend riechen-den" *A. anguicida* L. fliehen. Ein bis zwei Tropfen des Wurzelsaftes genügen, um das Tier zu betäuben, mehrere Tropfen töten es. ORFILA (*50*) (1818) prüfte die Wirkung der Wurzeln von *A. clematitis* L. an Hunden und stellte leichte Ent-zündungen der Organe fest, mit welchen sie in Berührung kamen. FRICKHINGER (*25*) beschrieb einen Selbstversuch mit der Wurzel von *A. clematitis* L. Ekel, Erbrechen, Schüttelfrost und vor Müdigkeit gelähmte Glieder waren die Folge. In neuerer Zeit prüften LALANNE und MATHOU (*37*) Extrakte von *A. eurystoma* DUCKARTRE.

Mit dem 19. Jahrhunderts wurden die ersten chemischen Untersuchungen an Aristolochiaarten durchgeführt. Nach BUCHOLZ (*4*) beschäftigte sich CHEVALLIER (*14*) (1820) mit der *A. serpentaria*. Er beschrieb erstmalig einen gelben Bitterstoff, den er neben einem ätherischen Öl und anderen indifferenten Stoffen isolierte. SOBRAL (*69*) fand in der *A. cymbifera* ein der „Quassia, Gentiana u. dgl. sehr ähnliches Bitter". Ebenso beschrieben BRANDES (*8*) (1835) und WITT-STEIN (*80*) (1836) eine gelbe Substanz, die sie aus *A. grandiflora* und *A. antihysterica* isolierten.

*Literaturverzeichnis: SS. 78—82.*

Über die chemische Natur dieses gelben Stoffes wurde bis dahin nichts ausgesagt. Man hatte vor allem auch nur uneinheitliche, amorphe Substanzen zur Verfügung. WINCKLER (79) machte sich als erster die sauren Eigenschaften des gelben Bitterstoffes zunutze und extrahierte die vorher mit Wasserdampf behandelten Wurzeln von *A. clematitis* mit verd. Alkali. Aus dieser Lösung fällte er durch Ansäuern mit verd. Schwefelsäure eine amorphe, gelbbraune Substanz.

FRICKHINGER (25) beschrieb einen aus der *A. clematitis* erhaltenen gelben Stoff von saurer Natur, den er als „Aristolochiagelb" bezeichnete; WALZ (77) erhielt ebenfalls aus der *A. clematitis* eine gelbe Substanz, die er „Aristolochiabitter" nannte und für die er die Formel $C_9H_{10}O_6$ aufstellte. PECKOLT (57) fand in den Wurzeln von *A. cymbifera genuina* MASTERS einen Bitterstoff, den er als „Aristolochin" bezeichnete. Ein gelber Bitterstoff wurde ferner von MAISCH (40) aus *A. fragrantissima* RUIZ et PAVON und von FERGUSON (21) aus *A. reticulata* isoliert. In der *A. indica* fanden DYMOCK und WARDEN (18) neben einem Alkaloid eine amorphe gelbe Säure.

POHL (61) (1892) hat erstmalig über die gelben Verbindungen eingehendere Untersuchungen angestellt. Er isolierte aus *A. clematitis*, *A. rotunda* und *A. longa* eine stickstoffhaltige saure Verbindung, die er als Aristolochin bezeichnete. Die von ihm ermittelte Bruttoformel war $C_{32}H_{22}O_{13}N_2$. POHL hat vor allem festgestellt, daß das Aristolochin für die Wirkung der Droge auf Warmblüter verantwortlich ist, während die „Schlangenwirksamkeit" einem ätherischen Öl zuzuschreiben ist. HESSE (31) (1895) isolierte aus *A. argentina* neben einem Alkaloid, das er „Aristolochin" nannte, drei gelbe Säuren:

Aristinsäure: Schmp. 260°, $C_{18}H_{13}O_7N$; 1,5% $OCH_3$ (ZEISEL); bei der Kalischmelze wurde $NH_3$ abgespalten.

Aristidinsäure: Schmp. 260°, $C_{16}H_{13}O_7N$; eine $OCH_3$-Gruppe.

Aristolsäure: Schmp. 260—270°, $C_{15}H_{11}O_7N$.

Er hielt das POHLsche „Aristolochin", für das er den Namen Aristolochiasäure vorschlug, für nahe verwandt mit seinen drei Säuren und glaubte, daß ihm die Formel $C_{17}H_{11}O_7N$ zuzuschreiben sei. Aus den Wurzeln der *A. sipho* isolierte CASTILLE (13) in 1%iger Ausbeute eine Substanz, die sich mit dem Aristolochin von POHL als identisch erwies und die er daher nach dem Vorschlag von HESSE als Aristolochiasäure bezeichnete.

Er konnte die von HESSE vorgeschlagene Formel $C_{17}H_{11}O_7N$ bestätigen und durch Molgewichtsbestimmung und Titration bestimmen, daß die Aristolochiasäure einbasisch ist. Durch Reduktion mit Zinkstaub und Eisessig wurde eine Verbindung erhalten, der am besten die Formel $C_{17}H_{13}O_4N$ entsprach. Durch Methylierung der Aristolochiasäure mit Dimethylsulfat und Alkali erhielt CASTILLE den Methylester [Schmp. 260—261° (Zers.)], der sich durch Alkalien in der Hitze angeblich leicht wieder verseifen ließ. Auf Grund der Analysen wurde angenommen, daß zwei Methylgruppen eingetreten seien. Bei der Kalischmelze entwich der Stickstoff in Form von Ammoniak, und aus dem

Rückstand isolierte Castille einen mit Brom fällbaren phenolischen Körper sowie eine Verbindung, die die Eigenschaften der Anthrachinone zeigte. Auf Grund dieser Ergebnisse nahm er an, daß das Grundgerüst der Aristolochiasäure ein Anthrachinonkern sei und der Stickstoff in tertiärer Bindung vorliegt.

### 2. Neuere Arbeiten zur Isolierung, Reindarstellung und Konstitutionsermittlung.

Krishnaswamy und Mitarbeiter (34, 35) isolierten aus der *A. indica* neben ätherischen Ölen, Fetten und Zuckern ein Alkaloid $C_{17}H_{19}O_3N$ und eine Säure $C_{17}H_{11}O_7N$, Schmp. 275° (Zers.), die sie Iso-aristolochiasäure nannten. Die Säure besitzt dieselbe Bruttoformel wie die Aristolochiasäure von Castille bzw. das Aristolochin von Pohl und zeigte auch ähnliche Eigenschaften. Da aber Pohl für seine Säure den Schmp. 215° angab, nahmen die Autoren an, daß ihre Säure mit der Pohlschen isomer sei. Sie stellten aktiven Wasserstoff fest, fanden keine Methoxyl- und Methylendioxygruppe, keine Reaktion mit Ketonreagenzien und keine Bildung eines Jodmethylates. Weder siedende 50%ige KOH noch Kochen mit Essigsäureanhydrid veränderten die Substanz. Mit Dimethylsulfat ließ sich ein neutrales Methylderivat $C_{18}H_{13}O_7N$, Schmp. 267° (Zers.), erhalten, das sich durch mehrstündiges Kochen mit 1-n-methanolischem KOH nicht verseifen ließ. Durch Einwirkung von Benzoylchlorid und Alkali wurde in schlechter Ausbeute ein neutraler Stoff der Formel $C_{24}H_{15}O_8N$ erhalten. Auf Grund dieser Ergebnisse vermuteten die Autoren, daß die Iso-aristolochiasäure keine Carboxylgruppe, sondern eine Enolgruppe enthält. Durch Oxydation der Iso-aristolochiasäure mit $H_2O_2$ ließ sich in geringer Ausbeute eine weiße, kristallisierte Dicarbonsäure $C_{16}H_{13}O_9N$ isolieren.

Eine ausführliche Zusammenfassung, besonders der älteren Literatur über Aristolochia, über die Drogen, die Inhaltsstoffe und ihre pharmakologische Wirksamkeit findet sich in einer neueren Arbeit von Rosenmund und Reichstein (66) (1943) und in zwei Veröffentlichungen von Pilarczyk (60). Die Schweizer Autoren isolierten auch aus den Wurzeln von *A. sipho* eine gelbe, kaliumbikarbonat-lösliche Substanz, die nach Umlösen aus Dioxan den Schmp. 274—278° (Zers.) hatte und deren Analysen auf $C_{17}H_{11}O_7N$ stimmten. Die Methoxylbestimmung gab einen Wert von 1,3%, während für eine $OCH_3$-Gruppe 9,09% berechnet waren. Es wurde angenommen, daß die geringe Menge $CH_3J$, welche bei der Methoxylbestimmung erhalten wurde, nicht, wie Pohl erklärte, von einer Verunreinigung stammt, sondern von einem leicht abspaltbaren N-Methyl. Durch Veresterung mit Diazomethan wurde ein Methylester erhalten, der nach Sublimation im Hochvakuum bei 280—282° (Zers.) schmolz und der sich auch durch mehrstündiges Kochen mit alkoholischer

KOH verseifen ließ. Die Oxydation der Aristolochiasäure mit $KMnO_4$ und Chromsäure brachte keine verwertbaren Ergebnisse.

Die Hydrierung des Methylesters mit $PtO_2$ in Eisessig lieferte eine Verbindung, für welche die Autoren auf Grund der Analysenergebnisse die Formel $C_{13}H_{18}O_4N + \frac{1}{2} H_2O$ annahmen. Durch Acetylierung dieses Hydrierungsproduktes erhielten sie eine bei 306—308° schmelzende kristallisierte Verbindung, für welche sie auf Grund der Analysen (ohne Acetylbestimmung) als Formel $C_{22}H_{15}O_6N$ bzw. $C_{22}H_{17}O_6N$ errechneten und das Vorliegen einer Diacetylverbindung vermuteten. Daraus schlossen ROSENMUND und REICHSTEIN auf zwei Hydroxygruppen im Molekül der hydrierten Aristolochiasäure. Zur Feststellung des Carbonsäurecharakters wurde die Aristolochiasäure durch Erhitzen mit Cu-Pulver in Chinolin decarboxyliert. Unter Abspaltung von 1 Mol $CO_2$ bildete sich eine Substanz $C_{16}H_{11}O_5N$, die bei 206—212° schmolz. Die Aristolochiasäure, der Methylester und das Decarboxylierungsprodukt reagierten nicht mit Ketonreagenzien und auch nicht mit Essigsäureanhydrid in Pyridin. Eine eindeutige Schlußfolgerung konnten die Autoren nach ihren Angaben nicht ziehen, sie erklärten aber, daß die Resultate darauf hindeuten, daß möglicherweise eine chinoide Gruppierung vorhanden ist, die durch Reduktion in ein empfindliches Phenol übergeht.

ROSENMUND und REICHSTEIN (66) schrieben ferner, daß sie, ohne einen direkten Beweis angeben zu können, der Vermutung Ausdruck geben möchten, daß das „Aristolochiagelb" von FRICKHINGER, das „Aristolochin" von POHL, die „Aristinsäure" von HESSE, die Aristolochiasäure von CASTILLE und die „Iso-aristolochiasäure" von KRISHNASWAMY alle ein und dieselbe Substanz darstellen, die sie nach HESSES Vorschlag als Aristolochiasäure bezeichneten. Durch Vergleich der bis dahin angegebenen Schmelzpunkte und Analysenwerte begründeten die beiden Autoren ihren Vorschlag.

In neuerer Zeit haben sich PAILER und Mitarbeiter (51, 52, 55, 56) mit der Untersuchung der Aristolochiasäure, die sie als Aristolochiasäure-I bezeichneten, und einer zweiten, ganz ähnlich gebauten Verbindung, der Aristolochiasäure-II, beschäftigt und die Konstitution dieser Verbindungen ermittelt.

Zusammenfassend sei nochmals kurz zitiert, welche Ergebnisse bis dahin bei der Bearbeitung der Aristolochiasäure-I erhalten worden waren: Eine einbasische Carbonsäure der Formel $C_{17}H_{11}O_7N$ mit einer N-Methyl-, keiner Methoxyl- und keiner Methylendioxygruppe. Die Verbindung enthält nur einen aktiven Wasserstoff. Es ließen sich weder acetylierbare Hydroxygruppen noch reaktive Ketogruppen nachweisen. Über die Funktion der restlichen fünf Sauerstoffe ließ sich nichts aussagen. Die Aristolochiasäure-I bzw. ihr Methylester ließen sich reduzieren, wobei sich die Anzahl der O-Atome von 7 auf 4 verminderte. Für die reduzierte Säure fand CASTILLE die Formel $C_{17}H_{13}O_4N$, für den hydrierten Ester fanden ROSENMUND und REICHSTEIN $C_{18}H_{11-13}O_4N$ bzw. für den reduktiv acetylierten Ester $C_{22}H_{15-17}O_6N$, was einem Diacetylprodukt der Säure $C_{17}H_{11-13}O_4N$ entsprechen würde.

Als Drogenmaterial wurden von PAILER und Mitarbeitern (*51, 52, 55, 56*) Rhizome und Wurzeln von *A. clematitis* L. verwendet. Die getrockneten und fein gemahlenen Wurzeln wurden in Petroläther und Äther entfettet und mit Alkohol extrahiert. Nach Aufarbeitung dieses Extraktes wurden die bicarbonatlöslichen Bestandteile abgetrennt, durch die Schwerlöslichkeit des K-Salzes die Aristolochiasäure-I angereichert und durch Umlösen aus Dimethylformamid und Alkohol gereinigt. Schmp. 281—286° (Zers.).

### 3. Konstitution der Aristolochiasäure-I.

Die Aristolochiasäure-I (XXVII, S. 71) gab mit Diazomethan einen Methylester (XXVIII), Schmp. 287—288°. Die Verbindung (*Formelübersicht 2*, S. 71) wurde auch analog den Angaben von ROSENMUND und REICHSTEIN mit Cu-Pulver in Chinolin decarboxyliert, und die so erhaltene Verbindung (XXIX) hatte die erwartete Bruttoformel $C_{16}H_{11}O_5N$ und Schmp. 212°. Damit war die Carbonsäurenatur der Aristolochiasäure, die gelegentlich angezweifelt worden war, bewiesen. Bei geringer Modifizierung der Methoxylbestimmungsmethode, wobei vor allem auf die geringe Löslichkeit der Aristolochiasäure Bedacht genommen wurde, konnten entgegen früheren Befunden für die Säure eindeutig ein und für den Methylester zwei Methoxyle festgestellt werden. Da später bei einer N-freien Abbauverbindung noch immer eine Methoxylgruppe nachweisbar war und die Kalischmelze nicht $NH_2 \cdot CH_3$, sondern $NH_3$ lieferte, schien das Vorhandensein eines Methoxyls und die Abwesenheit einer N-Methylgruppe bewiesen. Die Zinkstaubdestillation der Aristolochiasäure lieferte Phenanthren, das durch Schmelzpunkt, Mischprobe und UV-Spektrum identifiziert werden konnte. Die weitgehende Übereinstimmung der UV-Absorptionskurven der decarboxylierten Verbindung (XXIX) und eines anderen Abbauproduktes mit Phenanthrenderivaten und weitere Abbauergebnisse bestätigten, daß das Phenanthrengerüst schon im Naturstoff vorhanden war und nicht erst unter den drastischen Bedingungen der Zinkstaubdestillation gebildet wurde.

Die katalytische Hydrierung der Aristolochiasäure in Eisessig mit $PtO_2$ oder Pd-Tierkohle als Katalysator führte unter Aufnahme von 3 Molen Wasserstoff und Verlust von 3 Molen $H_2O$ zu einer Substanz $C_{17}H_{11}O_4N$ vom Schmp. 319° (XXX), die weder basische noch saure Eigenschaften zeigte. Dieselbe Verbindung wurde auch bei der Hydrierung des Methylesters der Aristolochiasäure unter Aufnahme von 3 Molen $H_2$ und Abspaltung von 2 Molen $H_2O$ und 1 Mol $CH_3OH$ erhalten. Daraus ergab sich der Schluß, daß in beiden Fällen, bei der Säure und beim Ester, eine Gruppe unter Aufnahme von 3 Molen Wasser-

*Formelübersicht 2.* Struktur der Aristolochiasäure-I.

stoff und Abspaltung von 2 Molen Wasser hydriert wird und daß diese neue Gruppe nun mit der Carboxylgruppe der Säure unter Abspaltung von Wasser und mit der Carbomethoxygruppe des Esters unter Abspaltung von $CH_3OH$ reagiert und die Verbindung (XXX) ergibt.

Bei der katalytischen Hydrierung der Decarboxy-aristolochiasäure-I war ebenfalls die Aufnahme von 3 Molen Wasserstoff feststellbar. Es wurden dabei nur 2 Mole $H_2O$ abgespalten und eine Verbindung $C_{16}H_{13}O_3N$ gebildet, Schmp. 170° (XXXI). Diese Verbindung war sehr labil und mußte unter besonderen Vorsichtsmaßnahmen isoliert werden. Sie war eindeutig, wenn auch schwach basisch und lieferte mit Essigsäureanhydrid ein Acetylderivat (XXXII), das auch bei der reduktiven Acetylierung der Aristolochiasäure-I erhalten wurde.

Diese Ergebnisse ließen sich nur mit dem Vorhandensein einer Nitrogruppe erklären, die bei der Reduktion mit der Carboxylgruppe bzw. der Estergruppe sofort das Lactam (XXX) der entsprechenden Aminosäure lieferte. In den IR-Spektren fanden sich tatsächlich die für eine Nitrogruppe charakteristischen Banden stark ausgebildet. Die Base (XXXI) zeigte die $NH_2$- und ihr Acetylderivat (XXXII) die CO—NH-Banden.

Nach diesen Ergebnissen war die Aufstellung der teilweisen Strukturformel (XXXIII) für die Aristolochiasäure möglich, wenn man alle C-Atome des Phenanthrens mit den bereits identifizierten Gruppen und den restlichen Wasserstoffen besetzte.

—OCH$_3$
—COOH
—NO$_2$
(CO$_2$)

(XXXIII.)

Es verblieben also noch ein C- und zwei O-Atome, die sich theoretisch in verschiedenen Gruppierungen formulieren ließen, von denen aber alle bis auf eine mit den bereits gewonnenen Erkenntnissen nicht in Einklang zu bringen waren.

Obwohl die Frage nach der Anwesenheit einer Methylendioxygruppe von früheren Bearbeitern verneint wurde, konnten diese PAILER, BELOHLAV und SIMONITSCH (51, 52) schließlich doch eindeutig feststellen. Die Bestimmung wurde nicht mit der klassischen Methode durchgeführt, da die Aristolochiasäure-I, aber auch die anderen in Frage kommenden Derivate bereits mit Schwefelsäure allein eine Farbreaktion, und zwar eine intensive Grünfärbung, zeigten. Es wurde daher die Methylendioxygruppe mit 80%iger $H_3PO_4$ gespalten und der gebildete Formaldehyd mit etwas Wasser abdestilliert. Das Destillat gab in 72%iger $H_2SO_4$ mit Chromotropsäure die für Formaldehyd charakteristische Violettfärbung. Dieses Ergebnis wurde auch dadurch bestätigt, daß sich

—OCH$_3$
—COOH
—NO$_2$
—O
        CH$_2$
—O

(XXXIV.)

bei einer Abbauverbindung, die das Phenanthrengerüst nicht mehr enthielt, die Methylendioxygruppe nach der klassischen Methode eindeutig nachweisen ließ. Damit waren die funktionellen Gruppen der Aristolochiasäure-I bestimmt und es ergab sich die Teilformel (XXXIV).

Zur genauen Festlegung des Sitzes der einzelnen Gruppen am Phenanthrengerüst wurde die decarboxylierte Aristolochiasäure-I (XXIX, S. 71) zu einer Diphensäure abgebaut, und zwar ergab die Oxydation der decarboxylierten Aristolochiasäure-I mit $H_2O_2$ in guter Ausbeute eine Dicarbonsäure (XXXV), die noch alle C-Atome des Ausgangsmaterials enthielt und in der sich noch die Methoxyl- und die Methylendioxygruppe bestimmen ließen. Die Abbausäure hatte die Bruttoformel $C_{16}H_{12}O_7$ und lieferte mit Diazomethan einen Dimethylester $C_{18}H_{16}O_7$. Sie enthielt nicht mehr die Nitrogruppe.

Dieses Abbauergebnis war ein weiterer Beweis für das Phenanthrengerüst des Naturstoffes, zeigte aber vor allem, daß die Nitrogruppe am Kohlenstoffatom 9 oder 10 haftet. Da die Aristolochiasäure-I und ihr Ester bei der Reduktion ein Lactam gaben, mußte also auch die COOH-Gruppe an einem geeigneten, benachbarten Kohlenstoffatom sitzen. Dafür kamen nur die C-Atome 1 bzw. 8 in Frage.

Bei der Behandlung der Abbausäure (XXXV) mit konzentrierter HCl im Rohr, bei Anwesenheit von Resorcin zur Bindung des Formaldehyds, wurden die Äthergruppen gespalten und eine Verbindung (XXXVI) $C_{13}H_8O_4$ erhalten, die, mit Diazomethan methyliert, einen Dimethyläther lieferte, welcher Lactoneigenschaften zeigte.

Zur weiteren Klärung der Konstitution dieses Dimethyläthers wurde in verdünnter Lauge unter Aufspaltung des Lactonringes gelöst und bei pH 8 mit $KMnO_4$ oxydiert. Als charakteristisches Oxydationsprodukt wurde o-Methoxyphthalsäure isoliert. Für die Konstitution des Dihydroxylactons ergaben sich damit nur mehr zwei Möglichkeiten (XXXVI) und (XXXVII, S. 71), von denen (XXXVI) die wahrscheinlichere war. Im anderen Fall (XXXVII) wäre bei der Ätherspaltung wahrscheinlich ein Dilacton entstanden.

Nun wurde das Lacton der 3,2',3'-Trihydroxy-2-biphenyl-carbonsäure (XXXVI) auf eindeutigem Weg synthetisiert und mit der Abbauverbindung identifiziert. 1,5,6-Trimethoxyphenanthren-carbonsäure-(10) (XXXVIII) wurde mit Natriumbichromat zum 1,5,6-Trimethoxyphenanthrenchinon-(9,10) (XXXIX) oxydiert, das dann bei weiterer Behandlung mit $H_2O_2$ 5,6,3'-Trimethoxy-2,2'-biphenyldicarbonsäure (XL) lieferte. Diese gab bei der Ätherspaltung mit HCl ein Lacton, das der beim Abbau des Naturstoffes erhaltenen Verbindung entsprach. Auch der daraus hergestellte Methyläther erwies sich mit dem aus der Abbauverbindung hergestellten identisch.

Damit ergab sich für die Diphensäure, welche bei der Oxydation der decarboxylierten Aristolochiasäure-I erhalten worden war, die Formel (XXXV). Die beiden Carboxylgruppen dieser Dicarbonsäure entsprachen den C-Atomen 9 und 10 des Phenanthrenringes im Naturstoff, von dem eines nach den bisher geschilderten Ergebnissen Träger der Nitrogruppe sein mußte. Da die Nitrogruppe bei der Reduktion

mit der Carboxylgruppe ein Lactam gab, kam also für die Carboxylgruppe, wie bereits erwähnt wurde, als Platz am Phenanthrengerüst nur
das C-Atom 1 oder 8 in Frage. C-Atom 8 ist mit einer Methoxylgruppe
besetzt, so daß die Carboxylgruppe nur mehr am C-Atom 1 haften konnte.
Die Aristolochiasäure-I ist somit eine 3,4-Methylendioxy-8-methoxy-10-
nitrophenanthren-carbonsäure-(1) (XXVII).

## 4. Konstitution der Aristolochiasäure-II.

Da die direkte Isolierung der Aristolochiasäure-II und ihre Reindarstellung große Schwierigkeiten bereitete, wurde das durch die Schwerlöslichkeit der K-Salze abgetrennte Gemisch der Aristolochiasäure-I
und der Aristolochiasäure-II mit Diazomethan methyliert und auf einer
$Al_2O_3$-Säule chromatographiert. Dabei ließen sich der Aristolochia-
säure-I- und -II-methylester gut trennen und außerdem konnten noch,
diesen beiden Zonen nachfolgend, drei weitere schmale Zonen festgestellt
werden, die drei, den Aristolochiasäure-I- und -II-estern sehr ähnlichen
Estern entsprechen dürften. Wegen der geringen Menge wurden sie
nicht weiter untersucht.

In ähnlicher Weise ließ sich auch das Gemisch der Aristolochiasäuren
decarboxylieren und dann auf einer $Al_2O_3$-Säule trennen. Diese Trennung
erfolgte sehr leicht und wesentlich besser als die der Methylester. Wie bei
der Chromatographie der Methylester wurden auch hier wieder 5 Zonen
erhalten, von denen die beiden Hauptmengen der decarboxylierten
Aristolochiasäure-II (Schmp. 174—175°) und der decarboxylierten
Aristolochiasäure-I (Schmp. 212°) entsprachen.

Der Aristolochiasäure-II-methylester vom Schmp. 274° ergab die
erwartete Bruttoformel $C_{17}H_{11}O_6N$, lieferte bei der Methoxylbestimmung
nach ZEISEL Werte für ein $OCH_3$ und nach der erwähnten Methode einen
positiven Test auf die Methylendioxygruppe. Das UV-Spektrum zeigte
gegenüber dem Spektrum des Aristolochiasäure-I-methylesters eine
hypsochrome Verschiebung, bedingt durch das Fehlen der Methoxylgruppe, und die für Phenanthrenderivate typische Bande bei 251 m$\mu$.
Nach diesen Befunden erschien es sehr wahrscheinlich, daß der
Aristolochiasäure-II die Konstitution einer Methylendioxy-nitro-
phenanthren-carbonsäure zukommt.

Der Aristolochiasäure-I-methylester war schwer verseifbar und auch
die Verseifung des Aristolochiasäure-II-methylesters machte Schwierigkeiten. Beim Kochen mit alkoholischer KOH wurden die Substanzen
tiefgreifend verändert und die so erhaltenen Verbindungen zeigten nicht
mehr die für das Phenanthrengerüst charakteristische UV-Absorption.

Da auf dem Weg über die Veresterung, Chromatographie und Verseifung die Darstellung einer für die Konstitutionsermittlung ausreichenden
Menge Aristolochiasäure-II zu verlustreich war, wurde der durch

*Literaturverzeichnis: SS. 78—82.*

Chromatographie erhaltene Aristolochiasäure-II-methylester und die decarboxylierte Aristolochiasäure-II, die aus dem Rohsäuregemisch hergestellt worden waren, für die weiteren Untersuchungen verwendet.

*Formelübersicht 3.* Struktur der Aristolochiasäure-II.

Bei der katalytischen Hydrierung des Aristolochiasäure-II-methylesters (XLI) *(Formelübersicht 3)* war wie beim -I-methylester die Aufnahme von 3 Molen Wasserstoff unter Austritt von 2 Molen $H_2O$ und 1 Mol $CH_3OH$ zu beobachten und es entstand ein Neutralstoff der Formel $C_{16}H_9O_3N$ (XLII). Die Hydrierung verlief also analog der des Aristolochiasäure-I-methylesters.

Oxydation des Aristolochiasäure-II-methylesters mit Perhydrol in alkalischer Tetrahydrofuranlösung ergab eine aromatische Methylendioxycarbonsäure, die als Trimethylester der Formel (XLIII) $C_{19}H_{16}O_8$ isoliert und charakterisiert wurde. Durch die Bildung der Biphenylcarbonsäure war auch bei der Aristolochiasäure-II die Stellung der Nitrogruppe auf 9 und 10 und damit, unter Berücksichtigung der Bildung des Lactams bei der Hydrierung, die der Carboxylgruppe auf 1 oder 8 festgelegt.

Der Aristolochiasäure-II-methylester wurde weiters mit HCl und Resorcin zur Spaltung der Methylendioxygruppe erhitzt und das so gewonnene Produkt, ohne Isolierung und Reindarstellung, mit $KMnO_4$ in alkalischer Lösung oxydiert. Als einziges Abbauprodukt wurde dabei Phthalsäure, isoliert als Anhydrid, erhalten. Damit war bewiesen, daß die Methylendioxygruppe und die Carboxylgruppe in der Aristolochiasäure-II am selben Ring haften, da andernfalls Hemimellithsäure hätte entstehen müssen. Damit war die Konstitution der Aristolochiasäure-II als die einer 3,$x$-Methylendioxy-10-nitrophenanthren-carbonsäure-(1) festgelegt.

Die genaue Stellung der Methylendioxygruppe wurde durch Abbau der decarboxylierten Aristolochiasäure (XLIV) ermittelt. Da diese aber über den gut zu reinigenden Aristolochiasäure-II-methylester zu schwer zugänglich war, wurde das ursprüngliche Gemisch der Säuren mit Kupferpulver in Chinolin decarboxyliert und die so gewonnenen Substanzen an Aluminiumoxyd chromatographiert. Es wurden wie bei der Auftrennung der Ester fünf Zonen erhalten, von denen die erste der decarboxylierten Aristolochiasäure-II entsprach. Das zweite breite Band ergab die schon beschriebene, decarboxylierte Aristolochiasäure-I, während die drei folgenden, rotorange gefärbten Zonen nicht weiter untersucht wurden.

Die decarboxylierte Aristolochiasäure-II (XLIV) ließ sich mit Pd-Tierkohle unter Aufnahme von 3 Molen Wasserstoff zum Amin (XLV) reduzieren, das sehr labil war. Das N-Acetylderivat (XLVI) wurde aus dem Amin und auch direkt aus der decarboxylierten Aristolochiasäure-II durch reduktive Acetylierung hergestellt. Pailer und Schleppnik (56) konnten diese Verbindung dann auch, ausgehend von der 3,4-Methylendioxyphenanthrencarbonsäure-10, mittels eines Säureazid-Abbaues bzw. eines modifizierten Schmidt-Abbaues synthetisieren. Das war ein weiterer Beweis der vorher referierten Abbauergebnisse.

Der oxydative Abbau der decarboxylierten Aristolochiasäure-II mit $H_2O_2$ in alkalischer Tetrahydrofuranlösung gab eine Methylendioxy-biphenyldicarbonsäure $C_{15}H_{10}O_6$ (XLVII), welche bei der Ätherspaltung das 3,4-Benzo-8-hydroxycumarin lieferte (XLVIII). Damit war die Abbausäure als 5,6-Methylendioxy-biphenyl-dicarbonsäure-(2,2') eindeutig charakterisiert. Die Aristolochiasäure-II ist somit die 3,4-Methylendioxy-10-nitrophenanthren-carbonsäure-(1) (XLIX) und unterscheidet sich von der Aristolochiasäure-I nur durch das Fehlen der Methoxylgruppe.

### 5. Weiteres Vorkommen von Aristolochiasäuren und Isolierung und Charakterisierung von strukturell verwandten Naturstoffen.

Tomita und Kura (72) isolierten aus *A. Kaempferi* Willd. und aus *A. debilis* Sieb. et Zucc. ein Gemisch von Aristolochiasäuren, in welchem die Aristolochiasäure-I in überwiegender Menge enthalten sein dürfte. Reine Aristolochiasäure-I wurde ferner von Coutts, Stenlake und Williams (16) aus *A. indica* und *A. reticulata* erhalten. Sie scheint in diesen Pflanzen als einzige Aristolochiasäure vorzukommen, da sie allein durch Umkristallisieren gereinigt werden konnte. Daneben wurde eine weitere Substanz mit konstitutioneller Beziehung zur Aristolochiasäure, das ,,Aristored" (L), aufgefunden. Dieser Verbindung wurde auf Grund der chemischen und spektroskopischen Befunde die Struktur eines 3,4-Methylendioxy-8,$x$,$y$-trimethoxy-phenanthrencarbonsäure-(1)-lactams zugeschrieben. Das Grundgerüst dieser Verbindung entspricht also dem Lactam, welches bei der Reduktion der Aristolochiasäure erhalten wurde (XXX, S. 71).

Aus *A. debilis* wurde auch von Tseng und Ku (74) Aristolochiasäure isoliert, sowie als Inhaltsstoff eine weitere gelbe, kristalline Substanz vom Schmp. 269° beschrieben. Die letztere Verbindung könnte möglicherweise mit der von Pailer und Mitarb. (51, 52, 55) aufgeklärten Aristolochiasäure-II identisch sein.

Es erscheint sehr wahrscheinlich, daß die beiden bisher aufgeklärten Aristolochiasäuren, aber möglicherweise auch noch andere, ähnliche Nitrokörper in einer Anzahl der vielen Arten der Familie Aristolochiaceae aufgefunden werden.

Neben den Aristolochiasäuren konnten Tomita und Kura (72) aus *A. Kaempferi* Willd. und *A. debilis* Sieb. et Zucc. eine quartäre Base der Aporphingruppe, das Magnoflorin (LI), isolieren. Dieses Alkaloid wurde auch von Pailer und Pruckmayr (54) in den Wurzeln von *A. clematitis* festgestellt. Die große Ähnlichkeit der Formeln der Aristolochiasäuren und dieser Base (Phenanthrengrundgerüst, Sitz des N am gleichen C des Ringsystems usw.) und das gemeinsame Vorkommen dieser Naturstoffe in den Drogen deutet auf einen engen biogenetischen Zusammenhang hin. Es ist wahrscheinlich, daß man auch die dem Magnoflorin entsprechende Aristolochiasäure in diesen Drogen finden wird. Auffällig ist auch die weitgehende Ähnlichkeit der Aristolochiasäure-I mit dem Stephanin (LII), einer natürlichen Aporphinbase, die bisher allerdings nicht aus Aristolochia-Arten isoliert werden konnte. Rein formal kann man sich die Bildung der Aristolochiasäuren durch Oxydation der entsprechenden Aporphinalkaloide vorstellen, ebenso aber umgekehrt die Entstehung der Basen durch Ringaufbau und Reduktion der Aristolochiasäuren. Verbindungen vom Typ des Aristoreds könnten bei diesen Umwandlungen Zwischen-, wahrscheinlicher aber Nebenprodukte sein.

Über die Biogenese dieser Aristolochiasäuren, sowie auch der anderen Naturstoffe mit einer Nitrogruppe, sind bisher keine Arbeiten erschienen.

(L.) Aristored.

(LI.) Magnoflorin.

(LII.) Stephanin.

### Literaturverzeichnis.

1. ALBERTI, C. G., B. ASERO, B. CAMERINO, R. SANNICOLÒ and A. VERCELLONE: The Synthesis of Chloromycetin. Chim. e ind. (Milano) **31**, 357 (1949) [Chem. Abstr. **46**, 7068 (1952)].
2. BENDAS, H. and E. D. BERGMANN: Configuration of Phenylserine and a Synthesis of Chloromycetin. Bull. Res. Council Israel **1**, 131 (1951) [Chem. Abstr. **46**, 7070 (1952)].
3. BESSMAN, S. P. and S. STEVENS: A Colorimetric Method for the Determination of Chloromycetin in Serum or Plasma. J. Lab. Clin. Med. **35**, 129 (1950) [Chem. Abstr. **44**, 3553 (1950)].
4. BUCHOLZ: Berl. Jahresber. **1807**, S. 129 [zitiert nach M. SPICA, Gazz. chim. ital. **17**, 313 (1887)].
5. BUSH, M. T. and A. GOTH: Flavicin; an Antibacterial Substance Produced by an *Aspergillus flavus*. J. Pharmacol. **78**, 164 (1943) [Chem. Abstr. **37**, 5101 (1943)].
6. BUSH, M. T., A. GOTH and H. L. DICKISON: Flavicin. II. An Antibacterial Substance Produced by an *Aspergillus flavus*. J. Pharmacol. **84**, 262 (1945) [Chem. Abstr. **40**, 108 (1946)].
7. BUSH, M. T., O. TOUSTER and J. E. BROCKMAN: The Production of β-Nitropropionic Acid by a Strain of *Aspergillus flavus*. J. Biol. Chem. **188**, 685 (1951).
8. BRANDES, R.: Aristolochiaceae — *Aristolochia grandiflora* GOMES. Buchners Repert. Pharmacie **50**, 365 (1835).
9. CARRIE, M. S.: Karakin, the Glucoside of *Corynocarpus laevigata*. J. Soc. Chem. Ind. (London) **53**, 288 T (1934) [Chem. Abstr. **29**, 173 (1935)].
10. CARTER, C. L.: Karakin, the Glucoside of *Corynocarpus laevigata*, and Hiptagenic Acid. J. Soc. Chem. Ind. (London) **62**, 238 (1943) [Chem. Abstr. **38**, 2629 (1944)].
11. CARTER, C. L. and W. J. MCCHESNEY: Hiptagenic Acid Identified as β-Nitropropionic Acid. Nature (London) **164**, 575 (1949).
12. CARTER, H. E., D. GOTTLIEB and H. W. ANDERSON: Chloromycetin and Streptomycetin. Science (Washington) **107**, 113 (1948).

*13.* CASTILLE, A.: Aristolochic Acid Extracted from *Aristolochia sipho.* J. pharm. Belg. **4**, 569 (1922) [Chem. Abstr. **16**, 3902 (1922)]; cf. Bull. acad. roy. méd. Belg. [5] **1**, 569 (1921) [Chem. Abstr. **16**, 1448 (1922)].

*14.* CHEVALLIER, A.: J. pharm. sci. access. **6**, 565 (1820).

*15.* CONTROULIS, J., M. C. REBSTOCK and H. M. CROOKS, Jr.: Chloramphenicol (Chloromycetin). V. Synthesis. J. Amer. Chem. Soc. **71**, 2463 (1949).

*16.* COUTTS, R. T., J. B. STENLAKE and W. D. WILLIAMS: The Chemistry of the *Aristolochia* Species. III. Aristolochic Acids and Related Substances from *Aristolochia reticulata* and *A. Indica.* J. Chem. Soc. (London) **1957**, 4120.

*17.* DÖLL, W.: A Simple Colorimetric Assay Method for Chloromycetin (Chloramphenicol, Leukomycin) in Aqueous Solution. Arzneimittelforsch. **5**, 97 (1955) [Chem. Abstr. **49**, 8565 (1955)].

*18.* DYMOCK, W. and C. J. H. WARDEN: Pharmac. J. and Trans. [3] **22**, 245 (1891).

*19.* EHRLICH, J., Q. R. BARTZ, R. M. SMITH, D. A. JOSLYN and P. R. BURKHOLDER: Chloromycetin, a New Antibiotic from a Soil Actinomycete. Science (Washington) **106**, 417 (1947).

*20.* ELPHIMOFF-FELKIN, I., H. FELKIN et Z. WELVART: Réductions sélectives au moyen de l'hydrure d'aluminium et de lithium. III. Nouvelle synthèse de la chloromycétine. C. R. hebd. Séances Acad. Sci. **234**, 1789 (1952).

*21.* FERGUSON, J. A.: Analysis of *Aristolochia reticulata* NUTTALL. Amer. J. Pharm. **59**, 481 (1887).

*22.* FODOR, G., I. TÓTH, E. KOVÁCS and J. KISS: Synthesis of Chloramphenicol. Izvest Akad. Nauk. SSSR. Otdel. Khim. Nauk **1955**, 441; Bull. Acad. Sci. USSR. Div. Chem. Sci. **1955**, 391 [Chem. Abstr. **50**, 6360 (1956)].

*23.* FORJAZ, P.: Spectrophotometric Techniques. Analytical Methods for Vitamin B$_{12}$, Chloromycetin, Rutin, and Sodium 4-Aminosalicylate. An. Azevedos (Lisbon) **1**, 163 (1949) [Chem. Abstr. **44**, 3675 (1950)].

*24.* FREEMAN, F. M.: Colorimetric Determination of Chloramphenicol. Analyst **81**, 299 (1956) [Chem. Abstr. **50**, 10612 (1956)].

*25.* FRICKHINGER, A.: Über *Aristolochia clematitis.* Buchners Repert. Pharmacie [3] **7**, 1 (1851).

*26.* FUNKE, A. et P. KORNMANN: Une nouvelle synthèse de la chloromycétine. C. R. hebd. Séances Acad. Sci. **233**, 1631 (1951).

*27.* GARCIA MADRID, H.: Chemical Determination of Chloramphenicol. Colegio farm. (Santiago Chile) **9**, No. 118, 1 (1952) [Chem. Abstr. **47**, 7949 (1953)].

*28.* GLAZKO, A. J., L. M. WOLF and W. A. DILL: Biochemical Studies on Chloramphenicol (Chloromycetin). I. Colorimetric Methods for the Determination of Chloramphenicol and Related Nitro Compounds. Arch. Biochemistry **23**, 411 (1949).

*29.* GORTER, K.: L'Hiptagine, glucoside nouveau retiré de l'*Hiptage Madablota Gaertn.* Bull. Jard. bot. Buitenzorg [3] **2**, 187 (1920).

*30.* HESS, G. B.: Polarographic Estimation of Chloramphenicol (Chloromycetin). Analyt. Chemistry **22**, 649 (1950).

*31.* HESSE, O.: Über die Wurzeln von *Aristolochia argentina.* Arch. Pharmaz. **233**, 684 (1895).

*32.* IKUMA, S.: Alkamines. VI. Synthesis of Chloramphenicol. J. Pharmac. Soc. Japan **73**, 284 (1953) [Chem. Abstr. **48**, 1996 (1954)].

*33.* KOLLONITSCH, J., A. HAJÓS, V. GABOR und M. KRAUT: Neue Synthesen des Chloramphenicols und deren stereochemischen Beziehungen. Experientia **10**, 458 (1954).

*34.* KRISHNASWAMY, P. R. and B. L. MANJUNATH: Chemical Examination of the Roots of *Aristolochia Indica* LINN. III. Isolation of the Alkaloid Aristolochine. J. Indian Chem. Soc. **14**, 39 (1937) [Chem. Abstr. **31**, 5101 (1937)].

*35.* KRISHNASWAMY, P. R., B. L. MANJUNATH and S. VENKATA RAO: Chemical Examination of the Roots of *Aristolochia Indica* LINN. I. J. Indian Chem. Soc. **12**, 476 (1935) [Chem. Abstr. **30**, 233 (1936)].

*36.* KUNZ, A. and C. S. HUDSON: Relations Between Rotatory Power and Structure in the Sugar Group. XV. Conversion of Lactose to Another Disaccharide, Neolactose. The Chloro-hepta-acetate and two Octa-acetates of Neolactose. J. Amer. Chem. Soc. **48**, 1978 (1926).

*37.* LALANNE, P. et TH. MATHOU: Une aristoloche médicinale de la Guadeloupe. Bull. sci. pharmacol. **41**, 460 (1934).

*38.* LONG, L. M. and H. D. TROUTMAN: Chloramphenicol (Chloromycetin). VI. A Synthetic Approach. J. Amer. Chem. Soc. **71**, 2469 (1949).

*39.* — — Chloramphenicol (Chloromycetin). VII. Synthesis through *p*-Nitroacetophenone. J. Amer. Chem. Soc. **71**, 2473 (1949).

*40.* MAISCH, J. M.: Materia medica of the Mexican Pharmacopeia. Guaco *Aristol. frangrantissimo* RUIZ et PAVON. Amer. J. Pharm. **57**, 601 (1885).

*41.* MARAL, R. et A. BLANDIN: Dosage biologique du chloramphénicol dans les liquides organiques. C. R. Séances Soc. Biol. **144**, 1381 (1950).

*42.* MORRIS, M. P., C. PAGÁN and H. E. WARMKE: Hiptagenic Acid, a Toxic Component of *Indigofera endecaphylla*. Science (Washington) **119**, 322 (1954).

*43.* MURRAY, A.: Apparatus medicaminum, 1. Bd., S. 502. Göttingen. 1793.

*44.* NAGAWA, M. and H. SHINDO: Determination of Chloramphenicol by Azotometry. J. Pharmac. Soc. Japan **76**, 99 (1956) [Chem. Abstr. **50**, 6747 (1956)].

*45.* NAKAMURA, S. and C. SHIMODA: An Antibiotic Substance, Oryzacidin, Produced by *Aspergillus oryzae*. V. Existence of $\beta$-Nitropropionic Acid in the Fermentation Broth. J. Agric. Chem. Soc. Japan **28**, 909 (1954) [Chem. Abstr. **50**, 15723 (1956)].

*46.* NEGORO, H.: Colorimetric Estimation of Chloromycetin by (2-Diethyl-aminoethyl)-1-naphthylamine. J. Pharmac. Soc. Japan **70**, 669 (1950) [Chem. Abstr. **45**, 4403 (1951)].

*47.* NOGUEIRA, A. L., A. S. ANTÃO and R. A. PAREIRA: Analytical Study of some Antibiotics. Preliminary Note. Rev. portug. farm. **1**, 10 (1951) [Chem. Abstr. **45**, 8074 (1951)].

*48.* ODA, M. and M. HIRANO: Colorimetric Determination of Chloramphenicol with the Tsuda Reagent. J. Pharmac. Soc. Japan **71**, 51 (1951) [Chem. Abstr. **45**, 5075 (1951)].

*49.* OLIVE, T. R.: Chloromycetin by Parke, Davis. Chem. Eng. **56**, 107 (1949) [Chem. Abstr. **44**, 1653 (1950)].

*50.* ORFILA, M. P.: Allgemeine Toxicologie, 3. Bd., S. 307. Berlin. 1818.

*51.* PAILER, M., L. BELOHLAV und E. SIMONITSCH: Zur Konstitution der Aristolochiasäuren. Monatsh. Chem. **86**, 676 (1955).

*52.* — — — Pflanzliche Naturstoffe mit einer Nitrogruppe. I. Die Konstitution der Aristolochiasäure. Monatsh. Chem. **87**, 249 (1956).

*53.* PAILER, M. und K. NOWOTNY: Über das Vorkommen von $\beta$-Nitropropionsäure in den Wurzeln des wohlriechenden Veilchens *(Viola odorata)*. Naturwiss. **45**, 419 (1958).

*54.* PAILER, M. und G. PRUCKMAYR: Über die basischen Inhaltsstoffe der *Aristolochia clematitis* L. Monatsh. Chem. **90**, 145 (1959).

*55.* PAILER, M. und A. SCHLEPPNIK: Pflanzliche Naturstoffe mit einer Nitrogruppe. II. Die Konstitution der Aristolochiasäure-II. Monatsh. Chem. **88**, 367 (1957).

*56.* — — Die Synthese eines Abbauproduktes der Aristolochiasäure-II, des 3,4-Methylendioxy-10-acetamidophenanthrens. Monatsh. Chem. **89**, 175 (1958).

*57.* PECKOLT, TH.: Pharmazeut. Rdsch., New York **11**, 181 (1893) (zitiert in Nr. 66).

58. PEPE, T. L.: Identification and Determination of Chloromycetin. Boll. chim. farm. (Milano) 88, 411 (1949) [Chem. Abstr. 44, 3066 (1950)].

59. PESEZ, M.: Identification of the Dichloroacetyl Group. Application to Chloramphenicol. Ann. pharm. franç. 9, 187 (1951) [Chem. Abstr. 45, 8203 (1951)].

60. PILARCZYK, W.: *Aristolochia clematitis* L., die Osterluzei. Planta medica 6, 258 (1958); 7, 73 (1959).

61. POHL, J.: Über das Aristolochin, einen giftigen Bestandteil der Aristolochiaarten. Arch. exp. Pathol. Pharmakol. 29, 282 (1892).

62. PUGA, R.: Color Reaction of Chloramphenicol Suitable for Quantitative Estimation. Rev. farm. (Buenos Aires) 93, 240 (1951) [Chem. Abstr. 46, 5260 (1952)].

63. RAISTRICK, H. and A. STÖSSL: Studies in the Biochemistry of Micro-organisms. 104. Metabolites of *Penicillium atrovenetum* G. SMITH: β-Nitropropionic Acid, a Major Metabolite. Biochemic. J. 68, 647 (1958).

64. REBSTOCK, M. C., H. M. CROOKS, Jr., J. CONTROULIS and Q. R. BARTZ: Chloramphenicol (Chloromycetin). IV. Chemical Studies. J. Amer. Chem. Soc. 71, 2458 (1949).

65. ROBLES, G. G. and L. UNZUETA R.: Qualitative and Quantitative Reactions of Chloramphenicol. Bol. soc. quím. Perú 18, 6 (1952) [Chem. Abstr. 47, 4553 (1953)].

66. ROSENMUND, H. und T. REICHSTEIN: Zur Kenntnis der Aristolochiasäure. Pharm. Acta Helv. 18, 243 (1943) [Chem. Abstr. 38, 1745 (1944)].

67. SANTI, R. and M. SEREMBE: Absorption, Distribution, and Elimination of Chloromycetin. Boll. soc. ital. biol. sper. 25, 1225 (1949) [Chem. Abstr. 44, 8521 (1950)].

68. SMITH, G. N. and C. S. WORREL: The Application of Paper-Partition Chromatography to the Analysis of Chloramphenicol (Chloromycetin) and Decomposition Products. Arch. Biochemistry 28, 1 (1950).

69. SOBRAL, T. R.: J. de Coimbra Nr. 36, I. Abt., S. 196.

70. TAGUCHI, T. and M. TOMOEDA: Chloramphenicol: A New Synthesis and its Stereochemical Findings. Pharm. Bull. (Japan) 2, 433 (1954) [Chem. Abstr. 50, 11970 (1956)].

71. TAMELEN, E. E. VAN: Structural Chemistry of Actinomycetes Antibiotics. Fortschr. Chem. organ. Naturstoffe 16, 90 (1958).

72. TOMITA, M. and S. KURA: Studies on the Ingredients of Aristolochiaceus Plants. I. The Ingredients of *A. caempferi* and *A. debilis*. J. Pharmac. Soc. Japan 77, 812 (1957) [Chem. Abstr. 51, 17963 (1957)].

73. TRUHAUT, R.: A Color Reaction of Chloromycetin. Ann. pharm. franç. 9, 347 (1951) [Chem. Abstr. 46, 1711 (1952)].

74. TSENG, K.-F. and Y.-T. KU: A Note on the Constituents of *Aristolochia debilis* SIEB. et ZUCC. Acta Chim. Sinica 23, 156 (1957) [Chem. Abstr. 52, 13188 (1958)].

75. VAÏSMAN, G. A. and M. D. KISLAYA: Determination of Syntomycin and *l*-Mycetin in Preparations and Medicinal Mixtures. Aptechnoe Delo 5, No. 4, 19 (1956) [Chem. Abstr. 51, 5369 (1957)].

76. VALSETH, A. and A. WICKSTRØM: Methods for the Application of Periodate Oxidation to the Analytical Control of Chloramphenicol. Medd. Norsk Farm. Selskap 17, 345 (1955) [Chem. Abstr. 50, 5979 (1956)].

77. WALZ, G. F.: Beitrag zur Kenntnis der chemischen Zusammensetzung der Aristolochieen, insbesondere die Untersuchung der *Aristolochia clematitis*. Jahrb. prakt. Pharmacie 24, 65 (1852).

*78.* Weigner, J.: Chloramphenicol. Chem. Prumysl. (Czechosl.) **1** (26), 115 (1951) [Chem. Abstr. **46**, 2522 (1952)].

*79.* Winckler, F. L.: Über die Bestandteile der *Aristolochia clematitis.* Jahrb. prakt. Pharmacie **19**, 71 (1849).

*80.* Wittstein, G. Ch.: Untersuchung der *Rad. Aristolochiae antihystericae* Mart. Buchners Repert. Pharmacie [2] **57**, 145 (1836).

*(Eingelaufen am 12. Oktober 1959.)*

# Dérivés guanidiques biologiques.

Par **Nguyen van Thoai** et **J. Roche**, Paris.

Les *abréviations* suivantes sont utilisées dans le présent article: ATP = acide adénosinetriphosphorique; ADP = acide adénosinediphosphorique; C$\sim$P = créatinephosphate; A$\sim$P = argininephosphate; Ty$\sim$P = taurocyaminephosphate; Gly$\sim$P = glycocyaminephosphate; HMP (1 ou 6) = Hexose (1 ou 6) monophosphate; HDP = hexosediphosphate; DPN = diphosphopyridine nucléotide; DPNH = diphosphopyridine nucléotide réduit.

## Introduction.

Depuis l'isolement de la créatine par Chevreul, en 1832, dans les extraits de viande (44) et son identification par Liebig, en 1847, l'attention des biochimistes a porté principalement, pendant longtemps, sur le rôle de l'arginine dans la production de l'urée (128), sur le cycle de l'uréo-genèse (134) et sur celui de la phosphocréatine (71) et de la phospho-arginine (168, 169) dans la contraction musculaire.

A la suite de la mise en évidence, d'une part des réactions de trans-amidination expliquant la convertibilité entre l'arginine et un certain nombre de dérivés guanidiques (28, 29, 33, 73, 205, 291, 292), et, d'autre part, du mécanisme complexe de la biosynthèse de l'arginine (34, 45, 46, 195, 196), l'importance du groupement guanidique dans la fixation et le transfert de l'azote organique est devenue beaucoup plus manifeste.

Par ailleurs, l'isolement de nouvelles substances dont certaines: la phosphotaurocyamine, la phosphoglycocyamine (280—282), la phospho-lombricine (270), jouent le même rôle de phosphagènes que le créatine- et l'argininephosphate (257), a mis en évidence la part importante que les dérivés guanidiques prennent dans la chimie du muscle. L'étude de leur biogenèse a montré en outre que leur rôle n'est pas seulement équivalent à l'arginine dans la fixation de l'azote du groupe amidinique, mais qu'ils sont capables de jouer également de rôle de régulateurs du métabolisme azoté (258).

## I. Structure et formation.

1. **Structure.** La guanidine (I), ainsi nommée parce qu'elle a été isolée parmi les produits d'oxydation de la guanine par Strecker en 1861 (247), et différents dérivés mono- (II), di- (III), (IV) ou tri-substitués (V) de celle-ci sont représentés chez les êtres vivants:

$$(\text{I.}) \quad HN=C \diagup \begin{matrix} NH_2 \\ NH_2 \end{matrix} \qquad (\text{II.}) \quad HN=C \diagup \begin{matrix} NH_2 \\ NH-R \end{matrix} \qquad (\text{III.}) \quad NH=C \diagup \begin{matrix} NH-R_1 \\ NH-R \end{matrix}$$

$$(\text{IV.}) \quad HN=C \diagup \begin{matrix} NH_2 \\ N(R)(R_1) \end{matrix} \qquad (\text{V.}) \quad HN=C \diagup \begin{matrix} N(R_1)(R_2) \\ NH-R \end{matrix}$$

On ne connait pas de dérivés tétrasubstitués. Les produits disubstitués peuvent être symétriques (III) comme l'argininephosphate ou asymétriques (IV) comme la diméthylguanidine; mais on ne rencontre jamais deux isomères de position à l'état naturel. Une seule forme cyclisée, représentée par la créatinine (XXXI, p. 87) existe dans les organismes vivants; les autres produits de cyclisation (glycocyamidine, anhydride argininosuccinique) ne sont pas d'origine biologique.

**2. Nomenclature.** On peut essayer de classer les composés biologiques par la nature des radicaux fixés à la guanidine et distinguer ainsi les catégories suivantes de corps (*258*):

1° ω-Guanido-α-amino-acides: l'arginine (VI) et la canavanine (VII).

$$HN=C \begin{matrix} NH_2 \\ \diagdown \\ NH-(CH_2)_3-CH-COOH \\ | \\ NH_2 \end{matrix}$$

(VI.) Arginine.

$$HN=C \begin{matrix} NH_2 \\ \diagdown \\ NH-O-(CH_2)_2-CH-COOH \\ | \\ NH_2 \end{matrix}$$

(VII.) Canavanine.

2° ω-Guanido-acides: la glycocyamine (VIII), la créatine (IX), les acides γ-guanidobutyrique (X), δ-guanidovalérianique (XI), α-céto-δ-guanidovalérianique (XII), α-hydroxy-δ-guanidovalérianique ou arginique (XIII), la taurocyamine (XIV), l'astérubine (XV).

$$HN-C \begin{matrix} NH_2 \\ \diagdown \\ NH-CH_2-COOH \end{matrix}$$

(VIII.) Glycocyamine.

$$HN=C \begin{matrix} NH_2 \\ \diagdown \\ N(CH_3)-CH_2-COOH \end{matrix}$$

(IX.) Créatine.

$$HN=C \begin{matrix} NH_2 \\ \diagdown \\ NH-(CH_2)_3-COOH \end{matrix}$$

(X.) Acide γ-guanidobutyrique.

$$HN=C \begin{matrix} NH_2 \\ \diagdown \\ NH-(CH_2)_4-COOH \end{matrix}$$

(XI.) Acide δ-guanidovalérianique.

$$HN=C \begin{matrix} NH_2 \\ \diagdown \\ NH-(CH_2)_3-CO-COOH \end{matrix}$$

(XII.) Acide α-céto-δ-guanidovalérianique.

$$HN=C \begin{matrix} NH_2 \\ \diagdown \\ NH-(CH_2)_3-CHOH-COOH \end{matrix}$$

(XIII.) Acide arginique.

$$HN=C \begin{matrix} NH_2 \\ \diagdown \\ NH-CH_2-CH_2-SO_3H \end{matrix}$$

(XIV.) Taurocyamine.

$$HN=C \begin{matrix} N(CH_3)_2 \\ \diagdown \\ NH-CH_2-CH_2-SO_3H \end{matrix}$$

(XV.) Astérubine.

3° Bases: la guanidine (XVI), l'hydroxyguanidine (XVII), la monométhylguanidine (XVIII), la diméthylguanidine (XIX), l'agmatine (XX), la méthylagmatine (XXI), la galégine (XXII), l'arcaine (XXIII), la streptidine (XXIV).

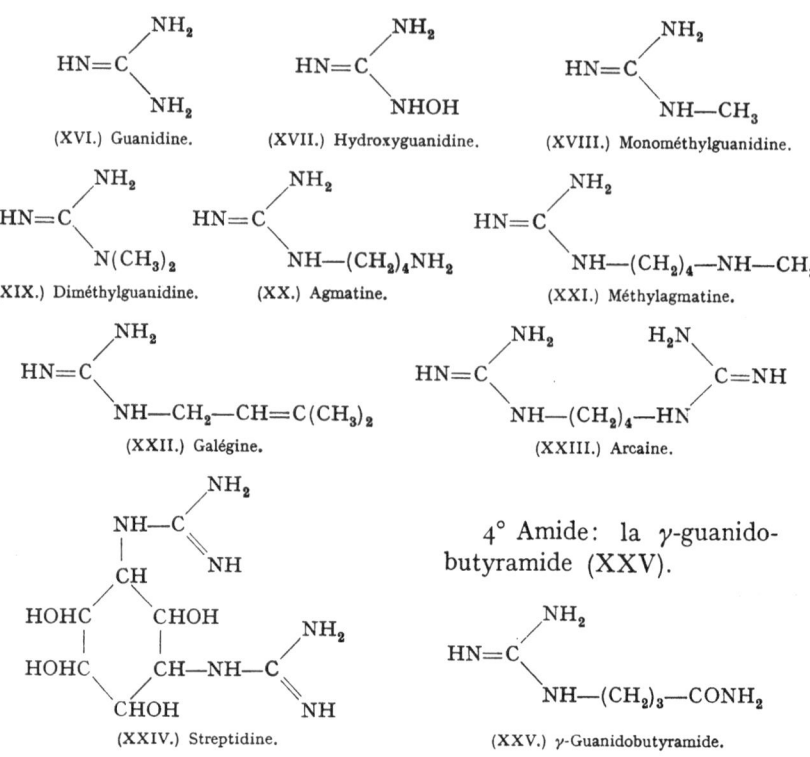

(XVI.) Guanidine.

(XVII.) Hydroxyguanidine.

(XVIII.) Monométhylguanidine.

(XIX.) Diméthylguanidine.

(XX.) Agmatine.

(XXI.) Méthylagmatine.

(XXII.) Galégine.

(XXIII.) Arcaine.

(XXIV.) Streptidine.

4° Amide: la γ-guanido-butyramide (XXV).

(XXV.) γ-Guanidobutyramide.

5° Diester: la lombricine ou guanidoéthylsérylphosphate (XXVI).

(XXVI.) Lombricine.

6° Produits de condensation de l'arginine: l'acide argininosuccinique (XXVII), l'acide canavaninosuccinique (XXVIII), l'octopine (XXIX) et la subérylarginine (XXX).

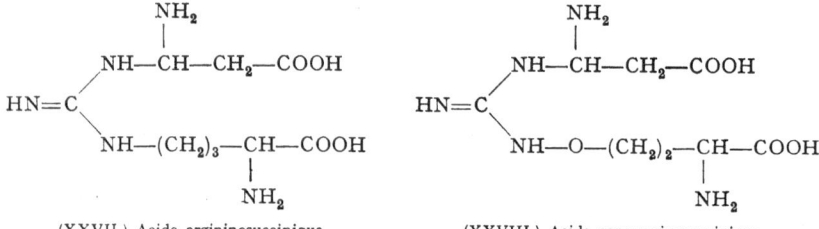

(XXVII.) Acide argininosuccinique.

(XXVIII.) Acide canavaninosuccinique.

$$HN=C\begin{cases}NH_2\\NH-(CH_2)_3-CH-COOH\\ \quad\quad\quad\quad\quad | \\ \quad\quad\quad\quad\quad NH \\ \quad\quad\quad\quad\quad | \\ \quad\quad\quad\quad CH_3-CH-COOH\end{cases}$$

(XXIX.) Octopine.

$$HN=C\begin{cases}NH_2\\NH-(CH_2)_3-CH-COOH\\ \quad\quad\quad\quad\quad\quad | \\ \quad\quad\quad\quad NH-OC-(CH_2)_6-COOH\end{cases}$$

(XXX.) Subérylarginine.

7° Produit de cyclisation: la créatinine (XXXI).

$$HN=C\begin{cases}NH-CO\\ \quad\quad\quad | \\ N-CH_2\\ | \\ CH_3\end{cases}$$

(XXXI.) Créatinine.

8° Produits de phosphorylation (phosphagènes): phosphorylglyco-cyamine (XXXII), phosphorylcréatine (XXXIII), phosphoryltauro-cyamine (XXXIV), phosphorylarginine (XXXV), phosphoryllombricine (XXXVI).

$$HN=C\begin{cases}NH\sim PO_3H_2\\NH-CH_2-COOH\end{cases}$$

(XXXII.) Phosphorylglycocyamine.

$$HN=C\begin{cases}NH\sim PO_3H_2\\N-CH_2-COOH\\ | \\ CH_3\end{cases}$$

(XXXIII.) Phosphorylcréatine.

$$HN=C\begin{cases}NH\sim PO_3H_2\\NH-CH_2-CH_2-SO_3H\end{cases}$$

(XXXIV.) Phosphoryltaurocyamine.

$$HN=C\begin{cases}NH\sim PO_3H_2\\NH-(CH_2)_3-CH-COOH\\ \quad\quad\quad\quad\quad\quad | \\ \quad\quad\quad\quad\quad NH_2\end{cases}$$

(XXXV.) Phosphorylarginine.

$$HN=C\begin{cases}NH\sim PO_3H_2\\NH-CH_2-CH_2-O-\underset{\underset{OH}{|}}{\overset{\overset{O}{||}}{P}}-O-CH_2-CH-COOH\\ \quad\quad\quad\quad\quad\quad\quad\quad\quad\quad\quad\quad\quad | \\ \quad\quad\quad\quad\quad\quad\quad\quad\quad\quad\quad\quad NH_2\end{cases}$$

(XXXVI.) Phosphoryllombricine.

3. **Formation par voie chimique.** D'un point de vue strictement chimique, la guanidine peut être considérée comme un dérivé à la fois amidique et imidique de l'acide carbonique. En fait, les méthodes générales de formation de ce corps et de ses dérivés sont pour la plupart basées sur l'action de la cyanamide, de l'urée, de l'isourée, de l'isothiourée ou de leurs dérivés de substitution.

Dessaignes, en 1854, faisant réagir l'urée sur la méthylamine a obtenu ce qu'il appelait la «méthyluramine» (53) et qui était en fait la méthylguanidine (247).

En vérification de l'hypothèse que la créatine est une combinaison de la sarcosine et de la cyanamide, l'emploi de celle-ci dans la synthèse des dérivés guanidiques a été appliqué pour la première fois à la préparation de la glycocyamine (247):

$$CH_2N_2 + C_2H_5O_2N \rightarrow C_3H_7O_2N_3$$

Cyanamide.        Glycine.        Glycocyamine.

Elle conduit à la créatine à partir de la sarcosine (287). La présence d'ammoniac favorise la réaction (248). L'emploi de la cyanamide, plus économique sur le plan industriel, ne fournit toutefois pas un rendement comparable à celui obtenu avec les dérivés de l'urée.

A l'échelle de laboratoire, les méthodes les plus efficaces sont basées sur l'emploi des dérivés de l'isourée, en particulier des dérivés O- et S-méthylés ou éthylés (XXXVII et XXXVIII), lesquels réagissent avec une grande facilité à froid, en présence d'un grand excès d'ammoniac, avec les acides aminés et les amines pour donner naissance aux guanidines substituées correspondantes (231, 238, 37, 176). Le rendement moyen obtenu par cette méthode est de l'ordre de 70—80%.

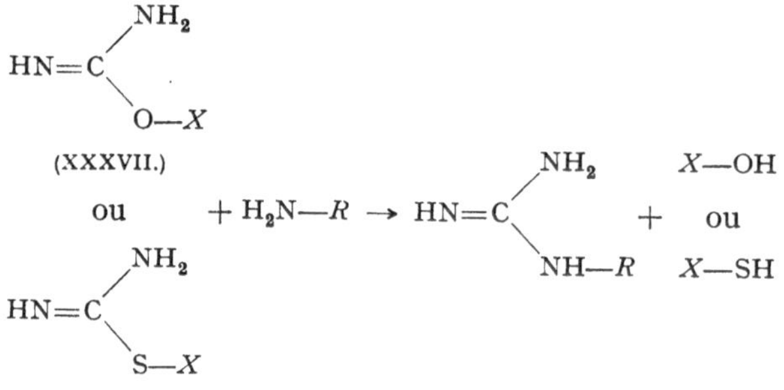

$$HN=C \begin{smallmatrix} NH_2 \\ \\ O-X \end{smallmatrix}$$

(XXXVII.)

ou        $+ H_2N-R \rightarrow HN=C \begin{smallmatrix} NH_2 \\ \\ NH-R \end{smallmatrix} + \begin{smallmatrix} X-OH \\ ou \\ X-SH \end{smallmatrix}$

$$HN=C \begin{smallmatrix} NH_2 \\ \\ S-X \end{smallmatrix}$$

(XXXVIII.)

La préparation des dérivés $\omega$-guanidiques des acides diaminés s'opère en bloquant le groupement $\alpha$-aminé soit par benzoylation préalable (77), soit, plus simplement, par formation d'un complexe cuivrique (246). On a préparé ainsi la canavanine à partir de la canaline (XXXIX) (136):

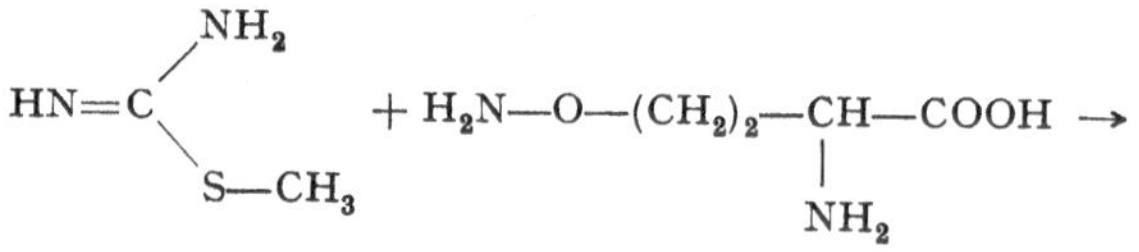

$$HN=C \begin{smallmatrix} NH_2 \\ \\ S-CH_3 \end{smallmatrix} + H_2N-O-(CH_2)_2-CH-COOH \rightarrow$$
$$\phantom{HN=C + H_2N-O-(CH_2)_2-CH-} | \phantom{COOH}$$
$$\phantom{HN=C + H_2N-O-(CH_2)_2-CH-} NH_2$$

(XXXIX.) Canaline.

$$\rightarrow \text{HN}{=}\text{C}\underset{\text{NH}-\text{O}-(\text{CH}_2)_2-\underset{|}{\text{CH}}-\text{COOH}}{\overset{\text{NH}_2}{<}} \qquad + \text{HSCH}_3$$
$$\text{NH}_2$$

(VII.) Canavanine.

La préparation des guanidines disubstituées symétriques ou tri-substituées est réalisée à partir des dérivés de la cyanamide, de l'isourée ou de l'isothiourée. L'action de la diméthylcyanamide (XL) sur la taurine (XLI) (5) ou sur la cystamine (XLII) (10) avec oxydation ultérieure (233) de la tétraméthyl-diguanidylcystamine (XLIII) permet d'obtenir ainsi la diméthylamidino-taurine ou astérubine (XV) (9, 13):

$$\text{NCN(CH}_3)_2 + \text{H}_2\text{N}-\text{CH}_2-\text{CH}_2-\text{SO}_3\text{H} \rightarrow \text{HN}{=}\text{C}\underset{\text{NH}-\text{CH}_2-\text{CH}_2-\text{SO}_3\text{H}}{\overset{\text{N(CH}_3)_2}{<}}$$

(XL.) Diméthylcyanamide.     (XLI.) Taurine.             (XV.) Astérubine.

$$\begin{array}{c}\text{S}-\text{CH}_2-\text{CH}_2-\text{NH}_2 \\ | \\ \text{S}-\text{CH}_2-\text{CH}_2-\text{NH}_2\end{array} \quad + 2\,\text{NCN(CH}_3)_2 \rightarrow$$

(XLII.) Cystamine.

$$\begin{array}{c}\text{S}-\text{CH}_2-\text{CH}_2-\text{NH}-\text{C}\overset{\text{N(CH}_3)_2}{\underset{\text{NH}}{<}} \\ | \\ \text{S}-\text{CH}_2-\text{CH}_2-\text{NH}-\text{C}\overset{\text{N(CH}_3)_2}{\underset{\text{NH}}{<}}\end{array} \rightarrow$$

(XLIII.)

$$\xrightarrow{\text{Ba(OH)}_2} 2\,\text{HN}{=}\text{C}\underset{\text{NH}-\text{CH}_2-\text{CH}_2-\text{SO}_3\text{H}}{\overset{\text{N(CH}_3)_2}{<}}$$

(XV.) Astérubine.

De leur côté, les dérivés N-phosphorylés de la S-méthyl ou de l'O-méthyl isourée permettent de préparer les guanidines N-phosphorylées symétriques.

En traitant le phosphorylphénylester de la S-méthylisothiourée (XLIV) par l'oxyde de mercure en présence de triéthylamine et d'éthanol, le mercaptide s'élimine et la cyanamide formée réagit sur le benzylester de la sarcosine (XLV) pour donner la diphénylphosphoryl-cyanamide correspondante (XLVI); celle-ci donne, par chauffage à 110° dans l'alcool isobutylique le N-diphényl-phosphoryl-créatine-benzylester (XLVII), qui, par hydrogénation successive en présence de charbon palladium et d'oxyde de platine fournit le créatinephosphate (XXXIII) (49). La méthode est extrêmement laborieuse et d'un rendement dérisoire.

$$(C_6H_5O)_2P(O)N=C\begin{array}{c} NH_2 \\ \backslash \\ S-CH_3 \end{array} \quad + \quad HN-CH_2-COO-CH_2-C_6H_5 \xrightarrow{HgO} $$

(XLIV.) S-Méthylisothiourée.                (XLV.) Sarcosine-benzylester.

$$\rightarrow [(C_6H_5)_2P(O)\overline{N}-C\equiv N](CH_3)_2\overset{\oplus}{H_2N}-CH_2-COO-CH_2-C_6H_5 \xrightarrow[110°]{isobutanol}$$

(XLVI.)

$$\rightarrow (C_6H_5)_2P(O)N=C\begin{array}{c} NH_2 \\ \diagdown \\ N(CH_3)-CH_2-COO-CH_2-C_6H_5 \end{array} \xrightarrow[2\,H]{2\,H}$$

(XLVII.) N-Diphényl-phosphoryl-créatine-benzylester.

$$\rightarrow HN=C\begin{array}{c} NH{\sim}PO_3H_2 \\ \diagdown \\ N(CH_3)-CH_2-COOH \end{array}$$

(XXXIII.) Créatinephosphate.

La technique inaugurée par Zeile et Fawaz (302) pour la préparation du créatinephosphate a été généralisée depuis en vue de la synthèse d'autres phosphagènes: créatinephosphate (65, 193), taurocyamine- (282), glycocyamine- (68, 282), arginine-phosphate (256).

Elle consiste à faire réagir l'oxychlorure de phosphore en milieu fortement alcalin sur le dérivé guanidique. Le phosphagène est purifié grâce au sel de baryum ou de calcium, qui est soluble dans l'eau et insoluble dans l'alcool aqueux. Le passage sur résine cationique chargée en ions $NH_4^+$, $Na^+$ ou $Li^+$ permet d'obtenir les sels cristallisés correspondants. L'action de l'oxychlorure de phosphore détermine parfois la formation de dérivés polyphosphorylés (256) dont les sels alcalino-terreux sont moins solubles, et qui peuvent être éliminés par passage sur résine (282, 256).

4. Biogenèse. Sur le plan biologique, la formation de dérivés guanidiques s'opère selon plusieurs mécanismes, que nous avons déjà exposés en détail par ailleurs (258). Il suffit de rappeler ici qu'ils sont de trois ordres:

a. Formation par amidination successive: cas de l'arginine. Contrairement à ce qui a été longtemps admis en ce qui concerne la fixation directe successivement de $CO_2$ et $NH_3$ sur l'ornithine, fixation théoriquement difficile en raison de la valeur élevée (pK = 10) du groupement $\omega$-aminé de celle-ci (142, 144, 145), la formation de l'arginine comporte de nombreuses étapes intermédiaires. Le première consiste en la formation du carbamylphosphate (XLIX), en présence de $CO_2$, $NH_3$ et d'acide adénosinetriphosphorique

$$O=C\begin{array}{c} NH_2 \\ \diagdown \\ O{\sim}PO_3^= \end{array}$$

(XLIX.)
Carbamylphosphate.

(ATP) (*110, 111*). Le carbamylphosphate transfert ensuite le groupement uréinique à l'ornithine en donnant naissance à la citrulline, réaction entièrement réversible et qui n'est, d'ailleurs, pas spécifique de l'ornithine.

$$O=C\begin{array}{c} NH_2 \\ \\ O{\sim}PO_3^= \end{array} + H_2N{-}R \underset{ATP}{\overset{ADP}{\rightleftharpoons}} O=C\begin{array}{c} NH_2 \\ \\ NH{-}R \end{array}$$

(XLIX.)

$$R = -(CH_2)_3-CH-COOH \; (110) \qquad R = -CH_2-CH_2-COOH \; (84)$$
$$\qquad\qquad\quad |$$
$$\qquad\qquad NH_2$$

$$R = -CH-COOH \qquad\quad (111) \qquad R = -CH-COOH \qquad\quad (70)$$
$$\quad\;\; | \qquad\qquad\qquad\qquad\qquad\qquad | $$
$$\quad CH_2-COOH \qquad\qquad\qquad\qquad\quad C_2H_5$$

La citrulline (L), réagissant sous sa forme isourée, se condense ensuite, réversiblement et spécifiquement, avec l'acide aspartique (LI), en présence d'ATP, pour former l'acide argininosuccinique (XXVII) (*198—203, 190*):

$$HN=C\begin{array}{c} OH \\ \\ NH-(CH_2)_3-CH-COOH \\ | \\ NH_2 \end{array} + H_2N-CH-CH_2-COOH \qquad \overset{ATP + Mg^{++}}{=\!=\!=\!=\!=}$$

(L.) Citrulline.      COOH (LI.) Acide aspartique.

$$\rightleftharpoons HN=C\begin{array}{c} NH-CH-CH_2-COOH \\ | \\ COOH \\ \\ NH-(CH_2)_3-CH-COOH \\ | \\ NH_2 \end{array}$$

(XXVII.) Acide argininosuccinique.

Ce dernier composé se scinde ensuite en arginine et en acide fumarique (LII), selon une réaction réversible (*197, 289*) qui se produit aussi avec l'acide canavaninosuccinique (XXVIII, p. 86) (*295*):

$$HN=C\begin{array}{c} COOH \\ | \\ NH-CH-CH_2-COOH \\ \\ NH-R \end{array} \rightleftharpoons HN=C\begin{array}{c} NH_2 \\ \\ NH-R \end{array} + HOOC-CH=CH-COOH$$

(LII.) Acide fumarique.

$$R = -(CH_2)_3-CH-COOH \qquad ou \qquad -O-(CH_2)_2-CH-COOH$$
$$\qquad\qquad\quad | \qquad\qquad\qquad\qquad\qquad\qquad\qquad | $$
$$\qquad\qquad NH_2 \qquad\qquad\qquad\qquad\qquad\qquad\quad NH_2$$

*b. Formation par transformation d'autres composés guanidiques.* C'est la dégradation ou la condensation de l'arginine avec d'autres corps qui fournissent le plus grand nombre de dérivés guanidiques. La désamination oxydative de l'arginine donne naissance à l'acide $\alpha$-céto-$\delta$-guanido-valérianique (XII) qui, par oxydation secondaire, spontanée, forme l'acide $\gamma$-guanidobutyrique (X) (*249, 278*). La même désamination, suivie de processus d'hydrogénation encore mal connus, conduit également à la formation d'acide arginique ou acide $\alpha$-hydroxy-$\delta$-guanidovalérianique (XIII) (*173*) et à l'acide $\delta$-guanidovalérianique (XI) (*36*):

$$HN=C\begin{array}{c} NH_2 \\ \\ NH-(CH_2)_3-CH-COOH \\ | \\ NH_2 \end{array} \xrightarrow[-NH_3]{+H_2O}$$

(VI.) Arginine.

$$\rightarrow HN=C\begin{array}{c} NH_2 \\ \\ NH-(CH_2)_3-CO-COOH \end{array} \qquad \rightarrow HN=C\begin{array}{c} NH_2 \\ \\ NH-R \end{array}$$

(XII.)

$$R=-(CH_2)_3-COOH \text{ ou } -(CH_2)_3-CHOH-COOH \text{ ou } -(CH_2)_4-COOH$$
(X.)            (XIII.)            (XI.)

La décarboxylation de l'arginine conduit à l'agmatine (XX) (*74*), qui, sans doute par méthylation secondaire, conduit à la méthyl-agmatine (*88*) (XXI):

$$HN=C\begin{array}{c} NH_2 \\ \\ NH-(CH_2)_3-CH-COOH \\ | \\ NH_2 \end{array} \xrightarrow{-CO_2} HN=C\begin{array}{c} NH_2 \\ \\ NH-(CH_2)_4-NH_2 \end{array} \rightarrow$$

(VI.) Arginine.           (XX.) Agmatine.

$$\xrightarrow{+CH_3} HN=C\begin{array}{c} NH_2 \\ \\ NH-(CH_2)_4-NHCH_3 \end{array}$$

(XXI.) Méthylagmatine.

Un autre type particulier de décarboxylation a été observé seulement en présence d'oxygène. Il conduit à la formation d'amide $\gamma$-guanido-butyrique (XXV) (*260, 261*) qui peut, à son tour, être hydrolysée enzymatiquement en acide $\gamma$-guanidobutyrique (X) ou transformée en acide guanidobutyro-hydroxamique (LIII) en présence d'hydroxyl-amine (*264, 267*):

(XXV.) Amide γ-guanidobutyrique.

(VI.) Arginine.

(LIII.) Acide guanidobutyro-hydroxamique.

(X.) Acide guanidobutyrique.

La canavanine peut subir la même décarboxylation oxydative produisant une amide et un acide hydroxamique, correspondant, sans doute à l'acide guanidoxy-propionique (*265*).

Enfin l'arginine peut se condenser ou s'estérifier par son groupement α-aminé, pour donner naissance à de nouveaux produits. La condensation réductive de l'arginine avec l'acide pyruvique (LIV) en présence de DPNH conduit à l'octopine (XXIX) (*271, 272*):

(LIV.) Acide pyruvique.

(VI.) Arginine.

(XXIX.) Octopine.

Le groupe α-aminé de l'arginine peut être également estérifié avec l'acide subérique; on obtient alors la subérylarginine (XXX, p. 87) (*299*).

De son côté la canavanine (VII) fournit par hydrolyse l'hydroxy-guanidine (XVII) (*114*) et par dismutation hydrogénante, la guanidine (XVI) (*117*). Dans les deux cas il se forme de l'homosérine (XVIa):

$$HN=C\begin{cases} NH_2 \\ NH-O-(CH_2)_2-CH-COOH \\ \qquad\qquad\qquad\quad NH_2 \end{cases}$$

(VII.) Canavanine.

$+ H_2O$     $+ 2 H$

$$HN=C\begin{cases} NH_2 \\ NHOH \end{cases}$$

(XVII.) Hydroxyguanidine.

$$HN=C\begin{cases} NH_2 \\ NH_2 \end{cases} + HOH_2C-CH_2-CH-COOH$$
$$\qquad\qquad\qquad\qquad\qquad\qquad\qquad\qquad NH_2$$

(XVI.) Guanidine.     (XVIa.) Homosérine.

*c. Formation par transamidination.* Le transfert enzymatique du groupe amidinique sur une amine ou un acide aminé constitue un autre mode de formation des dérivés guanidiques. La facile réversibilité de la réaction (73) indiquée par la position de l'équilibre (constante d'équilibre: 1,1) (204), ainsi que la large spécificité des transamidinases (205) permettent d'envisager une interconvertibilité pratiquement illimitée des différentes guanidines biologiques entre elles *(Tableau I)*.

$$HN=C\begin{cases} NH_2 \\ NH-R \end{cases} + H_2N-R' \rightleftharpoons HN=C\begin{cases} NH_2 \\ NH-R' \end{cases} + H_2N-R$$

Donateur d'amidine.    Accepteur d'amidine.    Guanidine formée.

Tableau I. Spécificité des réactions de transamidination.

| Donateur d'amidine | Accepteur d'amidine | Dérivé guanidique formé | Références |
|---|---|---|---|
| Arginine | Glycine | Glycocyamine | (33) |
| » | Ornithine | Arginine | (205, 291, 292) |
| » | Canaline | Canavanine | (33) |
| » | Hydroxylamine | Hydroxyguanidine | (293) |
| » | Acide γ-aminobutyrique | Acide γ-guanidobutyrique | (192, 258) |
| » | Acide δ-amino-valérianique | Acide δ-guanido-valérianique | (258) |
| » | β-Alanine | Acide guanidopropionique | (192, 258) |
| Glycocyamine | Ornithine | Arginine | (73, 205) |
| » | Glycine | Glycocyamine | (205, 293) |
| » | Canaline | Canavanine | (205, 291) |
| » | Hydroxylamine | Hydroxyguanidine | (293) |
| Canavanine | Glycine | Glycocyamine | (205, 291) |
| » | Ornithine | Arginine | (205, 291, 294, 294a) |
| » | Canaline | Canavanine | (205) |

En fait, il a été établi que la biogenèse des dérivés guanidiques ne relève pas toujours uniquement de ces facteurs; en tout état de cause, l'amine ou l'acide aminé correspondant à une guanidine biologique n'en constitue pas forcément le précurseur immédiat, prêt à lui donner naissance par transamidination (258). Il est intéressant de noter également qu'on ne connait pas jusqu'à présent d'exemple de transfert de groupe amidinique sur un groupement aminé secondaire. La formation des guanidines disubstituées symétriques (phosphagènes), asymétriques (créatine) ou trisubstituées (phosphocréatine) s'opère par fixation secondaire d'un radical phosphoryle ou méthyle sur la guanidine monosubstituée correspondante. Ainsi, la créatine se forme par méthylation enzymatique de la glycocyamine (27, 31, 32):

$$
HN=C \begin{cases} NH_2 \\ NH-CH_2-COOH \end{cases} \xrightarrow[\substack{\text{choline} \\ \text{bétaine} \\ \text{méthionine}}]{+ CH_3} HN=C \begin{cases} NH_2 \\ N(CH_3)-CH_2-COOH \end{cases}
$$

(VIII.) Glycocyamine.      (IX.) Créatine.

Les guanidophosphates naturels ou phosphagènes (58, 59) sont synthétisés par transphosphorylation de la guanidine correspondante (158), réaction entièrement réversible (141) mais d'une spécificité stricte (62, 64, 257, 174) (Tableau 2).

$$
HN=C \begin{cases} NH_2 \\ NRR' \end{cases} + ATP \underset{pH = 6,8-7,2}{\overset{pH = 8,8-9,1}{\rightleftharpoons}} HN=C \begin{cases} NH \sim PO_3^= \\ NRR' \end{cases} + ADP
$$

Accepteur de phosphoryle.      Phosphagène formé.

Tableau 2. Spécificité des transphosphorylations.

| Accepteur de $-PO_3^=$ | Phosphagène formé | Enzyme | Références |
|---|---|---|---|
| Créatine | Créatinephosphate | Créatinephosphokinase (Vertébrés) | (158, 21, 252, 135) |
| Arginine | Argininephosphate | Argininephosphokinase (Crustacés, Mollusques) | (159, 253, 174) |
| Taurocyamine | Taurocyaminephosphate | Taurocyaminephosphokinase (Arenicola marina) | (257, 194) |
| Glycocyamine | Glycocyaminephosphate | Glycocyaminephosphokinase (Nereis diversicolor) | (257) |

L'un des phosphagènes ainsi formés, le créatinephosphate est susceptible de se cycliser en donnant naissance à la créatinine. La réaction spontanée, lente (35), est accélérée par un enzyme en présence d'un accepteur de phosphate, l'hexose-1 (ou-6) phosphate (40, 41, 258):

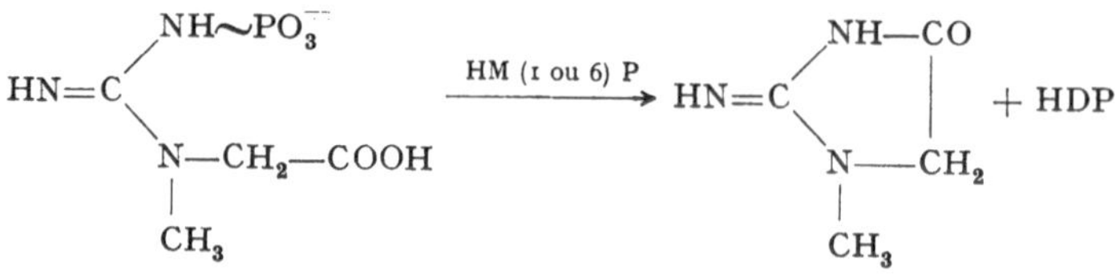

$$HN=C\begin{array}{c} NH\sim PO_3^{--} \\ \\ N-CH_2-COOH \\ | \\ CH_3 \end{array} \xrightarrow{\text{HM (1 ou 6) P}} HN=C\begin{array}{c} NH-CO \\ | \\ N-CH_2 \\ | \\ CH_3 \end{array} + HDP$$

(XXXIII.) Phosphorylcréatine.                          (XXXI.) Créatinine.

Les différents mécanismes indiqués ci-dessus n'expliquent toutefois pas la biogenèse de l'ensemble des guanidines biologiques connues; aussi l'existence d'autres voies de l'anabolisme de certaines d'entre elles doit-elle être prévue (258).

## II. Propriétés générales.

1. **Basicité.** Un caractère général des guanidines biologiques est leur forte basicité, qui, dans le cas de la guanidine, atteint celle de l'hydroxyde de sodium.

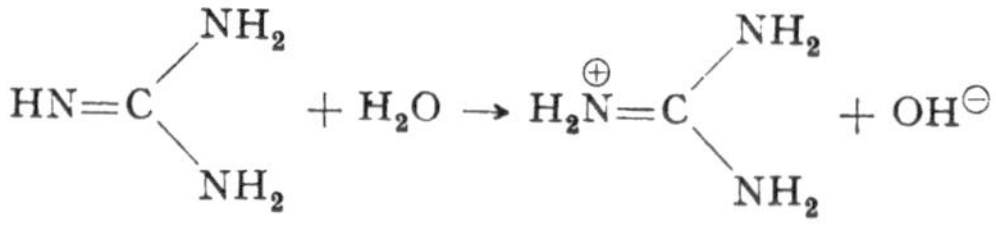

$$HN=C\begin{array}{c} NH_2 \\ \\ NH_2 \end{array} + H_2O \rightarrow H_2\overset{\oplus}{N}=C\begin{array}{c} NH_2 \\ \\ NH_2 \end{array} + OH^{\ominus}$$

La méthylguanidine, la diméthylguanidine, l'hydroxyguanidine sont également très basiques (pK = 14). Cette basicité est évidemment diminuée lorsque le substituant apporte des radicaux acides: carboxyle, phosphoryle, sulfonyle. Ainsi le pK du groupement guanidique de l'arginine, guanidine également monosubstituée, n'est pas proche de 14, mais compris entre 12,5 à 13,2 (47). Avec la créatine, où le groupement carboxyle est très proche du groupement guanidique, $pK_{ca}$ est encore plus bas (11,02). La méthylation du groupement guanidinium exalte l'acidité du groupement carboxyle ($pK_a$ = 2,62 à 30°) (39). Le caractère acide prédomine nettement dans les dérivés sulfonés (taurocyamine) et les phosphagènes; avec l'argininephosphate et le créatinephosphate, par exemple, $pK_a$ est égal à 1,96 (63). La basicité du groupement guanidinium est, par contre, renforcée lorsque le substituant apporte des groupements basiques supplémentaires: groupements aminés de l'agmatine et de la méthylagmatine ou groupements guanidiques de l'arcaine et de la streptidine.

2. **Formation des sels.** Cette basicité, renforcée ou atténuée par divers substituants, explique la facilité plus ou moins grande avec laquelle, les dérivés guanidiques donnent naissance à des sels.

D'une façon générale, ils forment des sels facilement cristallisables et peu solubles avec les acides picrique, flavianique, phosphotungstique, phosphomolybdique, chloroplatinique, chloroaurique. Mais la créatine, où pK$_{ca}$ n'est pas élevé, ne donne pas de combinaisons insolubles avec les acides phosphotungstique et phosphomolybdique et la plupart de ses sels, simples ou doubles, sont solubles. Il en est de même pour la taurocyamine.

Les dérivés guanidiques possédant des groupements acides donnent des sels de cuivre et d'argent cristallisables. Les sels alcalino-terreux (Ba, Ca) des guanidophosphates, plus ou moins solubles dans l'eau, sont tous cristallisables dans l'alcool aqueux.

**3. Stabilité et hydrolyse.** La structure parfaitement symétrique du groupement guanidinium et les possibilités de résonance qui en découlent (formules *A, B, C*) et qui sont démontrées à la fois par diffraction aux rayons X et le spectre Raman, lui confèrent une très grande stabilité (*47*). Lorsque cette structure est perturbée, le groupement guanidinium devient moins stable. Dans l'eau, la guanidine s'hydrolyse en urée, puis en NH$_3$,

cyanate et carbonate d'ammonium. Les guanidines monosubstituées, traitées par la baryte à chaud donnent de l'urée et de l'ammoniaque, et, en présence de soude à chaud, de l'ammoniaque et l'amine correspondante. La créatine donne ainsi naissance à la méthylhydantoine (LV) (*182*) et à la sarcosine (LVI) (*146, 147*):

(IX.) Créatine.      (LV.) Méthylhydantoine.

$$\rightarrow\ CO(NH_2)_2 + CH_3NH{-}CH_2{-}COOH$$

(LVI.) Sarcosine.

La structure des groupements guanidylés et phosphorylés explique l'instabilité particulière des phosphagènes. La molécule de ces derniers devient en effet le siège d'une opposition de résonance entre ces deux groupements, ce qui entraine une réduction des formes de résonance de la molécule du phosphagène, donc de sa stabilité (*113*). Avec la phosphocréatine où l'azote aminé est entièrement substitué, l'annulation de toute possibilité de résonance fait de ce corps le plus labile de tous les phosphagènes (*184*). Lorsqu'on compare l'énergie libre d'hydrolyse — $\Delta$G de ces derniers avec leur résistance aux acides on observe qu'elles vont effectivement de pair.

| Phosphagène | $-\Delta G$ calculé cal/mole | % d'hydrolyse ClH 0,1 N 30$^m$ à 30° |
|---|---|---|
| Créatinephosphate .............. | 11.000 (*113*) | 47,5 (*194*) |
| Argininephosphate .............. | 8.000 (*113*) | 25,4 (*194*) |
| Taurocyaminephosphate ......... | 7.600 (*194*) | 14,6 (*194*) |

Ce sont les monoanions (pH: 1 à 3,5) qui seraient les plus instables, les dianions et les molécules non chargées le seraient moins (*72, 79*). L'hydrolyse est considérablement accélérée par le molybdate dans le cas du créatinephosphate (*72*); elle est retardée dans le cas de l'arginine-phosphate (*157*), du taurocyamine- et du glycocyamine-phosphate (*79*).

**4. Cyclisation.** Avec le premier phosphagène, la libération de l'ion phosphorique et de la créatine est en outre, accompagnée, comme nous l'avons vu plus haut, de la cyclisation de celle-ci en créatinine, réaction peu intense en milieu faiblement acide mais devenant pratiquement totale en présence de molybdate (*169*):

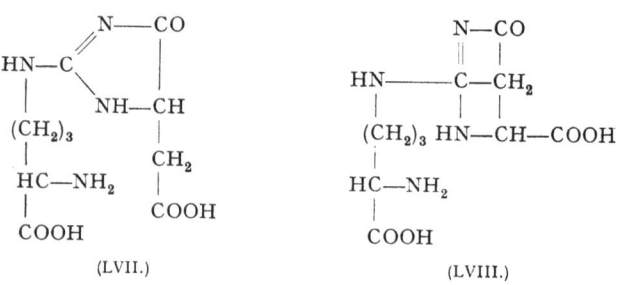

(XXXIII.) Phosphorylcréatine.  (XXXI.) Créatinine.

Cette cyclisation se produit également avec la créatine (*146*), mais dans des conditions plus drastiques. De même, la glycocyamine, traitée par les acides forts à 110° pendant quelques heures, se transforme en glycocyamidine (LVI):

(VIII.) Glycocyamine.  (LVI.) Glycocyamidine.

L'acide argininosuccinique (XXVII, p. 86) peut également s'anhydriser à chaud à pH = 3,2 en formes cyclisées (LVII et LVIII) encore indéterminées (*203*).

(LVII.)  (LVIII.)

5. **Oxydation.** En dehors des réactions d'hydrolyse et de cyclisation qui viennent d'être examinées, les dérivés guanidiques peuvent être oxydés plus ou moins profondément. En dehors de $CO_2$ et $NH_3$ libérés, les termes extrêmes de l'oxydation sont constitués par l'urée, la guanidine, la méthylguanidine (*53, 156, 50, 189*). Les produits intermédiaires sont très variés: avec l'arginine on obtient à côté de plusieurs dérivés inconnus, l'acide γ-guanidobutyrique (*221*), l'acide α-céto-δ-guanidovalérianique, l'acide guanidopropionique, le β-guanidopropanol, le guanidobutane (*221*); avec l'hirudonine, qui est présumée être la méthylagmatine, on peut identifier: l'agmatine, l'acide γ-guanidobutyrique, l'acide guanido-propionique, la guanidine (*213*). La formation de la méthylguanidine à partir de la créatine se fait avec une facilité extrême; il y aurait formation intermédiaire d'acide guanidoglyoxylique (LIX) (*67*):

$$HN=C\underset{N(CH_3)-CH_2-COOH}{\overset{NH_2}{<}} \rightarrow HN=C\underset{N(CH_3)-CO-COOH}{\overset{NH_2}{<}} \rightarrow$$

(IX.) Créatine.     (LIX.) Acide guanidoglyoxylique.

$$\rightarrow HN=C\underset{NH-CH_3}{\overset{NH_2}{<}}$$

(XVIII.)

L'oxydation facile des dérivés guanidiques en monométhyl- et diméthyl-guanidine explique que l'existence de ces dernières a souvent été mise en doute (*85*).

Les oxydants les plus divers peuvent être employés: oxyde et sels mercuriques, oxyde de cuivre en solution ammoniacale, peroxyde d'hydrogène, sulfate de fer, permanganate de calcium etc.

6. **Dégradation biochimique.** Sur le plan biochimique, la dégradation hydrolytique des dérivés guanidiques, étudiée en détail par ailleurs (*217*), peut se faire par trois mécanismes différents:

*a. Déguanidylation,* qui libère l'hydroxyguanidine et l'homosérine à partir de la canavanine. La réaction, qui ne s'observe qu'avec ce dérivé (voir page 94), est catalysée par des préparations de *Pseudomonas* du sol (*114*).

*b. Désamidination totale,* hydrolyse qui donne naissance à l'urée (LXI) et au dérivé aminé ou oxyaminé correspondant. C'est le mécanisme le plus général de dégradation des dérivés guanidiques. La réaction est catalysée soit par l'arginase (*129*), de spécificité assez restreinte, soit par l'hétéroarginase (*229*), dont l'action s'étend à la plupart des dérivés guanidiques: acides $\omega$-guanidiques comme les acides $\gamma$-guanido-butyrique (*283*), $\delta$-guanidovalérianique (*124*), guanidopropionique et les dérivés guanidiques des acides $\alpha$-aminés: glycine, alanine, *DL*-sérine, *DL*-isosérine, *DL*-thréonine, *DL*-méthionine, *DL*-cystine, *DL*-tyrosine, acide aspartique, phénylalanine (*219, 221*).

$$HN=C\begin{smallmatrix}NH_2\\NH-R\end{smallmatrix} + H_2O \rightarrow O=C\begin{smallmatrix}NH_2\\NH_2\end{smallmatrix} + H_2N-R$$

(LXI.)

L'arginase, comme l'hétéroarginase, n'hydrolyse pas les dérivés guanidiques dépourvus de groupement carboxyle tels que l'agmatine, l'arcaine. Il existe, par contre, dans certaines bactéries des enzymes capables d'hydrolyser ces bases ainsi que l'oxybutylguanidine (*180*), le 1-guanido-2-(*p*-hydroxyphényl)-éthane (*303*). En outre des extraits de *Streptomyces* hydrolysent la streptidine et la streptomycine (*262*).

*c. Désamidination partielle.* C'est une véritable désimination qui détache une molécule d'ammoniac et transforme le dérivé guanidique en composé uréinique. La réaction, catalysée par des désiminases (*95*), plus ou moins spécifiques, ne s'observe pas avec tous les dérivés guanidiques.

$$HN=C\begin{smallmatrix}NH_2\\NH-R\end{smallmatrix} + H_2O \rightarrow O=C\begin{smallmatrix}NH_2\\NH-R\end{smallmatrix} + NH_3$$

L'arginine (*13, 121, 232, 186, 191*), la canavanine (*118*), l'arcaine (*148, 149, 151*), l'agmatine (*150*) donnent ainsi naissance aux dérivés carbamylés correspondants. Dans le cas de d'arginine, la citrulline (L) formée subit à son tour non une hydrolyse (*13*), mais une phosphorolyse ou une arsénolyse (*122, 187, 188, 232, 126*); la première réaction est réversible, la seconde ne l'est pas:

$$O=C\begin{smallmatrix}NH_2\\NH-(CH_2)_3-CH-COOH\\ \qquad\qquad\quad | \\ \qquad\qquad\quad NH_2\end{smallmatrix}$$

(L.) Citrulline.

$$\xrightarrow{PO_4 + ADP}_{ATP}$$
$$\xrightarrow{AsO_4 + H_2O}$$

$$\underset{\rightarrow}{\rightleftarrows} H_2N-(CH_2)_3-CH-COOH + CO_2 + NH_3$$
$$\qquad\qquad\qquad\quad | $$
$$\qquad\qquad\qquad\quad NH_2$$

Une désimination analogue à celles qui viennent d'être décrites s'observe également avec la créatinine. Celle-ci perd une molécule d'ammoniac et se dégrade en méthylhydantoine (LV) (*2, 3, 254*):

$$HN=C \begin{array}{c} NH\text{------}CO \\ | \\ N(CH_3)\text{---}CH_2 \end{array} \quad \xrightarrow{-NH_3} \quad O=C \begin{array}{c} NH\text{------}CO \\ | \\ N(CH_3)\text{---}CH_2 \end{array}$$

(XXXI.) Créatinine.          (LV.) Méthylhydantoine.

En dehors des réactions d'hydrolyse qui viennent d'être examinées et des réactions de transamidination citées plus haut — et qui toutes intéressent le groupement guanidique — on ne connait pas d'autres réactions générales communes à tous les dérivés guanidiques. Des processus particuliers de dégradation portent en revanche sur le radical lié au groupement guanidique.

Ainsi la γ-guanidobutyramide et la lombricine peuvent être toutes les deux hydrolysées, pour donner naissance, la première à l'ammoniac et à l'acide γ-guanidobutyrique (*259, 267*), la seconde au guanido-éthanol (LXII) et à la sérine (LXIII) (*208*):

$$HN=C \begin{array}{c} NH_2 \\ \diagdown \\ NH\text{---}CH_2\text{---}CH_2\text{---}O\text{---}\overset{\overset{O}{\|}}{\underset{OH}{P}}\text{---}O\text{---}CH_2\text{---}\underset{NH_2}{CH}\text{---}COOH \end{array} + 2H_2O \rightarrow$$

(XXVI.) Lombricine.

$$\rightarrow HN=C \begin{array}{c} NH_2 \\ \diagdown \\ NH\text{---}CH_2\text{---}CH_2OH \end{array} + HOH_2C\text{---}\underset{NH_2}{CH}\text{---}COOH + PO_4H_3$$

(LXII.) Guanidoéthanol.          (LXIII.) Sérine.

Quant aux phosphagènes, il ne semble pas exister d'enzymes hydrolysant le groupement phosphorique; ce dernier peut, comme on l'a vu plus haut, être transféré à l'acide adénosine-diphosphorique pour reconstituer le nucléotide triphosphorique.

Deux réactions de dismutation intéressent l'µne la canavanine, l'autre l'octopine. La première peut être observée chez *Streptococcus faecalis* et *Str. equinus*; elle consiste en une hydrogénation du dérivé oxyguanidique en guanidine et homosérine (*117*). La seconde est au contraire une réaction de déshydrogénation, suivie d'une hydrolyse, sans doute spontanée. L'octopine est dégradée en arginine et en acide pyruvique qui, en présence de la lacticodéhydrogénase est réduit à l'état d'acide lactique (LXIV) (*273a*):

$$HN=C\overset{\displaystyle NH_2}{\underset{\displaystyle NH-(CH_2)_3-CH-COOH}{\Big\backslash}}$$

$$\xrightarrow[\text{+ lacticodéhydrogénase}]{\text{enzyme + DPN}}$$

$$\underset{\underset{CH_3-CH-COOH}{\overset{|}{NH}}}{\overset{|}{}}$$

(XXIX.) Octopine.

$$\rightarrow HN=C\overset{\displaystyle NH_2}{\underset{\displaystyle NH-(CH_2)_3-CH-COOH}{\Big\backslash}} \quad + \quad CH_3-CHOH-COOH$$

(LXIV.) Acide lactique.

$$\underset{NH_2}{\overset{|}{}}$$

(VI.) Arginine.

Enfin l'arginine peut, comme il a été dit plus haut, être soit désaminée oxydativement en acide $\alpha$-céto-$\delta$-guanido-valérianique et secondairement en acide $\gamma$-guanidobutyrique, soit décarboxylée en agmatine, soit décarboxylée oxydativement en $\gamma$-guanidobutyramide.

On peut en outre penser que des mécanismes d'oxydation enzymatique encore inconnus existent, ce qui expliquerait l'existence de la guanidine, de la méthylguanidine, de la diméthylguanidine, dans les milieux de prolifération bactérienne et chez certains invertébrés.

# III. Méthodes générales d'analyse et de préparation.

Une revue générale de cette question a été faite récemment par ROBIN (*209*); aussi n'entrerons-nous pas dans les détails des méthodes et nous contenterons-nous d'en indiquer les principes généraux.

**1. Réactions d'identification et de dosage.** *a. Réaction au diacétyle.* La réaction la plus anciennement connue est la coloration rouge avec fluorescence verte que donnent les dérivés guanidiques avec le diacétyle en milieu alcalin (*86*). Cette réaction est sensibilisée et stabilisée par addition d'$\alpha$-naphtol (*22*); elle est plus sensible avec les guanidines disubstituées qu'avec les monosubstituées (*60, 170, 223*). On a élaboré à partir de cette réaction une méthode commode de dosage (*226*) et l'on peut remplacer le diacétyle par l'acétylbenzoyle, dont l'excès est éliminé par l'hydroxylamine (*140*).

*b. Réaction à l'$\alpha$-naphtol.* Une autre réaction très connue des guanidiques est la coloration rouge qu'ils donnent avec l'$\alpha$-naphtol en présence d'hypochlorite alcalin (*227*), d'hypobromite (*296*), de N-bromosuccinimide (*230*) ou de 1,3-dibromo-5,5-diméthylhydantoine (*251*). L'$\alpha$-naphtol lui-même peut être remplacé par la 8-hydroxyquinoléine (*228*), la 7-chlorooxine (*109*), l'acide $\alpha$-naphtolsulfonique (*133*). Mais les méthodes de dosage les plus courantes sont basées sur l'utilisation du réactif $\alpha$-naphtol-hypobromite (*112, 57, 284, 162*).

*c. Réaction au nitroprussiate ferrique.* Le mélange nitroprussiate de sodium-ferricyanure de potassium en milieu très alcalin donne une coloration rouge avec la guanidine et ses dérivés mono- et di-substitués asymétriques mais non avec les dérivés disubstitués symétriques et trisubstitués (*296*).

Le réactif précédent peut être remplacé avantageusement par le pentacyano-ammonioferrate trisodique $(FeCy_5NH_3)Na_3$ qui, en milieu alcalin, donne avec les guanidines monosubstituées et disubstituées asymétriques, des colorations rouges variées, et, en milieu neutre ou légèrement acide, réagit spécifiquement avec le groupement guanidoxy (hydroxyguanidine, canavanine etc.) (*69*, *24*).

*d. Réaction à la ninhydrine.* On a montré récemment que la ninhydrine, en milieu très alcalin, se dégrade en O-carboxyphénylglyoxal, lequel forme avec les dérivés guanidiques des combinaisons hautement fluorescentes, permettant de déceler la présence de traces de ces composés. Cette réaction diffère de celle de formation des composés d'addition entre la ninhydrine et les dérivés guanidiques (*48*).

Les réactions colorées qui viennent d'être rapportées sont communes aux différentes guanidines mono- et di-substituées asymétriques; elles ne sont pas applicables à l'analyse des dérivés disubstitués symétriques comme les phosphagènes, les acides arginino- ou canavanino-succinique, l'astérubine. Il convient d'appliquer d'autres méthodes à ces corps et à d'autres présentant des propriétés particulières. Ainsi le picrate alcalin jaune se laisse réduire en picramate rouge orangé par la créatinine, la glycocyamidine (*107*, *105*), l'anhydride argininosuccinique (*203*) obtenus par cyclisation des acides correspondants. Le 3,5-dinitrobenzoate donne une coloration rouge violacée avec la créatinine (*142*) et la glycocyamidine (*25*). La dinitrophénylhydrazine permet d'identifier et d'isoler l'acide $\alpha$-céto-$\delta$-guanido-valérianique (*249*, *278*) et la ninhydrine les dérivés guanidiques possédant un groupement aminé supplémentaire: arginine, agmatine, lombricine, canavanine acides arginino- ou canavanino-succinique.

Pendant longtemps, l'arginine a été, en outre, dosée soit par précipitation sous forme de flavianate (*130*, *286*) soit par dosage des produits de son hydrolyse, dosage au xanthydrol de l'urée formée sous l'action arginasique (*108*, *30*) de l'ammoniac (*98*) ou du $CO_2$ libérés (*99*) par l'action de l'uréase sur l'urée. Ces méthodes reposant sur l'action arginasique sont précises, mais relativement laborieuses et celles basées sur la colorimétrie de diverses réactions sont aujourd'hui plus employées. Elles présentent l'avantage de pouvoir être appliquées à l'échelle microanalytique.

Les phosphagènes ne peuvent être identifiés ou dosés qu'en déterminant le phosphore ou le dérivé guanidique libérés par hydrolyse ménagée. En dehors du dosage du phosphore sous forme de phosphate ammoniacomagnésien, les différentes techniques usuelles comportent l'inconvénient majeur d'introduire du molybdate en milieu acide, ce qui provoque une hydrolyse plus ou moins forte des phosphagènes en particulier du créatinephosphate (*71*, *161*, *26*, *73a*). Cet inconvénient peut être évité en dosant, après hydrolyse le dérivé guanidique, par l'une des méthodes indiquées plus haut, réalisable en milieu alcalin.

## 2. Chromatographie sur papier et sur colonne. Electrophorèse.

Les différentes réactions qui viennent d'être signalées ne sont pas spécifiques; communes à la guanidine et à ses différents dérivés, elles ne sont applicables qu'après séparation de ces corps. La chromatographie sur papier permet de réaliser celle-ci. On dispose d'un jeu très varié de solvants permettant de développer également des chromatogrammes en vue de l'analyse ou de la préparation des dérivés guanidiques. Ces derniers sont révélés par différents réactifs: picrate alcalin (*52*, *164*, *16*) et diacétyle alcalin (*16*) pour la créatine et la glycocyamine; $\alpha$-naphtol-hypobromite alcalin (*268, 1, 218, 223*), nitroprussiate-ferricyanure (*223, 241*), diacétyle-α-naphtol (*223, 226*) pour l'ensemble des guanidines monosubstituées, pentacyanoammonioferrate (*69*) pour les oxyguanidiques. A la chromatographie sur papier peut être associée efficacement la chromatoélectrophorèse, permettant de réduire le temps de migration et une séparation plus nette (*154*).

La chromatographie sur papier a permis au cours des dernières années, d'identifier dans les tissus et les liquides biologiques la plupart des dérivés guanidiques nouveaux ou connus auparavant seulement comme des corps de synthèse: taurocyamine (*280, 277*) et taurocyamine-phosphate (*280, 281*), glycocyamine-phosphate (*280, 281*), lombricine et lombricine-phosphate (*270*), hirudonine (*214*), acides γ-guanido-butyrique (*279, 266*) et δ-guanidovalérianique (*263*), γ-guanidobutyr-amide (*260, 261*), acide argininosuccinique (*289, 51*).

L'emploi des *isotopes*, dans ce domaine comme dans d'autres, permet de suivre efficacement, le métabolisme des dérivés guanidiques. C'est ainsi qu'ont été mises en évidence les réactions de transamidination (*28, 244*), leur réversibilité (*96, 205, 292*), l'incorporation de $CO_2$ (*66, 206*) et de $NH_3$ (*301*) dans le groupement guanidique de l'arginine ainsi que les différentes étapes de la biosynthèse de celle-ci (*81—83, 202, 163*), la formation des groupements guanidiques chez la streptomycine (*100, 101, 183*), la dégradation de l'arginine en acide arginique (*173*) ou en acide γ-guanidobutyrique ou en acide δ-guanidovalérianique (*36*).

La chromatographie sur *colonnes de résines* permet également une séparation satisfaisante des dérivés guanidiques. En raison précisément du caractère plus ou moins basique de ces derniers, les résines peuvent être utilisées sur une gamme très large.

Sauf les phosphagènes, aucun de ces corps n'est fixé sur les résines anioniques (Dowex 1 ou 2). Parmi les résines cationiques, les unes, comme la permutite ou la Dowex 50 (Cl⁻), fixent la presque totalité des dérivés guanidiques, sauf la tauro-cyamine; d'autres, comme l'amberlite IR 120 (Na⁺) ou amberlite IRC 50, ne fixent que les corps ayant un $pK_{ca}$ supérieur à 12 (arginine, arcaine).

L'emploi combiné de ces différentes résines, tamponnées ou non, et le jeu des élutions permet la fixation et la séparation d'un très grand nombre de guanidines biologiques, aussi bien pour leur analyse que pour leur isolement.

On a dosé ainsi la glycocyamine et l'arginine (*55, 56, 242*), la créatine et la créatinine (*17*), l'octopine (*271*). On a isolé de cette manière la glycocyamine et la taurocyamine (*269*), leurs phosphagènes (*282*), la lombricine et son dérivé phosphorylé (*270*), l'octopine (*125*), l'acide α-céto-δ-guanidovalérianique (*165*), l'hirudonine (*214*), le créatine-phosphate (*193*) et l'argininephosphate (*256*).

*Bibliographie: pp. 107—121.*

## IV. Répartition des dérivés guanidiques.

L'un des caractères les plus remarquables de la biologie des dérivés guanidiques est la diversité de leur répartition. L'examen du *Tableau 3* (p. 106) montre que la plupart des produits connus se rencontrent chez des Invertébrés ou plus précisément chez des groupes restreints d'Invertébrés et qu'un tiers d'entre eux peuvent être considérés comme spécifiques de groupes différents d'organismes vivants. Certains d'entre eux ne se rencontrent que chez les végétaux tels la galégine, la streptidine, la canavanine. Celle-ci comme l'acide canavaninosuccinique, en raison de leurs relations structurales avec l'arginine et l'acide argininosuccinique peuvent se former dans les milieux de réaction enzymatique, en présence de transamidinases animales ou d'un enzyme animal ou végétal de condensation de l'arginine et de l'acide fumarique. En fait, et comme la galégine et la streptidine, la canavanine n'a été, jusqu'à présent, rencontrée que chez des groupes restreints de végétaux.

Une deuxième série de corps, également spécifiques de certains organismes, est constituée par les phosphagènes. En dehors du créatine-phosphate, qu'on rencontre également chez les Vertébrés et certains Invertébrés, les autres phosphagènes se trouvent diversement répartis chez les Invertébrés. L'argininephosphate est le plus répandu chez la plupart de ceux-ci, tandis que les phosphoryl-glycocyamine, -tauro-cyamine, -lombricine se rencontrent seulement ou surtout chez les différents groupes de vers.

Une troisième série de corps: astérubine, lombricine, arcaïne, n'a été mise en évidence que chez certains Invertébrés. On peut y ajouter la taurocyamine, qui sert d'accepteur de phosphate chez les Polychètes sédentaires et qui ne se rencontre qu'à l'état de traces, quoique d'une façon régulière, dans les urines des Mammifères (*277*). D'ailleurs ces différents corps ne se rencontrent pas chez tous les Invertébrés: ils n'ont été étudiés ou identifiés que dans un petit nombre de cas.

La quatrième série de corps, constituée par la glycocyamine et ses dérivés, la créatine et la créatinine, semble être spécifique des animaux; elle est plus ou moins abondamment répartie chez les Vertébrés, les Invertébrés mais absente chez les végétaux. On peut ajouter en outre, qu'elle prédomine chez les Vertébrés, où les trois corps de la série sont présents d'une façon constante tandis qu'elle est seulement représentée d'une façon partielle et inconstante chez les Invertébrés. La glycocyamine, et surtout la créatine, quoique présentes du haut en bas de l'échelle animale, ne se trouvent que dans des groupes limités d'Invertébrés; quant à la créatinine, elle semble en être absente. La formation de la créatinine dans les cultures de levure renfermant du phosphate et supplémentées en glucose (*243*), constitue sans doute un artefact

Tableau 3. Répartition des dérivés guanidiques biologiques.

Le signe + indique les produits trouvés à l'état naturel, le signe ◯, ceux identifiés dans des milieux de réactions enzymatiques. Le signe + n'indique pas la présence chez tous les représentants des groupes désignés: vertébrés, invertébrés, végétaux, mais seulement dans des familles, genres ou espèces indiqués par les références.

| Dérivés guanidiques | Vertébrés | Invertébrés | Végétaux |
|---|---|---|---|
| Arginine | + (*89, 90, 237*) | + (*89, 90, 237*) | + (*89, 90, 237*) |
| Agmatine | + (*245, 127*) | + (*94, 105, 210, 12, 155, 75, 279*) | + (*177, 240, 62a, 241*) |
| Méthylagmatine | | + (*88*) | |
| γ-Guanidobutyramide | | ◯ (*274*) | ◯ (*260, 261, 264*) |
| Acide γ-guanidobutyrique | + (*266, 263, 102, 103*) | + (*278, 279, 212, 208, 103, 76, 154, 225*) | + (*177, 103, 178, 173*) |
| Ac. γ-guanidovalérianique | + (*263*) | | |
| Ac. α-céto-δ-guanido-valérianique | | + (*279*) | |
| Ac. α-hydroxy-δ-guanido-valérianique | | | ◯ (*173*) |
| Ac. argininosuccinique | + (*15, 298, 207*) | | + (*51, 289*) |
| Octopine | | + (*12, 105, 104, 172*) | ◯ (*183a*) |
| Subérylarginine | + (*299*) | | |
| Guanidine | + (*139, 85*) | + (*11, 116, 152, 153, 85*) | + (*234, 235, 85, 179*) |
| Hydroxyguanidine | | | ◯ (*114*) |
| Méthylguanidine | + (*85*) | + (*85*) | + (*85*) |
| Diméthylguanidine | + (*85*) | + | |
| Glycocyamine | + (*91, 56, 297*) | + (*268, 269, 6, 280, 281*) | |
| Créatine | + (*97, 53, 54*) | + (*167, 181, 78, 208, 223, 224*) | |
| Créatinine | + (*146, 149, 97*) | | ◯ (*243*) |
| Taurocyamine | + (*277, 266*) | + (*280, 281, 269, 211, 7*) | |
| Astérubine | | + (*4*) | |
| Lombricine | | + (*270, 23*) | |
| Arcaine | | + (*137, 138, 215, 210, 280, 216*) | |
| Phosphocréatine | + (*71, 72, 168, 169*) | + (*174, 92, 93, 222, 63*) | |
| Phosphoarginine | | + (*168, 166, 239, 63*) | |
| Phosphoglycocyamine | | + (*280, 281, 80, 92, 93, 63*) | |
| Phosphotaurocyamine | | + (*280, 281, 80, 92, 93, 63*) | |
| Phospholombricine | | + (*270, 23*) | |
| Galégine | | | + (*255*) |
| Streptidine | | | + (*288*) |
| Canavanine | ◯ (*291*) | | + (*69*) |
| Ac. canavaninosuccinique | ◯ (*290*) | | ◯ (*290*) |

réactionnel ou une déviation de la réaction enzymatique, dénué en tout cas de signification biologique précise.

La dernière série de corps renfermant l'arginine et ses produits de dégradation ou de condensation, présente une répartition très diverse. L'arginine est universelle, comme certains de ses produits de dégradation: l'agmatine, l'acide γ-guanidobutyrique. On peut penser que d'autres intermédiaires du métabolisme: l'acide α-céto-δ-guanidovalérianique, l'acide argininosuccinique, peuvent se former également dans la majorité des organismes vivants; toutefois leur participation active au déroulement normal du métabolisme peut empêcher leur accumulation et gêner leur identification. Par contre, la présence de certains des produits de transformation de l'arginine doit relever d'un type particulier du métabolisme et leur répartition doit réfléter des différences profondes de métabolisme. Tel est le cas de l'acide guanidovalérianique, de l'acide arginique, de l'octopine, de la γ-guanidobutyramide et peut-être aussi celui de la subérylarginine, de la guanidine, de l'hydroxyguanidine, de la mono- et de la di-méthylguanidine.

En l'état actuel de nos connaissances sur la répartition des guanidines biologiques, il est prématuré d'essayer de tirer dès maintenant des conclusions sur la signification biologique de celles-ci. Les données que nous possédons sont trop fragmentaires et parfois encore incertaines; les méthodes anciennes employées pour l'identification de ces corps, trop brutales, ont, en effet, provoqué souvent des transformations spontanées et conduit à l'identification à partir des milieux biologiques des corps qui n'en sont pas des constituants naturels.

## Conclusions.

La conclusion la plus remarquable qui doit être retenue au terme de cette étude est la diversité des radicaux fixés au groupement guanidique, dont résulte la multiplicité des dérivés étudiés dans la nature. Ce fait confère aux guanidines substituées une signification exceptionnelle dans le cadre du métabolisme azoté. Il traduit leur importance dans la fixation et dans le transfert de l'azote organique. Il explique la variété des réactions de biogenèse et de dégradation auxquelles participent les dérivés guanidiques et illustre les rapports multiples reliant ces corps à divers domaines du métabolisme (258).

### Bibliographie.

1. ACHER, R. et C. CROCKER: Réactions colorées spécifiques de l'arginine et de la tyrosine réalisées après chromatographie sur papier. Biochim. Biophys. Acta 9, 704 (1952).
2. ACKERMANN, D.: Über den fermentativen Abbau des Kreatinins. Z. Biol. 62, 208 (1913).

3. ACKERMANN, D.: Über den fermentativen Abbau des Kreatinins. II. Z. Biol. **63**, 78 (1913).

4. — Asterubin, eine schwefelhaltige Guanidinverbindung der belebten Natur. Z. physiol. Chem. (Hoppe-Seyler) **232**, 206 (1935).

5. — Synthese des Asterubins. Z. physiol. Chem. (Hoppe-Seyler) **234**, 208 (1935).

6. — Über das Vorkommen von Glykocyamin, Cholin, Lysin, Leucin, Tyrosin und α-Alanin in dem Meereswurm *Nereis virens*. Z. physiol. Chem. (Hoppe-Seyler) **299**, 186 (1955).

7. — Über das Vorkommen von Homarin, Taurocyamin, Cholin, Lysin und anderen Aminosäuren sowie Bernsteinsäure in dem Meereswurm *Arenicola marina*. Z. physiol. Chem. (Hoppe-Seyler) **302**, 80 (1955).

8. ACKERMANN, D., R. ENGELAND und FR. KUTSCHER: Die Synthese der δ-Guanidinovaleriansäure. Z. Biol. **57**, 179 (1911).

9. ACKERMANN, D. und H. A. HEINSEN: Über die physiologische Wirkung des Asterubins und anderer, zum Teil neu dargestellter, schwefelhaltiger Guanidinderivate. Z. physiol. Chem. (Hoppe-Seyler) **235**, 115 (1935).

10. — — Darstellung des Tetramethyldiguanylcystamins. Vgl. Nr. **9**, S. 121.

11. ACKERMANN, D., F. HOLTZ und H. REINWEIN: Über das Vorkommen von Methyladenin, Dimethylhistamin, Guanidin, Betain und Eledonin bei *Geodia gigas*. Z. Biol. **82**, 278 (1924).

12. ACKERMANN, D. und M. MOHR: Über das Vorkommen von Octopin, Agmatin und Arginin in der Octopodenart *Eledone moschata*. Z. physiol. Chem. (Hoppe-Seyler) **250**, 249 (1937).

13. ACKERMANN, D. und E. MÜLLER: Zweite Synthese des Asterubins. Z. physiol. Chem. (Hoppe-Seyler) **235**, 233 (1935).

14. AKAMATSU, S. and T. SEKINE: Hydrolysis of Arginine by *Streptococcus faecalis*. J. Biochem. (Japan) **38**, 349 (1951).

15. ALLAN, J. D., D. C. CUSWORTH, C. E. DENT and V. K. WILSON: A Disease, Probably Hereditary, Characterised by Severe Mental Deficiency and a Constant Gross Abnormality of Aminoacid Metabolism. Lancet **1958**, 182.

16. AMES, S. R. and H. A. RISLEY: Determination of Creatine, Creatinine, and Related Compounds in Urine by Means of Paper Chromatography. Proc. Soc. exp. Biol. Med. **69**, 267 (1948).

17. ANDERSON, D. R., C. M. WILLIAMS, G. M. KRISE and R. M. DOWBEN: Determination of Creatine in Biological Fluids. Biochemic. J. **67**, 258 (1957).

18. BALDWIN, E. and D. M. NEEDHAM: The Phosphorus Distribution in Resting Fly Muscle. J. Physiol. **80**, 221 (1933).

19. — — A Contribution to the Comparative Biochemistry of Muscular and Electrical Tissues. Proc. Roy. Soc. (London), Ser. B **122**, 197 (1937).

20. BALDWIN, E. and W. H. YUDKIN: The Annelid Phosphagen: with a Note on Phosphagen in Echinodermata and Protochordata. Proc. Roy. Soc. (London), Ser. B **136**, 614 (1949/50).

21. BANGA, I.: ATP-Creatine Phosphopherase. Stud. Inst. Med. Chem. Univ. Szeged **3**, 59 (1943).

22. BARRITT, M. M.: The Intensification of the Voges-Proskauer Reaction by the Addition of α-Naphthol. J. Pathol. Bacteriol. **42**, 441 (1936).

23. BEATTY, I. M., D. I. MAGRATH and A. H. ENNOR: Occurrence of *D*-Serine in Lombricine. Nature (London) **183**, 591 (1959).

24. BELL, E. A.: Canavanine and Related Compounds in Leguminosae. Biochemic. J. **70**, 617 (1958).

25. BENEDICT, S. R. and J. A. BEHRE: Some Applications of a New Color Reaction for Creatinine. J. Biol. Chem. **114**, 515 (1936).

26. BERENBLUM, I. and E. CHAIN: An Improved Method for the Colorimetric Determination of Phosphate. Biochemic. J. **32**, 295 (1938).

27. BLOCH, K. and R. SCHOENHEIMER: Studies in Protein Metabolism. XI. The Metabolic Relation of Creatine and Creatinine Studied with Isotopic Nitrogen. J. Biol. Chem. **131**, 111 (1939).

28. — — The Biological Origin of the Amidine Group in Creatine. J. Biol. Chem. **134**, 785 (1940).

29. — — The Biological Precursors of Creatine. J. Biol. Chem. **138**, 167 (1941).

30. BONOT, A. et T. CAHN: Dosage de l'arginine par la méthode de Jansen modifiée. Bull. soc. chim. biol. (Paris) **9**, 1001 (1927).

31. BORSOOK, H. and J. W. DUBNOFF: The Formation of Creatine from Glycocyamine in the Liver. J. Biol. Chem. **132**, 559 (1940).

32. — — Creatine Formation in Liver and in Kidney. J. Biol. Chem. **134**, 635 (1940).

33. — — The Formation of Glycocyamine in Animal Tissues. J. Biol. Chem. **138**, 389 (1941).

34. — — The Conversion of Citrulline to Arginine in Kidney. J. Biol. Chem. **141**, 717 (1941).

35. — — The Hydrolysis of Phosphocreatine and the Origin of Urinary Creatinine. J. Biol. Chem. **168**, 493 (1947).

36. BOULANGER, P. et R. OSTEUX: Aspects du métabolisme de l'arginine chez le Rat blanc. Colloque intern. biochim. comparée acides aminés basiques, Concarneau, 1959.

37. BRAND, E. and F. C. BRAND: Guanidoacetic Acid (Glycocyamine). In: Organic Syntheses, Vol. 22, p. 59. London: Chapman & Hall; New York: J. Wiley. 1942.

38. BRAND, E. and B. KASSELL: Photometric Determination of Arginine. J. Biol. Chem. **145**, 359 (1942).

39. CANNAN, R. K. and A. SHORE: The Creatine-Creatinine Equilibrium. The Apparent Dissociation Constants of Creatine and Creatinine. Biochemic. J. **22**, 920 (1928).

40. CAPUTTO, R.: On the Biological Transformation: Creatine → Creatinine. Arch. Biochem. Biophys. **52**, 280 (1954).

41. CAPUTTO, R. and M. P. CARPENTER: Enzymatic Conversion of Creatine to Creatinine. Federat. Proc. (Amer. Soc. exp. Biol.) **14**, 189 (1955).

42. CERIOTTI, G. and L. SPANDRIO: An Improved Method for the Microdetermination of Arginine by Use of 8-Hydroxyquinoline. Biochemic. J. **66**, 603 (1957).

43. CHAMBERS, E. L. and T. J. MENDE: Alterations of the Inorganic Phosphate and Arginine Phosphate Content in Sea Urchin Eggs Following Fertilization. Exptl. Cell. Res. **5**, 508 (1953).

44. CHEVREUL, E.: Recherches sur la composition chimique du bouillon de viande (extrait d'un rapport sur le bouillon de la compagnie hollandaise fait à l'Académie des Sciences par Chevreul le 19 mars 1822). J. Pharm. **21**, 231 (1835).

45. COHEN, P. P. and M. HAYANO: Urea Synthesis by Liver Homogenates. J. Biol. Chem. **166**, 251 (1946).

46. — — The Conversion of Citrulline to Arginine (Transimination) by Tissue Slices and Homogenates. J. Biol. Chem. **166**, 239 (1946).

47. COHN, E. J. and J. T. EDSALL: Proteins, Amino Acids and Peptides as Ions and Dipolar Ions, pp. 132/33. New York: Reinhold Publ. 1943.

48. CONN, R. B., Jr. and R. B. DAVIS: Green Fluorescence of Guanidinium Compounds with Ninhydrin. Nature (London) **183**, 1053 (1959).

49. Cramer, F. und A. Vollmar: Zur Chemie der „energiereichen Phosphate". VI. Synthesen von Kreatin- und Glykocyaminphosphorsäure. Chem. Ber. 92, 392 (1959).
50. Dakin, H. D.: The Formation of Glyoxylic Acid. J. Biol. Chem. 1, 271 (1906).
51. Davison, D. C. and W. H. Elliott: Enzymic Reaction Between Arginine and Fumarate in Plant and Animal Tissues. Nature (London) 169, 313 (1952).
52. Dent, C. E.: The Amino-aciduria in Fanconi Syndrome. A Study Making Extensive Use of Techniques Based on Paper Partition Chromatography. Biochemic. J. 41, 240 (1947).
53. Dessaignes, V.: Recherche sur quelques produits de transformation de la créatine. C. R. hebd. Séances Acad. Sci. 38, 839 (1854).
54. Dittrich, E.: Über Methyltaurin und die Bildung von Methyltaurocyamin . und Taurocyamin. J. prakt. Chem. [N. F.] 18, 63 (1878).
55. Dubnoff, J. W.: A Micromethod for the Determination of Arginine. J. Biol. Chem. 141, 711 (1941).
56. Dubnoff, J. W. and H. Borsook: A Micromethod for the Determination of Glycocyamine in Biological Fluids and Tissue Extracts. J. Biol. Chem. 138, 381 (1941).
57. Dumazert, C. et R. Poggi: Sur le microdosage colorimétrique de l'arginine et de divers dérivés guanidiques monosubstitués. Application aux hydrolysats protéiques. Bull. soc. chim. biol. (Paris) 21, 1381 (1939).
58. Eggleton, P. and G. P. Eggleton: The Inorganic Phosphate and a Labile Form of Organic Phosphate in the Gastrocnemius of the Frog. Biochemic. J. 21, 190 (1927).
59. — — Further Observations on Phosphagen. J. Physiol. 65, 15 (1928).
60. Eggleton, P., S. R. Elsden and N. Gough: The Estimation of Creatine and of Diacetyl. Biochemic. J. 37, 526 (1943).
61. Elderfield, R. C. and M. Green: Synthesis of Some Guanidino Amino Acids from Cyanogen Bromide. J. Organ. Chem. (USA) 17, 442 (1952).
62. Elödi, P. and E. Szörényi: Properties of Crystalline Arginine Phospherase Isolated from Crustacean Muscle. Acta Physiol. Hungar. 9, 367 (1956).
62a. Engeland, R. und Fr. Kutscher: Über eine zweite wirksame Secalebase. Zbl. Physiol. 24, 479 (1910).
63. Ennor, A. H. and J. F. Morrison: Biochemistry of Phosphagens and Related Guanidines. Physiol. Revs. 38, 631 (1958).
64. Ennor, A. H., H. Rosenberg and M. D. Armstrong: Specificity of Creatine Phosphokinase. Nature (London) 175, 120 (1955).
65. Ennor, A. H. and L. A. Stocken: The Preparation of Sodium Phosphocreatine. Biochemic. J. 43, 190 (1948).
66. Evans, E. A., Jr. and L. Slotin: The Utilization of Carbon Dioxide in the Synthesis of α-Ketoglutaric Acid. J. Biol. Chem. 136, 301 (1940).
67. Ewins, A. J.: Note on the Isolation of Methylguanidine by the Silver Method. Biochemic. J. 10, 103 (1916).
68. Fawaz, G. and K. Seraidarian: Phosphoglycocyamine. J. Biol. Chem. 165, 97 (1946).
69. Fearon, W. R. and E. A. Bell: Canavanine: Detection and Occurrence in Colutea arborescens. Biochemic. J. 59, 221 (1955).
70. Fink, R. M., C. McGaughey, R. E. Cline and K. Fink: Metabolism of Intermediate Pyrimidine Reduction Products in vitro. J. Biol. Chem. 218, 1 (1956).
71. Fiske, C. H. and Y. Subbarow: The Nature of the "Inorganic Phosphorus" in Voluntary Muscle. Science (Washington) 65, 401 (1927).
72. — — Phosphocreatine. J. Biol. Chem. 81, 629 (1929).

*73.* FULD, M.: Studies on Transamidinase. Federat. Proc. (Amer. Soc. exp. Biol.) **13**, 215 (1954).

*73a.* FURCHGOTT, R. F. and T. DE GUBAREFF: The Determination of Inorganic Phosphate and Creatine Phosphate in Tissue Extracts. J. Biol. Chem. **223**, 377 (1956).

*74.* GALE, E. F.: The Bacterial Amino Acid Decarboxylases. Adv. Enzymology **6**, 1 (1946).

*75.* GARCIA, I. et F. MIRANDA: Sur les dérivés guanidiques des Spongiaires. C. R. Séances Soc. Biol. **148**, 1187 (1954).

*76.* GARCIA, I., J. ROCHE et M. TIXIER: Sur le métabolisme de la *L*-arginine chez les Insectes. I. Bull. soc. chim. biol. (Paris) **38**, 1423 (1956).

*77.* GREENSTEIN, J. P.: A Synthesis of Homoarginine. J. Organ. Chem. (USA) **2**, 480 (1937).

*78.* GREENWALD, I.: The Presence of Creatine in the Testes of Various Invertebrates. The Preparation of Creatine Phosphoric Acid from Fish Testes. J. Biol. Chem. **162**, 239 (1946).

*79.* GRIFFITHS, D. E.: cité par ENNOR et MORRISON (*63*).

*80.* GRIFFITHS, D. E., J. F. MORRISON and A. H. ENNOR: The Distribution of Guanidines, Phosphagens and *N*-Amidino Phosphokinases in Echinoids. Biochemic. J. **65**, 612 (1957).

*81.* GRISOLIA, S., R. H. BURRIS and P. P. COHEN: Carbon Dioxide and Ammonia Fixation in the Biosynthesis of Citrulline. J. Biol. Chem. **191**, 203 (1951).

*82.* — — — Fate of Deuterio-labeled Carbamyl Glutamate in Citrulline Biosynthesis. J. Biol. Chem. **210**, 761 (1954).

*83.* GRISOLIA, S. and P. P. COHEN: Catalytic Rôle of Glutamate Derivatives in Citrulline Biosynthesis. J. Biol. Chem. **204**, 753 (1953).

*84.* GRISOLIA, S., D. P. WALLACH and H. J. GRADY: Carbamyl Transfer with Mammalian and Bacterial Enzymic Preparations. Biochim. Biophys. Acta **17**, 150 (1955).

*85.* GUGGENHEIM, M.: Die biogenen Amine und ihre Bedeutung für die Physiologie und Pathologie des pflanzlichen und tierischen Stoffwechsels. 4. Aufl. Basel: S. Karger. 1951.

*86.* HAGIHARA, F.: Paper Microelectrophoresis. V. Detection of Enzymic Decarboxylation Product of Amino Acids. J. pharmac. Soc. Japan **76**, 613 (1956) [Chem. Abstr. **50**, 14850 (1956)].

*87.* HARDEN, A. and D. NORRIS: The Diacetyl Reaction for Proteins. J. Physiol. **42**, 332 (1911).

*88.* HAUROWITZ, F. und H. WAELSCH: Über die chemische Zusammensetzung der Qualle *Velella spirans*. Z. physiol. Chem. (Hoppe-Seyler) **161**, 300 (1926).

*89.* HEDIN, S. G.: Über die Bildung von Arginin aus Proteinkörpern. Z. physiol. Chem. (Hoppe-Seyler) **21**, 155 (1895).

*90.* — Einige Bemerkungen über die basischen Spaltungsprodukte des Elastins. Z. physiol. Chem. (Hoppe-Seyler) **25**, 344 (1898).

*91.* HOBERMAN, H. D.: The Determination of Guanidoacetic Acid and Arginine in Human Urine and Serum. J. Biol. Chem. **167**, 721 (1947).

*92.* HOBSON, G. E. and K. R. REES: The Annelid Phosphagens. Biochemic. J. **61**, 549 (1955).

*93.* — — The Annelid Phosphokinases. Biochemic. J. **65**, 305 (1957).

*94.* HOLTZ, F.: Über das Vorkommen des Agmatins bei mehreren Tieren. Z. Biol. **81**, 65 (1924).

*95.* HORN, F.: Über den Abbau des Arginins zu Citrullin durch *Bacillus pyocyaneus*. Z. physiol. Chem. (Hoppe-Seyler) **216**, 244 (1933).

96. HORNER, W. H., I. SIEGEL and J. BRUTON: The Synthesis of Arginine from Guanidoacetic Acid. J. Biol. Chem. **220**, 861 (1956).

97. HUNTER, A.: Creatine and Creatinine. London: Longmans, Green. 1928.

98. HUNTER, A. and J. A. DAUPHINEE: The Arginase Method for the Determination of Arginine and its Use in the Analysis of Proteins. J. Biol. Chem. **85**, 627 (1929).

99. HUNTER, A. and J. B. PETTIGREW: A Manometric Method for the Enzymatic Determination of Arginine. Enzymologia **1**, 341 (1936/37).

100. HUNTER, G. D., M. HERBERT and D. J. D. HOCKENHULL: Actinomycete Metabolism: Origin of the Guanidine Groups in Streptomycin. Biochemic. J. **58**, 249 (1954).

101. HUNTER, G. D. and D. J. D. HOCKENHULL: Actinomycete Metabolism. Incorporation of $^{14}$C-Labelled Compounds into Streptomycin. Biochemic. J. **59**, 268 (1955).

102. IRREVERRE, F. and R. L. EVANS: Isolation of $\gamma$-Guanidinobutyric Acid from Calf Brain. J. Biol. Chem. **234**, 1438 (1959).

103. IRREVERRE, F., R. L. EVANS, A. R. HAYDEN and R. SILBER: Occurrence of $\gamma$-Guanidinobutyric Acid. Nature (London) **180**, 704 (1957).

104. IRVIN, J. L.: Further Studies on Octopine. J. Biol. Chem. **123**, LXII (1938).

105. IRVIN, J. L. and D. W. WILSON: Studies on Octopine. I. The Synthesis and Titration Curve of Octopine. J. Biol. Chem. **127**, 555 (1939).

106. JAFFÉ, M.: Untersuchungen über die Entstehung des Kreatins im Organismus. Z. physiol. Chem. (Hoppe-Seyler) **48**, 430 (1906).

107. — Über den Niederschlag, welchen Pikrinsäure in normalem Harn erzeugt, und über eine neue Reaktion des Kreatinins. Z. physiol. Chem. (Hoppe-Seyler) **10**, 391 (1886).

108. JANSEN, B. C. P.: La transformation d'arginine en créatine dans l'organisme animal. Arch. néerl. physiol. **1**, 618 (1916).

109. JANUS, J. W.: Determination of Arginine. Nature (London) **177**, 529 (1956).

110. JONES, M. E., L. SPECTOR et F. LIPMANN: Carbamylphosphate. Rapport 3e Congrès intern. Biochimie, Bruxelles 1955, p. 67.

111. — — — Carbamyl Phosphate, the Carbamyl Donor in Enzymatic Citrulline Synthesis. J. Amer. Chem. Soc. **77**, 819 (1955).

112. JORPES, E. and S. THORÉN: The Use of the Sakaguchi Reaction for the Quantitative Determination of Arginine. Biochemic. J. **26**, 1504 (1932).

113. KALCKAR, H. M.: Nature of Energetic Coupling in Biological Syntheses. Chem. Revs. **28**, 71 (1941).

114. KALYANKAR, G. D., M. IKAWA and E. E. SNELL: The Enzymatic Cleavage of Canavanine to Homoserine and Hydroxyguanidine. J. Biol. Chem. **233**, 1175 (1958).

115. KAPFHAMMER, J. und H. MÜLLER: Guanidosäuren und Guanidopeptide. Z. physiol. Chem. (Hoppe-Seyler) **225**, 1 (1934).

116. KIESEL, A.: Über die stickstoffhaltigen Substanzen in reifenden Roggenähren. Z. physiol. Chem. (Hoppe-Seyler) **135**, 61 (1924).

117. KIHARA, H., J. M. PRESCOTT and E. E. SNELL: The Bacterial Cleavage of Canavanine to Homoserine and Guanidine. J. Biol. Chem. **217**, 497 (1955).

118. KIHARA, H. and E. E. SNELL: The Enzymatic Cleavage of Canavanine to o-Ureidoserine and Ammonia. J. Biol. Chem. **226**, 485 (1957).

119. KITAGAWA, M. and A. TAKANI: Studies on a Diaminoacid, Canavanin. IV. The Constitution of Canavanin and Canalin. J. Biochem. (Japan) **23**, 181 (1936).

120. KITAGAWA, M. and T. TOMIYAMA: A New Amino-Compound in the Jack Bean and a Correspond i g New Ferment. J. Biochem. (Japan) **11**, 265 (1930).

*121.* KNIVETT, V. A.: Phosphorylation Coupled with Anaerobic Breakdown of Citrulline. Biochemic. J. **55**, x (1953).

*122.* — Bacterial Decomposition of Arginine. Colloque intern. biochim. comparée acides aminés basiques, Concarneau, 1959.

*123.* KNORR, A.: Zur Kenntnis der Iminoester. II. Mitt. Der Mechanismus der Amidinbildung. Ber. dtsch. chem. Ges. **50**, 229 (1917).

*124.* KOBAYASHI, G.: Specificity of Heteroarginase. Biochem. Soc. (Japan) **19**, 85 (1947).

*125.* KOJIMA, Y. and H. KUSAKABE: Isolation of Natural Substances by Ion-exchange Resins. III. A New Simple Method for the Isolation of Octopine and Taurine from Squid Muscles and the Estimation of their Amounts. J. Sci. Research Inst. (Japan) **49**, 132 (1955) [Chem. Abstr. **49**, 13533 (1955)].

*126.* KORZENOVSKY, M. and C. H. WERKMAN: Conversion of Citrulline to Ornithine by Cell-free Extracts of *Streptococcus lactis*. Arch. Biochem. Biophys. **46**, 174 (1953).

*127.* KOSSEL, A.: Über das Agmatin. Z. physiol. Chem. (Hoppe-Seyler) **66**, 257 (1910).

*128.* KOSSEL, A. und H. D. DAKIN: Weitere Untersuchungen über fermentative Harnstoffbildung. Z. physiol. Chem. (Hoppe-Seyler) **42**, 181 (1904).

*129.* — — Über die Arginase. Z. physiol. Chem. (Hoppe-Seyler) **41**, 321 (1904).

*130.* KOSSEL, A. und R. E. GROSS: Über die Darstellung und quantitative Bestimmung des Arginins. Z. physiol. Chem. (Hoppe-Seyler) **135**, 167 (1924).

*131.* KOSSEL, A. und FR. KUTSCHER: Beiträge zur Kenntnis der Eiweißkörper. Z. physiol. Chem. (Hoppe-Seyler) **31**, 165 (1900/01).

*132.* KOSSEL, A. und W. STAUDT: Über die quantitative Bestimmung von Arginin und Histidin. Z. physiol. Chem. (Hoppe-Seyler) **156**, 270 (1926).

*133.* KRAUT, H., E. v. SCHRADER-BEIELSTEIN und M. WEBER: Eine Modifikation der Argininbestimmung nach Sakaguchi. Z. physiol. Chem. (Hoppe-Seyler) **286**, 248 (1951).

*134.* KREBS, H. A. und K. HENSELEIT: Untersuchungen über die Harnstoffe im Tierkörper. Z. physiol. Chem. (Hoppe-Seyler) **210**, 33 (1932).

*135.* KUBY, S. A., L. NODA and H. A. LARDY: Adenosinetriphosphate-Creatine Transphosphorylase. I. Isolation of the Crystalline Enzyme from Rabbit Muscle. J. Biol. Chem. **209**, 191 (1954).

*136.* KURTZ, A. C.: Use of Copper(II) Ion in Masking ʌ-Amino Groups of Amino Acids. J. Biol. Chem. **180**, 1253 (1949).

*137.* KUTSCHER, FR. und D. ACKERMANN: Über das Arcain. Z. physiol. Chem. (Hoppe-Seyler) **203**, 132 (1931).

*138.* KUTSCHER, FR., D. ACKERMANN und O. FLÖSSNER: Über das Arcain, eine bisher unbekannte tierische Base. Z. physiol. Chem. (Hoppe-Seyler) **199**, 273 (1931).

*139.* KUTSCHER, FR. und J. OTORI: Der Nachweis des Guanidins unter den bei der Selbstverdauung des Pankreas entstehenden Körpern. Z. physiol. Chem. (Hoppe-Seyler) **43**, 93 (1904).

*140.* LANG, K.: Über den Mechanismus der Diacetylreaktion von Guanidinen, ihre Umformung und Anwendung zur kolorimetrischen Bestimmung von Kreatin und Arginin. Z. physiol. Chem. (Hoppe-Seyler) **208**, 273 (1932).

*141.* LEHMANN, H.: Über die Umesterung des Adenylsäuresystems mit Phosphagenen. Biochem. Z. **286**, 336 (1936).

*142.* LEHNARTZ, E.: Über die Bestimmung des Kreatins im Muskel mit der Dinitrobenzoat-Methode. Z. physiol. Chem. (Hoppe-Seyler) **271**, 265 (1941).

*143.* Leuthardt, F. et R. Brunner: Réaction entre l'acétyl-acétanilide et les amines primaires. Helv. Chim. Acta **30**, 958 (1947).

*144.* — — Méthode de dosage des uréido-acides. Dosage dans l'urine du chat après injection de la *d,l*-phénylalanine. Helv. Chim. Acta **30**, 966 (1947).

*145.* Leuthardt, F. und B. Glasson: Über die biologische Harnstoffbildung. IV. Mitt. Helv. Chim. Acta **25**, 630 (1942).

*146.* Liebig, J.: Recherches de chimie animale. C. R. hebd. Séances Acad. Sci. **24**, 69 et 195 (1847).

*147.* — Über die Bestandteile der Flüssigkeiten des Fleisches. Ann. Chem. Pharm. **62**, 257 (1847).

*147a.* Linneweh, F.: Über den fermentativen Abbau des Kreatinins. Z. Biol. **90**, 109 (1930).

*148.* — Über die Spaltung des Arcains durch Mikroorganismen. Z. physiol. Chem. (Hoppe-Seyler) **200**, 115 (1931).

*149.* — Über die Spaltung des Arcains durch Mikroorganismen. II. Mitt. Z. physiol. Chem. (Hoppe-Seyler) **202**, 1 (1931).

*150.* — Über die Spaltung des Arcains durch Mikroorganismen. III. Mitt. Der biologische Abbau des Agmatins zum Carbaminyl-Putrescin. Z. physiol. Chem. (Hoppe-Seyler) **205**, 126 (1932).

*151.* — Über die Resistenz des Guanidins und alkylierter Guanidine gegenüber bakterieller Guanidodesimidase und Arginase. Z. physiol. Chem. (Hoppe-Seyler) **207**, 152 (1932).

*152.* Lipmann, E. O. v.: Über einige organische Bestandteile des Rübensaftes. Ber. dtsch. chem. Ges. **20**, 3201 (1887).

*153.* — Über stickstoffhaltige Bestandteile aus Rübensäften. Ber. dtsch. chem. Ges. **29**, 2645 (1896).

*154.* Lissitzky, S., I. Garcia et J. Roche: Sur la caractérisation de dérivés guanidiques d'origine biologique par électrophorèse et chromatographie sur papier (chromatoélectrophorèse). Experientia **10**, 379 (1954).

*155.* — — — Sur les dérivés guanidiques du muscle de Scorpion, *Androctonus australis* L. C. R. Séances Soc. Biol. **148**, 436 (1954).

*156.* Loew, O.: Kupferoxyd-Ammoniak als Oxydationsmittel. J. prakt. Chem. **126**, 298 (1874).

*157.* Lohmann, K.: Über die Isolierung verschiedener natürlicher Phosphorsäure-verbindungen und die Frage ihrer Einheitlichkeit. Biochem. Z. **194**, 306 (1928).

*158.* — Über die enzymatische Aufspaltung der Kreatinphosphorsäure; zugleich ein Beitrag zum Chemismus der Muskelkontraktion. Biochem. Z. **271**, 264 (1934).

*159.* — Über die Aufspaltung der Adenylpyrophosphorsäure und Argininphosphor-säure in Krebsmuskulatur. Biochem. Z. **282**, 109 (1935).

*160.* — Untersuchungen an Oktopusmuskulatur. Isolierung und enzymatisches Verhalten von Adenylpyrophosphorsäure und Argininphosphorsäure. Biochem. Z. **286**, 28 (1936).

*161.* Lowry, O. H. and J. A. Lopez: The Determination of Inorganic Phosphate in the Presence of Labile Phosphate Esters. J. Biol. Chem. **162**, 421 (1946).

*162.* Macpherson, H. T.: Modified Procedures for the Colorimetric Estimation of Arginine and Histidine. Biochemic. J. **36**, 59 (1942).

*162a.* Major, R. H. and C. J. Weber: Possible Increase of Guanidine in Blood of Certains Persons with Hypertension. Arch. int. Med. **40**, 891 (1927).

*163.* Marshall, M., R. L. Metzenberg and P. P. Cohen: Purification of Carbamyl Phosphate Synthetase from Frog Liver. J. Biol. Chem. **233**, 102 (1958).

*164.* MAW, G. A.: The Detection of Creatine and Creatinine by Partition Chromatography. Biochemic. J. **43**, 139 (1948).

*164a.* MAYEDA, H.: Über die Extraktivstoffe aus den Schließmuskeln von *Pecten* (Patinopecten) *yessoensis* JAY. Acta Schol. Med. Univ. Kyoto **18**, 218 (1936).

*165.* MEISTER, A.: The α-Keto Analogues of Arginine, Ornithine, and Lysine. J. Biol. Chem. **206**, 577 (1954).

*166.* MENDE, T. J. and E. L. CHAMBERS: The Occurrence of Arginine Phosphate in Echinoderm Eggs. Arch. Biochem. Biophys. **45**, 105 (1953).

*167.* MEYERHOF, O.: Über die Verbreitung der Argininphosphorsäure in der Muskulatur der Wirbellosen. Arch. Sci. Biol. (Napoli) **12**, 536 (1928).

*168.* MEYERHOF, O. und K. LOHMANN: Über die natürlichen Guanidinphosphorsäuren (Phosphagene in der quergestreiften Muskulatur). I. Mitt. Das physiologische Verhalten der Phosphagene. Biochem. Z. **196**, 22 (1928).

*169.* — — Über die natürlichen Guanidinphosphorsäuren (Phosphagene in der quergestreiften Muskulatur). II. Mitt. Die physikalisch-chemischen Eigenschaften der Guanidinophosphorsäuren. Biochem. Z. **196**, 49 (1928).

*170.* MOLD, J. D., J. M. LADINO and E. J. SCHANTZ: The Sakaguchi and Biacetyl Reactions for the Identification of Alkyl Guanidines. J. Amer. Chem. Soc. **75**, 6321 (1953).

*171.* MØLLER, V.: Distribution of Amino Acid Decarboxylases in Enterobacteriaceae. Acta Pathol. Microbiol. Scand. **35**, 259 (1954).

*172.* MOORE, E. and D. W. WILSON: Nitrogenous Extractives of Scallop Muscle. I. The Isolation and a Study of the Structure of Octopine. J. Biol. Chem. **119**, 573 (1937).

*173.* MOREL, G. et H. DURANTON: Le métabolisme de l'arginine par les tissus végétaux. Bull. soc. chim. biol. (Paris) **40**, 2155 (1958).

*174.* MORRISON, J. F., D. E. GRIFFITHS and A. H. ENNOR: Biochemical Evolution on Position of the Tunicates. Nature (London) **178**, 359 (1956).

*175.* — — — The Purification and Properties of Arginine Phosphokinase. Biochemic. J. **65**, 143 (1957).

*176.* MOURGUE, M.: Sur la préparation de divers acides α-guanidiques et de dérivés guanidiques d'amines et d'amino-alcools. Bull. soc. chim. France **15**, 181 (1948).

*177.* MOURGUE, M., R. BARET et R. DOKHAN: Sur la présence de dérivés guanidiques dans les graines de ricin *(Ricinus communis)*. C. R. Séances Soc. Biol. **147**, 1449 (1953).

*178.* MOURGUE, M. et R. DOKHAN: Les dérivés d'oxydation de l'arginine chez les végétaux. C. R. Séances Soc. Biol. **148**, 1434 (1954).

*179.* MÜLLER, E. und K. ARMBRUST: Über einige stickstoffhaltige Bestandteile der Sojabohne. Z. physiol. Chem. (Hoppe-Seyler) **263**, 41 (1940).

*180.* NAKAMURA, S.: Über die Spezifität der Arginase. J. Biochem. (Japan) **36**, 243 (1944).

*181.* NEEDHAM, D. M., J. NEEDHAM, E. BALDWIN and J. YUDKIN: A Comparative Study of the Phosphagens, with some Remarks on the Origin of Vertebrates. Proc. Roy. Soc. (London), Ser. B **110**, 260 (1932).

*182.* NEUBAUER, C.: Über Kreatinin und Kreatin. III. Ann. Chem. Pharm. **137**, 288 (1866).

*183.* NUMEROF, P., M. GORDON, A. VIRGONA and E. O'BRIEN: Biosynthesis of Streptomycin. I. Studies with C-14-Labeled Glycine and Acetate. J. Amer. Chem. Soc. **76**, 1341 (1954).

*183a.* OBATA, Y and M. IIMORI: Biosynthesis of Octopine. J. Chem. Soc. Japan, Pure Chem. Sect. **73**, 832 (1952) [Chem. Abstr. **47**, 6093 (1953)].

*184.* Oesper, P.: Sources of the High Energy Content in Energy-rich Phosphates. Arch. Biochem. **27**, 255 (1950).

*185.* Ogasawara, K., S. Abe, I. Ito, M. Yoneda, N. Asano, T. Takatori and S. Watanabe: Antibiological Studies on Ekiri. II. Production of Toxic Amines by Strains of *Shigella dysenteriae*. Nisshin Igaku **40**, 99 (1953) [Chem. Abstr. **48**, 13806 (1954)].

*186.* Oginsky, E. L. and R. F. Gehrig: The Arginine Dehydrolase System of *Streptococcus faecalis*. I. Identification of Citrulline as an Intermediate. J. Biol. Chem. **198**, 791 (1952).

*187.* — — The Arginine Dehydrolase System of *Streptococcus faecalis*. II. Properties of Arginine Desimidase. J. Biol. Chem. **198**, 799 (1952).

*188.* — — The Arginine Dehydrolase System of *Streptococcus faecalis*. III. The Decomposition of Citrulline. J. Biol. Chem. **204**, 721 (1953).

*189.* Orglmeister, G.: Über die Bestimmung des Arginins mit Permanganat. Beitr. chem. Physiol. Pathol. **7**, 21 (1905).

*190.* Petrack, B. and S. Ratner: Biosynthesis of Urea. VII. Reversible Formation of Argininosuccinic Acid. J. Biol. Chem. **233**, 1494 (1958).

*191.* Petrack, B., L. Sullivan and S. Ratner: Behavior of Purified Arginine Desiminase from *S. faecalis*. Arch. Biochem. Biophys. **69**, 186 (1957).

*192.* Pisano, J. J., C. Mitoma and S. Udenfriend: Biosynthesis of γ-Guanidino-butyric Acid from γ-Aminobutyric Acid and Arginine. Nature (London) **180**, 1125 (1957).

*193.* Pradel, L. A., N. v. Thiem, P. Pin et N. v. Thoai: Préparation du créatine-phosphate de sodium. Bull. soc. chim. biol. (Paris) **41**, 519 (1959).

*194.* Pradel, L. A. et N. v. Thoai: Étude de la taurocyaminephosphokinase. Colloque intern. biochim. comparée acides aminés basiques, Concarneau, 1959.

*195.* Ratner, S.: The Enzymatic Mechanism of Arginine Formation from Citrulline. J. Biol. Chem. **170**, 761 (1947).

*196.* — Urea Synthesis and Metabolism of Arginine and Citrulline. Adv. Enzymology **15**, 319 (1954).

*197.* Ratner, S., W. P. Anslow, Jr. and B. Petrack: Biosynthesis of Urea. VI. Enzymatic Cleavage of Argininosuccinic Acid to Arginine and Fumaric Acid. J. Biol. Chem. **204**, 115 (1953).

*198.* Ratner, S. and A. Pappas: Biosynthesis of Urea. I. Enzymatic Mechanism of Arginine Synthesis from Citrulline. J. Biol. Chem. **179**, 1183 (1949).

*199.* — — Biosynthesis of Urea. II. Arginine Synthesis from Citrulline in Liver Homogenates. J. Biol. Chem. **179**, 1199 (1949).

*200.* Ratner, S. and B. Petrack: Biosynthesis of Urea. III. Further Studies on Arginine Synthesis from Citrulline. J. Biol. Chem. **191**, 693 (1951).

*201.* — — Biosynthesis of Urea. IV. Further Studies on Condensation in Arginine Synthesis from Citrulline. J. Biol. Chem. **200**, 161 (1953).

*202.* — — The Mechanism of Arginine Synthesis from Citrulline in Kidney. J. Biol. Chem. **200**, 175 (1953).

*203.* Ratner, S., B. Petrack and O. Rochovansky: Biosynthesis of Urea. V. Isolation and Properties of Argininosuccinic Acid. J. Biol. Chem. **204**, 95 (1953).

*204.* Ratner, S. and O. Rochovansky: Biosynthesis of Guanidinoacetic Acid. I. Purification and Properties of Transamidinase. Arch. Biochem. Biophys. **63**, 277 (1956).

*205.* — — Biosynthesis of Guanidinoacetic Acid. II. Mechanism of Amidine Group Transfer. Arch. Biochem. Biophys. **63**, 296 (1956).

206. RITTENBERG, D. and H. WAELSCH: The Source of Carbon for Urea Formation. J. Biol. Chem. **136**, 799 (1940).

207. ROBERTS, E. and G. H. TISHKOFF: Distribution of Free Amino Acids in Mouse Epidermis in Various Phases of Growth as Determined by Paper Partition Chromatography. Science (Washington) **109**, 14 (1949).

208. ROBIN, Y.: Répartition et métabolisme des guanidines monosubstituées d'origine animale. Thèse Doct., Univ. Paris, 1954.

209. — Méthodes générales d'analyse et de préparation des dérivés guanidiques. Application à l'étude de leur répartition. Colloque intern. biochim. comparée acides aminés basiques, Concarneau, 1959.

210. ROBIN, Y., L. A. PRADEL, N. v. THOAI et J. ROCHE: Sur le phosphagène d'*Arca noae* L. et sur la signification physiologique de l'arcaïne chez quelques Invertébrés. C. R. Séances Soc. Biol. **153**, 21 (1959).

211. ROBIN, Y. et J. ROCHE: Sur la présence de taurocyamine (guanidotaurine) chez des Coelentérés et des Spongiaires. C. R. Séances Soc. Biol. **148**, 1783 (1954).

212. ROBIN, Y. et N. v. THOAI: Métabolisme oxydatif de la *L*-arginine chez la Limnée, *Limnaea stagnalis* L. I. Oxydation par la *L*-aminoacideoxydase. C. R. Séances Soc. Biol. **151**, 2093 (1957).

213. — — Travaux non publiés.

214. ROBIN, Y., N. v. THOAI et L. A. PRADEL: Métabolisme des dérivés guanidylés. VII. Sur une nouvelle guanidine monosubstituée biologique: l'hirudonine. Biochim. Biophys. Acta **24**, 381 (1957).

215. ROBIN, Y., N. v. THOAI, L. A. PRADEL et J. ROCHE: Sur les constituants guanidiques des œufs d'*Audouinia tentaculata* MTG. C. R. Séances Soc. Biol. **150**, 1892 (1956).

216. ROBIN, Y., N. v. THOAI et J. ROCHE: Sur la présence d'arcaïne chez la Sangsue, *Hirudo medicinalis* L. C. R. Séances Soc. Biol. **151**, 2015 (1957).

217. ROCHE, J.: Dégradation hydrolytique de l'arginine et des dérivés guanidiques. Colloque intern. biochim. comparée acides aminés basiques, Concarneau, 1959.

218. ROCHE, J., W. FÉLIX, Y. ROBIN et N. v. THOAI: Chromatographie sur papier des dérivés guanidiques monosubstitués. C. R. hebd. Séances Acad. Sci. **233**, 1688 (1951).

219. ROCHE, J., H. GIRARD, G. LACOMBE et M. MOURGUE: Sur la dégradation de la glycocyamine par *Pseudomonas ovalis*. Glycocyaminase et argininedihydrolase bactériennes. Biochim. Biophys. Acta **2**, 414 (1948).

220. ROCHE, J., M. MOURGUE et R. BARET: Sur la spécificité de l'arginase hépatique et de l'hétéroarginase intestinale. III. Bull. soc. chim. biol. (Paris) **36**, 511 (1954).

221. ROCHE, J., M. MOURGUE et R. DOKHAN: Sur la formation de dérivés guanidiques par oxydation de l'arginine. Bull. soc. chim. biol. (Paris) **37**, 55 (1955).

222. ROCHE, J. et Y. ROBIN: Sur les phosphagènes des Éponges. C. R. Séances Soc. Biol. **148**, 1541 (1954).

223. ROCHE, J., N. v. THOAI et J. L. HATT: Métabolisme des dérivés guanidylés. III. Analyse chromatographique des dérivés guanidylés. Biochim. Biophys. Acta **14**, 71 (1954).

224. ROCHE, J., N. v. THOAI et Y. ROBIN: Sur la présence de créatine chez les Invertébrés et sa signification biologique. Biochim. Biophys. Acta **24**, 514 (1957).

225. ROCHE, J., N. v. THOAI, I. GARCIA et Y. ROBIN: Sur la nature et la répartition des guanidines monosubstituées dans les tissus des Invertébrés. Présence de dérivés guanidiques nouveaux chez des Annélides. C. R. Séances Soc. Biol. **146**, 1902 (1952).

226. Rosenberg, H., A. H. Ennor and J. F. Morrison: The Estimation of Arginine. Biochemic. J. **63**, 153 (1956).
227. Sakaguchi, S.: Über eine neue Farbenreaktion von Protein und Arginin. J. Biochem. (Japan) **5**, 25 (1925).
228. — A New Method for the Colorimetric Determination of Arginine. J. Biochem. (Japan) **37**, 231 (1950).
229. Sano, M.: Studien über die Arginase. J. Biochem. (Japan) **33**, 467 (1941).
230. Satake, K. and J. M. Luck: The Spectrophotometric Determination of Arginine by the Sakaguchi Reaction. Bull. soc. chim. biol. (Paris) **40**, 1743 (1958).
231. Schenck, M. und H. Kirchhof: Über Äthylendiguanidin. Z. physiol. Chem. (Hoppe-Seyler) **155**, 306 (1926).
232. Schmidt, G. C., M. A. Logan and A. A. Tytell: The Degradation of Arginine by *Clostridium perfringens* (BP6K). J. Biol. Chem. **198**, 771 (1952).
233. Schöberl, A.: Die Oxydation von Disulfiden zu Sulfonsäuren mit Wasserstoffsuperoxyd. Eine neue Synthese von Taurin. Z. physiol. Chem. (Hoppe-Seyler) **216**, 193 (1933).
234. Schulze, E.: Über basische Stickstoffverbindungen aus den Samen von *Vicia sativa* und *Pisum sativum*. Z. physiol. Chem. (Hoppe-Seyler) **15**, 140 (1891).
235. — Über das Vorkommen von Guanidin im Pflanzenorganismus. Ber. dtsch. chem. Ges. **25**, 658 (1892).
236. — Zum Nachweis des Guanidins. Ber. dtsch. chem. Ges. **25**, 661 (1892).
237. Schulze, E. und E. Steiger: Über das Arginin. Z. physiol. Chem. (Hoppe-Seyler) **11**, 43 (1887).
238. Schütte, E.: Darstellung von Guanidosäuren und Guanidopeptiden. Z. physiol. Chem. (Hoppe-Seyler) **279**, 52 (1943).
239. Schütze, W.: Untersuchungen über die chemische Zusammensetzung der Arthropoden-Muskulatur. Zool. Jahrb. Abt. allg. Zool. Physiol. **51**, 505 (1932).
240. Shibuya, S. and S. Makizumi: Biochemical Studies on Guanidine Compounds. VIII. Distribution of Extractable Guanidine Bodies in Animals and Plants. J. Japan. Biochem. Soc. **25**, 210 (1953).
241. Shibuya, S., S. Mikami, G. Tsuchibashi and T. Tagawa: The Determination of Guanidine Compounds in Some Organisms by Paper Chromatography. J. Biochem. (Japan) Proc. 23rd Meeting Biochem Soc. **39**, 22 (1952).
242. Sims, E. A. H.: Microdetermination of Glycocyamine and Arginine by Means of a Synthetic Ion Exchange Resin for Chromatographic Separation. J. Biol. Chem. **158**, 239 (1945).
243. Soda, T., A. Yoshida and A. Oikawa: Creatinine Formation from Creatine by Yeast. J. Biochem. (Japan) **40**, 421 (1953).
244. Stetten, D., Jr. and B. Bloom: The Metabolism of the Amidine Group of Arginine in the Intact Rat. J. Biol. Chem. **220**, 723 (1956).
245. Steudel, H. und K. Suzuki: Zur Histochemie der Spermatogenese. Z. physiol. Chem. (Hoppe-Seyler) **127**, 1 (1923).
246. Stevens, C. M. and J. A. Bush: New Syntheses of $\alpha$-Amino-$\varepsilon$-guanidino-*n*-caproic Acid (Homoarginine) and its Possible Conversion in vivo into Lysine. J. Biol. Chem. **183**, 139 (1950).
247. Strecker, A.: Étude sur la guanine. C. R. hebd. Séances Acad. Sci. **52**, 1210 (1861).
248. — Note. Jahresber. Fortschr. Chem. **21**, 685 (1868).
249. Stumpf, P. K. and D. E. Green: *l*-Amino Acid Oxidase of *Proteus vulgaris*. J. Biol. Chem. **153**, 387 (1944).

*250.* Suzuki, T. and S. Muraoka: New Guanidyl Derivatives and Amino Acids in the Extract of Shellfish, *Cristaria plicata.* J. pharmac. Soc. Japan **74**, 171 (1954) [Chem. Abstr. **48**, 5386 (1954)].

*251.* Szilágyi, I. and I. Szabó: Microchemical Method for the Determination of Sakaguchi-positive Antibiotics. Nature (London) **181**, 52 (1958).

*252.* Szörényi, E. T. et R. G. Degtyar: Phosphocréatine adénosinetriphospho-phérase de muscle et sa réaction avec l'actomyosine. Ukrain. Biokhim. Zhur **20**, 234 (1948).

*253.* Szörényi, E. T., P. P. Dvornikova et R. G. Degtyar: Isolement à l'état cristallisé et description de quelques propriétés de l'enzyme arginine phospho-phérase. C. R. (Doklady) Acad. Sci. (USSR) **67**, 341 (1949).

*254.* Szulmajster, J.: Bacterial Degradation of Creatinine. II. Creatinine Des-imidase. Biochem. Biophys. Acta **30**, 154 (1958).

*255.* Tanret, G.: Sur la galégine, alcaloïde retiré du *Galega officinalis.* Bull. soc. chim. France **15**, 613 (1914).

*256.* Thiem, N. v., N. v. Thoai et J. Roche: Synthèse de l'argininephosphate. Colloque intern. biochim. comparée acides aminés basiques, Concarneau, 1959.

*257.* Thoai, N. v.: Sur la taurocyamine et la glycocyamine phosphokinase. Bull. soc. chim. biol. (Paris) **39**, 197 (1957).

*258.* — Les dérivés guanidiques. Leur rôle biologique. Colloque intern. biochim. comparée acides aminés basiques, Concarneau, 1959.

*259.* Thoai, N. v. et T. T. An: Sur une nouvelle amidase spécifique: la guanido-butyramidase. C. R. Séances Soc. Biol. **150**, 1722 (1956).

*260.* Thoai, N. v., J. L. Hatt et T. T. An: Sur un nouveau type de dégradation enzymatique de l'arginine: l'oxydation en guanidobutyramide. Biochim. Biophys. Acta **18**, 589 (1955).

*261.* — — — Métabolisme des dérivés guanidylés. V. Oxydation enzymatique de l'arginine en guanidobutyramide. Biochim. Biophys. Acta **22**, 116 (1956).

*262.* Thoai, N. v., J. L. Hatt, T. T. An et J. Roche: Métabolisme des dérivés guanidylés. VI. Dégradation des dérivés guanidiques chez *Streptomyces griseus* (Waksman). Biochim. Biophys. Acta **22**, 337 (1956).

*263.* Thoai, N. v. et G. Lacombe: Sur la présence de l'acide δ-guanido-*n*-valérianique dans les urines humaines. Biochim. Biophys. Acta **29**, 437 (1958).

*264.* Thoai, N. v. et A. Olomucki: Étude de l'argininedécarboxyoxydase. Colloque intern. biochim. comparée acides aminés basiques, Concarneau, 1959.

*265.* — — Travaux non publiés.

*266.* Thoai, N. v., A. Olomucki, Y. Robin, L. A. Pradel et J. Roche: Sur la présence de nombreux dérivés carbamylés et guanidiques dans les urines et leur signification biologique. C. R. Séances Soc. Biol. **150**, 2160 (1956).

*267.* Thoai, N. v., A. Olomucki, F. Thome-Beau et N. k. Dinh: Sur la γ-guanido-butyramide. Colloque intern. biochim. comparée acides aminés basiques, Concarneau, 1959.

*268.* Thoai, N. v. et Y. Robin: Méthylation de l'acide guanidoacétique et répartition de la créatine chez les Invertébrés marins. C. R. hebd. Séances Acad. Sci. **232**, 452 (1951).

*269.* — — Métabolisme des dérivés guanidylés. II. Isolement de la guanidotaurine (taurocyamine) et de l'acide guanidoacétique (glycocyamine) des vers marins. Biochim. Biophys. Acta **13**, 533 (1954).

*270.* — — Métabolisme des dérivés guanidylés. IV. Sur une nouvelle guanidine monosubstituée biologique: l'ester guanidoéthylsérylphosphorique (lombricine) et le phosphagène correspondant. Biochim. Biophys. Acta **14**, 76 (1954).

*271.* Thoai, N. v. et Y. Robin: Métabolisme des dérivés guanidylés. VIII. Bio-synthèse de l'octopine et répartition de l'enzyme l'opérant chez les Invertébrés. Biochim. Biophys. Acta **35,** 446 (1959).

*272.* — — Sur la biogénèse de l'octopine dans différents tissus de *Pecten maximus* L. Bull. soc. chim. biol. (Paris) **41,** 735 (1959).

*273.* — — Métabolisme de l'octopine; ses relations avec la glycolyse musculaire. Colloque intern. biochim. comparée acides aminés basiques, Concarneau, 1959.

*273 a.* — — Travaux non publiés.

*274.* Thoai, N. v., Y. Robin et L. A. Pradel: Métabolisme oxydatif de la *L*-arginine chez la Limnée, *Limnaea stagnalis* L. II. Oxydation en guanidobutyramide. C. R. Séances Soc. Biol. **151,** 2097 (1957).

*275.* Thoai, N. v. et J. Roche: Biochimie du groupement guanidique. Exposés annu. biochim. méd. **18,** 165 (1956).

*276.* — — Dérivés guanidophosphoriques et types nouveaux de phosphagènes. Biokhimiya (USSR) **22,** 319 (1957).

*277.* Thoai, N. v., J. Roche et A. Olomucki: Sur la présence de la taurocyamine (guanidotaurine) dans l'urine de rat et sa signification biochimique dans l'excrétion azotée. Biochim. Biophys. Acta **14,** 448 (1954).

*278.* Thoai, N. v., J. Roche et Y. Robin: Sur l'oxydation de la *L*(+)arginine par une *L*-aminoacideoxydase présente chez les Invertébrés marins (étapes de la réaction et produits formés). C. R. hebd. Séances Acad. Sci. **235,** 832 (1952).

*279.* — — — Métabolisme des dérivés guanidylés. I. Dégradation de l'arginine chez les Invertébrés marins. Biochim. Biophys. Acta **11,** 403 (1953).

*280.* Thoai, N. v., J. Roche, Y. Robin et N. v. Thiem: Sur la présence de la glycocyamine (acide guanidylacétique) de la taurocyamine (guanidyltaurine) et des phosphagènes correspondants dans les muscles des vers marins. Biochim. Biophys. Acta **11,** 593 (1953).

*281.* — — — Sur deux nouveaux phosphagènes: la phosphotaurocyamine et la phosphoglycocyamine. C. R. Séances Soc. Biol. **147,** 1241 (1953).

*282.* Thoai, N. v. et N. v. Thiem: Sels d'ammonium de phosphoglycocyamine et de phosphotaurocyamine, phosphagènes de vers marins. Isolement et synthèse. Bull. soc. chim. biol. (Paris) **39,** 355 (1957).

*283.* Thomas, K.: Über die Herkunft des Kreatins im tierischen Organismus. I. Das Verhalten der Arginase zur $\gamma$-Guanidylbuttersäure und $\varepsilon$-Guanidyl-capronsäure. Z. physiol. Chem. (Hoppe-Seyler) **88,** 465 (1913).

*284.* Thomas, L. E., J. K. Ingalls and J. M. Luck: The Determination of Arginine in the Presence of other Amino Acids by Means of the Sakaguchi Reaction. J. Biol. Chem. **129,** 263 (1939).

*285.* Tomiyama, T.: The Apparent Dissociation Constants of Canavanine and Canaline. J. Biol. Chem. **111,** 45 (1935).

*286.* Vickery, H. B.: The Determination of Arginine by Means of Flavianic Acid. J. Biol. Chem. **132,** 325 (1940).

*286 a.* Voges, O. und B. Proskauer: Beitrag zur Ernährungsphysiologie und zur Differentialdiagnose der Bakterien der hämorrhagischen Septicämie. Z. Hyg. Infektionskrankh. **28,** 20 (1898).

*287.* Volhard, J.: Über die Synthese des Kreatins. Sitzungsber. Königl. Bayer. Akad. Wissensch. **2,** 472 (1868).

*288.* Waksman, S. A. and H. B. Woodruff: Streptothricin, a New Selective Bacteriostatic and Bactericidal Agent, Particularly Active Against Gram-Negative Bacteria. Proc. Soc. exp. Biol. Med. **49,** 207 (1942).

289. WALKER, J. B.: Arginosuccinic Acid from *Chlorella*. Proc. Nat. Acad. Sci. (USA) **38**, 561 (1952).

290. — An Enzymatic Reaction Between Canavanine and Fumarate. J. Biol. Chem. **204**, 139 (1953).

291. — Biosynthesis of Arginine from Canavanine and Ornithine in Kidney. J. Biol. Chem. **218**, 549 (1956).

292. — Arginine-Ornithine Transamidination in Kidney. J. Biol. Chem. **221**, 771 (1956).

293. — Studies on the Mechanism of Action of Kidney Transamidinase. J. Biol. Chem. **224**, 57 (1957).

294. — Role for Pancreas in Biosynthesis of Creatine. Proc. Soc. exp. Biol. Med. **98**, 7 (1958).

294a. — Further Studies on the Mechanism of Transamidinase Action: Transamidination in the *Streptomyces griseus*. J. Biol. Chem. **231**, 1 (1958).

295. WALKER, J. B. and J. MYERS: The Formation of Arginosuccinic Acid from Arginine and Fumarate. J. Biol. Chem. **203**, 143 (1953).

296. WEBER, C. J.: A Modification of Sakaguchi's Reaction for the Quantitative Determination of Arginine. J. Biol. Chem. **86**, 217 (1930).

297. — The Presence of Glycocyamine in Urine. J. Biol. Chem. **109**, XCVI (1935).

298. WESTALL, R. G.: Argininosuccinicaciduria. Identification of the Metabolic Defect in a Newly Described Form of Mental Deficiency. Congrès intern. biochimie, Vienne, 1958. Résumé commun., p. 168.

299. WIELAND, H. und R. ALLES: Über den Giftstoff der Kröte. Ber. dtsch. chem. Ges. **55**, 1789 (1922).

300. WINTERSTEIN, E.: Über einige Bestandteile des Emmentaler Käses. II. Mitt. Z. physiol. Chem. (Hoppe-Seyler) **41**, 485 (1904).

300a. — Über die Bestandteile des Emmentaler Käses. V. Mitt. Z. physiol. Chem. (Hoppe-Seyler) **105**, 25 (1919).

301. WU, H. and D. RITTENBERG: Metabolism of *L*-Aspartic Acid. J. Biol. Chem. **179**, 847 (1949).

302. ZEILE, K. und G. FAWAZ: Synthese der natürlichen Kreatinphosphorsäure. Z. physiol. Chem. (Hoppe-Seyler) **256**, 193 (1938).

303. ZELLER, E. A., L. S. VAN ORDEN and W. VÖGTLI: Enzymology of Mycobacteria. VII. Degradation of Guanidine Derivatives. J. Biol. Chem. **209**, 429 (1954).

*(Reçu le 14 Décembre 1959.)*

# Naturally Derived *iso*Thiocyanates (Mustard Oils) and Their Parent Glucosides.

By **Anders Kjær**, Copenhagen.

## Contents.

# I. Introduction.

The *iso*thiocyanate producing glucosides constitute a well-defined and unique class of natural products which occur in a large variety of higher plants, belonging to a relatively small number of botanical families. They are characterized by the ability to undergo enzymic hydrolysis to *iso*thiocyanates (mustard oils), hydrogen sulphate and D-glucose. Invariably, the latter has been encountered as the sugar moiety of the more than thirty individual compounds so far recorded, justifying the designation of the latter as glucosides.

On account of the conspicuously pungent properties of many mustard oils this group of glucosides has attracted scientific interest for hundreds of years. The time period prior to about 1850, with its fragmentary concepts of chemical structure, was followed by a period of considerable progress, culminating around 1900 with the proposal of a general chemical structure of these glucosides. After almost five decades of little activity, this province of natural product chemistry has enjoyed a considerable revival since the end of the Second World War. Recently, the subject covered in this essay has been reviewed by various authors (*10, 37, 46, 57, 58, 60, 62, 157*). However, a more comprehensive treatise, including data for the individual representatives of this class of compounds, may be of help in future studies.

Particular emphasis will be placed on recent developments, whereas only scant attention will be given to the analytical aspects of the subject which have recently been surveyed elsewhere (*145*). Possible trends for future activities will be indicated.

# II. Historical Development.

References to higher plants as sources of volatile, pungent principles date back to antiquity. Mustard, and allied species, have been employed as condiments and remedies for centuries, and numerous speculations concerning the active compounds appeared in the literature, long before modern organic chemistry was founded a hundred years ago.

A valuable key to the oldest literature is the monograph by GILDEMEISTER and HOFFMANN (*36*).

About 1830, various investigators demonstrated that the production of volatile mustard oil from seed material required the presence of water. A decade later, the enzymic character of the reaction was envisaged by BUSSY (*8*) who isolated the parent substrate from black mustard seeds (*Brassica nigra* KOCH) in form of a crystalline potassium salt of an acid, termed "acide myronique" (myron = balsam), which has subsequently become known as *sinigrin*. An auxiliary compound, known today as an enzyme (myrosinase), was termed "myrosyne" (syn = with). As early as in 1831, ROBIQUET and BOUTRON (*113*) isolated from the seeds of white mustard (*Sinapis alba* L.) a crystalline, sulphur-containing constituent which later was named *sinalbin* and recognized as a mustard-oil producing glucoside. Only a single addition was made to the list of crystalline glucosides prior to the present post-war period, viz. *glucocheirolin*, a compound isolated by SCHNEIDER and SCHÜTZ (*128*) in 1913 from wall-flower seeds. Today, the number of crystalline glucosides of this type exceeds ten, in addition to half as many which have been characterized as crystalline acetates. The existence of a considerable number of further glucosides is indicated by current knowledge of their enzymic fission products. The latter comprise a variety of *iso*thiocyanates, $XNCS$, which, for the sake of the present discussion, will be divided into groups according to their structural types. Up to 1952, when interest in this field was revived, eight mustard oils of established structure had been recorded. Today, the number runs close to thirty, and several further additions may be expected in the near future.

A reliable survey covering the literature to about 1930 was presented by SCHNEIDER (*124*).

Clearly, this progress is a result of the development of modern analytical tools. The discovery within a few years of about four times as many glucosides and mustard oils as known earlier illustrates the immense influence of new methods such as paper chromatography on the study of natural products.

The traditional interest in the *iso*thiocyanate glucosides, virtually limited to their applications as condiments and remedies in folk medicine, has recently been deepened by several biological effects of the glucosides or rather their enzymic fission products.

## III. Parent Glucosides.

### 1. General Properties.

All glucosides discussed in this Chapter are of the same general chemical character and hence possess similar properties. The nine *iso-*

thiocyanate-producing glucosides thus far described in the crystalline state are listed in *Table 1*, p. 157, which also includes references to the original botanical sources of the individual compounds.

It will be noted that, apart from the long-established trivial names sinigrin and sinalbin and the lately introduced designation progoitrin (*42*)\*, the glucosides have been consistently named by adding the prefix "gluco" to an appropriate part of the Latin name of the botanical species in which the compound was first recognized. This arbitrary nomenclature often leads to rather unwieldy designations, which might conceivably be rationalized by an alternative nomenclature based on a generic trivial name for the molecular entity common to all glucosides, preceded by the systematic chemical name of the side-chain of the individual compound.

All glucosides of this class contain a sulphuric acid residue and are accordingly isolated and handled as salts; they contain mostly potassium as the cation, because of the abundant presence of this element in plant tissues. Sinalbin, the classical glucoside of white mustard, is unique in the sense that it contains sinapine (I), a rather widely distributed quaternary base, as its cationic moiety (cf. *124*).

$$CH_3O \diagdown$$
$$HO - \langle\rangle - CH{=}CH - COOCH_2CH_2\overset{+}{N}(CH_3)_3$$
$$CH_3O \diagup$$

(I.) Sinapine.

The excellent crystallization properties of a few glucosides such as sinigrin, sinalbin, glucocapparin and glucoiberin, are in marked contrast to the generally experienced difficulties in inducing the other purified mustard oil glucosides to crystallize. This explains, why only nine out of more than thirty glucosides have hitherto been obtained as crystals. Recently, ETTLINGER and LUNDEEN (*23, 24*) have shown that certain tetramethylammonium salts possess good crystallization properties.

In a number of instances, acetylation of amorphous *iso*thiocyanate glucosides was advantageously employed for characterization purposes. The well described crystalline glucoside acetates are listed in *Table 2*, p. 158. Usually, four acetyl groups are introduced, all located in the glucose moiety. In one instance, viz. that of glucorapiferin (= progoitrin), a fifth acetyl enters a secondary hydroxyl group in the side-chain (*141*), whereas glucoconringiin, possessing a tertiary hydroxyl function in the aglucone, is not acetylated in this moiety under the conditions employed (*82*).

---

\* Independently, the name "glukorapiferin", constructed in accord with the general practice within this field, was proposed for the same glucoside by other authors (*138*) but, unfortunately, it seems to have received less recognition than the rather unorthodox designation "progoitrin".

In addition to the acetyl derivatives listed in Table 2, several non-homogeneous preparations of other glucoside acetates have been reported (*151*). The acetylated products of glucoberteroin (*151*) and glucomatronalin (*151*) have been omitted from Table 2 because the infrared spectra (*142*), which display only minor differences in such glucoside acetates, are the sole documentation available for these derivatives. As seen from Tables 1 and 2 (pp. 157, 158), most glucosides and their acetates crystallize from water or aqueous alcohols as monohydrates.

In accordance with their structure, the crystalline glucosides are colourless and water-soluble. In aqueous solution they all exhibit *levo*-rotation. Apart from glucoiberin which possesses an asymmetric sulph-oxide-group in its side-chain, all known glucosides display rotations of roughly the same magnitude suggesting $\beta$-glucoside character*.

## 2. Distribution in Plant Tissues.

The distribution of the various *iso*thiocyanate glucosides within the vegetable kingdom will be discussed in Chapter VI, p. 154. It should be emphasized that the occurrence of more than one glucoside in a given botanical species is the rule rather than the exception. As many as eight individual glucosides have been clearly distinguished in a single seed specimen. Within the glucoside-containing species the parent compounds often seem to be distributed over the entire plant. According to GUIGNARD (*44*), the glucosides are diffusively present in parenchymal tissues, especially in the bark. In seeds, the embryos constitute the site of accumulation. Only few systematic studies have been undertaken to establish the variation of the glucoside content as a function of the stage of growth, or of environmental factors such as climate, soil composition, etc. It is interesting in this connexion that STAHMANN et al. (*144*) found 2-phenylethyl *iso*thiocyanate, the aglucone of gluco-nasturtiin, to be the predominant mustard oil enzymically liberated from roots of black mustard (*Brassica nigra* KOCH); the seeds represent the classical source of sinigrin which yields allyl *iso*thiocyanate upon enzymic fission. Further, DELAVEAU (*16*) has studied the total and relative amounts of the individual glucosides in *Alliaria officinalis* during its growth cycle and noticed considerable variation. Numerous (mostly unpublished) observations from the author's laboratory clearly indicate significant quantitative, and frequently also qualitative, changes in the glucoside pattern of a given species, from the roots to green organs and to seeds (cf. Table 5, p. 161). Likewise, considerable variations were observed occasionally when the glucoside contents in botanically

---

* The rotatory dispersion curves through the wavelength region 295–600 m$\mu$ of a number of representative glucosides from the author's laboratory have kindly been determined by Dr. W. KLYNE, Postgraduate Medical School, University of London. They all exhibit negative plain curves through the entire range, i. e. curves devoid of anomalous dispersion (no Cotton effect).

identical seed specimens of different provenance were compared. Hence, it is imperative, here as in most provinces of natural product chemistry, to define the botanical source very carefully.

Most glucoside studies have been conducted with seeds which are usually rich in the desired compounds. Seeds are often more easily available than the fresh plants and the isolation procedures are simpler. Moreover, seasonal independence is another convenience when working with seeds.

From the above it would appear that much remains to be learned about the location of mustard oil glucosides in various tissues. Clarification of such problems may prove most helpful in attempts to elucidate the biogenesis and metabolic pathway of this class of natural products.

### 3. Detection, Isolation, Separation and Determination.

#### a. Paper Chromatography.

Until a few years ago, all references to the occurrence of *iso*thiocyanate glucosides in higher plants were based on the detection of mustard oils, liberated upon enzymic hydrolysis, whereas no analytical procedure existed for the detection of the glucosides themselves. As in most other areas of biochemistry, paper chromatography has been put to good service in the class under discussion.

SCHULTZ et al. (*38, 131, 133, 139*) developed a method for paper chromatography of the genuine glucosides, applicable to crude extracts of plant material, with the use of various solvent systems (*n*-butanol : acetic acid : water, collodine : water etc.) and with ammoniacal silver nitrate as a spray reagent. This method, combined with chromatography of certain derivatives of the corresponding *iso*thiocyanates (p. 139), has afforded the experimental basis for practically all the progress in this field during the last decade; it has helped to establish the glucoside patterns of numerous plant species (*38, 131, 133, 139, 151*) and yielded many new glucosides. The ease of performance and the minimum requirement of material render paper chromatography well suited also for chemotaxonomic studies (*87*).

In view of the large number of well-known, genuine *iso*thiocyanate glucosides, paper chromatography alone does not provide sufficient evidence for the identification of an individual spot but it provides valuable assistance in this task. Evidently, a need does exist for developing of still better differential-diagnostic assays. At present, caution is recommended when assigning definite glucosides to a given plant solely on paperchromatographic evidence; several erroneous statements have appeared in the literature on this account.

The paperchromatographic technique has proved to be very useful also in purifying extracts as well as in synthetic studies.

## b. Isolation and Separation Methods.

The procedure of extraction and purification of *iso*thiocyanate glucosides involves disintegration of plant tissues in such a way that enzymic hydrolysis be prevented or minimized, extraction of the glucosides with water or aqueous alcohols, removal of impurities, crystallization and purification.

The classical isolation of sinigrin from black mustard seed (*124*) was improved by STOLL and SEEBECK (*146*), who employed fresh horse-radish as a source. Three kg. of fresh roots, containing 60—70% water, afforded 10.6 g. of analytically pure sinigrin. Well-crystallized glucoiberin was isolated from seeds of *Iberis amara* L. by a very simple procedure (*136*). In most instances, however, particularly when the plant extracts contained a mixture of glucosides, the traditional procedures were unsuccessful.

Fortunately, modern methods of *ion-exchange* have made the old-known glucosides, as well as a series of new compounds, more easily accessible in crystalline or highly purified form (*Table 1*, p. 157). SCHULTZ et al. (*38*, *137*) studied the applicability of various anion exchange resins (Lewatit MI, Amberlite IR-400 and Amberlite IR-4 B) in the isolation of mustard oil glucosides and found that the latter, due to their character of substituted sulphates, can be retained quantitatively on the resins and thus be freed from impurities, such as sinapine, and other cationic or neutral contaminants (carbohydrates, etc.). Subsequent elution with KOH or sulphate solutions afforded a highly purified glucoside which in several cases crystallized. A very useful modification, introduced by the same authors (*38*, *135*, *136*, *138*, *140*, *141*, *151*) utilizes the ion-exchange properties of acid-washed ("anionotropic") alumina. This exchanger has the additional ability of retaining coloured and other impurities. In a few instances, fractional elution of glucoside-loaded alumina columns has allowed the resolution of a glucoside mixture into its components (*151*), although this selectivity seems to be rather limited. Much experience in the author's laboratory has confirmed the broad applicability of anionotropic alumina in the purification and isolation of mustard oil glucosides; we consider this as the tool of choice. Clearly, ion-exchange resins are well suited also for introducing other cations, such as tetramethylammonium (*24*) or rubidium (*61*), into the glucosides.

Lead acetate, a commonly used reagent in the purification of glucosides (*149*) and other plant products also removes some impurities from crude extracts of mustard oil glucosides prior to ion-exchange (*138*, *151*), although prolonged contact with lead-containing reagents may lead to losses of thioglucosides as was demonstrated long ago (*128*). In the author's laboratory, lead acetate precipitations have been used extensively, with favourable results (*67*, *75*, *80*, *83*, *90*).

Other procedures include electrophoresis (*130*) and partition chromatography on cellulose powder (*38*, *132*). Neither of these proceed entirely satisfactorily, though the latter technique, when applied to acetylated glucosides, has afforded purified (*138*) or homogeneous (*140*) preparations. Isolated examples of the use of countercurrent distribution (*67*) and adsorption chromatography in 80% ethanol (*42*) for the separation of chemically similar glucosides suggest broader applicability.

To summarize, the method of isolation and purification should be selected in each case with due regard to the nature and amount of contaminants. Mostly, the available methods will prove satisfactory for the preparation of glucoside fractions free of extraneous matters. A great need exists, however, for efficient preparative procedures which would resolve complex mustard oil glucoside mixtures.

### c. Quantitative Determination.

Only a few systematic studies have hitherto been undertaken to determine the quantities of glucosides present in plant material. Customarily, the contents have been evaluated indirectly, by estimating the enzymically produced *iso*thiocyanates; for this purpose several methods exist [cf. (*145*)]. Repeated observations indicate, however, that the enzymic fission rarely affords quantitative yields of mustard oils. Hence, an analytical procedure, based on cleavage of the glucosides in strong sulphuric acid and followed by a colourimetric glucose determination, using the anthrone reagent (*38*, *134*), was helpful. A feature that detracts from the usefulness of this method is the necessity of removing, prior to analysis, all disturbing impurities, such as free sugars, glycosides, etc., by paper chromatography or ion exchange.

On the basis of an extensive series of paper chromatograms of *iso*-thiocyanate glucosides from higher plants, and also from literature references concerning the quantities of mustard oils liberated by enzymic hydrolysis, it can now be claimed that the total and relative glucoside contents are subject to considerable variation. The amounts range from traces to several per cent of the dry weight. As in many phytochemical comparisons, doubt may arise as to whether or not a trace glucoside should be counted as a characteristic constituent of a plant. This, of course, is entirely dependent on the sensitivity of the analytical method. Furthermore, the glucoside pattern of even a single organ of a given species may show considerable variation depending on environmental factors. Pertinent quantitative studies have received much less attention so far than did qualitative aspects.

### 4. Chemical Structure.

### a. Earlier Formulation.

Studies of the chemistry of sinigrin and sinalbin, the classical mustard oil glucosides, commenced several years before A. W. HOFMANN (*49*) established in 1868 the true structure of the *iso*thiocyanates and their isomeric relationship to the thiocyanates. Of particular importance in this connexion is a contribution by WILL and KÖRNER (*155*) (1863). They established the correct elementary composition of sinigrin and demonstrated the formation of free sulphur and allyl cyanide, in addition to allyl *iso*thiocyanate, glucose and sulphate, upon enzymic decomposition of sinigrin. Furthermore, they studied the cleavage of sinigrin with silver nitrate, resulting in precipitation of a glucose-free silver salt which on decomposition with hydrogen sulphide afforded elementary sulphur and allyl cyanide. The latter was established also as a minor constituent of allyl mustard oil of natural origin. WILL and KÖRNER concluded that sinigrin, in addition to glucose and sulphate, contained the elements of sulphur and allyl cyanide, arranged in such a manner as to allow for the simultaneous formation of allyl *iso*thiocyanate, allyl nitrile and sulphur, upon enzymic hydrolysis or chemical cleavage.

HOFMANN (*51*) was the first to report on the presence of 0.32–0.56% of carbon disulphide in naturally derived, as well as synthetic allyl *iso*-thiocyanate. The origin of this contamination has been discussed by CHALLENGER (*10*).

On this background it may surprise that GADAMER (*30*) in 1897 ventured to propose for sinigrin and sinalbin the structures (II) which remained virtually unchallenged until a few years ago and still appear in most elementary textbooks.

$$R\text{---}N\text{=}C \overset{\displaystyle S\text{---}C_6H_{11}O_5}{\underset{\displaystyle O\text{---}SO_2\text{---}O^-\ \ X^+}{}}$$

(II.)

(IIa.) $R = CH_2{=}CH\text{--}CH_2$, $X = $ K. Sinigrin.
(IIb.) $R = (p)HOC_6H_4CH_2$, $X = $ Sinapine. Sinalbin (GADAMER).

Structure (II) makes it difficult to explain the formation of nitriles. On the other hand, it satisfactorily accounts for the enzymic hydrolysis to *iso*thiocyanates, $R\text{---}N{=}C{=}S$, glucose and sulphate. Additional evidence given by GADAMER for structure (II) included the reaction of sinigrin with silver nitrate (*30*). When one equivalent of the latter was used, glucose was detached from the glucoside, and a silver mercaptide appeared, suggesting the original location of the sugar moiety in a thioglucosidic linkage. In case of an additional equivalent of silver nitrate, potassium was exchanged by silver to give a crystalline compound, $(C_3H_5NCS)(Ag)(SO_4Ag)$. Information on the molecular site of the sulphate-grouping was sought in the behaviour of sinigrin towards barium hydroxide. Whereas the glucoside was unaffected by boiling barium chloride, the hydroxide caused instantaneous precipitation of barium sulphate, indicating the presence in sinigrin of sulphuric acid in an ester linkage.

Analogous reasoning led GADAMER to propose the structure (IIb) for sinalbin, the parent glucoside of white mustard (*Sinapis alba* L.) (*30*), which furnishes 4-hydroxybenzyl *iso*thiocyanate upon enzymic hydrolysis, as rendered likely by SALKOWSKI (*114*) and confirmed in our laboratory (*94*). An unusual feature of the sinalbin structure is its content of the base sinapine (I, p. 125), the choline ester of sinapic acid, which GADAMER has identified as 3,5-dimethoxy-4-hydroxycinnamic acid (*31*).

A welcome corroboration of the presence of a thioglycoside linkage in sinigrin originated from studies by SCHNEIDER and WREDE (*129*) who showed that treatment of the glucoside with potassium methoxide yielded 1-thio-*D*-glucose, isolated as the silver salt. This observation was later extended by the same research group, with the result that sinigrin, and other analogous glucosides as well, are 1-β-*D*-thio-glucosides (*125*).

The same authors (*129*) demonstrated the formation also of "mero-sinigrin", for which a cyclic structure was proposed, during treatment of sinigrin with potassium methoxide. The strongly *dextro*rotatory compound yielded a triacetate and was obviously formed by elimination of one molecule of sulphuric acid from sinigrin.

In spite of the early recognized nitrile formation already mentioned, and several observations of the production of varying amounts of organic cyanides during both enzymic hydrolysis and chemical fission of other glucosides (cf. *10*), the GADAMER structure was not seriously questioned until a few years ago. This appears even more astonishing in view of the formation of other recognized by-products during the enzymic cleavage, such as free sulphur and carbon disulphide, the appearance of which is not easily reconcilable with GADAMER's formulation. It is only fair to point out that these difficulties were not ignored by GADAMER but he assumed that the positive evidence mentioned was sufficiently convincing to justify his proposed structure.

A thorough, critical discussion of these developments has been presented by CHALLENGER (*10*) in a recent monograph, that also contains speculations on the formation of the various by-products in the light of present-day knowledge. Especially, the many instances of concomitant nitrile formation during enzymic hydrolysis of thioglucosides are reviewed.

### b. Revised Structures.

In 1956, ETTLINGER and LUNDEEN (*23*) published an important communication in which structure (III) was convincingly established as a correct expression for sinigrin and sinalbin.

$$R-C\overset{\textstyle S-C_6H_{11}O_5}{\underset{\textstyle N-O-SO_2-O^-\ X^+}{\Big\langle}}$$

(III.)

(IIIa.) $R = CH_2=CHCH_2$, $X = $ K. Sinigrin.
(IIIb.) $R = (p)HOC_6H_4CH_2$, $X = $ Sinapine. Sinalbin.
(IIIc.) $R = (p)HOC_6H_4CH_2$, $X = (CH_3)_4N^+$.

The revised structure differs significantly from (II) by having the side-chain attached to carbon, rather than nitrogen, which is accommodated in an oxime-like arrangement. Accordingly, the glucosides may be interpreted as substituted thioimino acid esters or, rather, *iso*thiohydroxamic acids.

Conclusive support for the revised structure was provided by the following reactions: (i) Hydrogenolysis of sinigrin and the tetramethylammonium salt (IIIc) with Raney nickel furnished, respectively, *n*-butylamine and tyramine; (ii) acid hydrolysis of the same glucosides afforded, respectively, vinylacetic and *p*-hydroxyphenylacetic acid; (iii) the acid fission was accompanied in both instances by formation of hydroxylamine in 50–90% yields. None of these results are compatible with the GADAMER structure. In addition, the American authors presented evidence for sinigrin being a $\beta$-1-thio-D-glucopyranoside by desulphurizing its tetraacetate to 1,5-anhydro-D-glucitol tetraacetate. The only structural detail which still remains to be settled is the configuration around the C=N-double bond. Indirect evidence, quoted below, appears to support the *anti*-configuration (III) of the side-chain $R$ and the $OSO_2O^-$-grouping (23).

In the light of current knowledge, merosinigrin, the transformation product of sinigrin mentioned, almost certainly possesses the structure and conformation (IIId).

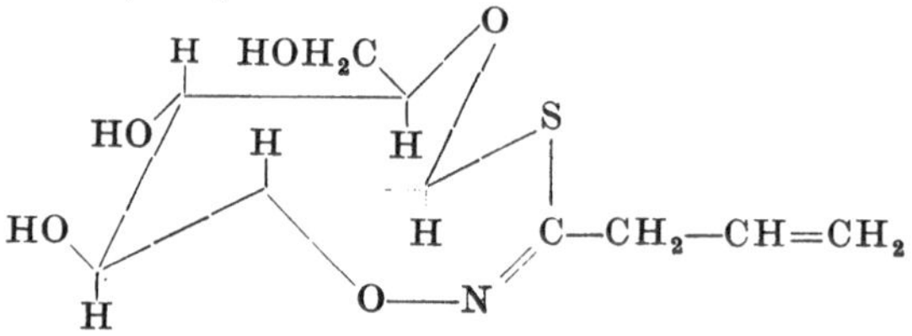

(IIId.) Merosinigrin.

Structure (III) makes the frequently observed nitrile and sulphur formation more understandable. As pointed out by ETTLINGER and LUNDEEN (23), thiohydroxamic acids have formerly been demonstrated (9, 150) to undergo facile decomposition to nitriles and elementary sulphur. Thus, *iso*thiocyanates ($R'NCS$), mainly of the aromatic type, add hydroxylamine to give compounds of the type $R'NHCSNHOH$ which readily decompose in the following manner:

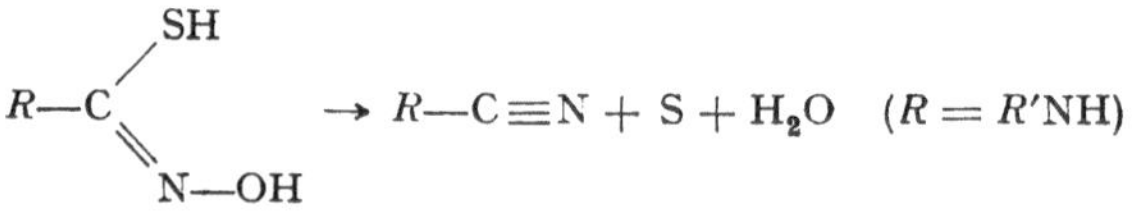

$$R{-}C\underset{N{-}OH}{\overset{SH}{\big\langle}} \rightarrow R{-}C\equiv N + S + H_2O \quad (R = R'NH)$$

Similar cleavage of the silver mercaptides, or intermediates in the enzymic hydrolysis, proceeding concurrently with the *iso*thiocyanate formation, may be responsible for the observed formation of the by-products.

The new formulation (III) implies that an intramolecular rearrangement takes place both during enzymic hydrolysis of glucosides and, occasionally, by nucleophilic displacement of the metal from the silver mercaptides. The analogy of this reaction with the well-known Lossen rearrangement of hydroxamic acids has been pointed out by the American authors (23).

This mechanism lends support to the suggested configuration around the C=N-bond, in view of the generally recognized *anti* configuration of the migrating group in such rearrangements. Moreover, there is some evidence available (24, 67, 86), though not as yet conclusive, to indicate that the rearrangement proceeds with retention of the configuration of the migrating group as expected for a reaction of this kind. The hydrolysis of the glucoside is probably unique in the sense that it involves an enzyme-initiated, intramolecular, nucleophilic displacement; hence, it is of considerable interest. A more detailed study of the factors influencing the relative amounts of *iso*thiocyanates and nitriles produced in the enzymic reaction would be desirable.

In this connexion, the recent demonstration by GMELIN and VIRTANEN (40) of an enzyme occurring in the seeds and fresh plants of *Thlaspi arvense* L. and *Lepidium ruderale* L. deserves attention: it cleaves the glucosides sinigrin and glucotropaeolin present to allyl and benzyl thiocyanate, respectively. Apparently, no concomitant production of *iso*thiocyanates takes place in these plants, whereas ordinary garden cress (*Lepidium sativum* L.) gives rise to a mixture of benzyl thiocyanate and benzyl *iso*thiocyanate when reacting with the plant's own enzyme system. Other species have been listed by the same authors (39) as additional sources of thiocyanates. Possibly, an unknown factor governing the course of enzymic attack is operative. Indeed, it was questioned by SCHMIDT (122) many years ago whether allyl *iso*thiocyanate was the primary reaction product of enzymic sinigrin hydrolysis, considering the facile isomerization of allyl thiocyanate to the corresponding *iso*-thiocyanate. Enzymic cleavage experiments conducted at 0° by the same author afforded, however, only traces of the rhodanide and did not support the assumption that the latter is the first reaction product.

In conclusion, at the present time three pathways seem to exist for the enzymic attack on this class of glucosides: (i) A course involving intramolecular rearrangement to *iso*thiocyanates; (ii) an alternative route, predominant in certain plants, that leads to thiocyanates, obviously by rearrangement; and (iii) the formation of nitriles and elementary sulphur with no change in the carbon skeleton.

Differences in the enzymic systems as well as in the initial sites of attack are likely to be responsible for this multiplicity of the observed end products. It would be important to clarify the detailed mechanism of this remarkable enzyme reaction.

Shortly after the publication of structure (III, p. 131), SCHULTZ and WAGNER (*142*) expressed doubt as to its correctness, because they failed to detect the expected C=N stretching mode in the infrared spectra of various glucoside acetates, and were unable to hydrogenate the same group catalytically by means of Raney nickel. Neither of these arguments seems, however, to affect the validity of (III). There are numerous cases on record (*26*) of substituted oximes, whose C=N stretching bands are very weak or absent. In fact, a complete set of infrared spectra of the crystalline tetraacetates listed in Table 2 (p. 158), recorded in the author's laboratory (*64*), invariably display a weak, but consistent band at $\sim 1640$ cm.$^{-1}$, the expected position for the C=N stretching mode. The failure to saturate the C=N linkage under the conditions employed can hardly surprise considering the sluggish reaction of a concurrently studied, synthetic model glucoside acetate and the insufficient analytical assays employed.

There are good reasons to believe that all glucosides encountered thus far in nature possess the same general structure as that of sinigrin, sinalbin and glucotropaeolin, with the individual features residing solely in the side-chains. Thus, various other glucosides [progoitrin (*23*), gluco-malcolmiin (*90*), glucohirsutin (*66*), glucoerypestrin (*78*), glucocameli-nin (*83*), glucoalyssin (*75*), glucocapparin (*74*), glucocinringiin (*41*), etc.] which have been subjected to degradation, all afforded at least two of the fission products: glucose, sulphate and hydroxylamine, the typical fragments of structure (III).

## c. Synthesis.

Shortly after the announcement of the glucoside structure (III), ETTLINGER and LUNDEEN (*24*) reported on the first successful synthesis of an *iso*thiocyanate glucoside, viz. that of glucotropaeolin. This noteworthy achievement provides an important argument for the correctness of (III).

The synthesis *(Chart 1)* proceeded from magnesium dithiophenyl-acetate (IV), obtained from benzylmagnesium chloride and carbon disulphide, which upon treatment with hydroxylamine yielded phenyl-aceto-thiohydroxamic acid (V). Interaction of the latter with aceto-bromoglucose afforded S-β-D-1-(tetraacetyl-glucopyranosyl)-phenylaceto-thiohydroximic acid (VI), which could be converted into the tetraacetyl-glucotropaeolate ion upon treatment with sulphur trioxide in pyridine. The ion was isolated as the potassium salt, which proved identical with a salt of natural origin (Table 2, p. 158), and as the tetramethylammonium salt (VII) which was then deacetylated to give glucotropaeolin as a salt of the same base (VIII); the latter was indistinguishable from a sample prepared by ion-exchange of the potassium salt that originated from natural sources.

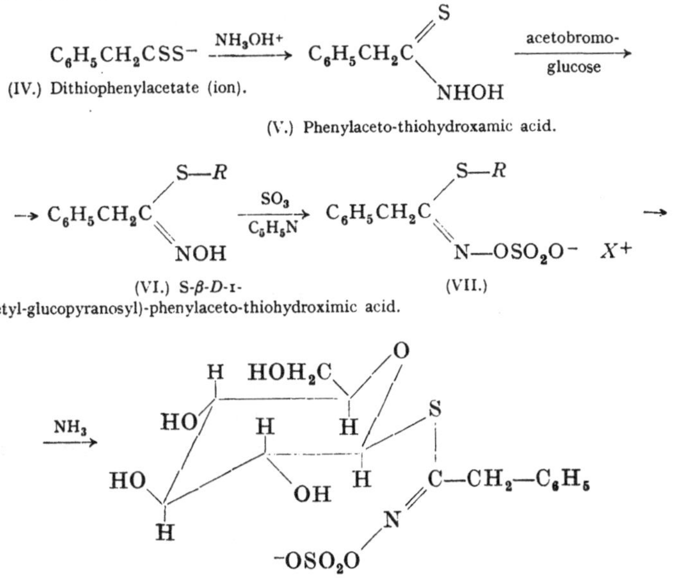

$$C_6H_5CH_2CSS^- \xrightarrow{NH_2OH^+} C_6H_5CH_2C \overset{\displaystyle S}{\underset{\displaystyle NHOH}{\big<}} \xrightarrow[\text{glucose}]{\text{acetobromo-}}$$

(IV.) Dithiophenylacetate (ion).

(V.) Phenylaceto-thiohydroxamic acid.

$$\rightarrow C_6H_5CH_2C \overset{\displaystyle S-R}{\underset{\displaystyle NOH}{\big<}} \xrightarrow[C_5H_5N]{SO_3} C_6H_5CH_2C \overset{\displaystyle S-R}{\underset{\displaystyle N-OSO_2O^- \ X^+}{\big<}} \rightarrow$$

(VI.) S-$\beta$-D-I-
(Tetraacetyl-glucopyranosyl)-phenylaceto-thiohydroximic acid.

(VII.)

(VIII.) Glucotropaeolin (ion).

Chart 1. Synthesis of Glucotropaeolin.
$R =$ Tetraacetyl $\beta$-D-I-glucopyranosyl, $X = (CH_3)_4N$.

The glucotropaeolate ion (VIII) is formulated above in its supposedly favoured conformation, with all substituents of the pyranose ring located in equatorial positions.

## 5. Individual Glucosides.

In addition to the natural glucosides which have been obtained in crystalline form, either as the genuine compounds (Table 1, p. 157) or as acetates (Table 2, p. 158), a considerable number exists, whose structures can be inferred from the chemical nature of the derived mustard oils. Most of the parent compounds have been assigned names in accordance with the general usage in this field. The complete series of glucosides for which the derived *iso*thiocyanates have so far been chemically established appears in *Table 3*, p. 159. For the sake of completeness, the glucosides listed in Tables 1 and 2 are retabulated in Table 3, together with their aglucones. A number of glucosides with unknown or only partially clarified side-chains, such as gluco-caulorapin (*138, 151*) and glucomatronalin (*142, 151*), have not been included in the Table.

A more detailed discussion of the individual *iso*thiocyanates (cf. Table 3) will be presented on p. 141. It has become an increasingly common practice to denote the more complex *iso*thiocyanates by trivial names derived from the corresponding glucoside designations by omitting the prefix gluco- (e. g. arabin, iberin, berteroin, etc.).

The natural sources for the individual glucosides (Table 3) will be treated on p. 154. As an outcome of paperchromatographic scanning of an extensive collection of botanical taxa for *iso*thiocyanate glucosides in the author's laboratory and elsewhere, it can be predicted that several additions to those listed in Table 3 will be forthcoming.

# IV. Enzymic Hydrolysis.

## 1. Distribution of Myrosinase.

The recognition of *iso*thiocyanate production as a result of enzymic hydrolysis (*8*), paired with the unique chemical structure of the substrates and the multiplicity of the reaction products, has stimulated early interest in myrosin, particularly its anatomical localisation and its distribution in the vegetable kingdom. The practical interest, associated with the manufacture of table mustard, should not be overlooked in this connexion.

In contrast to the glucosides, which are distributed diffusely throughout the parenchymal tissues, the enzyme is accumulated in particular cells (idioblasts), as was first demonstrated for Crucifers by GUIGNARD (*44*). The same author later showed a similar occurrence of myrosin in species of the families *Capparidaceae, Resedaceae, Tropaeolaceae* and *Limnantha-ceae* (*45*). It appears that myrosin-containing cells are present in all tissues of *iso*thiocyanate glucoside-producing plants and can be histo-chemically distinguished by staining with Millon's reagent, iodine, or orcinol in hydrochloric acid (*143*). PECHE (*106*) employed a sinigrin solution, saturated with barium chloride, to locate myrosin whereby a precipitate of barium sulphate formed inside the enzyme-containing cells. In accordance with the distribution of substrate and enzyme within the plant, no hydrolytic fission takes place until the tissues are disintegrated and the reactants brought into contact.

In an early, important contribution by LEPAGE (*99*) it was demonstrated that a large variety of Crucifer seeds contained myrosin. Since then, numerous references have appeared in the literature to myrosin-containing species, and a list of these has been published by SCHMALFUSS and MÜLLER (*117*). There are good reasons for believing that the presence of myrosin in plant species is usually connected with the glucoside content. However, the same enzyme system seems to be operative in the hydrolysis of all glucosides of this type.

## 2. Properties of Myrosinase.

A more detailed study of myrosin was initiated in 1926 by NEUBERG and WAGNER (*104*) as part of a broader investigation of sulphatases. They adopted the term "myrosinase" instead of "myrosin", coined in analogy with "sulphatase". Directions were also presented for the preparation, from white mustard seeds, of a cell-free, active enzyme solution, which has been employed in most subsequent work. An alternative procedure for the preparation of myrosinase, in form of a dry powder, was described by BRAECKE (*7*).

Surprisingly enough, few systematic studies have as yet been reported on the purification of myrosinase; and most authors have been concerned with rate studies of the enzymic fission. NEUBERG and WAGNER (*104*) clearly demonstrated that myrosinase is distinctly different from all known plant and animal sulphatases and possesses a high degree of specificity that is limited to the natural glucosides discussed. The existence in myrosinase of two entities, one liberating glucose in a fast reaction and another acting as a slow sulphatase, was suggested by VON EULER and ERIKSON (*25*). This interpretation was strengthened by SANDBERG and HOLLY (*115*), yet with the modification that both components seemed to start their action simultaneously. Whereas a quantitative yield of sulphuric acid was obtained, the yields of glucose never exceeded 66% of the theoretically possible value. STAHMANN et al. (*144*) have demonstrated that myrosinase at equilibrium had produced about 80% of the possible amount of allyl *iso*thiocyanate.

In contrast to these investigations, NAGASHIMA and UCHIYAMA (*103a*) have recently provided good experimental evidence to support the conception of myrosinase as a single enzyme that hydrolyses the thio-glucosidic linkage. The subsequent intramolecular rearrangement is visualized as a spontaneous, non-enzymic reaction. The same authors have observed inhibition of myrosinase by SH-inhibitors and a remarkable activation by ascorbic acid.

Several investigators have tested the stability of myrosinase by varying pH and temperature [cf. e. g. (*153*, *103a*)]. It appears that myrosinase has a pH optimum at 6.5–7.5 and a temperature optimum between 30° and 40°.

It is obvious that detailed studies of the enzyme and its action, in the light of our recent knowledge concerning the chemical constitution of the substrates, are greatly needed. The diversity of the products suggests that enzymic and non-enzymic conversions may proceed concomitantly and at various rates, depending on the conditions. Moreover, the participation of unknown co-factors or inhibitors in the enzymic hydrolysis is suggested by the recent finding of thiocyanates as ultimate reaction products in certain plants (*39*, *40*).

It is noteworthy that REESE et al. (*112*), as a result of screening 300 microorganisms, have recently reported the discovery of three fungi of the *Aspergillus versicolor* group that produce a specific thioglucosidase, for which the term *sinigrinase* was introduced (not to be confused with the same name used earlier for myrosinase). In one of these fungi, an *Aspergillus sydowi* strain, it was shown that sinigrinase represents a constitutive enzyme. Though distinctly different from myrosinase, the fungus enzyme acts on sinigrin to yield the same products as myrosinase. Sinigrinase effects virtually complete glucose formation from sinigrin but attacks phenyl-$\beta$-$D$-thioglucoside as little as does myrosinase. The relative activities of the two enzymes on various *iso*thiocyanate glucosides are different and so are the respective stabilities and inhibition patterns.

It is of interest in this connexion that GOODMAN et al. (*41a*) have recently demonstrated the ability of myrosinase to hydrolyze a series of synthetic $\beta$-$D$-thioglucopyranosides which contain thiol-substituted purines, pyrimidines, pyridazines or aromatic rings as aglycones. In this respect, myrosinase resembles a remarkable new enzyme detected by the same authors as a constituent of mammalian tissues of every living species investigated. The natural substrates for the latter enzyme are still obscure.

Evidently, several unanswered problems would be clarified by a more detailed knowledge of myrosinase and its mode of action.

## V. Naturally Derived *iso*Thiocyanates.

### 1. General Properties.

We have referred repeatedly to *iso*thiocyanates which result from enzymic hydrolysis of naturally occurring glucosides. The formulae of the thirty known mustard oils of natural derivation with clarified structures are listed in *Table 3*, p. 159. In the present Chapter the individual *iso*thiocyanates will be discussed.

The ability to liberate pungent principles upon disintegration is a conspicuous property of some plants which have been used extensively as potherbs, condiments or remedies. Evidently, the production of *iso*thiocyanate will often be the first indication of the presence of the corresponding glucoside. Accordingly, much pertinent information has originated from studies of enzymically produced mustard oils.

Notwithstanding the early recognition of such non-volatile *iso*thiocyanates as cheirolin, erysolin and *p*-hydroxybenzyl mustard oil, there has been a general tendency to consider the natural *iso*thiocyanates as "essential oils", i. e. steam-volatile compounds. This should be discontinued, however, because about half of all known natural mustard oils are not steam-distillable (cf. Table 3). Most of them are liquids, and it is a common practice to characterize them by their thiourea derivatives obtained by means of ammonia or amines.

The spectroscopic properties of *iso*thiocyanates, both in the ultra-violet and infrared regions, have recently been thoroughly studied (*147*).

When applied to the tongue, all mustard oils cause a sharp and burning sensation. Their odours, though mostly pungent, display characteristic individual differences which often are helpful in the detection and classification of mustard oils. Certain *iso*thiocyanates, that undergo rapid intramolecular cyclization, give rise to a transient biting taste, followed by a sensation of bitterness. Like most synthetic mustard oils, those of natural origin show vesicant and frequently also lachrymatory properties.

## 2. Detection, Isolation, Separation, and Determination.

### a. Chromatographic Methods.

Besides taste and smell, various chemical assays have been employed to detect *iso*thiocyanates in plant materials (*145*). While most earlier procedures were limited to the detection of steam-volatile mustard oils, paper chromatography has provided a more general and efficient analytical tool. It is also useful for the tentative identification of individual *iso*thiocyanates. In the author's laboratory, a method was developed several years ago for the separation of thioureas by chromatography on paper (*93*). Transformation of natural *iso*thiocyanates into thiourea derivatives, followed by paper chromatography (*72*), has been found very useful in the detection, isolation and identification of many new natural mustard oils. Originally, the chromatographic method had been developed for steam-volatile mustard oils (*72*), but it has since been extended to the study of non-volatile representatives.

Water-saturated chloroform has proved to be particularly useful for paper chromatography of thioureas (*93*), but other solvent systems also have been employed, such as ethyl acetate : water and 2-butanone : water (*56*); pyridine : amyl alcohol : water and heptane : 90% formic acid : *n*-butanol (*73*, *85*); *n*-butanol : : ethanol : water (*74*, *75*), formamide : chloroform (*65*), and aromatic hydrocarbons : : water, eventually with varying additions of ethanol, acetone or ethyl acetate (*13*). Grote's reagent, a modified nitroprusside reagent, has been extensively used for spraying, whereby the thioureas appear as blue spots (*93*). Ammoniacal silver solutions are also useful (*12*, *131*) for locating thiourea spots on paper. When not too volatile, the *iso*thiocyanates themselves can be chromatographed, with silver nitrate as a suitable spray reagent (76, 109). A Roumanian research group has reported on paper chromatography of various aromatic *iso*thiocyanates, after conversion of the latter into thiosemicarbazides by means of 2,4-dinitrophenyl-hydrazine (27). Alternatively, the same authors chromatographed simple thioureas in form of their yellow bismuth acetate complexes (28).

The powerful new tool of gas chromatography is a promising supplement to the paperchromatographic technique for the separation and identification of mustard oils (88).

## b. Isolation.

The procedures selected for the isolation of *iso*thiocyanates from enzymic hydrolysates depend, of course, on the chemical character. Provided the mustard oils are sufficiently stable and volatile, steam-distillation still affords a convenient method. Furthermore, steam-volatility of one or more components in a complex mixture of *iso*thio-cyanates may help in identifying individual constituents. From the aqueous distillates the substance can be isolated either by extraction, or by conversion directly into a thiourea derivative (references in Table 3, p. 159). When extraction is required, certain solvents, such as ether, do not interfere with the enzymic reaction and can be employed for continuous removal of the mustard oils during hydrolysis (94); other solvents, e. g. chloroform, may damage the enzyme system.

## c. Separation.

As mentioned, most glucoside-containing plant materials afford a mixture of *iso*thiocyanates upon enzymic hydrolysis. This raises the important problem of separating the individual constituents on a preparative scale in order to secure sufficient material for structural studies. Occasionally, the material available will allow fractional distillation in vacuo of the mustard oil mixture (20, 89, 116), but often-times thiourea derivatives will have to be prepared. Distribution between partly miscible solvents has been successfully applied to *iso*thio-cyanates (119, 120) and thioureas (67, 71, 85, 103). A similar method, followed by column chromatography on alumina, has proved useful also for the separation of free mustard oils (109). In many instances, however, the solubility properties of the individual thiourea derivatives are sufficiently different to permit fractional crystallization [e. g. (90, 110)].

No systematic studies have yet been undertaken to develop general methods for the separation on a preparative scale of *iso*thiocyanates or their derivatives. All workers active in the field would welcome progress along these lines.

### d. Quantitative Determination.

A detailed discussion of the quantitative analysis of mustard oils falls outside the scope of the present paper. Special interest is attached to this problem in pharmacy and food industry, and an extensive pertinent literature exists (3, 145). Evidently, the methods must be based on a preceding liberation of the isothiocyanates from the glucosides and hence depend on the character and amount of accompanying by-products or contaminants. The frequent occurrence of isothiocyanate mixtures in plants has often been disregarded.

Within the last decade, spectrophotometric assays have been increasingly used. In the author's laboratory, a method, taking advantage of the intense band shown by thioureas at ∼ 240 mµ, has given very satisfactory results (72).

The determination in plant extracts and milk samples of that particular group of mustard oils which undergo cyclization to biologically potent 2-oxazolidinethiones represents a problem of particular interest. The methods used for this purpose are based on the spectrophotometric evaluation of the heterocyclic ring (4, 97, 154).

### 3. Chemical Structure.

The following discussion includes a brief outline of the evidence on which the chemical structures are based, and also some references to the most important botanical sources. A more complete survey of the plants investigated appears in *Table 5*, p. 161. Only mustard oils of well-authenticated chemical structures, as listed in Table 3, will be considered.

### a. Saturated Alkyl isoThiocyanates.

**Methyl isothiocyanate.** This simple mustard oil, $CH_3NCS$, has not been definitely identified in any species of the *Cruciferae* but appears to be widely distributed in form of the glucoside *glucocapparin* throughout the *Capparidaceae* (74, 78). For the purpose of identification it was transformed into 1-methylthiourea (84).

**Ethyl isothiocyanate.** Seeds of the North-American crucifer *Lepidium Menziesii* DC. represent, thus far, the sole recorded source of this simple compound derived from the hypothetical glucoside "glucolepidiin". Conversion into 1-ethylthiourea has proved its structure (56).

**isoPropyl isothiocyanate.** The first recorded source of this mustard oil was the Indian plant *Putranjiva Roxburghii* WALL. (110)—a remarkable finding since this compound is the sole well-authenticated mustard oil ound in *Euphorbiaceae*. The same isothiocyanate, deriving from the parent glucoside *glucoputranjivin*, was isolated from seeds of the crucifer *Lunaria biennis* MNCH. (*69*). It is rather widely distributed in *Cruciferae*, occurring e. g. in species of *Cochlearia*, *Lunaria* and *Sisymbrium*. Furthermore, the species *Tropaeolum peregrinum* (*Tropaeolaceae*) afforded *iso*propyl mustard oil upon enzymic hydrolysis. The structural evidence is based on the conversion into 1-*iso*propylthiourea (*69*).

**(+)-2-Butyl *iso*thiocyanate** (IX) represents one of the "classical" mustard oils, recognized by HOFMANN (*50*) in 1870, as part of distillates of *Cochlearia officinalis* L.; it was further studied by GADAMER (*33*). More recent investigations have established its occurrence in form of the glucoside *glucocochlearin*, in species of the cruciferous genera, *Cardamine*, *Cochlearia*, *Draba*, *Lunaria*, *Sisymbrium*, etc. (cf. Table 5, p. 161). Other families, such as *Euphorbiaceae Phytolaccaceae* and *Tropaeolaceae*, also include species which afford, on enzymic fission, (+)-2-butyl *iso*thiocyanate. The latter is frequently accompanied by varying amounts of *iso*propyl *iso*thiocyanate, possibly indicating a common biogenetic pathway of the two compounds.

By chemical correlation, the *dextro*rotatory, natural 2-butyl *iso*-thiocyanate has been shown to possess the absolute configuration (IX) (*86*) which is identical with that prevailing around the $\beta$-carbon atom of natural *iso*leucine (X).

<br>

NCS

$$H_3C \text{------} H$$

$$C_2H_5$$

(IX.) (+)-2-Butyl *iso*thiocyanate.

COOH

$$H_2N \text{------} H$$

$$H_3C \text{------} H$$

$$C_2H_5$$

(X.) L-*Iso*leucine.

**Methyl 4-*iso*thiocyanatobutyrate (Erypestrin).** The detection in the seeds of *Erysimum rupestre* DC. (and other species of the same genus) of an alkali-labile glucoside, *glucoerypestrin*, was followed by a study of the corresponding mustard oil, *erypestrin*, in the author's laboratory (*78*). On treatment with ammonia, aniline and 1-naphthylamine, the *iso*thio-cyanate afforded crystalline thioureas that were indistinguishable from those prepared from synthetic methyl 4-*iso*thiocyanatobutyrate, $CH_3OOCCH_2CH_2CH_2NCS$. No other sources than *Erysimum* spp. have been reported thus far for glucoerypestrin which seems to be invariably accompanied by glucocheirolin.

b. Unsaturated Alkyl isoThiocyanates.

**Allyl *iso*thiocyanate.** This is the volatile mustard oil par excellence, $CH_2=CHCH_2NCS$, formed by enzymic fission of sinigrin. Seeds of black mustard (*Brassica nigra* KOCH) serve as the traditional source for this *iso*thiocyanate but other materials, such as horse-radish root (*146*), may serve equally well.

Numerous references exist to the production of allyl mustard oil upon disintegration of various botanical species, but the older literature should be accepted with reservation because of meagre documentation in many instances. A list of well-established and of doubtful sinigrin sources is available (*58*). Attention should be drawn to the recently observed enzymic production in *Thlaspi arvense* L. seeds of allyl thiocyanate rather than the mustard oil (*40*).

**3-Butenyl *iso*thiocyanate.** A rather contradictory literature exists on the occurrence in rape seeds (*Brassica napus* L.) of a glucoside, *gluconapin*, that affords on enzymic cleavage an unsaturated $C_5$-*iso*-thiocyanate. The identity of the latter as 3-butenyl *iso*thiocyanate, $CH_2=CHCH_2CH_2NCS$, was established by synthesis independently by ETTLINGER and HODGKINS (*20*), and KJÆR et al. (*71*). Some additional sources exist for this mustard oil, e. g. *Alyssum, Brassica, Cardamine* and *Isatis* spp. (cf. Table 5, p. 161).

**4-Pentenyl *iso*thiocyanate.** This higher homologue, $CH_2=$ $=CHCH_2CH_2CH_2NCS$, has been identified as a minor constituent of the volatile mustard oil fraction of rape seed, originating, by enzymic hydrolysis, from the glucoside, *glucobrassicanapin* (*89*). The structure is based on infrared evidence and on the desulphuration of the 1-naphthylthiourea derivative (XI) to the urea derivative (XII), which upon catalytic hydrogenation afforded the saturated urea (XIII), identical with a sample synthesized from *n*-pentylamine and 1-naphthyl *iso*-cyanate.

$$CH_2=CH—CH_2—CH_2—CH_2—NHCSNH—C_{10}H_7$$

(XI.)  $\downarrow$ AgNO$_3$

$$CH_2=CH—CH_2—CH_2—CH_2—NHCONH—C_{10}H_7$$

(XII.)  $\downarrow$ H$_2$/Pt

$$CH_3—CH_2—CH_2—CH_2—CH_2—NHCONH—C_{10}H_7$$

(XIII.) Pentyl-(1-naphthyl)-urea.

$\uparrow$

$$CH_3—CH_2—CH_2—CH_2—CH_2—NH_2 + C_{10}H_7NCO$$

Synthesis of 4-pentenyl *iso*thiocyanate confirmed the identity of the mustard oil mentioned (*89*). In addition to *Brassica* spp., the 4-pentenyl compound seems to occur in members of *Alyssum* and possibly other Crucifers as well.

*c. ω-Methylthioalkyl* iso*Thiocyanates and Related Sulphoxides and Sulphones.*

Under this heading an interesting group of natural products will be discussed, having in common the skeleton, $CH_3S(CH_2)_nNCS$, in which a double bond may be present in the alkyl chain and the sulphide atom may carry oxygen. Mustard oils of this type occur abundantly in Crucifers, and there is much interest in the problem of their biogenesis. The discovery within the last decade of ten new individual *iso*thiocyanates of this general structure was preceded by the structural elucidation of cheirolin, $CH_3SO_2(CH_2)_3NCS$ (*123*), and erysolin, $CH_3SO_2(CH_2)_4NCS$ (*126*), by SCHNEIDER et al. fifty years ago.

**3-Methylthiopropyl *iso*thiocyanate (Ibervirin).** Seeds of the Crucifer *Iberis sempervirens* L. were shown in this laboratory to afford two steam-volatile *iso*thiocyanates subsequent to enzymic hydrolysis. The corresponding thiourea mixture was subjected to countercurrent distribution, resulting in the separation of two pure fractions (*85*). One of these proved to be 1-(3-methylthiopropyl)-thiourea upon comparison with a synthetic specimen, whereas the other, minor constituent was the homologous 1-(4-methylthiobutyl)-thiourea. This result indicates the enzymic production of 3-methylthiopropyl *iso*thiocyanate, $CH_3SCH_2CH_2CH_2NCS$, *(ibervirin)* from a parent glucoside, named *glucoibervirin*. The botanical distribution of the latter is still unknown but appears to be rather limited. Glucoibervirin is present in green parts of *Cheiranthus cheiri* L. (wallflower) and in *Iberis amara* L. roots (unpublished).

**3-Methylsulphinylpropyl *iso*thiocyanate (Iberin).** The foregoing mustard oil is closely related to *iberin*, derivable from the glucoside *glucoiberin*, which was isolated by SCHULTZ and GMELIN (*136*) in crystalline form from the seeds of *Iberis amara* L. They presented evidence, later confirmed in the writer's laboratory (*75*), for the corresponding, optically active *iso*thiocyanate to possess the structure $CH_3SO(CH_2)_3NCS$. The sulphoxide group is very probably formed by in vivo oxidation of gluco-ibervirin. Among possible botanical sources of glucoiberin, the green parts of certain cabbage species deserve special interest. Thus, PROCHÁZKA et al. (*109*) found iberin in the press juice of Brussels sprouts, for example.

**3-Methylsulphonylpropyl *iso*thiocyanate (Cheirolin).** A still higher oxidation level than in iberin is present in the well-known mustard oil of wallflower seeds (*Cheiranthus cheiri* L.), for which SCHNEIDER (*123*) proved the structure $CH_3SO_2CH_2CH_2CH_2NCS$, derivable from the glucoside *glucocheirolin*. The latter has been encountered in several species of the cruciferous genera *Cheiranthus*, *Erysimum* and *Malcolmia*. It is most likely related biogenetically to the structurally similar mustard oils ibervirin and iberin (see above). Recently, BACHELARD et al. (*4a*) have isolated cheirolin from enzymically hydrolyzed extracts of fruits (1.2 g. per kg. dry weight) and fresh leaves (0.4 g. per kg. wet weight) of the common weed, *Rapistrum rugosum* (L.) ALL. A convenient synthesis of cheirolin has been developed in the author's laboratory (*92*).

**4-Methylthiobutyl *iso*thiocyanate (Erucin).** This sulphide mustard oil, $CH_3SCH_2CH_2CH_2CH_2NCS$, was first found in enzymically hydrolyzed glucoside extracts of *Eruca sativa* MILL., and its structure was established by comparison with synthetic preparations (*73*). The parent glucoside, *glucoerucin*, seems to be rather widespread throughout the Crucifer family. Thus, erucin was isolated by DELAVEAU (*17*) from *Diplotaxis tenuifolia* (L.) DC. roots and, as pointed out above, in this laboratory from *Iberis sempervirens* L. (*85*). Furthermore, there is strong paper-chromatographic evidence for the presence of glucoerucin in certain species of the genera *Brassica, Cheiranthus, Farsetia, Hesperis, Iberis, Matthiola* and *Vesicaria* (*73*).

**4-Methylsulphinylbutyl *iso*thiocyanate (Sulphoraphane).** It was found only recently that this next higher homologue, $CH_3SO(CH_2)_4NCS$, of iberin, synthesized by SCHMID and KARRER (*121*) more than ten years ago, is a natural product. In the author's laboratory a rather wide-spread occurrence of sulphoraphane has been noted (*66*), e. g. in species of *Brassica, Eruca* and *Iberis*. PROCHÁZKA (*108*) was able to isolate partially racemized sulphoraphane from the leaves of *Lepidium draba* L. According to the same author, sulphoraphane, together with iberin, is also present in the fresh juice of certain cabbage varieties (*109*). No name has yet been suggested for the parent glucoside, but *glucoraphanin* would be a logical term.

**4-Methylsulphinyl-3-butenyl *iso*thiocyanate (Sulphoraphene).** SCHMID and KARRER (*119*) concluded in 1948 that radish seeds (*Raphanus sativus* L.) produce, upon enzymic hydrolysis, the strongly *levo*rotatory mustard oil, sulphoraphene, $CH_3SOCH=CHCH_2CH_2NCS$, accompanied by the corresponding nitrile (*120*). It seems, that sulphoraphene is the first natural product whose optical activity is due solely to the presence of asymmetric sulphur. GMELIN (*38*) has introduced the designation *glucoraphenin* for the parent glucoside, which probably is also a constituent of *Matthiola bicornis* DC. seeds (*138*). The recent finding by PROCHÁZKA (*107*) of sulphoraphene in the press juice of *Plantago major* L. is remarkable, since it is the first recorded case of the presence of a mustard oil in *Plantaginacea*.

**4-Methylsulphonylbutyl *iso*thiocyanate (Erysolin).** This higher homologue, $CH_3SO_2(CH_2)_4NCS$, of cheirolin was reported by SCHNEIDER and KAUFMANN (126) as a constituent of enzymically hydrolyzed seed extracts of *Erysimum Perofskianum* FISCH. et MAY. Although this mustard oil, conceivably arising from the glucoside *glucoerysolin*, fits well into the present series, it has not been possible in the author's laboratory or elsewhere to prove its presence in the seeds mentioned or in any other species investigated. Hence, its occurrence may be dependent on special conditions of growth. A practical synthesis of erysolin was reported by KJÆR and CONTI (70).

**5-Methylthiopentyl *iso*thiocyanate (Berteroin).** It was shown in the author's laboratory, that seeds of the cruciferous weed *Berteroa incana* (L.) DC. furnish enzymically a volatile mustard oil, *berteroin*, whose 5-methylthiopentyl *iso*thiocyanate structure, $CH_3S(CH_2)_5NCS$, was established by comparison with suitable thiourea derivatives and by synthesis (91). The progenitor, *glucoberteroin*, was shown to be present also in seeds of *Lunaria rediviva* L., as well as in several *Alyssum* species. Usually, the glucoside is accompanied by varying amounts of the corresponding sulphoxide (see below).

**5-Methylsulphinylpentyl *iso*thiocyanate (Alyssin).** This sulphoxide mustard oil was first isolated in the author's laboratory from seeds of *Alyssum argenteum* VITM., following enzymic hydrolysis. Its structure, $CH_3SO(CH_2)_5NCS$, was established by transformation of the corresponding benzylthiourea derivative (XIV) into the benzylurea compound (XV). The optical activity of the latter disappeared on reduction with zinc and acid which yielded the sulphide-urea (XVI); this indicates the original sulphoxide-grouping as the sole centre of asymmetry in alyssin. Compound (XVI) was proved to be 1-(5-methylthiopentyl)-benzylurea by comparison with an authentic sample, produced by desulphurization of the corresponding thiourea (XVII); the latter had been prepared from berteroin (XVIII) and benzylamine (75).

$$CH_3SO(CH_2)_5NHCSNHCH_2C_6H_5 \xrightarrow{Ag^+} CH_3SO(CH_2)_5NHCONHCH_2C_6H_5$$
$$\text{(XIV.)} \qquad\qquad\qquad\qquad\qquad\qquad \text{(XV.)}$$

$$CH_3S(CH_2)_5NHCONHCH_2C_6H_5 \xleftarrow{\;\;\big|\;Zn/H^+\;} $$
$$\text{(XVI.) 1-(5-Methylthiopentyl)-benzylurea.} \quad\big|Ag^+$$

$$CH_3S(CH_2)_5NCS + C_6H_5CH_2NH_2 \rightarrow CH_3S(CH_2)_5NHCSNHCH_2C_6H_5$$
$$\text{(XVIII.) Berteroin.} \qquad\qquad\qquad\qquad \text{(XVII.)}$$

The progenitor, *glucoalyssin*, was encountered in several *Alyssum* and *Berteroa* spp. (75) and may well be related biogenetically to glucoberteroin. SCHULTZ and WAGNER (140) isolated glucoalyssin, in form of the crystalline tetraacetate, from seeds of an *Alyssum* species and proposed (though on meagre evidence) the above alyssin structure.

**8-Methylsulphinyloctyl** *iso***thiocyanate (Hirsutin).** The class of naturally derived sulphoxide *iso*thiocyanates was extended when it was observed in this laboratory that enzymic hydrolysis of seed extracts of the crucifer *Arabis hirsuta* (L.) Scop. afforded *hirsutin*, for which the structure $CH_3SO(CH_2)_8NCS$ was established upon degradation of the corresponding phenylthiourea derivative to 1-(8-methylthiooctyl)-3-phenylurea (*66*) by a procedure similar to that outlined for alyssin. The glucoside, *glucohirsutin*, was not further characterized and its natural distribution is still unknown.

**9-Methylsulphinylnonyl** *iso***thiocyanate (Arabin).** This next-higher homologue of hirsutin was encountered as the enzymic hydrolysis product of *glucoarabin*, present in seeds of the common crucifer *Arabis alpina* L. The arabin structure, $CH_3SO(CH_2)_9NCS$, was suggested on the basis of customary degradation methods (*77*), and was later unequivocally proved by partial synthesis (*65*). Although presumably present in several *Arabis* species, the exact distribution of glucoarabin remains to be clarified.

**10-Methylsulphinyldecyl** *iso***thiocyanate (Camelinin).** This highest known member of the sulphoxide mustard oil series appears in enzymically cleaved seed extracts from *Camelina sativa* (L.) Crantz that contain *glucocamelinin*. Its structure, $CH_3SO(CH_2)_{10}NCS$, was established in the author's laboratory (*83*) by a conventional scheme of degradation. The parent glucoside is present also in other *Camelina* species (*83*), but otherwise its occurrence in the vegetable kingdom is unknown.

*Stereochemistry.*

The homologous series of optically active ω-methylsulphinylalkyl *iso*thiocyanates of natural extraction, $CH_3SO(CH_2)_nNCS$ ($n = 3$, 4, 5, 8, 9 and 10), discussed above, raises the question as to the relative and absolute steric configurations. Convincing evidence is on hand that all members of this series possess the same steric configuration around the asymmetric sulphoxide grouping. Thus, an extensive series of analogous derivatives, listed in Table 4, p. 160, shows comparable rotations, with regard to sign as well as magnitude. Furthermore, rotatory dispersion studies of a selected number of these derivatives over a rather broad wavelength region (*96*) lend even stronger support to our assumption.

Absolute configurations, however, have not yet been established. The problem requires the stereochemical correlation of any representative of the mustard oil series with (+)-*S*-methyl-*L*-cysteine *S*-oxide, the sole sulphoxide of secured absolute configuration. The latter was established by HINE and ROGERS (*48*) by X-ray analysis.

### d. Aromatic iso Thiocyanates.

**Benzyl *iso*thiocyanate.** This structurally simple, aromatic mustard oil, $C_6H_5CH_2NCS$, was first recognized by GADAMER in 1899 as an enzymic hydrolysis product of *glucotropaeolin*, present in seeds and fresh parts of *Tropaeolum majus* L. (*34*) and in *Lepidium sativum* L. seeds (*35*). Since then, several other sources of benzyl mustard oil have been found, e. g. *Coronopus didymus* (L.) SM. (*101*) and *Lepidium* spp. Even more interesting is the frequent appearance of benzyl *iso*thiocyanate in species belonging to other families than *Cruciferae*. Thus, it has been encountered in *Caricaceae, Moringaceae, Salvadoraceae, Tropaeolaceae* and, possibly, *Phytolaccaceae* (cf. *Table 6*, p. 167). There is, furthermore, some support for its occurrence in *Jatropha multifida* L. *(Euphorbiaceae)* (*29*).

It is noteworthy that benzyl cyanide has been observed repeatedly as a product formed during fission of the parent glucoside (*6, 10*). Accordingly, SCHULTZ and GMELIN (*135*) suggested a possible relationship between glucotropaeolin and an unknown growth factor of *Lepidium sativum* L. which may plausibly be phenylacetic acid, formed by hydrolysis of benzyl cyanide. An alternative enzymic cleavage of glucotropaeolin in the seeds of *L. ruderale* L. and *L. sativum* L., viz. to benzyl thiocyanate, was recently reported (*40*). Further breakdown of the latter may be responsible for the benzyl mercaptan production observed by FORSS (*28a*) in macerates of the crucifer *Coronopus didymus*. Under different circumstances the same plant afforded mixtures of benzyl cyanide and benzyl *iso*thiocyanate (*28a, 101*).

**$p$-Hydroxybenzyl *iso*thiocyanate.** The non-volatile mustard oil formed during enzymic hydrolysis of the "classical" glucoside *sinalbin* was studied by SALKOWSKI (*114*) who suggested its structure, $(p)HOC_6H_4CH_2NCS$. His conclusion was confirmed in the author's laboratory by comparing the phenylthiourea derivative with an authentic sample of 1-(4-hydroxybenzyl)-3-phenylthiourea. The labile character of the initially formed mustard oil calls for special precautions during isolation (*94*). Thus, WILL and LAUBENHEIMER (*156*) have demonstrated the facile formation of thiocyanate upon treatment of the mustard oil with hot ammonia or alkali. Though abundantly present in white mustard seeds (*Sinapis alba* L.), sinalbin does not appear to be a glucoside of very wide distribution. There is good evidence for the presence of the glucosidic anion of sinalbin also in seeds of *Sinapis arvensis* L., *Lepidium* *campestre* (L.) R. Br. and in some species of the genera *Aubrietia*, *Brassica* and *Bunias*.

**$p$-Methoxybenzyl *iso*thiocyanate (Aubrietin).** In view of the common occurrence of O-methylated phenolic compounds in plants, it was not unexpected to find $p$-methoxybenzyl mustard oil, $CH_3OC_6H_4CH_2NCS$, as an enzymic fission product of *glucoaubrietin*, in fresh parts and seeds of various *Aubrietia* spp. *(Cruciferae)*. Its structure was secured by comparison of thiourea derivatives with authentic samples (*81*). Thus far, *Aubrietia* spp. represent the sole botanical source of aubrietin, easily recognizable by its taste which, besides being pungent, is reminiscent of anis.

**$m$-Methoxybenzyl *iso*thiocyanate (Limnanthin).** More surprising was the finding by Ettlinger and Lundeen (*22*) of $m$-methoxybenzyl mustard oil, $CH_3OC_6H_4CH_2NCS$, as an enzymic product of an unnamed glucoside in the seeds of the North-American plant *Limnanthes douglasii* R. Br., belonging to the small family of *Limnanthaceae*. No other sources have yet been recorded. The structural proof was provided by synthesis. As pointed out by the American authors, the limnanthin structure deserves special interest considering the rare occurrence of $m$-disubstituted aromatic compounds in nature.

**2-Phenylethyl *iso*thiocyanate.** This next higher homologue, $C_6H_5CH_2CH_2NCS$, of benzyl mustard oil, liberated from *gluconasturtiin*, was first encountered by Gadamer (*35*) in fresh parts of the Crucifers *Nasturtium officinale* R. Br. and *Barbarea praecox* R. Br., but has since been recognized as a mustard oil of extensive distribution. In fact, there is strong evidence that it is the predominant volatile *iso*thiocyanate in root tissues of several *Brassica* species, such as turnip, black and white mustard, as well as cabbage (*144*). Horse-radish also has a high content of this aromatic *iso*thiocyanate (*144*). André and Delaveau (*2*) found it to be a minor constituent of the volatile mustard oil fraction of rape seed cake. Outside the family *Cruciferae*, 2-phenylethyl *iso*thiocyanate has been observed in root tissues of various *Resedaceae* (*5*).

**3-Benzoyloxypropyl *iso*thiocyanate (Malcolmiin).** A rather unexpected aromatic mustard oil was discovered in the author's laboratory when studying seed extracts of the Crucifer *Malcolmia maritima* (L.) R. Br. Upon enzymic cleavage of *glucomalcolmiin*, a mustard oil with absorption characteristics indicative of aromatic character was obtained. Its thiourea derivative, $C_{11}H_{14}O_2N_2S$, yielded (in alkali) benzoic acid in accordance with the infrared evidence of an ester linkage. The alcoholic entity of the latter was established as 3-hydroxypropyl *iso*thiocyanate, $HOCH_2CH_2CH_2NCS$, by its spontaneous cyclization to a heterocyclic compound, which was identified with synthetic tetrahydro-1,3-oxazine-2-thione (76). Subsequently, the malcolmiin structure was confirmed by synthesis (90).

Since the genus *Malcolmia* is taxonomically closely related to *Cheiranthus*, the traditional botanical source of cheirolin, it is not surprising to find the latter accompanying malcolmiin in seed hydrolysates of *M. maritima* (76, 90). The further distribution of glucomalcolmiin is unknown.

e. Hydroxy-substituted isoThiocyanates.

There exist some natural *iso*thiocyanate glucosides whose aliphatic side-chains are substituted by a β-hydroxy group. The corresponding *iso*thiocyanates undergo a presumably spontaneous, intramolecular cyclization to 2-oxazolidinethiones, which are the recognizable end products of the enzymic reaction. This conclusion has been reached as a result of recent studies of several representatives of this type (see below).

**2-Hydroxy*iso*butyl *iso*thiocyanate.** More than twenty years ago, HOPKINS (52) isolated from enzymically cleaved seed extracts of the weed *Conringia orientalis* L. (DUMORT) [more correctly designated as *C. orientalis* (L.) ANDRZ. and synonymous with *Erysimum orientale* MILL.] a compound to which the structure (XIX) was attributed; it was corrected by ETTLINGER (18) to the isomeric form (XX) on the basis of infrared evidence. HOPKINS suggested as the likely pathway in the formation of (XIX) the initial formation of β-methallyl *iso*thiocyanate (XXI), followed by hydration of the double bond and cyclization. That this reaction sequence cannot be responsible for the production of (XX) became apparent, however, when a synthetic specimen of (XXI), studied in the author's laboratory, displayed no tendency to cyclize (95).

(XIX.)      (XX.)      (XXI.)      (XXII.)

β-Methallyl *iso*thiocyanate.    2-Hydroxy*iso*butyl *iso*thiocyanate.

More recently, the parent glucoside, *glucoconringiin*, was isolated as a crystalline acetate from seeds of *C. orientalis* by KJÆR et al. (82), who demonstrated spectroscopically that 2-hydroxy*iso*butyl *iso*thiocyanate (XXII) is the initial product of the enzymic hydrolysis and subsequently cyclizes to give the HOPKINS compound (XX). Similar conclusions were reached by SCHULTZ and WAGNER (141). Likewise, the presence of glucoconringiin in various species of the genus *Cochlearia* was proved, and 5,5-dimethyl-2-oxazolidinethione was isolated from the seeds of *C. officinalis* L. (*82*); its further distribution is unknown.

**2-Hydroxy-3-butenyl *iso*thiocyanate.** In 1949, ASTWOOD et al. (*4*) isolated from yellow turnip and various *Brassica* seeds an antithyroid factor, which was identified as (—)-5-vinyl-2-oxazolidinethione (XXIII), a structure later confirmed synthetically by ETTLINGER (*19*). The same investigators pointed to an *iso*thiocyanate glucoside as a plausible progenitor of the heterocyclic compound (*43*)—a suggestion confirmed by studies in the author's laboratory (*82*) and elsewhere (*141*). These indicated one of the stereoisomeric 2-hydroxy-3-butenyl *iso*thiocyanates (XXIV) to be the first reaction product.

(XXIII.) Goitrin.      (XXIV.) 2-Hydroxy-3-butenyl *iso*thiocyanate.

GREER (*42*) succeeded in isolating from rutabaga seeds, the genuine glucoside, *progoitrin*, which is identical with *glucorapiferin*, obtained as the crystalline pentaacetate from seeds of a rape variety [SCHULTZ and WAGNER (*141*)].

In this laboratory, the absolute configuration of goitrin (XXIII), and hence of the parent mustard oil (XXIV) as well as the side-chain of progoitrin, was recently established by chemical correlation with configurationally known compounds (*68*); see the formulas (XXV) and (XXVI).

(XXV.) Goitrin.      (XXVI.) 2-Hydroxy-3-butenyl *iso*thiocyanate.

Progoitrin occurs largely in *Brassica* spp., predominantly in seed materials, including rapeseed oil meal (2–4 mg. goitrin per g. of meal) (*111*). This occurrence is interesting, because progoitrin is a potential precursor of goitrin. In contrast to seeds, fresh cabbage contains only very small amounts of goitrin (*1*).

**2-Hydroxy-2-phenylethyl *iso*thiocyanate.** On evidence, similar in nature to that presented for the last two hydroxy-substituted mustard oils, it was established in this laboratory that seed material of various species of the cruciferous genus *Barbarea* afforded on enzymic hydrolysis a hydroxy-substituted *iso*thiocyanate, that underwent spontaneous cyclization to a 2-oxazolidinethione. Seed extracts from *B. vulgaris* R. BR., containing the parent glucoside *glucobarbarin* in addition to gluconasturtiin, were enzymically cleaved to a mixture of 2-phenylethyl mustard oil, originating from the latter glucoside, and (—)-5-phenyl-2-oxazolidine-thione (barbarin) (XXVII), the cyclization product of primarily formed 2-hydroxy-2-phenylethyl mustard oil (XXVIII) (79).

$$H_2C\text{----}NH$$
$$C_6H_5\text{---}HC\diagdown\quad C=S$$
$$O$$
$$\leftarrow$$
$$H_2C\text{----}N$$
$$C_6H_5\text{---}HC\diagdown\quad C=S$$
$$OH$$

(XXVII.) Barbarin.          (XXVIII.) 2-Hydroxy-2-phenylethyl *iso*thiocyanate.

A stereospecific synthesis of both enantiomers of (XXVII), starting from configurationally known compounds, has established the absolute configurations of barbarin (XXIX) and its mustard oil precursor (XXX) (79, 80).

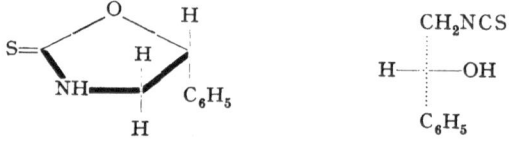

(XXIX.) Barbarin.          (XXX.) 2-Hydroxy-2-phenylethyl *iso*thiocyanate.

In addition to *Barbarea* species, leaves and inflorescenses of *Reseda luteola* L. (dyer's weed, weld) yielded barbarin on enzymic treatment. Seeds of the same plant, as well as other *Reseda* species, are additional sources of barbarin (80).

**2-Hydroxy-*iso*propyl *iso*thiocyanate.** A β-hydroxy-substituted, bran-ched, aliphatic mustard oil represents the newest addition to the *iso*thiocyanates that undergo cyclization. The glucoside *glucosisymbrin* was reported from this laboratory as a constituent of seeds of the Crucifer *Sisymbrium austriacum* JACQ. It was characterized by its

$$CH_3\text{---}HC\text{----}NH$$
$$H_2C\diagdown\quad C=S$$
$$O$$

$$CH_3\text{---}HC\text{----}N$$
$$H_2C\diagdown\quad C=S$$
$$OH$$

(XXXI.) Sisymbrin.          (XXXII.) 2-Hydroxy-*iso*propyl *iso*thiocyanate.

ability to undergo enzymic decomposition to *dextro*rotatory 4-methyl-2-oxazolidinethione (sisymbrin) (XXXI), produced by cyclization of the initially liberated 2-hydroxy-*iso*propyl *iso*thiocyanate (XXXII) (67).

A stereospecific synthesis starting from alanine of known configuration has given full information on the absolute configuration of the ring compound. Accordingly, the configurations (XXXIII) and (XXXIV) were established for the naturally occurring sisymbrin and its parent mustard oil (67).

(XXXIII.) Sisymbrin.   (XXXIV.) 2-Hydroxy-*iso*propyl *iso*thiocyanate.

In the same paper (67), the possible biogenetic relationship of (XXXIV) and the non-hydroxylated *iso*propyl mustard oil, formerly encountered in plants, was discussed as well as its formal similarity with $\beta$-hydroxy-*iso*butyric acid of the sesquiterpene lactone arctiopicrin, as well as with (—)-$\beta$-amino-*iso*butyric acid, isolated from bulbs of *Iris tingitana*.

### *f. iso Thiocyanates of Doubtful Structure.*

In addition to the naturally derived mustard oils of established structure, a number of other *iso*thiocyanates have been reported without sufficient documentation to permit conclusive formulations. The existence of still others is questionable, because their isolation could not be repeated.

The first category includes the compound, $CH_3CH_2CH_2CH_2SCH=CHCH_2CH_2NCS$, which Heiduschka and Zwergal (47) claimed to be present in fresh radish root. The same authors postulated the presence in horse-radish of 3-phenylpropyl *iso*thiocyanate.

Quite incompatible with the physical and chemical properties, Zwergal (159) formulated a component from kohlrabi seeds as 4,4-dimethyl-5-vinyl-2-oxazolidine-thione, a structure which was later revised by Schultz and Wagner (141) to $CH_3SOCH=CHCH_2CH_2CH_2NCS$, apparently again with no other evidence than its resemblance to the corresponding nitrile, $CH_3SOCH=CHCH_2CH_2CH_2CN$; the latter represents Zwergal's original proposal (158). The parent glucoside, isolated as a non-homogeneous tetraacetate from kohlrabi seeds, was named gluco-caulorapin (141, 151).

From the seeds of *Hesperis matronalis* L., Wagner (151) isolated gluco-matronalin as a crystalline hexa- or heptaacetate, suggesting the presence of two or three acetylable groupings in the otherwise unknown, derivable mustard oil.

An interesting observation deserving repetition was made by Puntambekar (110) who claimed, in addition to *iso*propyl and *sec*-butyl *iso*thiocyanates, the isolation of phenyl mustard oil (as its thiourea derivative) from seed kernels of the Indian *Euphorbia* species *Putranjiva Roxburghii* Wall.

As already pointed out the presence of numerous unidentified glucoside spots in paper chromatograms obtained from a large collection of plant species indicates that additional mustard oils will be forthcoming in the future.

## VI. Botanical Distribution of isoThiocyanate Glucosides.

It is remarkable that practically all investigated plants of the nearly cosmopolitan family *Cruciferae* contain one or more *iso*thiocyanate-producing glucosides. Although only a minor fraction of a total of about 1500 species, that belong to this family, has as yet been subjected to analysis, it appears likely that most Crucifers, in some or all parts of the plant, contain such glucosides. In *Table 5* (p. 161) an attempt has been made to present a rather complete survey of all those Crucifer species in which either *iso*thiocyanates or parent glucosides have been found up to the end of 1959. The Table does not include numerous other plants, studied in the author's laboratory, whose analyses are unpublished. In general, Table 5 contains the chemical identity of the *iso*thiocyanates or their glucosides. In some cases, however, and particularly when the evidence was exclusively of paperchromatographic nature, it was preferable to list the reported number of constituents rather than chemical structures, even when these were suggested in the original paper.

On the whole, the number of observed individual *iso*thiocyanate glucosides is now so large that chemical identifications based solely on paper chromatography are more than questionable. Furthermore, the number of the observed glucosides may vary considerably, depending on the sensitivity of the analytical method and the concentration of the extracts. It was noted repeatedly in the author's laboratory that plant materials, generally regarded as containing only one or two glucosides, on closer examination displayed a notable number of additional trace glucosides.

Where provided in the original text, or where no doubt seems to exist, botanical authors' names appear in Table 5. It should be stressed at this point that botanical designations should be as unambiguous as the proof of the chemical structure, when new chemical compounds are related to botanical sources. To illustrate: the labelling of more than half of 300 seed specimens, collected by us from botanical gardens and other sources all over the world, proved to be incorrect on closer inspection.

Although preponderant in *Cruciferae*, *iso*thiocyanate glucosides are by no means confined to this family. It may not surprise that species of other families, such as *Capparidaceae*, *Moringaceae* and *Resedaceae*, which belong to the same order *(Rhoeadales)* as *Cruciferae*, may occasionally contain the same glucosides. Quite unexpected, however, was the isolated appearance of *iso*thiocyanate glucosides in some species of plant families which are taxonomically remote from *Rhoeadales*, such as *Caricaceae, Euphorbiaceae, Limnanthaceae, Phytolaccaceae, Plantaginaceae, Salvadoraceae* and *Tropaeolaceae*. *Table 6* (p. 167) presents a survey of the reported glucoside-containing species which do not belong to *Cruciferae*.

Attempts to correlate chemical structure of plant constituents with the botanical character have become more frequent during recent years, and often rewarding. Such chemotaxonomic approaches have also been attempted (*87*) in the glucoside group discussed and will probably be extended. It should not be forgotten, however, that the structure of a selected type of compounds is only one of numerous factors that contribute to biological individualism. Nevertheless, further comparative studies of the distribution of mustard oil glucosides, when conducted with due criticism, may provide new and interesting results concerning the chemistry of natural products as well as botanical taxonomy.

## VII. Biological Properties of *iso*Thiocyanates and Their Parent Glucosides.

In connexion with the taxonomic and phylogenetic implications of comparative studies of *iso*thiocyanates and glucosidic progenitors in individual botanical species, biogenesis is a challenging problem. No systematic studies have as yet been undertaken to elucidate the anabolic or catabolic pathway of these compounds in plants. The conspicuous structural regularities, however, make such investigations most desirable. Thus, the formal similarity of several side-chains of the glucosides with those of some amino acids occurring in proteins is striking and would suggest a possible common pathway in biogenesis (*23, 59*). Other products, however, belong to a type not ordinarily encountered in nature, such as the long-chain methylthioalkyl derivatives. Another interesting feature is the recognition of the glucosides as hydroxylamine derivatives (*23*). Bound hydroxylamine groups have been extensively discussed in the literature in connexion with the nitrogen metabolism of higher plants. The frequent appearance of homologous mustard oils in extracts is another notable characteristic of potential interest for biogenetic considerations. As a whole, the detailed knowledge of an extensive series of formally simple glucosides with well-established structures calls for thorough biogenetical investigations.

From time to time, several biological actions have been ascribed to mustard oils or their simple conversion products, but a detailed discussion of this subject falls outside the scope of the present survey. It suffices to draw attention to the long recognized antibacterial and fungiostatic property of most mustard oils, on which an extensive literature exists.

Recently, McKay et al. (*102*) published the results of bacteriostatic assays of a large series of *iso*thiocyanates.

Much interest is being devoted to some *iso*thiocyanates originating from weeds, as factors responsible for a "scorched" or "burnt" flavour of cream and butter. McDowall et al. (*101*), however, were unable to confirm such suggestions in case of the land-cress taint in New Zealand dairy products (cf. *28a*).

The goitrogenic effect of several oxazolidinethiones, particularly those of plant origin, has led to a variety of biological studies to assess the possible importance of such compounds in animal feeding and in endemic goiter. In connexion with such problems, BACHELARD et al. (*4a*) have recently isolated cheirolin from ripe fruits and fresh leaves of *Rapistrum rugosum* (L.) ALL., a widely distributed weed in Australian pastures, and showed that cheirolin is goitrogenic in rats. Rumen liquor effected in vitro conversion of cheirolin into the less goitrogenic 1,3-bis-(3-methylsulphonylpropyl)-2-thiourea. Moreover, certain toxic effects observed occasionally in animals after feeding materials that contained excessive amounts of *iso*thiocyanates, have called for much attention.

An interesting biological effect has been attributed to the genuine glucosides rather than the free mustard oils, viz. that of stimulating the feeding of certain insects. According to THORSTEINSON (*148*), two oligophagous insects [*Plutella maculipennis* (CURT.) and *Pieris brassicae* (L.)] could be induced to feed on leaves which they ordinarily refuse, by painting the leaves with sinigrin or sinalbin solutions, but not with allyl *iso*thiocyanate. The importance of such observations is obvious, considering the immense, world-wide damages caused by host-specific insects to crops of important *Cruciferae*, such as cabbage, turnip and rape. These, and many other examples may characterize the *iso*thiocyanate glucosides and their enzymic fission products as natural products of considerable and increasing interest to a diverse group of modern scientists.

# VIII. Tables.

Table 1. Crystalline isoThiocyanate Glucosides, Known at the End of 1959.

| Glucoside | Cation | Seed source | M. p. | Rotation in water $[\alpha]_D$ | Temp. (°) | c | References |
|---|---|---|---|---|---|---|---|
| Glucocapparin | K | Cleome spinosa JACQ. | 208–210° | − 28.6° | 25 | 2.2 | (74) |
| Sinigrin[a] | K | Brassica nigra KOCH | 127° | − 17.6° | 27 | | (124) |
| Glucoiberin[a] | K | Iberis amara L. | 142–144° (dec.) | − 55.3° | 20 | 4.9 | (136) |
| Glucocheirolin[a] | K | Cheiranthus cheiri L. | 158–160° c | − 21.6° | 27 | 4.9 | (128) |
| Sinalbin[b] | Sinapine | Sinapis alba L. | 83–84° | − 8.4° | | | (124) |
| Glucotropaeolin | $(CH_3)_4N^+$ | Tropaeolum majus L. | 188–189° | − 16.7° | 28 | | (24) |
| Gluconasturtiin[d] | K | Nasturtium officinale R. BR. | 163° | − 21° | | | (11) |
| Glucoconringiin | K | Conringia orientalis (L.) ANDRZ. | 168° (dec.) | − 10.9° | 21 | 3.7 | (41) |
| Progoitrin | Na | Brassica rapa L. | 145° | − 22.3° | | | (42) |

[a] Monohydrate.

[b] Pentahydrate. An anhydrous preparation was reported (124) to have the m. p. 138.5–140°. The anhydrous tetramethyl-ammonium salt of the sinalbin anion (tetramethylammonium glucosinalbate), m. p. 191–192° (dec.); $[\alpha]_D^{35} − 19°$ ($H_2O$), was described by ETTLINGER and LUNDEEN (23).

[c] A sample prepared in this laboratory decomposed over a much broader range (61).

[d] Non-analyzed preparation of unstated purity.

Table 2. Crystalline *iso*Thiocyanate Glucoside Tetraacetates, Known at the End of 1959.

| Parent glucoside | M. p.[a] | Rotation in water | | | References |
| --- | --- | --- | --- | --- | --- |
| | | $[\alpha]_D$ | Temp. (°) | c | |
| Glucocapparin[b] .......... | 209–210° | − 31.0° | 25 | 1.8 | (74) |
| Glucoputranjivin .......... | 180° | − 15.4° | 23 | 1.2 | (61) |
| Glucocochlearin .......... | 193–194° | | | | (138) |
| Sinigrin ................ | 192–193° [c] | − 18.6° | 24 | 3.6 | (82) |
| Glucoiberin[d] ............ | 145–147° | − 16.6° | 23 | 2.0 | (61) |
| Glucocheirolin .......... | 193–194° | − 18.0° | 24 | 1.0 | (61) |
| Glucoraphenin[e] .......... | 155–156° | − 23.5° | 27 | 1.5 | (61) |
| Glucoalyssin ............ | 158° | − 10.9° | 20 | 1.7 | (140) |
| Glucoerypestrin[f] .......... | 188–190° | − 18.0° | 23 | 1.7 | (78) |
| Glucotropaeolin[g] .......... | 186–187° | − 20.0° | 25 | 1.0 | (63) |
| Glucoconringiin[h] .......... | 152° | − 5.3° | 25 | 5.3 | (82) |
| Glucorapiferin[i] (= Progoitrin) | 178–180° | − 9.5° | 25 | 2.6 | (42, 138, 141) |

[a] All glucoside acetates decompose on heating; the decomposition temperatures are much dependent on the rate of heating.

[b] If not otherwise stated the glucoside acetates are crystalline potassium salts.

[c] SCHULTZ and WAGNER (138) reported the m. p. 195–196° (dec.); no rotation data were given.

[d] Monohydrate. A preparation, with no reference to its water content, was reported by WAGNER (138, 151); m. p. 148–149°, $[\alpha]_D^{20}$ − 12.2° (c 1.4, $H_2O$).

[e] Monohydrate. The name glucoraphenin was introduced by GMELIN (38) for the parent (hypothetical) glucoside. WAGNER (151) reported the m. p. 157–159°, but no rotation data, for a tetraacetate sample containing less than 1 $H_2O$.

[f] Monohydrate.

[g] Monohydrate. Other reported data: m. p. 197–198° (151) and 187–189° (dec.) (138); no rotations were given. The anhydrous tetramethylammonium salt was described by ETTLINGER and LUNDEEN (24), m. p. 182–183° (dec.), $[\alpha]_D^{28}$ − 18.9° ($H_2O$).

[h] SCHULTZ and WAGNER (141) reported m. p. 160° (dec.), $[\alpha]_D^{23}$ − 2° (c 2.3, $H_2O$).

[i] Pentaacetate.

Table 3.  *iso*Thiocyanate Glucosides with Established Side-chains, Known at the End of 1959.

| No. | Parent glucoside | $R$ of the derived *iso*thiocyanate, $R-NCS$ | References |
|---|---|---|---|
| 1 | Glucocapparin[a, b] | $CH_3$[j] | (74, 84) |
| 2 | Glucolepidiin[d] | $CH_3CH_2$[j] | (56) |
| 3 | Glucoputranjivin[b] | $CH_3CH(CH_3)$[j] | (69, 110) |
| 4 | Glucocochlearin[b] | $(+)\text{-}CH_3CH_2\overset{*}{C}H(CH_3)$[j] | (33, 50, 86) |
| 5 | Sinigrin[a, b] | $CH_2=CHCH_2$[j] | (124, 155) |
| 6 | Gluconapin | $CH_2=CHCH_2CH_2$[j] | (20, 71) |
| 7 | Glucobrassicanapin | $CH_2=CHCH_2CH_2CH_2$[j] | (89) |
| 8 | Glucoibervirin | $CH_3SCH_2CH_2CH_2$[j] | (85) |
| 9 | Glucoiberin[a, b] | $CH_3SOCH_2CH_2CH_2$ | (75, 136) |
| 10 | Glucocheirolin | $CH_3SO_2CH_2CH_2CH_2$ | (123) |
| 11 | Glucoerucin | $CH_3SCH_2CH_2CH_2CH_2$[j] | (73) |
| 12 | (Glucoraphanin)[d] | $CH_3SOCH_2CH_2CH_2CH_2$ | (108) |
| 13 | (Glucoraphenin)[b] | $CH_3SOCH=CHCH_2CH_2$ | (119) |
| 14 | Glucoerysoline[e] | $CH_3SO_2CH_2CH_2CH_2CH_2$ | (126) |
| 15 | Glucoberteroin | $CH_3SCH_2CH_2CH_2CH_2CH_2$[j] | (91) |
| 16 | Glucoalyssin[b] | $CH_3SOCH_2CH_2CH_2CH_2CH_2$ | (75, 140) |
| 17 | Glucohirsutin | $CH_3SO(CH_2)_8$ | (66) |
| 18 | Glucoarabin | $CH_3SO(CH_2)_9$ | (77) |
| 19 | Glucocamelinin | $CH_3SO(CH_2)_{10}$ | (83) |
| 20 | Glucotropaeolin[a, b] | $C_6H_5CH_2$[j] | (34, 35) |
| 21 | Sinalbin[a] | $(p)HOC_6H_4CH_2$ | (94, 114) |
| 22 | Glucoaubrietin | $(p)CH_3OC_6H_4CH_2$[j] | (81) |
| 23 | (Glucolimnanthin)[d, l] | $(m)CH_3OC_6H_4CH_2$ | (22) |
| 24 | Gluconasturtiin[a] | $C_6H_5CH_2CH_2$[j] | (32, 35) |
| 25 | Glucoerypestrin[b] | $CH_3OOCCH_2CH_2CH_2$[k] | (78) |
| 26 | Glucomalcolmiin | $C_6H_5COOCH_2CH_2CH_2$[k] | (76, 90) |
| 27 | Progoitrin[a] (Glucorapiferin[c]) | $CH_2=CH\overset{*}{C}HOHCH_2$[f] | (42, 82, 141) |
| 28 | Glucoconringiin[a, b] | $(CH_3)_2C(OH)CH_2$[g] | (41, 82, 141) |
| 29 | Glucobarbarin | $C_6H_5\overset{*}{C}HOHCH_2$[h] | (79, 80) |
| 30 | Glucosisymbrin | $HOCH_2\overset{*}{C}H(CH_3)$[i] | (67) |

[a] Known in crystalline form, cf. Table 1, p. 157.

[b] Known as crystalline tetraacetate, cf. Table 2, p. 158.

[c] Known as crystalline pentaacetate, cf. Table 2.

[d] This name was not proposed in the original paper.

[e] The occurrence of this glucoside could not be verified in the author's laboratory.

[f] Cyclizes to (—)-5-vinyl-2-oxazolidinethione.

[g] Cyclizes to 5,5-dimethyl-2-oxazolidinethione.

[h] Cyclizes to (—)-5-phenyl-2-oxazolidinethione.

[i] Cyclizes to (+)-4-methyl-2-oxazolidinethione.

[j] Volatile with steam.

[k] Volatile with steam but may decompose.

[l] Name proposed in agreement with Dr. M. G. ETTLINGER.

Table 4. Molecular Rotations of Sulphoxide *iso*Thiocyanates of Natural Origin, Some Derivatives and Related Compounds.

| Formula | M. p. | $[M]_D{}^a$ | Solvent | Conc. g./100 ml. | References |
|---|---|---|---|---|---|
| $CH_3SO(CH_2)_3NH_2$ ............ | 27° | — 143° | EtOH | 1.0 | (55, 54) |
| $CH_3SO(CH_2)_3NCS$ (iberin) ...... | | — 120° | EtOH | | (55) |
| $CH_3SO(CH_2)_3NHCSNHC_6H_5$ .... | 137° | — 138° | EtOH | 1.9 | (75) |
| $CH_3SO(CH_2)_3NHCSNHCH_2C_6H_5$. | 108° | — 138° | EtOH | 2.0 | (75) |
| | | | | | |
| $CH_3SOCH=CH(CH_2)_2NCS$ (sulphoraphene) ........... | | — 238° | EtOH | 1.4 | (119) |
| $CH_3SOCH=CH(CH_2)_2NHCSNH_2{}^b$ | 219° | — 138° | $H_2O$ | 1.1 | (119) |
| $CH_3SOCH= =CH(CH_2)_2NHCSNHC_6H_5$ .. | 121° | — 280° | $CHCl_3$ | 1.0 | (119) |
| | | | | | |
| $CH_3SO(CH_2)_4NH_2$ ............ | | — 169° | MeOH | 0.4 | (121) |
| $CH_3SO(CH_2)_4NCS$ (sulphoraphane) ................ | | — 140° | $CHCl_3$ | 1.2 | (121) |
| | | — 117° | $CHCl_3$ | 2.1 | (108) |
| $CH_3SO(CH_2)_4NHCSNHC_6H_5$ .... | 145° | — 146°$^c$ | EtOH | 1.8 | (108) |
| | | | | | |
| $CH_3SO(CH_2)_5NCS$ (alyssin)$^d$ .... | | — 132° | $CHCl_3$ | 4.0 | (75) |
| $CH_3SO(CH_2)_5NHCSNH_2$........ | 106° | — 167° | $H_2O$ | 2.0 | (75) |
| $CH_3SO(CH_2)_5NHCSNHC_6H_5$ .... | 126° | — 176° | EtOH | 2.1 | (75) |
| $CH_3SO(CH_2)_5NHCONHC_6H_5$ .... | 102° | — 147° | EtOH | 0.8 | (75) |
| $CH_3SO(CH_2)_5NHCSNHCH_2C_6H_5$. | 104° | — 175° | EtOH | 1.2 | (75) |
| $CH_3SO(CH_2)_5NHCONHCH_2C_6H_5$. | 106° | — 182° | EtOH | 0.9 | (75) |
| $CH_3SO(CH_2)_8NHCSNH_2$........ | 88° | — 190° | EtOH | 1.0 | (66) |
| $CH_3SO(CH_2)_8NHCSNHC_6H_5$ .... | 139° | — 148° | $CHCl_3$ | 0.8 | (66) |
| $CH_3SO(CH_2)_8NHCONHC_6H_5$ .... | 121° | — 186° | EtOH | 1.3 | (66) |
| | | | | | |
| $CH_3SO(CH_2)_9NHCSNH_2$........ | 104° | — 175° | EtOH | 2.1 | (77) |
| $CH_3SO(CH_2)_9NHCSNHC_6H_5$ .... | 122° | — 200° | EtOH | 0.6 | (77) |
| $CH_3SO(CH_2)_9NHCSNHCH_2C_6H_5$. | 119° | — 226° | MeOH | 0.4 | (77) |
| $CH_3SO(CH_2)_9NHCONHCH_2C_6H_5$. | 121° | — 190° | EtOH | 1.0 | (77) |
| | | | | | |
| $CH_3SO(CH_2)_{10}NHCSNH_2$ ....... | 92° | — 181° | EtOH | 1.0 | (83) |
| $CH_3SO(CH_2)_{10}NHCSNHC_6H_5$ ... | 139° | — 163° | EtOH | 0.4 | (83) |
| $CH_3SO(CH_2)_{10}NHCONHCH_2C_6H_5$ | 114° | — 190° | EtOH | 1.2 | (83) |

$^a$ All rotations reported were determined at 15–25°.

$^b$ The author found that the reported compound (119) can hardly be authentic sulphoraphenethiourea.

$^c$ The value reported is a result of a microdetermination carried out in the author's laboratory with a specimen of natural provenance, kindly furnished by Dr. Ž. Procházka.

$^d$ Non-analyzed specimen.

Table 5. The Occurrence of *iso*Thiocyanates or Their Parent Glucosides
in Species of the Family *Cruciferae*.

R, root material; G, green parts; S, seeds. **Bold** figures in the column "Compounds"
refer to the individual mustard oils listed in Table 3, p. 159; the other figures in
the same column indicate the number of unspecified mustard oils observed. I and
PC in the column "Evidence" denote isolation and paper chromatography,
respectively.

| Plant | Part | Compounds | Evidence | References |
|-------|------|-----------|----------|-----------|
| *Aethionema cordifolia* DC. ... | G | **1** | PC | (*38, 133*) |
| *A. pulchella* ................ | S | 2 | PC | (*38, 133*) |
| *A. saxatile* ................ | S | 2 | PC | (*139, 151*) |
| | | | | |
| *Alliaria officinalis* ANDRZ. ... | R | **5** | I | (*152*) |
| | | **5, 20, (24)** | PC | (*14*) |
| | G | **5** | I, PC | (*38, 151*) |
| | | **5, 20** | PC | (*14*) |
| | S | **5** | I, PC | (*14, 38, 152*) |
| | | | | |
| *Alyssum alpestre* ........... | S | **1** | PC | (*38*) |
| *A. Arduini* ................ | S | 2 | PC | (*38, 133*) |
| *A. arenarium*.............. | S | **1**, 2 | PC | (*38, 133*) |
| *A. argenteum* .............. | S | **1** | PC | (*38*) |
| *A. argenteum* ALL. ......... | S | 2, **15, 16** | PC, I | (*139, 140, 151*) |
| *A. argenteum* (ALL.) VITM.... | S | **16, (15)** | I | (*75*) |
| *A. Benthami* .............. | S | **5** | PC | (*139, 151*) |
| *A. Bornmuelleri* HAUSKNECHT | S | **1, 6** | PC | (*38, 75, 133*) |
| | G | **1** | PC | (*139, 151*) |
| *A. Borzaeanum* NYÁR. ...... | S | **15, 16** | PC | (*75*) |
| *A. calycinum* L............ | S | **1** | PC | (*38*) |
| | G | **1** | PC | (*139, 151*) |
| *A. corymbosum* BOISS........ | S | **3, 6, 15, (16)** | PC | (*38, 75, 133*) |
| *A. maritimum* LAM......... | S | **16** | PC | (*75*) |
| *A. montanum* L............ | S | **1, 6, 15, 16** | PC | (*38, 75*) |
| *A. orientale* ARD. .......... | S | **6, 7, 16** | PC | (*75*) |
| *A. ovirense* A. KERN. ....... | S | **15, 16** | PC | (*75*) |
| *A. pedemontanum* .......... | S | 2 | PC | (*38*) |
| *A. rostratum* STEV. ......... | S | **1** | PC | (*139, 151*) |
| *A. saxatile* L. .............. | S | **3** | PC | (*139, 151*) |
| | S | **6, 7, 15, 16** | PC | (*75*) |
| | G | **3** | PC | (*14*) |
| *A. saxatile* L. var. *citrinum* . | S | **6, 15, 16** | PC | (*75*) |
| *A. sinuatum* L.............. | S | **7, 15, 16** | PC | (*75*) |
| *A. Wulfenianum* ........... | S | **1**, 2 | PC | (*38, 133*) |
| | | | | |
| *Arabis albida* .............. | S | **1** | PC | (*38*) |
| *A. alpestris* ............... | S | **3** | PC | (*38, 133*) |
| *A. alpina* L. .............. | S | **1**, 2 | PC | (*38, 139, 151*) |
| | S | **18** | I | (*77*) |
| | G | **3** | PC | (*139, 151*) |

(Table 5, continued.)

| Plant | Part | Compounds | Evidence | References |
|---|---|---|---|---|
| Arabis aubrietoides BOISS. ... | S | 1 | PC | (38) |
| | G | 4 | PC | (139, 151) |
| A. bellidifolia JACQ.......... | S | 2 | PC | (38, 133) |
| | G | 6 | PC | (139, 151) |
| A. colinsii FERNALD ....... | S | 6 | PC | (87) |
| A. corymbiflora VEST....... | G | 2 | PC | (139, 151) |
| A. hirsuta SCOP............. | G | 2 | PC | (139, 151) |
| | S | 17 | I | (66) |
| A. holboellii HORNEM........ | G | 5 | PC | (139, 151) |
| A. procurrens WALDST. et KIT. | S | 3 | PC | (38, 133) |
| | G | 2 | PC | (139, 151) |
| A. pumila JACQ............. | G | 4 | PC | (139, 151) |
| A. retrofracta (GRAHAM) RYD- | | | | |
| BERG ................... | S | 5 | PC | (87) |
| A. rochinensis .............. | S | 3 | PC | (38, 133) |
| A. rosea DC................ | S | 3, 4 | PC | (38, 133) |
| | G | 1 | PC | (139, 151) |
| A. soyeri RENT. ............ | S | 1 | PC | (38, 133) |
| Armoracia lapathifolia GILIB.. | R | 5 | I | (146, 152) |
| | | 5, 24 | I, PC | (14, 47, 144) |
| | G | 5, 24 | PC | (14) |
| Aubrietia columnifera GUSSM. | G | 3, 4, 22 | PC | (81, 139, 151) |
| A. deltoidea DC............. | G | 22 | I | (81) |
| | S | 21, 22 | PC | (81) |
| A. erubescens GRISEB....... | S | 21, 22 | PC | (81) |
| A. hybrida hort............. | G | 22 | PC | (81) |
| A. intermedia HELDR. et ORPH. | S | 21, 22 | PC | (81) |
| Barbaraea arcuata (OPIZ.) | | | | |
| REICHB. ................ | S | 1 (?), 24 | PC | (72) |
| B. intermedia BOR. ......... | S | 24 | PC | (72) |
| B. praecox R. BR. .......... | S | 24 | I | (32, 35) |
| B. stricta FRIES ............ | G | 1 | PC | (139, 151) |
| | S | 1 | PC | (139, 151) |
| B. vulgaris R. BR........... | G | 1 | PC | (14, 38, 131) |
| | R | 5, 24 | PC | (14) |
| | S | 24, 29 | I | (79) |
| Berteroa incana (L.) DC. .... | G | 2, 3, 4 | PC | (15, 38, 139, 151) |
| | S | 2, 3 | PC | (38, 139, 151) |
| | S | 15, 16 | I | (91) |
| | R | 1 | PC | (15) |
| Biscutella auriculata L....... | G | 2 | PC | (139, 151) |
| | S | 2 | PC | (38, 133) |
| B. laevigata L. ............. | G | 2 | PC | (139, 151) |
| | S | 1, 2 | PC | (38, 133, 139, 151) |

*(Table 5, continued.)*

| Plant | Part | Compounds | Evidence | References |
|---|---|---|---|---|
| *Brassica cernua* THUNB. . . . . . | S | ? | | *(152)* |
| *B. dichotoma* ROXB. . . . . . . . | S | ? | | *(152)* |
| *B. glauca* ROXB. . . . . . . . . . | S | ? | | *(152)* |
| *B. integrifolia* O. E. SCHULZ . | S | ? | | *(152)* |
| *B. juncea* CZERN. et COSS. . . . | S | 5 | I | *(118)* |
| | R | 24 | PC | *(15)* |
| | G | 5, 6, 7, 24 | PC | *(15)* |
| *B. napus* L. . . . . . . . . . . . . . . | S | 6, 7, 9 (or 12), 21, 24, 27 | I, PC | *(89)* |
| | R | 27 | I | *(4)* |
| *B. nigra* KOCH . . . . . . . . . . . . | R | 5, 24 | PC, I | *(14, 144)* |
| | G | 5 (24) | I | *(14)* |
| | S | 5 (24) | I | *(152)* |
| *B. oleracea* L. var. *capitata* L. | R | 24 | I | *(144)* |
| *B. oleracea* var. *hort.* . . . . . . . | S | 5, 6, 7, 24 | PC | *(13, 53)* |
| | S | 5, 27 | I | *(4, 127)* |
| | G | 5, 6 | PC | *(53)* |
| | G | (5, 6, 8), 27 | I | *(1, 100)* |
| | G | 9, 12 | I | *(109)* |
| *B. pseudojuncea* . . . . . . . . . . . | S | 5, 6 | PC | *(53)* |
| *B. rapa* L. . . . . . . . . . . . . . . . | S | (6, 7, 24), 27 | PC, I | *(4, 13)* |
| | R | 27, 24 | I | *(4, 144)* |
| *Bunias erucago* L. . . . . . . . . . | S | 2, 3 | PC | *(38, 133, 139, 151)* |
| *B. orientalis* L. . . . . . . . . . . . | S | 2, 3 | PC | *(38, 133, 139, 151)* |
| | G | 2 | PC | *(38, 131)* |
| | R | 1, 2 | PC | *(15, 131)* |
| *Cakile maritima* SCOP. . . . . . . . | S | 5 | PC | *(72)* |
| | S | 7 | PC | *(14)* |
| *Camelina dentata* (WILLD.) PERS. . . . . . . . . . . . . . . . . . . | S | 19 | PC | *(83)* |
| *C. microcarpa* ANDRZ. . . . . . . . | S | 19 | PC | *(83)* |
| *C. sativa* CRANTZ . . . . . . . . . . | S | 19 | I | *(83)* |
| *Capsella bursa pastoris* (L.) MEDIC. . . . . . . . . . . . . . . . | S | 5 | I | *(152)* |
| *Cardamine amara* L. . . . . . . . . | S | 4 | PC | *(38, 131)* |
| | G | 4 | I | *(152)* |
| *C. flexuosa* WITH. . . . . . . . . . | G | 1 | PC | *(139, 151)* |
| *C. graeca* L. . . . . . . . . . . . . . . | S | 6, 20 | PC | *(72)* |
| *C. hirsuta* L. . . . . . . . . . . . . . | G | ? | I | *(152)* |
| *C. pratensis* L. . . . . . . . . . . . . | R | 2 or 3 | PC | *(38)* |
| | G | 4, 2 | I, PC | *(38, 152)* |
| | S | 4 | PC | *(38, 131)* |

*(Table 5, continued.)*

| Plant | Part | Compounds | Evidence | References |
|---|---|---|---|---|
| *Cheiranthus Allioni* hort. . . . . | S | 3 | PC | *(139, 151)* |
| *C. cheiri* L. . . . . . . . . . . . . . . | S | 10 | I | *(123)* |
| *C. Kewensis* . . . . . . . . . . . . . . | S | 2 | PC | *(139, 151)* |
| *Cochlearia anglica* (L.) Asch. et Grb. . . . . . . . . . . . . . . . . | G | 3 | I | *(69)* |
| | S | 3, 4, 28 | PC | *(82)* |
| *C. danica* L. . . . . . . . . . . . . . | S | 3, 4, 28 | PC | *(82)* |
| *C. officinalis* L. . . . . . . . . . . | S | 3, 4, 28 | I | *(82)* |
| *Conringia orientalis* (L.) Dumort . . . . . . . . . . . . . . . | S | 28 | I | *(82, 141)* |
| *Coronopus didymus* (L.) Sm. . | G | 20 | I | *(101)* |
| *Crambe maritima* L. . . . . . . . . . | S | 5, 2 | PC | *(38, 72)* |
| | G | 1 | PC | *(139, 151)* |
| *Dentaria digitata* Lamarck. . . | G | 1 | PC | *(139, 151)* |
| *D. enneaphylla* L. . . . . . . . . . . | G | 3 | PC | *(139, 151)* |
| *Diplotaxis erucoides* (L.) DC.. | S | 5 | PC | *(38, 133)* |
| *D. muralis* (L.) DC. . . . . . . . . | R | 5 | PC | *(38)* |
| | G | 5 | PC | *(38, 131)* |
| | S | 5 | PC | *(72, 131)* |
| *D. tenuifolia* (L.) DC. . . . . . . | R, G | 11 | I | *(17)* |
| *Draba aizoides* . . . . . . . . . . . . . | S | 2 | PC | *(38, 133)* |
| *D. aizoon* Wallenb. . . . . . . . . | S | 1 | PC | *(139, 151)* |
| *D. borealis* DC. . . . . . . . . . . . . | S | 4, 20 (?) | PC | *(72)* |
| *D. Haynaldi* Stur. . . . . . . . . | S | 1 | PC | *(139, 151)* |
| *D. incana* L. . . . . . . . . . . . . . . | S | (5, 6), 2 | PC | *(38, 72, 133)* |
| | G | 1 | PC | *(139, 151)* |
| *D. norwegica* Gunn. . . . . . . . . . | S | 3 | PC | *(38, 133)* |
| | G | 2 | PC | *(139, 151)* |
| *D. pyrenaica* . . . . . . . . . . . . . . | S | 3 | PC | *(38, 133)* |
| *D. repens* . . . . . . . . . . . . . . . . . | S | 2 | PC | *(38, 133)* |
| *Eruca sativa* Mill. . . . . . . . . . . | G | 11 | PC | *(15)* |
| | S | 11, 2 | I, PC | *(38, 73)* |
| *E. sativa* Lmk. . . . . . . . . . . . . . | S | 2 | PC | *(139, 151)* |
| | G | 2 | PC | *(139, 151)* |
| *Erucastrum gallicum* (Willd.) O. E. Schulz . . . . . . . . . . . | S | 5, 6 | PC | *(72)* |
| *E. Pollichii* Sch. et Sp. . . . . . | G | 1, 24 | PC | *(14, 139, 151)* |
| *Erysimum alpinum* . . . . . . . . . | S | 1 | PC | *(38)* |
| *E. arcansanum* Nutt. . . . . . . . | S | 10 | PC, I | *(123, 131)* |

*(Table 5, continued.)*

| Plant | Part | Compounds | Evidence | References |
|---|---|---|---|---|
| *Erysimum cheiranthoides* L. ... | S | 5 | PC | *(72)* |
| | G | 4 | PC | *(139, 151)* |
| *E. helveticum* .............. | S | 2 | PC | *(38, 133)* |
| *E. nanum* Boiss. et Hohen.. | S | 10 | PC | *(131)* |
| *E. ochroleucum* DC. ........ | S | 10, 25 | PC | *(78)* |
| *E. pachycarpum* hort. fil. et Thoms. ................ | S | 2, 5 | PC | *(38, 133, 139, 151)* |
| | G | 2 | PC | *(139, 151)* |
| *E. Perofskianum* Fisch. et May.................... | S | 1 (?), 5 | PC | *(72, 139, 151)* |
| | S | 14 | I | *(126)* |
| *E. pumilum* DC............. | S | 1, 3, (10, 25) | PC | *(38, 78, 133, 139, 151)* |
| | G | 2 | PC | *(139, 151)* |
| *E. rupestre* DC.............. | S | 3, (10, 25) | PC, I | *(78, 139, 151)* |
| *E. strictissimum* L. ........ | S | 3 | PC | *(139, 151)* |
| *E. vincoleucum* ............ | S | 2 | PC | *(38, 133)* |
| *Eutrema wasabi* Maxim. ...... | R, G | 4, 5 | I | *(103)* |
| *Farsetia clypeata* ........... | R | 4 or 5, 11 | PC | *(38, 73)* |
| | G | 1 | PC | *(38)* |
| | S | 1 | PC | *(38)* |
| *Hesperis matronalis* L. ...... | G | 1–2, 5, 3 | PC | *(38, 139, 151)* |
| | S | 3–4, 5, 4, 11 | PC | *(38, 73, 139, 142, 151)* |
| *Hutchinsia alpina* R. Br. .... | G | 2 | PC | *(139, 151)* |
| | S | 24 | PC | *(72)* |
| *Iberis amara* L. ............ | R | 9, 2 | PC | *(15, 38)* |
| | G | 9 | PC | *(38)* |
| | S | 9 | I | *(136)* |
| | G | 9 | PC | *(38)* |
| *I. sempervirens* L. .......... | S | 8, 11 | I | *(85)* |
| | G | 5 | PC | *(14, 15)* |
| *I. umbellata* Dunnetti L. ... | S | 2 | PC | *(139, 151)* |
| *Isatis tinctoria* L............ | S | 2, 6 | PC | *(38, 72, 133, 139, 151)* |
| | G | 1 | PC | *(139, 151)* |
| *Kernera saxatilis* Medicus ... | S | 2 | PC | *(38, 133)* |
| | G | 3, 4 | PC | *(139, 151)* |
| *Lepidium campestre* R. Br. .. | S | 21, 2 | PC | *(38, 133, 139, 151)* |
| | G | 21 | PC | *(38)* |
| | R | 2 | PC | *(15)* |
| *L. densiflorum* Schrad. ..... | S | 20 | PC | *(72)* |
| *L. Draba* L................ | G | 3, 12 | PC | *(38, 108, 131, 139, 151)* |

(Table 5, continued.)

| Plant | Part | Compounds | Evidence | References |
|---|---|---|---|---|
| *Lepidium graminifolium* L. ... | R, G | 20 | PC | (*15*) |
| L. *latifolium* L.............. | S | 2 | PC | (*38, 133*) |
| | G | 2, 8 | PC | (*14, 38, 133*) |
| L. *Menziesii* DC. ........... | S | 2 | I | (*56*) |
| L. *ruderale* L. .............. | S | 20 | PC, I | (*38, 40, 133*) |
| | G | 20 | PC | (*15, 38, 133*) |
| | R | 20 | PC | (*15*) |
| L. *sativum* L. .............. | S | 20 | I | (*35, 40*) |
| | G | 20 | PC | (*38*) |
| L. *virginicum* L............. | S | 20 | PC | (*72*) |
| *Lunaria annua* L. .......... | S | 3, 4, 3 | PC | (*14, 72*) |
| L. *biennis* MNCH. (= *L. annua* L.) .................... | S | 3, 15 | I, PC | (*69, 91*) |
| L. *rediviva* L. .............. | S | 15 | PC | (*91*) |
| *Malcolmia maritima* R. BR. ... | S | 10, 16 | I | (*76, 90*) |
| *Matthiola annua* R. BR. ..... | S | 1, 2, **3**, 11 | PC | (*38, 69, 73, 139, 151*) |
| | G | 2, 24 | PC | (*14, 38*) |
| M. *bicornis* DC. ............ | S | 13 | I | (*138*) |
| M. *fenestralis* (L.) R. BR. ... | S | 1 (?) | PC | (*72*) |
| M. *incana* R. BR. .......... | S | 1 | PC | (*139, 151*) |
| *Nasturtium officinale* R. BR. . | S | 2, 24 | PC | (*38, 72, 133*) |
| | G | 2, 24 | PC, I | (*14, 35, 38, 131*) |
| *Parrya menziesii*............ | S | 1, 3 | PC | (*38, 133, 139, 151*) |
| *Raphanus raphanistrum* L.... | G | 3 | PC | (*38, 152*) |
| R. *sativus* L. var. *alba* ...... | R | 13 | I | (*119*) |
| | S | 1–2, 1 (?), 3 | PC | (*38, 72, 133*) |
| R. *sativus* L. var. *nigra* ..... | R | 3 | PC | (*139, 151*) |
| | S | 2 | PC | (*139, 151*) |
| R. *sativus* L. var. *radicula* PERS. | S | 5 (trace) | PC | (*72*) |
| *Rapistrum perenne* ALL. ..... | G | 1, 6 | PC | (*72, 139, 151*) |
| R. *rugosum* (L.) ALL. ....... | G, S | 10 | I | (*4 a*) |
| *Rorippa amphibia* (L.) BESS. | S | 2 | PC | (*38*) |
| R. *silvestris* (L.) BESS. ...... | R | 2 | PC | (*15*) |
| *Schiverekia Dörfleri* ......... | G | 2 | PC | (*139, 151*) |
| *Sinapis alba* L.............. | R | 24 | I | (*144*) |
| | S | 21 | I | (*94, 114*) |
| S. *arvensis* L. .............. | S | 3, 2, 5 | PC, I | (*38, 133, 152*) |
| | G | ? | | (*152*) |

(*Table 5, continued.*)

| Plant | Part | Compounds | Evidence | References |
|-------|------|-----------|----------|------------|
| *Sisymbrium austriacum* JACQ. | S | 3, 30 | PC, I | (*67*) |
| *S. cheiranthoides* ET. et W... | S | ? | I | (*99*) |
| *S. Loeselii* L. ............. | G | 1 | PC | (*38*) |
| *S. officinalis* (L.) SCOP. ..... | S | (?), 1 | I, PC | (*38, 99, 131*) |
| | G | 1 | PC | (*38, 131*) |
| *S. sophia* L. ............... | S | 5 | PC | (*72*) |
| | G | 2 | PC | (*15*) |
| *S. strictissimum* L........... | S | 2, (3, 4) | PC | (*38, 72, 133*) |
| *Thlaspi alpestre* L. .......... | S | 1, 2 | PC | (*38, 133, 139, 151*) |
| *T. arvense* L................ | S | 1, 5 | PC | (*38, 40, 72, 133, 152*) |
| *T. perfoliatum* L........... | S | 3 | PC | (*139, 151*) |
| | R, G | 2 | PC | (*15*) |
| *T. rotundifolium* GAUDIN .... | G | 1 | PC | (*139, 151*) |
| *Vesicaria graeca* RENT....... | G | 3 | PC | (*139, 151*) |
| *V. sinuata* POIR. ........... | S | 11 | PC | (*73*) |
| *V. utriculata* L. ............ | S | 11 | PC | (*73*) |
| *Vogelia paniculata* HORNEM. ... | G | 1 | PC | (*139, 151*) |

Table 6. The Occurrence of *iso*Thiocyanates or Their Parent Glucosides in Families Other than *Cruciferae*.

The symbols are those used in Table 5, p. 161.

| Plant | Part | Compounds | Evidence | References |
|-------|------|-----------|----------|------------|
| 1. *Capparidaceae* | | | | |
| *Capparis spinosa* L. ..... | S | 1 | I | (*63, 152*) |
| | G | 1 | PC | (*15*) |
| *Cleome arabica* L. ....... | S | 1 + unknown | PC | (*72, 84*) |
| *C. arborea* BSS. ......... | S | 1 + unknown | PC | (*84*) |
| *C. gigantea* L. .......... | S | 1 + unknown | PC | (*84*) |
| *C. graveolens* RAFIN...... | S | 1 + unknown | PC | (*84*) |
| *C. monophylla* L. ....... | S | 1 + unknown | PC | (*84*) |
| *C. speciosissima* DEPPE... | S | 1 + unknown | PC | (*84*) |
| *C. spinosa* JACQ......... | S | 2 (1 + unknown) | PC, I | (*84, 139, 151*) |
| *C. viscosa* L. ........... | S | 1 + unknown | PC | (*84, 152*) |
| *C. trachysperma* (TORR. et GRAY) PAX et K. HOFFM. ......... | S | 1 + unknown | PC· | (*84*) |
| *Gynandropsis gynandra* (L.) BRIQ............ | S | 1 | PC | (*72, 84*) |
| *G. pentaphylla* DC. ....... | S | unknown | I | (*152*) |

(*Table 6, continued.*)

| Plant | Part | Compounds | Evidence | References |
|---|---|---|---|---|
| *2. Caricaceae* | | | | |
| Carica papaya L. . . . . . . . | S | **20** | I | (*21*) |
| *3. Euphorbiaceae* | | | | |
| Jatropha multifida L. . . . . | Latex | **20** (?) | I (?) | (*29*) |
| Putranjiva Roxburghii | | | | |
| WALL. . . . . . . . . . . . . . | S | **3, 4**, unknown | I | (*110*) |
| *4. Limnanthaceae* | | | | |
| Limnanthes douglasii R. | | | | |
| BR. . . . . . . . . . . . . . . . | S | **23** | I | (*22*) |
| *5. Moringaceae* | | | | |
| Moringa pterygosperma | | | | |
| GAERTN. . . . . . . . . . . | R | "**20**" | I | (*98*) |
| *6. Phytolaccaceae* | | | | |
| Codonocarpus cotinifolius | | | | |
| (DESF.) . . . . . . . . . . . . . | G | **4, (20?)** | I | (*6*) |
| *7. Plantaginaceae* | | | | |
| Plantago major L. (?) . . . | G | **13** | PC | (*107*) |
| *8. Resedaceae* | | | | |
| Reseda alba L. . . . . . . . . . | R | **24** | PC | (*15*) |
| R. lutea L. . . . . . . . . . . . . | R | (?), **20** | PC | (*14, 152*) |
| | G | (?) | I | (*152*) |
| R. luteola L. . . . . . . . . . . . | R, G, S | (?) | | (*124*) |
| | R | **2, 24** | PC | (*14, 38*) |
| | R, G | **1** | PC | (*131*) |
| | G, S | **29** | I, PC | (*80*) |
| R. odorata L. . . . . . . . . . . | R | **24** | I | (*5*) |
| | S | **1, 2, 1** (?) | PC | (*72, 139, 151*) |
| *9. Salvadoraceae* | | | | |
| Salvadora oleoides DEN. . . | S | **20** | I | (*105*) |
| *10. Tropaeolaceae* | | | | |
| Tropaeolum majus L. . . . . | S | **20** | I | (*34*) |
| | G | **20** | I | (*34*) |
| T. peregrinum (canariense) | S | **3, 4** | PC | (*72*) |

### References.

*1.* ALTAMURA, M. R., L. LONG, Jr. and T. HASSELSTROM: Goitrin from Fresh Cabbage. J. Biol. Chem. **234**, 1847 (1959).

*2.* ANDRÉ, É. et P. DELAVEAU: Recherches chimiques sur la composition des essences sulfurées des graines de colza. Oléagineux **9**, 591 (1954).

*3.* ANDRÉ, É. et M. MAILLE: Contribution à l'étude des essences sulfurées de graines de Crucifères: IV. Dosage de l'essence fournie par les graines de *Brassica Juncea* CZERNY et COSSON. Ann. Agron. (Paris) **2**, 442 (1951).

*4.* ASTWOOD, E. B., M. A. GREER and M. G. ETTLINGER: *l*-5-Vinyl-2-thio-oxazolidone, an Antithyroid Compound from Yellow Turnip and from *Brassica* Seeds. J. Biol. Chem. **181**, 121 (1949).

*4a.* BACHELARD, H. S. and V. M. TRIKOJUS: Plant Thioglycosides and the Problem of Endemic Goitre in Australia. Nature (London) **185**, 80 (1960).

*5.* BERTRAM, J. und H. WALBAUM: Über das Resedawurzelöl. J. prakt. Chem. [2] **50**, 555 (1894).

*6.* BOTTOMLEY, W. and D. E. WHITE: The Chemistry of Western Australian Plants. II. The Essential Oil of *Codonocarpus cotinifolius* (DESF.). Roy. Australian Chem. Inst. J. Proc. **17**, 31 (1950) [Chem. Abstr. **45**, 820 (1951)].

*7.* BRAECKE, M.: Die Verwendung der hydrolysierenden Fermente zur Auffindung und zum Studium von Glucosiden. J. pharm. Belgique **10**, 463 (1928).

*8.* BUSSY, A.: Sur la formation de l'huile essentielle de moutarde. J. Pharmac. Chim. **26**, 39 (1840).

*9.* CAMBI, L.: Su gli acidi tiodrossamici. Atti reale accad. Lincei, Rend. classe sci. fis., mat. e nat. [5] **18**, I, 687 (1909).

*10.* CHALLENGER, F.: The Natural Mustard Oil Glucosides and their Related *iso*Thiocyanates and Nitriles. In: Aspects of the Organic Chemistry of Sulphur, p. 115. London: Butterworths. 1959.

*11.* DAS, B. R., P. A. KURUP and P. L. N. RAO: Antibiotic Principle from *Moringa Pterygosperma*. Part VII. Antibacterial Activity and Chemical Structure of Compounds Related to Pterygospermin. Ind. J. med. Res. **45**, 191 (1957).

*12.* DELAVEAU, P.: Recherches sur les sénevols des graines de *Brassica*, à l'aide de la chromatographie sur papier. I. Graines de *B. nigra* et *B. juncea*. Ann. pharm. franç. **14**, 765 (1956).

*13.* — Recherches sur les sénevols des graines de *Brassica*, à l'aide de la chromatographie sur papier. II. Graines de *B. Rapa*, *B. Napus* et *B. oleracea*. Ann. pharm. franç. **14**, 770 (1956).

*14.* — Sur la multiplicité des hétérosides à sénevol et leur relation avec la physiologie et la taxinomie des Crucifères. Bull. soc. bot. France **104**, 148 (1957).

*15.* — Multiplicité des hétérosides à sénevol chez les Crucifères; leur relation avec la physiologie et la taxinomie (2e note). Bull. soc. bot. France **105**, 224 (1958).

*16.* — Variations de la teneur en hétérosides à sénevol de l'*Alliaria officinalis* L. au cours de la végétation. C. R. hebd. Séances Acad. Sci. **246**, 1903 (1958).

*17.* DELAVEAU, P. et R. PARIS: Sur la composition chimique de l'essence de *Diplotaxis tenuifolia* (L.) DC. Ann. pharm. franç. **16**, 81 (1958).

*18.* ETTLINGER, M. G.: Infrared Spectra and Tautomerism of 2-Thiooxazolidone and Congeners. J. Amer. Chem. Soc. **72**, 4699 (1950).

*19.* — Synthesis of the Natural Antithyroid Factor *l*-5-Vinyl-2-thiooxazolidone. J. Amer. Chem. Soc. **72**, 4792 (1950).

*20.* ETTLINGER, M. G. and J. E. HODGKINS: The Mustard Oil of Rape Seed, Allyl-carbinyl Isothiocyanate, and Synthetic Isomers. J. Amer. Chem. Soc. **77**, 1831 (1955).

*21.* ETTLINGER, M. G. and J. E. HODGKINS: The Mustard Oil of Papaya Seed. J. Organ. Chem. (USA) **21**, 204 (1956).

*22.* ETTLINGER, M. G. and A. J. LUNDEEN: The Mustard Oil of *Limnanthes douglasii* Seed, *m*-Methoxybenzyl Isothiocyanate. J. Amer. Chem. Soc. **78**, 1952 (1956).

*23.* — — The Structures of Sinigrin and Sinalbin; an Enzymatic Rearrangement. J. Amer. Chem. Soc. **78**, 4172 (1956).

*24.* — — First Synthesis of a Mustard Oil Glucoside; the Enzymatic Lossen Rearrangement. J. Amer. Chem. Soc. **79**, 1764 (1957).

*25.* EULER, H. v. und S. E. ERIKSON: Zur Kenntnis der enzymatischen Spaltung des Sinigrins. Fermentforsch. **8**, 518 (1926).

*26.* FABIAN, J. et M. LEGRAND: Sur l'intensité d'absorption infrarouge du groupe imine. Bull. soc. chim. France **1956**, 1461.

*27.* FIŞEL, S., F. MODREANU şi A. CARPOV: Chromatografia pe hîrtie a isotiocianaţilor. Nota 1. Studii şi Cercetări Sti., Chim. (Fil. Iasi) **7**, 19 (1956).

*28.* — — — Chromatografia pe hîrtie a isotiocianaţilor. Nota 2. Studii şi Cercetări Sti., Chim. (Fil. Iasi) **8**, 277 (1957).

*28 a.* FORSS, D. A.: An Investigation of the Relation of the Essential Oils of *Coronopus didymus* to the Tainting of Butter. Austral. J. Appl. Sci. **2**, 396 (1951).

*29.* FREISE, F. W.: Ätherische Öle von brasilianischen Euphorbiaceen. Perfumery Essent. Oil Record **26**, 219 (1935) [Chem. Zbl. **1935** II, 2748].

*30.* GADAMER, J.: Über die Bestandteile der schwarzen und des weißen Senf-samens. Arch. Pharmaz. **235**, 44 (1897). — Über das Sinigrin. Ber. dtsch. chem. Ges. **30**, 2322 (1897). — Über das Sinalbin. Ber. dtsch. chem. Ges. **30**, 2327 (1897).

*31.* — Über Sinapinsäure. Ber. dtsch. chem. Ges. **30**, 2330 (1897).

*32.* — Über die ätherischen Öle und Glucoside einiger Kressenarten. Ber. dtsch. chem. Ges. **32**, 2335 (1899).

*33.* — Über das ätherische Öl von *Cochlearia officinalis*. Arch. Pharmaz. **237**, 92 (1899).

*34.* — Das ätherische Öl von *Tropaeolum majus*. Arch. Pharmaz. **237**, 111 (1899).

*35.* — Über ätherische Kressenöle und die ihnen zu Grunde liegenden Glukoside. Arch. Pharmaz. **237**, 507 (1899).

*36.* GILDEMEISTER, E. und F. HOFFMANN: Die ätherischen Öle. 3. Aufl. Bd. I, S. 145. Leipzig: Schimmel und Co. 1927.

*37.* — — Die ätherischen Öle. 4. Aufl. Bd. V, S. 151. Berlin: Akademie-Verlag. 1959.

*38.* GMELIN, R.: Präparative und analytische Versuche über Senfölglucoside. Dissert., Univ. Tübingen, 1954.

*39.* GMELIN, R. and A. I. VIRTANEN: Formation of Esters of Normal Thiocyanic Acid from Sulphur-containing Glucosides in Some *Cruciferae* Plants. Suomen Kemistilehti B **32**, 236 (1959).

*40.* — — A New Type of Enzymatic Cleavage of Mustard Oil Glucosides. Formation of Allylthiocyanate in *Thlaspi arvense* L. and Benzylthiocyanate in *Lepidium ruderale* L. and *Lepidium sativum* L. Acta Chem. Scand. **13**, 1474 (1959).

*41.* — — Preparation and Properties of Glucoconringiin, the Precursor of the Thyreostatic 5,5-Dimethyl-2-oxazolidinethione. Acta Chem. Scand. **13**, 1718 (1959).

*41 a.* GOODMAN, I., J. R. FOUTS, E. BRESNICK, R. MENEGAS and G. H. HITCHINGS: A Mammalian Thioglycosidase. Science (Washington) **130**, 450 (1959).

*42.* GREER, M. A.: Isolation from Rutabaga Seed of Progoitrin, the Precursor of the Naturally Occurring Antithyroid Compound, Goitrin (*L*-5-Vinyl-2-thioöxazolidone). J. Amer. Chem. Soc. **78**, 1260 (1956).

*43.* GREER, M. A., M. G. ETTLINGER and E. B. ASTWOOD: Dietary Factors in the Pathogenesis of Simple Goiter. J. Clin. Endocrinology **9**, 1069 (1949).

*44.* GUIGNARD, L.: Recherches sur la localisation des principes actifs des Crucifères. J. botanique **4**, 385 (1890).

*45.* — Recherches sur la localisation des principes actifs chez les Capparidées, Tropéolées, Limnanthées, Résédacées. J. botanique **7**, 345 (1893).

*46.* HEGNAUER, R.: Recente Onderzoekingen over de Zwavelverbindingen der Cruciferen. Pharmac. Weekbl. **90**, 577 (1955).

*47.* HEIDUSCHKA, A. und A. ZWERGAL: Beiträge zur Kenntnis der Geschmacksstoffe von Meerrettich und Rettich. J. prakt. Chem. [2] **132**, 201 (1931).

*48.* HINE, R. and D. ROGERS: The Crystal and Molecular Structure of (+)-*S*-Methyl-*L*-cysteine *S*-Oxide; a Standard of Absolute Configuration for Asymmetric Sulphur. Chem. and Ind. **1956**, 1428.

*49.* HOFMANN, A. W.: Über die dem Senföl entsprechenden Isomeren der Schwefelcyanwasserstoffäther. Ber. dtsch. chem. Ges. **1**, 169 (1868).

*50.* — Synthese des ätherischen Öls der *Cochlearia officinalis*. Ber. dtsch. chem. Ges. **7**, 508 (1874).

*51.* — Über Erkennung und Bestimmung kleiner Mengen von Schwefelkohlenstoff. Ber. dtsch. chem. Ges. **13**, 1732 (1880).

*52.* HOPKINS, C. Y.: A Sulphur-containing Substance from the Seed of *Conringia orientalis*. Canad. J. Res. **16**, 341 (1938).

*53.* JENSEN, K. A., J. CONTI and A. KJÆR: *iso*Thiocyanates. II. Volatile *iso*-Thiocyanates in Seeds and Roots of Various *Brassicae*. Acta Chem. Scand. **7**, 1267 (1953).

*54.* KARRER, P., N. J. ANTIA und R. SCHWYZER: Über zwei einfache, optisch aktive Sulfoxyde. Helv. Chim. Acta **34**, 1392 (1951).

*55.* KARRER, P., E. SCHEITLIN und H. SIEGRIST: Über Homologe des Sulphoraphans und über ω-Aminoalkyl-sulfoxyde. Helv. Chim. Acta **33**, 1237 (1950).

*56.* KJÆR, A.: *iso*Thiocyanates. IX. The Occurrence of Ethyl *iso*Thiocyanate in Nature. Acta Chem. Scand. **8**, 699 (1954).

*57.* — Naturligt Forekommende *iso*Thiocyanater (Sennepsolier). Dansk Tidsskr. Farmaci **30**, 117 (1956).

*58.* — Secondary Organic Sulfur Compounds of Plants. In: P. SCHWARZE, Handbuch der Pflanzenphysiologie, Bd. IX, S. 64. Berlin-Göttingen-Heidelberg: Springer-Verlag. 1958.

*59.* — Naturally Occurring *iso*Thiocyanates and their Possible Relationship with α-Amino Acids. Acta Chem. Scand. **8**, 1110 (1954).

*60.* — Naturally Occurring *iso*Thiocyanates and their Parent Glucosides. Suomen Kemistilehti A **32**, 53 (1959).

*61.* — *iso*Thiocyanates. XXXV. Miscellaneous *iso*Thiocyanate Glucoside Acetates. Acta Chem. Scand. **13**, 851 (1959).

*62.* — Naturally Occurring *iso*Thiocyanates and their Parent Glucosides. In: N. KHARASCH, Chemistry of Sulfur, Vol. I. London-New York: Pergamon Press. 1960.

*63.* — Unpublished results.

*64.* — Infra-red Spectra of *iso*Thiocyanate Glucosides. Acta Chem. Scand. (to be published).

65. KJÆR, A. and B. (W.) CHRISTENSEN: *iso*Thiocyanates. XXVI. Straight-Chain ω-Methylthioalkyl *iso*Thiocyanates and some Derivatives. Acta Chem. Scand. **11**, 1298 (1957).

66. — — *iso*Thiocyanates. XXX. Glucohirsutin, a New Naturally Occurring Glucoside Furnishing (—)-8-Methylsulphinyloctyl *iso*Thiocyanate on Enzymic Hydrolysis. Acta Chem. Scand. **12**, 833 (1958).

67. — — *iso*Thiocyanates. XXXVI. (+)-4-Methyl-2-oxazolidinethione, the Enzymic Hydrolysis Product of a Glucoside (Glucosisymbrin) in Seeds of *Sisymbrium austriacum* JACQ. Acta Chem. Scand. **13**, 1575 (1959).

68. KJÆR, A., B. W. CHRISTENSEN and S. E. HANSEN: *iso*Thiocyanates. XXXIV. The Absolute Configuration of (—)-5-Vinyl-2-oxazolidinethione (Goitrin) and its Glucosidic Progenitor (Progoitrin). Acta Chem. Scand. **13**, 144 (1959).

69. KJÆR, A. and J. CONTI: *iso*Thiocyanates. V. The Occurrence of *iso*Propyl *iso*Thiocyanate in Seeds and Fresh Plants of Various *Cruciferae*. Acta Chem. Scand. **7**, 1011 (1953).

70. — — *iso*Thiocyanates. VII. A Convenient Synthesis of Erysolin (δ-Methylsulphonylbutyl *iso*Thiocyanate). Acta Chem. Scand. **8**, 295 (1954).

71. KJÆR, A., J. CONTI and K. A. JENSEN: *iso*Thiocyanates. III. The Volatile *iso*Thiocyanates in Seeds of Rape (*Brassica napus* L.). Acta Chem. Scand. **7**, 1271 (1953).

72. KJÆR, A., J. CONTI and I. LARSEN: *iso*Thiocyanates. IV. A Systematic Investigation of the Occurrence and Chemical Nature of Volatile *iso*Thiocyanates in Seeds of Various Plants. Acta Chem. Scand. **7**, 1276 (1953).

73. KJÆR, A. and R. GMELIN: *iso*Thiocyanates. XI. 4-Methylthiobutyl *iso*Thiocyanate, a New Naturally Occurring Mustard Oil. Acta Chem. Scand. **9**, 542 (1955).

74. — — *iso*Thiocyanates. XVIII. Glucocapparin, a New Crystalline *iso*Thiocyanate Glucoside. Acta Chem. Scand. **10**, 335 (1956).

75. — — *iso*Thiocyanates. XIX. L-(—)-5-Methylsulphinylpentyl *iso*Thiocyanate, the Aglucone of a New Naturally Occurring Glucoside (Glucoalyssin). Acta Chem. Scand. **10**, 1100 (1956).

76. — — *iso*Thiocyanates. XXII. 3-Benzoyloxypropyl *iso*Thiocyanate, Present as a Glucoside (Glucomalcolmiin) in Seeds of *Malcolmia maritima* (L.) R. BR. Acta Chem. Scand. **10**, 1193 (1956).

77. — — *iso*Thiocyanates. XXIII. L-(—)-9-Methylsulphinylnonyl *iso*Thiocyanate, a New Mustard Oil Present as a Glucoside (Glucoarabin) in *Arabis* Species. Acta Chem. Scand. **10**, 1358 (1956).

78. — — *iso*Thiocyanates. XXV. Methyl 4-*iso*Thiocyanatobutyrate, a New Mustard Oil Present as a Glucoside (Glucoerypestrin) in *Erysimum* Species. Acta Chem. Scand. **11**, 577 (1957).

79. — — *iso*Thiocyanates. XXVIII. A New *iso*Thiocyanate Glucoside (Glucobarbarin) Furnishing (—)-5-Phenyl-2-oxazolidinethione upon Enzymic Hydrolysis. Acta Chem. Scand. **11**, 906 (1957).

80. — — *iso*Thiocyanates. XXXIII. An *iso*Thiocyanate Glucoside (Glucobarbarin) of *Reseda luteola* L. Acta Chem. Scand. **12**, 1693 (1958).

81. KJÆR, A., R. GMELIN and R. BOE JENSEN: *iso*Thiocyanates. XV. *p*-Methoxybenzyl *iso*Thiocyanate, a New Natural Mustard Oil Occurring as Glucoside (Glucoaubrietin) in *Aubrietia* Species. Acta Chem. Scand. **10**, 26 (1956).

82. — — — *iso*Thiocyanates. XVI. Glucoconringiin, the Natural Precursor of 5,5-Dimethyl-2-oxazolidinethione. Acta Chem. Scand. **10**, 432 (1956).

83. KJÆR, A., R. GMELIN and R. BOE JENSEN: isoThiocyanates. XXI. (—)-10-Methylsulphinyldecyl isoThiocyanate, a New Mustard Oil Present as a Glucoside (Glucocamelinin) in *Camelina* Species. Acta Chem. Scand. 10, 1614 (1956).

84. KJÆR, A., R. GMELIN and I. LARSEN: isoThiocyanates. XIII. Methyl iso-Thiocyanate, a New Naturally Occurring Mustard Oil, Present as Glucoside (Glucocapparin) in *Capparidaceae*. Acta Chem. Scand. 9, 857 (1955).

85. — — — isoThiocyanates. XII. 3-Methylthiopropyl isoThiocyanate (Ibervirin), a New Naturally Occurring Mustard Oil. Acta Chem. Scand. 9, 1143 (1955).

86. KJÆR, A. and S. E. HANSEN: isoThiocyanates. XXVII. The Absolute Configuration of Optically Active 2-Butylamine and 2-Butyl isoThiocyanate. Acta Chem. Scand. 11, 898 (1957).

87. — — isoThiocyanates. XXXI. The Distribution of Mustard Oil Glucosides in Some *Arabis* Species, a Chemotaxonomic Approach. Bot. Tidsskr. 54 374 (1958).

88. KJÆR, A. and AA. JART: isoThiocyanates. XXIX. Separation of Volatile isoThiocyanates by Gas Chromatography. Acta Chem. Scand. 11, 1423 (1957).

89. KJÆR, A. and R. BOE JENSEN: isoThiocyanates. XX. 4-Pentenyl isoThiocyanate, a New Mustard Oil Occurring as a Glucoside (Glucobrassicanapin) in Nature. Acta Chem. Scand. 10, 1365 (1956).

90. — — isoThiocyanates. XXII. Synthesis and Reactions of 3-Benzoyloxypropyl isoThiocyanate, Enzymically Liberated from Glucomalcolmiin, and some Compounds of Related Structure. Acta Chem. Scand. 12, 1746 (1958).

91. KJÆR, A., I. LARSEN and R. GMELIN: isoThiocyanates. XIV. 5-Methylthiopentyl isoThiocyanate, a New Mustard Oil Present in Nature as a Glucoside (Glucoberteroin). Acta Chem. Scand. 9, 1311 (1955).

92. KJÆR, A., F. MARCUS and J. CONTI: isoThiocyanates. VI. A Synthesis of Cheiroline (γ-Methylsulphonylpropyl isoThiocyanate). Acta Chem. Scand. 7, 1370 (1953).

93. KJÆR, A. and K. RUBINSTEIN: Paper Chromatography of Thioureas. Acta Chem. Scand. 7, 528 (1953).

94. — — isoThiocyanates. VIII. Synthesis of p-Hydroxybenzyl isoThiocyanate and Demonstration of its Presence in the Glucoside of White Mustard (*Sinapis alba* L.). Acta Chem. Scand. 8, 598 (1954).

95. KJÆR, A., K. RUBINSTEIN and K. A. JENSEN: Unsaturated Five-Carbon isoThiocyanates. Acta Chem. Scand. 7, 518 (1953).

96. KLYNE, W., J. DAY and A. KJÆR: isoThiocyanates. XXXVII. Rotatory Dispersion Studies of Optically Active Sulphoxides Derivable from isoThiocyanates of Natural Origin. Acta Chem. Scand. 14, 215 (1960).

97. KREULA, M. and M. KIESVAARA: Determination of L-5-Vinyl-2-thiooxazolidone from Plant Material and Milk. Acta Chem. Scand. 13, 1375 (1959).

98. KURUP, P. A. and P. L. N. RAO: Antibiotic Principle of *Moringa pterygosperma*. Part II. Chemical Nature of Pterygospermin. Indian J. med. Res. 42, 85 (1954).

99. LEPAGE, P. N.: Mémoire sur la formation de l'huile volatile dans les plantes antiscorbutiques sèches. J. chimie méd. [3], 2, 171 (1846).

100. LONG, L., Jr., R. C. CLAPP, G. P. DATEO, F. H. BISSETT and T. HASSELSTROM: Volatile Isothiocyanates of Fresh Cabbage. Abstr. Papers, 134th Meet. Amer. Chem. Soc., Washington 1958, 36 A. — R. C. CLAPP, L. LONG, Jr., G. P. DATEO, F. H. BISSETT and T. HASSELSTROM: The Volatile Isothiocyanates in Fresh Cabbage. J. Amer. Chem. Soc. 81, 6278 (1959).

101. McDOWALL, F. H., I. D. MORTON and A. K. R. McDOWELL: Land-cress Taint in Cream and Butter. New Zealand J. Sci. Technol., Sect. A 28, 305 (1947).

102. McKay, A. F., D. L. Garmaise, R. Gaudry, H. A. Baker, G. Y. Paris, R. W. Kay, G. E. Just and R. Schwartz: Bacteriostats. II. The Chemical and Bacteriostatic Properties of Isothiocyanates and their Derivatives. J. Amer. Chem. Soc. 81, 4328 (1959).

103. Nagashima, Z.: Studies of Wasabi (Eutrema wasabi Maxim.). Part 1. On Acrid Substances of Wasabi and Mustard (Black Mustard). J. Agric. Chem. Soc. Japan 28, 119 (1954).

103a. Nagashima, Z. and M. Uchiyama: Possibility that Myrosinase is a Single Enzyme and Mechanism of Decomposition of Mustard Oil Glucoside by Myrosinase. Bull. Agric. Chem. Soc. Japan 23, 555 (1959).

104. Neuberg, C. und J. Wagner: Über die Verschiedenheit der Sulfatase und Myrosinase. Biochem. Z. 174, 457 (1926).

105. Patel, C. K., S. N. Iyer, J. J. Sudborough und H. E. Watson: Über das Fett von Salvadora oleoides: Khakanfett. J. Indian Inst. Sci., Sect. A 9, 117 (1926) [Chem. Zbl. 1927 I, 465].

106. Peche, K.: Mikrochemischer Nachweis des Myrosins. Ber. dtsch. bot. Ges. 31, 458 (1913).

107. Procházka, Ž.: Chromatographic Proof of the Presence of Sulphoraphene in Plantain. Naturwiss. 46, 426 (1959).

108. — Isolation of Sulphoraphane from Hoary Cress (Lepidium draba L.). Collect. Czechoslov. Chem. Communs. 24, 2429 (1959).

109. Procházka, Ž., V. Šanda und L. Jirousek: Isothiocyanate im Wirsing- und Rosenkohl. Collect. Czechoslov. Chem. Communs. 24, 3606 (1959).

110. Puntambekar, S. V.: Mustard Oils and Mustard Oil Glucosides Occurring in the Seed Kernels of Putranjiva Roxburghii Wall. Proc. Indian Acad. Sci., Sect. A 32, 114 (1950).

111. Raciszewski, Z. M., E. Y. Spencer and L. W. Trevoy: Chemical Studies of a Goitrogenic Factor in Rapeseed Oilmeal. Canad. J. Technol. 33, 129 (1955).

112. Reese, E. T., R. C. Clapp and M. Mandels: A Thioglucosidase in Fungi. Arch. Biochem. Biophys. 75, 228 (1958).

113. Robiquet et Boutron: Sur la semence de moutarde. J. Pharmac. Chim. 17, 279 (1831).

114. Salkowski, H.: Über einige Derivate der p-Oxyphenylessigsäure und das ätherische Öl des weißen Senfs. Ber. dtsch. chem. Ges. 22, 2137 (1889).

115. Sandberg, M. and O. M. Holly: Note on Myrosin. J. Biol. Chem. 96, 443 (1932).

116. Schmalfuss, H.: Gewinnung und Erkennung der Senföle aus Raps. Forschungsdienst, Sonderheft 1, 37 (1935).

117. Schmalfuss, H. und H.-P. Müller: Gewinnung und Erkennung der Senföle aus Raps. Forschungsdienst 6, 83 (1938).

118. — — Über Senföle. III. Das flüchtige Senföl des Sarepta-Senf, Brassica juncea Czern. et Coss. Forschungsdienst 17, 205 (1944).

119. Schmid, H. und P. Karrer: Über Inhaltsstoffe des Rettichs. I. Über Sulphoraphen, ein Senföl aus Rettichsamen (Raphanus sativus L. var. alba). Helv. Chim. Acta 31, 1017 (1948).

120. — — Über Inhaltsstoffe des Rettichs. II. Optisch aktives 4-Methylsulfoxyd-buten-(3)-yl-cyanid als Spaltprodukt eines Glucosides aus den Samen von Raphanus sativus var. alba. Helv. Chim. Acta 31, 1087 (1948).

121. — — Synthese der racemischen und der optisch aktiven Formen des Sulphoraphans. Helv. Chim. Acta 31, 1497 (1948).

122. Schmidt, E.: Zur Kenntnis der Bildung des Allylsenföls. Ber. dtsch. chem. Ges. 10, 187 (1877).

*123.* SCHNEIDER, W.: Über Cheirolin, das Senföl des Goldlacksamens. Sein Abbau und Aufbau. Liebigs Ann. Chem. **375**, 207 (1910).

*124.* — Lauch- und Senföle. Senfölglucoside: In G. KLEIN, Handbuch der Pflanzenanalyse, Bd. III/2, S. 1063. Wien: J. Springer. 1932.

*125.* SCHNEIDER, W., H. FISCHER und W. SPECHT: Über Schwefel-Zucker und ihre Abkömmlinge, XV: Die Natur des Zuckers der Senföl-glucoside. Ber. dtsch. chem. Ges. **63**, 2787 (1930).

*126.* SCHNEIDER, W. und H. KAUFMANN: Untersuchungen über Senföle. II. Erysolin, ein Sulfonsenföl aus *Erysimum Perofskianum*. Liebigs Ann. Chem. **392**, 1 (1912).

*127.* SCHNEIDER, W. und W. LOHMANN: Untersuchungen über Senföle; das Cheirolin-glykosid. Ber. dtsch. chem. Ges. **45**, 2954 (1912).

*128.* SCHNEIDER, W. und L. A. SCHÜTZ: Untersuchungen über Senföl-glykoside. II. Glucocheirolin. Ber. dtsch. chem. Ges. **46**, 2634 (1913).

*129.* SCHNEIDER, W. und F. WREDE: Untersuchungen über Senföl-glykoside. V. Zur Konstitution des Sinigrins. Ber. dtsch. chem. Ges. **47**, 2225 (1914).

*130.* SCHULTZ, O.-E. und E. BARTHOLD: Die Bedeutung der Elektrophorese für die Reindarstellung des Senfölglykosides der Samen von *Lepidium sativum* L. (Kresse). Arzneimittel-Forsch. **2**, 532 (1952).

*131.* SCHULTZ, O.-E. und R. GMELIN: Papierchromatographie der Senfölglucosid-Drogen. Z. Naturforsch. **7** b, 500 (1952).

*132.* — — Die Reindarstellung des Glukosids von *Lepidium sativum* durch Säulenchromatographie mit Zellulosepulver. Arzneimittel-Forsch. **2**, 568 (1952).

*133.* — — Papierchromatographie senfölglucosidhaltiger Pflanzen. Neue Ergebnisse. Z. Naturforsch. **8** b, 151 (1953).

*134.* — — Quantitative Bestimmung von Senfölglucosiden mit dem Anthron-reagens. Z. Naturforsch. **9** b, 27 (1954).

*135.* — — Das Senfölglukosid von *Tropaeolum majus* L. (Kapuzinerkresse) und Beziehungen der Senfölglukoside zu den Wuchsstoffen. Arch. Pharmaz. **287/59**, 342 (1954).

*136.* — — Das Senfölglukosid „Glukoiberin" und der Bitterstoff „Ibamarin" von *Iberis amara* L. (Schleifenblume). Arch. Pharmaz. **287/59**, 404 (1954).

*137.* SCHULTZ, O.-E., R. GMELIN und A. KELLER: Reindarstellung von Senföl-glucosiden mit Hilfe der Ionenaustauscher. Z. Naturforsch. **8** b, 14 (1953).

*138.* SCHULTZ, O.-E. und W. WAGNER: Kristallisierte Azetylderivative von nicht oder schwer kristallisierenden Senfölglukosiden. Arch. Pharmaz. **288/60**, 525 (1955).

*139.* — — Trennung der Senfölglucoside durch absteigende Papierchromatographie. Z. Naturforsch. **11** b, 73 (1956).

*140.* — — Glucoalyssin, ein neues Senfölglucosid aus *Alyssum*-Arten. Z. Natur-forsch. **11** b, 417 (1956).

*141.* — — Senfölglukoside als genuine Muttersubstanzen von natürlich vor-kommenden antithyreoiden Stoffen. Arch. Pharmaz. **289/61**, 597 (1956).

*142.* — — Beitrag zur Struktur der Senfölglucoside. Arzneimittel-Forsch. **6**, 647 (1956).

*143.* SPATZIER, W.: Über das Auftreten und die physiologische Bedeutung des Myrosins in der Pflanze. Jahrb. wiss. Bot. **25**, 39 (1893).

*144.* STAHMANN, M. A., K. P. LINK and J. C. WALKER: Mustard Oils in Crucifers and their Relation to Resistance to Clubroot. J. Agric. Res. **67**, 49 (1943).

*145.* STOLL, A. und E. JUCKER: Senföle, Lauchöle und andere schwefelhaltige Pflanzenstoffe. In: K. PAECH und M. V. TRACEY, Moderne Methoden der Pflanzenanalyse, Bd. IV, S. 689. Berlin-Göttingen-Heidelberg: Springer-Verlag. 1955.

*146.* Stoll, A. und E. Seebeck: Die Isolierung von Sinigrin als genuine, kristallisierte Muttersubstanz des Meerrettichöls. Helv. Chim. Acta 31, 1432 (1948).

*147.* Svátek, E., R. Zahradník and A. Kjær: Absorption Spectra of Alkyl *iso*-Thiocyanates and N-Alkyl Monothiocarbamates. Acta Chem. Scand. 13, 442 (1959).

*148.* Thorsteinson, A. J.: The Chemotactic Responses that Determine Host Specificity in an Oligophagous Insect (*Plutella maculipennis* (Curt.) *Lepidoptera*). Canad. J. Zool. 31, 52 (1953).

*149.* Trim, A. R.: Glycosides as a General Group. In: K. Paech und M. V. Tracey, Moderne Methoden der Pflanzenanalyse, Bd. II, S. 295. Berlin-Göttingen-Heidelberg: Springer-Verlag. 1955.

*150.* Voltmer, L.: Über die Einwirkung von Hydroxylamin, Äthoxylamin und Benzyloxylamin auf Senföle. Ber. dtsch. chem. Ges. 24, 378 (1891).

*151.* Wagner, W.: Papierchromatographische Analyse der Senfölglucoside, präparative Darstellung ihrer Acetylderivative und ein Beitrag zu ihrer allgemeinen Struktur. Dissert., Univ. Tübingen, 1956.

*152.* Wehmer, C.: Die Pflanzenstoffe, 2. Aufl. Jena: G. Fischer. 1929.

*153.* Weis-Fogh, O.: Undersøgelser over Holdbarheden af sort Sennep. Dansk Tidsskr. Farmaci 28, 69, 117 (1954).

*154.* Wetter, L. R.: The Estimation of Substituted Thiooxazolidones in Rapeseed Meals. Canad. J. Biochem. Physiol. 35, 293 (1957).

*155.* Will, H. und W. Körner: Zur Kenntnis der Bildung des Senföls aus dem Samen des schwarzen Senf. Liebigs Ann. Chem. 125, 257 (1863).

*156.* Will, H. und A. Laubenheimer: Über das Glucosid des weißen Senfsamens. Liebigs Ann. Chem. 199, 150 (1879).

*157.* Zinner, G.: Die Senföle und ihre Glykoside. Dtsch. Apoth.-Zeitg. 98, 335 (1958).

*158.* Zwergal, A.: Beitrag zur Kenntnis der Inhaltsstoffe des Kohlrabis. Pharmazie 6, 245 (1951).

*159.* — Der *Brassica*-Faktor und andere antithyreoide Stoffe als die Ursache der Kropfnoxe. Pharmazie 7, 93 (1952).

(*Received, January 19, 1960.*)

# Die Farbstoffe im Gefieder der Vögel.

Von **OTTO VÖLKER**, Gießen.

Mit 6 Abbildungen.

## I. Einleitung.

Die Farbenerscheinungen im Gefieder der Vögel können auf sehr verschiedene Weise zustande kommen. Hier ist zu unterscheiden zwischen den sogenannten optischen oder Strukturfarben und den chemischen oder Pigmentfarben. Bei vielen Gefiederfärbungen kommt es zu einer innigen Verflechtung dieser beiden so verschiedenartigen farbenerzeugenden Prinzipien.

Wenden wir uns zunächst den optischen Farben zu, so ist ihr jeweiliges Auftreten grundsätzlich gebunden an eine gewisse morphologische

Struktur im Aufbau der Feder. Die einfachste derartige Strukturfarbe ist das Weiß pigmentfreier Federn. Sie kommt zustande durch die lufterfüllten Markzellen der Federäste, an denen das Licht total reflektiert wird. Eine andere sehr häufig auftretende optische Farbe ist das Strukturblau, ein Blau optisch trüber Medien. Die Wand modifizierter Markzellen wirkt hierbei als lichtzerstreuendes, also optisch trübes Medium, wobei die Farbe Blau resultiert. Stets trägt unterlagerndes dunkles Pigment (Melanin), indem es die Reststrahlen absorbiert, dazu bei, das Blau besonders intensiv hervortreten zu lassen. Ein Beispiel hierfür ist das schöne Blau der Flügeldeckfedern beim Eichelhäher *(Garrulus glandarius)*. Gesellt sich zu dem Strukturblau gelbes Pigment in der Rindenschicht der Feder, so erscheint durch diese Kombination die Feder grün; tritt ein rotes Pigment an die Stelle des gelben, so wirkt die Feder violett. Wieder andere optische Farben sind die weitverbreiteten, metallisch glänzenden Dünnblatt- oder Schillerfarben, die besonders eindrucksvoll im Gefieder der Kolibris *(Trochilidae)* und der Glanzstare *(Lamprocolius)* hervortreten. Die mit dem Lichteinfall sich ändernden Farben entstehen durch Interferenz an den Schillerradien der Federäste, in denen feinste Melaninblättchen das Interferenz erzeugende Medium darstellen.

Bei den vielen Modifikationen, welche die Strukturfarben in den verschiedenen Verwandtschaftsgruppen der Vögel erfahren, ist es für das unbewaffnete Auge nicht immer leicht, diese als solche zu erkennen. Hinzu kommt, daß in neuester Zeit bei gewissen Tauben *(Ptilinopodinae)* Schillerstrukturen gefunden wurden, die, unabhängig vom Lichteinfall, z. B. grünes Licht nach allen Richtungen reflektieren, so daß eine Wirkung erzielt wird wie bei einer Pigmentfarbe. In der Regel verschafft jedoch ein Blick ins Mikroskop oder ein Extraktionsversuch die wünschenswerte Klarheit.

Es kann im Rahmen dieses Artikels nicht näher auf die teilweise recht komplizierten physikalischen Verhältnisse eingegangen werden, vielmehr sei nur auf die neueren Arbeiten von Dorst *(15)*, Frank *(24)*, Mason *(47)*, W. J. Schmidt *(59—61)* sowie auf die zusammenfassende Darstellung von D. L. Fox *(21)* hingewiesen, welche die eingehende Analyse der optischen Farben der Vogelfeder zum Gegenstand haben.

Es wurde bereits auf die große Bedeutung dunkler Pigmente, der *Melanine*, beim Zustandekommen der Strukturfarben hingewiesen. Diesen Farbkörpern kommt jedoch weit darüber hinaus eine große Verbreitung im Vogelgefieder zu, denn nach der Häufigkeit ihres Auftretens beurteilt, stehen sie unter allen Pigmenten weitaus an erster Stelle. Es sind in granulärer Form abgelagerte Pigmente von hellbrauner bis schwarzer Farbe, die man aufgrund ihrer unterschiedlichen Löslichkeit in Alkalien in Eu- und Phaeomelanine einzuteilen pflegt. Ihre

*Literaturverzeichnis: SS. 218—222.*

Behandlung muß an dieser Stelle jedoch ebenfalls unterbleiben, da bis heute noch kein Melanin als einheitliche chemische Verbindung isoliert werden konnte. Man weiß nur, daß sie als hochmolekulare Körper außer Kohlenstoff, Wasserstoff und Sauerstoff auch Stickstoff enthalten und biochemisch als Endprodukte des Eiweißstoffwechsels zu betrachten sind.

Gegenstand der vorliegenden Darstellung sind vielmehr nur solche Pigmente, die man als *Diffusfarbstoffe* zu bezeichnen pflegt, da sie am Orte ihrer Ablagerung keine mikroskopischen Strukturen erkennen lassen.

## II. Carotinoide (Lipochrome).

Den Carotinoiden kommt als Federfarbstoffen nächst den Melaninen die größte Verbreitung zu. Ihr Anteil beim Zustandekommen der Färbung des Gesamtgefieders ist oft erheblich, z. B. *Oriolus*. Es handelt sich bei ihnen um Pigmente von gelber bis roter Farbe, die in der Hornsubstanz der Feder, dem Keratin, dilut verteilt sind. In geradezu idealer Weise sind diese labilen Farbkörper durch diesen Einschluß vor den zerstörenden Wirkungen des Sauerstoffes und des Lichtes geschützt, denn sonst wäre es nicht möglich, daß Vogelbälge nach hundert und mehr Jahren die lipochromatischen Färbungen noch in ihrer nahezu ursprünglichen Frische zeigen. Durch ihr kombiniertes Auftreten mit optischen Farben tragen sie vielfach zur besonderen Buntheit der Vogelgefieder bei.

Die Namen, die man den Lipochromen im biochemischen und zoologischen Schrifttum gegeben hat: Zooxanthin, Zoochlorin, Zooerythrin, Coriosulfurin u. a., haben heute kaum mehr als historische Bedeutung.

In der Mehrzahl der Fälle gelingt die Freilegung der Federcarotinoide durch kurzes Behandeln des Materials mit schwachem alkoholischem Alkali auf dem Wasserbad. Dabei zerfällt das Keratin der Feder rasch und es gelingt meist die quantitative Überführung der Farbstoffe ins Benzin. Die Benzinlösung wird dann in der üblichen Weise analysiert und zur weiteren Zerlegung der chromatographischen Adsorptionsanalyse unterworfen (*43, 100*).

Sämtliche Spektraldaten sind im Gittermeß-spektroskop nach LÖWE-SCHUMM (Zeiß, Jena) durch visuelle Beobachtung ermittelt.

### 1. Lutein (Xanthophyll).

Das Lutein (I, S. 186) ist zweifellos eines der am weitesten verbreiteten gelben Carotinoide. Neben Vogelarten, bei denen dieses Carotinoid als alleiniger Farbstoff vorkommt, gibt es solche, wo es von wechselnden Mengen anderer gelber Carotinoide begleitet ist. Auch in der Gesellschaft roter Carotinoide ist es häufig nachweisbar. In den gelben Federn der folgenden und vieler anderer Arten ist das Lutein das alleinige oder dominierende Pigment:

*Ampelion cucullatus,*
*Emberiza citrinella,*
*Emberiza icterica,*
*Icterus xanthornus,*

*Motacilla flava,*
*Oriolus oriolus,*
*Pitangus sulphuratus,*
*Ploceus cucullatus.*

Die in den Federn eines Vogels eingeschlossene Farbstoffmenge ist minimal. Sie beträgt bei einem finkengroßen Vogel 100—140 $\gamma$, bei einem Pirol etwa 1 mg Farbstoff, für Lutein berechnet. Die Isolierung des Luteins aus Federn gelang erstmals Völker (83). Als Ausgangsmaterial dienten hierzu zehn goldgelbe Bälge des afrikanischen Pirols *Oriolus auratus*. Nach Freilegung mit alkoholischer Natronlauge, chromatographischer Reinigung und Umkristallisieren, Schmp. 182° (*Abb. 1*, S. 183). Das Lutein hatte also seine Kristallisationsfähigkeit nicht eingebüßt trotz seiner Verteilung auf die enorme Oberfläche der Federn und sein über sechs Jahrzehnte langes Lagern in den Bälgen.

Lutein findet sich als Hauptfarbstoff in veresterter Form auch in der Epidermis der Fußhaut von Huhn, Ente und Gans, ferner in der Epidermis des Gänseschnabels. In Substanz konnte es auch aus dem Hühnereidotter und dem Fett des Huhnes dargestellt werden (*101*).

## 2. Kanarienxanthophyll.

Oft ist das Lutein der Federn von einem weiteren Xanthophyll begleitet, dessen mengenmäßiger Anteil bei den einzelnen Vogelarten stark schwankt. Es wurde „Kanarienxanthophyll" benannt, weil es als alleiniger Farbstoff zunächst in den gelben Federn des Kanarienvogels *(Serinus canaria)* gefunden wurde (*4*). Es haftet stärker als Lutein an Calciumcarbonat und unterscheidet sich weiter von diesem durch sein kürzerwelliges Spektrum. Die Lage der Absorptionsschwerpunkte im Benzin bei 472, 443, 418 m$\mu$ sowie der negative Ausfall der Salzsäurereaktion* sprechen für Taraxanthin. Von diesem unterscheidet es sich jedoch in charakteristischer Weise dadurch, daß die mit 25%iger Salzsäure unterschichtete ätherische Lösung des Farbstoffes keine Verschiebung der Absorptionsbanden zeigt, während bei Taraxanthin eine starke Verschiebung der Banden nach kürzeren Wellenlängen eintritt. Kanarienxanthophyll ist gegenüber Mineralsäuren viel beständiger als Taraxanthin, Violaxanthin und Xanthophyllepoxyd, die alle sehr säureempfindlich sind, an die es jedoch in seinen spektralen Eigenschaften so sehr erinnert. In den Federn der folgenden Arten ist das Kanarienxanthophyll das alleinige oder vorherrschende Pigment:

| | |
|---|---|
| *Carduelis carduelis,* | *Malaconotus cruentus,* |
| *Chloris chloris,* | *Serinus canaria,* |
| *Laniarius preussi,* | *Serinus serinus.* |

In Substanz konnte das Kanarienxanthophyll noch nicht gefaßt werden.

---

* Für eine Anzahl von Xanthophyllen (Epoxyde) ist die Blaufärbung charakteristisch — ebenso deren Intensität und zeitlicher Bestand —, die beim Schütteln einer ätherischen Lösung dieser Farbstoffe mit konzentrierter, wäßriger Salzsäure auftritt. Diese Reaktion, die auf der Bildung farbiger Salze beruht, pflegt man der Kürze halber als Salzsäurereaktion zu bezeichnen.

*Literaturverzeichnis: SS. 218—222.*

### 3. Zeaxanthin.

Verhältnismäßig selten ist das Zeaxanthin (II, S. 186) als Federfarbstoff anzutreffen. Bis jetzt fand es sich nur neben Lutein, von dem es in wechselnder Menge bei den einzelnen Vogelarten begleitet wird (*84*), so in den gelben Federn der folgenden Spezies:

| | |
|---|---|
| *Chrysoena victor,* | *Sericulus chrysocephalus,* |
| *Rupicola peruviana,* | *Xanthomelus ardens.* |

Zweifellos kommt dem Zeaxanthin eine größere Verbreitung in Federn zu, als es nach den bis jetzt vorliegenden Befunden den Anschein erweckt. Sein Vorkommen in untergeordneten Mengen neben relativ viel Lutein wird stets leicht zu übersehen sein, da es sich säulenchromatographisch vom Lutein nur schwer abtrennen läßt.

Auch in der roten Epidermis der Fußhaut von Haustaube und Ringeltaube (*Columba palumbus*) ist Zeaxanthin-Ester als dominierendes Carotinoid vorhanden.

### 4. Picofulvin.

In den gelbgrünen Federn vieler Spechte *(Picidae)* findet sich ein hypophasisches Carotinoid von auffallend kurzwelliger Absorption. Es wurde bereits von KRUKENBERG (*39, 40*) entdeckt, der es seiner Herkunft entsprechend „Picofulvin" nannte, da er ihm in der Klasse der Vögel sonst nirgends begegnete. Ferner war es KRUKENBERG bereits bekannt, daß es auch Spechte gibt, bei denen Picofulvin und Lutein in ein und demselben Individuum auftreten, jedoch nach Gefiederregionen getrennt. So enthalten bei *Chloronerpes yucatensis* die Federn der Bauchseite überwiegend Picofulvin, während die des Rückens im wesentlichen Lutein einschließen. Die Absorptionsschwerpunkte des Picofulvins liegen im Benzin bei 450, 423 m$\mu$ (sehr scharf) und stimmen mit denen des Flavoxanthins weitgehend überein, doch ergibt das Picofulvin keine Blaufärbung beim Schütteln seiner ätherischen Lösung mit 25%iger Salzsäure, die für Flavoxanthin charakteristisch ist. Eine Identität beider ist also auszuschließen.

Bemerkenswerterweise enthalten die intensiv gelbgrünen Bürzelfedern eines Singvogels, und zwar des Tanagriden *Ramphocelus icteronotus*, ebenfalls ein Carotinoid mit dem Spektrum des Picofulvins. Dieses ergibt zwar eine schwache Salzsäurereaktion, doch ist diese sehr unbeständig, während mit Flavoxanthin diese Farbreaktion intensiver ausfällt und erst nach einigen Minuten wieder verblaßt. Da ferner im Misch-Chromatogramm mit Flavoxanthin deutliche Zonenbildung auftritt, so können auch diese beiden spektroskopisch gleichen Carotinoide nicht miteinander identisch sein.

Die Federpigmente aus Spechten und Tanagriden sind also trotz ihrer mit Flavoxanthin nahezu übereinstimmenden Absorptionsspektren nicht mit diesem identisch. Ob eine Identität des Picofulvins mit Chrysanthemaxanthin vorliegt, mit dem es das Spektrum und die negative Salzsäurereaktion gemeinsam hat, wäre noch zu prüfen.

Einige Beispiele für untersuchte Spechte und Tangaren mögen hier folgen:

|  |  |
|---|---|
| *Chloronerpes yucatensis,* | *Picus viridis,* |
| *Hypoxanthus rivolii,* | *Ramphocelus icteronotus,* |
| *Picus canus,* | *Ramphocelus passerini.* |

Es verdient hier noch hervorgehoben zu werden, daß in den gelben und roten Federn der beiden *Ramphocelus*-Arten das Lutein völlig fehlt.

Hier sei auch darauf hingewiesen, daß alle gelben Lipochrome in Kombination mit dem optischen Strukturblau oder gemeinsam mit dunkeln Melaninen auftreten können, was stets den Eindruck einer grünen Gefiederfarbe hervorruft.

### 5. Astaxanthin bzw. Astacin.

Trotz der Untersuchung einer großen Zahl von Vogelarten mit rotem Gefieder ist es bisher nur in einigen Fällen gelungen, das Astaxanthin (III, S. 186) als Inhaltsstoff roter Federn mit Sicherheit nachzuweisen, und zwar bei:

|  |  |
|---|---|
| *Laniarius atrococcineus,* | *Laniarius erythrogaster,* |
| *Laniarius barbarus,* | *Ramphocelus brasilius.* |

Hiermit wurde das erste rote Carotinoid aus Vogelfedern chemisch charakterisiert (*83*). Die feuerroten Federn der *Laniarius*-Arten schließen nur rotes Lipochrom ein, das im wesentlichen aus freiem Astaxanthin besteht, denn man erhält mit Alkali unter strengem Ausschluß von Sauerstoff die für Astaxanthin charakteristische Farbreaktion, eine tiefe Blaufärbung (die sog. Astaxanthinreaktion), die bei Luftzutritt sofort wieder nach Rot umschlägt, wobei unter Dehydrierung Astacin (ein Tetraketon) gebildet wird.

Bei *Ramphocelus brasilius* hingegen ergibt diese Reaktion mehr violettschwarze Farbtöne, was darauf beruht, daß sich hier das Astaxanthin chromatographisch nicht mit genügender Schärfe von anderen roten Carotinoiden abtrennen läßt, die es in der Feder begleiten. Das nach alkalischer Behandlung entstandene Astacin ist jedoch an seinem Spektrum. seiner Alkalilöslichkeit sowie an seinem Verhalten gegenüber Aluminiumoxyd zu identifizieren.

Die Isolierung des Astaxanthins aus Federn gelang erstmals VÖLKER (*89*). Als Ausgangsmaterial dienten die feuerroten Federn von 22 frisch gesammelten Bälgen von *Laniarius atrococcineus* (Südwest-Afrika). Um den Federn unter Vermeidung von Alkali das Pigment entziehen zu können, ist es erforderlich, diese nach VÖLKER in eine konzentrierte Lösung von Kaliumrhodanid für mehrere Tage einzulegen. Dann erst läßt sich den Federn mit Methanol die Hauptmenge des Pigments auf dem Wasserbad entziehen. Nach der üblichen Weiterverarbeitung gelingt es, das Astaxanthin in seiner typischen Kristallform zu erhalten; nach Umkristallisation, Schmp. 205° (*Abb. 2*). Auch das im Dehydrierungsversuch gebildete Astacin konnte in Substanz gefaßt werden.

Astaxanthin wurde auch isoliert aus den Papillen der roten Augenfelder des Jagdfasans (*Phasianus colchicus*), in denen es in größtenteils veresterter Form vorliegt (*4, 34*), und ebenfalls als Ester findet es sich in der tiefroten Epidermis der Fußhaut des wildlebenden Flamingos (*Phoenicopterus ruber*) neben anderen roten Carotinoiden. Auch hier gelingt seine kristallisierte Abscheidung als Astacin nach vorausgegangener alkalischer Behandlung (VÖLKER, unveröffentlicht).

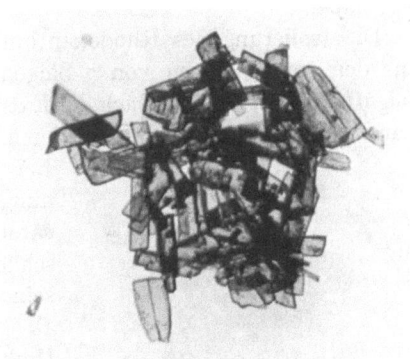

Abb. 1. Lutein aus Federn des Pirols *Oriolus auratus* (aus Methanol-Äther).

Überblickt man die gesicherten Nachweise von Astaxanthin bzw. Astacin im Tierreich, so muß man feststellen, daß dieses rote Carotinoid den Schwerpunkt seiner Verbreitung offensichtlich bei den Gliederfüßlern (*Arthropoda*) und den Fischen (*Pisces*) hat. Demgegenüber tritt sein Vorkommen im Gefieder der Vögel weit zurück. Die intensiv roten Federn vieler Vogelarten, die unter dem dringenden Verdacht eines Astaxanthin-Vorkommens untersucht wurden, brachten größtenteils nur negative Ergebnisse (s. S. 186).

### 6. Rhodoxanthin.

Das Rhodoxanthin (IV, S. 186), der rote Farbstoff der Eibenfrucht (*Taxus baccata*), wurde erstmals in den

Abb. 2. Astaxanthin aus Federn des Rotbauchwürgers *Laniarius atrococcineus* (aus Pyridin-Wasser).

scharlachroten Federn des Cotingiden *Phoenicircus nigricollis* (Ekuador) aufgefunden [VÖLKER (*84, 85*)]. Sein Vorkommen an dieser Stelle bedeutete zunächst eine Überraschung, weil dieses Carotinoid bis dahin im Tierkörper überhaupt noch nicht nachgewiesen war. Mit ihm wurde das zweite rote

Lipochrom aus Federn chemisch charakterisiert. Es zeigte sich weiter, daß gewisse lipochromatische Federregionen vieler Flaumfußtauben *(Ptilinopodinae)* (Südsee) ebenfalls dem Rhodoxanthin ihre Färbung verdanken. Dieser Farbstoff findet sich demnach bei Vogelarten, die weder geographisch noch stammesgeschichtlich etwas miteinander zu tun haben.

Die Isolierung des Rhodoxanthins gelang erstmals VÖLKER *(86, 87)* aus den roten Federn von 7 Bälgen der Flaumfußtaube *Megaloprepia magnifica* (Australien). Nach der Extraktion der Federn mit alkoholischer Lauge wurde der Farbstoff nach den üblichen Methoden weiter verarbeitet.

Bei der chromatographischen Abtrennung geringer Mengen gelber Begleitcarotinoide (Xanthophylle) schied sich schließlich das Rhodoxanthin in seiner charakteristischen Kristallform ab (vgl. *Abb. 3*). Schmelzpunkt nach Umkristallisieren: 216°. Auch in diesem Falle hatte das Carotinoid seine Kristallisationsfähigkeit nicht eingebüßt, obgleich die Bälge ein Durchschnittsalter von 40 Jahren hatten.

Abb. 3. Rhodoxanthin aus Federn der Flaumfußtaube *Megaloprepia magnifica* (aus Benzol-Methanol).

In der Gruppe der Flaumfußtauben, die mit zu den farbenprächtigsten Vögeln gehören, erhält man das Rhodoxanthin nicht nur aus roten Federn, sondern auch aus purpurroten, violetten und blauen Federregionen, welche meist die Kopfplatten, aber auch andere Gefiederteile schmücken (s. *Tabelle 1*).

Tabelle 1. Spektroskopische Untersuchung der roten bis blauen Lipochrome einiger Flaumfußtauben.

| Vogelart | Federfarbe | Absorptionsschwerpunkte in Benzin oder Hexan (mμ) | |
|---|---|---|---|
| *Megaloprepia magnifica*............... | rot | 523 | 489 |
| *Ptilinopus perousii*.................. | purpurrot | 522 | 488 |
| *Ptilinopus porphyreus* .............. | purpurrot | 519 | 487 |
| *Ptilinopus superbus* ................. | violett | 521 | 488 |
| *Ptilinopus hyogaster*................. | violett | 521 | 488 |
| *Ptilinopus monachus* ................ | blau | 518 | 486 |
| Rhodoxanthin aus *Taxus baccata*, reiner Farbstoff (zum Vergleich) ......... | — | 524 | 489 |

Beim Zustandekommen der purpurroten, violetten und blauen Federfarbe auf der Grundlage des Rhodoxanthins ist ein farbenerzeugender optischer Strukturfaktor, der sich dem roten Pigment überlagert und mit diesem die von Rot abweichenden Federfarben ergibt, auszuschließen, da ein Pressen oder Quetschen der Federn keine Änderung ihres Farbtones zur Folge hat. Auch für die Anwesenheit zusammengesetzter Verbindungen nach Art der dunklen Chromoproteide des Hummers (*Astacus gammarus*), bei denen das Astaxanthin mit Proteinen verknüpft ist, sprechen keine Beobachtungen. Hingegen läßt sich bei der Adsorption des Rhodoxanthins an geeigneten Adsorbentien die gesamte natürliche Farbenskala dieser Federn, von Rot über Violett bis Blau, nachahmen. Es ergibt sich eine Reihe von Adsorbatfarben, die für Rhodoxanthin spezifisch ist, wie sich beim Vergleich mit anderen roten Carotinoiden, wie Lycopin, Capsanthin und Astaxanthin, herausstellte. So haftet das Rhodoxanthin vor allem an wasserhaltigen Silikaten, besonders eindrucksvoll an Talkum aus Benzin-Benzol-Gemischen mit tief violetter Farbe, die jener der Federn völlig gleicht. Bei der Verwendung von Benzol als Lösungsmittel wird der Farbton der Adsorbate nach Blau verschoben. Tief purpurrot ist der Farbton aus Benzin an Kaolin und Bleicherde.

Überträgt man aus Analogiegründen diese Verhältnisse auf die Federn, indem man eine Adsorption des Rhodoxanthins an der stark verkieselten Hornmasse der Federn annimmt (*26, 87*), die Farben Purpurrot, Violett, Blau, demzufolge Adsorbatfarben des Rhodoxanthins, an der Feder darstellen, so erklären sich alle Erscheinungen an diesen Federn widerspruchsfrei, die zuvor unverständlich bleiben mußten. Es ist jedoch einzusehen, daß ein derartiger Vorgang der Farbstoffadsorption am Substrat der Feder sich naturgemäß nur während des Verhornungsprozesses im Rahmen der Federentwicklung abzuspielen vermag, und es darf daher nicht überraschen, wenn man diese Adsorptionseffekte durch äußeres Anfärben depigmentierter Federn mit einer Lösung von Rhodoxanthin nicht erreicht. Möglicherweise sind artliche Unterschiede in der Zusammensetzung der Feder-Keratine (*57*) die Ursache für die verschiedenfarbige Ausfärbung des Rhodoxanthins. Ob dem Rhodoxanthin als Federpigment bei anderen Arten eine weitere Verbreitung zukommt, ist vorläufig nicht abzusehen.

Es sei zum Abschluß der Besprechung der bisher in Vogelfedern nachgewiesenen und genauer bekannten gelben und roten Lipochrome besonders hervorgehoben, daß es sich hierbei durchweg um sauerstoffhaltige Carotinoide handelt, die durch den Besitz von Hydroxyl- und Ketogruppen ausgezeichnet sind (vgl. Tabelle 2 und Formeln I—IV, S. 186), während der in der Natur so weitverbreitete Kohlenwasserstoff Carotin hier so gut wie völlig fehlt. Selbst bei der Aufarbeitung des Lipochroms aus den Federn von 10 Pirolen ließ sich Carotin nur in Spuren nachweisen (*83*). Entsprechendes gilt ja bekanntlich auch für die Carotinoide des Hühnereidotters und des Hühnerfettes (*101*). Es herrscht demzufolge in der Klasse der Vögel eine selektive Spezifität in der Aufnahme carotinoider Stoffe, die auch bei einigen anderen Klassen der Wirbeltiere ausgeprägt ist. So finden sich bei den Fischen fast ausschließlich Xanthophylle, während bei den Säugetieren das Carotin bevorzugt abgelagert wird.

Tabelle 2. Lage der Absorptionsbanden der wichtigsten Carotinoide aus Federn in Benzin (Sdp. 70—80°).

| Vogelart | Absorptionsschwerpunkte (m$\mu$) | | Carotinoid |
|---|---|---|---|
| *Ramphocelus icteronotus* . . . . . . . . . . . | 450 | 423 | (Picofulvin) |
| *Ramphocelus passerini* . . . . . . . . . . . . | 450 | 423 | (Picofulvin) |
| *Picus viridis* . . . . . . . . . . . . . . . . . . . | 450 | 423 | Picofulvin |
| *Chloronerpes yucatensis* . . . . . . . . . . . | 450 | 423 | Picofulvin |
| *Serinus canaria* . . . . . . . . . . . . . . . . | 472 | 443 | Kanarienxanthophyll |
| *Carduelis carduelis* . . . . . . . . . . . . . . | 472 | 443 | Kanarienxanthophyll |
| *Oriolus auratus* . . . . . . . . . . . . . . . . . | 477 | 447 | Lutein, isoliert |
| *Emberiza citrinella* . . . . . . . . . . . . . . | 477 | 447 | Lutein |
| *Rupicola peruviana* . . . . . . . . . . . . . . | 482 | 451 | Zeaxanthin |
| *Chrysoena victor* . . . . . . . . . . . . . . . . | 482 | 451 | Zeaxanthin |
| *Megaloprepia magnifica* . . . . . . . . . . . | 523 | 489 | Rhodoxanthin, isoliert |
| *Phoenicircus nigricollis* . . . . . . . . . . . | 524 | 489 | Rhodoxanthin |
| *Laniarius atrococcineus* . . . . . . . . . . . | 472 | | Astaxanthin, isoliert |
| *Phasianus colchicus* (Papillen) . . . . . . | 472 | | Astaxanthin, isoliert |

(I.) Lutein, $C_{40}H_{56}O_2$.

(II.) Zeaxanthin, $C_{40}H_{56}O_2$.

(III.) Astaxanthin, $C_{40}H_{52}O_4$.

(IV.) Rhodoxanthin, $C_{40}H_{50}O_2$.

## 7. Rote Carotinoide unbekannter Struktur.

Die Betrachtung vieler Vogelbälge mit rotem Gefieder hat ergeben, daß unser Auge nicht befähigt ist festzustellen, ob die Rotfärbung auf der Anwesenheit von Astaxanthin, Rhodoxanthin oder anderen roten

Carotinoiden unbekannter Zusammensetzung beruht. Aus diesem Grunde war die biochemische Untersuchung eines möglichst umfassenden Materials geboten. Mit Absicht wurden dabei Vertreter ganz verschiedener systematischer Gruppen ausgewählt, doch kann von einer nur annähernd vollständigen Erfassung der hierbei wichtigen Objekte nicht die Rede sein. Das Ergebnis der Analysen zeigt, daß in allen diesen Fällen rote Pigmente vorliegen, die mit den bisher bekannten roten Carotinoiden keine Identifizierung gestatten (siehe auch *Tabelle 3*). Bei vielen Arten besitzen die roten Lipochrome keine sauren Eigenschaften, d. h. sie sind nach ihrer Freilegung mit alkoholischer Lauge quantitativ ins Benzin überführbar im Gegensatz zum Astacin, das alkalilöslich ist.

Tabelle 3. Rote Federcarotinoide unbekannter chemischer Struktur.

Vogelarten, bei denen noch eine saure Farbstoffkomponente vorhanden ist, sind durch ein + gekennzeichnet. Ein (+) bedeutet, daß ihr Verhalten gegenüber Aluminiumoxyd noch nicht geprüft wurde.

| Vogelart | Lage der Absorptionsbanden in Benzin (mμ) | |
|---|---|---|
| *Cardinalis cardinalis* + | 475 | |
| *Carduelis carduelis* + | ∼ 472 | ∼ 445 |
| *Chloronerpes yucatensis* (+) | ∼ 478 | ∼ 449 |
| *Cinnyris gutturalis* | ∼ 457 | ∼ 426 |
| *Cotinga cotinga* | 480 | |
| *Cymborhynchus macrorhynchus* (+) | ∼ 480 | ∼ 450 |
| *Dendrocopus major* (+) | ∼ 475 | ∼ 446 |
| *Guara rubra* | 477 | |
| *Haematoderus militaris* | 465 | |
| *Harpactes diardi* | 479 | |
| *Hypoxanthus rivolii* (+) | ∼ 480 | ∼ 450 |
| *Leistes militaris* | 476 | |
| *Loxia curvirostra* | ∼ 482 | ∼ 452 |
| *Paroaria cucullata* | ∼ 482 | ∼ 450 |
| *Pericrocotus miniatus* + | 470 | |
| *Pharomachrus mocinno* | 476 | |
| *Pharomachrus pavoninus* | 478 | |
| *Phoenicopterus antiquorum* + | 480 | |
| *Phoenicopterus ruber* + | 476 | |
| *Picus viridis* + | ∼ 480 | ∼ 450 |
| *Poecilothraupis igniventris* | 477 | |
| *Pyranga aestiva* | ∼ 498 | ∼ 465 |
| *Pyrocephalus rubineus* | 480 | |
| *Pyromelana franciscana* | 478 | |
| *Pyrrhula pyrrhula* | ∼ 476 | ∼ 447 |
| *Ramphocelus nigragularia* | 474 | |
| *Rupicola sanguinolenta* | ∼ 480 | ∼ 448 |
| *Spinus cucullatus* | 478 | |
| *Xipholena lamellipennis* | 465 | |
| *Xipholena punicea* | 465 | |

Diese Carotinoide adsorbieren an Aluminiumoxyd mit mehr oder minder intensiv roter Farbe und sind mit Benzin-Alkohol wieder quantitativ eluierbar, während Astacin fest daran haften bleibt. Im Spektrum ist in der Regel nur ein breites Absorptionsband vom Astacintypus erkennbar mit der Streifenmitte bei etwa 476 m$\mu$ im Benzin. Bei einigen Arten treten rote Carotinoide auf, die zwei unscharf begrenzte Absorptionsstreifen eben noch erkennen lassen (Tabelle 3). Auch bei der Verteilung zwischen Benzin und 90%igem Methanol verhalten sich diese Carotinoide nicht einheitlich, indem es epiphasische gibt (unverseifbare, z. B. bei *Guara*), während andere beide Schichten anfärben, und schließlich solche, die rein hypophasisches Verhalten zeigen.

Es gibt auch rote Lipochrome, die neben ihrem stets vorherrschenden, nicht sauren Pigmentanteil eine meist nur geringfügige Menge saures Pigment erkennen lassen, das an Aluminiumoxyd mit tiefer Farbe stark adsorbiert und mit den gebräuchlichen Lösungsmitteln nicht eluierbar ist. Möglicherweise handelt es sich hierbei um Astacin, doch ließen die stets minimalen Mengen noch in keinem Falle eine sichere Identifizierung mit diesem zu. Bedenken, daß diese roten Lipochrome durch das Herauslösen aus den Federn mit alkoholischer Lauge auf dem Wasserbad eine teilweise Veränderung erfahren haben könnten, erscheinen bei der bekannten Alkalistabilität der meisten Carotinoide wenig begründet. Sie werden vollends beseitigt durch die Möglichkeit, bei *Pharomachrus mocinno* (Quesal) den Federn das feuerrote Lipochrom mit Methanol bei Zimmertemperatur vollständig zu entziehen. Dabei stimmt die Analyse völlig überein mit jener des nach Alkalibehandlung gewonnenen Farbstoffes.

Alle Eigenschaften, die bis jetzt an diesen roten Lipochromen ermittelt werden konnten (*38, 71, 84*), sprechen dafür, daß es sich bei ihnen höchstwahrscheinlich um rot gefärbte Zersetzungsprodukte carotinoider Stoffe handelt, die dem Verlangen des Vogels nach Farbe noch durchaus genügen. Die intensive Rotfärbung der Vogelfeder wird eben mit verschiedenen Mitteln realisiert. Ob dabei rotgefärbte Zersetzungs- bzw. Umwandlungsprodukte von Carotinoiden oder intakte Farbstoffe, wie Astaxanthin und Rhodoxanthin, Verwendung finden, ist für die Entstehung der roten Farbwirkung offensichtlich ganz belanglos. In jedem Falle ist die Rotfärbung gesichert.

Einer besonderen Erwähnung bedürfen an dieser Stelle noch die eigenartigen dunklen lipochromatischen Federfarben einiger *Cotingiden* (Schmuckvögel). Es sei hier nur an das dunkle Purpurrot von *Xipholena punicea*, das Rotschwarz von *Xipholena lamellipennis*, das Violett von *Cotinga cotinga* oder *Cotinga ridgwayi* und das Veilchenblau von *Jodopleura isabellae* erinnert. Von diesen Pigmentierungen ist schon seit langem bekannt, daß sie alle bei gelindem Erhitzen einen Farbumschlag nach Rot erfahren. In der gleichen Richtung wirkt auch ein Behandeln dieser Federn mit Alkalien und Säuren (*30*). In neuester Zeit stellte dann VÖLKER (*84*) fest, daß dieser Farbumschlag nach Rot auch bei der rein mechanischen Beanspruchung solcher Federn, beim Quetschen oder Pressen, erfolgt, wobei stets die Farbe jener gelbroten, nicht näher charakterisierbaren Carotinoide sichtbar wird, die man auch bei der Extraktion dieser Federn erhält (Tabelle 3). Da andere Pigmente in diesen Federn fehlen, so handelt es sich bei diesem Vorgang offenbar um die Zerstörung einer Feinstruktur, die sich in der intakten Feder dem roten Pigment überlagert und im Zusammenwirken mit diesem die beachtliche Farbvertiefung nach Rotschwarz bzw. Violett hervorruft.

Neuerdings haben MATTERN (48) histologische und W. J. SCHMIDT (62) polarisations-optische Untersuchungen an einigen dieser Cotingiden-Federn angestellt, ohne daß es dabei gelungen wäre, zu einer übereinstimmenden Auffassung über die beim Farbumschlag sich vollziehenden Vorgänge zu gelangen.

## 8. Physiologische Grundlagen der Carotinoid-Ablagerung.

Bei der Frage nach der Herkunft der Lipochrome im Tierkörper sind grundsätzlich drei Möglichkeiten gegeben: direkte Übernahme mit der Nahrung aus Pflanzen oder aus Tieren, bei denen sie direkt oder indirekt pflanzlichen Ursprungs sind, Umbau aufgenommener Carotinoide und endlich Biosynthese. Für die Lipochrome des Hühnereidotters schien schon seit langem eine Beziehung zu bestehen zwischen der Art der Fütterung und der Dotterfarbe. Doch vermochte erst PALMER (52) in vorbildlichen Versuchen zu zeigen, daß der gesamte Carotinoidgehalt des Eidotters aus der Nahrung stammt und daß bei reichlichem Carotin- und Luteingehalt des Futters fast ausschließlich Lutein in den Dotter übergeht, während Carotin nur in sehr geringer Menge abgelagert wird.

Von besonderem Interesse war daher die Frage nach der Entstehung der gelben und roten Gefiederfarben der Vögel. Als Versuchsobjekt diente die gelbe Zuchtrasse des Kanarienvogels *(Serinus canaria)*, die sich infolge ihrer Domestikation weit besser an die jeweiligen Versuchsbedingungen anzupassen vermag als andere Vögel. Als Versuchstier ist der Kanarienvogel auch schon deshalb besonders reizvoll, weil in seinen Federn nicht Lutein, sondern Kanarienxanthophyll abgelagert wird, das vor allem in seinem Spektrum von jenem abweicht (S. 180). Zur Feststellung der Herkunft und der Spezifität der gelben Federfarbe wurden die Vögel zunächst carotinoidfrei ernährt. Dabei ergab sich die zu erwartende völlige Abhängigkeit der Federfarbe von der Nahrung, denn nach einiger Zeit trat nach wiederholter Rupfung und Mauser ein allmählicher Schwund der gelben Federfarbe ein, und man erhielt schließlich Federn, die praktisch frei von Farbstoff, also weiß waren (4, 25). Auch die gelbe Färbung von Fettgewebe und Dotter ging während der Versuchsdauer verloren.

Da die Analysen des normalen carotinoidhaltigen Futters ergeben hatten, daß die Hauptkomponente des Farbstoffs aus Lutein besteht, lag es nahe, dieses als Ausgangssubstanz für den Federfarbstoff anzusehen. Zur Prüfung dieser Annahme erhielten die weiß gewordenen Vögel reines Lutein als Futterzusatz mit dem Ergebnis, daß die neugebildeten Federn eine intensive gelbe Farbe annahmen. Die Untersuchung dieser Federn ergab das Vorliegen desselben Farbstoffes wie bei den zahlreichen vorher untersuchten Vögeln verschiedener Herkunft. Damit ist wohl eindeutig erwiesen, daß der gelbe Federfarbstoff des Kanarienvogels, das Kanarienxanthophyll, ein Umwandlungsprodukt des mit der Nahrung aufgenommenen Luteins ist (4).

Auch bei der Verabreichung von Zeaxanthin trat intensive Gelbfärbung der Federn ein. In diesem Falle enthielten die Federn, die etwas orangestichiger waren als bei Luteinfütterung, deutlich nachweisbare Mengen Zeaxanthin. Die Hauptmenge des Farbstoffes zeigte jedoch unscharfe Absorptionsbanden und bestand offenbar aus einem Gemisch von Zersetzungs- bzw. Umwandlungsprodukten. Die Fütterungsversuche zeigen, daß in ganz spezifischer Weise nur Lutein, das im Futter vorherrschende Carotinoid, als Ausgangssubstanz für das Kanarienxanthophyll normal ernährter Vögel in Frage kommt. Die Verfütterung anderer Carotinoide, die in der Nahrung vorkommen können, wie Violaxanthin, Lycopin und Carotin, war ohne jeden Einfluß auf die Federfarbe. Dabei ging die Indifferenz gegen Carotin sogar so weit, daß Vögel, die neben ihrer carotinoidfreien Grundkost dauernd reichliche Gaben Carotin erhielten, nach Rupfung und Mauser allmählich völlig weiß wurden, sich also genau so verhielten wie bei carotinoidfreier Kost. Während sich das Ausbleiben der Federfärbung bei der Verfütterung von Violaxanthin durch dessen schwere Resorbierbarkeit und große Säureempfindlichkeit erklären läßt, ist das negative Ergebnis mit Carotin und Lycopin bemerkenswert.

Die Fütterungsversuche geben keinen Anhaltspunkt dafür, daß beim Vogel eine Neubildung von Carotinoiden stattfindet, wohl aber zeigen sie, daß eine Übernahme und teilweise Umwandlung (oxydative Veränderung) verfütterter Carotinoide möglich ist, bei der eine Änderung in der Lage der Absorptionsbanden auftritt, während die carotinoiden Eigenschaften erhalten bleiben. Sehr beachtlich ist es, daß es nur sauerstoffhaltige Carotinoide sind, die hierbei vom Vogel übernommen werden.

Es ist schon seit langem bekannt, daß auch Capsanthin, der Farbstoff der Paprikaschote *(Capsicum annuum)*, beim gelben Kanarienvogel Orangefärbung seiner Federn hervorruft, wobei allerdings bemerkt werden muß, daß dieses Carotinoid in den Federn wildlebender Vogelarten bisher noch nicht nachgewiesen werden konnte und es daher fraglich erscheint, ob es in der Natur als Pigmentbildner eine Rolle spielt. Erst neuerdings konnte gezeigt werden, daß Rhodoxanthin, der Farbstoff der Eibenfrucht, deutliche Rotfärbung des Gefieders bewirkt [VÖLKER (90, 91)].

*Literaturverzeichnis: SS. 218—222.*

Die lipochromatischen Federfarben der Vögel bilden in ihrer Gesamtheit eine lückenlose Farbenskala vom hellsten Gelb bis zum dunkelsten Rot. Dies steht in einem sehr auffallenden Gegensatz zu dem in der üblichen Nahrung eintönigen Angebot an Carotinoiden, die zur Federfärbung geeignet sind. Daher erscheint die Annahme berechtigt, daß in viel weitgehenderem Maße, als dies bis jetzt durch Fütterungsversuche erwiesen werden konnte, spezifische Veränderungen der Nahrungscarotinoide eintreten müssen. Denn erst dann wird die oft von Art zu Art wechselnde und für die Spezies charakteristische Lipochromfarbe verständlich. Daß bei dieser im Vogel sich vollziehenden Umwandlung dem Lutein eine ganz überragende Bedeutung zukommt, steht außer Zweifel. Das tatsächliche Überangebot an Lutein in praktisch allen Nahrungsbestandteilen ist hierbei der ausschlaggebende Faktor. Demgegenüber benötigt der Vogel zur Deckung seines Lipochrombedarfes nur geringe Luteinmengen. In 10 g Rübsamen (Brassica rapa), einem wichtigen Bestandteil des üblichen Körnermischfutters, sind etwa 175 γ Lutein enthalten, das ist reichlich die Farbstoffmenge, die in sämtlichen gelben Federn eines finkengroßen Vogels eingeschlossen ist.

Die schönsten Beispiele für Umwandlungsprodukte des mit der Nahrung aufgenommenen Luteins sind wohl die intensiv roten Lipochrome einiger Körnerfresser, von denen genannt seien: *Amadina fasciata, Cardinalis cardinalis, Ramphocelus brasilius* und *Spinus cucullatus*. Da wir die Zusammensetzung der Nahrung während des Käfiglebens genau kennen und diese Vögel ihre roten Lipochrome auch in der Gefangenschaft reproduzieren (im Gegensatz zu anderen Arten, die dazu nicht befähigt sind, vgl. S. 194), so haben wir hier besonders überzeugende Beispiele dafür, wie aus dem Lutein des Futters rote Lipochrome entstehen. Ihre Bildung ohne diese Luteinzufuhr ist jedenfalls ganz unwahrscheinlich.

Unter den gelben Lipochromen muß auch das Picofulvin der Spechte *(Picidae)* und der Tangaren *(Ramphocelus)* als ein Umwandlungsprodukt des Luteins gelten. Dafür spricht das völlige Fehlen des Luteins in den gelben Federn des Grünspechtes *(Picus viridis)* und der Tangare *Ramphocelus icteronotus*, die nur Picofulvin beinhalten, obwohl diese Vögel doch zweifelsohne viel Lutein mit ihrer Nahrung aufnehmen. Ferner spricht zugunsten dieser Annahme das gemeinsame Vorkommen von Lutein und Picofulvin in den Federn des Spechtes *Chloronerpes yucatensis*, bei dem außerdem noch einige „Zwischenprodukte" anwesend sind, welche gleichsam spektroskopisch den Übergang von Lutein in Picofulvin erkennen lassen (88). Und schließlich entsteht aus Lutein auch bei der Einwirkung von starker Salzsäure ein Produkt mit dem Spektrum des Picofulvins.

Das Astaxanthin in den Federn von *Laniarius* (Rotbauchwürger) dürften diese Vögel ebenfalls nicht aus der Nahrung beziehen, sondern höchstwahrscheinlich durch Umwandlung gelber Nahrungscarotinoide selbst bereiten. Diese Annahme wird nunmehr durch das Auftreten einer gelbbäuchigen Mutante von *Laniarius atrococcineus* fast zur Gewißheit. Diese seltene Abart besitzt eben nicht mehr die Fähigkeit, gelbes Lipochrom in rotes umzuwandeln, und lagert daher nur gelbes Kanarienxanthophyll ab. Eine allerdings weniger wahrscheinliche Alternative wäre die, daß bei einem gleichzeitigen Angebot von gelben und roten Nahrungscarotinoiden die Mutante nur auf das gelbe anspricht, während sie das ausschließliche Selektionsvermögen für rotes — das der Norm entspräche — eingebüßt hat. Für eine endogene Bildung des Astaxanthins sprechen ferner die Erfahrungen der Tiergärtner, wonach die rote Federfarbe von *Laniarius* auch im Käfigleben ohne nennenswerte Intensitätsverminderung reproduziert wird, und zwar bei einer Kost, die keine roten Carotinoide enthält. Das gleiche gilt auch für das Astaxanthin in den Augenfeldern des Jagdfasans *(Phasianus)*.

Es sei in diesem Zusammenhang daran erinnert, daß in den Eiern der Heuschrecke *Locusta migratoria* während der Embryonalentwicklung eine Umwandlung von $\beta$-Carotin in Astaxanthin nachgewiesen ist, was sich im Schwund des ursprünglich in ihnen enthaltenen $\beta$-Carotins und im Auftreten von Astaxanthin klar zu erkennen gibt [GOODWIN (27)].

Die Möglichkeiten der Entstehung von Astaxanthin aus gelben Carotinoiden sind demnach, wie die erwähnten Beispiele zeigen, mehrfach realisiert. Hingegen liegen für eine Synthese des Astaxanthins im Tierkörper aus ungefärbten Vorstufen keinerlei Anhaltspunkte vor. Wenn schließlich der Flamingo *(Phoenicopterus)* und der Rote Sichler *(Guara)* nur rotes Lipochrom (ohne Beimengung von gelbem) in ihren Federn führen, so kann dies ebenfalls nur im Sinne einer Umwandlung gelber Nahrungscarotinoide in rote zu verstehen sein.

Die Beantwortung der Frage nach der Herkunft des Rhodoxanthins bei Flaumfußtauben *(Ptilinopodinae)* und Schmuckvögeln *(Cotingidae)* dürfte in derselben Richtung liegen. So verlockend die Vorstellung auch sein mag, daß diese Vögel als ausgesprochene Früchtefresser das Rhodoxanthin lediglich aus diesen beziehen und es in ihren Federn in unverändertem Zustand wieder ablagern, so ist auch hier mit der Möglichkeit einer oxydativen Veränderung gelber Nahrungscarotinoide zu rechnen, deren Endprodukt in diesem Falle das Rhodoxanthin ist. Über den Carotinoidgehalt der Nahrung dieser Vögel liegen zwar keine Angaben vor, doch kann kaum ein Zweifel darüber bestehen, daß in ihr Lutein und sicher auch Zeaxanthin quantitativ stark vorherrschen. Ob daneben auch Rhodoxanthin in einer zur Pigmentierung ausreichenden Menge vorhanden ist, entzieht sich zunächst jeder Beurteilung.

Wenn es andererseits gelingt, durch Verfütterung von Rhodoxanthin beim während des Käfiglebens ausgeblichenen Kreuzschnabel *(Loxia)*

*Literaturverzeichnis: SS. 218—222.*

intensive Rotfärbung seiner Federn zu erzielen [VÖLKER (91)], wenn ferner beim Flamingo (Phoenicopterus) mit Astaxanthin die Intensivierung der Rotfärbung seines Gefieders erreicht wird [Fox (22)] und sich die gebotenen Carotinoide im wesentlichen unverändert in den Federn wieder nachweisen lassen, so steht dies mit der obigen Auffassung nicht im Widerspruch. Beide Versuche besagen lediglich, daß es möglich ist, definierte rote Carotinoide in den Federn dieser Arten zur Ablagerung zu bringen, was deren intensive Rotfärbung zur Folge hat. Sie sagen jedoch nichts darüber aus, wie im Freileben dieser Vögel der Vorgang der Rotpigmentierung sich vollzieht. Daß dieses Geschehen dort anders verläuft als im Experiment, beweist die Analyse des natürlichen Materials. Denn würden Flamingo und Kreuzschnabel in der Freiheit Astaxanthin bzw. Rhodoxanthin in nennenswerter Menge zu sich nehmen, so müßten diese Carotinoide nach den jetzt vorliegenden Erfahrungen auch in den Federn der Wildvögel nachweisbar sein. Dies ist jedoch nicht der Fall.

Die Möglichkeiten des oxydativen Umbaues gelber Nahrungscarotinoide (Lutein, Zeaxanthin) sind in der Klasse der Vögel offenbar sehr groß. In der Mehrzahl der Fälle führt dieser Vorgang zu undefinierbaren, intensiv rot gefärbten Zersetzungs- oder Umwandlungsprodukten carotinoider Stoffe, wie man solche in den roten Federn vieler Vögel antrifft (Tabelle 3, S. 187). In einer Minderzahl jedoch entstehen dabei wohldefinierte Pigmente, wie das gelegentliche Auftreten von Astaxanthin und Rhodoxanthin zeigt, die hier als Endprodukte einer Umwandlung gelber Nahrungscarotinoide aufzufassen sind.

Gemeinsam ist allen diesen Umwandlungsprodukten ihre örtliche Beschränkung auf ganz bestimmte Bezirke des Vogelorganismus. Während unverändertes Lutein nicht nur im Integument, sondern auch in allen fettreichen Geweben und im Eidotter auftritt, sind Kanarienxanthophyll, Picofulvin, Astaxanthin und wohl auch Rhodoxanthin stets nur im Integument zu finden. Und auch beim Huhn ist das Vorkommen von Astaxanthin nur auf die Ölkugeln der Retina beschränkt. Es sind dies Gegebenheiten, die vermuten lassen, daß diese Stoffe am Orte ihres Auftretens auch entstanden sind. Mit dem Nachweis der autochthonen Bildung des roten Lipochroms in den Kopffedern des Stieglitzes (Carduelis) gelang es, eine experimentelle Stütze für diese Auffassung zu finden. Reicht man diesem Vogel anstelle der normalen Kost eine stark rhodoxanthinhaltige und rupft ihm die Schwingen mit ihrem gelben Lipochrom, so wird dieses durch ein ziegelrotes ersetzt. Damit ist gezeigt, daß die Papillen der Schwungfedern auch zur Aufnahme eines roten, im Futter gebotenen Lipochroms befähigt sind und nicht etwa nur auf gelbes anzusprechen vermögen. Weiterhin läßt sich hieraus schließen, daß der freilebende Stieglitz bei einem nennenswerten Angebot von „federfähigen" roten Carotinoiden keine gelbe, sondern eine rote

Flügelbinde haben müßte. In den Federpapillen der roten Gesichtsmaske hingegen ist der Ort zu suchen, wo Wirkstoffe (Fermente oder Hormone) gelbes Nahrungscarotinoid zu einem erheblichen Teil in rotes umwandeln [VÖLKER (93)].

Oft verhalten sich nächstverwandte Vogelformen (Rassen einer Art) hinsichtlich der Qualität ihrer Federlipochrome gänzlich verschieden: *Rupicola p. peruviana* enthält in den sattgelben Federn ein Gemisch von Lutein und Zeaxanthin, das der Vogel aus der Nahrung bezieht und in seinen Federn unverändert wieder ablagert. Hingegen führen die tiefroten Federn von *Rupicola p. sanguinolenta* nur rote, nicht näher identifizierbare Carotinoide ohne Beimengung gelber. Bei den sicherlich gleichen Nahrungsansprüchen dieser beiden Rassen kann nur ihre verschiedene Reaktionsnorm gegenüber den Nahrungscarotinoiden für diesen unterschiedlichen Befund verantwortlich zu machen sein.

Neben den Vogelarten, deren rote Gefiederfarbe auch während des Käfiglebens stets in voller Schönheit reproduziert wird (einige Beispiele hierfür sind oben genannt, vgl. S. 191), gibt es auch solche, die dazu nicht oder nur sehr unvollkommen in der Lage sind, z. B. *Loxia curvirostra*, *Pyromelana franciscana*, *Carduelis cannabina*, *Carpodacus erythrinus*, *Carduelis flammea*, *Guara rubra* u. a. Diese Vogelarten verblassen in der Regel bereits nach der ersten Mauser und tragen dann ein mehr oder minder unscheinbares Gefieder von gelblicher oder grünlicher Farbe. Trotz zahlreicher, meist vergeblicher Versuche, mit Carotinoid- und Hormongaben den früheren Zustand wieder herzustellen, sind wir auch heute nur ungenügend über die Gründe für das Ausbleiben der roten Gefiederfarbe unterrichtet. Ein hoher Carotinoidgehalt der Nahrung ist bei diesen Vögeln jedenfalls nicht die alleinige Voraussetzung für die Reproduktion ihrer roten Lipochrome. Diese erfolgt offenbar nur dann, wenn die der betreffenden Art gemäßen optimalen Umweltbedingungen erfüllt sind.

## III. Die gelben und roten Federfarbstoffe der Papageien.

In der Ordnung der Papageien *(Psittaci)* sind gelbe und rote Federfarbstoffe sehr weit verbreitet. Nach ihrem Äußeren zu urteilen, ist zumindest ihre Ähnlichkeit mit den in Federn vorkommenden Carotinoiden verblüffend, so daß man kaum Zweifel hegt, an eine nahe Verwandtschaft oder gar Identität mit diesen Fettfarbstoffen zu denken. Auch gleicht der Verteilungszustand innerhalb der Feder dem der Carotinoide, indem das Keratin stets diffus von ihnen angefärbt ist, eine morphologische Struktur dieser Pigmente also nie beobachtet werden kann; und vollends ist ihnen allen wie den Lipochromen die Eigenschaft gemeinsam, beim Behandeln mit konzentrierter Schwefelsäure einen Farbumschlag nach Blau zu geben (79).

Da die Freilegung der untersuchten Papageienfarbstoffe ausnahmslos erhebliche Schwierigkeiten bereitet, wobei Veränderungen der nativen Pigmente sehr wohl möglich sein können, seien zunächst die Ergebnisse der direkten spektroskopischen Messungen hier wiedergegeben. Zu diesem Zweck werden die gelben und grünen Federn (die roten sind hierzu ungeeignet) vor den Spalt eines Gittermeßspektroskops gebracht und deren Absorption im durch- oder auffallenden Lichte gemessen (s. *Tabelle 4*).

Tabelle 4. Direkte spektroskopische Prüfung gelber und grüner Papageienfedern im durchfallenden bzw. reflektierten Licht.

Die angegebenen Wellenlängen sind die Schwerpunkte der Absorption.

| Vogelart | Lage der Absorptions-maxima (m$\mu$) | | Bemerkungen |
|---|---|---|---|
| *Lorius salvadorii* | $\sim 472$ | $\sim 440$ | |
| *Pionopsitta pyrilia* | $\sim 472$ | $\sim 443$ | |
| *Melopsittacus undulatus* | $\sim 468$ | $\sim 445$ | unscharfe Absorptionsbanden |
| *Amazona leucocephala* | $\sim 473$ | — | |
| *Platycercus eximius* | $\sim 480$ | — | |
| *Oriolus oriolus* (Lutein) | 487 | 457 | zum Vergleich |

Dagegen sind bei den folgenden Arten keine Absorptionsbanden erkennbar: *Caica leucogaster*, *Amazona oratrix*, *Poicephalus rüppelli*, *Aratinga solstitialis*, *Ara ararauna*, *Conurus guarouba*, *Conurus pyrocephalus*, *Urochroma cingulata*, *Pyrrhura vittata* und *Cacatua moluccensis*.

Läßt die Mehrzahl der Pigmente der angeführten Arten Absorptionsbanden vermissen, so ist dies noch kein endgültiger Beweis für das Fehlen derselben, da, wie die Erfahrung zeigt, wenig stark ausgeprägte Banden, die nahe dem blauen Ende des Spektrums liegen, dieser Art von subjektiver Beobachtung leicht entgehen können. Wie man jedoch aus Tabelle 4 ersieht, gibt es auch eine Anzahl Spezies, deren gelbe Farbstoffe zwar unscharfe, aber dennoch mit hinreichender Genauigkeit meßbare Absorptionsstreifen im kurzwelligen Bereich des Spektrums erkennen lassen. Jeweils zwei Banden im Sichtbaren scheinen die· Regel zu sein. Ihre nach dem kurzwelligen Ende hin verschobene Lage wird am deutlichsten beim Vergleich mit dem Absorptionsverhalten bekannter Federcarotinoide. So absorbieren z. B. die gelben Steuerfedern von *Oriolus*, die Lutein enthalten, im Durchschnitt für jeden Streifen etwa um 15 m$\mu$ langwelliger (Tabelle 4). Gibt sich schon hierin eine recht beachtliche Verschiedenheit gegenüber den Federcarotinoiden zu erkennen, so werden die trennenden Merkmale noch markanter bei dem Bestreben, die gelben und roten Federpigmente der Papageien in möglichst unveränderter Form in Lösung zu bringen. Alle Versuche, die Farbstoffe mit schwacher alkoholischer Lauge — der bewährten Freilegungsmethode für Federcarotinoide — den Federn zu entziehen, befriedigen nicht.

In den meisten Fällen erleiden dabei die Pigmente irgendeine Veränderung oder sie gehen zugrunde, da eine Überführung von Pigment in organische Lösungsmittel nicht gelingt. Gerade die Empfindlichkeit gegen Alkalien ist ein recht bezeichnender Unterschied gegenüber den Carotinoiden, die durch eine ausgesprochene Alkalibeständigkeit gekennzeichnet sind. Auch die anderwärts bewährte Methode der Vorbehandlung der Federn mit konzentrierter Kaliumrhodanidlösung hilft hier nicht weiter. Indessen gelingt es, den gelben Federfarbstoff des Wellensittichs nach Behandlung der Federn mit Pyridin-Salzsäure in Benzin überzuführen (74). Unter den roten Papageienfedern erweisen sich einige der Einwirkung von alkoholischer Lauge etwas zugänglicher, zumindest gelingt es, aus ihnen gelb gefärbte Benzinlösungen zu erhalten, deren Absorptionen zusammen mit denen des Wellensittichs in *Tabelle 5* zusammengestellt sind. Bei einer ganzen Reihe gelber Pigmente, auch bei solchen, die im nativen Zustande Absorptionserscheinungen erkennen lassen (vgl. Tabelle 4), gestattet bis jetzt keine der angewandten Methoden, Lösungen zu erhalten, die zur Messung der Absorption geeignet wären.

KRUKENBERG (39) kannte bereits die Schwierigkeiten, die der Extraktion dieser Pigmente im Wege stehen. Auch waren ihm eine Reihe von Eigenschaften bekannt, die sie von denen bei anderen Vogelarten auftretenden unterscheiden. Er schlug deshalb für den gelben Farbstoff der Papageienfedern den Namen ,,Psittacofulvin" vor, während er den roten ,,Araroth" nannte. Auf eine weitere Verwendung dieser Bezeichnungen wird hier verzichtet, da bei der Kompliziertheit der Pigmentierungsverhältnisse bei den Papageien (s. a. Kapitel IV, S. 199) heute nicht mehr mit Sicherheit festgestellt werden kann, was KRUKENBERG in Händen hatte.

Tabelle 5. Spektroskopische Prüfung der Benzinlösungen gelber und roter Federpigmente von Papageien.

(Die angegebenen Wellenlängen sind die Absorptionsmaxima.)

| Vogelart | Absorptionsbanden in Benzin (m$\mu$) | | Bemerkungen |
|---|---|---|---|
| Melopsittacus undulatus... | 440 | 420 | gelbe Federn, gelbe Lösung |
| Ara chloroptera .......... | 456 | 432 | rote Federn, gelbgrüne Lösung |
| Eclectus polychlorus ...... | 456 | 432 | rote Federn, gelbe Lösung |
| Eos cyanogenys .......... | 458 | 432 | rote Federn, gelbgrüne Lösung |
| Lorius salvadorii ........ | 458 | 432 | rote Federn, gelbgrüne Lösung |
| Alisterus amboinensis..... | 458 | 432 | rote Federn, gelbgrüne Lösung |
| Domicella garrula ........ | 460 | 434 | rote Federn, gelbe Lösung |
| Lutein aus Pirolfedern ... | 477 | 447 | } zum Vergleich |
| Picofulvin aus Spechtfedern .............. | 450 | 423 | |

Die Lage der Absorptionsbanden gestattet keine spektroskopische Identifizierung mit den bis jetzt bekannten Carotinoiden pflanzlicher oder tierischer Herkunft. Die recht kurzwelligen Absorptionen sind auch im Benzin sehr auffällig. Stellt man ihnen zum Vergleich eine Lutein-lösung gegenüber (Tabelle 5), so beträgt die Differenz auch hier wie im Federkeratin im Durchschnitt etwas über 15 m$\mu$ für jede Bande. (Das hinsichtlich seines Absorptionsspektrums ähnliche Picofulvin hat die typischen Eigenschaften eines Xanthophylls und gehört schon deshalb nicht hierher.) Bemerkenswert ist ferner, daß man beim Aufarbeiten der roten Federn der in Tabelle 5 genannten Papageienspezies stets gelbe Lösungen erhält. Dies hängt offenbar damit zusammen, daß im Verlauf der Extraktion die „Rotkomponente" dieser Pigmente Zersetzung erleidet und der gelbe Anteil allein übrigbleibt bzw. sich in Benzin überführen läßt.

Man hat diese gelben und roten Federfarbstoffe bisher schlechthin den Lipochromen zugeordnet. Dafür sprach die diffuse Verteilung in der Federsubstanz und weiter die Blaufärbung, die sie alle mit konzentrierter Schwefelsäure ergeben. Was nun diese Farbreaktion betrifft, so ist hier vor allem die große Geschwindigkeit ihres Ablaufs sehr bemerkenswert. Stets ist ein leuchtendes und tiefes Blau zunächst die Endstufe der Reaktion. Bei den Federcarotinoiden dagegen erfolgt der Farbumschlag mit konzentrierter Schwefelsäure ganz allmählich. In der Regel sind alle Stufen der durchlaufenen Farbskala erkennbar, bis schließlich ein mehr oder minder farbstarkes Blaugrün das vorläufige Ende des Farbenspiels anzeigt. Wie man seit den Untersuchungen von KUHN und WINTERSTEIN (44) weiß, ist diese Reaktion lediglich das Kennzeichen für das Vorhandensein von stark ungesättigten Verbindungen von mindestens sechs konjugierten Kohlenstoff-Doppelbindungen. Dem Vorgang kommt also nur die Bedeutung einer Gruppenreaktion zu, deren Bereich über den der Carotinoide hinausgeht.

Aus diesem Grund wäre es auch verfehlt, allein beim positiven Ausfall dieser Reaktion stets auf das Vorliegen von Carotinoiden, d. h. Fettfarbstoffen im üblichen Sinne, zu schließen oder gar darüber hinaus noch weitergehende Schlüsse auf die Natur der fraglichen Pigmente zu ziehen. Nach Ausweis dieser Reaktion ist unseren Pigmenten allen gemeinsam ihre stark ungesättigte Natur, und mit dieser, ihrer einzig gesicherten Eigenschaft (Polyene) müssen wir uns vorläufig begnügen. Das Fehlen einer morphologischen Struktur ist ebenfalls eine Eigenschaft, die außer bei Lipochromen noch bei vielen anderen Pigmenten verbreitet ist.

Die Frage nach der endogenen oder exogenen Herkunft dieser Farbstoffe, denen man nur in den Federn begegnet, kann auch hier nur durch Fütterungsversuche beantwortet werden. Sie ist von besonderem Interesse auch schon deshalb, weil die Papageien im Eidotter, in Körperfett und Leber Lutein zu speichern vermögen. Fütterungsversuche am gelben Wellensittich *(Melopsittacus undulatus)*, die im Prinzip in derselben Weise wie beim Kanarienvogel durchgeführt wurden (vgl. S. 189), ergaben keinerlei Anhaltspunkte für eine Abhängigkeit der Federpigmentbildung vom Carotinoidgehalt der Nahrung [VÖLKER (74)]. Weder gelang es, durch carotinoidfreie Fütterung ein allmähliches Weißwerden der nachwachsenden Federn zu erreichen, noch hatte die Verabreichung des zur Federfärbung geeigneten farbstarken Capsanthins in diesem Falle irgendeinen Einfluß auf die Pigmentierung der sprossenden Federn. Auch bei einer durch Thyroxin ausgelösten Totalmauser wachsen die gelben und grünen Federn des Wellensittichs mit unveränderter Farbe und Intensität nach wie unter normalen Bedingungen, während Thyroxingaben bei Lipochromträgern (Kanarienvogel, Grünfink, Gimpel u. a.) einen nahezu völligen Ausfall der Carotinoide zur Folge haben, was sich im Extremfall in einer Depigmentierung der Federn durch den Entzug des lipoiden Substrats (Fett) auswirkt (16, 37).

Die Entstehung der Federfarbstoffe der Papageien erfolgt somit völlig unabhängig vom Lutein- und Fettgehalt der lipoiden Gewebe. Dafür spricht auch die normale Histogenese dieser Pigmente: In der Federanlage der Papageien ist kein lipoides Substrat mit Sudan III oder Osmiumsäure nachweisbar (14, 16). Dagegen findet sich Fett in reichlicher Menge in den Federkeimen gelber Finkenvögel, wo es nach unseren derzeitigen Kenntnissen die Voraussetzung für die Lipochromeinwanderung in die Feder darstellt.

Die gelben Federfarbstoffe einiger Papageien des Amazonasgebietes sollen auf Störungen des Federwachstums mit einer höchst eigenartigen Umfärbung reagieren.

WALLACE (94) hat zuerst über diesen von den Eingeborenen geübten künstlichen Umfärbungsprozeß berichtet, und später beschrieben diesen Vorgang unabhängig von ihm A. B. MEYER (50) und KOCH-GRÜNBERG (36). Das Verfahren der künstlichen Farbänderung besteht im Prinzip darin, daß nach dem Ausrupfen grüner Federn und dem Einreiben der Rupfstelle mit dem Hautsekret eines Frosches oder einer Kröte die Federn mit gelber bis roter Farbe nachwachsen. Und werden auch diese Federn ausgerupft, so sollen ohne weiteren Eingriff neue mit derselben Farbe nachwachsen. Ein solcher, wie es scheint, komplizierter Vorgang ist in seinem Ablauf kaum zu begreifen. Vielleicht wird das Verständnis für Erscheinungen der genannten Art größer, wenn wir über die chemische Natur dieser Federpigmente genauer unterrichtet sind. Sie haben mit den Carotinoiden nur die Farben gemeinsam. In allen übrigen Eigenschaften — Absorptionsverhalten, Löslichkeit, Entstehung im Organismus, Histogenese — weichen sie von diesen jedoch erheblich ab.

In morphologisch-anatomischer Hinsicht sind die Papageien recht einheitlich und sehr spezialisiert. Beides kommt auch in der isolierten Stellung im System zum Ausdruck. Zweifellos sind sie auch bei der Erzeugung ihrer gelben und roten Federpigmente ihre eigenen, noch unbekannten Wege gegangen.

*Literaturverzeichnis: SS. 218—222.*

## IV. Fluoreszierende, gelbe Federfarbstoffe.

### 1. Bei Papageien.

Viele Papageien *(Psittaci)* sind dadurch ausgezeichnet, daß gewisse Regionen ihres Gefieders im filtrierten ultravioletten Licht (Schwarz-UV-Filter UG 2 von Schott & Gen., Jena) eine prachtvolle gelbe bzw. grüne Fluoreszenz zeigen [VÖLKER (75)]. Wie die Untersuchung eines großen Materials ergeben hat, ist diese merkwürdige Eigenschaft sporadisch bei Papageien aller geographischen Zonen anzutreffen. Der Schwerpunkt der Verbreitung liegt jedoch zweifellos bei den Papageien der australischen Region und Indiens, die überdies durch die Brillanz und Leuchtkraft ihrer Fluoreszenzen alle übrigen übertreffen. Lebhafte Fluoreszenzerscheinungen sind festgestellt an den Federn von Vertretern der folgenden Gattungen:

Altweltliche: *Cacatua, Licmetis, Leptolophus, Calyptorhynchus, Callocephalon, Polytelis, Micropsitta, Platycercus, Alisterus, Barnardius, Psephotus, Melopsittacus, Neophema, Cyanorhamphus, Geoffroyus, Prioniturus, Tanygnathus, Palaeornis, Loriculus* und *Opopsitta.* Afrikanische: *Agapornis.* Neuweltliche: *Bolborhynchus* und *Caica.* Daneben gibt es eine große Anzahl von Gattungen, denen diese Eigenschaft offensichtlich fehlt.

Träger der gelben bis grünen Fluoreszenz ist in allen Fällen blaß-gelbes, in der Feder diffus verteiltes Pigment, das in seinem Vorkommen nur auf die Konturfedern beschränkt ist. Sehr wahrscheinlich liegen den verschiedenen Fluoreszenzfarben mehrere, einander chemisch nahe verwandte Körper zugrunde. Von einer Benennung dieser Farbstoffklasse sei vorläufig abgesehen, solange über deren chemische Natur nur wenig bekannt ist.

Es erweist sich als vorteilhaft, die bei den verschiedenen Arten von goldgelb bis grün variierende Fluoreszenzfarbe drei Fluoreszenztypen zuzuordnen, die natürlich der großen Mannigfaltigkeit der Erscheinungen nicht gerecht werden, vielmehr nur der ungefähren Abgrenzung dienen sollen.

Typ I, goldgelbe Fluoreszenz: *Cacatua.*
Typ II, schwefelgelbe Fluoreszenz: *Melopsittacus.*
Typ III, grüne Fluoreszenz: *Palaeornis.*

Außer der Farbe ist auch die Intensität des Fluoreszenzlichtes bei den einzelnen Arten recht verschieden. Das intensivste Aufleuchten ist wohl den Federn vieler Arten der Gattung *Platycercus* eigen. Fluoreszierendes Pigment kann, wenn es sich ohne Beimengungen anderer Farbstoffe in der Feder findet (Haubenfeder von *Cacatua sulfurea*), an seiner eigentümlich schwefel- oder fahlgelben Färbung auch bei Tageslicht als solches erkannt werden. Tritt es dagegen neben Blaustruktur und Melaninen auf *(Platycercus)* oder liegt ein Gemisch mehrerer gelber Pigmente innerhalb desselben Federbezirkes vor *(Barnardius zonarius semitorquatus)*, so kann nur die Prüfung im filtrierten UV-Licht Aufschluß über das Vorhandensein dieses Farbstoffes geben. Wie weitgehend sich das

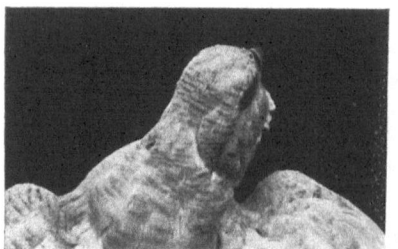

Abb. 4. *Melopsittacus undulatus* ad. (Wellensittich). a) Im Tageslicht; b) im filtrierten UV-Licht intensives schwefelgelbes Aufleuchten der Federn der Kopfplatte.

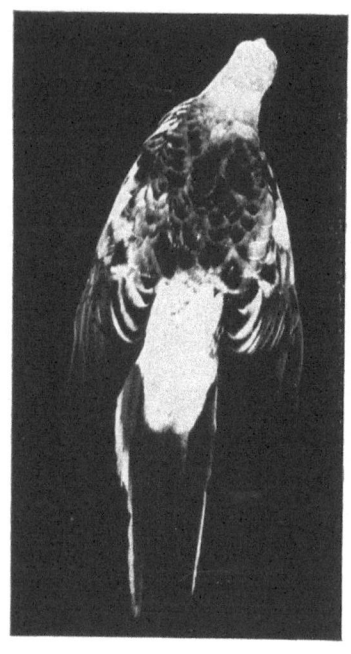

Abb. 5. *Platycercus adscitus palliceps* ad. (Blaßkopfrosella). a) Im Tageslicht; b) im filtrierten UV-Licht leuchten nur die Federn an Kopf, Hals und Bürzel intensiv grüngelb auf.

Aussehen vieler Papageien im UV-Licht durch die Anwesenheit von fluoreszierendem Pigment verändert, mögen die *Abb. 4* und *5* veranschaulichen.

Allen fluoreszierenden Federn oder Federabschnitten ist die Eigenschaft gemeinsam, beim Versetzen mit konzentrierter Schwefelsäure

*Literaturverzeichnis: SS. 218—222.*

entweder keinen Farbumschlag oder einen solchen nach Braun bzw. Rot zu zeigen. Hierin unterscheiden sie sich chemisch grundsätzlich von den bei den Papageien sehr weit verbreiteten gelben und roten Pigmenten (S. 194), mit denen sie häufig vergesellschaftet sind und die mit der Säure den bekannten Farbumschlag nach Blau ergeben.

Infolge ihrer schweren Löslichkeit und der minimalen in den Federn vorhandenen Mengen sind die fluoreszierenden Pigmente bis jetzt der chemischen Analyse schwer zugänglich. Auffallend ist ihre Resistenz gegenüber verdünnten Mineralsäuren, in denen fein zerschnittene Federn oft wochenlang ihre Fluoreszenz unverändert beibehalten. Dagegen zerstören Alkalien allmählich den Farbstoff. Nach längerer Behandlung der Federn mit 1%iger Sodalösung oder nach dem Andauen mit stark salzsaurem Pepsin gelingt es, den Federn mit siedendem Alkohol etwas Farbstoff zu entziehen. Doch müssen alle diese Versuche zur Freilegung als unbefriedigend betrachtet werden, da der Farbstoff dann nicht mehr die Eigenschaften des nativen Zustandes besitzt. Aus diesem Grunde ist auch die weitere Untersuchung des Pigments innerhalb der Feder anderen Methoden zunächst vorzuziehen.

Die Federn mit fluoreszierendem Farbstoff lassen im auf- oder durchfallenden Licht bei der Beobachtung mit dem Gittermeßspektroskop im sichtbaren Bereich des Spektrums keine Absorptionsbanden erkennen. Es ist von Interesse, die spektrale Zusammensetzung des Fluoreszenzlichtes zu untersuchen. Zu diesem Zweck bringt man die zu prüfende Feder in das mit einem Kondensor konzentrierte UV-Licht der Analysenlampe und beobachtet mit dem Spektroskop in Richtung der Lichtquelle. Das Fluoreszenzspektrum im Sichtbaren erstreckt sich von Rot bis Blau, läßt jedoch im grünen Bereich deutlich zwei oder drei dunkle Streifen, die Minima der Emission, erkennen. In *Tabelle 6* sind diese Streifen von einigen der untersuchten Federn verzeichnet. Am deutlichsten ausgeprägt sind die Streifen bei *Cacatua sulfurea*, ihre Breite beträgt jeweils etwa 15 m$\mu$ bei einer Spaltbreite von 0,44 mm. Die photographische Aufnahme dieses Fluoreszenzspektrums mit einem Steinheil-Spektrographen läßt die Streifen ebenfalls erkennen.

Tabelle 6. Emissionsminima der Fluoreszenzspektren einiger Papageienfedern.

| Vogelart und Fluoreszenztyp | | Minima der Emission (m$\mu$) | | |
|---|---|---|---|---|
| *Cacatua sulfurea* | I | 540 | 505 | |
| *Cacatua citrinocristata* | I | ~ 540 | ~ 505 | |
| *Leptolophus novaehollandiae* | I | ~ 540 | ~ 505 | |
| *Melopsittacus undulatus* | II | ~ 542 | ~ 508 | ~ 473 |
| *Platycercus adscitus palliceps* | III | | ~ 514 | ~ 479 |
| *Alisterus cyanopygius* | III | | | ~ 478 |

Wie aus der Tabelle 6 hervorgeht, lassen die Fluoreszenzspektren der drei Typen in bezug auf die Lage ihrer Emissionsminima deutliche

Unterschiede erkennen. Auf Grund dieses Verhaltens ist zu erwarten, daß den verschiedenen Fluoreszenzfarben verschiedene chemische Individuen zugrunde liegen. Die Fluoreszenzspektren dieser Federn haben mit denen anderer fluoreszierender Naturfarbstoffe nichts gemeinsam.

In organischen Lösungsmitteln bleibt die Fluoreszenz der Federn unverändert erhalten. Auch verdünnte Säuren sind ohne Einfluß darauf. Bringt man dagegen die Federn in verdünnte Laugen, so schwindet die Fluoreszenz innerhalb weniger Sekunden völlig. Fügt man Säure bis zur neutralen Reaktion hinzu, so tritt die Fluoreszenz plötzlich mit gleicher Intensität und Farbe wieder auf. Die Federfluoreszenz ist also vom pH abhängig. Mit dem Schwinden der Fluoreszenz im alkalischen Medium ist gleichzeitig ein allmähliches Aufhellen der Federfarbe verbunden, die ebenso wie die Fluoreszenz im Neutralen bzw. Sauren wiederkehrt.

Um die mögliche Identität mit anderen in der Natur verbreiteten fluoreszierenden Farbstoffen festzustellen, ist es erforderlich, diese mit den fluoreszierenden Federfarbstoffen zu vergleichen.

So sind nach COCKAYNE (7) die Farbstoffe einiger Schmetterlinge durch eine gelbe Fluoreszenz des nativen Farbstoffes im filtrierten UV-Licht ausgezeichnet, und in der Tat erinnern diese lebhaft gelben Fluoreszenzen — etwa des Papilioniden *Ornithoptera rubicollis* — sehr an die der Federn. Der Fluoreszenzbereich dieser chemisch noch unbekannten Pigmente ist jedoch von dem der Federfarbstoffe verschieden, indem ihre Fluoreszenz in verdünnten Mineralsäuren stets verschwindet, während die der Federn darin unbegrenzt erhalten bleibt. Eine Identität mit diesen Pigmenten ist schon deshalb auszuschließen. Auch Lactoflavin und Xanthopterin, die ebenfalls zum Vergleich mit herangezogen wurden, kommen für eine Identität nicht in Frage, da die fluoreszierenden Federfarbstoffe in ihrer unter sich gleichen pH-Abhängigkeit weder mit Lactoflavin noch mit Xanthopterin übereinstimmen.

Das auffallendste Merkmal dieser Federpigmentklasse ist ihre lebhafte Gelbfluoreszenz, die sie von allen bisher bekannten Farbstoffen der Vögel scharf unterscheidet. Als spezifische Federfarbstoffe sind sie in ihrem Auftreten stets nur an die Feder gebunden, da man nach ihnen anderen Ortes im Organismus der Papageien (Eischale, Eidotter, Organfett, Schuppen) vergeblich fahndet. Diese Farbstoffe sind endogener Natur. Fütterungsversuche am gelben Wellensittich haben jedenfalls gezeigt, daß kein Zusammenhang besteht zwischen dem Carotinoidgehalt des Futters und der Entstehung des fluoreszierenden Pigments (74). Auch eine Thyroxin-Mauser ist ohne Einfluß auf die Wiederkehr des fluoreszierenden Pigments der Kopfplatte (16). Schließlich dürfte auch der Lebensraum der Arten ebenfalls ohne Bedeutung sein, da die Bewohner der Grassteppe *(Platycercus)* in gleichem Maße wie die Bewohner des Urwaldes *(Cacatua)* fluoreszierende Federfarbstoffe hervorzubringen vermögen.

Zweifellos hat bei den Papageien die Differenzierung des Pigmentstoffwechsels einen Höhepunkt erreicht, denn wir finden Carotinoide

*Literaturverzeichnis: SS. 218—222.*

im Eidotter und im Fett, Melanine sind als Federfarben ebenso weit verbreitet wie die große Gruppe der ungesättigten gelben und roten Pigmente (Polyene), und zu diesen treten bei vielen Arten noch die fluoreszierenden Farbstoffe hinzu, so daß bis jetzt Pigmente aus vier Farbstoffklassen bei der Färbung der Papageien am Werke sind. Als die unscheinbarsten von allen verleihen jedoch die letzteren ihren Trägern auch im filtrierten UV-Licht ein prachtvolles Aussehen.

### 2. Bei anderen Arten.

Die Sichtung eines größeren Materials hat ergeben, daß die bei den Papageien aufgefundenen Federfluoreszenzen bei anderen Vögeln nicht auftreten. Insbesondere überrascht es, daß zahlreiche blaßgelbe Federpigmente, wie sich solche bei vielen Vogelarten finden, zwar die Farbe mit denen der Papageien zum Verwechseln gemeinsam haben, im UV-Licht jedoch keinerlei Fluoreszenzreaktion zeigen. Nur in der Gruppe der Paradiesvögel *(Paradisaeidae)* konnten einige Arten gefunden werden, deren strohgelbe Federn sich durch eine intensive fahlgelbe Fluoreszenz auszeichnen [VÖLKER (unveröffentlicht)]. Es sind dies: *Diphyllodes magnificus, Paradisea minor finschi* und *Schlegelia respublica*. Die in den Federn dieser Arten gefundenen fluoreszierenden Farbkörper zeigen untereinander große Ähnlichkeit. Sie ergeben mit konzentrierter Schwefelsäure keine Farbreaktion und sind in organischen Lösungsmitteln unlöslich, so daß es sich vermutlich in allen Fällen um den gleichen Farbstoff handelt. Eine pH-Abhängigkeit der Fluoreszenz ist nicht ausgeprägt. Eine Verwandtschaft oder Identität mit den fluoreszierenden Federfarbstoffen der Papageien, von denen sie sich auch im Farbton der Fluoreszenz unterscheiden, ist daher auszuschließen.

Ein recht helles Leuchten im UV-Licht zeigen auch die Dunen einiger Vogelküken. So fluoreszieren die hellen Dunen vom Hühnchen weißlich, die von der Taube gelblich, die vom Birkhuhn *(Lyrurus tetrix)* und der Chinesischen Zwergwachtel *(Excalfactoria chinensis)* intensiv grün. Im Gegensatz zu den fluoreszierenden Pigmenten der Papageien und Paradiesvögel sind diese Dunenfarbstoffe jedoch recht lichtunbeständig, so daß schon aus diesem Grunde eine Identität mit jenen auszuschließen ist. Die zur Verfügung stehenden minimalen Federmengen von Paradiesvögeln und Dunen ließen eine genauere chemische Untersuchung indessen noch nicht zu.

## V. Pyrrolfarbstoffe.

Über das Vorkommen von Porphyrinen in der Natur hat LEMBERG *(44a)* in diesen *Fortschritten* zusammenfassend berichtet.

### 1. Koproporphyrin.

DERRIEN und TURCHINI *(13)* konnten das Vorkommen von Porphyrin in Federn wahrscheinlich machen durch die Beobachtung einer roten

Fluoreszenz an den Federkielen junger Tauben, und bald darauf fand DERRIEN (9) an den vor Licht geschützten Federregionen von Eulen *(Striges)* und Nachtschwalben *(Caprimulgi)* dieselbe Erscheinung. Diesmal gestattete ihm das Material, den die Rotfluoreszenz bedingenden Farbstoff zu extrahieren, den er auf Grund seiner spektralen Eigenschaften für identisch hielt mit Protoporphyrin. Diese Mitteilungen DERRIENS sind kritiklos von der Literatur übernommen worden, trotz des Fehlens analytischer Daten. Zur Gewinnung eines möglichst großen Überblickes über die Verbreitung der Porphyrineinlagerung in den Federn der verschiedensten Arten war die Untersuchung eines recht großen Materials erforderlich. Da sich herausstellte, daß die in den Federn vorhandenen Porphyrinmengen in den meisten Fällen äußerst gering sind, diente die Anwendung der empfindlichen Fluoreszenzreaktion als erste Orientierung. Es wurde daher in allen Fällen, wo deutliche Rotfluoreszenz im filtrierten UV-Licht auftrat, auf die Anwesenheit von Porphyrin geschlossen. Die Empfindlichkeitsgrenze dieser Reaktion liegt bei 1 : 5 000 000.

Die rote Fluoreszenz beschränkt sich grundsätzlich nur auf Federn und Federabschnitte, die durch ihre Stellung innerhalb des Gefieders weitgehend vor der Einwirkung des Lichtes geschützt sind. Geringe, durch die rote Fluoreszenz deutlich nachweisbare Porphyrinmengen finden sich bei einer Anzahl von Arten, die den folgenden Ordnungen angehören: *Columbae, Ralli, Rhinocheti, Eurypygae, Anseres, Accipitres, Cuculi, Halcyones, Trogones* und *Passeres.* Es ist jedoch zu bemerken, daß innerhalb dieser Ordnungen Porphyrin nur sporadisch auftritt, also keineswegs bei allen Familien, Gattungen und Arten einer Ordnung. Eine Ausnahme macht die zur Ordnung der *Cuculi* gehörende Familie der *Musophagidae* (Turacos). Hier tritt Rotfluoreszenz in allen Gattungen auf, und zwar deutlich intensiver bei turacinführenden als bei turacinfreien Formen. Die Fähigkeit, freies Porphyrin abzulagern, ist demnach der ganzen Familie eigen. Die Kiele und die basalen Abschnitte der Federfahne sind die Träger der Fluoreszenz. Zur Extraktion sind die Farbstoffmengen aber bei weitem nicht ausreichend, so daß eine genauere Untersuchung des Porphyrins in diesen Fällen nicht möglich ist.

Größere, zur Extraktion ausreichende Porphyrinmengen sind dagegen in den Federn der *Caprimulgi, Striges* und *Otides* vorhanden, so daß die Gesamtzahl der Ordnungen mit Porphyrinfedern nunmehr 13 beträgt. Bei sämtlichen Vertretern dieser zuletzt genannten 3 Ordnungen ist die Eigenschaft der roten Federfluoreszenz ausnahmslos anzutreffen.

Bei den Nachtschwalben *(Caprimulgi)* ist der Porphyringehalt nicht weiter auffällig und erst im UV-Licht oder durch Extraktion nachweisbar. Bei den Eulen *(Striges)* kommt jedoch in einigen Fällen durch reichliche Porphyrineinlagerung schwache Rosafärbung der Federn zustande. So sind rosafarbene Axillaren geradezu ein systematisches

Merkmal aller *Ninox*-Formen der Salomon-Inseln, was auch in der Namengebung seinen Niederschlag findet: *Ninox roseoaxillaris.* Von HEINROTH (*33*) und MEISE (*49*) wurde eine Rosafärbung auch an der Sumpfohreule *(Asio flammeus)* beschrieben, die ebenfalls auf Porphyrin beruht [VÖLKER (*77*)]. Von allen bisher untersuchten Fällen besitzen jedoch die Federn der Trappen *(Otides)* den weitaus höchsten Porphyringehalt. Er ist hier so beträchtlich, daß er den basalen Federabschnitten und Dunen jene rosenrote Färbung verleiht, die ein Charakteristikum der ganzen Ordnung darstellt. Es ist das Verdienst von HEINROTH (*31, 32*), als erster auf diese eigentümlich rote Farbe der Trappenfedern und auf deren große Lichtempfindlichkeit mit allem Nachdruck hingewiesen zu haben.

Den 13 Ordnungen, bei denen Rotfluoreszenz an Teilen des Gefieders nachgewiesen ist, steht eine größere Anzahl gegenüber, die diese Eigenschaft völlig vermissen lassen. Es sind dies: *Struthiones, Apteryges, Galli, Pterocletes, Cariamae, Grues, Laro-Limicolae, Alcae, Colymbi, Podicipedes, Sphenisci, Tubinares, Steganopodes, Phoenicopteri, Gressores, Psittaci, Coraciae, Meropes, Upupae, Macrochires* und *Picidae.* Da geringe Porphyrinmengen, wie die Erfahrung zeigt, durch Lichteinwirkung und andere Umstände allmählich schwinden können, wurde zur Erhärtung der an Bälgen gemachten Befunde auch frisches Material untersucht. Von mausernden Vögeln: Wendehals, Fasan, Haushuhn, Birkhuhn; von Nestjungen: Wendehals, Grünspecht, Wellensittich. In Übereinstimmung mit den Eigenschaften des Balgmaterials konnte hier nirgends, auch nicht andeutungsweise, Rotfluoreszenz beobachtet werden. Dies ist besonders bei den jungen Höhlenbrütern recht bemerkenswert.

Porphyrin kann allen Federn und Federelementen eingelagert sein. Der Farbstoff ist dabei in der Hornsubstanz der Feder diffus verteilt und liegt, wofür die Rotfluoreszenz spricht, in echter Lösung vor, da Porphyrin nur in gelöstem Zustand fluoresziert. Daneben beobachtet man bei den Trappen in beachtlicher Menge porphyrinhaltigen Puder, im Sinne der Definition von SCHÜZ (*64*), also kleine, der Feder von außen aufgelagerte Hornpartikelchen. Merkwürdigerweise fluoresziert hier das Porphyrin erst beim Behandeln mit Alkali oder Säure. Möglicherweise liegt in diesem Falle das Porphyrin in einer anderen Form vor. Vielleicht ist aber auch die Dichte der Ablagerung für das Fehlen der spontanen Fluoreszenz verantwortlich zu machen. Wie bereits erwähnt, finden sich die durch Porphyrin bedingten Federfärbungen stets an verborgenen, vor Lichtzutritt geschützten Teilen des Gefieders. Um so mehr überrascht es daher, daß die Genickfedern der Schopftrappe, *Lophotis r. ruficrista*, lediglich einem von außen als Puder aufgelagerten Porphyrin ihre schöne weinrote Färbung verdanken. Es ist dies der einzige bisher bekannte Fall, wo durch beträchtliche Anhäufung von Porphyrin eine Schmuckfarbe zustande kommt [VÖLKER (*76*)]. Welchen Umständen allerdings der lichtunbeständige Farbstoff an dieser exponierten Stelle — am lebenden Vogel und in der Sonne Afrikas — seine Erhaltung verdankt, ist noch völlig unklar.

Bei der üblichen Beschränkung des Porphyrins auf Gefiederregionen, die weitgehend vor Licht geschützt sind, gewinnt man den Eindruck, als sei seine Verteilung lediglich eine Folge der Lichteinwirkung. Dies trifft bei gewissen Jungvögeln, z. B. bei der Blaumeise, auch zweifellos zu, denn bei derartigen Höhlenbrütern leuchtet, falls sie noch nicht am Tageslicht waren, die gesamte Oberfläche des Gefieders im UV-Licht blutrot auf. Dagegen lassen sprossende Federn von Trappen und Nachtschwalben deutlich erkennen, daß deren distale Enden keine Fluoreszenz aufweisen und auch kein Porphyrin enthalten, obwohl sie, noch völlig vor Licht geschützt, im fluoreszierenden Bereich der Nachbarfedern stehen. Die Porphyrin-einlagerung dürfte somit in diesen Fällen nicht regellos über die ganze Feder verteilt, sondern nur auf deren basale Zone beschränkt sein. Weiter sprechen dafür alle die Fälle, bei denen im lichtgeschützten Bereich entweder nur die Basaldune oder nur der Schaft fluoreszieren. Die Autonomie der einzelnen Bezirke einer Feder, die sich bei jeder Pigmentierung äußert, gilt demnach auch für die Porphyrine.

Bei keiner Ordnung ist die Menge des in den Federn abgelagerten Porphyrins solchen Schwankungen unterworfen wie bei den Eulen (Striges). Hier sind Fälle, wo rote Fluoreszenz eben noch festgestellt werden kann (Schnee-Eule), durch alle Übergänge mit solchen verbunden, bei denen durch reichliche Ablagerung zarte Rosafärbung im Tageslicht resultiert (Ninox-Arten). Ganz allgemein zeigt sich bei den Eulen, daß dunkel pigmentierte Formen mehr Porphyrin einlagern als helle, und dieses Prinzip gilt selbst innerhalb derselben Großart, z. B. bei der Schleiereule, deren Gesamtpigmentierung bedeutendem Wechsel unterworfen ist. Die Porphyrinpigmentierung geht hier also parallel der übrigen Pigmentierungstendenz des Vogels. Ähnliche Verhältnisse zeigen die Trappen, bei denen dunkel pigmentierte Formen im allgemeinen einen höheren Porphyringehalt aufzuweisen haben als helle. Freilich gibt es auch Fälle, bei denen der Feder trotz weitestgehender Reduktion anderer Pigmente die Fähigkeit mäßiger Porphyrineinlagerung erhalten geblieben ist, wie dies deutlich die weißen, praktisch melaninfreien Federn von *Otis tarda, Heterotetrax rüppelli* und *Chlamydotis undulata* erkennen lassen.

Porphyrine beteiligen sich nie an der Entstehung eines Zeichnungsmusters der Feder. Sie können entweder allein in Federn oder Federabschnitten auftreten oder, wie dies meist der Fall ist, neben anderen Pigmenten (Melaninen), die sie dann gleichmäßig überlagern.

Beobachtet man porphyrinreiche, weinrote Trappenfedern bei auf- oder durchfallendem Licht im Spektroskop, so erkennt man vier deutliche

*Literaturverzeichnis: SS. 218—222.*

Absorptionsbanden, deren Schwerpunkte z. B. bei den folgenden Wellenlängen liegen:

*Lophotis r. ruficrista* 623, 576, 537, 505 mμ.

*Lissotis melanogaster* 622, 574, 537, 504 mμ.

Dieses vierbandige „alkalische Spektrum" ist das Kennzeichen für die Anwesenheit eines freien Porphyrins. Zur genauen Ermittlung der Lage der Absorptionsbanden ist es erforderlich, diese in Salzsäure und Äther zu messen. Zu diesem Zweck wird der Feder mit verdünnter Salzsäure das Porphyrin entzogen, das sich beim Abstumpfen der Säure mit Natriumacetat quantitativ in Äther überführen läßt. Die spektroskopische Prüfung der Lösung, die Verteilung des Farbstoffes zwischen 0,1%iger Salzsäure und Chloroform sowie die Salzsäurezahl sprechen eindeutig für Koproporphyrin, und dasselbe Verhalten zeigen die aus den Federn der Eulen und Nachtschwalben gewonnenen Porphyrinlösungen (s. *Tabelle 7*). Das Vorliegen von Koproporphyrin in allen untersuchten Fällen ist also bemerkenswert.

Tabelle 7. Spektroskopische Messung der Salzsäure- und Ätherlösungen einiger Federporphyrine von Trappen, Eulen und einer Nachtschwalbe. Die angegebenen Wellenlängen sind die Absorptionsmaxima. Die Hauptbanden der Ätherlösung sind kursiv gedruckt (mμ).

| Vogelart | Spektrum in 25%iger Salzsäure | | | Spektrum in Äther | | | | | |
|---|---|---|---|---|---|---|---|---|---|
| | I | II | III | I | II | III | IV | V | VI |
| *Lophotis r. ruficrista* ... | 593,4 | 573,1 | 550,4 | *623,6* | 597,4 | 577,3 | *568,3* | *528,7* | *497,1* |
| *Lissotis melanogaster* .. | 593,0 | 572,0 | 550,0 | *623,2* | 596,4 | 576,0 | *568,1* | *529,4* | *497,1* |
| *Chlamydotis undulata*.. | 592,5 | 574,1 | 549,5 | *623,0* | 596,9 | 576,2 | *568,2* | *529,5* | *497,8* |
| *Heterotetrax rüppelli* ... | 593,4 | 573,5 | 551,3 | *623,1* | 596,5 | 576,0 | *568,2* | *528,5* | *498,1* |
| *Asio flammeus*........ | 593,5 | 573,5 | 550,5 | *623,2* | 597,1 | 576,1 | *568,3* | *529,5* | *497,7* |
| *Asio otus* ........... | 593,5 | 574,5 | 551,0 | *623,0* | 596,8 | 576,8 | *568,2* | *529,3* | *497,1* |
| *Athene noctua* ........ | 593,5 | 574,1 | 550,6 | *623,6* | 597,1 | 576,4 | *568,2* | *528,1* | *497,0* |
| *Ketupa flavipes* ....... | 593,5 | * | 550,5 | ** | | | | | |
| *Podargus papuensis* ... (Nachtschwalbe) | 593,8 | * | 550,8 | *623,4* | 597,5 | 576,4 | *568,2* | *529,1* | *497,1* |

\* Die Bande II des „sauren Spektrums" ist nur bei Lösungen höherer Konzentration hinreichend genau ablesbar.

\*\* Wegen Materialmangel konnte nur das „saure Spektrum" gemessen werden.

Eine Ausnahme machte bis jetzt nur ein Waldkauz *(Strix aluco)*, der spektroskopisch neben Koproporphyrin auch Protoporphyrin erkennen ließ. Im Gegensatz zu diesen Feststellungen steht die Angabe von DERRIEN (9, 10), nach der in den Federn von Eulen und Nachtschwalben in überwiegender Menge Protoporphyrin vorkommen soll. Sein Befund ist um so auffallender, als die in Tabelle 7 angeführten Spektren in allen Einzelheiten übereinstimmen mit reinem Koproporphyrin, das zum Vergleich herangezogen wurde.

Zur Isolierung des Federporphyrins in Form seines Methylesters dienten die Federn dreier afrikanischer Trappenarten: *Lophotis r. ruficrista, Lophotis r. gindiana* und *Lissotis melanogaster.* Sie boten als Ausgangsmaterial günstige Bedingungen und wurden nach den bekannten Methoden getrennt aufgearbeitet. Eine Kristallisation der veresterten Porphyrine konnte allerdings erst erzielt werden, nachdem diese durch zweimalige chromatographische Adsorption an Aluminiumoxyd weitgehend gereinigt worden waren. Die Ester kristallisierten in allen drei Fällen aus Chloroform-Methanol in Form schöner, sternförmig angeordneter, prismatischer Nadeln *(Abb. 6)*, die für Koproporphyrin-

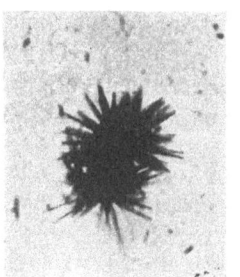 

a                          b

Abb. 6. Koproporphyrin-„III"-tetramethylester aus Trappenfedern. a) Von *Lophotis r. ruficrista* (aus Methanol-Chloroform); b) von *Lissotis melanogaster* (aus Methanol-Chloroform).

„III"-ester charakteristisch sind. Nach Umkristallisieren Schmp. der drei Präparate: 172° (Präp. 1), 152° bzw. 175° (Präp. 2) und 152° bzw. 176° (Präp. 3) (Doppelschmelzpunkte). Im Misch-Schmp. mit syntheti-

(V.) Protoporphyrin, $C_{34}H_{34}O_4N_4$.

*Literaturverzeichnis: SS. 218—222.*

CH$_2$—COOH

(VI.) Koproporphyrin „III", C$_{36}$H$_{38}$O$_8$N$_4$.

(VII.) Turacin (Uroporphyrin-„III"-Cu-Komplexsalz), C$_{40}$H$_{36}$O$_{16}$N$_4$Cu.

$R_1 = CH_2—CH_2—COOH$
$R_2 = CH_2—COOH$

schem Koproporphyrin-„III"-tetramethylester trat keine Depression ein. Auch stimmte die Elementaranalyse von Präparat 1, das in dazu ausreichender Menge anfiel, ebenfalls auf den Tetramethylester eines Koproporphyrins [VÖLKER (77, 78)]. Die aus den Federn der drei Trappen isolierten Porphyrine sind also identisch mit Koproporphyrin „III" (Formel VI). Damit ist erstmals aus Federn die Kristallisation eines freien Porphyrins in Form seines Methylesters gelungen. Die Farbstoffausbeute beträgt pro Vogel (der nahezu die Größe eines Haushuhnes erreicht) etwa 12 mg Ester. Das Porphyrin der Eulenfedern ließ sich papierchromatographisch ebenfalls mit Koproporphyrin „III" identifizieren (nach brieflicher Mitteilung von T. K. WITH).

In diesem Zusammenhange verdient darauf hingewiesen zu werden, daß die so weitverbreitete Fleckenfarbe vieler Vogeleier stets durch Protoporphyrin (V), die eisenfreie Stammverbindung des Hämins, hervorgerufen wird. Dieses Porphyrin wurde zuerst von H. FISCHER und KÖGL (*19*) aus Kiebitz- und Möweneierschalen als Ooporphyrin isoliert und ist neuerdings auch aus dunkelbraun gefärbten, ungefleckten Hühnereischalen, die lebhaft rot fluoreszieren, in Substanz erhalten worden (*65*).

Tabelle 8. Porphyrine in Eischalen und Federn der Vögel.

m. = Schalen mit Flecken; o. = Schalen ohne Flecken; (+) = Anwesenheit von Porphyrin bis jetzt nur durch die Fluoreszenzreaktion nachgewiesen.

| Ordnung | Protoporphyrin in Eischalen | | Koproporphyrin in Federn |
|---|---|---|---|
| *Otides*, Trappen.......................... | m. + | | + isoliert |
| *Caprimulgi*, Nachtschwalben ............. | m. + | | + |
| *Striges*, Eulen ......................... | o. (+) | | + |
| *Accipitres*, Raubvögel .................. | m. + | | (+) |
| *Columbae*, Tauben ...................... | o. + | | (+) |
| *Ralli*, Rallen .......................... | m. + | | (+) |
| *Anseres*, Entenvögel..................... | o. + | | (+) |
| *Cuculi*, Kuckucksartige ................. | m. + | | (+) |
| *Passeres*, Sperlingsvögel ................ | m. + | | (+) |
| *Laro-Limicolae*, Möwen und Watvögel .... | m. + | isoliert | — |
| *Galli*, Hühnervögel...................... | m. + | isoliert | — |
| *Jacanae*, Blatthühnchen ................. | m. + | | — |
| *Grues*, Kraniche ........................ | m. + | | — |
| *Alcae*, Alken .......................... | m. + | | — |
| *Colymbi*, Seetaucher ................... | m. + | | — |
| *Pici*, Spechtartige ..................... | o. — | | — |
| *Psittaci*, Papageien..................... | o. — | | — |
| *Macrochires*, Seglerartige................ | o. — | | — |

Da das Protoporphyrin, wie die *Tabelle 8* zeigt, bis jetzt bei Vogelarten, die 14 Ordnungen angehören, als Eischalenpigment spektroskopisch nachgewiesen ist [VÖLKER (*80*)], darf man wohl schließen, daß diesem Porphyrin eine ganz ausschließliche Verbreitung als Fleckenfarbstoff zukommt. Man kann, wie die Erfahrung zeigt, überall, wo Schalenflecke auftreten, auf die Anwesenheit dieses Farbstoffes schließen. Hierin machen auch die Trappen und die Nachtschwalben keine Ausnahme, in deren Federn sich Koproporphyrin findet. Das Vorkommen von Koproporphyrin in Federn und von Protoporphyrin in Eischalen ist für die Vögel also charakteristisch.

*Literaturverzeichnis: SS. 218—222.*

Es ist jedoch zu bemerken, daß die Zahl der Vogelordnungen mit porphyrinhaltigen Eischalen in Wirklichkeit weit höher liegen dürfte, als es in Tabelle 8 den Anschein erweckt, da es außer den hier genannten noch eine Anzahl von Ordnungen gibt, bei denen gefleckte Eier vorkommen, die vermutlich ebenfalls protoporphyrinhaltig sind. Weiterhin muß darauf hingewiesen werden, daß die Porphyrinvorkommen bei den einzelnen Ordnungen in quantitativer Hinsicht in sehr weiten Grenzen schwanken, was in Tabelle 8 nur mangelhaft zum Ausdruck kommt. In minimaler Menge kann Protoporphyrin auch Bestandteil weißer Schalen sein, z. B. bei den Eulen, Enten, Tauben und Hühnern. Demgegenüber ist das Auftreten von Porphyrin in Federn eine viel weniger weit verbreitete Erscheinung.

Völlig porphyrinfrei sind auch in frischem Zustand die weißen Eischalen der Spechte, Segler, Papageien und sicher mancher anderen Gruppe. Wie bei den Federn, so stehen auch bei den Eischalen Fälle von starker Porphyrineinlagerung solchen gegenüber, wo dieser Farbstoff völlig fehlt.

### 2. Turacin.

Das Turacin, wie der rote Farbstoff in den Schwungfedern vieler Helmvögel *(Turacos)* oder Bananenfresser *(Musophagiden)* heißt, hat schon früh das lebhafte Interesse der Forschung erregt. Der Grund hierfür ist zweifellos in einer Reihe von Eigenschaften zu suchen, die diesem Farbstoff eine Sonderstellung unter den Federpigmenten einräumen. Es darf deshalb nicht wundernehmen, wenn kein anderes Pigment in der Klasse der Vögel eine so vielseitige und gründliche Bearbeitung erfahren hat wie dieses. Von den zahlreichen, das Turacin behandelnden Untersuchungen seien nur diejenigen herausgestellt, die wesentliche Etappen auf dem Wege zu seiner Erforschung bedeuten.

Auf das berühmte, offenbar durch Wasser bedingte Abfärben der roten Turacofedern hat zuerst VERREAUX (73) aufmerksam gemacht, ein Verhalten, das später seine Erklärung fand, als die chemischen Eigenschaften des Turacins bekannt wurden. Der aufsehenerregende Nachweis des Kupfers im Molekül des Turacins ist von CHURCH (6) erbracht worden, dem es im weiteren Verlauf seiner grundlegenden Studien auch gelang, das Turacin in amorphem Zustand zu gewinnen. Die Analyse seines Produktes ergab Werte, die, wie wir heute durch die Untersuchungen von H. FISCHER wissen, der von der Theorie geforderten Zusammensetzung für das Turacin sehr nahekamen. Damit fällt CHURCH das Verdienst zu, mit dem Turacin das erste chemisch definierte Federpigment in fast reiner Form in Händen gehabt zu haben. Er stellte weiterhin fest, daß aus dem Turacin nach dem Abspalten des Kupfers mit konzentrierter Schwefelsäure ein Stoff entsteht mit dem Absorptionsspektrum des „Hämatoporphyrins", also eines eisenfreien Abkömmlings des Blutfarbstoffes. Mit dieser wichtigen Feststellung war die Zugehörigkeit des Turacins zur Reihe der Blutfarbstoffderivate erkannt.

In einer klassischen Untersuchung vermochten später H. FISCHER und HILGER (*17, 18*) das im Turacin komplex gebundene Kupfer mit Natriumamalgam abzuspalten und auf dem Umwege über die Leukoverbindung und die Reoxydation mit Luft den kristallisierten Ester eines Uroporphyrins darzustellen, den sie für Uroporphyrin „I" hielten. Unter Beibehaltung der Methode von FISCHER untersuchte dann RIMINGTON (*54*) das Turacin der Federn von 11 Turacus-Arten und fand bei der Decarboxylierung des Uroporphyrins zum Koproporphyrin in allen Fällen überraschenderweise nur Koproporphyrin „III", was besagt, daß auch im Turacin ein Uroporphyrin vom Typ „III" vorliegen muß. NICHOLAS und RIMINGTON (*51*) gelang es schließlich, den aus Turacin nach Abspaltung des Kupfers erhaltenen Uroporphyrin-„III"-ester nach chromatographischer Reinigung zur Kristallisation zu bringen (Schmp. 264°); sein Abbau lieferte ebenfalls Koproporphyrin „III". Damit ist eindeutig bewiesen, daß im Turacin das Kupferkomplexsalz des Uroporphyrins „III" vorliegt (Formel VII, S. 209).

Nach einer vereinfachten Methode gewinnt neuerdings WITH (*96, 97*)˙ das Uroporphyrin „III" aus dem Turacin, indem er die turacinhaltigen Federabschnitte mit konzentrierter Schwefelsäure tränkt und nach einer Stunde mit so viel absolutem Methanol übergießt, daß der Schwefelsäuregehalt des Gemisches 5% beträgt. Nach längerem Stehen wird mit wäßrigem Natriumacetat neutralisiert und der Ester mit Chloroform extrahiert. Seine Trennung und Reinigung wird dann nach NICHOLAS und RIMINGTON (*51*) durchgeführt.

Im Tierreich ist das Turacin bisher nur aus der Klasse der Vögel bekanntgeworden, und selbst hier ist sein Vorkommen etwas Einmaliges, da es nur in den roten Schwungfedern der *Turacos* (Helmvögel) als Schmuckfarbe auftritt. Man muß STRESEMANN (*70*) zustimmen, der das Turacin als einen der rein afrikanischen Gruppe der *Musophagiden* eigentümlichen Federfarbstoff charakterisiert. Auch hinsichtlich der Menge des in den Federn gespeicherten Farbstoffes ist es ein Sonderfall, denn es gelingt, aus einer einzigen Schwungfeder mit Ammoniak eine sehr farbstarke Lösung zu bereiten und aus ihr mit Eisessig das Turacin auszuflocken. Mit einem Gehalt von etwa 5% Porphyrin sind nach WITH (*96*) die roten Turaco-Federn das bisher bekannte porphyrinreichste Material überhaupt und die einzige Quelle für die Darstellung von reinem Uroporphyrin „III".

Relativ einfach ist der Nachweis des Turacins. Spektroskopisch erkennt man es leicht an der Lage seiner Absorptionsbanden (sogen.

*Literaturverzeichnis: SS. 218—222.*

„metallisches Spektrum" = Spektrum eines an Metall gebundenen Porphyrins), die im Keratin der Feder bei 582 und 541 m$\mu$, in verdünntem Ammoniak bei 562 und 526 m$\mu$ liegen. Man erkennt es ferner an der intensiv roten Porphyrinfluoreszenz im UV-Licht, die beim Behandeln der Federn mit konzentrierter Schwefelsäure infolge der Abspaltung des Kupfers auftritt. Und schließlich ist das Kupfer des Turacins an der grünen Flammenfärbung beim Verbrennen der Federn zu erkennen.

Uroporphyrinkupfersalz ist in Alkalien ungemein leicht löslich, so daß schon Spuren genügen, um das Pigment mit weinroter Farbe in Lösung zu bringen. Durch diesen Umstand wird das Abfärben der Turacos auch verständlich. Es darf daher nicht überraschen, wenn das Badewasser der Vögel infolge seines von Kotbeimengungen herrührenden Ammoniakgehaltes den Federn das Turacin allmählich entzieht und dadurch eine rötliche Färbung annimmt. Diese Verhältnisse wurden von KRUMBIEGEL (*41, 42*) quantitativ untersucht. Nach seinen Feststellungen besitzt Ammoniak in einer Verdünnung von 1 : 2 000 000 noch abfärbende Wirkung.

### 3. Pyrrolfarbstoffe im Stoffwechsel der Vögel.

Die wichtige Ermittlung von RIMINGTON (*54*), NICHOLAS (*51*) und WITH (*96*), nach der auch das Uroporphyrin der Turacusvögel dem Typ „III" angehört (Uroporphyrin „III" als Kupferkomplex), ergänzt die Vorstellung, die man bisher von der Biogenese der Porphyrine bei den Vögeln hatte, in klärender Weise, denn nunmehr lassen sich alle aus adulten Vögeln isolierten Porphyrine, das Protoporphyrin der Eischalen, das Koproporphyrin „III" der Trappen- und Eulenfedern und schließlich das Uroporphyrin „III" der Turacos, auf ihren gemeinsamen Grundfarbstoff, das Ätioporphyrin „III", zurückführen. Die Entstehung von Porphyrinen des Typs „III" ist in der Klasse der Vögel nach den vorliegenden Erfahrungen eine auffallend einseitige. Ist es doch bisher noch in keinem Falle gelungen, Porphyrine des Typs „I" vom Typ „III" abzutrennen und in Substanz zu fassen. Auch lassen die bei der Aufarbeitung der Porphyrine der Vögel gemachten Erfahrungen ein geringfügiges Vorkommen von Vertretern des Typs „I" neben solchen vom Typ „III" praktisch ausschließen. Lediglich während der Embryonalentwicklung ist bei den Vögeln die Fähigkeit zur gleichzeitigen Synthese von Porphyrinen beider Isomerenreihen noch vorhanden, wie die Isolierung von Koproporphyrin „I" aus dem Hühnerembryo zeigt [SCHØNHEYDER (*63*)]. Dieses ist dort gleichzeitig mit dem Auftreten von normalem Hämoglobin (Porphyrin Typ „III") nachweisbar. Demgegenüber ist für den adulten Säuger und Menschen der Dualismus der Porphyrine (nach H. FISCHER) die Regel, worunter man die grundsätzliche

Fähigkeit des Organismus zu verstehen hat, nebeneinander isomere Porphyrine vom Typ „I" und „III" zu erzeugen.

Porphyrine sind auch ein Bestandteil der normalen Exkrete vieler Vögel. Im Kot der Lachmöwe findet man Protoporphyrin, in den Fäzes der Turacos zuweilen Turacin (18), und in den normalen Ausscheidungen der Trappen und Eulen ist regelmäßig ein Gemisch von Proto- und Koproporphyrin nachweisbar (77). Demgegenüber überrascht das ausschließliche Vorkommen von Protoporphyrin in den Eischalen, wie das von Koproporphyrin in den Federn vieler Vögel. Dies mag in der Selektionsfähigkeit der Feder seine Ursache haben, möglicherweise aber auch darin begründet sein, daß nur das an Ort und Stelle gebildete Porphyrin jeweils zur Ablagerung gelangt. Dies würde allerdings voraussetzen, daß im Uterus nur Protoporphyrin, in der Federanlage dagegen nur Koproporphyrin gebildet wird. Für die letztere Möglichkeit spricht sehr eine Beobachtung von BORST (2), nach der die Synthese des Turacins erst in der Federanlage erfolgt. Solange wir über den Porphyringehalt von Blut und inneren Organen dieser Vögel noch nichts wissen, wäre es verfrüht, hier einen festen Standpunkt zu beziehen.

Die relativ starke Pigmentierung vieler Eischalen durch Pyrrolfarbstoffe — Protoporphyrin meist in Gesellschaft von Oocyan, einem Gallenfarbstoff — ist ohne weiteres verständlich. Diese Pigmente haben hier zweifellos die Aufgabe, die Eier der Freibrüter der Färbung ihrer Umgebung weitgehendst anzupassen (Tarnung). Indessen bereitet das Auftreten von Porphyrin in den Federn dem Verständnis einige Schwierigkeiten, da der Einlagerung des Farbstoffes hier kaum pigmentbildende Bedeutung zukommt. Offenbar stellt dieses Prinzip nur einen Versuch zur Pigmentierung der Feder dar. Dabei sind jene vereinzelten Fälle gewissermaßen als Ausnahmen zu werten, in denen auf dieser Grundlage dennoch Schmuckfarben entstehen, wie bei den weinroten Schopffedern einiger afrikanischer Trappen, z. B. bei *Lophotis r. ruficrista* und dieser nahestehenden Formen. Den durch Porphyrin rosenrot gefärbten Dunen der Trappen kommt wohl ebenfalls eine biologische Bedeutung zu, da nach SIEWERT (66) durch die Gefiederwandlung bei der Hochbalz des Großtrappen *(Otis tarda)* „die rosenrote Farbe der Dunen sichtbar wird". Völlig anders liegen dagegen die Verhältnisse bei den *Turacos*. Hier erfährt der lichtunbeständige Porphyrinring durch den Eintritt des Kupfers eine Stabilisierung, so daß auf diese Weise eine recht lichtechte und weithin sichtbare Schmuckfarbe zustande kommt.
Es hat nicht an Versuchen gefehlt, den Porphyrinstoffwechsel der Vögel, insbesondere den der Turacos, mit dem in pathologischen Fällen beträchtlich vermehrten Auftreten von Porphyrinen beim Säuger und Menschen zu vergleichen (H. FISCHER, DERRIEN). Dabei erblickte man

in der Feder ein Gebilde, das geeignet erschien, eine erhebliche Menge des „lichtgiftigen" Porphyrins (das erst bei der Belichtung giftig wirkt) in sich aufzunehmen und auf diese Weise aus dem Stoffwechsel zu eliminieren. Hierzu ist jedoch zu bemerken, daß die Feder nur während der kurzen Dauer ihres jährlichen Wachstums (Mauser) zu dieser Aufnahme fähig ist, da eine Einlagerung von Pigment in die fertige, verhornte Feder völlig ausgeschlossen ist. Die mögliche Funktion der Feder als Exkretions-organ für Porphyrine kann also nur mit dieser zeitlichen Einschränkung angenommen werden.

Die mannigfachen Fälle der Porphyrineinlagerung in Federn lassen sich vorläufig unter keinerlei einheitlichen Gesichtspunkten betrachten. Man findet diese Erscheinung in den verschiedensten systematischen Kategorien ebenso wie bei Vögeln der verschiedensten Lebensweise und Ernährung. Die Voraussetzungen für die (endogene) Entstehung von Porphyrin sind offensichtlich bei Pflanzenfressern (Trappen) und Fleisch-fressern (Eulen) die gleichen. Auch ist die nächtliche Lebensweise der Nachtschwalben und Eulen, die DERRIEN (10) für die Erhaltung des in ihren Federn gespeicherten Porphyrins verantwortlich macht, nicht von entscheidender Bedeutung, da andererseits die Trappen als ausgesprochene Tagvögel diese Erscheinung am intensivsten erkennen lassen. Das oft selbst innerhalb einer Ordnung verblüffend sporadische Auftreten von Porphyrin ist nicht ohne Analogie in der Tierreihe. So besitzen nach den Untersuchungen von H. FISCHER (20) unter den Muscheln nur einige wenige Arten die Fähigkeit, in ihren Schalen Porphyrin abzulagern, und der Nachweis von Porphyrin in den Igelstacheln [DERRIEN (10)] ist bis heute wohl der einzige bekannte Fall eines normalen Porphyrin-vorkommens im Integument der Säuger geblieben.

## VI. Chemisch ungeklärte Federpigmente.

Mit den bis jetzt bekannten, in der vorliegenden Darstellung charakterisierten Pigmenten sind die Möglichkeiten, die zur Färbung des Vogelgefieders beitragen, noch nicht erschöpft. Viele Pigmente sind jetzt als Vertreter der beiden wichtigsten Farbstoffklassen, der Carotinoide und der Pyrrolfarbstoffe, erkannt und ihre weite Verbreitung sichergestellt, doch harrt noch eine große Zahl von Pigmenten der Erforschung. Zunächst wären hier die Melanine, die verbreitetsten unter allen Pigmenten, zu nennen, die aber, wie eingangs erwähnt, als chemische Individuen vorläufig nicht charakterisierbar sind. Es sei jedoch in diesem Zusammenhang auf die Arbeit von FRANK (24) verwiesen, welche die vielfache Beteiligung der Melanine beim Zustandekommen der Feder-färbung aufzeigt.

Von den bis jetzt noch wenig charakterisierten diffusen Federpigmenten, deren Namen man oft im Schrifttum begegnet, seien einige
markante Beispiele herausgegriffen. Es sei jedoch bemerkt, daß die
chemischen Eigenschaften all dieser Pigmente von den bisher besprochenen
abweichen, so daß sie mit diesen nicht identisch sein können. Ferner
hat es durchaus den Anschein, daß es sich hier um Pigmente handelt,
denen keine größere Verbreitung zukommt, die vielmehr nur sporadisch
bei einzelnen Vogelarten auftreten.

Man findet:

*Gelbe Pigmente* bei: *Catarrhactes chrysocome* (Stirn). *Prionops alberti* (Kopf).

*Grüne Pigmente* bei: *Ithaginis sinensis* (Körper), „Phasianoverdin". *Turacus*-
Arten, einige (Körper), „Turacoverdin". *Parra variabilis* (Schwingen) und *Somateria
mollissima* (Kopf), „Zooprasinin".

*Rote Pigmente* bei: *Ithaginis sinensis* (Körper), „Phasianorubin". *Cicinnurus
regius* (Körper), „Zoorubin". *Rhodonessa caryophyllacea* (Kopf). *Megacephalon
maleo* (Körper).

Unbekannt ist auch noch die chemische Natur roter Pigmente, welche den
weißen Federregionen vieler Säger *(Mergus)*, Möwen *(Larus)* und Seeschwalben
*(Sterna)* einen rosenroten Anhauch während der Brutzeit verleihen, der aber nach
dem Tode des Vogels alsbald ausbleicht. Über den Einfärbungsvorgang als solchen
ist kürzlich bekanntgeworden, daß beim Pelikan *(Pelecanus onocrotalus)* das rote
Sekret der Bürzeldrüse beim Putzen der Federn in diese eingerieben wird, was
zu einer nachträglichen äußerlichen Einfärbung der fertigen (verhornten) Feder
führt [STEGMANN (68)]. Eine solche „Schminkfärbung" hält jedoch naturgemäß
nur so lange an, wie die Bürzeldrüse durch Lipochrom rot gefärbtes Sekret liefert.
Durch eine sehr auffällige Rostfarbe sind Teile des Gefieders von *Gypaëtus barbatus*
(Bartgeier), *Grus grus* (Kranich) und *Bubulcus ibis* (Kuhreiher) ausgezeichnet.
Die Farbe haftet den Federn dieser Arten offenbar nur äußerlich an und zeichnet
sich durch einen hohen Eisengehalt aus.

# VII. Rückblick.

Unsere jetzigen Kenntnisse von den Pigmenten im Gefieder der
Vögel lassen eine große Mannigfaltigkeit erkennen hinsichtlich der
Struktur und der Klassenzugehörigkeit der hierbei beteiligten Farbstoffe.
Faßt man die Fülle aller chemischen Individuen ins Auge, die farben-
erzeugend in der Feder wirken, so kann man darin das stoffliche Gegen-
stück zur morphologischen Struktur der Feder erblicken, die ja bekanntlich
das differenzierteste Hautgebilde der Wirbeltiere darstellt. Durch ihre
weite, an keine systematischen Grenzen gebundene Verbreitung heben
sich die drei großen Farbstoffklassen der *Melanine*, der *Carotinoide* und
der *Tetrapyrrolfarbstoffe* heraus, die als Stoffe endogener und exogener
Herkunft im Dienste der Pigmentierung des Vogels stehen. Daneben
gibt es aber noch einige kleinere Pigmentgruppen von mehr lokaler

*Literaturverzeichnis: SS. 218—222.*

Verbreitung, die offenbar die Funktion der großen übernehmen. Beispiele hierfür sind die gelben und roten Federfarbstoffe der Papageien und die ebenfalls nur sehr vereinzelt auftretenden fluoreszierenden gelben Pigmente bei Papageien und Paradiesvögeln. Es ist zu erwarten, daß ihre Zahl mit dem Fortschreiten der Forschung sich noch etwas erhöhen wird.

Geradezu erstaunlich ist bei vielen Federn die völlige Übereinstimmung ihrer Farbe, obwohl die sie erzeugenden Pigmente chemisch nichts miteinander zu tun haben. Man gewinnt daher den Eindruck, daß dem biologischen Bedürfnis nach einer ganz bestimmten Färbung vielfach mit den verschiedensten stofflichen Hilfsmitteln entsprochen wird. So entsteht *Gelbfärbung* in der Regel durch die Einlagerung von Lutein. Die gleiche Farbwirkung — von sicherlich derselben biologischen Bedeutung — wird bei den Papageien mit einem Pigment erreicht, das nur die Farbe und die ungesättigte Natur mit dem Lutein gemeinsam hat. Oder es können fluoreszierende Pigmente bei Papageien und Paradiesvögeln Gelbfärbung hervorrufen. Entsprechend kann die *Rotfärbung* der Feder ebenfalls auf ganz verschiedener stofflicher Grundlage erfolgen. Im allgemeinen sind es im Organismus entstandene, rot gefärbte Zersetzungsprodukte pflanzlicher Carotinoide, die hierzu dienen, und nur in einigen wenigen Ausnahmefällen übernehmen wohlcharakterisierte rote Carotinoide, wie Astaxanthin und Rhodoxanthin, diese Aufgabe. Wiederum sind es die Papageien, die bei der Rotfärbung ihrer Federn ihre eigenen Wege gegangen sind. Bei den Trappen und den Eulen bedingt die Einlagerung von Koproporphyrin einen Ansatz zur Rotfärbung von meist verborgenen Gefiederregionen, während schließlich bei den Turacos das Kupfersalz des Uroporphyrins zur Erzielung einer recht auffallend roten Farbe in den Schwungfedern dient. Auch bei der Pigmentierung der Vogeleischalen sind entsprechende Fälle von gleicher Farbwirkung auf verschiedener stofflicher Grundlage bekannt (Parallelentwicklung der Pigmentfarben).

Die Frage nach der stofflichen Verwandtschaft, also nach der gemeinsamen Basis aller Pigmentierungen, die in der Frühzeit der Pigmentforschung vielfach diskutiert wurde, ist zurückgetreten zugunsten einer ausgesprochen stoffwechsel-physiologischen Fragestellung. Hier gilt es, einerseits zu ermitteln, welche Pigmente auf dem Wege des tierischen Intermediärstoffwechsels entstehen, während andererseits die Aufgabe besteht, den Veränderungen nachzugehen, die sich an nicht körpereigenen Farbstoffen (Carotinoiden) in oft artspezifischer Weise im Organismus vollziehen, ehe sie als Pigmente Verwendung finden. Der Versuch, zur Klärung dieser Fragen beizutragen, dürfte eine lohnende Aufgabe künftiger biochemischer und physiologischer Forschung sein.

## Literaturverzeichnis.

*1.* BIEDERMANN, W.: Vergleichende Physiologie des Integumentes der Wirbeltiere. 3. Teil. Erg. Biologie 3, 354 (1928).

*2.* BORST, M.: Morphologie der Porphyrine. Naturwiss. 18, 1038 (1930).

*3.* BORST, M. und H. KÖNIGSDÖRFFER: Untersuchungen über Porphyrie. Leipzig: S. Hirzel. 1929.

*4.* BROCKMANN, H. und O. VÖLKER: Der gelbe Federfarbstoff des Kanarienvogels (*Serinus c. canaria* [L.]) und das Vorkommen von Carotinoiden bei Vögeln. Z. physiol. Chem. (Hoppe-Seyler) 224, 193 (1934).

*5.* BRUGSCH, J.: Porphyrine. Bedeutung, Stoffwechsel, Untersuchungsverfahren beim gesunden und kranken Menschen. Leipzig: J. A. Barth. 1952.

*6.* CHURCH, A. H.: Researches on Turacin, an Animal Pigment containing Copper. Philos. Trans. Roy. Soc. (London) 159, II, 627 (1870); 183, 511 (1893).

*7.* COCKAYNE, E. A.: The Distribution of Fluorescent Pigments in *Lepidoptera*. Trans. Entomol. Soc. London 1924, 1.

*8.* DANCKWORTT, P. W.: Lumineszenz-Analyse im filtrierten ultravioletten Licht. Leipzig: Akad. Verlagsges., 5. Aufl. 1949.

*9.* DERRIEN, E.: Sur la biologie des porphyrines naturelles. Bull. soc. chim. biol. (Paris) 8, 218 (1926).

*10.* DERRIEN, E. et CH. BENOIT: [Sur les porphyrines des phanères de certains vertébrés homéothermes]. Bull. soc. chim. France [4], 45, 689 (1929).

*11.* — — Quelques faits nouveaux pour l'histoire naturelle des porphyrines cuprifères. Bull. soc. chim. France [4], 45, 391 (1929).

*12.* DERRIEN, E. et J. TURCHINI: Sur les fluorescences rouges de certains tissus ou secreta animaux en lumière ultraparaviolette. C. R. Séances Soc. Biol. 92, 1028 (1925).

*13.* — — Nouvelles observations de fluorescences rouges chez les animaux. C. R. Séances Soc. Biol. 92, 1030 (1925).

*14.* DESSELBERGER, H.: Über das Lipochrom der Vogelfeder. J. Ornithol. 78, 328 (1930).

*15.* DORST, J.: Recherches sur la structure des plumes des *Trochilidés*. Mém. Muséum Nat. (Paris) N. S. (A) Zoologie I, 3, p. 125 (1951).

*16.* DRIESEN, H.-H.: Untersuchungen über die Einwanderung diffuser Pigmente in die Federanlage, insbesondere beim Wellensittich (*Melopsittacus undulatus* [SHAW]). Z. Zellforschg. 39, 121 (1953).

*17.* FISCHER, H. und J. HILGER: Zur Kenntnis der natürlichen Porphyrine. II. Über das Turacin. Z. physiol. Chem. (Hoppe-Seyler) 128, 167 (1923).

*18.* — — Zur Kenntnis der natürlichen Porphyrine. 8. Mitt. Über das Vorkommen von Uroporphyrin (als Kupfersalz, Turacin) in den Turakusvögeln und den Nachweis von Koproporphyrin in der Hefe. Z. physiol. Chem. (Hoppe-Seyler) 138, 49 (1924).

*19.* FISCHER, H. und F. KÖGL: Zur Kenntnis der natürlichen Porphyrine. IX. Über Ooporphyrin aus Kiebitzeierschalen und seine Beziehungen zum Blutfarbstoff. Z. physiol. Chem. (Hoppe-Seyler) 138, 262 (1924).

*20.* FISCHER, H. und H. ORTH: Die Chemie des Pyrrols. 2. Band. Pyrrolfarbstoffe. 1. Hälfte. Leipzig: Akad. Verlagsges. 1937.

*21.* FOX, D. L.: Animal Biochromes and Structural Colours. Cambridge: University Press. 1953.

*22.* — Astaxanthin in the American Flamingo. Nature (London) 175, 942 (1955).

*23.* Fox, H. M.: Farben im Tierreich. Endeavour **14,** Nr. 53, 40 (1955).

*24.* Frank, F.: Die Färbung der Vogelfeder durch Pigment und Struktur. J. Ornithol. **87,** 426 (1939).

*25.* Giersberg, H. und R. Stadie: Zur Entstehung der gelben und roten Gefiederfarben der Vögel. Z. vergl. Physiol. **18,** 696 (1933).

*26.* Gonnermann, M.: Zur Biologie der Kieselsäure und der Tonerde in den Vogelfedern. Z. physiol. Chem. (Hoppe-Seyler) **102,** 78 (1918).

*27.* Goodwin, T. W.: The Comparative Biochemistry of the Carotenoids. London: Chapman & Hall. 1952.

*28.* Görnitz, K.: Versuch einer Klassifikation der häufigsten Federfärbungen. J. Ornithol. **71,** 127 (1923).

*29.* Görnitz, K. und B. Rensch: Über die violette Färbung der Vogelfedern. J. Ornithol. **72,** 113 (1924).

*30.* Götz, W. H.: Über die Pigmentfarben der Vogelfedern. Verh. Ornithol. Ges. Bayern **16,** 193 (1925).

*31.* Heinroth, O.: [Die rote Färbung der Dunenbasis bei *Otis tarda*]. J. Ornithol. **74,** 563 (1926).

*32.* — Die Farbe der Trappendaunen. Sitz.-Ber. naturf. Freunde Berlin **1926,** 34.

*33.* Heinroth, O. und M. Heinroth: Die Vögel Mitteleuropas. 4 Bände. Berlin: Hugo Bermühler. 1924—1931.

*34.* Karrer, P. und E. Jucker: Carotinoide. Basel: Birkhäuser. 1948.

*35.* Koch, E. L.: Zur Frage der Beeinflußbarkeit der Gefiederfarben der Vögel. Z. wiss. Zoologie, Abt. A **152,** 27 (**1939**).

*36.* Koch-Grünberg, Th.: Zwei Jahre unter den Indianern [Brasiliens]. 1. Bd., SS. 84, 286. Berlin: 1909.

*37.* Krätzig, H.: Histologische Untersuchungen zur Frage der Struktur- und Farbveränderungen an Federn nach künstlicher (Thyroxin-)Mauser. Arch. Entwickl. Mech. **137,** 86 (1937).

*38.* Kritzler, H.: Carotenoids in the Display and Eclipse Plumages of Bishop Birds. Physiol. Zoölogy **16,** 241 (1943).

*39.* Krukenberg, C. F. W.: Die Farbstoffe der Federn. 1. bis 4. Mitt. Vergl.-physiol. Studien 1. und 2. Reihe. Heidelberg: Carl Winter. 1881—1882.

*40.* — Grundzüge einer vergleichenden Physiologie der Farbstoffe und der Farben. Vergl.-physiol. Vorträge. III. Heidelberg: Carl Winter. 1884.

*41.* Krumbiegel, I.: Versuche über das Abfärben des Turacins. J. Ornithol. **73,** 440 (1925).

*42.* — Versuche über das Abfärben der *Musophagiden*. Biol. Zbl. **45,** 735 (1925).

*43.* Kuhn, R. und H. Brockmann: Bestimmung von Carotinoiden. Z. physiol. Chem. (Hoppe-Seyler) **206,** 41 (1932).

*44.* Kuhn, R. und A. Winterstein: Über konjugierte Doppelbindungen. I—IV. Helv. Chim. Acta **11,** 87, 116, 123, 144 (1928).

*44a.* Lemberg, R.: Porphyrins in Nature. Fortschr. Chem. organ. Naturstoffe **11,** 299 (1954).

*45.* Lönnberg, E.: Zur Kenntnis der „Lipochrome" der Vögel. Ark. Zool. **21** A, No. 11 (1930).

*46.* — The Occurrence and Importance of Carotenoid Substances in Birds. Proc. 8th Internat. Ornithol. Congr. Oxford 1934, p. 410 (1938).

*47.* Mason, C. W.: Structural Colors in Feathers. II. J. Physic. Chem. **27,** 401 (1923).

*48.* Mattern, I.: Zur Histologie und Histochemie der lipochromatischen Federn einiger *Cotingiden* (Schmuckvögel). Z. Zellforschg. **45,** 96 (1956).

*49.* MEISE W.: Alte Sumpfohreule, *Asio f. flammeus* (PONTOPP.), mit Rosafärbung. Mitt. Ver. sächs. Ornithol. **5**, 89 (1936).

*50.* MEYER, A. B.: Über den Xanthochroismus der Papageien. Sitz.-Ber. Akad. Wissensch. Berlin **1882**, XXIV, 517.

*51.* NICHOLAS, R. E. H. and C. RIMINGTON: Isolation of Unequivocal Uroporphyrin III. A Further Study of Turacin. Biochemic. J. **50**, 194 (1951).

*52.* PALMER, L. S.: Carotinoids and Related Pigments. New York: Chemical Catalogue Co. 1922.

*53.* RENSCH, B.: Die Farbaberrationen der Vögel. J. Ornithol. **73**, 514 (1925).

*54.* RIMINGTON, C.: A Reinvestigation of Turacin, the Copper Porphyrin Pigment of Certain Birds Belonging to the *Musophagidae*. Proc. Roy. Soc. (London) B **127**, 106 (1939).

*55.* — Über das Vorkommen von Koproporphyrin III. Z. physiol. Chem. (Hoppe-Seyler) **259**, 45 (1939).

*56.* — Die Porphyrine. Endeavour **14**, Nr. 55, 126 (1955).

*57.* SCHENCK, E. G.: Über das Keratin der Federn. Z. physiol. Chem. (Hoppe-Seyler) **211**, 160 (1932).

*58.* SCHERESCHEWSKY, H.: Einige Beiträge zum Problem der Verfärbung des Gefieders beim Gimpel. Arch. Entwickl. Mech. **115**, 110 (1929).

*59.* SCHMIDT, W. J.: Altes und Neues über Strukturfarben im Tierreich. Gießener Naturwiss. Vorträge, Heft 6. Gießen: W. Schmitz. 1949.

*60.* — Wie entstehen die Schillerfarben der Federn? Naturwiss. **39**, 313 (1952).

*61.* — Über die Buckelreflektoren der grünen Federn von Flaumfußtauben (*Megaloprepia, Ptilinopus*). Ber. Oberhess. Ges. Natur- u. Heilkde. N. F. Naturwiss. Abt. **25**, 93 (1952).

*62.* — Polarisationsoptik und Farbenerscheinungen der lipochromführenden Federäste von *Xipholena lamellipennis*. Z. Zellforschg. **45**, 152 (1956).

*63.* SCHØNHEYDER, F.: The Formation of Coproporphyrin I and Hemoglobin During Embryonic Life. J. Biol. Chem. **123**, 491 (1938).

*64.* SCHÜZ, E.: Beitrag zur Kenntnis der Puderbildung bei den Vögeln. J. Ornithol. **75**, 86 (1927).

*65.* SCHWARZ, L., W. DECKERT und H. KETELS: Über Ooporphyrin in den Schalen von Hühner- und anderen Eiern und seine quantitative Bestimmung. Z. physiol. Chem. (Hoppe-Seyler) **312**, 37 (1958).

*66.* SIEWERT, H.: Die Balz des Großtrappen. Z. Jagdkunde **1**, 7 (1939).

*67.* STADIE, R.: Ein Beitrag zur hormonalen Beeinflussung der Gefiederfarben. Z. wiss. Zool. A **151**, 445 (1938).

*68.* STEGMANN, B.: Über die Herkunft des flüchtigen rosenroten Federpigments. J. Ornithol. **97**, 204 (1956).

*69.* STEINER, H.: Vererbungsstudien am Wellensittich, *Melopsittacus undulatus* (SHAW). Arch. Klaus-Stiftg. Vererbungsforsch. **7**, 37 (1932).

*70.* STRESEMANN, E.: Aves. In: KÜKENTHAL-KRUMBACH, Handbuch der Zoologie, VII, 2. Hälfte. Berlin u. Leipzig: W. de Gruyter & Co. 1927—1934.

*71.* TEST, F. H.: The Nature of the Red, Yellow, and Orange Pigments in Woodpeckers of the Genus *Colaptes*. Univ. Calif. Publ. Zoölogy **46**, 371 (1942).

*72.* VANNOTTI, A.: Porphyrine und Porphyrinkrankheiten. Berlin: J. Springer. 1937.

*73.* VERREAUX, J.: Observations on the Colouring-matter of the Wing-Feathers of *Turacoes*. Proc. Zool. Soc. London **1871**, 40.

*74.* Völker, O.: Über den gelben Federfarbstoff des Wellensittichs (*Melopsittacus undulatus* [Shaw]). J. Ornithol. **84**, 618 (1936).

*75.* — Über fluoreszierende, gelbe Federpigmente bei Papageien, eine neue Klasse von Federfarbstoffen. J. Ornithol. **85**, 136 (1937).

*76.* — Ein eigenartiges Prinzip der Federpigmentierung. Ornithol. Monatsber. **46**, 107 (1938).

*77.* — Porphyrin in Vogelfedern. J. Ornithol. **86**, 436 (1938).

*78.* — Zur Kenntnis des Porphyrins in Vogelfedern. Z. physiol. Chem. (Hoppe-Seyler) **258**, 1 (1939).

*79.* — Die gelben und roten Federfarbstoffe der Papageien. Biol. Zbl. **62**, 8 (1942).

*80.* — Über das Vorkommen von Protoporphyrin in den Eischalen der Vögel. Z. physiol. Chem. (Hoppe-Seyler) **273**, 277 (1942).

*81.* — Die stofflichen Grundlagen der Pigmentierung der Vögel. Biol. Zbl. **64**, 184 (1944).

*82.* — Über die Struktur der bei Vögeln vorkommenden Porphyrine. Z. Naturforsch. **2** b, 316 (1947).

*83.* — Die Isolierung von Lutein aus Vogelfedern und der Nachweis von Astaxanthin im Gefieder einiger Vögel. Z. physiol. Chem. (Hoppe-Seyler) **288**, 20 (1951).

*84.* — Die Lipochrome in den Federn der *Cotingiden*. J. Ornithol. **93**, 122 (1952).

*85.* — Über das Vorkommen von Rhodoxanthin im Gefieder einiger Vögel. Z. physiol. Chem. (Hoppe-Seyler) **290**, 223 (1952).

*86.* — Die Isolierung von Rhodoxanthin aus den Federn einer Flaumfußtaube. Z. physiol. Chem. (Hoppe-Seyler) **292**, 75 (1953).

*87.* — Das Farbkleid der Flaumfußtauben (*Ptilinopodinae*). J. Ornithol. **94**, 263 (1953).

*88.* — Über Umwandlungsprodukte pflanzlicher Carotinoide in den Federn von Tangaren, Spechten und Paradiesvögeln. J. Ornithol. **95**, 124 (1954).

*89.* — Die Isolierung von Astaxanthin aus den Federn des Rotbauchwürgers *Laniarius atrococcineus*. J. Ornithol. **96**, 50 (1955).

*90.* — Die experimentelle Rotfärbung der Vogelfeder mit Rhodoxanthin, dem Arillus-Farbstoff der Eibe (*Taxus baccata*). J. Ornithol. **96**, 54 (1955).

*91.* — Die experimentelle Rotfärbung des Gefieders beim Fichtenkreuzschnabel (*Loxia curvirostra*). J. Ornithol. **98**, 210 (1957).

*92.* — Die Rotfärbung der Flamingos (*Phoenicopterus*) im Freileben und in der Gefangenschaft. J. Ornithol. **99**, 209 (1958).

*93.* — Über die autochthone Entstehung des roten Lipochroms in den Kopffedern des Stieglitzes (*Carduelis carduelis*). J. Ornithol. **99**, 422 (1958).

*94.* Wallace, A. R.: A Narrative of Travels on the Amazon and Rio Negro. London. 1853, p. 294.

*95.* Weber, H.: Bewirkung des Farbwechsels bei männlichen Kreuzschnäbeln. J. Ornithol. **94**, 342 (1953).

*96.* With, T. K.: Uroporphyrin from Turacin: a Simplified Method. Nature (London) **179**, 824 (1957).

*97.* — Pure Unequivocal Uroporphyrin III, Simplified Method of Preparation from *Turaco* Feathers. Scand. J. Clin. and Lab. Invest. **9**, 398 (1957).

*98.* Zechmeister, L.: Carotinoide. Ein biochemischer Bericht über pflanzliche und tierische Polyenfarbstoffe. Berlin: J. Springer. 1934.

*99.* — Die Carotinoide im tierischen Stoffwechsel. Erg. Physiol. **39**, 117 (1937).

*100.* Zechmeister, L. und L. v. Cholnoky: Die chromatographische Adsorptionsmethode. 2. Aufl. Wien: J. Springer. 1938.

*101.* ZECHMEISTER, L. und P. TUZSON: Isolierung des Lipochroms aus Hühner-
    und Pferdefett. Z. physiol. Chem. (Hoppe-Seyler) **225**, 189 (1934).
*102.* — — Zur Kenntnis der selektiven Aufnahme von Carotinoiden im Tierkörper.
    Z. physiol. Chem. (Hoppe-Seyler) **234**, 235 (1935).
*103.* ZIEGLER-GÜNDER, I.: Pterine: Pigmente und Wirkstoffe im Tierreich. Biol.
    Revs. **31**, 313 (1956).

*(Eingelaufen am 5. Oktober 1959.)*

# *Cis-trans* Isomeric Carotenoid Pigments.

By **L. ZECHMEISTER**, Pasadena, California.

With 49 Figures.

**Contents.**

*Acknowledgement.* The writer wishes to thank Mrs. L. ZECHMEISTER and Dr. WARREN V. BUSH for their help. Drs. BUSH and F. J. PEȚRACEK have kindly read the proofs.

# I. Introduction.

As is well known, the carotenoid pigments belong to the class of polyenes and represent the most unsaturated mass products of biosynthesis. Several carotenoid pigments are extremely widespread in the vegetable and animal kingdoms. Although the name "poly-ene" would a priori include any compound whose molecules contain more than two carbon-carbon double bonds, the accepted use of the term is restricted to long, aliphatic, conjugated double bond systems.

The carbon skeleton of the carotenoids is highly branched and composed of isoprenic building blocks. Most naturally occurring carotenoids contain 40 carbon atoms, corresponding formalistically to 8 isoprene units. The number of conjugated C=C double bonds varies from 7 to 15 and is 10 or 11 in many common carotenoids. By synthesis, analogous systems up to 19 conjugated double bonds have been obtained (*132*).

The plant carotenoids show intense yellow, orange, red, or violet color. They are accompanied in the tissue by "colorless carotenoids" such as phytofluene (*311, 296, 241, 212*) in which only 5 double bonds are conjugated. Phytofluene (p. 316) is practically colorless but fluoresces intensely in ultraviolet light. The lowest member of this series is phytoene (p. 316) (*214, 220, 219*) with 3 conjugated double bonds; it shows neither color nor fluorescence. The two colorless compounds mentioned have the same carbon skeleton as the colored ones and can be dehydrogenated in vitro to carotenoid pigments (*297, 285, 146*).

It is a general feature of carotenoid pigments that the chromophore does not extend over the whole main carbon chain, although it occupies the largest, middle section of the molecule. The latter carries two characteristic "terminal groups" that have either aliphatic or hydro-aromatic or (exceptionally) aromatic structure. In a number of common carotenoids $\alpha$- or $\beta$-ionone rings are found at the molecule ends.

Since the two terminal groups may be either identical or different, both symmetrical and non-symmetrical carotenoid structures are known.

The number of the more or less well-characterized, naturally occurring carotenoids is about 80 to which several total-synthetic products which do not occur in plants have recently been added.

From the viewpoint of functional groups the carotenoid pigments include hydrocarbons, alcohols, aldehydes, ketones, carboxylic acids, epoxides, and combinations of these types. The oxygen function is located in terminal groups and not in the multiconjugated system. Evidently, this feature as well as the presence of characteristically arranged methyl side-chains is determined by the path of biosynthesis.

In some instances a $C_{40}$-carotenoid undergoes partial oxidative cleavage in the plant to give lower-molecular pigments carrying either aldehyde or carboxyl groups at the point(s) of attack. Examples: $\beta$-citraurin ($C_{30}$) (*317*) and $\beta$-apo-8'-carotenal ($C_{30}$) (*268a*); for some $C_{20}$- and $C_{24}$-dicarboxylic acids cf. Chapter X, p. 320.

In most $C_{40}$-carotenoids such as $\alpha$-carotene, $\beta$-carotene and their hydroxylated derivatives, etc., the hydroaromatic rings and the middle section of the molecule are held together by means of single bonds; this defines a *normal* (cyclohexenyl) structure (V). In less frequent instances, double bonds play this part and then the term *retro* (cyclohexylidene) structure (*197*) is used (VI, p. 228).

A detailed discussion of the biosynthesis, occurrence, isolation, structural clarification and in vitro synthesis of the carotenoids lies outside the scope of the present stereochemical survey, and the reader is referred to pertinent monographs (*277, 135, 78, 119*). Some in vitro conversions of natural carotenoids have been recently surveyed in this Series (*285*).

The structural formulas of some important carotenoids are listed on pp. 227–229; several other structures will appear later.

The conventional manner of writing carotenoid structures is shown in the formulas (I)–(IV), while in (V) and (VI) only carbon skeletons appear. Finally, the abbreviated symbols (VII)–(XXIV) are restricted to the terminal groups, and the two dots represent the following grouping:

## Nomenclature.

As shown in the *β*-carotene formula (I), the C-atoms of the main chain are numbered 1–15 and 1'–15', beginning at the carbons which carry *gem.* methyl groups (KARRER's nomenclature). In the presence of two different hydroaromatic rings, the *β*-ionone ring carries the unprimed numbers [example, (II)]; and when one of the terminal groups is an open chain, the latter is primed [example, (III)]. In the case of fully aliphatic carotenoids the number 1 and 1' will be assigned to the same carbon atoms as in the corresponding cyclic structures [example, (IV)].

The latter arrangement, of course, does not comply with assigning No. 1 to the first carbon of the longest chain according to the Geneva Nomenclature. Earlier, some authors began the numbering at the first carbons of the aliphatic chromophore (*99*). This system is no more in use except in the case of terminal aromatic groups (*71*). Our earlier proposal to assign italicized numbers to double bonds has likewise been abandoned (cf. *278*).

The term "stereoisomeric set" includes all possible *cis-trans* forms of a given carotenoid, and each stereoisomer is a "member" of the set. A "stereoisomeric (or *cis-trans*-isomeric) equilibrium mixture" is present in a solution after a rearranging treatment has been applied to an all-*trans* compound or any other member of the set. A *cis* isomer which is encountered in substantial amounts is termed a "main" or "preferred" isomer.

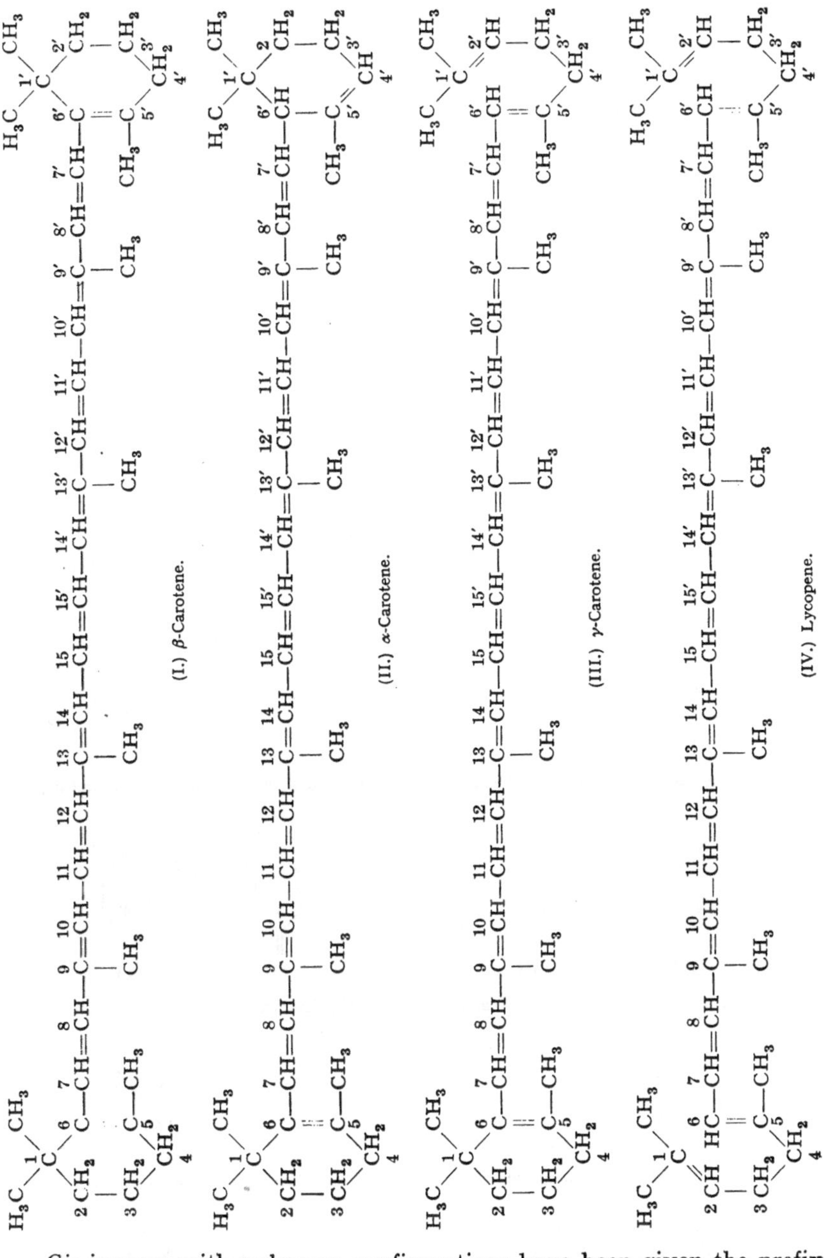

(I.) β-Carotene.

(II.) α-Carotene.

(III.) γ-Carotene.

(IV.) Lycopene.

*Cis* isomers with unknown configurations have been given the prefix "neo". In our laboratory neo A, neo B, etc. also mean that the *cis* isomer appears below the corresponding all-*trans* zone on the chromatographic column, while neo U (U for ultra), neo V, etc. indicate a location above

15*

(V.) β-Carotene *(normal)*.

(VI.) *retro*-Dehydrocarotene.

(VII.) β-Carotene.

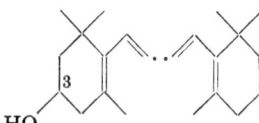

(VIII.) Cryptoxanthin (3-hydroxy-β-carotene).

(IX.) Isocryptoxanthin (4-hydroxy-β-carotene).

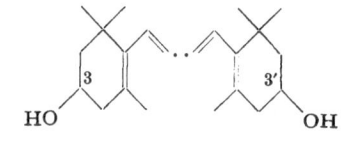

(X.) Zeaxanthin (3,3'-dihydroxy-β-carotene).

(XI.) Isozeaxanthin (4,4'-dihydroxy-β-carotene).

(XII.) Canthaxanthin (4,4'-diketo-β-carotene).

(XIII.) β-Carotene monoepoxide.

(XIV.) β-Carotene diepoxide.

(XV.) 3,4-Dehydro-β-carotene.

(XVI.) 3,4,3',4'-Bisdehydro-β-carotene.

*References, pp. 334—349.*

(XVII.) α-Carotene.

(XVIII.) α-Cryptoxanthin (3′-hydroxy-α-carotene).

(XIX.) Lutein (3,3′-dihydroxy-α-carotene).

(XX.) 3,4-Dehydro-α-carotene.

(XXI.) γ-Carotene.

(XXII.) Capsanthin (*27*; but see p. 310).

(XXIII.) Lycopene.

(XXIV.) Rhodoviolascin (spirilloxanthin) (*9, 122*).

the all-*trans* compound. Such differentiation does not always appear in older papers.

## Some Historical Remarks on Stereoisomerism of Polyenes.

In the following paragraphs only a few milestones in the history of *cis-trans* isomerism will be mentioned.

In 1819 a French pharmacist, POUTET (*215*), shook olive oil with a reagent that had been obtained by dissolving mercury in excess nitric acid and observed the solidification of the oil. POUTET was, of course, unaware of the fact that he had performed a *cis → trans* rearrangement of some unsaturated fatty acids about half a century before the foundation of modern stereochemistry was laid by VAN T'HOFF, LEBEL and others.

As is well known, VAN T'HOFF developed the basic, still valid theory of geometrical isomerism for ethylenic compounds (1875). The terms "*cis*" and "*trans*" were introduced by BAEYER much later (*7*).

Considerable time had elapsed before the successful stereochemical study of simple ethylenes could be extended to polyenes. The situation as it appeared in 1930 was correctly outlined by WITTIG and WIEMER (*270*) as follows:

„Daß die *cis-trans* Isomerie mit steigender Zahl der konjugierten Doppelbindungen in den Hintergrund tritt und schließlich ganz verschwindet, obwohl
die Zahl der Raumisomeren rasch zunehmen müßte, ist dann auf die wachsende
Beweglichkeit der Valenzelektronen zurückzuführen, die die Isomerisationen zu
den energieärmsten, stabilsten Lagen erleichtert. Ein lehrreiches Beispiel hierfür
sind die von R. KUHN dargestellten Diphenyl-polyene $C_6H_5(CH=CH)_xC_6H_5$, deren
Farbe sich mit wachsender Kettenlänge vertieft. Während die niederen Glieder
der Reihe mit einer und mit zwei Doppelbindungen *cis-trans*-Isomere bilden, sind
die höherkonjugierten Derivate nur in einer Raumform zu isolieren.''

This interpretation of the few then available data was, however,
not accepted by KUHN and WINTERSTEIN (*157*):

„Wir sind geneigt, die sterische Einheitlichkeit der höheren Diphenyl-polyene
wenigstens teilweise den zur Synthese angewandten Methoden zuzuschreiben,
ohne die Existenzfähigkeit entsprechender *cis*-Formen grundsätzlich zu bezweifeln.''

KUHN (*151*) has also drawn the attention to the natural product
bixin as an example of *cis-trans* stereoisomeric polyenes (p. 321).

Subsequent developments have shown that long conjugated aliphatic
systems undergo *trans* → *cis* rearrangements even more easily than do
isolated *trans* double bonds.

The broad field of *cis-trans* isomeric $C_{40}$-carotenoids was opened up
by GILLAM's pioneer studies in 1935–1936. GILLAM and EL RIDI (*75, 76*)
observed that, after *repeated* adsorptions on alumina columns, β-carotene
separated into two zones, an upper one containing unchanged starting
material and a lower one, termed pseudo-α-carotene, a new product.
According to GILLAM this isomerization process is reversible and leads
to the equilibrium, β-carotene ⇄ ψ-α-carotene. α-Carotene (*77*) behaves
similarly and forms the equilibrium, α-carotene ⇄ neo-α-carotene.

These changes were originally attributed to an action of the adsorbent
on the carotene: ". . . the analytical process itself affects the substances
which it is designed only to separate." It was soon shown, however,
by the writer in collaboration with TUZSON that the phenomenon is
independent of the adsorption (*318, 319*). Indeed, it takes place
spontaneously when a carotenoid solution is kept at room temperature
for several hours or days. In the case of the tomato pigment lycopene,
a continuous displacement of the spectral maxima towards shorter
wavelengths can be followed in the visual spectroscope, i. e. without
the use of an adsorbent (*320*).

It came to light that the lengthy operations, also including evaporation
of solutions, were responsible for the heterogeneity discovered by GILLAM;
this was confirmed by CARTER and GILLAM (*23*).

As expected, the reaction rates increased at higher temperature
(refluxing). A further, particularly strong promoting effect was observed
even at room temperature upon the addition of catalytic amounts of
iodine, in light [writer and TUZSON (*319, 320*; cf. *211, 298*)].

*References, pp. 334—349.*

Independently, STRAIN (*240*) had noticed (1938) that several leaf xanthophylls were reversibly altered by heat or some other agents and yielded complex chromatograms. While characteristic spectral data were reported, the thermal effect and the alleged adsorbent effect were not yet clearly differentiated in these early studies (cf. *291*).

As a theoretical interpretation of the isomerization GILLAM considered at first both geometrical isomerism and migration of a terminal double bond out of conjugation (*76, 77*). The latter assumption seemed reasonable at that time since, in contrast to $\beta$-carotene molecules which possess eleven conjugated double bonds, the color and spectrum of $\psi$-$\alpha$-carotene corresponded to the presence of only ten such bonds.

When studying the behavior of the main red pepper pigment, the polyene-ketone capsanthin *(ex Capsicum annuum)*, CHOLNOKY and the writer (*288*) reported in 1937 on some isomerization phenomena quite similar to those described by GILLAM. As possible explanations double bond migration, *cis-trans* rearrangement and keto-enol isomerism were considered.

When more experimental data had become available, it could be claimed with increasing emphasis that all conversions mentioned were *trans* → *cis* rearrangements (*320, 290, 289*). Valuable information was gained by studying the behavior of $\beta$-carotenone whose chromophore is blocked at both ends by conjugated carbonyl groups. Although this structure excludes double bond migration, the observed isomerization phenomena were the same as in the case of carotenoid hydrocarbons (*290*). Furthermore, as was pointed out later by HUNTER et al. (*95*), since the optical activity was maintained during the conversion, $\alpha$-carotene → neo-$\alpha$-carotene (*77*), no movement of the terminal 5,6-double bond could have taken place. The final decision in favor of geometrical isomerism was made on the basis of the high number of reversibly formed pigments, e. g. about a dozen in the case of $\beta$-carotene [POLGÁR and the writer (*209*)]. No other theory would allow for this richness of forms. Although this development encountered considerable scepticism at first [KARRER and RUTSCHMANN (*136*)], the basic concepts mentioned now appear to have been generally accepted.

A new epoch in polyene stereochemistry was initiated in 1950 when KARRER and EUGSTER (*130*) and, almost simultaneously, INHOFFEN et al. (*98*) reported the first total syntheses of $\beta$-carotene. In the last phase of these and several other syntheses certain *cis* intermediates appeared whose configurations follow beyond doubt from the very reaction sequence (cf. Chapter VI, p. 274).

In the course of the short history of polyene stereochemistry it has become increasingly clear that the carotenoid molecule is "morphologically sensitive" to spatial variations (stereomutation) in the sense that a single *trans* → *cis* shift may drastically alter the overall shape of the molecule. This feature is not shared by some other types of compounds which are rich in steric forms; thus, the general pattern of a sugar molecule is but little modified by epimerization.

Survey articles: CROMBIE (*36*), MACKINNEY (*175*), GOODWIN (*80*, *78*), WYMAN (*272*), BRODE (*19*), AMES (*5*), writer (*278*, *284*). Bibliography of papers published by our research group (*286*).

## II. Number and Types of *cis* Carotenoids. Steric Hindrance.

As mentioned, the carotenoids and related multiconjugated systems constitute a unique case among low-molecular weight substances because of the great number of possible geometrical isomers.

In order to calculate the total number of possible *cis-trans* isomers (*301,156*) the polyenes may be divided into "unsymmetrical" and "symmetrical" types; in the first type the two halves of the molecule are dissimilar but in the second type they are identical. If the conjugated system contains $n$ aliphatic double bonds, the number of possible stereoisomers $N$ will be:

For unsymmetrical systems: $N = 2^n$.
For symmetrical systems, $n$ odd: $N = 2^{(n-1)/2} \cdot (2^{(n-1)/2} + 1)$.
For symmetrical systems, $n$ even: $N = 2^{(n/2)-1} \cdot (2^{(n/2)} + 1)$.

Some pertinent values are listed in *Table 1*.

Table 1. Total Number $N$ of *cis-trans* Isomers Calculated for Polyenes Containing $n$ Sterically Effective Double Bonds.

| $n$ | $N$ (symmetrical molecules) | $N$ (unsymmetrical molecules) | $n$ | $N$ (symmetrical molecules) | $N$ (unsymmetrical molecules) |
|---|---|---|---|---|---|
| 1 | 2 | 2 | 7 | 72 | 128 |
| 2 | 3 | 4 | 8 | 136 | 256 |
| 3 | 6 | 8 | 9 | 272 | 512 |
| 4 | 10 | 16 | 10 | 528 | 1024 |
| 5 | 20 | 32 | 11 | 1056 | 2048 |
| 6 | 36 | 64 | 12 | 2080 | 4096 |

As will be shown in Chapter V (p. 264), any ordinary (all-*trans*) carotenoid can be converted by various treatments into a (quasi-) equilibrium mixture of *cis-trans* isomers. Theoretically, such a mixture should contain all possible spatial forms. Furthermore, when isolated from the mixture and submitted again to rearrangement, each stereoisomer should yield the same equilibrium. This postulate has been verified experimentally. Indeed, it must be fulfilled before an observed pigment can be accepted as a member of a given stereoisomeric set.

At present, there still exists a wide gap between the number of calculated and observed spatial forms of carotenoids. Experiment has shown that each carotenoid yields only a very limited number of *cis* isomers, mostly from two to a dozen. In no case did more than two or three main *cis* pigments appear whose respective quantities ·amounted to at least 10% of the stereoisomeric mixture.

Although the differences between the thermodynamic behavior of the many possible isomers do not seem to be excessive, the action of a selective mechanism is manifest whose nature is but partly understood. It is hoped that the configurations of an increasing number of *cis* isomers will be clarified in the future. Then it should become possible to work out a theory concerning the relationship between a given configuration, its relative stability and the probability of its appearance in a *cis-trans* isomeric equilibrium. At present the following points (a)–(c) can be made as an attempt to narrow the gap between the number of calculated and observed spatial forms of carotenoids.

(a) The composition of equilibrium mixtures is influenced by the chemical structure in the sense that it will be more complex in a non-symmetrical than in an analogous symmetrical set. In the latter instance each double bond (except the central one) finds its stereochemically equal counterpart in the other half of the molecule and thus the total number of possible *cis* configurations is reduced. The probability of formation of a given preferred *cis* double bond increases by a factor of two. The observed composition of the *cis-trans* isomeric mixture is simpler in some symmetrical than in comparable unsymmetrical sets.

Examples (*309, 209*). The relative amounts of the *cis* isomers were found as follows: α-Carotene set (unsymmetrical), neo U : neo V : neo W : neo B : minor *cis* isomers = 15 : 3 : 16 : 13 : 3; β-carotene set (symmetrical), neo U : neo B : minor isomers = 22 : 25 : 5. The unsymmetrical all-*trans*-γ-carotene yields a very complex equilibrium mixture but in the case of the symmetrical lycopene practically only a single *cis* isomer (neo A) is present, to the extent of 50% of the total pigment.

(b) A second restricting factor is the spectroscopically established feature that the rearrangement of all-*trans* carotenoids yields mainly mono*cis* and di*cis* (but not poly*cis*) forms (cf. $\lambda_{max}$ shifts, p. 294). This is understandable as follows. Suppose that the equilibrium ratio, mono-*cis*/all-*trans* is of the order of magnitude $^1/_{10}$. If it is assumed for the sake of simplicity that roughly the same change in free energy accompanies a *trans* → *cis* rotation about each double bond, then this ratio, e. g. for a tetra*cis* form, would amount to only $^1/_{10\,000}$. In other words, the isomer would not be detectable by any current analytical method (*300*).

That indeed the probability of the appearance of a given configuration decreases rapidly with the increasing number of *cis* double bonds, was demonstrated by PETRACEK and the writer (*303*) in the following, unusually

large-scale experiment; it has shown that the order of magnitude
of poly*cis* lycopenes in thermally obtained equilibria must be less than
1 part in 3 million parts of the total pigment.

A benzene solution of 30 g. of pure, crystalline all-*trans*-lycopene was refluxed
for 30 min. and submitted to careful chromatographic fractionation. None of
the pigment fractions showed upon iodine catalysis the migration of the spectral
maxima typical for poly*cis* forms (Fig. 17, p. 284). This test turned, however, positive,
when 40 μg. of prolycopene was introduced. Furthermore, as little as 10 μg. of
added poly*cis* lycopene when washed through the chromatographic column was
detected easily in the first liter of the filtrate.

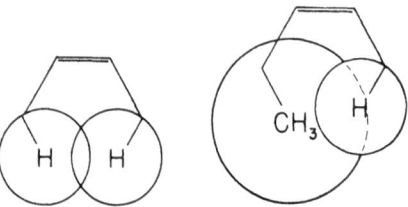

Fig. 1. Overlapping of hydrogen atoms in —CH—CH=CH—CH— and of hydrogen and methyl in
—CH—CH=CH—CCH₃ with *cis* configuration; according to PAULING (*201*). [From: Fortschr. Chem.
organ. Naturstoffe 3, 203 (1939).]

(c) In order to evaluate a third and important restricting factor,
the double bonds of a carotenoid are divided into two types, termed,
respectively, "unhindered" and "hindered" ones. This distinction was
proposed by PAULING (*201*):

A *cis* double bond located in a polyene chain will have the orientation,

$$\begin{array}{ccc} H & & H \\ \diagdown & & \diagup \\ & C=C & \\ \diagup & & \diagdown \\ =C & & C= \\ \diagdown & & \diagup \\ & X \quad X' & \end{array}$$

In carotenoid molecules $X$ and $X'$ represent either hydrogen atoms
or methyl groups. When both $X$ and $X'$ are hydrogens the formation
of the *cis* isomer by rotation about a *trans* double bond does not encounter
any significant steric hindrance. This, however, does happen when $X$
represents a hydrogen and $X'$ a methyl *(Fig. 1)*. Then a *trans → cis*
rearrangement is opposed by a considerable strain, and a significant
energy barrier must be overcome.

Inspection of the β-carotene formula (I, p. 227), for example, will
show that the molecule contains both types of double bonds, viz. out
of the nine aliphatic such bonds only five are unhindered, namely the
four that carry methyl groups (positions 9, 13, 13', 9') and the central
double bond (position 15).

When an all-*trans* carotenoid is subjected to stereoisomerization, the formation of unhindered and hindered *cis* isomers will compete with each other and the former type will be given preference. Recent work has shown that the presence of a hindered *cis* double bond in a polyene molecule degrades the spectral curve in the visible region in the sense that the extinction values and the degree of fine structure decrease substantially; thus, the fundamental band flattens (cf. Fig. 17, p. 284). Degraded spectra did, however, as a rule not appear on *trans → cis* rearrangements and hence no hindered *cis* isomers were formed. *Table 2* shows the practical importance of this simplification: the calculated

Table 2. Calculated Number of the Sterically Unhindered and Total *cis-trans* Isomeric Forms of Certain $C_{40}$-Carotenoids.

| Name | Symmetrical (s) or unsymmetrical (u) structure | Number of aliphatic, conjugated C=C double bonds | Number of theoretically possible steric forms | |
|---|---|---|---|---|
| | | | unhindered | total |
| α-Carotene . . . . . . . . | u | 9 | 32 | 512 |
| β-Carotene . . . . . . . . | s | 9 | 20 | 272 |
| γ-Carotene . . . . . . . . | u | 10 | 64 | 1024 |
| Lycopene . . . . . . . . . | s | 11 | 72 | 1056 |
| Lycoxanthin . . . . . . | u | 11 | 128 | 2048 |
| Cryptoxanthin . . . . . | u | 9 | 32 | 512 |
| Zeaxanthin. . . . . . . . | s | 9 | 20 | 272 |
| Lutein . . . . . . . . . . . . | u | 9 | 32 | 512 |
| Astacin . . . . . . . . . . | s | 9 | 20 | 272 |
| Capsanthin . . . . . . . . | u | 9 | 32 | 512 |
| β-Carotene mono-epoxide . . . . . . . . . | u | 9 | 32 | 512 |
| Violaxanthin (zeaxanthin diepoxide) | s | 9 | 20 | 272 |

number of unhindered *cis* isomers amounts to only 6–7% of that of the total possible isomers.

In the case of β-carotene only 20 unhindered isomers were expected and 12 were observed (*209*)—a reasonably good agreement. *Fig. 2* (pp. 236 and 237) represents the skeleton models of the 20 unhindered β-carotenes.

To summarize, several factors have been helpful in the interpretation of the existing gap between the calculated and found number of isomers, viz. structural features, the low probability of the formation of poly*cis* forms, and the sterically hindered nature of some double bonds.

Two decades ago, when PAULING (*201*) developed his theory of hindered and unhindered double bonds, the only available way for the preparation of *cis* carotenoids was direct rearrangement. Subsequently, several authors (among them the present writer) more or less tacitly interpreted his "rule" as being prohibitive for the existence of hindered

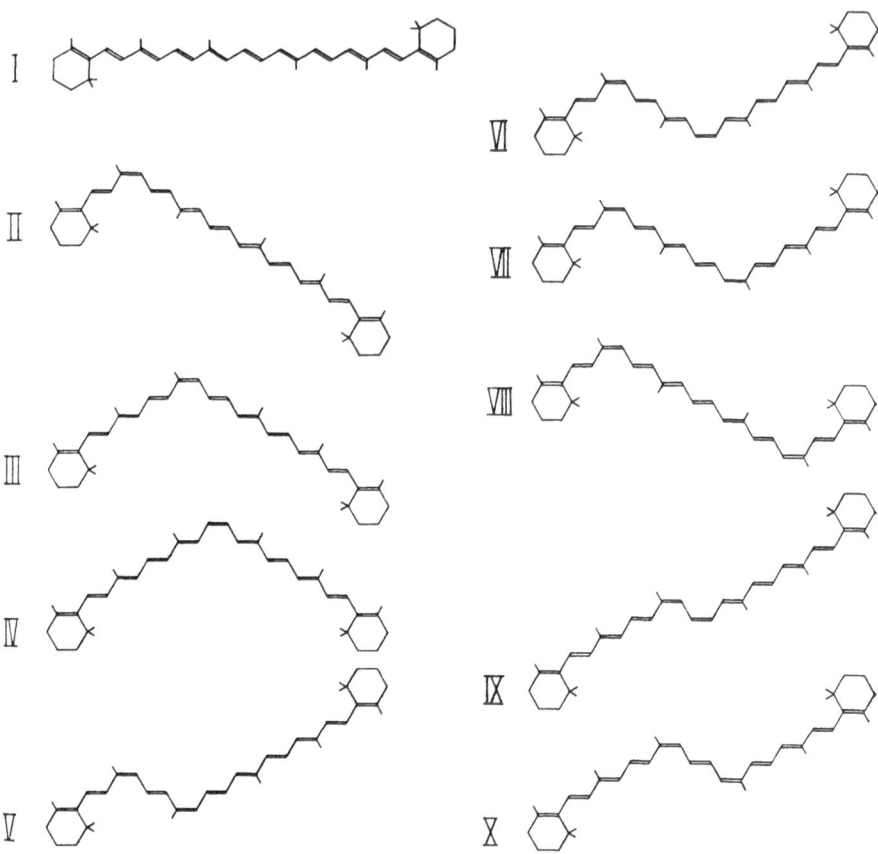

Fig. 2. Skeleton models of the twenty possible unhindered stereoisomers of β-carotene: all-*trans*-β-carotene, three mono*cis*-β-carotenes, six di*cis*-β-carotenes, six tri*cis*-β-carotenes, three tetra*cis*-β-carotenes, and one penta*cis*-β-carotene (*278*). [From: Chem. Rev. **34,** 267 (1944).]

*cis* forms in the $C_{40}$-carotenoid series. It can now be stated as the outcome of a controversy (*138, 203, 58, 60, 71, 74*) that while it is indeed difficult or impossible to overcome the energy barrier in the direct process, *trans* → hindered *cis*, this obstacle can be circumvented by following an entirely different route, namely total synthesis. As will be shown in Chapter VI (p. 274) a triple bond is built up at the site of the prospective hindered *cis* double bond and is then catalytically semi-hydrogenated in a stereospecific manner to give a *cis* olefin. Thus, a hindered *cis* carotenoid is obtained in good yield.

The first clear exception from the "rule" was reported by OROSHNIK, KARMAS and MEBANE (*197*) in the *retro*-vitamin A methylether set. Soon after the same research group synthesized hindered *cis* isomers of several isoprenic dienes, trienes, tetraenes and pentaenes of the $C_{20}$-series (*197a*).

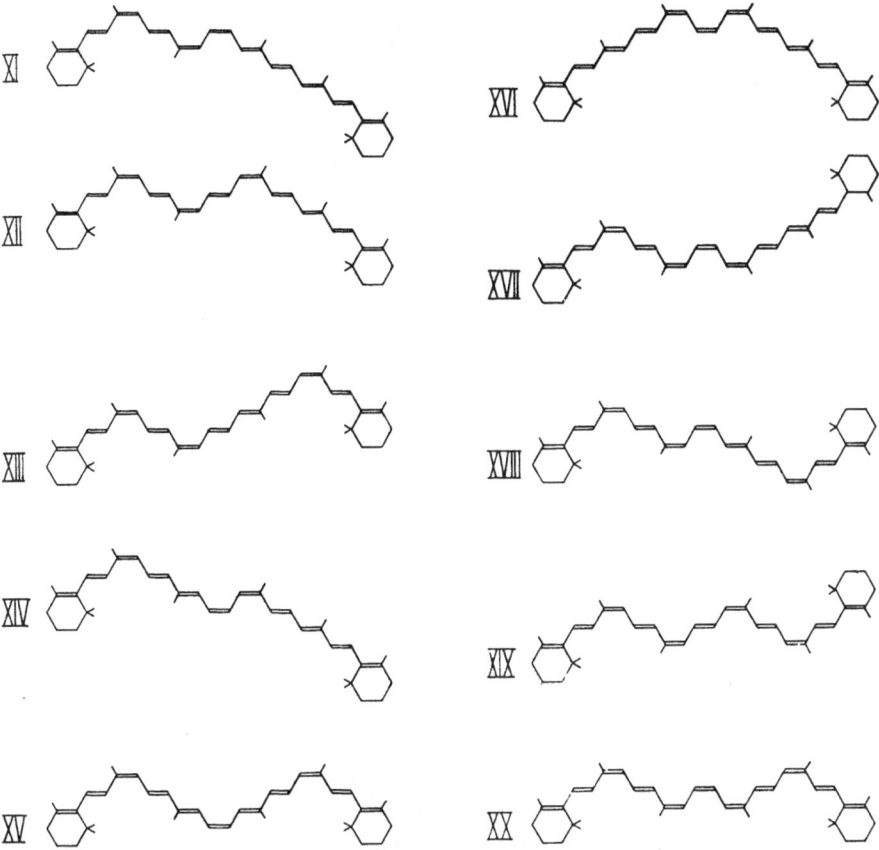

Fig. 2.

KARRER et al. were successful in synthesizing hindered *cis* forms of β-caro-tene, lycopene, and 1,18-diphenyl-3,7,12,16-tetramethyl-octadecanonaene (*60, 71, 72, 74, 170*). Even such hindered *cis* isomers were prepared that carried phenyl, instead of methyl, side-chains (*324*) (p. 320).

A slightly different type of steric conflict between a H-atom and a CH₃-group does not seem to have been realized with carotenoid pigments; it would be caused by a *s-trans* conformation at the 6,7-bond (cf. *197a, 18, 262*):

*s-cis* (6,7)
(interference of the 8-hydrogen with the 5-methyl).

*s-trans* (6,7)
(interference of the 8-hydrogen with one of the 1-methyls).

In case the chromophoric carbon skeleton is unbranched, e. g. in diphenyl-polyenes, the moderate steric hindrance operative between *ortho* H-atoms of the benzene ring and some H-atoms of the aliphatic chain does not prevent the preparation of hindered forms by direct rearrangement (*299*).

# III. Some Properties of *cis* Carotenoids.

## 1. Relative Stabilities.

An all-*trans* carotenoid, because of its coplanar chromophore and nearly perfect resonance, is generally expected to surpass in stability its *cis* isomers. Nevertheless it should be pointed out that simple characterizations such as "stable" or "labile" have lost much of their meaning in the class of multiconjugated systems. Because the degree of the double bond character is subject to considerable variation along the chain, the extent of lability depends not only on the number but also on the location of *cis* double bonds, hence on the overall shape of the molecule.

*Sterically Unhindered cis Forms.* Various stereoisomerizing treatments (especially thermal as compared to photochemical) may affect individual configurations quite differently. For instance, during the (theoretical) transition, all-*trans* → mono*cis* → di*cis* → poly*cis*, the thermostability in solution first decreases and then increases again when the bent molecule gradually straightens out and assumes a compact overall shape. This behavior is contrasted with the increasing photochemical sensitivity in the presence of iodine during the above process.

Considerable variations in the stereostability have also been noted during crystallization. All-*trans* pigments and certain natural and artificial *cis* forms crystallize easily and without change. In other instances, however, either no crystallization takes place or spatial rearrangement competes with the formation of the crystal grating. Thus, neo-$\beta$-carotene B (GILLAM's $\psi$-$\alpha$-carotene, p. 230) crystallizes easily but (in our hands) even fresh solutions of the crystals contained some all-*trans* form and afforded a complex chromatogram. Still more sensitive is the configuration of the main lycopene isomer, neo A, that yields exclusively all-*trans* crystals (*319*). The closely related, synthetic central-mono*cis*-lycopene rearranges to the all-*trans* pigment when heated for a short time during recrystallization (*110*).

Recent studies into the behavior of several central-mono*cis* carotenoids have shown that this highly bent molecular shape causes strong photo-sensitivity (pp. 269, 293).

Several mono*cis* carotenoids of the neo U type which according to the spectra (p. 296) must possess a peripherally located *cis* double bond and a slightly bent molecular shape, are distinguished by a high degree of thermostability. They remain unchanged when the solution is refluxed and crystallize rapidly on cooling.

Examples: neo-α-carotene U and neo-β-carotene U (*309, 209*).

*Sterically Hindered cis Forms.* As we have seen, these isomers show, once formed, considerable photo- and thermostability. Some of them are well crystallized and can be heated almost to their melting points without suffering rearrangements.

Examples. 11,11'-Dicis-β-carotene (*cis*-β-carotene B) (*61, 108*); *cis*-β-carotene C (*61*); 5-*cis*-1,18-diphenyl-3,7,12,16-tetramethyl-octadecanonaene and its 5,13-di*cis*isomer (*71*); several *cis*1,3,7,12,16,18-hexaphenyl-octadecanonaenes (*324*). The hindered *cis* lycopenes b, c and c' are much more labile and could not be crystallized (*74*).

All known sterically hindered *cis* carotenoids are extremely sensitive to iodine, in light, and the stereochemical equilibrium is reached at much higher rates than in the case of an all-*trans* or unhindered *cis* isomer.

*Polycis Forms.* The behavior of the naturally occurring poly*cis* carotenoids will be discussed in Chapter VII, p. 280.

## 2. Melting Points.

The situation is simple with reference to solubilities and melting points, since in each stereochemical set the least soluble and highest melting member is the all-*trans* compound. The melting point depression caused by a *trans* → *cis* rotation is considerable and amounts to 20–120°. The effect is not proportional to the number of *cis* double bonds. Neither hindered nor poly*cis* isomers show any particular behavior in this respect.

In some instances unsharp melting points are observed because of partial rearrangement in the fused or sintered state; some isomers are converted rapidly into the corresponding all-*trans* compounds and show the phenomenon of double melting points.

Examples. (a) Central-mono*cis*-lycopene melts at about 105°, then solidifies and melts again at higher temperature (*110*). (b) Central-mono*cis*-3,4,3',4'-bis-dehydro-β-carotene melts between 135° and 140°, and for the second time at 193° (m. p. of the all-*trans* form, 196–197°) (*105, 115*). (c) 5,13-Dicis-1,18-diphenyl-3,7,12,16-tetramethyl-octadecanonaene, a hindered isomer, becomes considerably darker at 155° and melts at 210° which is the m. p. of the all-*trans* form (*60*).

## 3. Rotatory Power.

In the presence of an asymmetric carbon atom, *trans* → *cis* rotations involve changes in the optical activity.

Thus, according to Strain's early observations (*240*) and those made by our group (*290*), the *levo* rotation of natural zeaxanthin is inverted to *dextro* by heating or iodine catalysis. The resulting solution contains a strongly dextrorotatory isomer. It was also found ($316$) that all-*trans*-gazaniaxanthin, whose rotatory power was unmeasurably small, afforded with iodine a stereoisomeric mixture showing $[\alpha]_D^{20} = + 160°$. After chromatography, some weakly adsorbed *cis* forms had the value $+ 220°$ that decreased to $+ 155°$ when the catalytic treatment was repeated.

Such experiments should be evaluated with caution, especially when the rotatory power decreases, since racemization and/or irreversible cleavage may interfere. It should be demonstrated that, upon re-isomerization of a *cis* form an all-*trans* sample is obtained that shows the initial optical activity.

Table 3. Examples of the Influence of the Configuration on the Rotatory Power.

| Stereoisomeric set | Solvent | Steric form | $[\alpha]_{Cd}^{25}$ or $[\alpha]_C^{20}$ | References |
|---|---|---|---|---|
| α-Carotene | chloroform | All-*trans* | $+ 359°$ | ($309$) |
|  |  | Neo U | $+ 221°$ |  |
| Gazaniaxanthin | petroleum ether | All-*trans* | $\pm 0°$ | ($316$) |
|  |  | Neo-Group I | $+ 220°$ |  |
| Zeaxanthin | chloroform | All-*trans* | $- 42.5°$ | ($290$) |
|  |  | Neo A | $+ 120°$ |  |
|  |  | Neo B | $\pm 0°$ |  |
| Capsanthin | benzene | All-*trans* | $\pm 0°$ | ($289$) |
|  |  | Neo A | $+ 89°$ |  |
|  |  | Neo B | $+ 21°$ |  |
|  |  | Neo C | $+ 27°$ |  |
| Capsanthin dipalmitate | petroleum ether | All-*trans* | $- 30°$ | ($289$) |
|  |  | Neo A | $- 22°$ |  |
|  |  | Neo B | $- 20°$ |  |
| Capsorubin | benzene | All-*trans* | $\pm 0°$ | ($289$) |
|  |  | Neo A | $- 134°$ |  |
|  |  | Neo B | $- 69°$ |  |
| Capsorubin dipalmitate | petroleum ether | All-*trans* | $\pm 0°$ | ($289$) |
|  |  | Neo A | $- 75°$ |  |
|  |  | Neo B | $- 15°$ |  |

*Table 3* refers to crystalline products. We note that in some instances the rotatory power of one steric form by far surpasses that of all others. On the basis of spectral readings such isomers seem to be mono*cis* compounds containing the *cis* double bond at or near the center. It is still unknown which particular configuration type is responsible in general for a substantial increase of the molecular asymmetry. Both theoretical and experimental problems would merit closer attention in this interesting field.

## 4. Relative Adsorption Affinities.

It has been established empirically, in the absence of a pertinent theory, that the relative adsorption affinities of stereoisomeric carotenoids are so sharply dependent on the configuration that the resolution of even complicated mixtures can be realized easily. Indeed, the systematic use of chromatography has made possible substantial progress in the polyene field (surveys, *283, 279, 280, 278, 163*). In considerable contrast, classical methods such as fractional crystallization may be satisfactory in the field of simple olefins.

Column chromatographic techniques have been applied with advantage in countless instances to the resolution of *cis-trans* isomeric mixtures and isolation of stereochemically pure, individual *cis* carotenoids. While the reader is referred to some monographs (*287, 283, 243, 163, 135*), only a few practical hints can be offered at this point.

Frequently used adsorbents are, calcium hydroxide (+ Celite), calcium carbonate, alumina, magnesia, zinc carbonate; and developers, hexane, petroleum ether, benzene, hexane-benzene mixtures, and hexane containing a polar solvent [cf. (*283*)].

In our laboratory, a series of hexane-acetone mixtures are kept ready for use, ranging from 99.5 : 0.5 to 80 : 20. When a new substance is to be studied, first pure hexane, then hexane with a low acetone content is tried out as a developer. The acetone content is increased gradually until optimum resolution takes place.

At the end of the developing process we extrude the column and excise the individual zones; fractional washings are often less reliable. One can, however, allow weakly adsorbed components such as carotenes and phytofluene, to pass into the filtrate, and separate by cutting the more strongly adsorbed pigments. In order to detect fluorescent impurities, inspection of the chromatogram by means of a portable ultraviolet lamp, under a black cloth, is recommended.

Each individual pigment is eluted with a solvent more polar than the developer such as hexane-acetone 1 : 1. After addition of water, the acetone can be removed by using LeRosen's automatic device (*164*).

For relative eluting strengths of solvents and $R_f$ values of *cis* β-carotenes, see Bickoff (*12*). — Combination of column chromatography and countercurrent distribution: Curl (*38–40*). — Paper chromatography of carotenoids: Grangaud and Garcia (*85*), Bauer (*10*), Lederer (*163*). This method has been scarcely used for separating stereoisomeric polyenes.

Table 8 (p. 266) lists a few data concerning relative positions of stereoisomeric carotenoids on the column.

Unfortunately, the theory is lagging behind practical experimentation in this field, perhaps because very little is known about the orientation of the individual carotenoid isomers on an active surface or about the strength of the adsorption forces as a function of the molecular form. Nobody has yet been able either to predict or to explain why a *trans* → *cis* rotation, i. e. the bending of a rod-like, all-*trans* carotenoid molecule may either increase or decrease the adsorption affinity with the result that in a chromatogram certain *cis* zones appear above but others below the unchanged portion of the all-*trans* pigment.

Evidently, this dual effect has a character quite different from the all-*trans* → *cis* spectral shift which invariably weakens the color of carotenoids. Under the influence of spatial rearrangements adsorbability and light extinction may change either in a parallel or in an opposite sense.

Let us now endeavor to summarize some aspects of the relationship between adsorption behavior and geometrical isomerism. When evaluating the following paragraphs it should be kept in mind that any statement on adsorption forces is valid only for a given system, consisting of substance, adsorbent, and developer. By changing the adsorbent or developer or both, the chromatogram may be altered substantially.

We assume that not every section of an adsorbed organic molecule will be equally responsible for the fixation process but certain "anchoring groups" will play the decisive part (*280, 283, 279*). In favorable instances the anchoring group can be identified experimentally by demonstrating that modification or elimination of that particular section of the molecule causes an unusually drastic change in the adsorption behavior (cf. some statements on dihydroxy-carotenes and their esters on p. 243). Within a complex molecule two or more potential anchoring groups may compete for the active spots of the adsorbent, and it seems that in principle more than one orientation of the molecule is possible on the surface. Consequently, certain changes in the chromatographic system may induce even an inversion of the top-to-bottom sequence of two substances.

As a characteristic example, the behavior of the pair, lycopene (IV, no rings, 11 conjugated and 2 isolated double bonds, no OH-group) and cryptoxanthin (VIII, 2 rings, 11 conjugated double bonds, 1 OH-group) may be mentioned. On benzene-developed alumina or calcium carbonate columns the sequence is, cryptoxanthin (top) and lycopene (bottom) which indicates that the hydroxyl group is responsible for the stronger fixation of cryptoxanthin; its effect overrules that of the greater number of double bonds present in lycopene. The top-to-bottom sequence is inverted, however, on lime where the hydroxyl does not seem to function as the (main) anchoring group (*165*). Some other impressive inversions were reported by STRAIN (*244–246*). It would be well worth while to investigate how far inversion phenomena can be realized pertaining to the chromatographic sequence of several *cis* forms of a given carotenoid.

Let us now compare the changes in adsorption affinities that are induced, respectively, by stereochemical and by (reasonably chosen) structural alterations of the molecule. Experiments have shown that structural and stereochemical factors are here of equal order of magnitude.

To illustrate, it was rather surprising to observe that the difference between the adsorption affinities of all-*trans*-β-carotene (VII, 11 conjugated double bonds, top position) and all-*trans*-α-carotene (XVII, 10 such bonds, bottom position) can be overruled by bending suitably the α-carotene molecule. After a mixture of the two all-*trans* carotenes had been subjected to a treatment with iodine and chromatography, one of the *cis* α-carotenes (neo U) appeared *above* the all-*trans*-β-carotene zone (*309*).

The top-to-bottom sequence was: neo-β V, neo-α U, all-*trans*-β, neo-α V, neo-β B, neo-β E, neo-α W, neo-β F, all-*trans*-α, and neo-α B.

Were the effect of the structural difference between β- and α-carotene of a higher order of magnitude than that of configurational differences, then all observed members of the β-carotene set would be expected to appear in a top section of the column followed by the well-separated family of stereoisomeric α-carotenes.

Although the adsorption forces are profoundly influenced by the number of *cis* double bonds in a polyene molecule, the dependence is not a simple one, because the effect is also a function of the position of these bonds.

In the subclass of the carotenoid hydrocarbons, the entirely aliphatic lycopene yields exclusively such *cis* isomers that show weakened adsorption affinities and appear below the all-*trans* zone (*319, 320, 300*). A treatment of either of the three hydroaromatic carotenes, α-, β-, or γ-, results, however, in the formation of isomers some of which preceed and others follow the all-*trans* zone in the chromatogram (*209, 309, 310*). As we will see later in connection with some spectral phenomena, a *trans* → *cis* rotation in the middle section of the carotene molecule decreases the adsorption affinity, and this effect is especially strong in the case of a central-mono*cis* carotene. In contrast, the adsorption affinity of a peripheral mono*cis* carotene may surpass that of the all-*trans* form.

Upon the introduction of one hydroxyl group the described chromatographic behavior of α- or β-carotene remains essentially unaltered in the sense that some *cis* forms of, e. g., 3-hydroxy-β-carotene (VIII, p. 228) appear above but others below the all-*trans* zone (*320, 298, 25, 22*). An impressive change takes place, however, when a second OH-group is introduced: each *cis* zone is then found above the all-*trans* pigment. The best known representatives of this type are lutein (3,3′-dihydroxy-α-carotene) (XIX) and zeaxanthin (3,3′-dihydroxy-β-carotene) (X) (*240, 320, 290*). Possibly, the reduced distance between the two OH-groups in the bent *cis* molecules is responsible for this behavior; and perhaps both hydroxyls are able to participate in the anchoring process. Upon esterification, the "excess" adsorption affinity disappears and each *cis* isomer is retained below the all-*trans* zone.

Examples. Physaliene = zeaxanthin dipalmitate, and esters of the dihydroxy-ketones capsanthin and capsorubin (*290, 289*). It may be mentioned that cantha-xanthin (4.4′-diketo-β-carotene) yields *cis* isomers with weakened adsorption affinities (*70*). The *cis* β-carotene epoxides are adsorbed in part below and in part above the all-*trans* zone (*259*).

16*

Much more uniform is the chromatographic behavior in the presence of several *cis* double bonds: each known poly*cis* carotenoid is adsorbed far below the corresponding all-*trans*, mono*cis* and di*cis* forms. Inversely, when a poly*cis* compound is submitted to *cis* → *trans* rearrangement under mild conditions, the stepwise formed isomers (with rare exceptions) display stronger adsorbabilities than that of the starting material (*180*, cf. p. 286).

## IV. *Cis-trans* Isomerism and UV Spectra.

### 1. Some Remarks on the Spectra of all-*trans* Carotenoids.

The intense color of carotenoids is caused by the strong extinction in the region, 400–500 m$\mu$, where the spectral curve shows a massive band. If the latter exhibits vibrational fine structure, in most instances three maxima or two maxima and a shoulder appear, whereby the middle peak has the highest intensity.

As is well known, the spectral properties of an all-*trans* polyene are determined, first of all, by the number of conjugated double bonds. A second factor is the shape of the carbon chain that carries the chromophore. This is well illustrated by the considerable difference between the lycopene and the $\beta$-carotene spectra (Figs. 6 and 4, pp. 248, 246). Both pigments possess eleven conjugated double bonds; however, the lycopene chromophore is entirely aliphatic while the two ends of the $\beta$-carotene chromophore reach into non-planar cyclohexene rings. The transition, $\beta$-carotene → lycopene, has a strong bathochromic effect.

The presence in carotenoid chromophores of four characteristically located methyl side-chains also influences the spectrum. Thus, the $\lambda_{max}$ of INHOFFEN's synthetic lower $\beta$-carotene homolog, 13,13'-bis-desmethyl-$\beta$-carotene (*100*) from which the two middle CH$_3$-groups are missing, is located at 10 m$\mu$ shorter waves than that of $\beta$-carotene. In contrast, the introduction of 2- and 2'-methyl groups displaces the $\beta$-carotene maxima only slightly, because the chromophore is not affected (*100*).

Interesting is the marked spectral difference between YAMAGUCHI's renieratene and isorenieratene (*273–275*); these natural products differ only with respect to their aromatically bound methyl groups that interfere sterically with the main aliphatic chain (p. 319).

Inspection of the extinction curves of numerous all-*trans* carotenoids has shown that the fundamental band may or may not possess vibrational fine structure; this is especially sharp in the case of *retro* carotenoids. However, the individual differences are also considerable in the *normal* series. Aliphatic or semi-aliphatic pigments such as lycopene or $\gamma$-carotene show much more pronounced fine structures in the main band than do the hydroaromatic $\alpha$- and $\beta$-carotenes. On lengthening one side of the β-carotene chromophore by a conjugated carbonyl group or a conjugated ring double bond the fine structure is reduced considerably and may disappear altogether when both sides of the system are affected.

Examples. 4-Keto-β-carotene, 4,4′-diketo-β-carotene; 3,4-dehydro-β-carotene, and 3,4,3′,4′-bisdehydro-β-carotene (*205, 127, 285, 105, 115, 304*). The curves of the dehydro compounds mentioned are in accordance with the partial loss of (the slight) fine structure during the transition, vitamin $A_1 \rightarrow$ 3,4-dehydrovitamin $A_1$ ($A_2$).

For theoretical interpretations of vibrational fine structures in polyenes see PLATT (*207*), DALE (*41, 42*), and H. KUHN (*149*).

## 2. Spectral Effect of *trans* → *cis* Isomerization in the Visible Region.

The profound influence of the molecular shape on the selective absorption of light has secured a preponderant role to spectroscopic readings in the study of *cis-trans* isomeric polyenes.

It was found that when an all-*trans* carotenoid is converted into a mixture of *cis-trans* isomers (p. 264) the color intensity decreases. At relatively high pigment concentrations the effect may be followed by the naked eye and at low concentrations in the visual spectroscope.

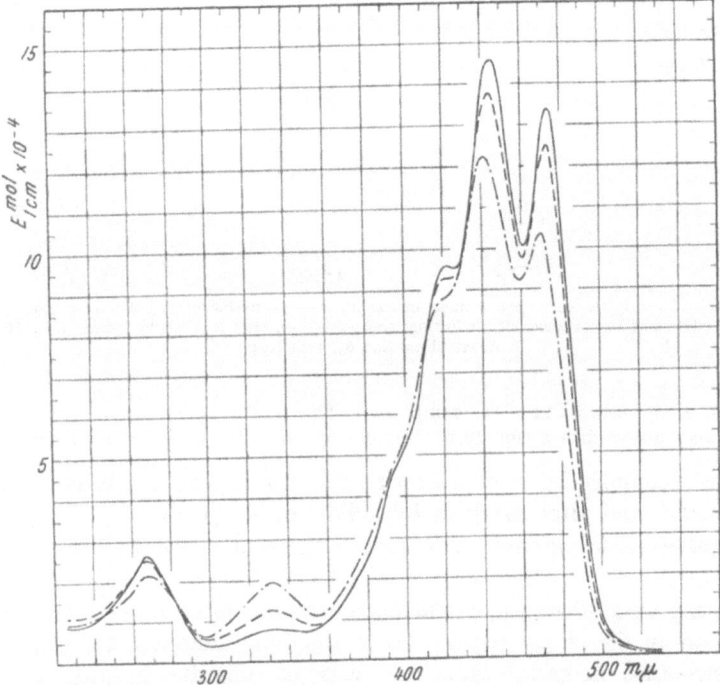

Fig. 3. Molecular extinction curves of α-carotene, in hexane: ————, fresh solution of the all-*trans* compound; —————, mixture of stereoisomers after refluxing; and —·—·—, after catalysis by iodine (*308*). [From: J. Amer. Chem. Soc. 65, 1522 (1943).]

In the visual spectroscope, first the position of the maxima is determined and then a drop of dilute iodine solution is introduced (the light required for the catalysis is furnished by the apparatus itself). The bands become moderately blurred and migrate towards shorter wavelengths in 1–2 minutes; the new positions are then determined. When starting from a *cis* pigment, migration in the opposite direction takes place. Thus, the observer is able to differentiate, within minutes,

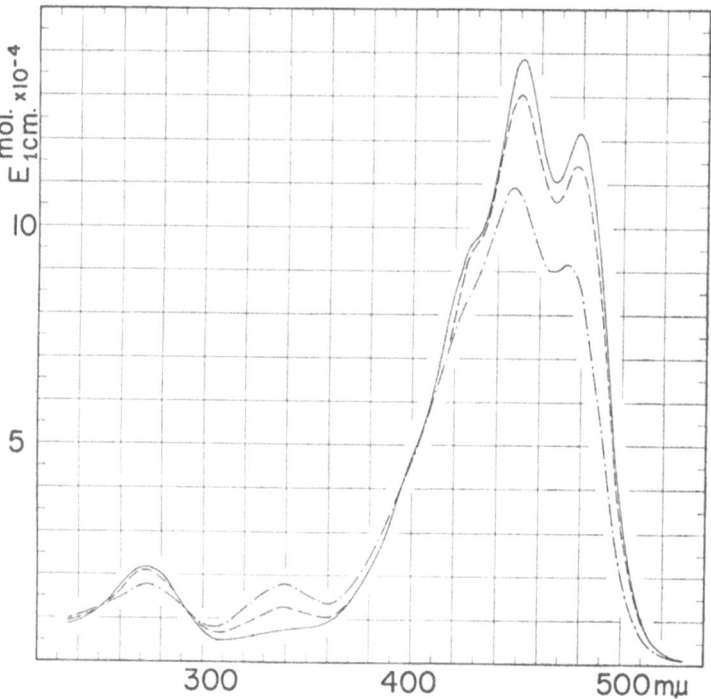

Fig. 4. Molecular extinction curves of β-carotene, in hexane: ————, fresh solution of the all-*trans* compound; — — —, mixture of stereoisomers after refluxing; and — · — · —, after catalysis by iodine (*308*). [From: J. Amer. Chem. Soc. 65, 1522 (1943).]

between an all-*trans* and a *cis* compound. The final wavelength position of the bands may also indicate the stereochemical set to which a *cis* isomer belongs.

It is recommended to characterize a polyene by recording its maxima both before and after catalysis by iodine, even when no further stereochemical work is planned. The same statement is valid for extinction curves.

During an all-*trans* → *cis* rearrangement the spectral curve is altered as follows in the visible region: The extinction values and the degree of fine structure decrease while the maxima migrate towards shorter wavelengths *(Figs. 3–6)*. Since the resulting equilibrium mixture usually contains 40–60% unchanged all-*trans* carotenoid, the spectral difference between the latter and the main *cis* isomer(s) evidently surpasses the observed shift. This can be demonstrated in somewhat larger scale experiments by chromatographic resolution of the equilibrium mixture and determination of the spectral curve of each individual *cis* isomer. Such curves undergo the following changes upon iodine catalysis: The

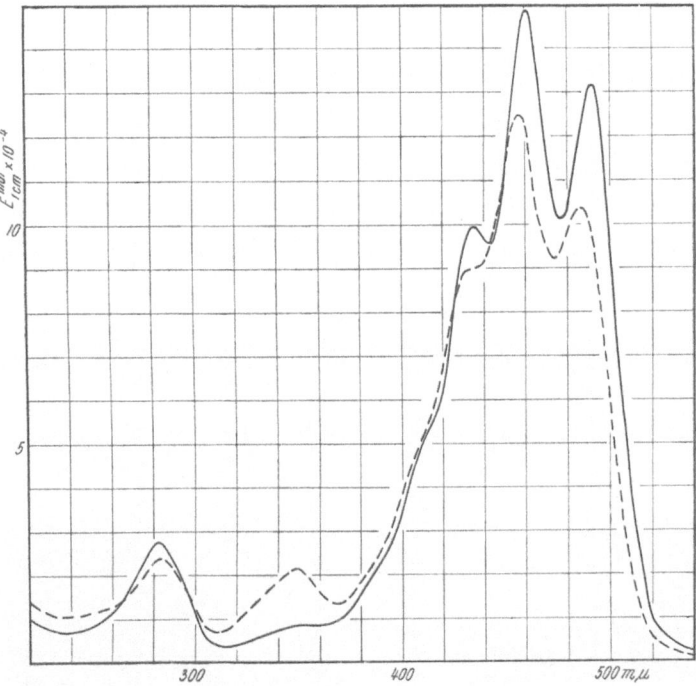

Fig. 5. Molecular extinction curves of γ-carotene, in hexane: ————, fresh solution of the all-*trans* compound — — —, and mixture of stereoisomers after catalysis by iodine (*205 a*).

extinction values at the maxima as well as the degree of fine structure increase and the maxima migrate towards longer wavelengths. Finally, the spectrum of the same equilibrium mixture is obtained as on interaction of the all-*trans* compound and iodine.

The wavelength difference, in m$\mu$, between the location of $\lambda_{max}$ of an all-*trans* compound and that of a *cis* isomer is termed the "$\lambda_{max}$ shift".

The validity of the statement that an all-*trans* carotenoid shows longer wavelength maxima in the visible region than any of its *cis* isomers can best be illustrated in the lycopene set of which about 40 spectroscopically well-defined members are known [Table 14, p. 289; (*180*)]. Some of them were prepared by rearranging the all-*trans* pigment, others by partial *cis* → *trans* isomerization of a poly*cis* lycopene, again others by total synthesis, while some were isolated from plants. Without a single exception each *cis* isomer exhibited maxima at shorter wavelengths than those of all-*trans*-lycopene.

Although the spectral behavior of diphenylpolyenes is very similar (*173, 306, 46*), this "rule" does not apply to dienes and some simple unbranched

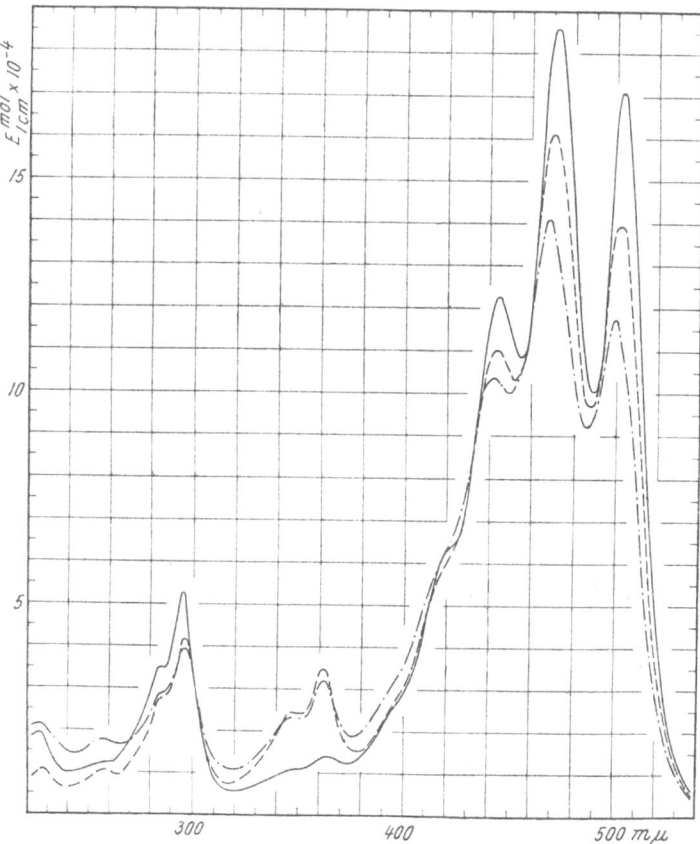

Fig. 6. Molecular extinction curves of lycopene, in hexane: ————, fresh solution of the all-*trans* compound; — — —, mixture of stereoisomers after refluxing in darkness for 45 min.; and — · — · —, after catalysis by iodine (*300*). [From: J. Amer. Chem. Soc. 65, 1940 (1943).]

polyenes [Nayler and Whiting (*196*), Crombie, Harper and Smith (*37*)]. Furthermore, according to Holme, Jones and Whiting (*93*), the approximately coplanar all-*cis*-deca-2,4,6,8-tetraene shows longer wavelength maxima than the all-*trans* form. In our opinion (*302*) these observations should not disturb the present .interpretation of the carotenoid spectra which is based on a very great number of experimental facts. In the future, we hope, a more exact theory will cover all available spectroscopie observations.

While according to our concepts the spectral phenomena described are caused by spatial rearrangements about double bonds, an essentially different interpretation was proposed by SIMPSON (*235*). This author seems to assume that the formation of stereoisomeric carotenoids generally involves *cis* and *trans* configurations around single bonds of the aliphatic conjugated system ["*s-cis*" and "*s-trans*"; cf. (*193*)]. From the viewpoint of the organic chemist various experimental data would be in disagreement with such a theory.

(a) Iodine, the most successful agent in pertinent experiments, is a well-known catalyst that affects rearrangements especially around aliphatic double bonds. Most main *cis* forms furnished by means of iodine have been identified with those obtained by thermal rearrangements. (b) If some single bonds in carotenoids were *cis*, isomers might exist with very weak absorption in the fundamental region; such isomers have never been observed (cf. p. 259). (c) The presence of *cis* double bonds follows from infrared readings (p. 294). (d) Recently, numerous *cis* carotenoids have been prepared by partial reduction of triple bonds (p. 274). Although no *s*-isomerism could be involved here, the synthetic *cis* carotenoids show striking similarity with those obtained by rearrangement of all-*trans* forms; in certain instances identity has been established. (e) In the case of some simple polyenes, such as diphenylbutadiene, the number of the observed *cis* forms is identical with that calculated on the basis of *cis-trans* isomerism around double bonds. (f) According to the best available evidence, the contribution of *s-cis* conformations in acrolein and some methyl-substituted $\alpha,\beta$-unsaturated aldehydes is of low order of magnitude (1–2%) (*261a, 69*).

The above comments refer to generalizations and should, of course, not imply that *s-cis* conformations are excluded from all aliphatic polyene systems. We refer, for example, to observations reported by WEEDON et al. (*2, 267*) concerning acyclic conjugated polyene diketones with bulky end groups. Some of their data could well be explained by the presence of *s-cis* bonds. Connected problems have been reviewed by WAIGHT and ERSKINE (*262*); see also pp. 326 and 291.

### 3. Spectral Effect of *trans* → *cis* Isomerization in the Near Ultraviolet Region: the *cis*-Peak.

It was observed in 1943 in collaboration with POLGÁR (*308*) that, when an all-*trans* carotenoid undergoes stereoisomerization, involving the described changes in the visible region, simultaneously a new maximum, the so-called "*cis*-peak" grows out in the near-ultraviolet region, somewhere between 320 and 380 m$\mu$ (Figs. 3–11, pp. 245–251).

The *cis*-peak effect can be demonstrated, for example, by adding catalytic amounts of iodine and illuminating the solution or keeping it in diffuse daylight. As *Fig. 7* shows, no *cis*-peak appeared when a solution was kept in darkness for an hour, but a light impulse as short as 5 seconds did produce a sizeable effect (*211*).

The appearance of the peak had been observed by earlier authors, after long standing of carotene solutions, but the reversible nature of this spectral change was overlooked and the phenomenon was ascribed to oxidation—a reasonable assumption at a time when nothing was known about stereoisomeric polyenes (*185, 189, 190*). In some other instances *cis*-peaks were represented as a feature of all-*trans* curves; evidently, non-intended partial stereoisomerization was responsible for such effects [cf. e. g. (*141, 60*)].

The best media for the observation of *cis*-peaks are non-polar solvents such as hexane, cyclohexane, or benzene. In carbon disulfide (which

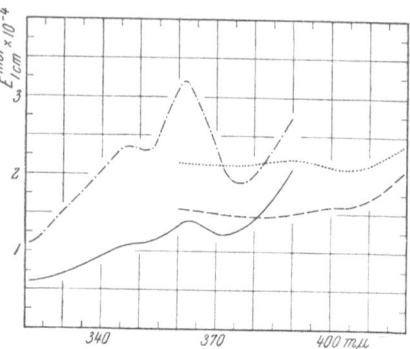

Fig. 8. Molecular extinction curves in the *cis*-peak region, in hexane and carbon disulfide solutions: ——————, all-*trans*-lycopene, in hexane; and — · — · —, mixture of stereoisomers after catalysis by iodine; — — —, all-*trans*-lycopene, in $CS_2$; and · · · · · · · , after catalysis by iodine (*310*). [From: J. Amer. Chem. Soc. 67, 108 (1945).]

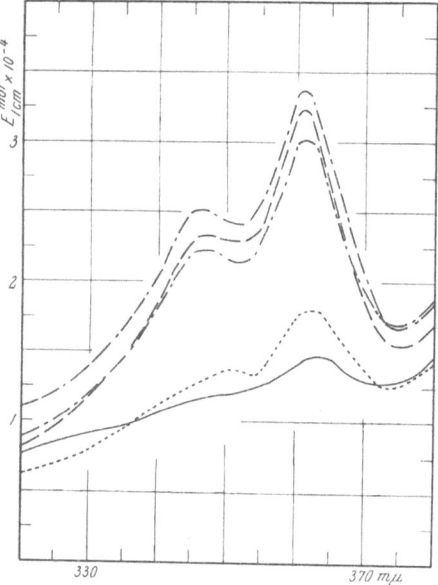

Fig. 7. Influence of illumination on the development of the *cis*-peak in an iodine-catalyzed solution of lycopene, in hexane: · · · · · · , after 0 sec.; — — —, after 5 sec.; — — · — · —, after 30 sec.; — · — · — , after 15 min. illumination; and ——————, before the addition of iodine, without illumination (*211*). [From: J. Amer. Chem. Soc. 66, 186 (1944).] ·

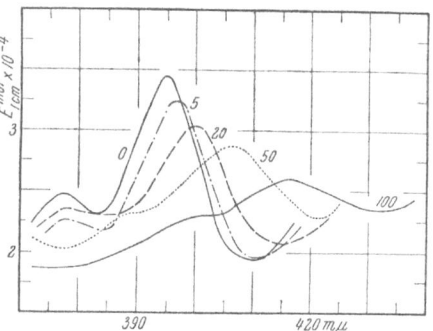

Fig. 9. Alteration of the molecular extinction coefficient in the *cis*-peak region of an iodine-catalyzed rhodoviolascin (spirilloxanthin) solution, in benzene, upon the addition of carbon disulfide (the figures on the curves indicate % $CS_2$ in the solvent) (*208*). [From: Arch. Biochemistry 5, 243 (1944).]

decreases or destroys the fine structure also in the fundamental band) the curves are flat, although elevated, in the *cis*-peak region *(Figs. 8—9)* (*310*).

*References, pp. 334—349.*

An extensive study of *cis*-peaks exhibited by stereoisomeric mixtures and by chromatographically pure, individual *cis* isomers has revealed the following regularities:

(a) The curves of all-*trans* carotenoids are flat in the *cis*-peak region but most *cis* isomers obtainable by direct rearrangement show *cis*-peaks (*Table 4*, next page).

(b) The *cis*-peaks of all observable members of a given stereochemical set are located at the same wavelength. Usually, some of the individual *cis*-peaks are higher and others are lower than

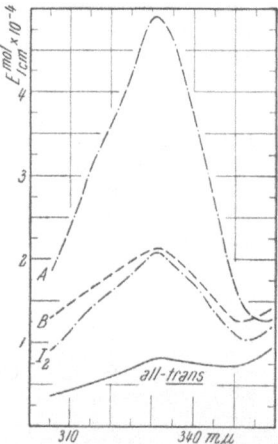

Fig. 10. Molecular extinction curves of some lutein stereoisomers in the *cis*-peak region, in hexane. $I_2$ indicates the equilibrium mixture obtained upon catalysis by iodine (*309*). [From: J. Amer. Chem. Soc. **66**, 137 (1944).]

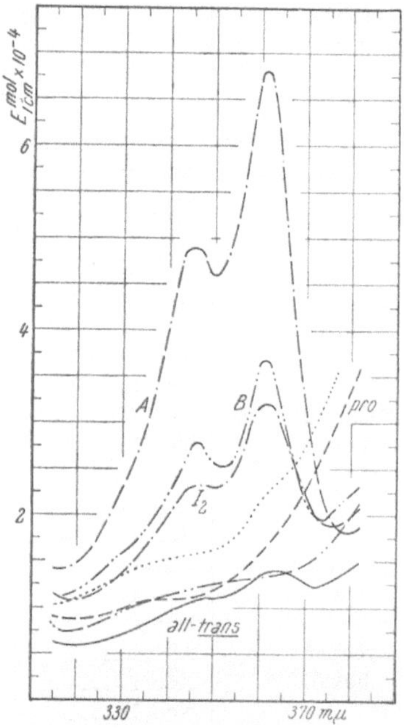

Fig. 11. Molecular extinction curves of some lycopene stereoisomers in the *cis*-peak region, in hexane. $I_2$ indicates the equilibrium mixture obtained upon catalysis by iodine, and — ·· — ·· —, an unnamed crystallizable stereoisomer (*309*). [From: J. Amer. Chem. Soc. **66**, 137 (1944).]

that of the stereoisomeric equilibrium mixture; on catalysis by iodine the former peaks decrease and the latter increase to reach the equilibrium value *(Figs. 10–11)*.

(c) In each stereochemical set only a few of the observed stereoisomers have high *cis*-peaks and these are mainly responsible for the peak observed in the stereochemical equilibrium mixture; in some instances a single *cis* form must be credited for the major part of the effect (*Table 5*, next page).

(d) In the curves of the $C_{40}$-carotenoids, possessing about 10—11 conjugated double bonds, the *cis*-peak has a well-defined location: the wavelength difference between its position and the longest wavelength maximum of the all-*trans* compound is practically a constant, viz. 142 m$\mu$

Table 4. Examples of the *cis*-Peak Intensities in someStereochemical Sets (within the respective sets the stereoisomers are listed in the sequence of decreasing adsorption affinities).

| Stereoisomeric set | Member of the set | Difference between the longest wavelength maximum of the *cis* isomer and that of the all-*trans* form (mμ) | Molecular extinction coefficient at *cis*-peak $E_{1\ cm.}^{mol.} \times 10^{-4}$ | Difference of $E_{1\ cm.}^{mol.} \times 10^{-4}$ for member of the set and all-*trans* form at *cis*-peak | References |
|---|---|---|---|---|---|
| α-Carotene (in hexane) | Neo U .... | 5.5 | 1.2 | 0.4 | (308) |
| | Neo V .... | 11.5 | 1.1 | 0.3 | |
| | Neo W.... | 6.5 | 1.6 | 0.8 | |
| | Neo X .... | 13.5 | 2.7 | 1.9 | |
| | All-*trans*... | 0 | 0.8 | 0 | |
| | Neo A .... | 8.5 | 3.8 | 3.0 | |
| | Neo B .... | 10.5 | 3.8 | 3.0 | |
| | Neo C .... | 4.5 | 4.5 | 3.7 | |
| β-Carotene (in hexane) | Neo U .... | 5 | 1.3 | 0.5 | (308) |
| | Neo V .... | 13.5 | 0.8 | 0 | |
| | All-*trans*... | 0 | 0.8 | 0 | |
| | Neo B .... | 10.5 | 3.4 | 2.6 | |
| | Neo E .... | 8.5 | 3.4 | 2.6 | |
| γ-Carotene (in hexane) | All-*trans*... | 0 | 0.95 | 0 | (300, 310) |
| | Neo U .... | 5 | 1.4 | 0.4 | |
| | Pro-γ ..... | 31 | 1.3 | 0.35 | |
| Lycopene (in hexane) | All-*trans*... | 0 | 1.4 | 0 | (300, 308) |
| | Neo A .... | 5 | 6.8 | 5.4 | |
| | Neo B .... | 8 | 3.7 | 2.3 | |
| | Unnamed crystalline isomer ... | 28 | 1.3 | — 0.1 | |
| | Prolycopene | 34 | 1.6 | 0.2 | |
| | Another poly*cis* form ..... | 38.5 | 2.2 | 0.8 | |

Table 5. Individual Stereoisomers Responsible for the Major Part of the *cis*-Peak Effect Observed in Equilibrium Mixtures upon Iodine Catalysis of All-*trans* Compounds.

| Stereoisomeric set | Name of *cis* form | Approximate percentage of the *cis* form in the equilibrium | Approximate percentage of the total *cis*-peak effect caused by the *cis* form | References |
|---|---|---|---|---|
| α-Carotene .......... | Neo B | 13 | 55 | (309) |
| β-Carotene .......... | Neo B | 25 | 75 | (209) |
| Lycopene .......... | Neo A | 30—40 | 95 | (300) |
| Lutein .............. | Neo A | 17 | 70 | (320) |
| Cryptoxanthin ....... | Neo A | 23 | 60 | (298) |
| Zeaxanthin.......... | Neo A + B | 30 | 90 | (298) |
| Capsanthin.......... | Neo A | 20 | 80 | (211) |

($\pm$ 2 m$\mu$) in hexane solution and a little larger (about 145–146 m$\mu$) in benzene. This statement is valid for a large number of carotenoids whose molecules contain, respectively, aliphatic, hydroaromatic, or

Table 6. Examples of the Position of the *cis*-Peak in Various Stereoisomeric Sets.

| Name | Number of conjugated C=C double bonds | Position of *cis*-peak (m$\mu$) | Distance between *cis*-peak and | | References |
|---|---|---|---|---|---|
| | | | longest wave length max. | $\lambda_{max}$ | |
| | | | of the all-*trans* form (m$\mu$) | | |
| In hexane or petroleum ether solution: | | | | | |
| Rhodoviolascin (spirilloxanthin)............. | 13 | 384 | 143 | 109 | (208) |
| 3,4,3′,4′-Bisdehydro-$\beta$-carotene............. | 13 | 368 | — | 103 | (105) |
| Lycopene............... | 11 | 362 | 141 | 111 | (308, 110) |
| Canthaxanthin......... | 11 + 2 C=O | 356 | — | 110 | (70) |
| Capsanthin ............ | 10 + 1 C=O | 355 | 143 | 109 | (211) |
| 16,16′-*Homo*-$\beta$-carotene*. | 12 | 355 | 141 | 111 | (97) |
| 3,4-Dehydro-$\beta$-carotene.. | 12 | 352 | — | 109 | (115) |
| 4-Hydroxy-$\gamma$-carotene... | 11 | 350 | 141 | 110 | (22) |
| $\gamma$-Carotene............. | 11 | 349 | 143 | 112 | (308) |
| $\beta$-Carotene............. | 11 | 338 | 142 | 114 | (308, 99) |
| Cryptoxanthin ......... | 11 | 339 | 141 | 113 | (298, 113) |
| Physaliene............. | 11 | 338 | 141 | 113 | (308) |
| Isozeaxanthin.......... | 11 | 337 | 141 | 114 | (116) |
| Zeaxanthin ............ | 11 | 336 | 144 | 116 | (112, 118) |
| Neurosporene .......... | 9 | 332 | 136 | 118 | (297) |
| $\alpha$-Carotene ............ | 10 | 331 | 143 | 114.5 | (308, 106) |
| Lutein ................ | 10 | 331 | 143 | 114.5 | (308) |
| Antheraxanthin** ...... | 10 | 331 | 141 | 115 | (254) |
| 5,6-Dihydro-$\beta$-carotene .. | 10 | 331 | 143 | 114 | (308) |
| 4-Hydroxy-$\alpha$-carotene... | 10 | 330 | 144 | 115 | (22) |
| $\beta$-Carotene monoepoxide. | 10 | 330 | 145 | 116 | (259) |
| 5,6-Dihydro-$\alpha$-carotene.. | 9 | 329 | 140 | 109 | (308) |
| $\beta$-Carotene diepoxide ... | 9 | 328 | 142 | 111 | (259) |
| Dimethylcrocetin....... | 7 + 2 C=O | 314 | 134 | 108 | (102) |
| $\zeta$-Carotene............. | 7 | 296 | 129 | 104 | (297) |
| Diphenyloctatetraene ... | 4 + 2 C$_6$H$_5$ | 283 | 111 | 89 | (306) |
| Diphenylhexatriene..... | 3 + 2 C$_6$H$_5$ | 268 | 101 | 83 | (173) |
| *retro*-Dihydro-C$_{19}$-aldehyde .......... | 5 + 1 C=O | 267 | 135 | 114 | (105) |
| Phytofluene ........... | 5 | 260 | 107 | 88 | (204) |
| 2,7-Dimethyl-octa-2,4,6-triene-1,8-dial........ | 3 + 2 C=O | 234 | 102 | 85 | (96) |

  \* In ether.
  \*\* In alcohol.

*(Table 6, continued.)*

| Name | Number of conjugated C=C double bonds | Position of *cis*-peak (m$\mu$) | Distance between *cis*-peak and | | References |
|---|---|---|---|---|---|
| | | | longest wave length max. | $\lambda_{max}$ | |
| | | | of the all-*trans* form (m$\mu$) | | |

In benzene solution:

| Name | Number of conjugated C=C double bonds | Position of *cis*-peak (m$\mu$) | longest wave length max. | $\lambda_{max}$ | References |
|---|---|---|---|---|---|
| Rhodoviolascin (spirillo-xanthin)............... | 13 | 395 | 151 | 115 | (*208*) |
| 1,18-Diphenyl-3,7,12,16-tetramethyl-octa-decanonaene ......... | 9 + 2 C$_6$H$_5$ | 378 | 142 | 109 | (*60*) |
| Canthaxanthin ......... | 11 + 2 C=O | 366 | — | 114 | (*70*) |
| Methylbixin .......... | 9 + 2 C=O | 363 | 145 | 112 | (*295, 96*) |
| Capsanthin ........... | 10 + 1 C=O | 362 | (146) | 122 | (*211*) |
| Renieratene............ | 9 + 2 C$_6$H$_5$ | 362 | 145 | 114 | (*273—275*) |
| Cryptoxanthin ......... | 11 | 348 | 145 | 115 | (*298, 25*) |
| Zeaxanthin ........... | 11 | 348 | 144 | 114 | (*298*) |

aromatic terminal groups. In the case of shorter conjugated systems the distance mentioned is markedly reduced *(Table 6)*.

(e) The spectral curves of the naturally occurring poly*cis* compounds such as prolycopene (p. 284) are as flat in the *cis*-peak region as those of all-*trans* carotenoids; a peak appears, however, on catalysis by iodine.

(f) The presence of a hindered *cis* double bond does not exclude the presence of a *cis*-peak. (Examples: synthetic, hindered *cis* lycopenes, pp. 313, 318.)

(g) No distinct *cis*-peak was observed in the case of *cis retro* carotenoids which possess a long conjugated system (e. g. *retro*-dehydrocarotene, *Fig. 12*), possibly because the peak was "covered up" by the very broad main band (*321*). Certain *retro* compounds with shorter chromophores may show the peak (*105*).

(h) Some of the observed *cis*-peaks show fine structure but others do not. Although such differences cannot be interpreted at the present time, the following empirical statements can be made. If the molecule of a C$_{40}$-carotenoid (hydrocarbon or alcohol) contains two hydroaromatic end groups, the *cis*-peak is void of fine structure ($\alpha$-carotene, $\beta$-carotene, 3,4-dehydro-$\beta$-carotene, 3,4,3',4'-bisdehydro-$\beta$-carotene, 16,16'-*homo*-$\beta$-carotene; cryptoxanthin, zeaxanthin, canthaxanthin, etc.). Nor does fine structure appear in the semi-aliphatic sets of $\gamma$-carotene and capsanthin but it is observed in several entirely aliphatic sets (lycopene, rhodoviolascin, phytofluene, etc.). $\beta$-Carotene diepoxide does show fine structure in the *cis*-peak region, in contrast to the monoepoxide. When the terminal groups are unmethylated aromatic rings, fine structure in the *cis*-peak region appears (1,18-diphenyl- or -dinaphthyl-3,7,12,16-tetramethyl-octadecanonaene; aromatic polyene azines, etc.); this is not the case, however, in the presence of several aromatically bound $CH_3$-groups (renieratene, isorenieratene).

(i) The presence or absence of fine structure in the fundamental band does not determine the corresponding situation in the *cis*-peak region.

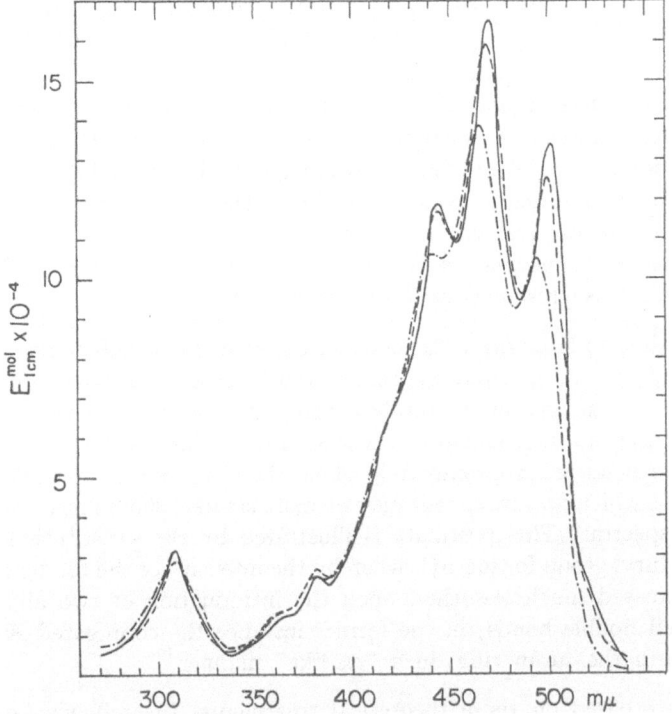

Fig. 12. Molecular extinction curves of *retro*-dehydrocarotene, in hexane: ————, fresh solution of the all-*trans* compound; — — —, mixture of stereoisomers after refluxing for 1 hour, in darkness; and — · — · —, after catalysis by iodine, in light (*321*). [From: J. Amer. Chem. Soc. **75**, 5341 (1953).]

(j) Whenever the *cis*-peak region exhibits fine structure, the curve has a characteristic shape; it shows a lower maximum (or shoulder) and a higher maximum, the latter being located nearer the longer wavelengths.

## 4. A Theoretical Interpretation of Spectral Phenomena, Especially of the *cis*-Peak Effect.

This Section will be restricted to a few features that are of immediate interest to the organic chemist and illustrate the close relationship between light absorption and the over-all shape of the polyene molecule.

For recent surveys cf. CROMBIE (*36*), WYMAN (*272*), H. KUHN (*148–150*), BRAUDE and WAIGHT (*18*). (The latter authors term the fundamental and the *cis*-peak bands, respectively, "full-chromophore" and "half-chromophore" bands.) Molecular orbital methods will not be considered here. They were applied to polyenes by COULSON (*35*); several possible theoretical approaches were outlined by MULLIKEN (*194*). The free-electron molecular-orbital method was surveyed recently by PLATT (*207*) and discussed in this Series by H. KUHN (*148, 149*).

The writer would recommend caution in the theoretical interpretation of minor bands because of possible spontaneous stereoisomerization in solutions and the resulting appearance of minor *cis* bands which are not immediately recognizable as such.

In 1939 there appeared several theoretical treatments that have profoundly influenced further research. We mention PAULING's contributions (*201, 202*) and the extensive paper by LEWIS and CALVIN (*168*) on the color of organic compounds in which the bands of "partial oscillation" are also treated. An important theoretical study of electronic transitions in molecular spectra of conjugated polyenes including carotenoids was presented by MULLIKEN (*192*).

According to MULLIKEN the shape of the conjugated polyene molecule strongly affects the distribution of absorption intensity. The more elongated the molecule, the greater the total intensity and the greater the preponderance of intensity in the longest wavelength transition; such $N \rightarrow V_1$ transitions are polarized approximately along the long axis of the molecule. "Molecules which for any reason may be more *cis*-like, should have weaker $N \rightarrow V_1$ spectra." This postulate is illustrated by the substitution of a terminal furyl group for methyl, whereby the intensity of the fundamental band increased much less than upon the introduction of two aliphatic conjugated double bonds; in the former instance the conjugated system is bent into the furan ring, in a *"cis*-like" manner.

In part based on these theoretical treatments, the following quasi-classical interpretation of our spectroscopic observations was presented in 1943 by PAULING, LEROSEN, SCHROEDER, POLGÁR, and the writer (*300*).

The three regions in carotenoid spectra which are of practical importance are: (a) the region of extraordinarily strong extinction in the visible region (fundamental or $K$ band or $\lambda_1$-band), (b) the *cis*-peak region in the near-ultraviolet (first overtone band or $\lambda_2$-band), and (c) a region in the farther ultraviolet (second overtone or $\lambda_3$-band). These bands correspond to transitions from the normal electronic state to excited states. The fundamental band results from the transition 0 → 1, the *cis*-peak from 0 → 2, and the second overtone from 0 → 3. (For example, the three corresponding maxima of lycopene in hexane solution are located at 471, 361, and 296 m$\mu$; Fig. 6, p. 248.) These electronic levels may be discussed in terms of the conventional structure ....⟋⟍⟋⟍⟋⟍⟋⟍⟋⟍⟋····, and a great number of ionic structures, such as

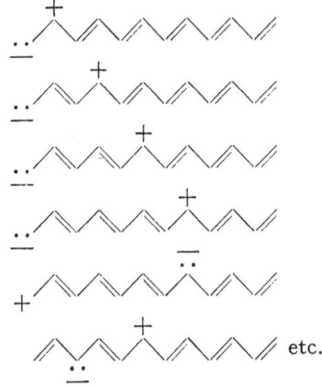

etc.

The conventional structure of alternating double and single bonds makes a most important contribution to the normal state, whereas the ionic structures contribute in the main to the excited states.

The three transitions mentioned correspond to oscillation of the electric charge along the unsaturated chain. Following LEWIS and CALVIN (*168*) we may compare these with the classical modes of vibration of mobile "unsaturation" electrons of the conjugated system along the chain. The observed bands can be correlated with the following manners of oscillation (*300*).

(a) Fundamental band: The electrons tend to concentrate first near one end and then near the other end of the chain; this simple oscillation would, according to classical electromagnetic theory, result from the absorption of light of the proper frequency because of interaction of the electric vector of the light and the regularly reversing electric dipole moment of the molecule. This mode of oscillation of the charge corresponds to the transition o → 1.

(b) Band in the *cis*-peak region: This results from the oscillation of the electrons from the two ends of the conjugated system toward the middle and from the middle toward the two ends (transition o → 2).

(c) Next band in the farther ultraviolet: This involves concentration of the electrons alternately in the first and third and the second and fourth quarters of the conjugated system (o → 3).

The intensity of an absorption band is proportional to the square of the corresponding dipole moment, and hence essentially to the square of the length of the system (*201, 192, 193*). The maximum intensity

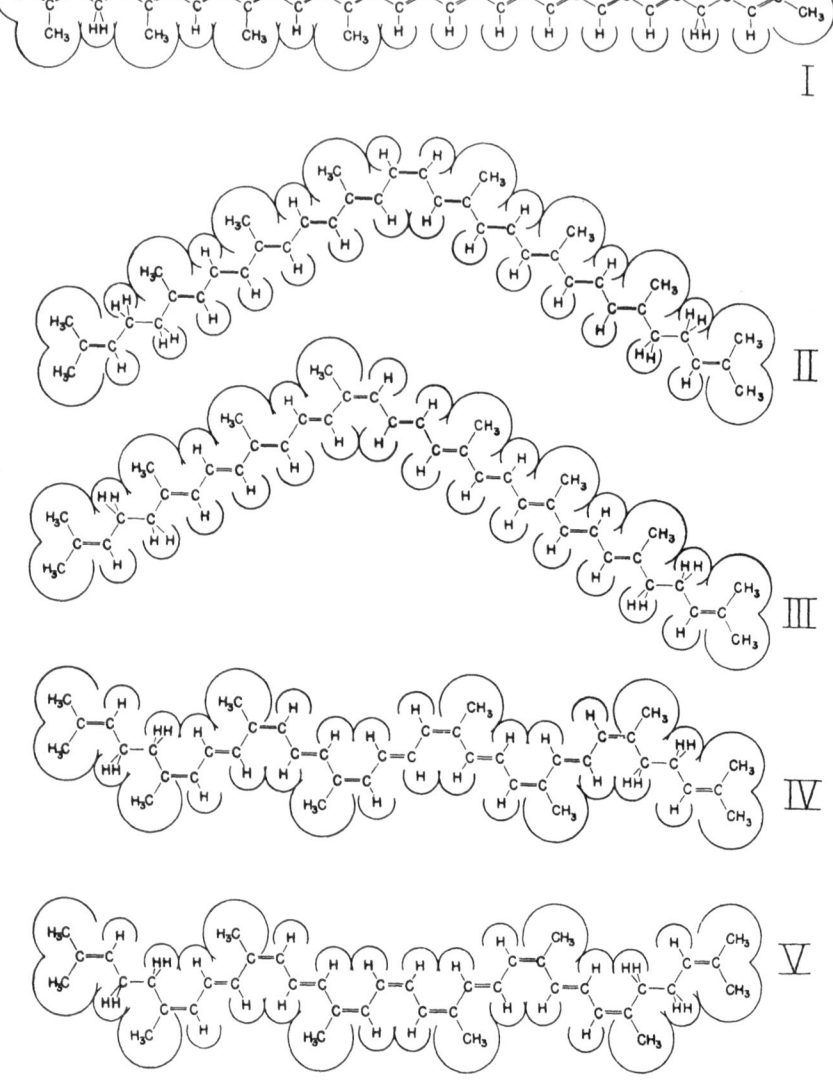

Fig. 13. Models of some lycopene stereoisomers: I, all-*trans* form; II, central-mono*cis*-lycopene; III, next-to-central-mono*cis*-lycopene; IV, a hexa*cis*-lycopene; and V, a hepta*cis*-lycopene in which all sterically unhindered double bonds are in the *cis* configuration. The bond angles, relative bond distances and van der Waals radii correspond approximately to the accepted values (*300*). [From: J. Amer. Chem. Soc. **65**, 1940 (1943).]

of the fundamental band is shown by the all-*trans* molecule (Model I, above).

*References, pp. 334—349.*

All stereoisomers which have a vertical plane of symmetry, such as Models II and V (Fig. 13) have a distance between the ends of the conjugated system smaller than the all-*trans* isomer by the factor cos $\alpha$, with $\alpha = 27° 22'$ (carbon bond angle along the chain $= 125° 16'$). The fundamental band for these isomers should then be approximately 80% as intense for the all-*trans* form since $\cos^2 \alpha = 0.80$.

According to BRAUDE and WAIGHT (*18*) one should calculate rather with the value 120°; hence, $\cos^2 30° = 0.75$.

The intensities for all mono*cis* isomers should lie between the extremes of the all-*trans* and central-mono*cis* forms. (In comparing intensities, essentially the ratio of the extinguished areas rather than the height of peaks should be considered.)

The fact that this agrees with observation provides support for the assumption that the single bonds of the conjugated system have essentially the *trans* configuration. If some single bonds also were *cis*, stereoisomers might exist with very small absorption in the fundamental region (cf. p. 249).

The nature of the $0 \to 2$ oscillation, which was discussed by MULLIKEN (*192*) is such that it gives rise to no dipole moments and to no absorption band for the all-*trans* molecule or any other molecule with a center of symmetry. In contrast, a central-mono*cis* form (such as Model II, Fig. 13) does have a dipole moment for the transition $0 \to 2$, because of its bent shape (*300*). This dipole moment is perpendicular to the straight line, connecting the ends of the conjugated system, instead of parallel to it*. This isomer would show the strongest *cis*-peak of all. Certain other isomers such as Model III (Fig. 13) would also have *cis*-peaks but of smaller intensity. As a rough approximation the intensity of the *cis*-peak can be taken proportional to the square of the distance between the center of the conjugated system and the mid-point of the straight line between its two ends. Accordingly, only a few steric forms can have *cis*-peaks approaching the intensity of Model II.

It should be emphasized that the above discussion is necessarily of qualitative nature in the sense that in a given stereoisomeric set it has not yet been possible to predict on theoretical grounds the absolute value of the maximum possible (or any other) *cis*-peak intensity. Luckily,

---

* Quite recently, this has been confirmed experimentally by ECKERT and H. KUHN (*57 a*) who investigated the dichroism obtained by stretching a polyethylene film that contained central-mono*cis*-$\beta$-carotene in solid solution. The pigment molecules were oriented preferentially parallel to the direction of stretching. It was found that the sign of the dichroism was positive in the main band but negative at *cis*-peak. The linearly polarized light was more strongly absorbed when the electric vector was parallel to the direction of stretching than when it was perpendicular to it. *(Added in Proof.)*

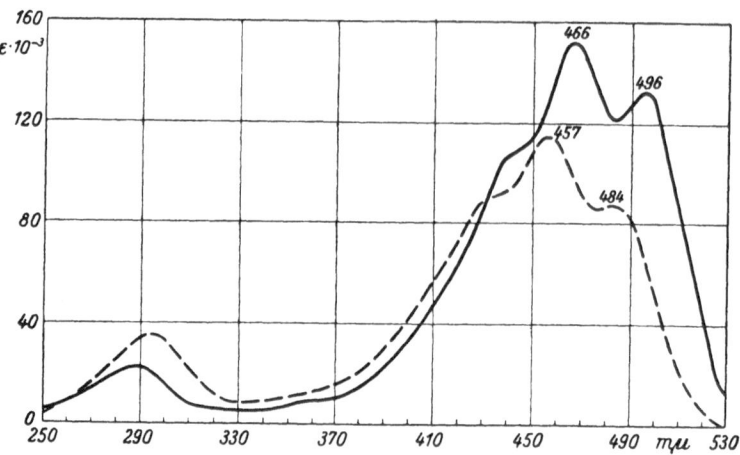

Fig. 14. Molecular extinction curves of 16,16'-*homo-β*-carotene, in ether: — — —, middle-di*cis* form and ————, all-*trans* form; according to Inhoffen et al. *(97)*. [From: Liebigs Ann. Chem. **573**, 1 (1951).]

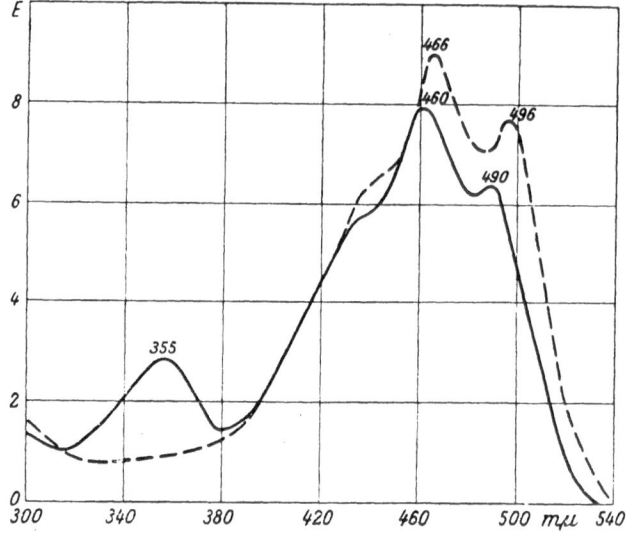

Fig. 15. Molecular extinction curves of middle-mono*cis*-16,16'-*homo-β*-carotene, in ether: ————, fresh solution, and — — —, after exposure to diffuse daylight for 5 hours; according to Inhoffen et al. *(97)*. [From: Liebigs Ann. Chem. **573**, 1 (1951).]

as will be shown later, this gap has been filled by measuring the *cis*-peaks of total-synthetic central-mono*cis*-carotenoids (p. 292).

Let us now, in a mental experiment, start from an all-*trans* carotenoid and bend the molecule in the center by forming there a *cis* double bond, and then either slide the *cis* configuration along the chain or introduce more *cis* double bonds. The initially very high *cis*-peak will decrease in both instances and disappear as the molecule straightens out. These postulates are in good agreement with numerous experimental data, especially with the absence of *cis*-peaks from the curves of poly*cis* carotenoids.

A clear experimental demonstration of the modification of the *cis*-peak during the process, di*cis* (XXV) → middle-mono*cis* (XXVI) → all-*trans* (XXVII), was given recently by INHOFFEN (97). The absolute configuration of the di*cis* compound used was established by total synthesis and, as expected, did not cause any *cis*-peak. On the stepwise rearrangement of the 15,15'-di*cis* form of the 16,16'-$\beta$-carotene homolog $C_{42}H_{58}$ first an intense *cis*-peak appeared which disappeared during the second step *(Figs. 14–15)*. Such well-defined, stepwise isomerization experiments might well become important tools in future stereochemical research.

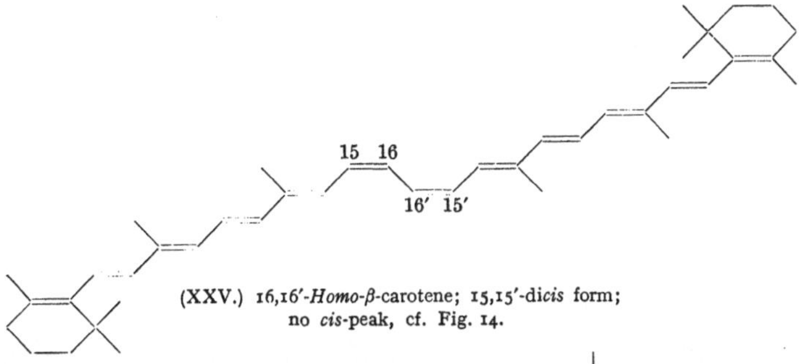

15 16
16' 15'

(XXV.) 16,16'-*Homo-$\beta$*-carotene; 15,15'-di*cis* form; no *cis*-peak, cf. Fig. 14.

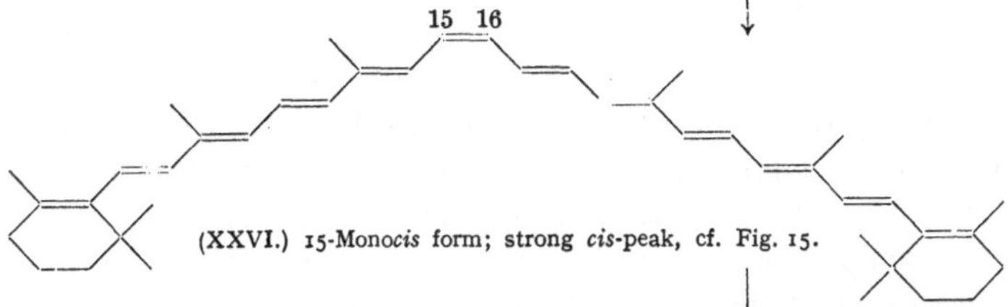

15 16

(XXVI.) 15-Mono*cis* form; strong *cis*-peak, cf. Fig. 15.

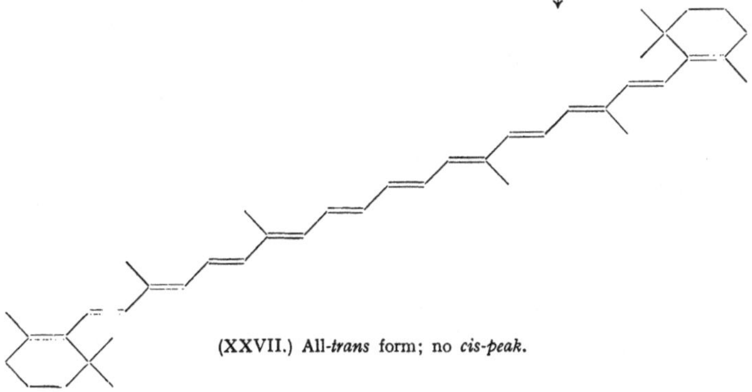

(XXVII.) All-*trans* form; no *cis-peak*.

Recently, the empirical relationships of the minor bands, including the *cis*-peak, in polyene spectra have been subjected to a detailed analysis by DALE (*41*) who has stated the following simple rule: In case of an aliphatic polyene containing *n* conjugated double bonds, the wavelength maximum of any minor band ($\lambda_s$) will lie very close to that of the main band of a corresponding polyene with *n/s* conjugated double bonds. This rule seems to apply also to poly*cis* carotenoids and numerous other polyenes.

The observed characteristics of certain *retro* carotenoid spectra (very sharp fine structure in the main band; great similarity between *trans* and *cis* extinction curves in the main band but lack of a conspicuous *cis*-peak effect; cf. Fig. 12, p. 255) as compared to those of *normal* carotenoids is explained by a spatial conflict between the *gem*-dimethyl group and the opposing chain-hydrogen in *normal* carotenoids whose cyclohexene ring is somewhat twisted. In contrast, the ring is planar and the whole conjugated system very nearly coplanar in *retro* carotenoids (*41*).

DALE (*42*) has presented a theoretical basis of the above findings by analyzing the free-electron model of polyenes. Furthermore, he has shown that the minor bands are overtones also in polyenynes and, especially in dehydro-carotenoids, not due to isolated partial chromophores [connected papers (*43–45*)]. In our Series, some quantum-mechanical aspects of the polyene spectra, based on the free electron gas theory have recently been discussed by H. KUHN (*148, 149*). He presented a theoretical explanation of DALE's selection rule, according to which only the electrons of the outermost $\pi$ electron shell are excited by light in polyenes.

Considering the nature of the *cis*-peak effect it is hardly necessary to stress that this phenomenon is by no means restricted to carotenoid pigments. Since the intensity of the peak is determined by the bent molecular shape, the *cis*-peak must be considered as an important characteristic of all such aliphatic multiconjugated systems that can be subjected to geometrical isomerization. Accordingly, *cis*-peaks have been observed in the following classes of compounds (among others): Acidic carotenoids (*276*); lower-molecular weight isoprenic systems, e. g. methyl-bixin (*295*); vitamin A (*94*); retinene (*94*); non-isoprenic aliphatic systems with aromatic end groups: diphenylhexatriene (*173*), diphenylocta-tetraene (*306*); renieratene (*273, 275*); polyphenyl-polybutadienes (*57*); symmetrical azines (*46*); and entirely aliphatic, non-isoprenic systems: corticrocin (*59*), unbranched *cis*-polyenynes (cf. e. g. *99, 17*), etc. In contrast, when a *trans* → *cis* rearrangement does not alter the straight overall shape of the aliphatic system, no *cis*-peak appears (example, dibiphenylene-hexatriene) (*178*).

*Cis*-peaks have also been observed in the solid solution state (polystyrene films, p. 259) (*57 a*, *161*, *162*) and in the glassy state. $\beta$-Carotene "glasses" were obtained by ROSENBERG (*226*) by fusing the crystals and solidifying the melt. He found that the *trans* → *cis* rearrangement process was important in determining the photoconduction phenomena in the glass. The observed photoconductive excitation spectrum was very different from the optical extinction curve of all-*trans*-$\beta$-carotene solutions and showed maximum effect in the *cis*-peak region.

## 5. Spectra at Extremely Low Temperatures.

The first observation in this interesting field was made in 1935 by HAUSSER, KUHN and SEITZ (*90a*) who found that the UV spectra of $\alpha,\omega$-diphenylpolyenes, some polyene-carboxylic acids and lycopene underwent conspicuous changes on cooling their solutions in liquid nitrogen ($-185°$ to $-195°$): in the fundamental region the curves became much sharper, some new maxima appeared and the extinction values increased far beyond those observed at room temperature. While only all-*trans* compounds could be investigated at HAUSSER's time, quite recently WALD and his collaborators have published an important study into the spectral behavior of geometric isomers of retinene, vitamin A and some carotenoids [JURKOWITZ (*126a*), LOEB, BROWN and WALD (*170a*)].

On cooling with liquid nitrogen, the all-*trans*-$\beta$-carotene spectrum shows the following changes: the main maximum is shifted by 17.5 m$\mu$ towards longer wavelengths, the fine structure is accentuated with two new maxima appearing; the longest wavelength maximum becomes the highest. The intensity of the middle maximum is then 1.34 times higher than at 20°. The sterically unhindered central-mono*cis*-$\beta$-carotene (p. 296) behaves similarly; the *cis*-peak is displaced by 14 m$\mu$ towards longer wavelengths but its intensity is increased by only 9%.

Dramatic changes take place on cooling the sterically hindered 11,11'-di*cis*-$\beta$-carotene (p. 300): instead of the degraded spectrum extensive fine structure appears, the position of $\lambda_{max}$ is shifted by 49.5 m$\mu$ towards longer waves and its intensity becomes 2.9 times higher than that observed at room temperature. Very similar is the behavior of 11-*cis*-retinene. Spectroscopically speaking, it seems as if the cooling would "relieve" the hindrance caused by the conflict between a H-atom and $CH_3$-group at a hindered *cis* double bond or bonds.

All-*trans*-lycopene shows similar extensive spectral changes, analogous to those of $\beta$-carotene.

A theoretical discussion of these phenomena has been presented by WALD (*170a*).

# V. Preparation of *cis* Carotenoids by Direct Rearrangement of the All-*trans* Form.

*Cis* carotenoids have been obtained by stereoisomerization of all-*trans* pigments, by total synthesis, and by isolation from plants or animals.

In general, the rearrangement methods afford quasi-equilibria or steady-state mixtures that contain *cis* isomers and a substantial portion of unchanged all-*trans* pigment. The composition of the mixtures is revealed on chromatographic resolution followed by spectral characterization and estimation of the individual *cis* isomers. The result will depend on the nature of the treatment; thus, thermic and photochemical methods may afford markedly different equilibria.

Examples. (a) Very little neolutein B was obtained upon refluxing the all-*trans* solution but it appeared in substantial quantities upon catalysis by iodine, in light (*320*). (b) One of the *cis* β-carotene diepoxides was not contained in equilibrium mixtures prepared by refluxing or illumination in the absence of catalysts, but it was formed to the extent of 25% in the presence of iodine (*259*).

## 1. Thermal Methods of *cis-trans* Isomerization.

*(a) Spontaneous Rearrangements in Solution at Room Temperature.* This usually slow process starts on dissolution of carotenoid crystals or elution of adsorbates and proceeds at rates which depend on structure, configuration, solvent, temperature, and light.

Examples. (1) According to Carter and Gillam (*23*) the extent of stereo-isomerization of all-*trans*-β-carotene in benzene-petroleum ether amounted to 11% in 49 days at 20° but only to 3–4% in 90 days at — 2° (in darkness, protected from air). (2) At room temperature, the following fractions of all-*trans* carotenoids underwent stereoisomerization in benzene or light petroleum solution within a day, in part in diffuse daylight (*278, 319, 310, 289, 208*): α-, β-, γ-carotene, crypto-xanthin, and capsanthin, 1–2%; zeaxanthin, 4–5%; capsorubin, 8%; lycopene, 10%; rhodoviolascin (spirilloxanthin), 23%.

Table 7. Examples of the Spontaneous Rearrangement of *cis* Carotenoids in Benzene Solution.

(The neo-β-carotene B solutions were sealed in dark brown ampoules; the others were exposed to weak diffuse daylight.)

| Compound | Temperature | Duration | Fraction stereoisomerized (%) | References |
|---|---|---|---|---|
| Neo-β-carotene B . . . . . . . . | — 2° | 60 days | 5–6 | (*23*) |
| Neo-β-carotene B . . . . . . . . | 20° | 1 day | 5 | (*23*) |
| Neo-β-carotene B . . . . . . . . | 20° | 14 days | 20 | (*23*) |
| Neo-β-carotene B . . . . . . . . | 20° | 60 days | 47 | (*23*) |
| Neocapsanthin A or B . . . | 20° | 1 day | 50 | (*289*) |
| Neorhodoviolascin A . . . . . | 20° | 9 min. | 55 | (*208*) |

The photolabile central-mono*cis*-β-carotene remained unchanged for a long time at — 5°, in darkness (*101*).

When working with carotenes the spontaneous isomerization can be neglected in short experiments, but the lability of some pigments such as all-*trans*-rhodoviolascin is a disturbing factor; chromatography revealed that even fresh solutions of crystals contained appreciable amounts of *cis* isomers (*208*).

The few *cis* carotenoids whose spontaneous *cis* → *trans* rearrangement has been studied in some detail show substantial individual differences, also depending on the conditions (*Table 7*).

*(b) cis-trans Isomerization in Refluxed Solutions.* When a dilute petroleum ether, hexane, cyclohexane, or benzene solution of an all-*trans* carotenoid is refluxed (preferably in darkness or semi-darkness) the stereoisomeric equilibrium is reached within 10–60 minutes. *Table 8* gives some information on the composition of the resulting solutions; and in *Table 9* the behavior of all-*trans*-β-carotene is compared with that of its two main isomers (pp. 266, 268).

Under the conditions of refluxing many neo forms are thermolabile. SAVINOV (*230*), however, was able to prepare neo-β-carotene B in good yield by keeping all-*trans* solutions at 100° in sealed tubes.

Some poly*cis* carotenoids are as thermostable as the corresponding all-*trans* forms (p. 284).

*(c) cis-trans Isomerization by Melting Crystals.* The crystals are sealed in vacuo or under an inert gas, in a narrow tube which is kept in a bath a few degrees above the melting point for 1–15 min. The tube is then rapidly cooled in ice-water, the substance is dissolved in cold solvent and chromatographed immediately (*209, 309, 211, 166*). No true equilibria are expected under these conditions but more or less stable steady-state mixtures that usually contain also some irreversibly formed product. The partial cleavage process may be checked to a certain extent when the melting point is depressed by the addition of several parts of naphthalene (*298*).

The method is useful for the isolation of some otherwise unavailable steric forms, the more so since in the melt *cis* isomers may predominate over the *trans* form. In exceptional instances small amounts of hindered *cis* carotenoids have also been obtained.

Examples. (1) All-*trans*-β-carotene, at 190° for 15 min.; 33% of the recovered pigment remained unchanged and six *cis* forms were observed. (2) Neo-β-carotene U at 135° for 15 min.; ratio, unchanged neo U : all-*trans* : seven *cis* forms = 40 : 22 : 38. (3) All-*trans*-zeaxanthin, with 4 parts of naphthalene at 180° for 1 min.; unchanged all-*trans* : *cis* forms = 56 : 44.

Analogous rearrangements were realized by heating *cis* dimethylcrocetins (p. 329) in KBr discs above the melting points. After cooling, the IR spectrum of the all-*trans* compound was observed (*154*).

Table 8. Examples of all-*trans*→*cis* Isomerization of Carotenoids in Refluxed Solutions.
(In the column "Ratio of stereoisomers," the first figure designates the top zone in the chromatogram; the values referring to unchanged all-*trans* forms are italicized; trace amounts are neglected.)

| Type of compound | Name | Solvent | Duration (min.) | Observed number of *cis* forms | Extent of isomerization in the recovered pigment (%) | Ratio of stereoisomers in the recovered pigment | References |
|---|---|---|---|---|---|---|---|
| Hydrocarbons | α-Carotene | petr. ether b. p. 60–70° | 30 | 2 | 8 | 4 : 92 : 4 | (309) |
| | β-Carotene | petr. ether b. p. 60–70° | 60 | > 4 | 14 | 4 : 86 : 8 : 1 : 1 | (209) |
| | γ-Carotene | petr. ether b. p. 60–70° | 30 | ≫ 4 | 27 | 73 : 6 : 5 : 8 : 8 | (310) |
| | 5,6-Dihydro-α-carotene | petr. ether b. p. 60–70° | 30 | > 4 | 36 | 3 : 27 : 64 : 6 | (210) |
| | 5,6-Dihydro-β-carotene | petr. ether b. p. 60–70° | 30 | 3 | 11 | 2 : 89 : 9 | (210) |
| | 3,4-Dehydro-α-carotene | hexane | 45 | 3 | 10 | 2 : 90 : 5 : 3 | (127) |
| | *retro*-Dehydrocarotene | hexane | 60 | > 4 | 36 | 64 : 28 : 5 : 2 : 1 | (321) |
| | Neurosporene | hexane | 60 | 8 | 16 | 84 : 16 | (179) |
| Alcohols | Cryptoxanthin (3-hydroxy-β-carotene) | ligroin b. p. 120° | 30 | 2 | 38 | 62 : 32 : 6 | (290, 298) |

| | Compound | Solvent | | | | | Ref. |
|---|---|---|---|---|---|---|---|
| | 4-Hydroxy-α-carotene | hexane | 30 | 3 | 5 | 95 : 2 : 2 : 1 | (21) |
| | Zeaxanthin (3,3'-dihydroxy-β-carotene) | benzene | 30 | > 2 | 30 | 24 : 6 : 70 | (320, 298) |
| | Lutein (3,3'-dihydroxy-α-carotene) | benzene | 30 | > 2 | 13 | 13 : 87 | (320) |
| | 5,6-Dihydroxy-5,6-dihydro-lycopene | benzene | 30 | > 4 | 4 | 96 : 1 : 2 : 1 | (21) |
| Esters | Physaliene (zeaxanthin dipalmitate) | petr. ether b. p. 70–80° | 60 | 1 | 42 | 58 : 42 | (290) |
| | Capsanthin dipalmitate | petr. ether b. p. 70° | 30 | 2 | 36 | 64 : 36 | (289) |
| Dimethoxy-hydrocarbons | Rhodoviolascin (spirilloxanthin) | benzene | 30 | > 2 | 41 | 59 : 41 | (208) |
| Ketones | Canthaxanthin (4,4'-diketo-β-carotene) | benzene | 30 | 4 | 38 | 62 : 29 : 4 : 5 | (70) |
| Hydroxyketones | Capsanthin | benzene | 45 | 3 | 35 | 22 : 78 | (289, 211) |
| | Capsorubin | benzene | 15 | 2 | 20 | 20 : 80 | (289) |
| Epoxides | β-Carotene monoepoxide | hexane | 60 | 5 | 9 | 5 : 91 : 4 | (259) |
| | β-Carotene diepoxide | hexane | 60 | 1 | 2 | 98 : 2 | (259) |
| | Taraxanthin (trihydroxy-α-carotene monoepoxide?) | benzene | 30 | 3 | 17 | 17 : 83 | (320) |

Table 9. Composition of Stereoisomeric Mixtures Obtained from All-*trans*-β-carotene and its Main Isomers by 60-min. Refluxing of the Petroleum Ether Solution (b. p. 60—70°), in Weak, Diffuse Daylight (*209*).

| Starting material | Relative photometric values in % of the recovered pigment | | | |
|---|---|---|---|---|
| | Neo U | All-*trans* | Neo B | Other *cis* forms |
| Neo-β-carotene U ........ | 31 | 40 | 19 | 10 |
| All-*trans*-β-carotene ...... | 4 | 86 | 8 | 2 |
| Neo-β-carotene B ........ | 4 | 50 | 40 | 6 |

Stereoisomerization in the fused or sintered state is responsible for the phenomenon of double melting points (p. 239).

## 2. Photochemical *cis-trans* Isomerization in the Absence of Catalysts.

Each carotenoid solution studied so far was found to be more or less photosensitive, depending on the structure and configuration. As expected the wavelengths corresponding to the fundamental band are most effective.

In our laboratory the solutions are illuminated either by artificial light (Photoflood bulbs or Mazda fluorescent lamps) or exposed to intense sunlight ("insolation"). When necessary, they are protected by an inert gas and/or cooling.

As a rule, the photo-stereoisomerization competes with irreversible side-reactions. In part only slight structural alterations occur, with preservation of a strong chromophore, but partial cleavage to colorless (fluorescent) substances may also take place. The destruction is promoted by the presence of air and accidental catalysts. On overexposure complete bleaching may be observed.

An all-*trans* carotenoid must loose some of its color intensity during illumination, while in the case of mono- or di-*cis* isomers either a partial bleaching effect or the bathochromic effect of the *cis* → *trans* rearrangement may prevail. When a poly*cis* compound is insolated for a short time, the tremendous increase in color intensity predominates. Thus, very dilute prolycopene solutions that appear almost colorless to the naked eye turn intensely yellow while exposed to sunshine for a few minutes (*278, 300*).

Examples. (a) Upon a 45-min. insolation of petroleum ether solutions at 30° the following ratios of unchanged to stereoisomerized starting material were found in the recovered pigment. All-*trans*-β-carotene, 98 : 2; neo-β-carotene B, 5 : 95; neo-β-carotene U, 37 : 73 (*309*). (b) When a dilute benzene solution of all-*trans*-zeaxanthin was insolated for 15 min., $1/_5$ of the color intensity was lost. Composition of the recovered pigment, 48% unchanged all-*trans* form, 11% *cis* zeaxanthins, 12% minor pigments, and 29% crystallizable new carotenoid (*298*). (c) A dilute hexane solution of *retro*-dehydrocarotene was illuminated with two Mazda lamps from 60 cm. distance for 8 hours. Irreversible loss, 6%; *cis* forms present in the recovered pigment, 11% (*321*).

The behavior of the total-synthetic, photolabile central-mono*cis*-β-carotene (which does not occur in equilibria ex all-*trans*-β-carotene) was described by INHOFFEN et al. (*101*). In diffuse daylight, at room temperature, this isomer rearranged almost completely within an hour. The wavelengths corresponding to the main band were most effective. Since those above 550 mμ do not act, this isomer (as some other similar pigments) can be handled safely in red light. It is remarkable that all-*trans*-β-carotene was practically the sole product of this illumination. In contrast, the iodine-catalyzed equilibrium mixture, formed within 3 minutes, contained only half of the pigment in the all-*trans* configuration. The *cis* configuration remained unchanged during a 20 hours' stay in darkness, in the presence of iodine.

### 3. *cis-trans* Isomerization by Iodine Catalysis, in Light.

The catalytic action of iodine on carotenoid solutions at room temperature, in light, is the most frequently used method of stereo-isomerization and yields within minutes a quasi-equilibrium mixture which can be resolved chromatographically. The process being reversible, each member of the set (or a mixture of members) should afford, after elution and renewed catalysis, the same stereoisomeric mixture. This postulate turned out to be very helpful in practical experimentation:

(a) If the investigator is interested in the isolation of a certain (minor) *cis* form only, that zone is cut out and set aside, while the mixture of all other pigments is subjected to a second treatment with iodine. These operations may be repeated and the desired *cis* zones combined. (b) Any *cis* carotenoid, even a non-crystallizable minor isomer, can be estimated quantitatively as follows, provided that the corresponding all-*trans* pigment is available in crystalline form: After the *cis* isomer has been eluted from its adsorbate and treated with iodine, one determines the extinction at $\lambda_{max}$ of the resulting stereoisomeric mixture. This value is then compared with the molecular extinction coefficient at $\lambda_{max}$ of an iodine-catalyzed equilibrium mixture which was prepared from a weighed sample of the all-*trans* compound.

The exact mechanism of iodine-catalyzed photoisomerizations about ethylenic double bonds is not yet clearly understood and will not be discussed in detail.

It has been postulated repeatedly that halogen is added and a free-radical formed at an intermediate stage of the catalytic process [cf. e. g., KHARASCH (*145*); URUSHIBARA (*260*)]. It should be noted in this connection, however, that iodine is only one of several paramagnetic substances ($O_2$, NO, $NO_2$, $Br_2$, Na, K, etc.) which are effective catalysts. HARMAN and EYRING (*89*) proposed that magnetic

Table 10. Examples of All-*trans*→*cis* Isomerizations of Carotenoids, Catalyzed by Iodine, in Light.

(In the column "Ratio of steric forms" the first figure designates the top zone on the chromatographic column; data referring to unchanged all-*trans* forms are italicized; trace amounts are neglected.)

| Type of compound | Name | Observed number of *cis* forms | Extent of isomerization in the recovered pigment (%) | Ratio of steric forms in the recovered pigment | References |
|---|---|---|---|---|---|
| Hydro-carbons | Lycopene.......... | 2 | 43 | *57*:43 | (*320, 300*) |
| | α-Carotene ........ | 7 | 48 | 14:3:15:*52*:13:3 | (*309, 195*) |
| | β-Carotene.......... | 12 | 52 | 22:*48*:25:3:2 | (*209*) |
| | γ-Carotene.......... | > 5 | 47 | 3:*53*:7:14:16:7 | (*310*) |
| | 5,6-Dihydro-α-carotene........ | 5 | 44 | 4:20:*56*:16:4 | (*210*) |
| | 5,6-Dihydro-β-carotene ........ | 5 | 33 | 18:*67*:15 | (*210*) |
| | 3,4-Dehydro-α-carotene........ | 4 | 32 | 7:2:*68*:6:17 | (*127*) |
| | 3,4-Dehydro-β-carotene ........ | 4 | 51 | 22:*49*:26:3 | (*127*) |
| | *retro*-Dehydrocarotene | > 8 | 80 | 20:32:5:12:14:5:7:4:1 | (*321*) |
| | Neurosporene ....... | 6 | 56 | *44*:13:11:12:15:3:2 | (*179*) |
| | Phytofluene ........ | 1 | 10 | *90*:10 | (*204, 147*) |
| Alcohols | Cryptoxanthin (3-hydroxy-β-carotene) | 3 | 41 | 18:*59*:18:5 | (*290, 298*) |
| | 4-Hydroxy-α-carotene | 6 | 37 | 13:*63*:21:3 | (*21*) |
| | Zeaxanthin (3,3'-dihydroxy-β-carotene) | 3 | 34 | 10:21:3:*66* | (*320, 298*) |
| | Lutein (3,3'-dihydroxy-α-carotene) | · 2 | 40 | 17:23:*60* | (*320*) |
| | 5,6-Dihydroxy-5,6-dihydrolycopene ..... | 4 | 49 | 11:*51*:12:17:9 | (*21*) |
| Ethers | 4-Ethoxy-α-carotene . | 5 | 39 | 13:1:*61*:11:8:6 | (*21*) |
| | Rhodoviolascin (spirilloxanthin) .... | 2 | 40 | *60*:40 | (*208, 237*) |
| Esters | Physaliene (zea-xanthin dipalmitate) | 1 | 55 | *45*:55 | (*290*) |
| | Capsanthin dipalmitate ........ | 2 | 33 | *67*:33 | (*289*) |
| | Capsorubin dipalmitate........ | 2 | 25 | *75*:25 | (*289*) |
| Mono-ketones | 4-Keto-α-carotene ... | 4 | 48 | 11:*52*:19:12:6 | (*21*) |
| Diketones | Canthaxanthin (4,4'-diketo-β-carotene) .. | 6 | 38 | *62*:18:10:5:5 | (*70*) |

(*Table 10, continued.*)

| Type of compound | Name | Observed number of *cis* forms | Extent of isomerization in the recovered pigment (%) | Ratio of steric forms in the recovered pigment | References |
|---|---|---|---|---|---|
| Hydroxy-ketones | Capsanthin ........ | 3 | 31 | 15:11:5:69 | (*289, 211*) |
| | Capsorubin ......... | 2 | 32 | 32:68 | (*289*) |
| Epoxides | β-Carotene mono- | | | | |
| | epoxide .......... | 9 | 44 | 14:1:15:56:7:5:2 | (*259*) |
| | β-Carotene diepoxide . | 7 | 41 | 4:25:59:10:2 | (*259*) |
| | Taraxanthin (trihydr-oxy-α-carotene | | | | |
| | monoepoxide?) ..... | 3 | 44 | 44:56 | (*320, 62*) |

interactions are responsible for catalytically promoted *cis-trans* rearrangements. According to McCONNELL (*184*), when a doublet electronic state of a catalyst atom interacts with the singlet and triplet states of an isomer, two doublet states are formed whose minimum separation is determined by the strength of the chemical binding between catalyst and substance. It is assumed that even weak chemical interactions may account for the observed effect. These arguments are perhaps preferable to the older concept that iodine promotes *cis-trans* rearrangements by adding to the molecule and then "breaking" the double bond.

The history of iodine-catalyzed rearrangements about ethylenic double bonds can be traced back to ANSCHÜTZ (*6*) (1878) who obtained ethyl fumarate by the interaction of silver maleate and ethyl iodide, because his ethyl iodide was contaminated with free iodine; the latter rearranged catalytically the primary reaction product, viz. ethyl maleate. In 1929 KARRER et al. (*134*) were able to convert, by means of iodine, the natural product bixin into "stable bixin"; and they interpreted the reaction tentatively as a geometrical isomerization (p. 327). This was proved to be correct by KUHN and WINTERSTEIN (*157*).

In the field of the $C_{40}$-carotenoids iodine catalysis was introduced by TUZSON and the writer (*320*) as late as 1939, closely followed by similar studies in collaboration with CHOLNOKY and POLGÁR (*289, 290*). Stress was laid on the reversibility of the process—a feature of practical importance that had not been claimed earlier. It was also demonstrated that iodine does not rearrange carotenoids in darkness (Fig. 7, p. 250).

In our early experiments, conducted in diffuse laboratory daylight, the concentration of the pigment was roughly 0.1 mg. per ml. of hexane or benzene ($^1/_{5000}$ molar) and the weight of the iodine amounted to 1–2% of the pigment. Stereoisomeric quasi-equilibrium was reached within 15–60 min. but in some instances within a few minutes.

Example. Lycopene in benzene (*320*):

| Minutes of catalysis ......... | 0 | 5 | 30 | 90 | 180 |
|---|---|---|---|---|---|
| Ratio, all-*trans* : *cis* forms ..... | 100 : 0 | 82 : 18 | 71 : 29 | 57 : 43 | 53 : 47 |

It was recognized later that brief illumination by means of standardized artificial light is the method of choice: One or several well-filled and stoppered 25- or 50-ml. Pyrex volumetric flasks, each containing a hexane or benzene solution of 5–25 mg. of pigment and 1–2% iodine (pigment = 100%), are exposed for 5–45 min. to the light of two parallel Mazda fluorescent lamps (40 W, 3500°, white and yellow) from 60 cm. distance.

*Table 10* (p. 270) shows that the resulting stereoisomeric mixtures contained $^1/_3$–$^1/_2$ of the pigment in *cis* configurations, with a few exceptions for which some special structural features were responsible.

The stereoisomeric phytofluene mixture contained only 10% *cis* forms (*147*). In contrast, *retro*-dehydrocarotene isomerized to the extent of 80% (*321*). Indeed, in this instance the amount of one main mono*cis* isomer by far surpassed the unchanged portion of the all-*trans* compound. This feature is in accordance with the finding that the preparation of *retro*-dehydrocarotene, when carried out by the interaction of β-carotene and iodine, followed by a treatment with thiosulfate (*155*), leads mainly to *cis* forms, in contrast with most structural conversions of *normal* carotenoids (*321*).

The main factors that influence iodine-catalyzed rearrangements are, the concentrations of pigment and catalyst, the absolute quantity of the iodine, the pigment/iodine ratio, and the mode and duration of the illumination. In light, iodine does not act exclusively as a promoter of *cis-trans* rearrangements but also participates in irreversible side-reactions. Thus, some pigment may be lost, especially on over-exposure.

For α-carotene ZSCHEILE et al. (*326*) have estimated the irreversible losses. Further, they found that the effective radiation is that absorbed by the iodine rather than by the carotene, and that the free iodine gradually disappears from the solution in prolonged experiments. Of course, catalysis by iodine cannot succeed if either the solvent or the plant extract contains unsaturated impurities which add halogen at high rates and thus remove the catalyst from the system (*326*). A single chromatographic operation will correct such situations.

STRAIN (*242*) has reported that, under certain conditions, zeaxanthin yielded (in part) the same irreversible pigments with halogen which were obtained by means of acids (see below). The presence of a little organic base prevented this to happen.

The nature of a few irreversible side-reactions has been clarified: (a) 4-Hydroxy-α-carotene afforded small amounts of the corresponding ketone (*21, 22*). (b) Either of the two β-carotene epoxides was converted to a limited extent into furanoid oxides (*259*). In the case of another epoxide, trollixanthin, no *cis→trans* isomerization could be realized at all because of conversion into a furanoid oxide (*62*).

In spite of the possible complications that were rather emphasized above, iodine, when applied under proper conditions, is a very efficient tool in stereochemical research.

When working with carotenoid pigments it should also be considered that the degree of iodine sensitivity is a function of the number and type of *cis* double bonds. Especially sensitive are sterically hindered such bonds; these afford the equilibrium mixture at much higher rates than unhindered *cis* or *trans* double bonds. Very similar is the behavior of naturally occurring poly*cis* compounds (p. 284). Iodine sensitivity and degraded spectrum go together in many instances. Although this statement is also valid in the diphenylhexatriene and diphenyloctatetraene sets (*173, 306*), it does not apply to dienes. Certain isoprenic, hindered *cis* dienes of the $C_{20}$ series are resistant to iodine while some unhindered *cis* isomers respond to the catalyst normally (*197a*). In two polyene-azine sets *cis* $C=C$ bonds were found to be more sensitive to iodine than $C=N$ bonds but no pertinent generalization is yet possible (*46*). It would be well worth while to carry out systematic kinetic studies in the fields just mentioned.

### 4. cis-trans Isomerization by Acid Catalysis.

Although this method may be useful in certain instances, its application has been abandoned in our laboratory, because the spatial rearrangement was frequently accompanied or overruled by irreversible processes. It has been recognized long time ago that especially the xanthophylls are acid-sensitive, even in the presence of weak acids (*268, 160, 240, 242, 248, 247, 218*). A classical example of acid sensitivity is the behavior of KARRER's carotenoid epoxides: Even the acid traces in commercial chloroform convert them into furanoid oxides quantitatively, at room temperature, within a few seconds [KARRER (*128*)]. A similar $H^+$-catalysis causes the almost instantaneous elimination of allylic hydroxyl groups. Example: 4-hydroxy-$\alpha$-carotene $\rightarrow$ 3,4-dehydro-$\alpha$-carotene [BUSH et al. (*22*)].

Carotenes are much less acid-sensitive. When a petroleum ether solution of $\beta$-carotene was shaken with strong hydrochloric acid, about half of the pigment underwent stereoisomerization but very little was changed irreversibly. Application of strong hydriodic acid afforded 5,6-dihydro-$\beta$-carotene [POLGÁR et al. (*210*)].

### 5. cis-trans Isomerization by Contact with Active Surfaces.

As mentioned in connection with GILLAM's pioneer experiments (p. 230), the all-*trans* configuration of carotenoids is not affected by routine chromatography. According to MEUNIER et al. (*187*), however, this situation may be altered under special conditions. The conditions applied by these authors were static and hence essentially different from those existing in the chromatographic column: Very dilute pigment solutions, contained in conical tubes, were left in contact (without shaking) with a large excess of strongly active $Al_2O_3$ or $TiO_2$, for days and weeks, protected from light and air. A treatment of $\beta$-carotene resulted in the appearance of a marked *cis*-peak observed in the supernatant petroleum ether solution; and the intensity of the peak seemed to indicate a *trans* → *cis* shift in the middle section of the molecule. No spectral change was noticed in the absence of the adsorbent within a month. MnO₂ also caused the described spectral effect but this was soon overruled by the progressing cleavage of carotene to retinene (*186*).

The stereoisomerization of fucoxanthin is accelerated by contact with sugar or glass powder (p. 317).

Evidently, a more profound study of the behavior of carotenoids on strongly active surfaces is desirable.

A reported stereoisomerization of astaxanthin on alumina needs confirmation, since the pigment was also exposed to acetic acid (*84*). An attempt to stereo-isomerize β-carotene epoxides by MEUNIER's method did not afford *cis* compounds but furanoid oxides (*259*). For the influence of moist Celites on carotenoids cf. (*269*).

### 6. *cis-trans* Isomerization via Boron Trifluoride Complexes.

PRICE and MEISTER (*217*) predicted long ago that this reagent, when added to ethylenic double bonds, will afford on cleavage *cis-trans* equilibria; an alleged application of this principle to stilbene, however, turned out to be erroneous (*216, 217*). Recent experiments have shown that BF₃ etherate may well become a tool for the preparation of certain *cis* carotenes, under strictly defined conditions (*265, 266, 22, 285*).

After a 1-min. interaction of BF₃ and β-carotene the pigment mixture was composed of 36% unchanged all-*trans* form, 44% *cis* isomers, 1% foreign carotenoids, and 19% unidentified (in part colorless, fluorescent) substances. In the case of *retro*-dehydrocarotene profound structural changes took place within a minute.

### 7. Bio-stereoisomerization.

It is interesting to learn that ingested carotenoids may alter their configurations in the body. The factors directing such rearrangements are more complex than those dealt with in the above Sections. They may include thermal and pH effects, the action of unknown catalysts, the metabolism of intestinal bacteria, etc. Some observations available in this field will be mentioned in Chapter XI, p. 331.

## VI. Preparation of *cis* Carotenoids by Total Synthesis.

The most brilliant achievement in polyene chemistry during the last decade has been the total synthesis of certain naturally occurring carotenoids and of some analogs and homologs that have not yet been found in plants. The synthesis of β-carotene was announced in 1950 by KARRER and EUGSTER (*130*), almost simultaneously by INHOFFEN, BOHLMANN, BARTRAM and POMMER (*98*), and somewhat later by MILAS et al. (*188*). Since this breakthrough the new field has been broadened and deepened by the research groups of KARRER, INHOFFEN, ISLER,

WEEDON, and others [cf. also SURMATIS et al. (*251*)]. The pertinent literature till 1952 has been discussed in our Series by INHOFFEN and SIEMER (*107*); for a brief survey cf. KARRER (*129*). In 1957 ISLER and ZELLER's excellent survey has appeared (*119*). The present report, which of course cannot do justice to the authors active in this fascinating field, will be restricted to a few stereochemically important aspects of carotenoid synthesis.

The $C_{40}$-molecule was built up in most instances by joining three fragments, according to the schemes, $C_{10} + C_{20} + C_{10}$; $C_{14} + C_{12} + C_{14}$; $C_{15} + C_{10} + C_{15}$; $C_{16} + C_8 + C_{16}$; $C_{18} + C_4 + C_{18}$; or $C_{19} + C_2 + C_{19}$. In the last two cases diacetylene or acetylene units were incorporated into the chain by means of Grignard derivatives, lithium-organic compounds, etc.

As characteristic intermediates, $C_{40}$-glycols were thus obtained which contained ethylenic and acetylenic bonds but were void of a

Table 11. Examples of Total-synthetic *cis* Carotenoids.

| Stereoisomeric set | Configuration | References |
|---|---|---|
| β-Carotene | central-mono*cis* ............. | (*99, 101, 114*) |
| | 11,11'-di*cis* (hindered) ....... | (*61, 108*) |
| | *cis*-β-carotene C (hindered)... | (*61, 108*) |
| 3,4-Dehydro-β-carotene | central-mono*cis* ............. | (*115*) |
| 3,4,3',4'-Bisdehydro-β-carotene | central-mono*cis* ............. | (*115, 105*) |
| 13,13'-Bis-desmethyl-β-carotene $C_{38}H_{52}$ | central-mono*cis* ............. | (*100*) |
| 16,16'-*Homo*-β-carotene $C_{42}H_{58}$ | middle-mono*cis* (16-*cis*) ...... | (*97*) |
| | middle di*cis* (16,16'-di*cis*) .... | (*97*) |
| α-Carotene | central-mono*cis* ............. | (*106*) |
| Lycopene | central-mono*cis* ............. | (*110*) |
| | hindered *cis* forms b, c and c' | (*74*) |
| Cryptoxanthin | central-mono*cis* ............. | (*113*) |
| Zeaxanthin | central-mono*cis* ............. | (*112, 118*) |
| Isozeaxanthin | central-mono*cis* ............. | (*116*) |
| Echinenone | central-mono*cis* ............. | (*2 a*) |
| Canthaxanthin | central-mono*cis* ............. | (*116, 117, 323*) |
| Torularhodin methylester | central-mono*cis* ............. | (*109*) |
| Apo-β-carotenals | 15,15'-*cis* ................. | (*227, 228*) |
| 1,18-Diphenyl-3,7,12,16-tetramethyl-octadecanonaene | 5-*cis* (hindered) ............ | (*60, 71, 72*) |
| | 5,13-di*cis* (hindered) | (*60, 71, 72*) |
| | other hindered and unhindered *cis* isomers | |
| 1,3,7,12,16,18-Hexaphenyl-octadecanonaene | four hindered *cis* forms and from them (with iodine) several unhindered ones ... | (*324*) |
| "Naphthyl-carotene" | central(?)-mono*cis* ......... | (*170*) |

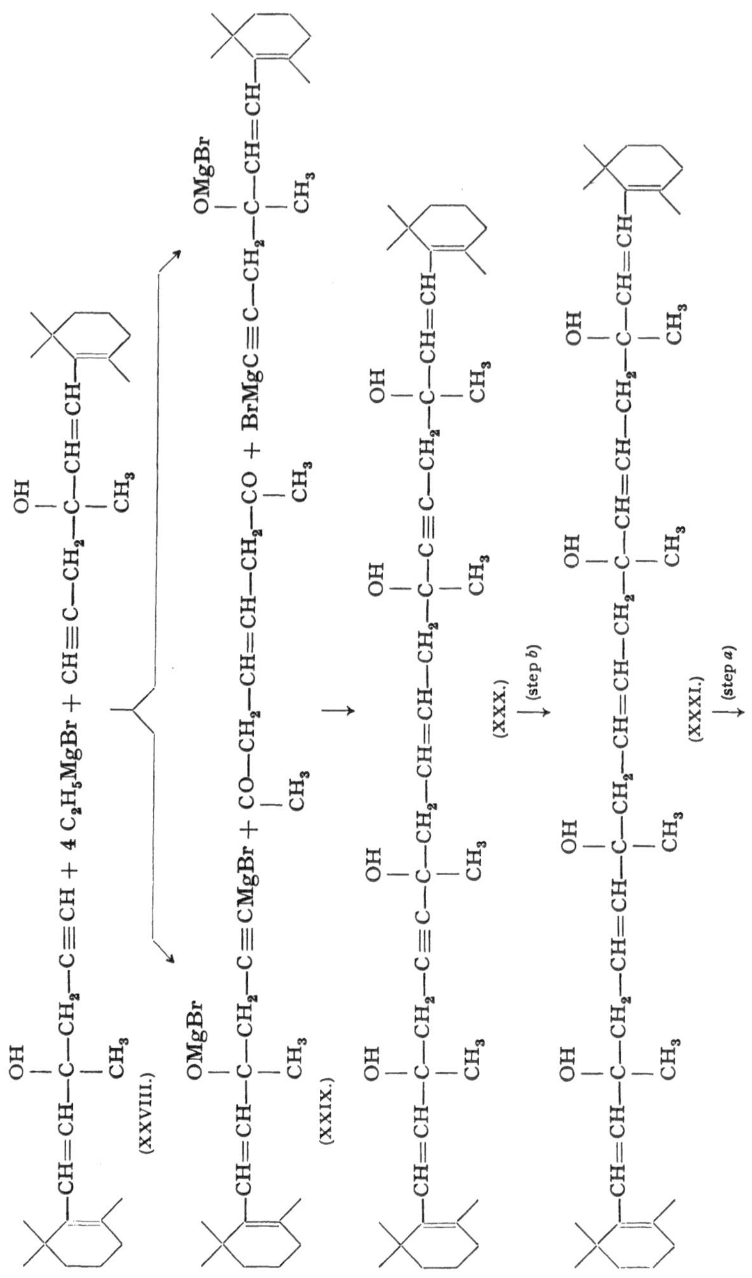


(I.) All-*trans*-β-carotene.

*Chart 1.* Synthesis of β-Carotene according to KARRER and EUGSTER (*131*).

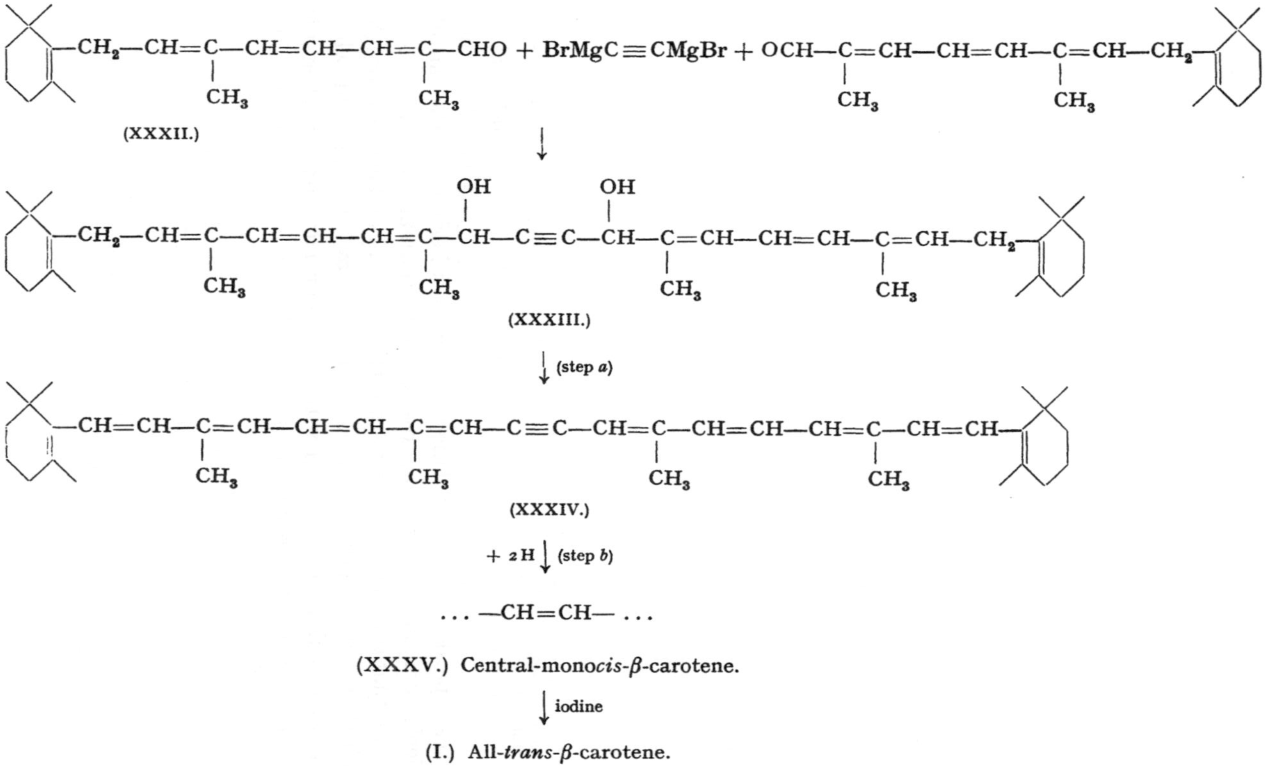

*Chart 2.* Synthesis of β-Carotene according to INHOFFEN et al. (99) (see also Models IV and I on p. 236).

complete conjugated system. Hence, at this stage the following further transformations were required: (a) elimination of the OH-groups by dehydration with formation of double bonds; and (b) partial reduction of the triple bond(s) to double bond(s). In some of the syntheses step (a), in others step (b) was carried out first *(Charts 1 and 2)*. When (a) preceeds (b), the reaction becomes stereochemically important because it is possible to conduct the partial saturation of the acetylenic bond(s) in a stereospecific manner *(198, 4, 67)*; surveys *(36, 20)*. A number of *cis* carotenoids have thus been obtained in which the location of the *cis* double bond(s) is established beyond doubt *(Table 11, p. 275)*.

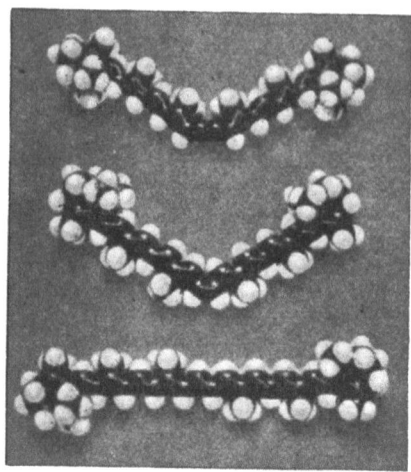

Fig. 16. Models of 15,15′-dehydro-β-carotene (top), central-mono*cis*-β-carotene (middle), and all-*trans*-β-carotene (bottom); according to Inhoffen et al. (99).
[From: Liebigs Ann. Chem. 570, 54 (1950).]

In some instances the synthesis yielded three intermediate pigments that contained, respectively, two triple bonds, two cumulene groups, and or triple bond and one cumulene group (besides other olefinic bonds). After chromatography, each of these compounds was submitted to partial saturation [see e. g. Garbers, Eugster and Karrer (71).]

A few remarks on the partial reduction of triple bonds follow. Sly's X-ray crystallographic study *(236)* of a typical intermediate of the β-carotene synthesis, to wit the central-acetylenic compound 15,15′-dehydro-β-carotene (formula XXXIV, p. 277) has shown that this molecule possesses axial symmetry and hence a straight over-all shape in the crystal. The situation is, however, different when dissolved molecules are in contact with an appropriate catalyst. Then this compound may react either in the straight form or in a form bent at the center *(Fig. 16) (99)*. Consequently, the configuration about the emerging double bond will depend on experimental conditions; for instance, LiAlH$_4$ affords an all-*trans* polyene but the Lindlar catalyst a *cis* form *(103)*.

The well-known Lindlar catalyst is prepared by deactivating Pd or Pd/CaCO$_3$ by means of hot lead acetate solution *(169)*. In some instances the presence of quinoline secures maximum efficiency and/or specificity. The interaction with carotenoids is best carried out in darkness or red light considering the photosensitivity of some *cis* compounds.

A comparison of total-synthetic *cis* carotenoids with our neo compounds has confirmed and extended numerous earlier observations concerning adsorption affinities, spectral curves, stabilities, $\lambda_{max}$ shift, *cis*-peak effect, and so forth. The synthetic studies have also shown, however, the availability of hindered *cis* isomers that failed to appear in our stereoisomeric equilibria and whose very existence was questioned at a certain stage of research (cf. p. 235).

No synthetic poly*cis* carotenoid with known configuration has been reported up to the present time and it would be desirable to carry out such theoretically important work in the near future.

## VII. Naturally Occurring *cis* and Poly*cis* Carotenoids.

### 1. Mono- and Di*cis* Carotenoids.

As we have seen, the overwhelming majority of plant carotenoids possess the all-*trans* configuration. To what extent the more energy-rich *cis* carotenoids appear in plants and animals is a question of biological and genetical interest. The relative scarcity of *cis* carotenoids in living tissue indicates the presence of some directing and protective mechanism that forms and preserves the all-*trans* configuration, even when the pigment is exposed to strong sunlight and heat; similar conditions cause *trans* → *cis* rearrangements in extracts. When a *cis* carotenoid has been formed, its configuration is likewise protected in the plant.

Examples. (a) Large quantities of *cis*-phytofluene do occur in the tomato but upon extraction rapid isomerization to the all-*trans* form takes place, especially in light (*204, 147*). (b) Although intact red pepper fruits contain (almost) exclusively all-*trans* pigments, exposure to sunshine of opened fruits results in stereo-isomerization (*182, 183*). (c) It has been shown by SAPERSTEIN and STARR (*229*) that some carotenoids of non-photosynthetic bacteria do not undergo *trans* → *cis* rearrangements under the influence of iodine and light, while attached to protein. (This should be further confirmed by showing that the catalyst is not used up by the complex.)

In regard to the natural occurrence of neo carotenoids that contain one or two *cis* double bonds, the literature appears to be in a state of confusion. Because of the spontaneous attainment of *trans-cis* equilibria, even at room temperature, more than one *cis* isomer that was claimed as a natural product might well have been formed during extraction, purification, storage, evaporation at elevated temperature, exposure to alkali, etc. Whether or not certain chromatographic *cis* zones contained "genuine" pigment remains open to criticism.

Although some authors have accepted the observed "occurrence" of *cis* forms at face value, others did recognize the situation correctly. According to PORTER and ZSCHEILE (*214*), the neo-$\beta$-carotene U found in canned tomatoes resulted from isomerization during the canning process. STRAIN (*240*) emphasized the necessity of extracting carotenoids under mild conditions. COOK (*34*) when reviewing algal pigments, has expressed doubts about the occurrence of some of the reported isomers. Similar statements are valid for some pigments in petals (28), in *Rhodospirillum* (123), in *Haematococcus* (82), in fossil sediments (261), etc.

It should also be remembered that previous storage conditions of plant materials may be responsible for the occurrence of some *cis* carotenoids in extracts. Thus, high temperatures applied during dehydration of alfalfa meal promotes *trans* → *cis* changes; and exposure of the meal to visible light decreases the neo-β-carotene B content while more neo U appears [Bickoff et al. (256, 239)].

The situation is much clearer when the observed *cis* carotenoid does not occur in any detectable amounts in the stereoisomeric equilibrium. Such an isomer would disappear rapidly upon the addition of iodine to the extract and illumination. This happens, for example, in the case of the poly*cis* compounds (see below) and of neo-γ-carotene P, a constituent of *Pyracantha* berries (305).

Likewise, a clear decision can be made when a native pigment, which, although a component of the equilibrium, occurs in the plant either as the sole or as the preponderant member of the set. For example, according to Tappi and Karrer (254), the main pigment of the pollen sacs of *Lilium candidum* is a crystallizable mono*cis*-antheraxanthin (zeaxanthin monoepoxide) which also occurs in *L. umbellatum*, *L. regale*, *L. Willmottiae unicolor*, *L. Maxwill*, and *L. mantchuricum* (133, 255). (The corresponding all-*trans* compound was isolated from some other species.) Along slightly different lines, Suzuki and Tsukida (252) have shown that neo-β-carotene B is a genuine constituent of *Osmanthus fragrans* flowers, because the fresh extract was free of neo-β-carotene U which is contained in amounts matching those of neo B at equilibrium.

When the interconversion rates of the stereoisomers are very low, then the probability of their natural occurrence increases. This may be assumed for the three stereoisomeric fucoxanthins, "unless a rapid conversion takes place immediately upon death of the cell" (248, 250).

Under less favorable circumstances it may still be possible to decide, as follows, whether or not an observed *cis* carotenoid is a natural product.

Two equal samples of the plant material are taken and one of them is mixed with pure all-*trans* pigment, in an amount similar to that present in the natural material. All extraction and purification operations are then carried out simultaneously with both samples, and the two final chromatograms are compared. If the column originating from the sample to which pigment had been added does not show an increase of the *cis* zones, a strong argument has been gained in favor of natural occurrence.

## 2. Poly*cis* Carotenoids.

The first representative of this class was detected as late as 1941 in some tomato varieties such as "tangerine tomatoes". The reason that had prevented its earlier detection is its absence from stereoisomeric equilibria ex all-*trans*-lycopene. The color of the ripe tangerine tomato is orange instead of red. Because the pigment substituted for ordinary lycopene, it was termed "prolycopene" (*301*). A few similar pigments have been isolated since. Most vegetable tissues are free of poly*cis* carotenoids, quite a number of plants contain trace amounts, but considerable quantities have rarely been found. More than 20 mg. of crystalline prolycopene was isolated from 1 kg. of fresh tangerine tomatoes (yield, 93%) [LE ROSEN, WENT, PAULING and the writer (*301*, *166*)].

A systematic search for naturally occurring poly*cis* pigments is still lacking. It will be noted that the relatively few higher plants listed in *Table 12* (p. 282) belong to ten families, and that poly*cis* pigments occur also in microorganisms. Although they are evidently wide-spread, poly*cis* carotenoids are not mass products of biosynthesis, in sharp contrast to all-*trans* forms.

With respect to the poly*cis* configuration of their carotenoids, even closely related plants may show considerable differences.

Example. The ripe berries of *Pyracantha angustifolia* harvested in California are rich in poly*cis* forms, but no such pigments have been found in *P. yunanensis* (California) or in *P. coccinia* (Europe) (*137*); the latter plants do produce all-*trans*-lycopene.

Although almost all known poly*cis* carotenoids are hydrocarbons (*Table 12*), poly*cis* α- or β-carotenes are unknown.

As we have seen, the conspicuous characteristics of poly*cis* carotenoids are the unusually weak adsorption affinity, the degraded fundamental band and the very high value of the $\lambda_{max}$ shift (p. 288). No carotenoid should be assigned, however, a poly*cis* configuration on this basis alone, rather the impressive spectral shift on iodine catalysis should be considered as decisive (Fig. 17, p. 284).

Very little is known about the biosynthesis of poly*cis* carotenoids; the suggestion that prolycopene may be a normal precursor of all-*trans*-lycopene in tomatoes does not seem to hold (*80*, *121*, *213*, *257*).

WENT and the writer (*322*) have considered this problem in the light of the chemical steering power of genes which is generally directed toward a single structural target. In some instances, however, a specially high degree of selectivity is required in order to build up a certain steric configuration. It is expected that, as in the case of the tomato, in many other instances one gene will be responsible for the synthesis of the polyene structure but another gene will steer the molecules to a predetermined configuration. This latter function may be important for the plant, because *cis-trans* configurations determine the overall molecular shape and because only certain geometrical forms may fit into a given biological system.

In general, the stereochemical status of plant polyenes, poly*cis* carotenoids included, is strongly influenced by both genetical and environmental factors.

Table 12. Examples of the Occurrence of Polycis Carotenoids in Nature.

| Compound | Plant family | Plant | Organ | References |
|---|---|---|---|---|
| Prolycopene* | Solanaceae | Lycopersicum esculentum (Mill.) var. "Tangerine tomato" | Fruit | (301, 166, 305) |
| | | var. "Golden Jubilee" | Fruit | (214, 213) |
| | | L. esculentum, L. hirsutum, L. pimpinellifolium L. (crossings, selections) | Fruit | (177) |
| | Pomoideae | Pyracantha (Cotoneaster) angustifolia Schneid. | Fruit | (314) |
| | Celastraceae | Evonymus fortunei Rehd. (Winter creeper) | Seed hulls | (294) |
| | Scrophulariaceae | Mimulus longiflorus Grant (Monkey flowers) | Flowers | (231) |
| | Palmae | Butia capitata Becc. | Fruit | (312, 315) |
| | | Butia eriospatha Becc. | Fruit | (312, 315) |
| Polycis-lycopenes I–VI | Algae | Chlorella vulgaris mutant | — | (30–32) |
| Polycis-lycopene I | Pomoideae | Pyracantha angustifolia Schneid. | Fruit | (305) |
| Polycis-lycopenes | Compositae | Calendula officinalis (Pot marigold) | Flowers | (79) |
| | Araceae | Arum maculatum (Cuckoo pint) | Fruit | (81) |
| | Rosoideae | Rosa canina, R. moyesii (Rose hips) | Fruit | (81) |
| | Cruciferae | Brassica rutabaga | Root | (125, 126) |
| | Solanaceae | Lycopersicum esculentum (strains, crosses, selections) | Fruit | (176, 213, 258, 179, 214) |
| Polycis-lycopenes (seven), Proneurosporene (protetra-hydrolycopene, neoneurosporene P, polycis-ψ-carotene, unidentified carotene I) | | | | |
| Pro-γ-carotene | Algae | Chlorella vulgaris mutant | — | (30–32) |
| | Palmae | Butia capitata Becc. | Fruit | (312, 315) |
| | | Butia eriospatha Becc. | Fruit | (312, 315) |
| | Pomoideae | Pyracantha angustifolia Schneid. | Fruit | (305, 314) |
| | Celastraceae | Evonymus fortunei Rehd. | Seed hulls | (294) |
| | Scrophulariaceae | Mimulus longiflorus Grant | Flowers | (231) |
| | Sulfur bacteria | Chlorobium spp. | — | (83) |
| Monohydroxy-pro-γ-carotene | Pomoideae | Pyracantha angustifolia Schneid. | Fruit | (314) |
| Polycis-cryptoxanthin | Aurantioideae | Citrus aurantium (Valencia orange) | Fruit juice | (40) |

* In our nomenclature "prolycopene" designates an individual polycis-lycopene, viz. the main pigment of the Tangerine tomato; we characterize some similar isomers as polycis-lycopenes I, etc., while Goodwin makes use of the term "prolycopenes I, ...".

Along genetic lines, by means of crossing and selection experiments, detailed studies of the tomato fruit have been reported by PORTER and ZSCHEILE *(214)*, ZSCHEILE and PORTER *(327)*, MACKINNEY and JENKINS *(176, 177)*, PORTER and LINCOLN *(213)*, TROMBLY and PORTER *(258)*, as well as by MANUNTA *(181)*; cf. also *(300, 121, 257)*. Surveys on connected subjects have been presented by MACKINNEY *(175)* and by GOODWIN *(80)*.

Turning now to environmental factors, light should be mentioned as of primary importance. This was shown by KUHN and WINTER-STEIN *(158)* in early experiments concerning crocetin (pp. 328, 329).

In our laboratory, SCHROEDER *(231)* made the interesting observation that when cut stems (with buds on) of "Monkey flowers" *(Mimulus longiflorus)* were placed in water and the flowers were allowed to open in diffuse room light, the petals produced substantial amounts of the poly*cis* pigments prolycopene and pro-γ-carotene. Conversely, from flowers developed on the bush, in the open, *cis* forms were missing and the extracts contained only the two corresponding all-*trans* carotenoids. Furthermore, when the flowering plant was exposed to weak sunshine only, the petals remained yellow but during a period of prolonged clear weather the usual orange coloration did appear. Such observations also demonstrate how profoundly *cis-trans* isomerism may enrich the variety of colors in petals and fruits.

Possibly, similar environmental factors were responsible for the isolation of either *cis-* or *trans*-trollixanthin from the petals of *Trollius europaeus* [EUGSTER and KARRER *(62)*].

In the class of the green algae CLAES *(30, 31)* has reported that when the *Chlorella* mutant "5/520" (obtained by X-ray irradiation) was cultivated in darkness, it produced poly*cis* carotenoids (prolycopene and protetrahydrolycopene). However, illumination of the dark cultures with blue light resulted in the decrease and disappearance of the poly*cis* isomers and in formation of all-*trans*-lycopene and all-*trans*-tetrahydro-lycopene instead. According to CLAES and NAKAYAMA *(32)* the photo-isomerization, poly*cis* → all-*trans*, in the algal cell can be realized even in red light, provided that chlorophyll is present. The same effect (which is inhibited by oxygen) can be demonstrated also in vitro, with petroleum ether solutions of prolycopene or protetrahydrolycopene. It would seem that, under certain conditions, the spatial configuration of chloro-plast carotenoids might be controlled by chlorophyll.

## Prolycopene.

Pure, crystalline prolycopene was described by LeROSEN and the writer *(166)*. It is sharply differentiated from neo-lycopenes by the thermostability of solutions and easy crystallizability. Its melting point,

111°, lies 64° below that of lycopene. Prolycopene is at least as thermo-stable as all-*trans*-lycopene. An impressive difference from the all-*trans* compound is in the color.

Under the microscope the crossing sections of individual prolycopene crystals are dull-brown, in contrast to the corresponding ruby-red color of all-*trans*-lycopene.

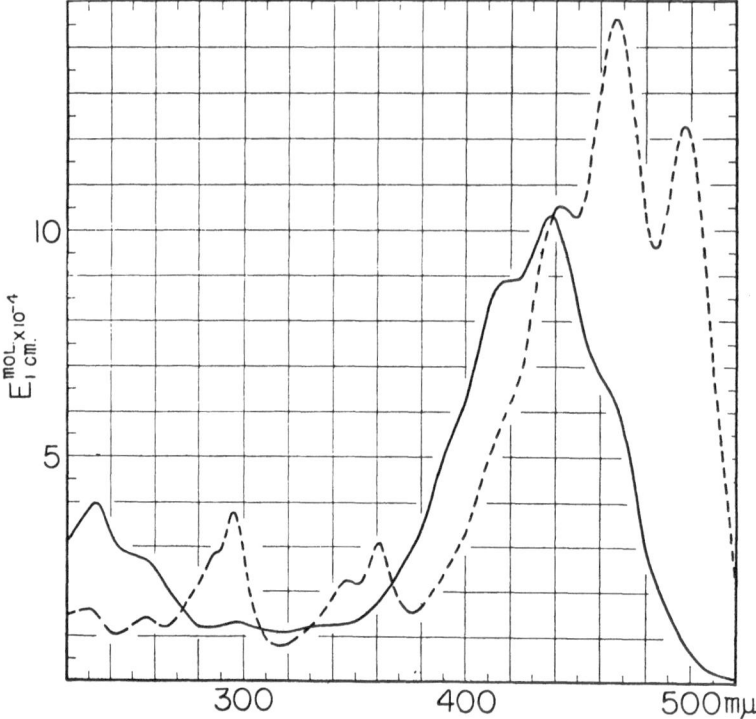

Fig. 17. Molecular extinction curves of prolycopene, in hexane: ————, fresh solution; and — — —, mixture of stereoisomers after catalysis by iodine *(305)*. [From: J. Amer. Chem. Soc. **69**, 1930 (1947).]

When measured in a simple colorimeter, the color intensity of prolycopene in hexane is about one half of that of lycopene.

The spectral maximum of prolycopene in hexane solution is located at 438 m$\mu$, i. e. at about 35 m$\mu$ shorter wavelengths than that of lycopene. The degraded curve shows much depressed extinction values and almost no fine structure *(Fig. 17)*. A *cis*-peak is absent and hence the molecule must have an overall straight form.

Prolycopene is highly photosensitive in the presence of iodine. Upon addition of the catalyst and exposure to light a striking change takes place almost instantaneously and the spectral bands of the well-known lycopene-neolycopenes equilibrium mixture appear (Fig. 17).

*References, pp. 334—349.*

Demonstration in the visual spectroscope: A hexane solution of prolycopene shows a blurred spectrum without distinct bands. Upon the addition of a few drops of a dilute hexane solution of iodine sharp bands appear within a minute; the light necessary for this rearrangement is supplied by the instrument. The same effect may be demonstrated simply by exposing the pigment solution in two small Petri dishes to the light of an ordinary electric bulb and adding a drop of iodine to one of the samples: the pale yellow color rapidly turns a deep orange.

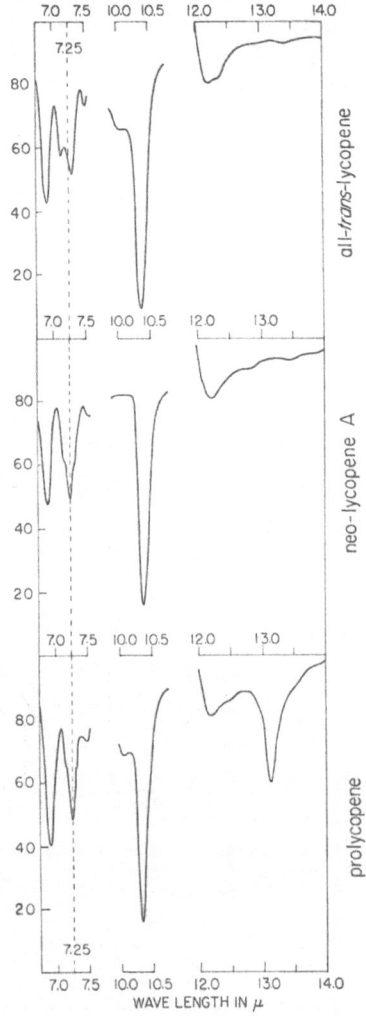

With reference to the configuration of prolycopene the following few comments can be made: The degraded character of the fundamental band could be explained a priori either by the presence of several unhindered *cis* double bonds or by that of at least one sterically hindered such bond (or both). When discussing similarly degraded fundamental bands of some total-synthetic, hindered mono- and di*cis* polyenes, GARBERS, EUGSTER and KARRER (*74, 72,* cf. *60, 71*) seem to have assumed that the molecules of the natural poly*cis* carotenoids possess one or perhaps two hindered *cis* double bonds rather than a larger number of unhindered such bonds. However, as has been pointed out repeatedly [DALE (*41*), LUNDE et al. (*174*)], representatives of the two series can be distinguished in the *cis*-peak region where the curves of the natural poly*cis* carotenoids are flat while GARBERS and KARRER's substances (hindered *cis* lycopenes, β-carotenes, and 1,18-diphenyl-3,7,12,16-tetramethyl-octadecanonaenes) show marked *cis*-peaks (Figs. 35, 36, 26 and 40, pp. 313, 299, 318). This feature has not been considered by KARRER's research group. In the lycopene set a further difference lies in the high thermostability and

Fig. 18. Stereochemically important sections taken from infrared curves of some *cis-trans* isomeric lycopenes: 1% CCl₄ solutions (1 mm. cell) in the 7.0—7.5 μ and 10.0—10.5 μ regions; 0.25% cyclohexane solutions (2 mm. cell) in the 12.0—14.0 μ region (*174*). [From: J. Amer. Chem. Soc. **77,** 1647 (1955).]

easy crystallization of prolycopene while the synthetic, hindered *cis* lycopenes b, c and c′ have not crystallized (*74*).

Another approach to the stereochemistry of prolycopene, viz. stepwise rearrangement was initiated some time ago (*166*). A prolycopene solution was treated with unusually small amounts of iodine and the resulting mixture chromatographed. The column showed several pigment zones located between those of unchanged prolycopene and the very little all-*trans*-lycopene formed. A hindered mono*cis* isomer should have yielded the

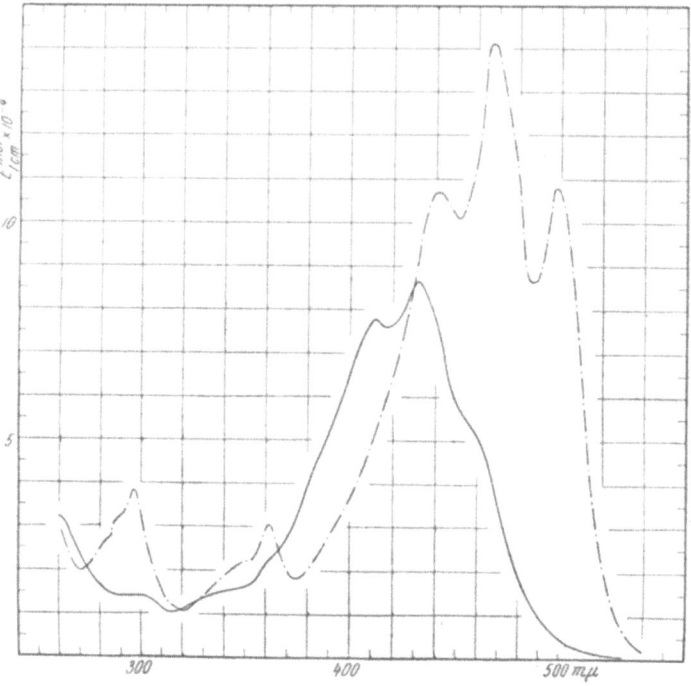

Fig. 19. Molecular extinction curves of a poly*cis*-lycopene (obtained by melting prolycopene crystals or by insolation of solutions), in hexane: ————, fresh solution; and — · — · —, mixture of stereoisomers after catalysis by iodine (*300*). [From: J. Amer. Chem. Soc. **65**, 1940 (1943).]

all-*trans* form as the primary, main rearrangement product. Furthermore, when prolycopene was kept just above its melting point for a few minutes, the subsequent chromatogram was also composed of a dozen colored zones.

Recently, in collaboration with MAGOON (*180*), some further stepwise isomerization experiments were conducted by exposing prolycopene to sunshine or artificial light, in the absence of catalysts. These conditions allowed the almost quantitative recovery of the pigment. While one half of the prolycopene remained unchanged and only little all-*trans*-lycopene appeared in the chromatogram, the latter contained relatively large amounts of thirteen "intermediate" *cis* lycopenes, seven of them crystallizable (Nos. 6, 8, 9, 10, 14, 18, and 21 in Table 14, p. 289). (One isomer was identical with a natural product, the "poly*cis*-lycopene III" ex *Pyracantha*; cf. Table 12, p. 282.) The respective spectra of the "intermediates" showed wide variety with reference to both *cis*-peak intensity and fine structure of the main band. On the column the pigment zones formed distinct groups, probably corresponding to mono-, di- and tri*cis* isomers. This whole picture can be explained exclusively by the poly*cis* nature of prolycopene. Evidently, only in this case can such intermediates appear, "en route" to the all-*trans* configuration, which are absent from the usual *cis-trans* isomeric equilibria.

This interpretation is also in accordance with pertinent infrared readings (*Fig. 18*, p. 285) [LUNDE et al. (*174*)]. The prolycopene spectrum includes both methylated and unmethylated *cis* double bonds with high intensities (7.25 $\mu$ and 13.15 $\mu$; cf. p. 295). In the course of the stepwise isomerization, prolycopene → all-*trans*-lycopene, both bands gradually decrease and finally disappear when the all-*trans* stage is reached (*180*).

In summary, we propose that the prolycopene molecule has a straight overall shape and contains four or five *cis* double bonds. All available data would be explained by a symmetrical penta*cis* model in which the central double bond and four other, sterically unhindered double bonds were *cis*. A confirmation of this working hypothesis could be gained either by the total synthesis of prolycopene or, possibly, by the steric clarification of some partially rearranged pigments. At present, the occurrence of one or two hindered *cis* double bonds in this poly*cis* molecule cannot be excluded.

During the stepwise isomerization a minor isomer (*Fig. 19*) [wrongly named "all-*cis*-lycopene" earlier by this writer (*300*)] appeared *below* the unchanged prolycopene zone. It contains probably one more *cis* double bond than prolycopene. A very similar pigment was observed on feeding prolycopene to chicks (p. 333).

### Further Polycis Lycopenes.

As mentioned on p. 281 the fruits of *Pyracantha angustifolia*, grown in Southern California, are unusually

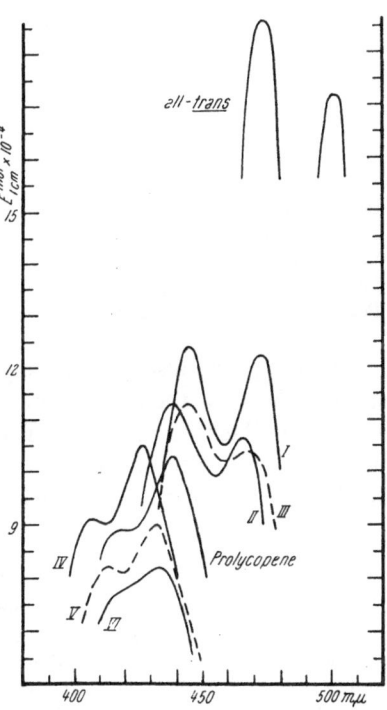

Fig. 20. Molecular extinction curves of all-*trans*-lycopene, prolycopene and poly*cis*-lycopenes I–VI at their main maxima, in hexane (*305*). [From: J. Amer. Chem. Soc. **69**, 1930 (1947).]

rich in poly*cis* hydrocarbons [PINCKARD and the writer (*305*)]. Besides prolycopene, six more similar isomers, the "poly*cis*-lycopenes I–VI" were isolated from the ripe berries, and "I"–"III" were crystallized.

Although 1 kg. of the fresh berries contained as little as 14 mg. of all-*trans*-lycopene, the following amounts of crystalline *cis* samples were obtained: 46 mg. of prolycopene, 5 mg. of poly*cis*-lycopenes I–III, 13 mg. of pro-γ-carotene (p. 291), and 3 mg. of neo-γ-carotene P (p. 309)—in all, 67 mg. of *cis* forms.

Some spectral data of the poly*cis*-lycopenes I–VI are presented in *Fig. 20* (p. 287) and *Table 13*. In Figs. 35–36 (p. 313) our poly*cis*-lycopenes "III" and "V" are compared with KARRER's synthetic *cis*-lycopenes b, c and c'.

Table 13. Spectral Data of Some Poly*cis* Lycopenes ex *Pyracantha* Listed in the Sequence of Decreasing Wavelengths at $\lambda_{max}$, and Compared with the All-*trans* and a Mono-*cis* Form (*305*).

| Name | Location of $\lambda_{max}$ in hexane (m$\mu$) | Difference in location of $\lambda_{max}$ from the all-*trans* form (m$\mu$) $\lambda_{max}$ shift | $E^{mol.}_{1\ cm.} \times 10^{-4}$ at $\lambda_{max}$ | Planimetrically measured relative extinguished area between 320 m$\mu$ and 560 m$\mu$ |
|---|---|---|---|---|
| All-*trans* ...... | 472–473 | 0 | 18.6 | 100 |
| Neo A........ | 465 | 7.5 | 12.2 | 82 |
| Poly*cis* I ..... | 444–445 | 28 | 12.3 | 76 |
| Poly*cis* III.... | 444–445 | 28 | 11.3 | 71 |
| Poly*cis* II..... | 441 | 31.5 | 11.4 | 70 |
| Prolycopene ... | 438 | 34.5 | 10.3 | 60 |
| Poly*cis* VI .... | 433 | 39.5 | 8.1 | 54 |
| Poly*cis* V ..... | 431–432 | 41 | 9.0 | 52 |
| Poly*cis* IV .... | 426 | 46.5 | 10.4 | 62 |

*Table 14* lists also several poly*cis* isomers reported by other research groups. Most of them have not yet been obtained in crystals, probably because of lack of material. The last pigment mentioned in Table 14 shows a $\lambda_{max}$ shift of 54.5 m$\mu$ and the shortest wavelength maximum (418 m$\mu$) ever recorded in the lycopene set [JOYCE (*125, 126*)].

### Proneurosporene (Protetrahydrolycopene, Neoneurosporene P).

The structure of the all-*trans* compound appears on p. 316.

Proneurosporene, a poly*cis*, form has been observed repeatedly in the fruit of *Lycopersicon* selections and described as "unidentified carotene I" [PORTER and ZSCHEILE (*214*)], "poly-*cis*-ψ-carotene" [MACKINNEY and JENKINS (*176*)], and "protetrahydrolycopene" [PORTER and LINCOLN (*213*); TROMBLY and PORTER (*258*)]. More recently, MAGOON and the writer (*179*) called it "neo-neurosporene P" (P for *Pyracantha*). The identity of all these preparations is very likely.

Table 14. Stereoisomeric Lycopenes Listed, where Possible, in the Sequence of Decreasing Adsorption Affinities (*180*).

Nos. 2–5 are in vitro stereoisomerization products (ex all-*trans*-lycopene); 3–5 are unpublished (the ext. coefficients of No. 2 are higher than reported earlier); 6–8, 12–16, 18–21, 23, 24 are in vitro isomerization products (ex prolycopene); 28–31 are in vivo isomerization products (chick); 33–35 were obtained by total synthesis; and the others were isolated from plants (including No. 18). The relative chromatographic positions of Nos. 11 and 12 are approximate, and those of 28–41 could not be determined by direct comparison. Spectra in hexane or light petroleum. *Italicized* numbers on the left designate crystalline products.

| Name of *cis-trans* isomer | $\lambda_{max}$ (m$\mu$) | $E^{mol.}_{1\,cm.} \times 10^{-4}$ | $\lambda_{max}$ (m$\mu$) | $E^{mol.}_{1\,cm.} \times 10^{-4}$ | *cis*-Peak | References |
|---|---|---|---|---|---|---|
| *1.* All-*trans*.......... | 472.5 | 18.6 | 504.5 | 17.2 | none | (*300, 320*) |
| 2. Neo A .......... | 465 | 15.1 | 496 | 12.9 | highest | (*300, 320*) |
| 3. Neo B .......... | 464.5 | 14.5 | 495.5 | 12.3 | high | (*300, 320*) |
| 4. Neo C .......... | 461 | 13.6 | 492 | 11.0 | high | (*300, 320*) |
| 5. Neo D .......... | 457.5 | 13.2 | 488.5 | 10.6 | medium | (*300, 320*) |
| 6. I A ............. | 468.5 | 15.2 | 499 | 12.8 | none | (*180*) |
| 7. I B ............. | 469 | 15.1 | 500 | .12.6 | none | (*180*) |
| 8. I C ............. | 464 | 13.8 | 495 | 11.2 | medium | (*180*) |
| 9. II A ............. | 466.5 | 13.0 | 495 | 9.8 | none | (*180*) |
| *10.* II B ............. | 444.5 | 12.6 | 471.5 | 12.7 | none | (*180*) |
| *11.* Poly*cis* I ........ | 444.5 | 12.3 | 472 | 12.4 | none | (*305*) |
| *12.* "Cryst. isomer" ... | 445.5 | 10.2 | 472 | 9.7 | none | (*300*) |
| 13. II C ............. | 439 | 11.6 | 464.5 | 11.8 | none | (*180*) |
| *14.* II D ............. | 439.5 | 11.0 | 465 | 11.8 | slight | (*180*) |
| 15. II E ............. | 439 | 10.8 | 464.5 | 10.3 | small | (*180*) |
| 16. II F ............. | 443 | 11.3 | 466 | 10.3 | none | (*180*) |
| *17.* Poly*cis* II ........ | 441.5 | 11.1 | 465.5 | 10.6 | none | (*305*) |
| *18.* Poly*cis* III ....... | 444 | 11.4 | 464 | 10.4 | none | (*305*) |
| 19. III B ............ | 437 | 10.5 | 463 | 8.8 | very slight | (*180*) |
| 20. III C ............ | 439.5 | 9.6 | 464.5 | 9.5 | small | (*180*) |
| *21.* III D ............ | 437 | 10.5 | — | — | none | (*180*) |
| *22.* Prolycopene ...... | 438 | 10.3 | — | — | none | (*301, 166*) |
| 23. IV B............. | 436 | 9.7 | 462 | 9.4 | small | (*180*) |
| 24. "All-*cis*" (abandoned term) ........ | 432 | 8.6 | — | — | none | (*300*) |
| 25. Poly*cis* IV ....... | 426 | 10.6 | — | — | none | (*305*) |
| 26. Poly*cis* V ....... | 431.5 | 8.9 | — | — | none | (*305*) |
| 27. Poly*cis* VI ....... | 433 | 8.1 | — | — | none | (*305*) |
| 28. Poly*cis* (ex chick) . | 446 | 11.0 | 472 | 10.4 | ? | (*48*) |
| 29. Poly*cis* (ex chick) . | 442 | 11.0 | 467 | 10.0 | ? | (*48*) |
| 30. Poly*cis* (ex chick) . | 440 | 11.0 | 464 | 9.9 | ? | (*48*) |
| 31. Poly*cis* (ex chick) . | 433 | 8.9 | — | — | ? | (*48*) |
| 32. Fraction "1 H" ... | 439 | — | 468 | — | — | (*79*) |
| 33. *cis*-Lycopene b.... | 433 | 7.4 | 457 | 6.9 | present | (*74*) |
| 34. *cis*-Lycopene c.... | 425 | 10.0 | 446 | 9.8 | present | (*74*) |
| 35. *cis*-Lycopene c' ... | 400–403 | 7.1 | 418–420 | 7.6 | ? | (*74*) |

(*Table 14, continued.*)

| Name of *cis-trans* isomer | $\lambda_{max}$ (m$\mu$) | $E_{1\ cm}^{mol.} \times 10^{-4}$ | $\lambda_{max}$ (m$\mu$) | $E_{1\ cm}^{mol.} \times 10^{-4}$ | *cis*-Peak | References |
|---|---|---|---|---|---|---|
| 36. Polycis-lycopene "44 : 66"* ........ | 444 | — | 466 | — | none | (*125, 126*) |
| 37. Polycis-lycopene "37" ............. | 437 | — | — | — | none | (*125, 126*) |
| 38. Polycis-lycopene "34" ............. | 434–435 | — | — | — | none | (*125, 126*) |
| 39. Polycis-lycopene "32 : 54"⁻........ | 432 | — | 454 | — | none | (*125, 126*) |
| 40. Polycis-lycopene "26" ............. | 426 | — | 440–445 | — | none | (*125, 126*) |
| 41. Polycis-lycopene "29" ............. | 429 | — | — | — | none | (*125, 126*) |
| 42. Polycis-lycopene "18" ............. | 418 | — | — | — | none | (*125, 126*) |

* "44–46" means that the maxima are located at 444 m$\mu$ and 466 m$\mu$ in petroleum ether solution.

Some tomato selections contained as much as 16 mg. of proneurosporene per kg., besides 26 mg. of prolycopene (*213*); only 3–4 mg./kg. was present, however, in some berries of *Pyracantha angustifolia* (*179*).

Proneurosporene is absent from stereoisomeric equilibrium mixtures. It does not crystallize but yields, when treated with iodine, the crystalline all-*trans* form (*179*) which is identical with HAXO's preparation ex *Neurospora crassa* (*91*).

Spectroscopically, proneurosporene represents an interesting type. Although its fundamental band shows considerable fine structure, with two peaks and two shoulders (Fig. 37, p. 314), the drastic increase of the extinguished area caused by iodine may well be compared with that observed for prolycopene. During this process the 461 m$\mu$ maximum doubles its intensity—a feature characteristic of several polycis carotenoids. The absence of a *cis*-peak excludes a bent molecular shape, and the fine structure in the main band excludes the presence of hindered *cis* double bonds. These data are in accordance with the infrared spectrum that does not contain the 13.15 $\mu$ band (unmethylated *cis* double bond), while the 7.25 $\mu$ band is present with marked intensity (methylated *cis* double bond). The central double bond is very probably *trans*, because of the absence of a band at 12.84 $\mu$ which appears in the case of central-mono*cis*-$\beta$-carotene but is missing in all-*trans* curves (*174*) (cf. p. 295). These considerations seem to exclude *cis* configurations around four or five of the conjugated double bonds. Hence, we tentatively propose that proneurosporene represents a poly(unhindered)-*cis* type which contains a *trans* CH—CH=CH—CH group in the center [MAGOON and the writer (*179*)].

Upon refluxing, a small fraction of proneurosporene ($\lambda_{max}$ at 432 m$\mu$) is converted into "neo R" whose maximum extinction (425 m$\mu$) indicates at least one more *cis* double bond than present in the starting material (Fig. 37, p. 314).

### Pro-γ-carotene.

This pigment was detected in some palm fruits (Table 12, p. 282) from which only 0.3 mg. of pure crystals were isolated per kg. of fresh material. A practical source of pro-γ-carotene is *Pyracantha angustifolia* whose ripe berries yielded 9 mg. of crystals per kg. [SCHROEDER and the writer (*314, 315*)].

Pro-γ-carotene forms brick-red glittering plates. Under the microscope the dull brownish-yellow crystals show orange colored crossings. To the naked eye the shade of dilute solutions is similar to that of β-carotene. With reference to thermostability, degraded spectrum, and the dramatic spectral shift upon iodine catalysis, pro-γ-carotene offers a perfect parallel to prolycopene. It yields crystalline γ-carotene when treated with iodine.

The melting point of pro-γ-carotene (135°) cannot be compared directly with that of the all-*trans* form since, for some reason, analytically pure all-*trans*-γ-carotene samples as obtained from various sources show sharp melting points anywhere between 131° and 178° (*315, 88, 313*). Furthermore, it has been observed repeatedly that at a certain stage of chromatographic development all-*trans*-γ-carotene formed twin-zones which eventually were washed together. One could speculate that in the two γ-carotenes the relative spatial positions of the long aliphatic chain and of the adjacent monomethyl group might be different with reference to the ring (*s-cis* and *s-trans*?).

The spectrum in the visible region shows two almost equal maxima which are divided by a shallow minimum (Fig. 31, p. 309); this rare feature differentiates pro-γ-carotene from prolycopene.

The relationship between pro-γ-carotene and neo-γ-carotene P will be discussed on p. 310.

## VIII. Some General Remarks on Configurational Assignments.

As we have seen, the spectral curve of a carotenoid is a function of the molecular shape and hence of the spatial configuration. Were the rules that govern this relationship fully known, it would be easy to assign configurations. Since, however, this is not the case, some of the following considerations and assignments have a tentative character and require confirmation by synthesis.

### 1. Stereoisomeric Types.

As *Table 15* shows, some simple spectroscopic observations can be made in the *cis*-peak and visible regions, before and after catalysis by iodine, that furnish rapid information on the type to which a given *cis* compound belongs. A priori, the following types could appear in stereo-isomeric equilibria:

Sterically unhindered forms: central-mono*cis*,
next-to-central-mono*cis*,
peripheral mono*cis*,
di- and tri*cis*,
poly*cis*.

Sterically hindered forms: mono-, di-, and poly*cis* (in part also containing unhindered *cis* double bonds).

Table 15. Configurational Types of *normal* Carotenoids as Indicated by Spectroscopic Changes upon Treatment with Iodine.

| Configuration | Shape of the curve in the *cis*-peak region | Change of the extinction upon iodine catalysis | | Typical example |
|---|---|---|---|---|
| | | in the *cis*-peak region | in the visible region | |
| All-*trans* | flat | increase | decrease | Fig. 4, p. 246 |
| Central-mono*cis* | high peak | decrease | increase | Fig. 23, p. 296 |
| Peripheral mono*cis* | moderate peak | slight increase | increase | Fig. 22, p. 296 |
| Poly*cis* | flat | increase | strong increase | Fig. 17, p. 284 |

As was pointed out earlier, the available *trans* → *cis* stereoisomerization methods are highly selective and afford only a few types: (a) With rare exceptions, no sterically hindered isomers appear at equilibrium, and hence no degraded spectra are observed. (b) The isomerized portion of the pigment consists mainly of mono- and di*cis* forms. (c) The incidence of those main *cis* forms that show high *cis*-peaks is remarkably important, in spite of the theoretical postulate that only a small fraction of the many possible spatial forms can have *cis*-peaks of high intensity [(*300*), cf. p. 259].

In the absence of a theory that would predict the absolute value of the maximum possible *cis*-peak intensity in a given stereoisomeric set, we assumed earlier that those neo forms that showed the highest *cis*-peaks in the set were central-mono*cis* compounds. However, most of these assignments had to be revised and substituted by next-to-central-*cis* configurations, after the true central-mono*cis* compounds had been synthesized (p. 275). The *cis*-peak intensities of the latter were 10–15% higher than those of the neo forms mentioned.

For rapid characterization of stereoisomers Inhoffen (*101*) proposed to use the quotient $Q =$ extinction at $\lambda_{max}$/extinction at *cis*-peak. This term is also applicable to trace amounts of pigments whose *cis*-peaks cannot be determined quantitatively. The $Q$ value is lowest for central-mono*cis* configurations. Examples: Central-mono*cis*-lycopene, 1.5; central-mono*cis* form of α-carotene, β-carotene, cryptoxanthin, zeaxanthin or isozeaxanthin, ∼ 1.7; next-to-central-mono*cis*-lycopene (neo A), 1.8; and peripheral mono*cis*-β-carotene (neo U), 10.8.

It can now be claimed that in the great majority of our *trans* ⇄ *cis* equilibria central-mono*cis* forms either did not appear at all or perhaps only in trace amounts. This finding might seem to contradict the statements that the activation energy for the *trans* → *cis* rotation about a centrally located double bond is less than for any other double bond in the system, and that this effect may be still further enhanced by the presence, in the middle section only, of the unbranched grouping =CH—CH=CH—CH= (*300*).

This situation was (in part) clarified by two unexpected properties of the synthetic central-mono*cis* isomers: (a) When compared to other *cis* forms, these compounds showed surprisingly low adsorption affinities and hence trace amounts might have been retained at unexpected locations and escaped detection on the column. (b) It was found [cf. e. g., INHOFFEN et al. (*101*)] that central-mono*cis* configuration involved photosensitivity, even in scattered daylight. Future experiments, to be conducted in red light, might still afford some representatives of this theoretically important type as products of direct isomerization. Central-mono*cis*-canthaxanthin was obtained in daylight (p. 302).

Configurational assignments become difficult when one or more sterically hindered *cis* double bonds are present. As indicated on p. 285, one cannot differentiate between poly*cis* and hindered mono*cis* types by comparison of the fundamental bands either before or after a treatment with iodine; both types show a spectacular increase of the extinguished area upon catalysis. A differentiation may be feasible, however, in the *cis*-peak region because the presence or absence of a *cis*-peak is not determined by the hindered or unhindered nature of *cis* double bonds but by the shape of the chromophore. Whereas the overall straight form of a poly*cis* molecule does exclude the presence of a *cis*-peak, a more or less marked peak may appear in the case of a bent molecule that contains hindered *cis* double bond(s) (cf. Figs. 35, 36, p. 313).

That a *cis*-peak and a degraded main band may occur in the same spectrum is also illustrated by the curve of a *cis-retro*-dehydrocarotene that was prepared by melting all-*trans* crystals [WALLCAVE et al. (*321*)].

The distinction of poly*cis* and hindered mono*cis* carotenoids by the method of stepwise isomerization was discussed on p. 287. We cannot differentiate, however, between poly*cis* compounds that contain, respectively, only unhindered, only hindered, or both hindered and unhindered *cis* double bonds, except perhaps in the IR regia.

## 2. Number and Location of *cis* Double Bonds.

Most attempts to rearrange all-*trans* carotenoids have yielded only one or two main *cis* forms whose interpretation as mono- or di*cis* isomers was reasonable considering the high probability of their formation.

Furthermore, numerous spectroscopic readings have revealed that in a number of instances $\lambda_{max}$ of the main neo form(s) is located at a 5 m$\mu$ ($\pm$ I m$\mu$) shorter wavelength than that of the corresponding all-*trans* compound (in hexane, petroleum ether, cyclohexane or benzene solution) (*278*). It is safe to assume that this consistently observed minimum value of the $\lambda_{max}$ shift indicates a mono*cis* configuration. Other main isomers which displayed a shift of $\sim$ 10 m$\mu$ were classified as di*cis* forms, etc. As the following discussion will show, the reliability of such assignments decreases rapidly when the $\lambda_{max}$ shift exceeds 10 m$\mu$.

If in a long, conjugated chain all (sterically unhindered) double bonds were energetically equivalent, then any single *trans* → *cis* rotation would have the same and additive spectral effect; the number of the *cis* double bonds present would simply follow from the position of $\lambda_{max}$. Evidently, this is not the case. As is well known, each double bond in the polyene molecule loses an individual fraction of its double bond character to the adjoining single bonds, and the amount thus lost increases from the ends toward the center of the conjugated system. Consequently, among the various ionic structures mentioned on p. 257, there are more which give partial single bond character to the double bonds at or near the center than near the ends [Coulson (*35*), cf. (*300*)]. The individual (unhindered) *cis* double bonds produce different $\lambda_{max}$ shifts and these are not necessarily additive.

The correctness of the above considerations was confirmed by recent total syntheses of the central-mono*cis* forms of $\alpha$-carotene, $\beta$-carotene, 3,4-dehydro-$\beta$-carotene, 3,4,3',4'-bisdehydro-$\beta$-carotene, cryptoxanthin, zeaxanthin, isozeaxanthin etc. (p. 275): In all these instances the $\lambda_{max}$ shift amounted to only 3 m$\mu$, and for lycopene to 2 m$\mu$ (*110*)—a clear demonstration of the decrease of the $\lambda_{max}$ shift value as we proceed from an end of the conjugated system towards the center.

Another spectral feature that is dependent on the number of *cis* double bonds is the degree of fine structure in the fundamental band. As mentioned before, this fine structure decreases upon any *trans* → *cis* rotation (p. 246). The contributions of individual *cis* double bonds to the flattening effect may well be additive but, unfortunately, no quantitative use can be made of this relationship at the present time.

### 3. Configuration and Infrared Spectrum.

The application of this powerful tool in the field of *cis-trans* isomeric carotenoids is still in an early stage. Most of the pertinent reports are concerned with shorter conjugated systems, e. g. *cis-cis*, *cis-trans*, and *trans-trans* dienes [Rasmussen et al. (*222, 224, 225*); Sheppard et al. (*234, 233*); Jackson et al. (*120*); Paschke et al. (*199*); Celmer et al. (*24*); Szasz et al. (*253*)].

A modern survey of this field has been presented by BRAUDE and WAIGHT (*18*); cf. also ALLAN et al. (*3*), and COLE (*33*).

LUNDE and the writer (*172*) have found in the diphenylpolyene series that the influence of the configuration on the IR curve is manifest in the following regions: 7.0–7.1 $\mu$ (in-plane vibration of CH being part of a *cis* C=C double bond); 12.84–12.95 $\mu$ (the analogous out-of-plane vibration); and 10.0–10.6 $\mu$ (out-of-plane vibration of CH in the corresponding *trans* grouping). In the class of the carotenoid pigments the stereochemically sensitive regions are, ~7.25 $\mu$, ~13 $\mu$, and 10.0–10.6 $\mu$ (example, *Fig. 21*).

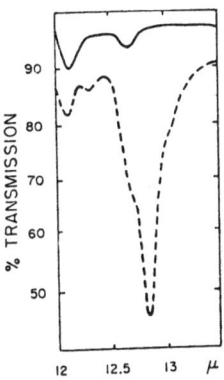

Fig. 21. Infrared spectral curves in cyclohexane (1 mm. NaCl cell): ———, 0.5% solution of all-*trans*-β-carotene; and ———, 1.0% solution of synthetic central-mono*cis*-β-carotene (*284*). [From: Experientia, **10**, 1 (1954).]

Because of the absence of aromatic vibrations, the IR curves of carotenes are simpler than those of diphenylpolyenes (*172*). On the other hand, the isoprenic distribution of methyl groups in the carotenoid molecule creates two types of olefinic bonds, conveniently termed "methylated" —(CH$_3$)C=CH— and "unmethylated" —CH=CH— double bonds. Thus, in the β-carotene molecule four aliphatic double bonds are methylated and five unmethylated. One of the latter occupies an exceptional position at the center, i. e. in the sole section of the molecule where neither of four consecutive carbon atoms carries a side-chain: —CH=CH—CH=CH—.

*Cis* configurations of methylated and of unmethylated double bonds cause different IR effects. Some pertinent data will be given below for the β-carotene set (p. 297) and a few other sets.

## IX. Configurational Assignments in Certain Stereoisomeric Sets.

In this Chapter an attempt will be made to summarize our knowledge concerning *cis* configurations proposed on the basis of stereoisomerization experiments and to coordinate them with the information gained by total synthesis. The reader will note that the available data are fragmentary.

### 1. Stereoisomeric Sets with Two Hydroaromatic Terminal Groups.

#### β-Carotene Set.

*Structure*, p. 227.

*Cis isomers obtained by rearrangement*, GILLAM and EL RIDI (*75, 76*); POLGÁR and the writer (*209, 308*).

*Cis isomers prepared by total synthesis*, INHOFFEN et al. (*99, 101*); ISLER et al. (*114, 108*); EUGSTER, GARBERS and KARRER (*61*).

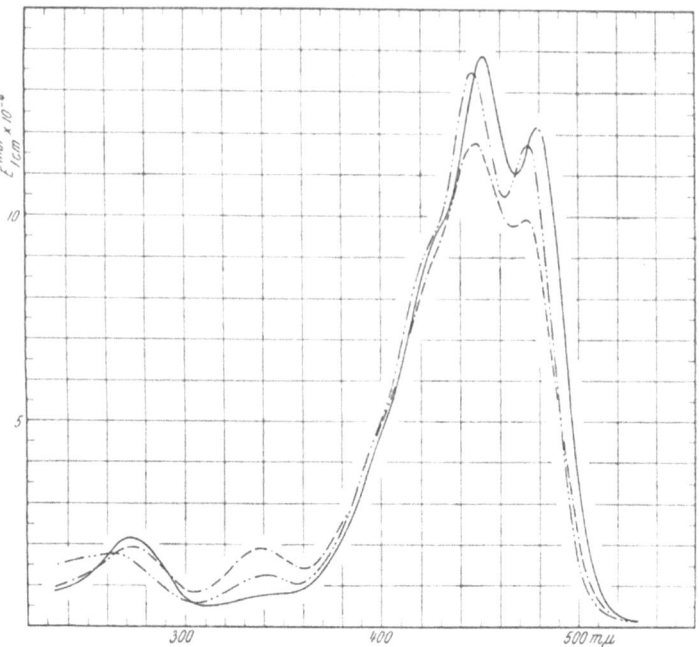

Fig. 22. Molecular extinction curves of neo-β-carotene U, compared with that of all-*trans*-β-carotene, in hexane: ————, β-carotene; — · · —, fresh solution of neo-β-carotene U; and — · —, mixture of stereoisomers obtained from neo U with iodine (*308*). [From: J. Amer. Chem. Soc. **65**, 1522 (1943).]

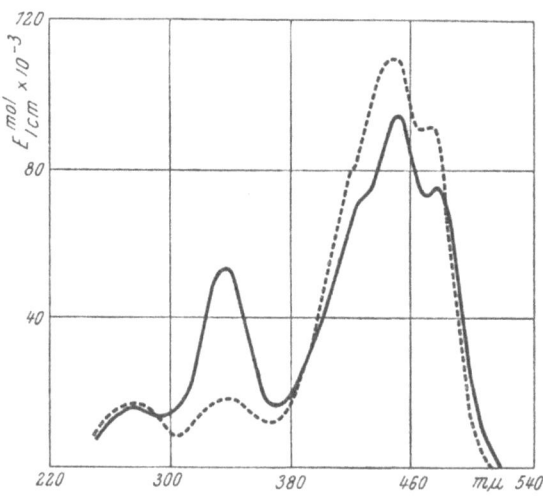

Fig. 23. Molecular extinction curves, in light petroleum: ————, fresh solution of synthetic central-mono*cis*-β-carotene; and — — —, mixture of stereoisomers after catalysis by iodine; according to Inhoffen et al. (*99*). [From: Liebigs Ann. Chem. **570**, 54 (1950).]

Stereoisomerization of all-*trans*-β-carotene has yielded 12 *cis* forms, among them 2 main isomers (20–25% each), viz. neo U (adsorbed above the all-*trans* pigment) and B (adsorbed below it). Considering the UV spectra *(Fig. 22)* neo U was interpreted as a peripheral 9-mono*cis*-β-carotene (Model II, p. 236; $\lambda_{max}$ shift, 5.5 m$\mu$, distinct but low *cis*-peak, considerable thermostability), while for neo B (GILLAM's ψ-α-carotene; $\lambda_{max}$ shift, 10.5 m$\mu$, high *cis*-peak, marked lability; partial rearrangement during crystallization) the 9,15-di*cis* configuration was tentatively proposed. Recently, these isomers were compared with INHOFFEN's synthetic central-mono*cis* compound (Model IV, p. 236; $\lambda_{max}$ shift, 3 m$\mu$) which has the highest *cis*-peak in this set, viz. $\varepsilon = 5.2 \times 10^4$ *(Fig. 23)*. It is crystalline but shows photolability (p. 293).

With reference to the IR spectra *(Fig. 24)* the following points can be made [LUNDE and the writer (*174*)]. Within the 10.0–10.6 $\mu$ region certain bands had been assigned by earlier authors (*234*) to out-of-plane vibrations of the two H-atoms in a *trans* C—CH=CH—C grouping. The latter group occurs five times in the all-*trans*-β-carotene, neo U, and neo B molecules but only four times in the central-mono*cis* isomer.

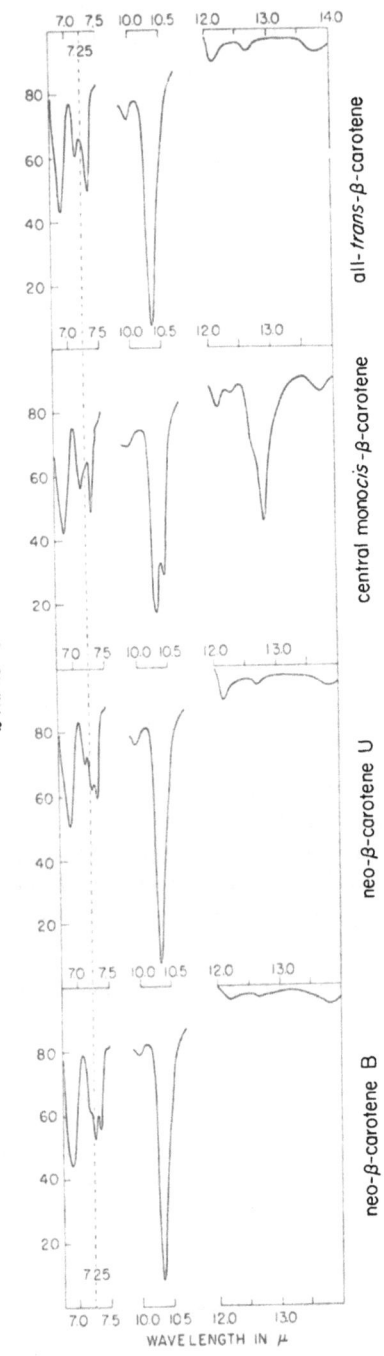

Fig. 24. Stereochemically important sections taken from infrared curves of some *cis-trans* isomeric β-carotenes: 1.0% CCl$_4$ solutions (1 mm. cell) in the 7.0–7.5 $\mu$ and 10.0–10.5 $\mu$ regions; and 0.5% cyclohexane solutions (1 mm. cell) in the 12.0–14.0 $\mu$ region, except for the central-mono*cis* isomer whose solubility in this solvent permitted the use of a 1.0% solution (*174*). [From: J Amer. Chem. Soc. **77**, 1647 (1955).]

Indeed, the 10.35 $\mu$ band appears in the latter curve with markedly lower intensity than in the three others. Furthermore, it has been known that conjugation of the mentioned group with a similar *cis* grouping, to form a *trans-cis* diene, causes the splitting of this band into a doublet or triplet (*120, 24, 233, 224, 267*). We did observe such a split when working with lower-molecular diphenylpolyenes but predicted that it might be obscured in more extended conjugated systems (*172*). In the central-mono*cis*-$\beta$-carotene curve, however, a doublet clearly appears (at 10.35 $\mu$ and 10.47 $\mu$, Fig. 24) as confirmed by ISLER et al. (*114*).

The neo-$\beta$-carotene U and B curves show a distinct band at 7.25 $\mu$ that as a rule is missing in central-mono*cis*- and all-*trans*-carotenoid spectra. We attribute this band which is located within the region assigned earlier (*223*) to deformation vibrations of methyl groups, to such vibrations in a *methylated cis* double bond.

This "rule" is not an absolute one. Thus, the all-*trans*-$\alpha$-carotene curve (p. 307) does show a very slight peak at 7.25 $\mu$ (interpretation on p. 307; cf. also p. 304).

Central-mono*cis*-$\beta$-carotene exhibits a strong band at 12.84 $\mu$ that is absent from the all-*trans*, neo U and neo B curves. This band has also been observed in the curves of the central-mono*cis* forms of methylbixin and dimethylcrocetin (pp. 325, 330) and indicates the presence of *unmethylated cis* double bonds. It is not expected to appear in the spectra of such carotenoids which either do not possess *cis* double bonds at all (all-*trans* forms) or contain methylated *cis* double bonds only. These findings confirm the 9-mono*cis* configuration of neo-$\beta$-carotene U which cannot contain a central *cis* double bond and must possess a methylated *cis* double bond. Hence, *cis* configurations at 7, 11, 15, 7', and 11' are excluded (numbering, p. 227). The 13-*cis* (and 13'-*cis*) configurations are excluded too because they would involve high *cis*-peaks (cf. Model III, p. 236); thus, the 9-mono*cis* configuration is vindicated.

The neo-$\beta$-carotene B curve shows the presence of a methylated *cis* double bond (7.25 $\mu$) and the absence of unmethylated such bonds (12.0—14.0 $\mu$). Hence, both *cis* double bonds of neo B must be methylated. Clearly, our earlier assumption (induced by the relatively high *cis*-peak) of the presence of a central-*cis* double bond had to be abandoned. *Cis* configurations about the other four unmethylated double bonds (7, 7', 11, and 11') are also excluded because they represent sterically hindered types which would be incompatible with the spontaneous formation of neo B from all-*trans*-$\beta$-carotene and with the non-degraded fundamental band. This leaves only the double bonds 9, 9', 13, and 13' available for *cis* assignments, i. e. the di*cis* configurations 9,9'; 9,13; 13,13'; and 9,13' (Models VIII, V, X, and VII on p. 236). Since the three former

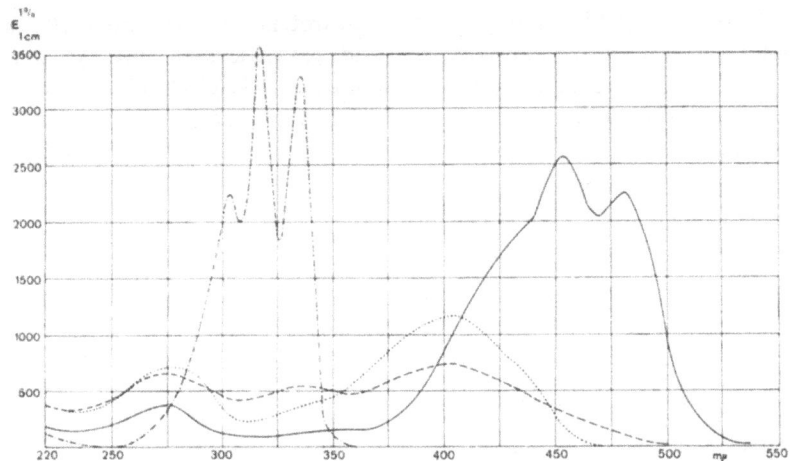

Fig. 25. Specific extinction curves, in light petroleum: — — —, synthetic (hindered) 11,11'-di*cis*-β-carotene; ———, all-*trans*-β-carotene; ····, 11,12;11',12'-bisdehydro-β-carotene; and — · — · —, 3,8-dimethyl-decatrien-(3,5,7)-diin-(1,9); according to Isler et al. (*108*). [From: Helv. Chim. Acta 40, 1256 (1957).]

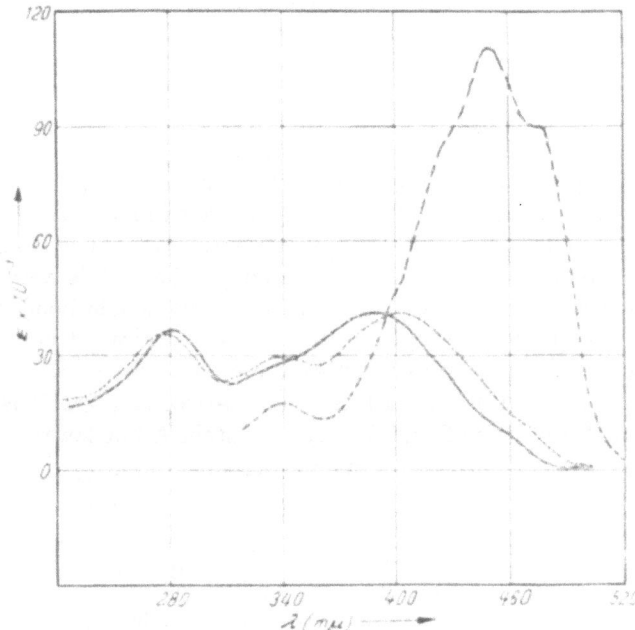

Fig. 26. Molecular extinction curves, in cyclohexane: — × — × —, synthetic (hindered) *cis*-β-carotene B; —o—o—, synthetic (hindered) *cis*-β-carotene C; and the highest curve, same after catalysis by iodine; according to Eugster, Garbers and Karrer (*61*). [From: Helv. Chim. Acta 36, 1378 (1953).]

configurations would involve slight or no *cis*-peaks, we propose that neo-β-carotene B is 9,13'-di*cis*-β-carotene (Model VII).

The sterically hindered 11,11′-di*cis*-β-carotene (*cis*-β-carotene B) was synthesized by EUGSTER, GARBERS and KARRER (*61*) and by ISLER et al. (*108*). As required by the configuration (XXXVI) this compound shows a degraded fundamental band (*Fig. 25*, p. 299).

In the IR region two unexplained bands appear (13.13 $\mu$, 13.50 $\mu$) (*108*) which do not occur in the spectra of the known unhindered *cis* β-carotenes; however, a strong 13.15 $\mu$ band is shown by prolycopene.

(XXXVI.) 11,11′-Di*cis*-β-carotene (both *cis* double bonds are sterically hindered).

The synthetic, crystalline "*cis*-β-carotene C" may possess, besides the sterically hindered 11- and 11′-*cis* double bonds, some unhindered such bonds adjacent to the hindered ones (*61*) (*Fig. 26*, p. 299).

### Cryptoxanthin Set.

*Structure*, 3-hydroxy-β-carotene, p. 228.
*Cis isomers obtained by rearrangement*, LEMMON and the writer (*298*); CHATTERJEE and the writer (*25*).
*Cis isomers prepared by total synthesis*, ISLER et al. (*113*).

Upon iodine catalysis of all-*trans*-cryptoxanthin, the mixture contained two main, crystallizable *cis* isomers, neo U and neo B (yield of either one, 18%; $\lambda_{max}$ shifts, 5 m$\mu$ and 4.5 m$\mu$). These two mono*cis* compounds are easily differentiated by their *cis*-peaks. In the case of neo U (adsorbed above the all-*trans* form) the moderate height of the peak is very similar to that of neo-β- or -α-carotene U, hence the *cis* double bond must be located peripherally. Because of the non-symmetrical structure, neo-cryptoxanthin U may represent either the 9-*cis* or the 9′-*cis* form. For neo B the molar extinction coefficient at *cis*-peak (4.5 × 10⁴) is almost identical with the value found for neozeaxanthin A but lower than that for ISLER's (amorphous) central-mono*cis*-cryptoxanthin (4.8 × 10⁴). We assign to neocryptoxanthin B the next-to-central-mono*cis* configuration (13- or 13′-*cis*).

In the minor isomer, neocryptoxanthin A, probably two double bonds have assumed *cis* configurations ($\lambda_{max}$ shift, 6.5 m$\mu$).

For the reported occurrence of a poly*cis* form cf. Table 12, p. 282.

### Zeaxanthin Set.

*Structure*, 3,3′-dihydroxy-β-carotene (p. 228).
*Cis isomers obtained by rearrangement*, CHOLNOKY, POLGÁR, TUZSON, LEMMON and the writer (*290, 320, 298*).
*Cis isomers prepared by total synthesis*, ISLER et al. (*112, 118*).

Both main, crystallizable isomers, neo A and neo B (yields, 10% and 20%), show a $\lambda_{max}$ shift of 5.5 m$\mu$. On the basis of the *cis*-peak intensities they had been assigned earlier the 15- and 13-mono*cis* configurations, respectively. In the IR region both *cis* curves (but not the all-*trans* curve) show, with very similar intensities, a band at 7.25 $\mu$ which is attributed to a methylated *cis* double bond.

The minor differences between the all-*trans* and the two *cis* curves in the 12.0–14.0 $\mu$ region could not possibly prove the presence of an unmethylated *cis* double bond. This is confirmed by the equal intensities of the *trans* double bond maxima at 10.35 $\mu$ in all three curves.

The proposed mono*cis* nature of the neozeaxanthins A and B is in accordance with the IR data; and the observed absence of unmethylated *cis* double bonds excludes *cis* configuration about the central double bond. Furthermore, the *cis*-peak of neo A, although the highest among all isomers obtained by rearrangement of all-*trans*-cryptoxanthin, is much lower than that of ISLER's synthetic central-mono*cis*-zeaxanthin. Hence, we propose that neo A is 13-mono*cis*-zeaxanthin (next-to-central-*cis* form). If so, then the only possible assignment for neo B is that of 9-mono*cis*-zeaxanthin, considering the postulate that its *cis* double bond must be methylated.

No IR data are available for the minor, crystalline di*cis* isomer neo C ($\lambda_{max}$ shift, 8.5 $\mu$); its considerable *cis*-peak indicates a bent molecular form.

### Isozeaxanthin Set.

*Structure*, 4,4′-dihydroxy-$\beta$-carotene (p. 228).
*Cis isomers prepared by total synthesis*, ISLER et al. (*116*).

The only reported stereoisomer belonging to this set is the synthetic, crystalline (meso or racemic) central-mono*cis*-isozeaxanthin that exists in two unclarified forms, a and b. The $\lambda_{max}$ shift amounts to 2–3 m$\mu$.

### Echinenone Set.

*Structure*, 4-keto-$\beta$-carotene.
*Cis isomer prepared by total synthesis*, AKHTAR and WEEDON (*2a*).

The central mono*cis* form ($\lambda_{max}$ at 454 m$\mu$) shows a $\lambda_{max}$ shift of 2 m$\mu$ (in petroleum ether).

### Canthaxanthin Set.

*Structure*, 4,4′-diketo-$\beta$-carotene (p. 228).
*Cis isomers obtained by rearrangement*, GANSSER and the writer (*70*).
*Cis isomers prepared by total synthesis*, ISLER et al. (*116, 323*).

This fungus pigment yielded on rearrangement the neo compounds A to F of which A, B, and C crystallized. They show decreased adsorption affinities. Although central-mono*cis* forms do not occur as a rule in

stereochemical equilibria, the minor isomer, neo C is identical with
ISLER's total-synthetic central-mono*cis*-canthaxanthin.   Whereas the
$\lambda_{max}$ shift of central-mono*cis*-$\beta$-carotene is only 3 m$\mu$, the corresponding
value in the canthaxanthin set is 4–5 m$\mu$ (in hexane).

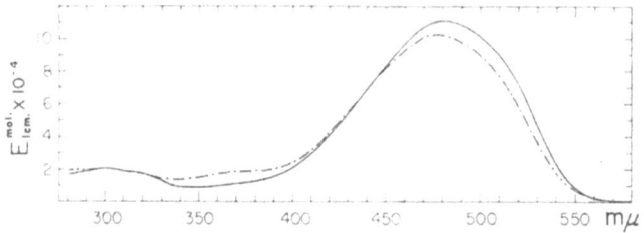

Fig. 27. Molecular extinction curves of canthaxanthin, in hexane: ————, all-*trans* compound; and
— · — · —, mixture of stereoisomers after catalysis by iodine (*70*). [From: Helv. Chim. Acta 40, 1757 (1957).]

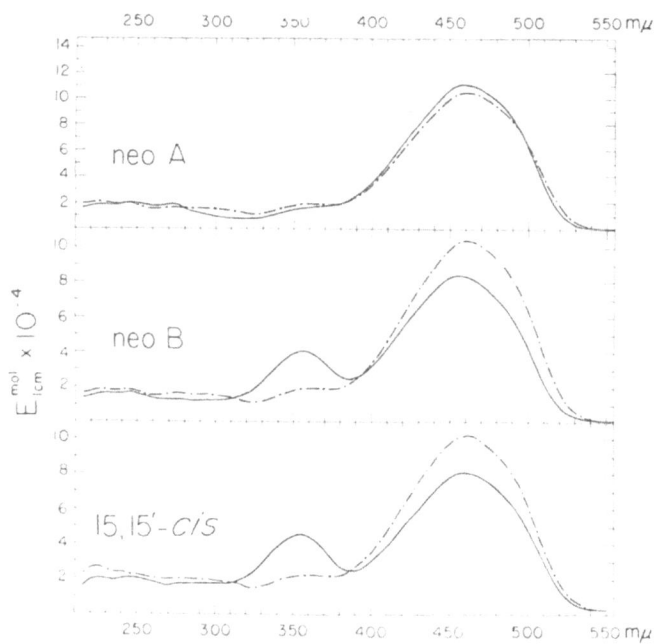

Fig. 28. Molecular extinction curves of some neocanthaxanthins: ————, neo forms A, B, and C (the
latter is identical with the synthetic 15,15′-*cis* form); and — · — · —, mixture of stereoisomers after catalysis
by iodine (*70*). [From: Helv. Chim. Acta 40, 1757 (1957).]

The two main neo forms A and B (yields, 18% and 10%) show the
respective $\lambda_{max}$ shifts, 7 m$\mu$ and 9 m$\mu$ in hexane and 6 m$\mu$ and 10 m$\mu$ in
benzene *(Figs. 27–28)*. Neo A represents a mono*cis* form and, considering
the flatness of its curve in the *cis*-peak region, we assign to it the 9-mono-

References, pp. 334—349.

*cis* configuration. The next-to-central-mono*cis* configuration (13-*cis*) is tentatively proposed for neo B in spite of its considerable $\lambda_{max}$ shift, because its *cis*-peak is but slightly inferior (by 13% in hexane, 8% in benzene) to that of the central-mono*cis* compound.

As expected, the IR spectrum shows a band at 12.84 $\mu$ (unmethylated *cis* double bond) only in the central-mono*cis* curve.

The intense bands at 13.15–13.20 $\mu$ in the neo A, B, and C curves must have some stereochemical significance that cannot be defined at the present time; no such band is present in the all-*trans* curve.

In the neocanthaxanthins D, E, and F more than one (probably two) *cis* double bonds are present ($\lambda_{max}$ shifts in hexane 14–18 m$\mu$, and in benzene 16–18 m$\mu$). Since the curves are flat in the *cis*-peak region, neither of these isomers can possess a bent molecular shape.

### β-*Carotene Monoepoxide Set.*

*Structure*, p. 228.
*Cis isomers obtained by rearrangement*, TSUKIDA and the writer (259).

Among the nine *cis* forms observed on iodine catalysis the neo-epoxides W (yield, 14%), T (15%) and B (7%) are preponderant. W and T show moderate *cis*-peaks; that of B is higher but not the highest in the set. The $\lambda_{max}$ shifts amount to 6–8 m$\mu$ in all three instances. It is difficult to assign configurations in this unsymmetrical set. We believe that neo T, the most abundant isomer, has a peripheral mono*cis* configuration (9- or 9'-*cis*) because of the flatness of its curve in the *cis*-peak region. The much reduced fine structure of the neo D and E curves in the fundamental band is a remarkable feature. The two curves are similar in this respect, although the $\lambda_{max}$ shifts are very different (5.5 m$\mu$ and 13 m$\mu$).

Some of the *cis* epoxides show in the 12.8–13.2 $\mu$ region strong bands whose evident stereochemical significance is unclear. Concerning the occurrence of the 7.25 $\mu$ band in the all-*trans* curve a tentative explanation is given on p. 307.

### β-*Carotene Diepoxide Set.*

*Structure*, p. 228.
*Cis isomers obtained by rearrangement*, TSUKIDA and the writer (259).

The all-*trans* form yielded seven *cis* isomers; the two main ones, neo U (25%) and neo C (10%), are mono*cis* forms ($\lambda_{max}$ shifts, 4 m$\mu$ and 5.5 m$\mu$). For neo U (adsorbed above the all-*trans* form; low *cis*-peak) the 9-*cis* configuration is proposed, while the extraordinarily high *cis*-peak of neo C ($\varepsilon = 6.8 \times 10^4$) would be indicative of the central-*cis* configuration (*Fig. 29*, p. 304). Nevertheless, because of the absence of unmethylated *cis* double bonds, the next-to-central-mono*cis* (13-*cis*) configuration is tentatively assigned to this isomer. No configuration can yet be proposed for the mono*cis* form neo D ($\lambda_{max}$ shift, 3.5 m$\mu$;

high *cis*-peak). The minor di*cis* form, neo V ($\lambda_{max}$ shift, 8.5 m$\mu$) must have a straight molecular shape because of the flatness of its curve in the *cis*-peak region.

Unexpectedly, the 7.25 $\mu$ band appeared in the IR spectra of the two all-*trans*-$\beta$-carotene epoxides. It is believed to be caused by

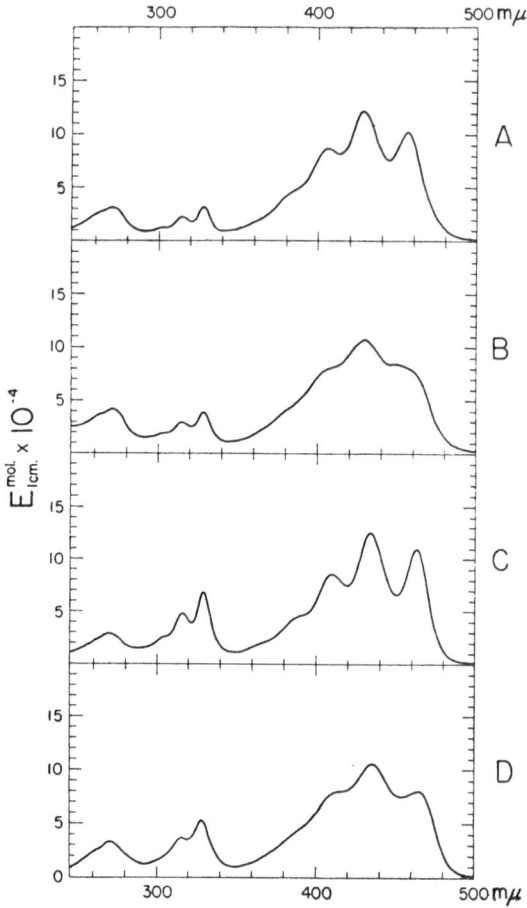

Fig. 29. Molecular extinction curves of the neo-$\beta$-carotene diepoxides A, B, C, and D, in hexane (*259*). [From: Arch. Biochem. Biophys. **74**, 408 (1958).]

deformation vibrations of $CH_3$ attached to the epoxide group. Hence, the situation is "*cis*-like" as at the $\alpha$-ionone end of the $\alpha$-carotene molecule where a methyl is connected with an isolated ring double bond (cf. p. 307). This explanation seems to be vindicated by the markedly higher intensity of the 7.25 $\mu$ band in the $\beta$-carotene diepoxide than in the monoepoxide curve.

*References, pp. 334—349.*

The strong bands located at 12.75 $\mu$ and 13.05 $\mu$ in the neo U and V curves, as well as the 13.05 $\mu$ band in the neo B curve, are missing in the spectra of the neo forms A, C, and D. Their evident stereochemical significance is unclear.

### 3,4-Dehydro-β-carotene and 3,4,3',4'-Bisdehydro-β-carotene Sets.

*Structures*, p. 228.

*Cis isomers obtained by rearrangement,* KARMAKAR and the writer (*127*).

*Cis isomers prepared by total synthesis,* ISLER et al. (*115*); INHOFFEN and RASPÉ (*105*).

All-*trans*-3,4-dehydro-β-carotene afforded on iodine catalysis the two main isomers neo U (yield 22%; adsorbed above the all-*trans* form; $\lambda_{max}$ shift, 7 m$\mu$; slight *cis*-peak) and neo A (27%; adsorbed below all-*trans*; $\lambda_{max}$ shift, 11 m$\mu$; considerable *cis*-peak). They were interpreted, respectively, as a peripheral mono*cis* and a di*cis* isomer, the latter having a bent molecular shape.

The total-synthetic central-mono*cis*-3,4-dehydro- and 3,4,3',4'-bis-dehydro-β-carotenes ($\lambda_{max}$ shifts, 4 m$\mu$) show as expected very high *cis*-peaks *(Fig. 30).*

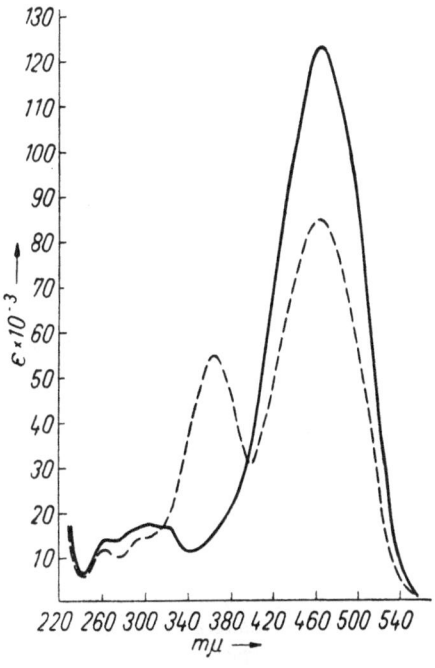

Fig. 30. Molecular extinction curves of 3,4,3',4'-bisdehydro-β-carotene, in light petroleum: ————, all-*trans* form; and — — —, central-mono*cis* form; according to INHOFFEN and RASPÉ (*105*). [From: Liebigs Ann. Chem. **594**, 165 (1955).]

### 16,16'-Homo-β-carotene Set.

*Structure*, p. 261; composition $C_{42}H_{58}$.

*Cis isomers prepared by total synthesis,* INHOFFEN et al. (*97*).

The configurations of the middle-mono*cis* (16-*cis*) and di*cis* (16,16'-di*cis*) isomers ($\lambda_{max}$ shifts, 6 m$\mu$ and 9 m$\mu$) are represented on p. 261, where the respective *cis*-peak regions are also discussed.

### 13,13'-Bis-desmethyl-β-carotene Set.

*Structure*, see next page; composition $C_{38}H_{52}$.

*Cis isomers prepared by total synthesis,* INHOFFEN, BOHLMANN and RUMMERT (*100*).

The main, crystalline *cis* form *a* shows a high *cis*-peak and a $\lambda_{max}$ shift of only 2 m$\mu$; it represents the central-mono*cis* compound. The curve of the minor isomer *b* is almost flat in the *cis*-peak region.

$$
\left[
\begin{array}{l}
\mathrm{CH_3} \quad \mathrm{CH_3} \\
\quad \diagdown \, \mathrm{C} \diagup \\
\quad \diagup \quad \diagdown \\
\mathrm{CH_2} \quad \mathrm{C-CH=CH-C=CH-CH=CH-\overset{13}{CH}=CH-CH=} \\
\;\; | \qquad | \qquad\qquad\qquad\; | \\
\mathrm{CH_2} \quad \mathrm{C-CH_3} \qquad\quad\; \mathrm{CH_3} \\
\quad \diagdown \, \mathrm{CH_2} \diagup
\end{array}
\right]_2
$$

(XXXVII.) 13,13′-Bis-desmethyl-β-carotene.

## *retro-Dehydrocarotene Set.*

*Structure*, p. 228.

*Cis isomers obtained by rearrangement*, WALLCAVE and the writer (*321, 264*).

Whereas *normal* carotenes, when submitted to iodine catalysis retain about one half of their molecules in the all-*trans* configuration, the corresponding value for this *retro* compound is as low as $^1/_6$. Among the observed eight *cis* forms which all appeared below the all-*trans* zone, two main, crystallizable isomers were preponderant, viz. neo A (yield, 28%; $\lambda_{max}$ shift, 4 m$\mu$) and neo D (12%; 7 m$\mu$); they represent a mono- and a di*cis* form. Their molecular shape is obscure since no definite *cis*-peaks are observable in this set (cf. pp. 254, 255).

When fused, all-*trans-retro*-dehydrocarotene was converted to the extent of 1% into neo J ($\lambda_{max}$ shift, 22 m$\mu$), an interesting hindered *cis* isomer with a degraded spectrum; the extinction at $\lambda_{max}$ amounted to only 63% of that of the all-*trans* compound.

## *α-Carotene Set.*

*Structure*, p. 227.

*Cis isomers obtained by rearrangement*, GILLAM, EL RIDI and KON (*77*); POLGÁR and the writer (*309*).

*Cis isomer prepared by total synthesis*, INHOFFEN, SCHWIETER and RASPÉ (*106*).

This structurally and sterically unsymmetrical carotenoid contains ten conjugated double bonds and hence two non-equivalent "middle" double bonds. Iodine catalysis afforded ten isomers, among them the three main forms, neo U (yield, 9%; adsorbed above the all-*trans* zone; crystalline; $\lambda_{max}$ shift, 5.5 m$\mu$), neo W (8%; crystalline; 6.5 m$\mu$), and neo B (13%; 10.5 m$\mu$); the latter represents a di*cis* compound. Like neo-β-carotene U, neo-α U is considerably thermostable. It shows a moderate *cis*-peak and is to be interpreted as a peripheral mono*cis* isomer, viz. 9-*cis*-α-carotene (the 9′-*cis* configuration cannot be excluded). Our assignment is confirmed by IR readings: a band appears at 7.25 $\mu$ (methylated *cis* double bond), while the 12.0–14.0 $\mu$ band is missing. This eliminates *cis* configurations about the unmethylated double bond in position 15 (those in 7, 11, 11′, and 7′ are hindered).

The *cis*-peak of neo-α-carotene B is much higher than that of neo U. Neo B was at first tentatively assigned the 13,9′- or 15,9′-di*cis* configuration. The IR data exclude, however, the presence of un-methylated *cis* double bonds and hence the 15,9'-di*cis* configuration. We believe that neo-α-carotene B is the 13,9'- (or possibly the 9,13'-) di*cis* isomer (cf. Model VII, p. 236).

The neo-α- and -β-carotenes B on one hand, and the two neo U forms on the other, seem to have analogous configurations; within the two corresponding pairs they show practically equal *cis*-peak intensities.

A special feature of all-*trans*-α-carotene is a slight but distinct maximum at 7.25 μ which can be explained as follows [LUNDE et al. (*174*)]: In general, this band is brought about by deformation vibrations of a methyl group attached to a *cis* double bonded carbon atom in an open chain. In α-carotene a methyl group is attached to an isolated double bond that is located in a cyclohexene ring and has "*cis*-like" character. Since the influence of such a double bond on the vibrations of the methyl group must be similar to that of an aliphatic *cis* double bond, a 7.25 μ band had to be expected to appear in the all-*trans*-α-carotene curve [cf. FARRAR et al. (*68*)].

Somewhat similar is the situation in case of the β-carotene epoxides (p. 304). It should also be mentioned that the inflection at 7.2 μ is much more distinct in the α-ionone than in the β-ionone curve (*232*). There are also other "*cis*-like" situations (cf. p. 256).

In the α-carotene set the intensities at 7.25 μ increase in the sequence, all-*trans* < neo U < neo B, indicating the respective presence of 0, 1, and 2 *aliphatic* methylated *cis* double bonds.

The crystalline 15,15'-mono*cis* form of (rac.) α-carotene ("central" with reference to the carbon skeleton but not to the conjugated system) was synthesized by INHOFFEN. Possibly, our minor isomer neo-α-caro-tene C (yield, 1%) is identical with this compound, except for the rotatory power (*cis*-peaks, 4.4 and 4.6 × 10⁴).

Neo-α-carotene X, a di- (or tri-) *cis* isomer ($λ_{max}$ shift, 13.5 mμ) is formed preponderantly upon mild heating of neo U solutions. It is reasonable to assume that the location of one of its *cis* double bonds is identical with that of neo U (9- or 9'-position).

No assignment can yet be proposed for neo-α-carotene W.

### x-Cryptoxanthin (Physoxanthin) Set.

*Structure*, 3'-hydroxy-α-carotene (p. 229).

*Cis isomers obtained by rearrangement*, CHOLNOKY et al. (*29*); cf. BODEA et al. (*15, 16*).

The three observed *cis* forms, neo U, neo A, and neo B, showed the respective $λ_{max}$ shifts, 4 mμ, 7 mμ, and 6 mμ. Neo-α-cryptoxanthin U adsorbs above the all-*trans* zone.

## Lutein Set.

*Structure*, 3,3′-dihydroxy-α-carotene (p. 229).
*Cis isomers obtained by rearrangement*, TUZSON, POLGÁR and the writer (*320, 309*)

In the presence of iodine two main, crystallizable *cis* isomers, neo A and neo B appeared, both above the all-*trans* zone; yields, 16% and 23%. The respective $\lambda_{max}$ shifts, 6 m$\mu$ and 5 m$\mu$, indicate mono*cis* configurations. The *cis*-peak of neo A ($\varepsilon = 4.9 \times 10^4$) surpasses slightly the highest value observed in the α-carotene set ($4.6 \times 10^4$) and almost reaches that for central-mono*cis*-β-carotene ($5.2 \times 10^4$); evidently, the *cis* double bond must be located in one of the two "middle" positions (13- or 15-*cis*). Neolutein B possesses a much lower *cis*-peak and could well represent the 9- or the 9′-*cis* compound (Fig. 10, p. 251).

## Trollixanthin Set.

*Structure*, see below.
*Cis isomer found in nature*, EUGSTER and KARRER (*62*).

Both a *cis* and an all-*trans* form were isolated from plants (m. p. 143 to 145°, and 199°; $\lambda_{max}$ shift, 2 m$\mu$). It has been impossible, however, to effect a *cis* → *trans* rearrangement because the starting material was either destroyed or remained unchanged during the treatment.

(XXXVIII.) Trollixanthin.

## 3,4-Dehydro-α-carotene Set.

*Structure*, p. 229.
*Cis isomers obtained by rearrangement*, KARMAKAR and the writer (*127*).

Upon iodine catalysis two main isomers appeared: neo U (yield, 8%; adsorbed above the all-*trans* zone; $\lambda_{max}$ shift 5 m$\mu$; practically no *cis*-peak), and neo B (18%; adsorbed below the all-*trans* form; $\lambda_{max}$ shift, 10 m$\mu$; considerable *cis*-peak). Evidently, a peripheral mono*cis* and a di*cis* isomer were formed; the latter must have a bent molecular shape.

## Taraxanthin Set.

*Structure* unknown, composition $C_{40}H_{56}O_4$; contains probably 2 hydroaromatic rings (trihydroxy-α-carotene monoepoxide?).
*Cis isomers obtained by rearrangement*, TUZSON and the writer (*320*).

This pigment yielded on catalysis by iodine the neotaraxanthins A–C, which all showed increased adsorbabilities. In $CS_2$, the $\lambda_{max}$ shifts were, 5 m$\mu$ (neo A) and 20 m$\mu$ (neo C); a very small value was observed for neo B.

## 2. Stereoisomeric Sets with One Hydroaromatic and One Aliphatic Terminal Group.

### γ-Carotene Set.

*Structure*, p. 227.

*Cis isomers found in nature*, PINCKARD and the writer (*305, 205a*). For pro-γ-carotene, see p. 291.

*Cis isomers obtained by rearrangement*, POLGÁR and the writer (*310*).

Our knowledge concerning the stereochemistry of this half-aliphatic, non-symmetrical set is unsatisfactory, in part because of the great number

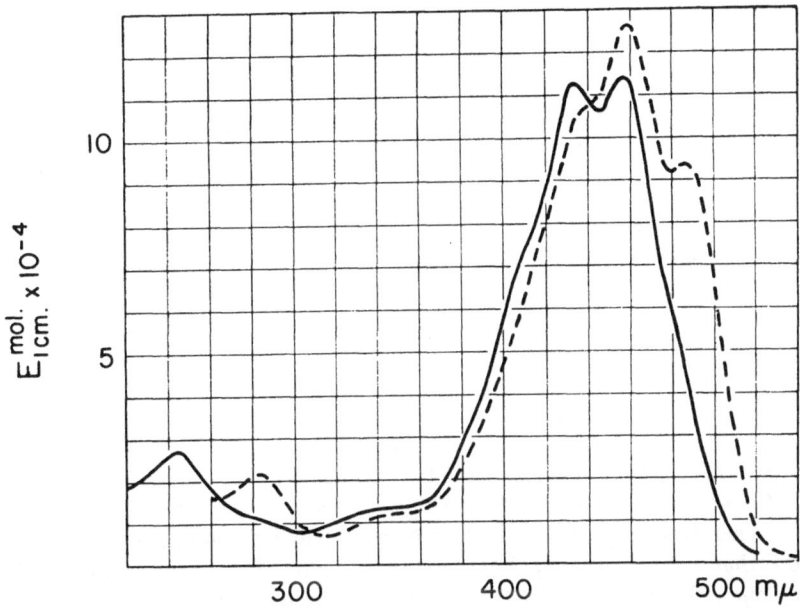

Fig. 31. Molecular extinction curves, in hexane: ———, pro-γ-carotene; and — — —, neo-γ-carotene P (*205a*).

of observed spatial forms, most of which either cannot be separated chromatographically or only after wasteful fractionations. The well-characterized isomer, neo-γ-carotene U (adsorbed above the all-*trans* form; $\lambda_{max}$ shift, 5 m$\mu$; low *cis*-peak) is certainly a peripheral mono*cis* compound (9- or 9'-*cis*). The *cis*-peak of the minor isomers neo H (obtained by refluxing solutions; $\lambda_{max}$ shift, 5 m$\mu$) is as high ($\varepsilon = 5.4 \times 10^4$) as that of central-mono*cis*-β-carotene ($5.2 \times 10^4$). This isomer could possibly be the central-mono*cis* form (15-*cis*).

Two crystalline isomers, neo-γ-carotene P and pro-γ-carotene (p. 291) are natural products; they are absent from stereochemical equilibria. Neo-γ-carotene P was isolated from the berries of *Pyracantha angustifolia*.

Its spectral curve is flat in the *cis*-peak region. The small $\lambda_{max}$ shift (3 m$\mu$) would be in accordance with a mono*cis* configuration. However, IR measurements seem to indicate some unclarified feature of the neo P molecule. We had stated earlier *(174)* that the curves of pro-$\gamma$-carotene and neo P are almost identical; this may have been caused by some experimental error. A reinvestigation (in $CS_2$) has now revealed that there is one important difference between the two IR spectra, viz. the presence of a 13.02–13.19 $\mu$ doublet band in the neo P curve. It then seems that neo-$\gamma$-carotene P contains both methylated and unmethylated *cis* double bond(s) while the latter type is absent from pro-$\gamma$-carotene [Lunde *(171)*].

It is interesting to note how profoundly this configurational difference influences the spectra in the visible and UV regions: in the fundamental band pro-$\gamma$-carotene shows two almost equal maxima (434, 457 m$\mu$), while the neo P curve contains a single main maximum (459 m$\mu$) *(Fig. 31, p. 309)*.

## Gazaniaxanthin Set.

*Structure*, unknown.
*Cis isomers obtained by rearrangement*, Schroeder and the writer *(316)*.

Like $\gamma$-carotene, gazaniaxanthin yielded on catalysis by iodine a complex chromatogram in which the *cis* zones "Group I" and "Group II" were stereochemically heterogeneous. The respective $\lambda_{max}$ shifts, 5.5 m$\mu$ and 8.5 m$\mu$, may indicate mono- and di*cis* configurations.

## Celaxanthin Set.

*Structure*, 3(?)-hydroxy-3′,4′-dehydro-$\gamma$-carotene.
*Cis isomers obtained by rearrangement*, LeRosen and the writer *(167)*.

When treated with iodine, this pigment yielded the neocelaxanthins A to C that were adsorbed below the all-*trans* zone; $\lambda_{max}$ shifts, 3 m$\mu$ (neo A), 6.5 m$\mu$ (B), and 3.5 m$\mu$ (C). A and C must be mono*cis* isomers.

## Capsanthin Set.

*Structure*, p. 229; but see R. Entschel and P. Karrer, Helv. Chim. Acta **43**, 89 (1960)*.
*Cis isomers obtained by rearrangement*, Cholnoky, Polgár and writer *(289, 211)*.

The fundamental band of this ketone shows almost no fine structure *(Fig. 32)*. Under the influence of iodine three isomers appeared. Ratio, all-*trans* : neo A : neo B : neo C = 66 : 23 : 7 : 4; neo A crystallized. The $\lambda_{max}$ shifts observed for A and B amounted to 6 m$\mu$ (in benzene), indicating mono*cis* compounds.

Considering the respective *cis*-peak intensities ($\varepsilon = 4.4 \times 10^4$ for neo A and 2.65 $\times$ 10$^4$ for neo B), the *cis* double bond must occupy a more central position in A than in B. The minor isomer, neocapsanthin C,

---

* *(Added in Proof.)*

represents a di*cis* form ($\lambda_{max}$ shift, 10.5 m$\mu$), and its low *cis*-peak indicates an almost straight molecular shape. The neo forms appeared below the *trans* zone.

*Capsanthin-dipalmitate* (*289*) yielded two *cis* isomers that showed weakened adsorbabilities and did not separate sharply from each other. Ratio, all-*trans* : *cis* forms = 65 : 35.

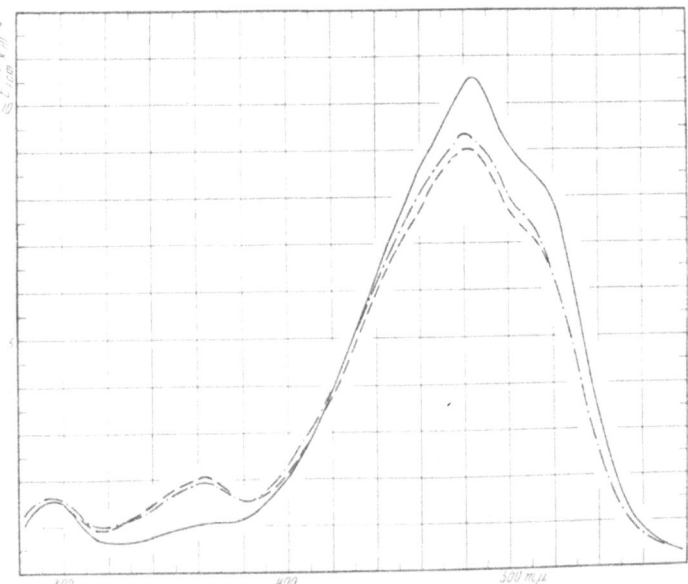

Fig. 32. Molecular extinction curves of capsanthin, in benzene: ————, fresh solution of the all-*trans* compound; — — —, mixture of stereoisomers after 45 min. refluxing; and — · — · —, after catalysis by iodine (*211*). [From: J. Amer. Chem. Soc. **66**, 186 (1944).]

### Torularhodin Set.

*Structure*, see below.
*Cis* isomer prepared by total synthesis, ISLER et al. (*109*).

Central-mono*cis*-torularhodin methylester showed a $\lambda_{max}$ shift of 2 m$\mu$.

(XXXIX.) Torularhodin.

### 3. Stereoisomeric Sets with Two Aliphatic Terminal Groups.

### Lycopene Set.

*Structure*, p. 227.
*Cis* isomers obtained by rearrangement, TUZSON et al. (*320*); LeROSEN, SCHROEDER, POLGÁR, PAULING and the writer (*300*).

*Cis isomers found in nature*, Tables 12 and 14 (pp. 282, 289).

*Cis isomers prepared by total synthesis*, ISLER et al. (*110*); GARBERS and KARRER (*74*).

This set occupies a unique position because about forty of its members are known (Table 14, p. 289). In contrast, only a single main isomer, neolycopene A (yield, 40–50%) appears upon rearrangement of the all-*trans* form. Considering the $\lambda_{\max}$ shift of 5 m$\mu$ and the extraordinarily high *cis*-peak *(Fig. 33)*, we had first assigned the central-mono*cis*

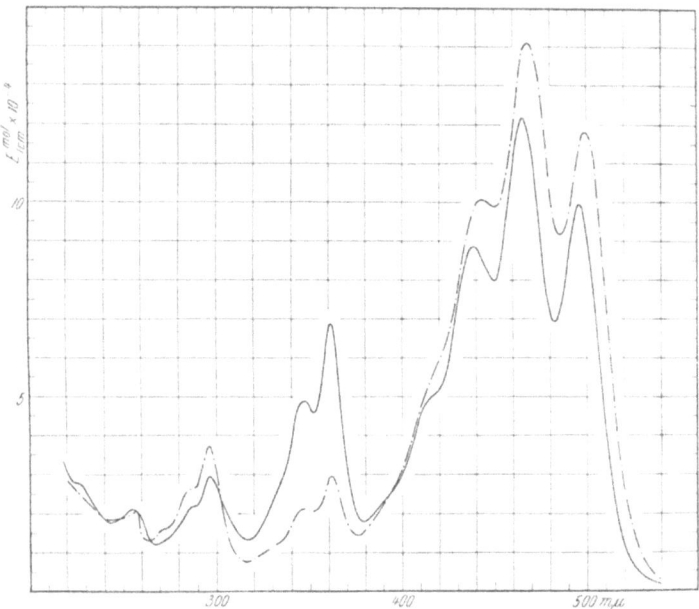

Fig. 33. Molecular extinction curves of neolycopene A: ————, fresh solution; and — · — · —, mixture of stereoisomers after catalysis by iodine (*300*). [From: J. Amer. Chem. Soc, 65, 1940 (1943).]

configuration (15-*cis*) to this isomer. Eventually, IR readings have shown that the 7.25 $\mu$ band, indicative of methylated *cis* double bonds, is present with considerable intensity. Hence, the assignment had to be changed to the next-to-central-mono*cis* configuration (13-*cis*) (*174*).

The correctness of this revision was confirmed by the synthesis of the true central-mono*cis*-lycopene (ISLER; KARRER) *(Fig. 34)*. This crystalline compound is clearly different from neolycopene A which can be obtained in solutions only. The *cis*-peak intensities are almost identical*. The $Q$ value (p. 292) is 1.5 for the synthetic product but 1.8 for neolycopene A.

---

* $\varepsilon = 6.0 \times 10^4$ for the synthetic product; our value, determined for neo A indirectly, is evidently too high ($6.8 \times 10^4$) (*300*).

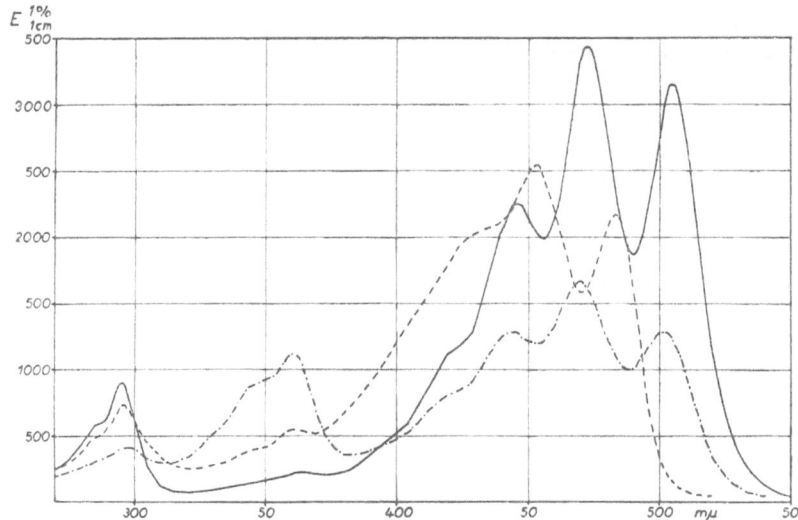

Fig. 34. Specific extinction curves, in light petroleum: — · — · —, synthetic central-mono*cis*-lycopene; ————, all-*trans*-lycopene; and — — —, 15,15′-dehydrolycopene; according to ISLER et al. (*110*). [From: Helv. Chim. Acta 39, 463 (1956).]

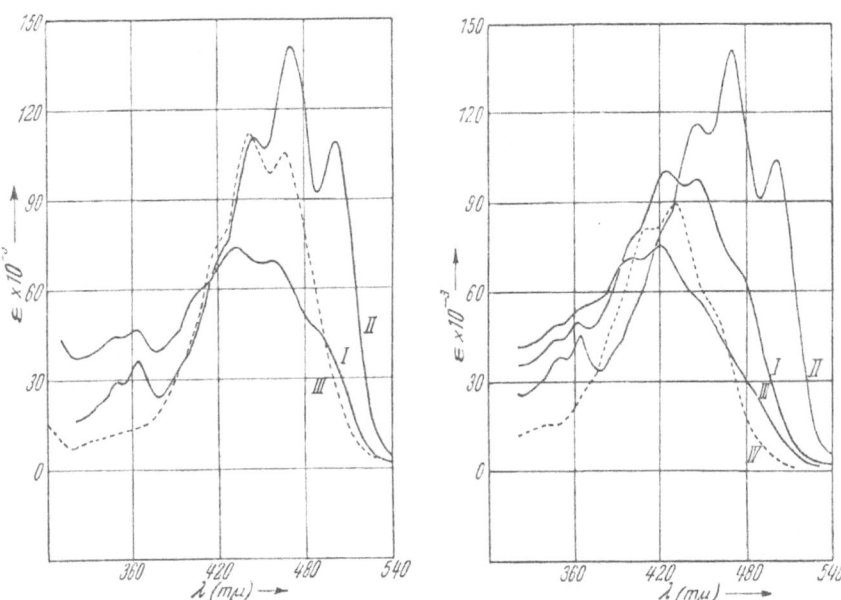

Fig. 35. Molecular extinction curves, in cyclohexane: Curve I, synthetic (hindered) *cis*-lycopene b; Curve II, the same, after catalysis by iodine; and Curve III, natural poly*cis*-lycopene "III", in hexane (*305*); according to GARBERS and KARRER (*74*). [From: Helv. Chim. Acta 36, 828 (1953).]

Fig. 36. Molecular extinction curves, in cyclohexane; Curve I, synthetic (hindered) *cis*-lycopene c; Curve II, the same, after catalysis by iodine; Curve III, synthetic (hindered) *cis*-lycopene c′; and Curve IV, natural poly*cis*-lycopene "V" in hexane (*305*); according to GARBERS and KARRER (*74*). [From: Helv. Chim. Acta 36, 828 (1953).]

The minor isomer, neolycopene B, shows a $\lambda_{max}$ shift of 8 m$\mu$ and a *cis*-peak of medium intensity. It seems to possess two *cis* double bonds, one of which has identical location with that of neolycopene A. Indeed, neo B is spontaneously formed from neo A when the solution is kept at room temperature.

Three hindered members of this set, the *cis*-lycopenes b, c, and c', were synthesized by KARRER et al. (*74*). Although they are relatively labile and not crystallizable, their preparation has proved that hindered *cis* lycopenes do exist and are characterized by degraded spectra. These compounds show *cis*-peaks (with fine structure) and must possess bent molecular form (*Figs. 35–36*, p. 313). Their *cis*-peaks differentiate them from our poly*cis* lycopenes ex *Pyracantha* (cf. p. 313).

Poly*cis*-lycopenes were discussed on pp. 283–286. The IR curve of prolycopene shows, with high intensities, the presence of both methylated and unmethylated *cis* double bonds. The poly*cis* character of prolycopene was confirmed in stepwise isomerization experiments (p. 286).

### Neurosporene (5,6;5',6'-Tetrahydrolycopene) Set.

Structure, p. 316.
*Cis isomer found in nature*, p. 288.
*Cis isomers obtained by rearrangement*, MAGOON and the writer (*179*).

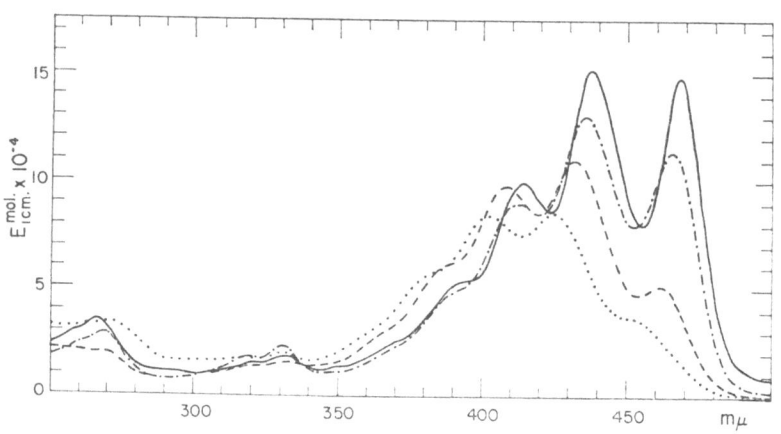

Fig. 37. Molecular extinction curves of stereoisomeric neurosporenes, in hexane: ————, all-*trans*; — — —, neo P (proneurosporene); · · · · ·, neo R; and — · — · —, iodine-catalyzed equilibrium mixture (*179*). [From: Arch. Biochem. Biophys. **68**, 263 (1957).]

Upon treatment of all-*trans*-neurosporene (ex *Neurospora crassa* or neo P, p. 288), six neo forms, A–F, were obtained, A–D in remarkably equal quantities. Ratios, unchanged all-*trans* : neo A : neo B : neo C : : neo D : neo E : neo F = 44 : 13 : 11 : 12 : 15 : 3 : 2. Although no configurational assignments can be made at the present time, mono*cis* forms are clearly preponderant ($\lambda_{max}$ shifts in the above sequence: 0, 1, 5, 5.5, 7, 9, and 14 m$\mu$) *(Fig. 37)*. For neo R cf. p. 291.

### *Phytofluene (7,8;11,12;8',7'-Hexahydrolycopene) Set.*

*Structure*, see next page.

*Cis isomers occurring in nature and obtained by rearrangement,* PETRACEK, KOE and the writer *(204, 147)*; WALLACE and PORTER *(263)*.

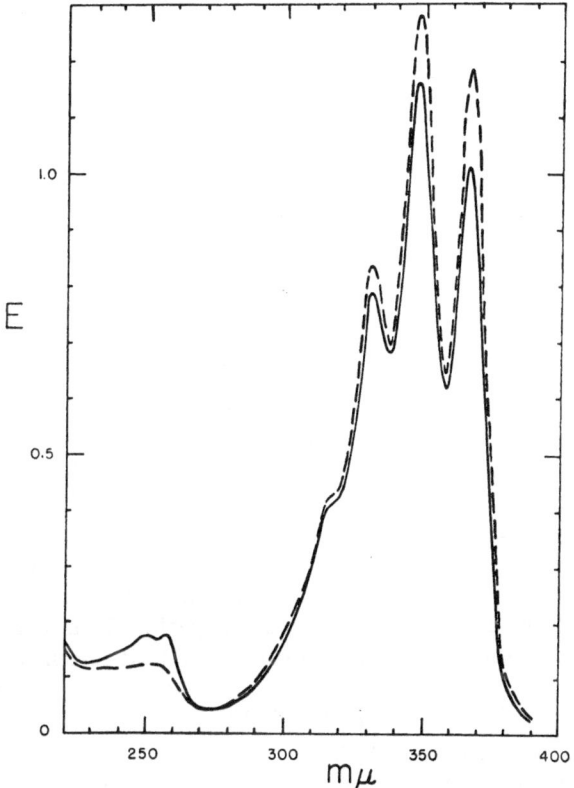

Fig. 38. Extinction curves of tomato phytofluene *(cis)*, in hexane: ————, fresh solution; and — — — after catalysis by iodine *(204)*. [From: J. Amer. Chem. Soc. 74, 184 (1952).]

In this nearly colorless set which contains only five conjugated double bonds, stereochemical work is difficult because of the oily nature and photosensitivity of the substance. The main phytofluene form in the tomatoes studied by our group was a mono*cis* phytofluene ($\lambda_{max}$ shift, 0.5–1 m$\mu$) *(Fig. 38)*. Considering the low *cis*-peak, the *cis* double bond cannot be located at the center of the phytofluene chromophore (position 13'); the hindered 11'-position is excluded by the spectrum. A 15'-*cis* configuration would probably account for all known facts.

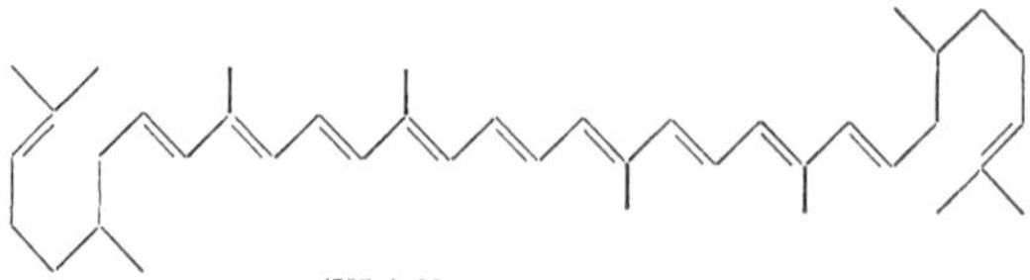

(XL.) Neurosporene *(63)*.*

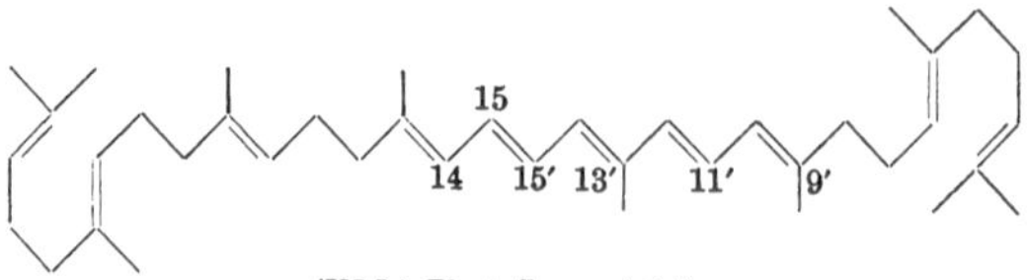

(XLI.) Phytofluene *(285)*.

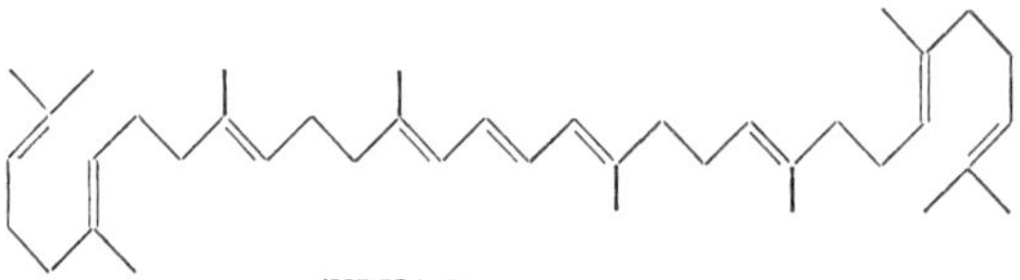

(XLII.) Phytoene *(220)*.

### *Rhodoviolascin (Spirilloxanthin) Set.*

*Structure*, p. 229 [BARBER, JACKMAN and WEEDON *(9)*; JENSEN *(122)*].
*Cis isomers obtained by rearrangement*, POLGÁR, VAN NIEL and the writer *(208)*.

The spectral curve of this dimethoxy compound appears in *Fig. 39*. The sole main stereoisomer in this set is neo A which is formed easily in about 40% yield. The all-*trans* configuration of natural rhodoviolascin is remarkably labile and even fresh solutions of the crystals contain appreciable amounts of neo A. The $\lambda_{max}$ shift (7 m$\mu$ in hexane or 10.5 m$\mu$ in benzene) seems to indicate a di*cis* compound whose molecules must possess a bent shape ($\varepsilon = 5.1 \times 10^4$ at *cis*-peak). It is believed that one of the *cis* double bonds has the central or next-to-central location.

### *Capsorubin Set.*

*Structure*, a fully aliphatic dihydroxy-diketone with nona-unsaturated dione chromophore [AHMAD and WEEDON *(2)*; WARREN and WEEDON *(267)*; cf. CHOLNOKY and SZABOLCS *(27)*]; but see R. ENTSCHEL and P. KARRER, Helv. Chim. Acta **43**, 89 (1960)**.
*Cis isomers obtained by rearrangement*, CHOLNOKY and the writer *(289)*; CHOLNOKY et al. *(26)*.

When treated with iodine, capsorubin yields two main isomers, the neo forms A and B, adsorbed above the all-*trans* zone. Ratio, all-*trans* : : *cis* forms = 70 : 30. Both neo A and B seem to be mono*cis* compounds ($\lambda_{max}$ shifts, 4 m$\mu$ and 3 m$\mu$).

---

* According to our numbering system, 5,6;5',6'-tetrahydrolycopene, but designated as the 6,7,6',7'-tetrahydro compound by EUGSTER et al. *(63)*.
** *(Added in Proof.)*

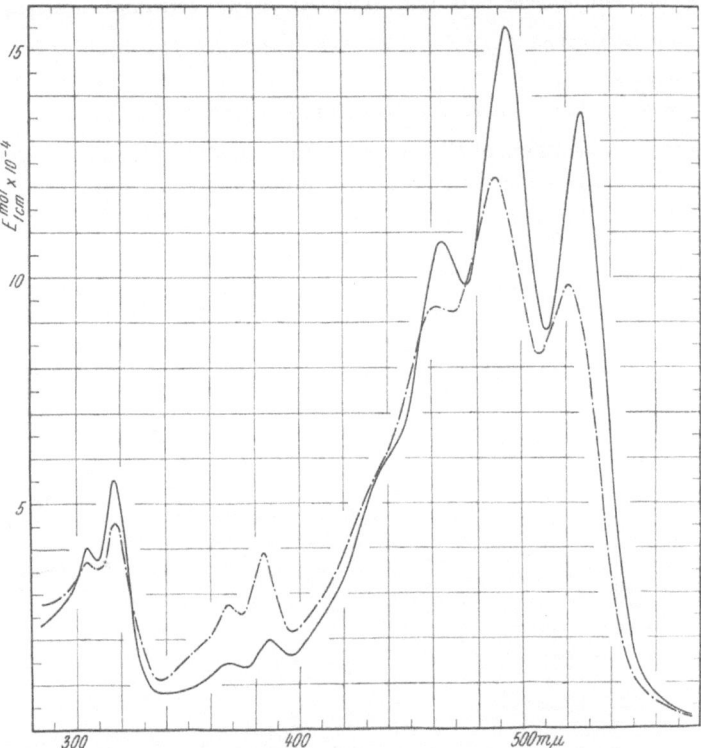

Fig. 39. Molecular extinction curves of rhodoviolascin, in hexane: ————, fresh solution of the all-*trans* compound; and —·—·—, mixture of stereoisomers after catalysis by iodine (*208*). [From: Arch. Biochemistry 5, 243 (1944).]

*Capsorubin dipalmitate* afforded two *cis* isomers that appeared below the all-*trans* zone and could not be sharply separated from one another ($\lambda_{\max}$ shifts, 5 m$\mu$ and 6 m$\mu$).

### Fucoxanthin Set.

*Structure* unknown, probably fully aliphatic; composition, $C_{40}H_{56}O_6$.
*Cis isomers found in nature and obtained by rearrangement*, STRAIN, MANNING and HARDIN (*248*, *250*).

This algal and diatom pigment yields two isomers, neo A and neo B which are adsorbed above the all-*trans* form ($\lambda_{\max}$ shifts in ethanol, 6–7 m$\mu$). The same stereoisomeric mixture (all-*trans* : *cis* forms = 90 : 10) obtains when diatoms are rapidly extracted. The thermal interconversion of the fucoxanthins is accelerated by light and by contacting petroleum ether solutions with sugar or powdered glass which strongly retain the pigment. Considering the HI-sensitivity, iodine catalysis is successful only in the presence of some organic base.

## 4. Stereoisomeric Sets with Two Aromatic Terminal Groups.

*1,18-Diphenyl-3,7,12,16-tetramethyl-octadecanonaene Set.*

*Structure*, see below.

*Cis isomers prepared by total synthesis,* GARBERS, EUGSTER and KARRER (*60, 71, 72*).

Several *cis* members of this set were synthesized by KARRER's research group. Although some hindered and unhindered *cis* forms could not

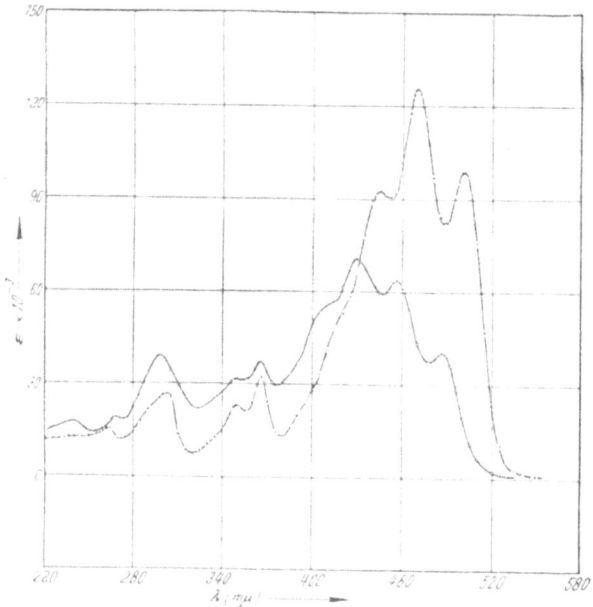

Fig. 40. Molecular extinction curves of synthetic (hindered) 5-*cis*-1,18-diphenyl-3,7,12,16-tetramethyl-octadecanonaene, in cyclohexane: —o—o—, fresh solution; and — ✕ — ✕ —, mixture of stereoisomers after catalysis by iodine; according to GARBERS, EUGSTER and KARRER (*72*). [From: Helv. Chim. Acta **36**, 562 (1953).]

yet be assigned configurations, the two following well-defined, hindered members of the set have theoretical importance. They are crystalline, stable and show degraded spectra (cf. p. 285).

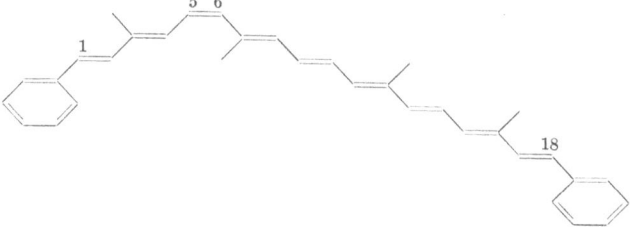

(XLIII.)  5-*cis*-1,18-Diphenyl-3,7,12,16-tetramethyl-octadecanonaene (the *cis* double bond is sterically hindered) (cf. *Fig. 40*).

*References, pp. 334—349.*

(XLIV.) 5,13-Di*cis* form of the same compound (both *cis* double bonds are sterically hindered).

## Renieratene and Isorenieratene. Sets.

*Structures*, see below.

*Mixture of cis isomers obtained by rearrangement*, YAMAGUCHI (273–275).

These sponge pigments are closely related to the synthetic polyenes just mentioned. Renieratene and isorenieratene show upon isomerization with iodine the expected spectral changes, including *cis*-peaks. No individual *cis* forms have been described up to the present time.

(XLV.) Renieratene.

(XLVI.) Isorenieratene.

## 1,3,7,12,16,18-Hexaphenyl-octadecanonaene Set.

*Structure*, see next page.

*Cis isomers prepared by total synthesis and by rearrangement*, ZIEGLER, EUGSTER and KARRER (324).

Four crystalline, hindered *cis* members of this set, the "cis forms I–IV" are known. They have degraded spectra and are iodine-sensitive, yielding, besides the all-*trans* form, a number of new *cis* isomers whose curves indicate the presence of unhindered *cis* double bonds. One of the isomerization products might well represent a second all-*trans* form that differs from the ordinary one by a crosswise orientation of phenyl side-chains (324).

(XLVII.) 1,3,7,12,16,18-Hexaphenyl-octadecanonaene.

## "Naphthyl-carotene" Set.

*Structure*, 1,18-di-β-naphthyl-3,7,12,16-tetramethyl-octadecanonaene.
*Cis isomer prepared by total synthesis*, LINNER, EUGSTER and KARRER (*170*).

A sterically unhindered mono*cis* form ($\lambda_{max}$ shift in $CS_2$, 4 m$\mu$) was isolated and showed remarkable thermostability (m. p. 225°). The extraordinarily high *cis*-peak indicates a central position of the *cis* double bond.

# X. Lower-molecular Weight Carotenoid-carboxylic Acids: Bixin and Crocetin.

The carbon skeletons of these stereochemically important pigments are identical with the middle section of the β-carotene molecule but the chromophore is terminated by conjugated carboxyl groups. Hence, it can be tentatively assumed that bixin, $CH_3OOC \cdot C_{22}H_{26} \cdot COOH$, and crocetin, $HOOC \cdot C_{18}H_{22} \cdot COOH$, are products of the bio-oxidative cleavage of primarily formed $C_{40}$-carotenoids.

## *Bixin Set.*

This monomethylester of a nonaene-dicarboxylic acid (XLVIII) occurs in the seed hulls of *Bixa orellana* in substantial amounts and constitutes the main pigment of industrial Orlean. The dimethylester is termed methylbixin and the free acid, norbixin. When methylated, the bixin molecule acquires symmetry whereby the calculated number of *cis-trans* forms decreases from 512 to 272 and that of the unhindered isomers from 32 to 20. Considering this simplification and the low solubility of bixin and norbixin, most stereochemical work has been carried out with methylbixin.

(XLVIII.) Bixin (all-*trans* form).

Historically, bixin was the first naturally occurring *cis* polyene. In 1923, HERZIG and FALTIS (*92*) in a single, unreproducible experiment obtained from Orlean a second bixin with higher melting point and longer wavelength spectrum. Presumably, an accidental catalyst had caused a *cis* → *trans* rearrangement during isolation. Six years later, KARRER et al. (*134*) tentatively assumed that the two bixins are in the relationship of geometrical isomerism. They showed that natural bixin can be converted into HERZIG's bixin by means of iodine of which,

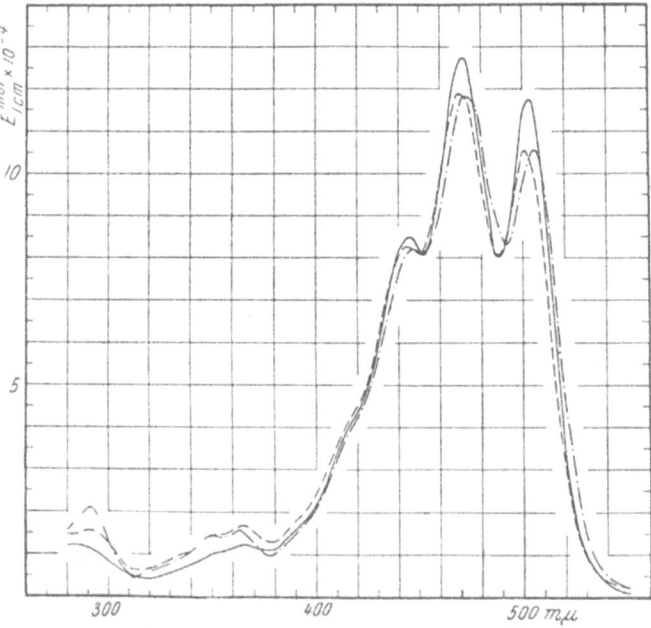

Fig. 41. Molecular extinction curves of all-*trans*-methylbixin, in benzene: ————, fresh solution; and —·—·—, mixture of stereoisomers after catalysis by iodine (*295*). [From: J. Amer. Chem. Soc. 66, 322 (1944).]

according to KUHN and WINTERSTEIN (*157*) catalytic amounts suffice. The structural identity of the two bixins has been demonstrated as follows [KUHN et al. (*152*, *153*)].

Since both pigments had afforded the same dihydro derivative, it became possible to convert the Orlean bixin into HERZIG's bixin by reduction and subsequent air oxidation (in the presence of piperidine): natural bixin → dihydrobixin → second bixin (*152*, *157*). An analogous transition was realized by methylation followed by saponification: natural bixin → natural methylbixin → second bixin.

In the literature, several names have been in use for the two bixins: Bixin = ordinary bixin = natural bixin = Orlean bixin = *cis*-bixin = α-bixin = labile bixin = = bixin II = lower melting bixin (m. p. 198°).—Second bixin = isobixin = *trans*-bixin = all-*trans*-bixin = β-bixin = stable bixin = bixin I = higher melting bixin (m. p. 220°).—Another "isobixin" described by VAN HASSELT (*90*) could not be reproduced by KARRER and TAKAHASHI (*140*); its homogeneity is doubtful.

Turning to stabilities, we may stress that as in the class of the $C_{40}$-carotenoids the simple terms "stable" and "labile" are unsatisfactory. The natural product, "labile" bixin, shows a high degree of thermo-stability and is also relatively photostable but is extremely sensitive to iodine, in light. We will designate the main pigment of the *Bixa* seeds as "natural bixin" and the dimethyl ester as "natural methyl-

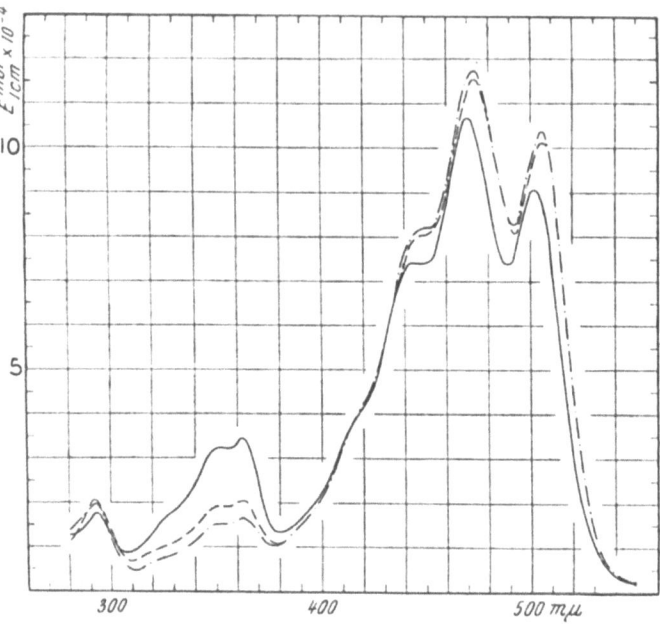

Fig. 42. Molecular extinction curves of natural methylbixin, in benzene: ————, fresh solution; — — —, mixture of stereoisomers after refluxing in darkness for 45 min.; and — · — · —, after catalysis by iodine *(295)*. [From: J. Amer. Chem. Soc. **66**, 322 (1944).]

bixin", although the latter does not seem to occur in plants. The pigment obtained by iodine catalysis is all-*trans*-bixin.

Strictly speaking, the name "natural bixin" is no longer a precise term because small quantities of the all-*trans* form have been isolated from *Aristolochia cymbifera* roots under conditions which exclude *cis-trans* rearrangement *(86)*.

Since earlier studies had been restricted to the two bixins mentioned and since the reversibility of the stereoisomerization had not been clearly claimed, ESCUE and the writer *(293, 295)* have re-investigated the bixin set. Upon spatial rearrangement three new *cis* forms, the neomethyl-bixins A, B, and C were isolated, A and C in crystalline form *(Figs. 41–44)*. A few years ago, total-synthetic all-*trans*-methylbixin was prepared by three research groups [AHMAD and WEEDON *(1)*, INHOFFEN and RASPÉ *(104)*, ISLER et al. *(111)*] and identified with the pigment that is obtained

from "natural" methylbixin by means of iodine. INHOFFEN also described
the crystalline central-mono*cis*-methylbixin (*Fig. 45*, p. 325) and found it
different from the neo compounds A–C. Its configuration is confirmed
by the IR spectrum, viz. absence of methylated *cis* double bonds (*174*).
Central-mono*cis*-methylbixin does not occur in stereoisomeric equilibrium
mixtures, pending experiments to be conducted in red light.

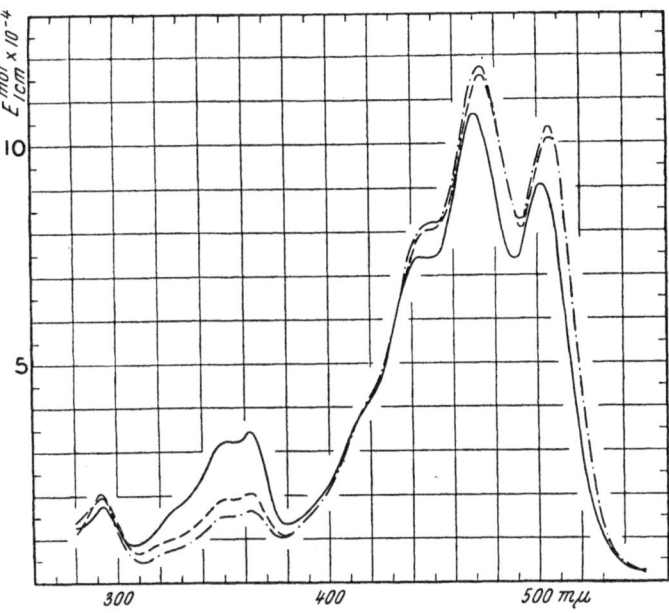

Fig. 43. Molecular extinction curves of neomethylbixin A, in benzene: ————, fresh solution; — — —,
mixture of stereoisomers after refluxing in darkness for 45 min.; and — · — · —, after catalysis by iodine (*295*).
[From: J. Amer. Chem. Soc. **66**, 322 (1944).]

Some characteristics of the five known crystalline members of the
methylbixin set appear in *Table 16*.

Table 16. Stereoisomeric Methylbixins (cf. *295, 104*).

| Name | Adsorbed above (a) or below (b) the all-*trans* form | Melting point | Position of $\lambda_{max}$ | | $\lambda_{max}$ shift (m$\mu$) | $\varepsilon$ at *cis*-peak | | $Q$ (p. 292) |
|---|---|---|---|---|---|---|---|---|
| | | | in benzene (m$\mu$) | in petr. ether (m$\mu$) | | in benzene | in petr. ether | |
| All-*trans*. | — | 220° | 475 | 458 | — | — | — | — |
| Natural . | a | 198° | 471 | — | 4 | $1.2 \times 10^4$ | — | 10.7 |
| Neo A .. | b | 190° | 470 | — | 5 | $3.5 \times 10^4$ | — | 3.1 |
| Neo C .. | b | 150–151° | 465. | — | 10 | $2.2 \times 10^4$ | — | 4.4 |
| Central-mono*cis* | b | 193–194° | — | 455 | 3 | — | $6.1 \times 10^4$ | 1.6 |

21*

Considering the $\lambda_{\max}$ shift values, neo C represents a dicis form but both natural methylbixin and neo A must be monocis compounds. On the basis of the high cis-peak we had assumed earlier that the cis double bond is located in the center of the neo A molecule. Since, however, the intensity of this peak is surpassed by that of Inhoffen's synthetic product, neomethylbixin A is now assigned the next-to-central-monocis

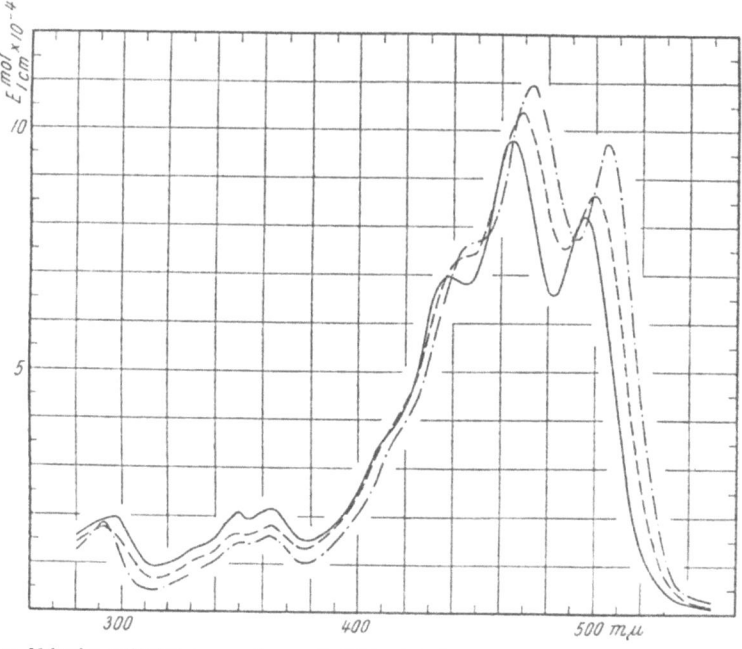

Fig. 44. Molecular extinction curves of neomethylbixin C, in benzene: ————, fresh solution; — — —, mixture of stereoisomers after refluxing in darkness for 45 min.; and — · — · —, after catalysis by iodine (295).
[From: J. Amer. Chem. Soc. 66, 322 (1944).]

configuration (8-cis). (The reported absence of bands both at $7.25\,\mu$ and $12.0–14.0\,\mu$ remains unexplained.)

This leaves the positions 2, 4, and 6 open for the cis double bond of natural methylbixin. Although the biochemically interesting configurational problem of natural methylbixin can still not be solved, as we will see a tentative assignment can be made.

In the cis-peak region, the spectral curve of this isomer is almost flat as foreseen for a peripheral monocis compound. A configuration of this "neo U type" is further indicated by the strong adsorption affinity that surpasses that of the all-trans pigment.

Natural methylbixin is not a component of the cis-trans equilibrium mixtures that are obtained from all-trans-methylbixin by refluxing, insolation, or iodine catalysis: in these instances the neo forms A–C appear. Neither is all-*trans* pigment formed in any marked amounts when natural methylbixin is insolated or refluxed. In the absence of iodine the thermo- and photostability of natural methylbixin is so high that the molecule rather undergoes further *trans → cis* rearrangement than to rotate its *cis* double bond into the *trans* configuration. Iodine

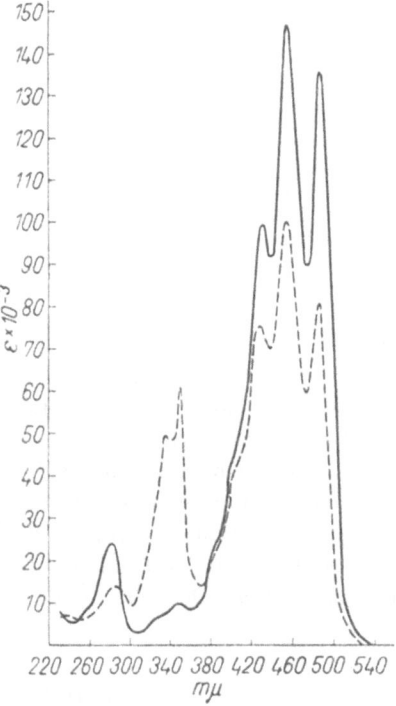

Fig. 45. Molecular extinction curves of methylbixin, in light petroleum: ————, all-*trans* form; and — — —, central-mono*cis* form; according to INHOFFEN and RASPÉ (*104*). [From: Liebigs Ann. Chem. **592**, 214 (1955).]

Fig. 46 (right). Stereochemically important sections taken from infrared curves of some *cis-trans* isomeric methylbixins: saturated CCl₄ solutions (1 mm. cell) in the 7.0—7.5 μ and 8.0—9.0 μ regions; mineral oil mulls in the 10.0—10.5 μ and 12.0—14.0 μ regions (*174*). [From: J. Amer. Chem. Soc. **77**, 1647 (1955).]

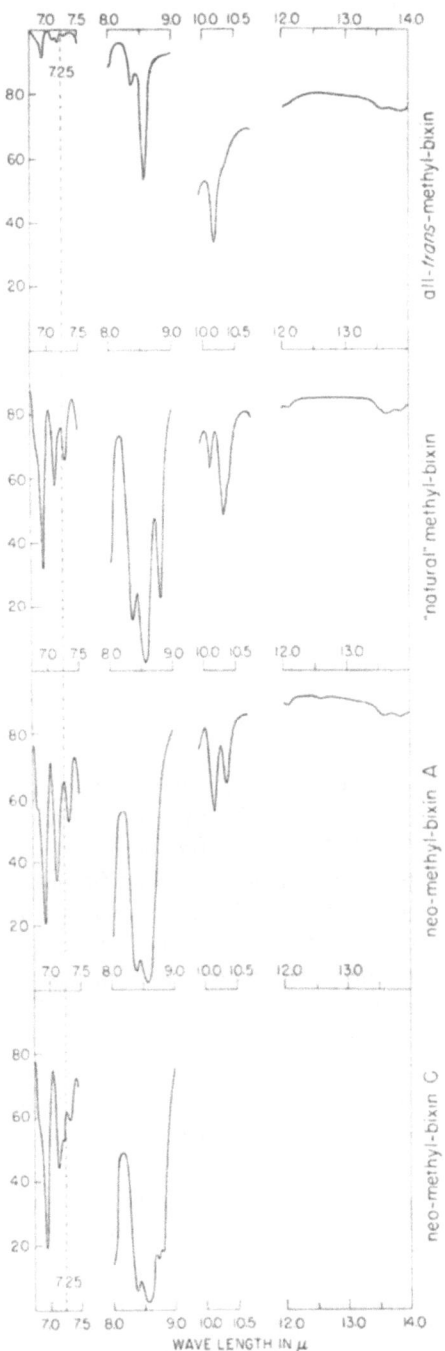

brings about the expected equilibrium mixture wherein natural methylbixin has disappeared. In contrast, the interconversion of natural
methylbixin and the di*cis* isomer neo C is easy. Therefore the location
of one of the *cis* double bonds in neo C is probably identical with that
of the natural pigment. At present the sole in vitro source of natural
methylbixin is neo C (ex all-*trans*) that yields it in substantial amounts
upon various treatments (in the absence of iodine).

If full arrows stand for easy interconversion and dotted arrows for
difficult or impossible ones, then we may write the following scheme,
assuming that iodine is absent:

All-*trans*-methylbixin ⇄ Neomethylbixin A

(mono*cis*)

Natural methylbixin ⇄ Neomethylbixin C
(mono*cis*)                        (di*cis*)

Typical examples of equilibrium mixtures (*295*):

All-*trans*-methylbixin: Refluxing; unchanged *trans* : neo A : minor isomers =
= 62 : 35 : 2.—Insolation; unchanged *trans* : neo A = 94 : 6.—Iodine, in daylight;
unchanged *trans* : neo A : other neo forms = 72 : 19 : 9.

Natural methylbixin: Refluxing; unchanged natural form : neo C (+ minor
isomers) = 74 : 26.—Insolation; unchanged natural form : minor isomers : neo C =
= 90 : 2 : 8.

Neomethylbixin C: Refluxing; natural form : all-*trans* : neo A : neo B : unchanged
neo C = 33 : 12 : 6 : 3 : 46.—Insolation; natural form : neo A : unchanged neo C =
= 21 : 2 : 77.

The peculiar behavior of natural methylbixin can be tentatively
explained by the influence of a conjugated carbonyl group on the *cis*
double bond, i. e. by the 2-*cis* configuration. Although this assignment
is in accordance with the flatness of the curve in the *cis*-peak region,
it necessitates further discussion because the 2-position is a hindered
one and the natural methylbixin curve is not degraded. This situation
might be explained by the special character of the $CH_3OOC—CH=CH—$
group which could assume a conformation essentially normal with
reference to the plane of the rest of the molecule.

The proposed 2-*cis* configuration is in accordance with the IR spectrum
(*Fig. 46*, p. 325) (*174*): (a) The 7.25 $\mu$ band and hence a methylated *cis*
double bond is absent; (b) a strong band appears at 8.00 $\mu$ that does not
occur in the all-*trans* spectrum. It may well represent a stretching C—O
vibration for which a conjugated *cis* double bond is responsible. (A more
remote such bond could not have this effect.)

Our configurational assignment to natural methylbixin, and hence to bixin
itself, is not in accordance with KARRER and SOLMSSEN's earlier observations (*139*)
who submitted the all-*trans* and natural bixins to a parallel treatment with permanganate. On the basis of the identity of some of the aldehydes obtained from

both sources and the steric difference between those two aldehydes that were obtained by cleavage at the C=C double bond third from the free carboxyl group, these authors tentatively concluded that bixin was the 6-*cis* compound. This interesting approach, which would require re-investigation, does show that the *cis* double bond cannot be located in the middle section of the natural bixin molecule.

If natural methylbixin is the proposed 2-*cis* form, then neo C must have one of the following eight di*cis* configurations: 2,4; 2,6; 2,8; 2,10; 2,8'; 2,6'; 2,4'; or 2,2'. Of these 2,6 and 2,6' are eliminated because of steric hindrance (the spectrum is not degraded); furthermore, 2,4; 2,4'; and 2,2' are excluded because they would contradict the marked *cis*-peak. Consequently, neo C may be either 2,8- or 2,10- or 2,8'-di*cis*-methylbixin.

The reader will note that the foregoing conclusions have tentative character and would require confirmation by total synthesis.

## Crocetin Set.

Crocetin (IL), a lower homolog of bixin, is the principal pigment of saffron *(Crocus sativus)* in which it occurs mainly in the form of the di-gentiobioseester crocin. Crocetin has also been observed in several other flowers and in some fruits (cf. *135*).

(IL.) Crocetin.

Crocetin is a heptaene-dicarboxylic acid. Its methyl- and dimethyl-ester are termed, respectively, methylcrocetin and dimethylcrocetin.

An earlier nomenclature (crocetin = $\alpha$-crocetin; monomethylester = $\beta$-crocetin; and dimethylester = $\gamma$-crocetin) has been abandoned.

The calculated number of stereoisomeric crocetins is 72, and that of the unhindered forms is 20. In contrast to the bixin molecule, the double bonds adjacent to the carbonyls are sterically unhindered. Only three members of this set are known. The all-*trans* and a *cis* form originate from plants while the central-mono*cis* isomer is a product of total synthesis; it can be rearranged to all-*trans*-crocetin. As in the bixin set, most of the stereochemical work has been carried out with the relatively easily soluble dimethylester.

KUHN and WINTERSTEIN (*158*, *159*) should be credited with the first observation about geometrical isomerism in the crocetin set. On trans-esterification of the native glucoside (from Spanish saffron) they isolated, besides the well-known all-*trans*-dimethylcrocetin (m. p. 222°) a second pigment (m. p. 141°) that was interpreted as a *cis* compound on the basis of the lower melting point, higher solubility, spectrum, and relatively greater lability. Its quantities in the stigmata are much inferior to those of the all-*trans* compound; and subsequently, the attempts to obtain *cis*-crocetin from various other saffron samples have failed altogether. Recently, it was possible, however, to repeat the isolation (cf. *96*). Perhaps working in red light would improve the yields.

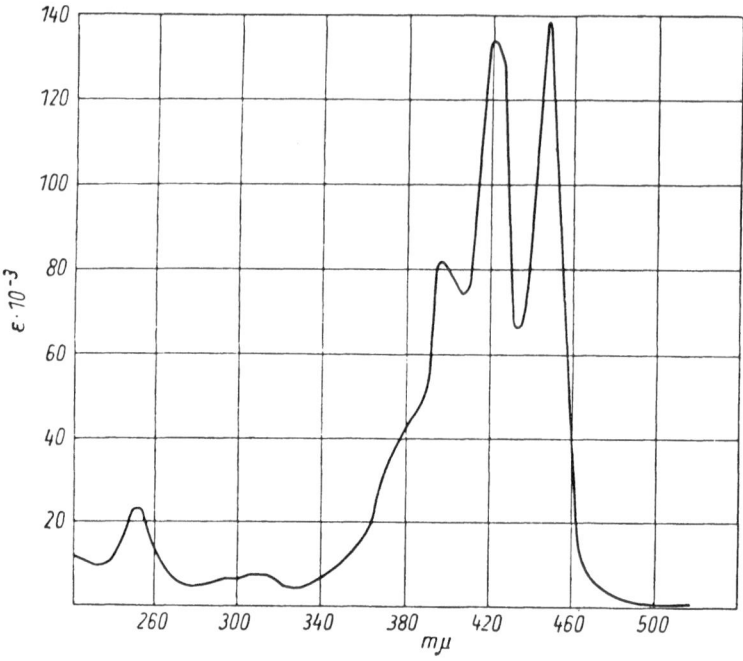

Fig. 47. Molecular extinction curve of all-*trans*-dimethylcrocetin, in light petroleum; according to INHOFFEN et al. (*102*). [From: Liebigs Ann. Chem. **580**, 7 (1953).]

The following names have been used for the two natural dimethylcrocetins (and analogous ones for the corresponding crocetins): Stable dimethylcrocetin = = *trans*-dimethylcrocetin = all-*trans*-dimethylcrocetin = dimethylcrocetin I = higher melting dimethylcrocetin (m. p. 222°).—Labile dimethylcrocetin = *cis*-dimethylcrocetin = natural labile dimethylcrocetin = natural *cis*-dimethylcrocetin = lower melting dimethylcrocetin (m. p. 141°).

The steric nature of the relationship between these two pigments was demonstrated also by chemical conversions: (a) Since the reduction of both dimethylcrocetins afforded the same dihydro compound, the following transition was realized (*152*): natural *cis*-dimethylcrocetin → dihydro derivative → all-*trans*-dimethylcrocetin. (b) Saponification of the *cis* ester in the heat furnished all-*trans*-crocetin.

According to KUHN and WINTERSTEIN (*158*) direct *cis → trans* rearrangement takes place upon melting crystals or adding to solutions trace amounts of iodine in diffuse daylight. The most impressive method is, however, illumination in the absence of catalysts. Although the configuration did not change when *cis* crystals were kept in darkness for 20 years (*154*), during the observation of a fresh solution in the visual spectroscope the bands migrated towards longer waves and soon reached the positions characteristic for all-*trans*-dimethylcrocetin. Under certain conditions of artificial illumination the observed half time of the

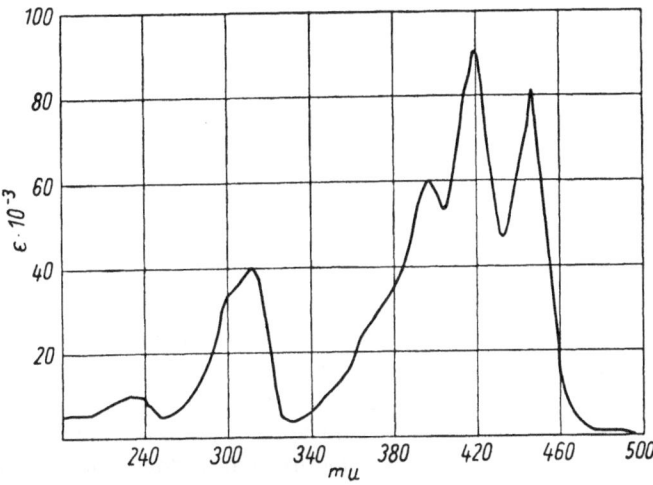

Fig. 48. Molecular extinction curve of central-mono*cis*-dimethylcrocetin, in light petroleum; according to INHOFFEN et al. (*102*). [From: Liebigs Ann. Chem. 580, 7 (1953).]

rearrangement was 6 minutes. The in vitro photosensitivity of natural *cis*-dimethylcrocetin by far surpasses that of the two other isomers.

The main spectral maxima of all-*trans*-dimethylcrocetin (in light petroleum) were found at 422, *448* mµ and at 422, *450* mµ (*102*, *111*). KUHN and WINTERSTEIN (*158*) had reported 420, *445.5* mµ for the all-*trans* compound and 416, *442* mµ for the *cis* form ex saffron. The small $\lambda_{max}$ shift indicates a mono*cis* configuration. A priori four locations can be considered for the *cis* double bond in natural *cis* dimethylcrocetin, namely the positions 2, 4, 6, and 8. Of these the non-degraded spectrum eliminates position 4 and the moderate *cis*-peak excludes the central position 8. Because of the new feature of extreme photosensitiveness, which could well be caused by a carbonyl group in conjugation with the *cis* C=C double bond, it seems reasonable to give preference to the 2-*cis* configuration, pending confirmation by synthesis.

Other authors have considered the 6-*cis* configuration (*154*).

All-*trans*-dimethylcrocetin (m. p. 146°) (*Fig. 47*, p. 328) was obtained by total synthesis and identified with the main saffron pigment [INHOFFEN et al. (*102*); ISLER et al. (*110, 111*)]. The last intermediate in the INHOFFEN synthesis was central-mono*cis*-dimethylcrocetin (also termed 8-*cis* or 8,8'-*cis*) (*Fig. 48*). Its main maxima lie at *420*, *446* mμ; and the value of the $\lambda_{max}$ shift (2 mμ) is smaller at the center of the chromophore than in the peripheries also in this relatively short conjugated system (cf. p. 294).

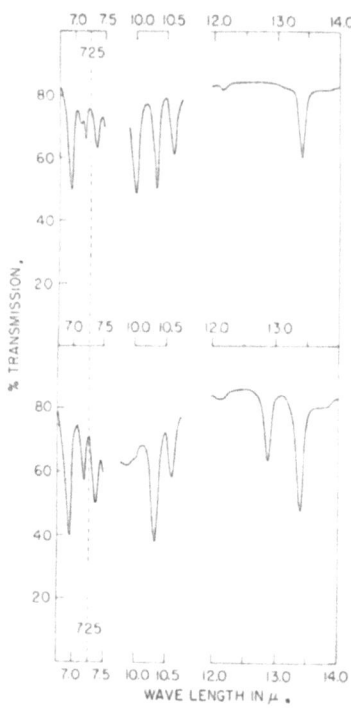

Fig. 49. Stereochemically important sections taken from infrared curves of all-*trans*-dimethylcrocetin (top) and central-mono*cis*-dimethylcrocetin (bottom): saturated CCl₄ solutions (1 mm. cell) in the 7.0-7.5 μ and 10.0—10.5 μ regions; mineral oil mulls in the 12.0—14.0 μ region (*174*). [From: J. Amer. Chem. Soc. 77, 1647 (1955).]

Comparative infrared measurements in the dimethylcrocetin set were reported by KUHN, INHOFFEN, STAAB and OTTING (*154*) and confirmed by LUNDE and the writer (*174*) (*Fig. 49*). The results agree with the conclusions given above. Among the mono*cis*-dimethylcrocetins only the central *cis* form has a symmetrical configuration. Accordingly, the IR curve of the *cis* compound (ex saffron) shows doublets or triplets at several wavelengths where singlets occur in the curve of the synthetic *cis* form. Furthermore, in case of the central-mono*cis*-dimethylcrocetin an intense band appears at 12.88 μ representing an unmethylated *cis* double bond (cf. p. 295). Because this maximum is missing in the curve of the natural *cis* isomer, the latter cannot have a *cis* configuration about the 4- or 8- double bond. As expected the 7.25 μ band, characteristic of methylated *cis* double bonds, is absent from the all-*trans* and central-mono*cis* spectra.

The properties of the synthetic and natural *cis* dimethylcrocetins were carefully compared by KUHN, INHOFFEN, STAAB and OTTING (*154*): Both substances show the phenomenon of double melting points (p. 239); the fundamental bands of the two isomers are very similar; the synthetic *cis* form is distinguished from the natural one (or from all-*trans*-dimethyl-crocetin) by its higher *cis*-peak, the IR spectrum, and differences in the Debye-Scherrer diagram.

*References, pp. 334—349.*

# XI. *Cis-trans* Isomerism and Provitamin A Effect of Carotenoids.

The chemical and biological relationships between carotene and vitamin A have been established by the classical investigations of STEENBOCK (*238*), MOORE (*191*), EULER (*65*), KARRER (*66*), and others. It is now well known that $\beta$-carotene whose molecules contain two unsubstituted $\beta$-ionone rings is partially converted in the body into vitamin A, according to the *schematic* expression, $C_{40}H_{56} \rightarrow 2\ C_{20}H_{29}OH$. Several other naturally occurring carotenoid pigments such as $\alpha$-carotene, $\gamma$-carotene, and cryptoxanthin undergo analogous cleavage but in these instances only one half of each molecule may yield the biologically active substance. Although the postulate that at least one unsubstituted $\beta$-ionone ring must be present in provitamins A has been rather generally accepted, there are some exceptions. For example, 2,2'-dimethyl-$\beta$-carotene is effective in the rat to the extent of one half of the $\beta$-carotene potency [EUGSTER, TRIVEDI and KARRER (*64*)].

Whereas earlier investigations into the relative strengths of various provitamins A had been restricted to the behavior of all-*trans* carotenoids, newer assays have demonstrated the profound dependence of the biopotency also on the spatial configuration, hence on the morphological features of the molecule. [Survey articles: (*281, 282, 129, 47*)]. The first observations were made in 1936–1937 by GILLAM et al. (*76, 77*) who found that their "pseudo-$\alpha$-carotene" (mainly neo-$\beta$-carotene B, p. 230) had a provitamin A potency "of the same order" as that of all-*trans*-$\beta$-carotene; furthermore, "neo-$\alpha$-carotene" displayed about 70% of the all-*trans*-$\beta$-carotene effect. A few years later KEMMERER and FRAPS (*142*) reported that "pseudo-$\alpha$-carotene" was half as potent as $\beta$-carotene [cf. also BEADLE and ZSCHEILE (*11*)].

Under the guidance of the late H. J. DEUEL, Jr. our group conducted an extended series of pertinent bioassays (1944–1952) in which either crystalline or chromatographically homogeneous samples were used (*48–56, 292, 307, 87*). *Table 17* (next page) contains a summary of our results and also some of the KARRER, INHOFFEN, and ISLER research groups.

In the course of these and other studies the following fundamental facts have emerged: (a) If an all-*trans* carotenoid shows provitamin A activity, then its *cis* forms are also active. (b) All-*trans* → *cis* rearrangements decrease the biopotency (with rare exceptions). (c) There is, in principle, no difference between the biological behavior of unhindered and hindered *cis* carotenoids. (d) The order of magnitude of the influence of stereoisomerization on the biopotency is the same as that of reasonably chosen structural conversions.

To illustrate point (d) let us start from all-*trans*-β-carotene and convert it into all-*trans*-α-carotene by moving a terminal double bond out of conjugation; the biopotency falls from 100% to 53%. This biological weakening is, however, surpassed when, without altering the conjugated β-carotene system, a peripheral double bond is rotated from the *trans* into the *cis* configuration; the resulting neo-β-carotene U shows only 38% of the initial activity.

Table 17. Relative Potencies of Some Stereoisomeric Provitamins A in the Rat.

| Stereoisomeric set | Configuration | Biopotency | | References |
|---|---|---|---|---|
| | | in % of that of the all-*trans* form of the set | in % of that of all-*trans*-β-carotene | |
| β-Carotene | all-*trans*............. | 100 | 100 | — |
| | central-mono*cis* (synth.) | 50–30* | 50–30* | (292) |
| | peripheral mono*cis* (neo U)............. | 38 | 38** | (54) |
| | di*cis* (neo B) .......... | 53 | 53 | (52) |
| | 11,11'-di*cis* (hindered) (synth.)............. | 30 | 30 | (108) |
| | "*cis* C" (hindered) (synth.) | 20–25 | 20–25 | (73) |
| β-Carotene homolog $C_{42}H_{58}$ | all-*trans* (synth.)....... | 100 | 20 | (51) |
| | middle-di*cis* (synth.) ... | 100 | 20 | (51) |
| α-Carotene | all-*trans*............. | 100 | 53 | (56) |
| | peripheral mono*cis* (neo U)............. | 25 | 13 | (56) |
| | di*cis* (neo B) .......... | 30 | 16 | (52) |
| γ-Carotene | all-*trans**** ............ | 100 | 43 | (53, 307, 281) |
| | group of *cis* forms..... | 60 | 16 | (50) |
| | neo P ............... | 70 | 19 | (50) |
| | pro-γ-carotene ......... | 100 | 43 | (53, 307) |
| Cryptoxanthin | all-*trans*............. | 100 | 57 | (87 a) |
| | peripheral mono*cis* (neo U)............. | 45 | 27 | (49) |
| | mono*cis* (neo A) ....... | 71 | 42 | (55) |

* In low doses, 50%. Otherwise, the experimental errors do not seem to exceed 10% of the values.

** Bahl et al. (8) found 18.5%.

*** Ex pro-γ-carotene, by iodine catalysis. All-*trans*-γ-carotene preparations ex tomato paste or *Pyracantha* berries showed only 26–27% of the β-carotene potency instead of 43%. For possible reasons see p. 291.

The conspicuous variation of the individual provitamin A potencies within a stereoisomeric set is evidently the result of numerous partial bioprocesses, whose outcome is the promotion of growth that can be measured in A-depleted young rats. Very little is known about the complicated mechanism of these processes. Contributing factors are, the extent of absorption, stepwise cleavage, destruction and excretion of the provitamin on one hand, and the rates of complexing and release from enzyme systems on the other. The oxidative destruction of pro-vitamins in the intestinal tract is also influenced by the presence or absence of antioxidants such as tocopherol.

Besides the mentioned growth tests, several other experimental approaches are useful in this field. Thus, according to JOHNSON and BAUMANN (*124*) different doses of stereoisomers are required in order to accumulate equal amounts of vitamin A in the liver and kidneys of the rat.

Such biologically equivalent daily doses are, 20 μg. of all-*trans*-β-carotene, 40 μg. of neo-β-carotene B, and 60 μg. of neo-β-carotene U. The ratios of these A-accumulating effects are, all-*trans* : neo B : neo U = 100 : 48 : 33, in remarkably good agreement with the ratios of the growth-promoting effects established in our laboratory, viz. 100 : 53 : 38.

As mentioned on p. 274, *cis-trans* isomeric carotenoids may undergo bio-stereoisomerization of which a few examples follow.

(a) KEMMERER and FRAPS observed (*144, 143*) that, upon the ingestion of neo-β-carotene U, half of the carotene derived from rat feces was present in the all-*trans* form. Further, after a Wesson oil solution of neo U had remained in the digestive tract for several hours the recovered pigment contained, besides unchanged neo U, substantial amounts of the neo B and all-*trans* isomers. Cf. also PAUL et al. (*200*). (b) Feeding all-*trans*-lycopene to carotenoid-depleted chickens resulted in the presence of some neo forms in the feces (*48*). (c) DEUEL et al. (*48*) administered pro-γ-carotene and prolycopene to chickens, whereupon the feces, gut washings and livers were investigated, with the result that 50–70% of the recovered pigment showed altered configuration. Interestingly, this bio-stereoisomerization did not lead directly from the starting material to the all-*trans* compound. Indeed, four poly*cis* lycopenes appeared, one of which possessed shorter wavelength maxima and probably contained more *cis* double bonds than prolycopene (p. 287).

When trying to interpret the observations reported above, it would be tempting to assume that a *cis* provitamin A must first rearrange to its all-*trans* form in the body before it may become a proper substrate for the enzymatic cleavage. Such oversimplification would, however, contradict the experimental evidence. We rather propose that the biological dividing line does not run between active *trans* and inactive *cis* compounds but between such isomers whose molecular shape fits into the enzyme system and those that do not satisfy this postulate. Characteristically, all-*trans*-γ-carotene and pro-γ-carotene show equal activities in the rat; and in the chick this *cis* form is even more effective (by 25%) than the all-*trans* isomer (*87*).

## References.

*1.* Ahmad, R. and B. C. L. Weedon: Carotenoids and Related Compounds. I. Total Synthesis of "all-*trans*"-Methylbixin and of a Diketone with the Capsorubin Chromophore. J. Chem. Soc. (London) **1953**, 3286.

*2.* — — Carotenoids and Related Compounds. IV. A New Synthesis of the Capsorubin Chromophore. J. Chem. Soc. (London) **1953**, 3815.

*2a.* Akhtar, M. and B. C. L. Weedon: Carotenoids and Related Compounds. VIII. Novel Syntheses of Echinenone and Canthaxanthin. J. Chem. Soc. (London) **1959**, 4058.

*3.* Allan, J. L. H., G. D. Meakins and M. C. Whiting: Researches on Acetylenic Compounds. L. The Infrared Absorption of Some Conjugated Ethylenic and Acetylenic Systems. J. Chem. Soc. (London) **1955**, 1874.

*4.* Ames, D. E. and R. Г. Bowman: Synthetic Long-chain Aliphatic Compound . VIII. The Preparati. of *cis*- and *trans*-Undec-9-enoic Acids. J. Chem. Soc. (London) **1952**, 677.

*5.* Ames, S. R.: Fat-soluble Vitamins. Annu. Rev. Biochem. **27**, 371 (1958).

*6.* Anschütz, R.: Über Fumar- und Maleinsäureäther. II. Ber. dtsch. chem. Ges. **12**, 2280 (1879).

*7.* Baeyer, A.: Über die Constitution des Benzols. Liebigs Ann. Chem. **245**, 103 (1888).

*8.* Bahl, A. N., J. C. Sadana and B. Ahmad: The Provitamin A Activities of Neo-β-carotene B and Neo-β-carotene U. J. Sci. Industr. Res. (India) **7 B**, 169 (1948).

*9.* Barber, M. S., L. M. Jackman and B. C. L. Weedon: The Structures of Spirilloxanthin and Related Carotenoids. Proc. Chem. Soc. (London) **1959**, 96.

*10.* Bauer, L.: Trennung der Karotinoide und Chlorophylle mit Hilfe der Papierchromatographie. Naturwiss. **39**, 88 (1952).

*11.* Beadle, B. W. and F. P. Zscheile: Studies on the Carotenoids. II. The Isomerization of β-Carotene and its Relation to Carotene Analysis. J. Biol. Chem. **144**, 21 (1942).

*12.* Bickoff, E. M.: Chromatographic Separation of β-Carotene Stereoisomers as a Function of Developing Solvent. Analyt. Chemistry **20**, 51 (1948).

*13.* Bickoff, E. M.. L. M. White, A. Bevenue and K. T. Williams: Isolation and Spectrophotometric Characterization of four Carotene Isomers. J. Assoc. Official Agric. Chem. **31**, 633 (1948).

*14.* Blankenhorn, D. H.: Carotenoids in Man. II. Fractions Obtained from Atherosclerotic and Normal Aortas, Serum, and Depot Fat by Separation on Alumina. J. Biol. Chem. **227**, 963 (1957).

*15.* Bodea, C. und E. Nicoara: Die Konstitution des Physoxanthins. Liebigs Ann. Chem. **609**, 181 (1957).

*16.* — — Zur Isolierung von α-Carotin, Physoxanthin und Lutein aus *Physalis Alkekengi*. Liebigs Ann. Chem. **622**, 188 (1959).

*17.* Bohlmann, F. und H. J. Mannhardt: Acetylenverbindungen im Pflanzenreich. Fortschr. Chem. organ. Naturstoffe **14**, 1 (1957).

*18.* Braude, E. A. and E. S. Waight: The Relationships between the Stereochemistry and Spectroscopic Properties of Organic Compounds. Progr. Stereochem. **1**, 126 (1954). — Cf. E. A. Braude: The Labile Stereochemistry of Conjugated Systems. Experientia **11**, 457 (1955).

*19.* Brode, W. R.: Steric Effects in Dyes. In: The Roger Adams Symposium, p. 8. New York: Wiley, and London: Chapman & Hall. 1955.

*20.* Burwell, R. L., Jr.: Stereochemistry and Heterogeneous Catalysis. Chem. Revs. **57**, 895 (1957).

*21.* Bush, W. V.: Thesis, California Institute of Technology, Pasadena, 1958.

*22.* Bush, W. V. and L. Zechmeister: On Some Cleavage Products of the Boron Trifluoride Complexes of α-Carotene, Lycopene, and γ-Carotene. J. Amer. Chem. Soc. **80**, 2991 (1958).

*23.* Carter, G. P. and A. E. Gillam: The Isomerization of Carotenes. III. Reconsideration of the Change β-Carotene to ψα-Carotene. Biochemic. J. **33**, 1325 (1939).

*24.* Celmer, W. D. and I. A. Solomons: Mycomycin. IV. Stereoisomeric 3,5-Diene Fatty Acid Esters. J. Amer. Chem. Soc. **75**, 3430 (1953).

*25.* Chatterjee, A. and L. Zechmeister: On Some Stereoisomeric Cryptoxanthins. J. Amer. Chem. Soc. **72**, 254 (1950).

*26.* Cholnoky, L., D. Szabó und J. Szabolcs: Untersuchungen über die Carotinoidfarbstoffe. II. Die Struktur des Capsanthins und des Capsorubins. Liebigs Ann. Chem. **606**, 194 (1957).

*27.* Cholnoky, L. und J. Szabolcs: Über die Struktur des Capsanthins und des Capsorubins. Naturwiss. **44**, 513 (1957).

*28.* — — Untersuchungen über die Carotinoidfarbstoffe. V. Über die Farbstoffe der Kelchblätter von *Physalis Alkekengi*. Liebigs Ann. Chem. **626**, 206 (1959).

*29.* Cholnoky, L., J. Szabolcs und E. Nagy: Untersuchungen über die Carotinoidfarbstoffe. IV. α-Kryptoxanthin. Liebigs Ann. Chem. **616**, 207 (1958).

*30.* Claes, H.: Biosynthese von Carotinoiden bei *Chlorella*. 2. Mitt. Tetrahydrolycopin und Lycopin. Z. Naturforsch. **11 b**, 260 (1956).

*31.* — Biosynthese von Carotinoiden bei *Chlorella*. III. Untersuchungen über die lichtabhängige Synthese von α- und β-Carotin und Xanthophyllen bei der *Chlorella*-Mutante 5/520. Z. Naturforsch. **12 b**, 401 (1957).

*32.* Claes, H. and T. O. M. Nakayama: Isomerization of Poly-*cis*-carotenes by Chlorophyll *in vivo* and *in vitro*. Nature (London) **183**, 1053 (1959).

*33.* Cole, A. R. H.: Infrared Spectra of Natural Products. Fortschr. Chem. organ. Naturstoffe **13**, 1 (1956).

*34.* Cook, A. H.: Algal Pigments and their Significance. Biol. Revs. **20**, 115 (1945).

*35.* Coulson, C. A.: The Electronic Structure of Some Polyenes and Aromatic Molecules. VII. Bonds of Fractional Order by the Molecular Orbital Method. Proc. Roy. Soc. (London), Ser. A **169**, 413 (1939).

*36.* Crombie, L.: Geometrical Isomerism about Carbon-Carbon Double Bonds. Quart. Revs. (Chem. Soc. London) **6**, 101 (1952).

*37.* Crombie, L., S. H. Harper and R. J. D. Smith: Stereochemical Studies of Olefinic Compounds. VI. Preparation of the Four Stereoisomeric Sorbyl Alcohols from a Single Precursor. J. Chem. Soc. (London) **1957**, 2754.

*38.* Curl, A. L.: Application of Countercurrent Distribution to Valencia Orange Juice Carotenoids. Agric. Food Chem. **1**, 456 (1953).

*39.* Curl, A. L. and G. F. Bailey: Orange Carotenoids. Polyoxygen Carotenoids of Valencia Orange Juice. Agric. Food Chem. **2**, 685 (1954).

*40.* — — Orange Carotenoids. I. Comparison of Carotenoids of Valencia Orange Peel and Pulp. Agric. Food Chem. **4**, 156 (1956).

*41.* Dale, J.: Empirical Relationships of the Minor Bands in the Absorption Spectra of Polyenes. Acta Chem. Scand. **8**, 1235 (1954).

*42.* — The Free-Electron Model, "Overtone" Bands, and Vibrational Structure in Absorption Spectra of Polyenes and Polyenynes. Acta Chem. Scand. **11**, 265 (1957).

*43.* — Infrared Absorption Spectra of *ortho*- and *para*-Linked Polyphenyls. Acta Chem. Scand. **11**, 640 (1957).

44. CURL, A. L.: Ultraviolet Absorption Spectra of *ortho*- and *para*-Linked Polyphenyls. Acta Chem. Scand. 11, 650 (1957).
45. — Ultraviolet Absorption Spectra of Chain Molecules Consisting of Alternating Benzene Rings and Ethylenic Bonds. Acta Chem. Scand. 11, 971 (1957).
46. DALE, J. and L. ZECHMEISTER: On the Stereochemistry of Azines: Cinnamalazine and Phenylpentadienalazine. J. Amer. Chem. Soc. 75, 2379 (1953).
47. DEUEL, H. J., Jr.: The Lipids. Vol. III: Biochemistry, p. 540. New York: Interscience Publ. 1957.
48. DEUEL, H. J., Jr., J. GANGULY, B. K. KOE and L. ZECHMEISTER: Stereoisomerization of the Poly*cis* Compounds, Pro-γ-carotene and Prolycopene in Chickens and Hens. Arch. Biochem. Biophys. 33, 143 (1951).
49. DEUEL, H. J., Jr., S. M. GREENBERG, E. STRAUB, T. FUKUI, A. CHATTERJEE and L. ZECHMEISTER: Stereochemical Configuration and Provitamin A Activity. VII. Neocryptoxanthin U. Arch. Biochemistry 23, 239 (1949).
50. DEUEL, H. J., Jr., C. HENDRICK, E. STRAUB, A. SANDOVAL, J. H. PINCKARD and L. ZECHMEISTER: Stereochemical Configuration and Provitamin A Activity. VI. Some *cis-trans* Isomers of γ-Carotene. Arch. Biochemistry 14, 97 (1947).
51. DEUEL, H. J., Jr., H. H. INHOFFEN, J. GANGULY, L. WALLCAVE and L. ZECHMEISTER: Stereochemical Configuration and Provitamin A Activity. XI. A Comparison of Synthetic All-*trans*- and Di*cis*-16,16'-homo-β-carotene $C_{42}H_{58}$ with All-*trans*-β-carotene in the Rat. Arch. Biochem. Biophys. 40, 352 (1952).
52. DEUEL, H. J., Jr., C. H. JOHNSTON, E. R. MESERVE, A. POLGÁR and L. ZECHMEISTER: Stereochemical Configuration and Provitamin A Activity. IV. Neo-α-carotene B and Neo-β-carotene B. Arch. Biochemistry 7, 247 (1945).
53. DEUEL, H. J., Jr., C. H. JOHNSTON, E. SUMNER, A. POLGÁR, W. A. SCHROEDER and L. ZECHMEISTER: Stereochemical Configuration and Provitamin A Activity. II. All-*trans*-γ-carotene and Pro-γ-carotene. Arch. Biochemistry 5, 365 (1944).
54. DEUEL, H. J., Jr., C. H. JOHNSTON, E. SUMNER, A. POLGÁR and L. ZECHMEISTER: Stereochemical Configuration and Provitamin A Activity. I. All-*trans*-β-carotene and Neo-β-carotene U. Arch. Biochemistry 5, 107 (1944).
55. DEUEL, H. J., Jr., E. R. MESERVE, A. SANDOVAL and L. ZECHMEISTER: Stereochemical Configuration and Provitamin A Activity. V. Neocryptoxanthin A. Arch. Biochemistry 10, 491 (1946).
56. DEUEL, H. J., Jr., E. SUMNER, C. H. JOHNSTON, A. POLGÁR and L. ZECHMEISTER: Stereochemical Configuration and Provitamin A Activity. III. All-*trans*-α-carotene and Neo-α-carotene U. Arch. Biochemistry 6, 157 (1945).
57. DREFAHL, G. und G. PLÖTNER: Untersuchungen über Stilbene. XX. Polyphenyl-polybutadiene. Chem. Ber. 91, 1285 (1958).
57a. ECKERT, R. und H. KUHN: Richtungen der Übergangsmomente der Absorptionsbanden von Polyenen, Cyaninen und Vitamin $B_{12}$ aus Dichroismus und Fluoreszenzpolarisation. Z. Elektrochem. (1960) (im Druck).
58. ELVIDGE, J. A., R. P. LINSTEAD and P. SIMS: Polyene Acids. III. A Reinvestigation of Karrer's β-Methylmuconic Acid. J. Chem. Soc. (London) 1951, 3398.
59. ERDTMAN, H.: Corticrocin, a Pigment from the Mycelium of a *Mycorrhiza* Fungus. II. Acta Chem. Scand. 2, 209 (1948).
60. EUGSTER, C. H., C. (F.) GARBERS und P. KARRER: Carotenoid-Synthesen. IX. Stereoisomere 1,18-Diphenyl-3,7,12,16-tetramethyl-octadeca-nonaene. Helv. Chim. Acta 35, 1179 (1952).
61. — — — Carotinoidsynthesen. XIII. Über zwei isomere *cis*-β-Carotine mit *cis*-Konfiguration an „behinderten" Doppelbindungen. Helv. Chim. Acta 36, 1378 (1953).

62. EUGSTER, C. H. und P. KARRER: Taraxanthin und Tarachrom sowie Beobachtungen über stereoisomere Trollixanthine. Helv. Chim. Acta 40, 69 (1957).

63. EUGSTER, C. H., E. LINNER, A. H. TRIVEDI und P. KARRER: Carotinoidsynthesen. XIX. Synthese eines 6,7,6',7'-Tetrahydro-lycopins und dessen Beziehung zum Neurosporin. Helv. Chim. Acta 39, 690 (1956).

64. EUGSTER, C. H., A. H. TRIVEDI und P. KARRER: Carotinoidsynthesen. XVII. Synthese des 2,2'-Dimethyl-β-carotins. Helv. Chim. Acta 38, 1359 (1955).

65. EULER, B. v., H. v. EULER und H. HELLSTRÖM: A-Vitaminwirkungen der Lipochrome. Biochem. Z. 203, 370 (1928).

66. EULER, B. v., H. v. EULER und P. KARRER: Zur Biochemie der Carotinoide. Helv. Chim. Acta 12, 278 (1929).

67. FARKAS, A. and L. FARKAS: The Mechanism of Hydrogenation Reactions and the Formation of Stereochemical Isomers. Trans. Faraday Soc. 33, 837 (1937).

68. FARRAR, K. R., J. C. HAMLET, H. B. HENBEST and E. R. H. JONES: Studies in the Polyene Series. XLIII. The Structure and Synthesis of Vitamin A₂ and Related Compounds. J. Chem. Soc. (London) 1952, 2657.

69. FORBES, W. F. and R. SHILTON: Electronic Spectra and Molecular Dimensions. III. Steric Effects in Methyl-substituted α,β-Unsaturated Aldehydes. J. Amer. Chem. Soc. 81, 786 (1959).

70. GANSSER, CH. und L. ZECHMEISTER: Über einige *cis*-Formen des Canthaxanthins. Helv. Chim. Acta 40, 1757 (1957).

71. GARBERS, C. F., C. H. EUGSTER und P. KARRER: Carotinoidsynthesen. X. Weitere stereoisomere 1,18-Diphenyl-3,7,12,16-tetramethyl-octadecanonaene. Zugleich ein Beitrag zu L. Pauling's Theorie der sterischen Hinderung bei *cis-trans*-isomeren Polyenen. Helv. Chim. Acta 35, 1850 (1952).

72. — — — Carotinoidsynthesen. XI. Weitere, mit 1,18-Diphenyl-3,7,12,16-tetramethyl-octadecanonaen verwandte Polyene. Helv. Chim. Acta 36, 562 (1953).

73. — — — Über die Vitamin-A-Wirkung des *cis*-β-Carotins C (mit behinderter Doppelbindung). Helv. Chim. Acta 37, 382 (1954).

74. GARBERS, C. F. und P. KARRER: Carotinoidsynthesen. XII. Bis-dehydrolycopine und totalsynthetische *cis*-Lycopine. Helv. Chim. Acta 36, 828 (1953).

75. GILLAM, A. E. and M. S. EL RIDI: Adsorption of Grass and Butter Carotenes on Alumina. Nature (London) 136, 914 (1935).

76. — — The Isomerization of Carotenes by Chromatographic Adsorption. I. *Pseudo*-α-carotene. Biochemic. J. 30, 1735 (1936).

77. GILLAM, A. E., M. S. EL RIDI and S. K. KON: The Isomerization of Carotenes by Chromatographic Adsorption. II. *Neo*-α-carotene. Biochemic. J. 31, 1605 (1937).

78. GOODWIN, T. W.: The Comparative Biochemistry of the Carotenoids. London: Chapman & Hall. 1952.

79. — Studies in Carotenogenesis. 13. The Carotenoids of the Flower Petals of *Calendula officinalis*. Biochemic. J. 58, 90 (1954).

80. — Carotenoids. Annu. Rev. Biochem. 24, 497 (1955).

81. — Studies in Carotenogenesis. 19. A Survey of the Polyenes in a Number of Ripe Berries. Biochemic. J. 62, 346 (1956).

82. GOODWIN, T. W. and M. JAMIKORN: Studies in Carotenogenesis. 11. Carotenoid Synthesis in the Alga *Haematococcus pluvialis*. Biochemic. J. 57, 376 (1954).

83. GOODWIN, T. W. and D. G. LAND: Studies in Carotenogenesis. 20. Carotenoids of Some Species of *Chlorobium*. Biochemic. J. 62, 553 (1956).

*84.* Grangaud, R. et P. Chardenot: Stéréoisomérisation *cis-trans* de l'asta-xanthine. C. R. hebd. Séances Acad. Sci. **242**, 1767 (1956).

*85.* Grangaud, R. et I. Garcia: Chromatographie de partage des caroténoïds. I. Séparation de l'astaxanthine. Bull. soc. chim. biol. (Paris) **34**, 754 (1952).

*86.* Green, A., C. H. Eugster und P. Karrer: Über Inhaltsstoffe der Wurzeln von *Aristolochia cymbitera* Mart. Helv. Chim. Acta **37**, 1717 (1954).

*87.* Greenberg, S. M., C. E. Calbert, J. H. Pinckard, H. J. Deuel, Jr. and L. Zechmeister: Stereochemical Configuration and Provitamin A Activity. IX. A Comparison of All-*trans*-γ-carotene and Pro-γ-carotene with All-*trans*-β-carotene in the Chick. Arch. Biochemistry **24**, 31 (1949).

*87a.* Greenberg, S. M., A. Chatterjee, C. E. Calbert, H. J. Deuel, Jr. and L. Zechmeister: A Comparison of the Provitamin A Activity of β-Carotene and Cryptoxanthin in the Chick. Arch. Biochemistry **25**, 61 (1950).

*88.* Haagen-Smit, A. J., J. H. Pinckard and L. Zechmeister: Contribution to the Structure of Pro-γ-carotene and Prolycopene Obtained from Various Sources. Arch. Biochemistry **26**, 358 (1950).

*89.* Harman, R. A. and H. Eyring: The Structure of Substituted Ethylenes and their Isomerization, Polymerization and "Peroxide Addition" Reactions. J. Chem. Physics **10**, 557 (1942).

*90.* Hasselt, J. F. B. van: Études sur la constitution de la bixine. Rec. trav. chim. Pays-Bas **30**, 1 (1911).

*90a.* Hausser, K. W., R. Kuhn und G. Seitz: Lichtabsorption und Doppelbindung. V. Über die Absorption von Verbindungen mit konjugierten Kohlenstoffdoppelbindungen bei tiefer Temperatur. Z. physik. Chem. B **29**, 391 (1935).

*91.* Haxo, F. (T.): Studies on the Carotenoid Pigments of *Neurospora*. I. Composition of the Pigment. Arch. Biochemistry **20**, 400 (1949).

*92.* Herzig, J. und F. Faltis: Zur Kenntnis des Bixins. Liebigs Ann. Chem. **431**, 40 (1923).

*93.* Holme, D., E. R. H. Jones and M. C. Whiting: The Synthesis of an All-*cis*-tetraene. Chem. and Ind. **1956**, 928.

*94.* Hubbard, R.: Geometrical Isomerization of Vitamin A, Retinene and Retinene Oxime. J. Amer. Chem. Soc. **78**, 4662 (1956).

*95.* Hunter, R. F., A. D. Scott and J. R. Edisbury: Palm Oil Carotenoids. 2. The Isolation of Lipoid Pigments from a West African Plantation Oil and Some Remarks on the Isomerization of Carotenoids. Biochemic. J. **36**, 697 (1942).

*96.* Inhoffen, H. H. und G. von der Bey: Synthesen in der Carotinoid-Reihe. XXVII. Stereoisomere C₁₀-Diole und Dialdehyde. Liebigs Ann. Chem. **583**, 100 (1953).

*97.* Inhoffen, H. H., F. Bohlmann, H.-J. Aldag, S. Bork und G. Leibner: Synthesen in der Carotinoid-Reihe. XXI. Kondensation von Carotinoidketonen und -aldehyden mit Diacetylen; zugleich eine weitere Synthese des β-Carotins. Liebigs Ann. Chem. **573**, 1 (1951).

*98.* Inhoffen, H. H., F. Bohlmann, K. Bartram und H. Pommer: Totalsynthese des β-Carotins. Chem.-Ztg. **74**, 285 (1950).

*99.* Inhoffen, H. H., F. Bohlmann, K. Bartram, G. Rummert und H. Pommer: Synthesen in der Carotinoid-Reihe. XV. Über die Darstellung von *trans* und von 9,9′-mono-*cis*-β-Carotin. Liebigs Ann. Chem. **570**, 54 (1950).

*100.* Inhoffen, H. H., F. Bohlmann und G. Rummert: Synthesen in der Carotinoid-Reihe. X. 7,7′-Bis-desmethyl-β-carotin. Liebigs Ann. Chem. **569**, 226 (1950).

*101.* — — — Synthesen in der Carotinoid-Reihe. XVIII. Über die Stereoisomerisierung des 9,9′-mono-*cis*-β-Carotins. Liebigs Ann. Chem. **571**, 75 (1951).

*102.* INHOFFEN, H. H., O. ISLER, G. VON DER BEY, G. RASPÉ, P. ZELLER und R. AHRENS: Synthesen in der Carotinoid-Reihe. XXVI. Totalsynthese des Crocetin-dimethylesters. Liebigs Ann. Chem. **580**, 7 (1953).

*103.* INHOFFEN, H. H., H.-J. KRAUSE und S. BORK: Synthesen in der Carotinoid-Reihe. XXIX. Liebigs Ann. Chem. **585**, 132 (1953).

*104.* INHOFFEN, H. H. und G. RASPÉ: Synthesen in der Carotinoid-Reihe. XXXI. Synthese des 10,10'-*cis*-Bixinmethylesters. Ein Beitrag zur Stereochemie der *cis-trans*-isomeren Bixine. Liebigs Ann. Chem. **592**, 214 (1955).

*105.* — — Synthesen in der Carotinoid-Reihe. XXXII. Totalsynthese des 3,4,3',4'-Bisdehydro-$\beta$-carotins. Liebigs Ann. Chem. **594**, 165 (1955).

*106.* INHOFFEN, H. H., U. SCHWIETER und G. RASPÉ: Synthesen in der Carotinoid-Reihe. XXX. Totalsynthese des *d,l*-$\alpha$-Carotins. Liebigs Ann. Chem. **588**, 117 (1954).

*107.* INHOFFEN, H. H. und H. SIEMER: Synthetische Chemie der Carotinoide. Fortschr. Chem. organ. Naturstoffe **9**, 1 (1952).

*108.* ISLER, O., L. H. CHOPARD-DIT-JEAN, M. MONTAVON, R. RÜEGG und P. ZELLER: Synthesen in der Carotinoid-Reihe. 12. Mitt. Synthese von 11,11'-Di-*cis*-$\beta$-carotin nach einem neuen Aufbauprinzip. Helv. Chim. Acta **40**, 1256 (1957).

*109.* ISLER, O., W. GUEX, R. RÜEGG, G. RYSER, G. SAUCY, U. SCHWIETER, M. WALTER und A. WINTERSTEIN: Synthesen in der Carotinoid-Reihe. 16. Mitt. Carotinoide vom Typus des Torularhodins. Helv. Chim. Acta **42**, 864 (1959).

*110.* ISLER, O., H. GUTMANN, H. LINDLAR, M. MONTAVON, R. RÜEGG, G. RYSER und P. ZELLER: Synthesen in der Carotinoid-Reihe. 6. Mitt. Synthese von Crocetindialdehyd und Lycopin. Helv. Chim. Acta **39**, 463 (1956).

*111.* ISLER, O., H. GUTMANN, M. MONTAVON, R. RÜEGG, G. RYSER und P. ZELLER: Synthesen in der Carotinoid-Reihe. 10. Mitt. Anwendung der Wittig-Reaktion zur Synthese von Estern des Bixins und Crocetins. Helv. Chim. Acta **40**, 1242 (1957).

*112.* ISLER, O., H. LINDLAR, M. MONTAVON, R. RÜEGG, G. SAUCY und P. ZELLER: Synthesen in der Carotinoid-Reihe. 7. Mitt. Totalsynthese von Zeaxanthin und Physalien. Helv. Chim. Acta **39**, 2041 (1956).

*113.* — — — — — — Synthesen in der Carotinoid-Reihe. 8. Mitt. Totalsynthese von Kryptoxanthin und eine weitere Synthese von Zeaxanthin. Helv. Chim. Acta **40**, 456 (1957).

*114.* ISLER, O., H. LINDLAR, M. MONTAVON, R. RÜEGG und P. ZELLER: Synthesen in der Carotinoid-Reihe. 1. Mitt. Die technische Synthese von $\beta$-Carotin. Helv. Chim. Acta **39**, 249 (1956).

*115.* — — — — — Synthesen in der Carotinoid-Reihe. 3. Mitt. Die Synthese von 3,4;3',4'-Bisdehydro-$\beta$-carotin und 3,4-Monodehydro-$\beta$-carotin. Helv. Chim. Acta **39**, 274 (1956).

*116.* — — — — — Synthesen in der Carotinoid-Reihe. 4. Mitt. Synthese von Isozeaxanthin. Helv. Chim. Acta **39**, 449 (1956).

*117.* ISLER, O., M. MONTAVON, R. RÜEGG, G. SAUCY und P. ZELLER: Synthese hydroxylhaltiger Carotinoide. Verh. naturforsch. Ges. Basel **67**, 379 (1956).

*118.* ISLER, O., M. MONTAVON, R. RÜEGG und P. ZELLER: Synthesen in der Carotinoid-Reihe. 9. Mitt. Neuer Aufbau symmetrischer Carotinoide. Liebigs Ann. Chem. **603**, 129 (1957).

*119.* ISLER, O. and P. ZELLER: Total Synthesis of Carotenoids. Vitamins and Horm. **15**, 31 (1957).

*120.* JACKSON, J. E., R. F. PASCHKE, W. TOLBERG, H. M. BOYD and D. H. WHEELER: Isomers of Linoleic Acid. Infrared and Ultraviolet Properties of Methyl Esters. J. Amer. Oil Chem. Soc. **29**, 229 (1952).

*121.* JENKINS, J. A. and G. MACKINNEY: Inheritance of Carotenoid Difference in the Tomato Hybrid Yellow × Tangerine. Genetics **38**, 107 (1953).

*122.* JENSEN, S. L.: A Note on the Constitutions of Spirilloxanthin and P-481. Acta Chem. Scand. **13**, 381 (1959).

*123.* JENSEN, S. L., G. COHEN-BAZIRE, T. O. M. NAKAYAMA and R. Y. STANIER: The Path of Carotenoid Synthesis in a Photosynthetic Bacterium. Biochim. Biophys. Acta **29**, 477 (1958).

*124.* JOHNSON, R. M. and C. A. BAUMANN: Storage and Distribution of Vitamin A in Rats Fed Certain Isomers of Carotene. Arch. Biochemistry **14**, 361 (1947).

*125.* JOYCE, A. E.: Some Polyenes from *Brassica rutabaga*. Nature (London) **173**, 311 (1954).

*126.* — Carotenoids of *Brassica napus*. J. Sci. Food Agric. **10**, 342 (1959).

*126a.* JURKOWITZ, L.: Photochemical Behaviour of Retinene. Nature (London) **184**, 614 (1959).

*127.* KARMAKAR, G. and L. ZECHMEISTER: On Some Dehydrogenation Products of α-Carotene, β-Carotene and Cryptoxanthin. J. Amer. Chem. Soc. **77**, 55 (1955).

*128.* KARRER, P.: Carotinoid-epoxyde und furanoide Oxyde von Carotinoid-farbstoffen. Fortschr. Chem. organ. Naturstoffe **5**, 1 (1948).

*129.* — Syntheses and Stereochemistry of Carotenoids. J. Sci. Industr. Res. (India), Ser. A **14**, 166 (1955).

*130.* KARRER, P. et C. H. EUGSTER: Synthèse totale du β-carotène. C. R. hebd. Séances Acad. Sci. **230**, 1920 (1950).

*131.* — — Synthese von Carotinoiden. II. Totalsynthese des β-Carotins. I. Helv. Chim. Acta **33**, 1172 (1950).

*132.* — — Carotinoidsynthesen. VIII. Synthese des Dodecapreno-β-carotins. Helv. Chim. Acta **34**, 1805 (1951).

*133.* KARRER, P., C. H. EUGSTER und M. FAUST: Über das Auftreten von Carotinoiden in Pollen und Staubbeuteln verschiedener Pflanzen. Helv. Chim. Acta **33**, 300 (1950).

*134.* KARRER, P., A. HELFENSTEIN, R. WIDMER und TH. B. VAN ITALLIE: Über Bixin. XIII. Mitt. über Pflanzenfarbstoffe. Helv. Chim. Acta **12**, 741 (1929).

*135.* KARRER, P. and E. JUCKER: Carotenoids. Translated and Revised by E. A. BRAUDE. New York: Elsevier Publ. Co. 1950.

*136.* KARRER, P. und J. RUTSCHMANN: Über Violaxanthin, Auroxanthin und andere Pigmente der Blüten von *Viola tricolor*. Helv. Chim. Acta **27**, 1684 (1944).

*137.* — — Carotinoide aus den Früchten von *Cotoneaster occidentalis* und *Pyracantha coccinia*. Helv. Chim. Acta **28**, 1528 (1945).

*138.* KARRER, P., R. SCHWYZER und A. NEUWIRTH: Oxydation von 4-Methyl-*o*-benzochinon zu *cis-cis*-β-Methylmuconsäure-anhydrid. Helv. Chim. Acta **31**, 1210 (1948).

*139.* KARRER, P. und U. SOLMSSEN: Stufenweiser Abbau des labilen und stabilen Bixins. Zur Stereochemie der Carotinoide. Helv. Chim. Acta **20**, 1396 (1937).

*140.* KARRER, P. und T. TAKAHASHI: Pflanzenfarbstoffe. XLVII. Über die Isomerieverhältnisse beim Bixin. Bemerkungen zu den Theorien über die Bildung von Carotinoidpigmenten in der Pflanze. Helv. Chim. Acta **16**, 287 (1933).

*141.* KARRER, P. und E. WÜRGLER: Absorptionsspektren einiger Carotinoide. Helv. Chim. Acta **26**, 116 (1943).

*142.* KEMMERER, A. R. and G. S. FRAPS: Constituents of Carotene Extracts of Plants. Ind. Eng. Chem., Analyt. Ed. **15**, 714 (1943).

*143.* KEMMERER, A. R. and G. S. FRAPS: Relative Value of Carotene in Vegetables for Growth of the White Rat. Arch. Biochemistry **8**, 197 (1945).

*144.* — — The Vitamin A Activity of neo-β-Carotene U and its Steric Rearrangement in the Digestive Tract of Rats. J. Biol. Chem. **161**, 305 (1945).

*145.* KHARASCH, M. S., J. V. MANSFIELD and F. R. MAYO: *cis-trans* Isomerization by Bromine Atoms. J. Amer. Chem. Soc. **59**, 1155 (1937).

*146.* KOE, B. K. and L. ZECHMEISTER: In Vitro Conversion of Phytofluene and Phytoene into Carotenoid Pigments. Arch. Biochem. Biophys. **41**, 236 (1952).

*147.* — — Preparation and Spectral Characteristics of all-*trans*- and a *cis*-Phytofluene. Arch. Biochem. Biophys. **46**, 100 (1953).

*148.* KUHN, H.: The Electron Gas Theory of the Color of Natural and Artificial Dyes: Problems and Principles. Fortschr. Chem. organ. Naturstoffe **16**, 169 (1958).

*149.* — The Electron Gas Theory of the Color of Natural and Artificial Dyes: Applications and Extensions. Fortschr. Chem. organ. Naturstoffe **17**, 404 (1959).

*150.* — Neuere Untersuchungen über das Elektronengasmodell organischer Farbstoffe. Angew. Chem. **71**, 93 (1959).

*151.* KUHN, R.: *Cis-trans*-Umlagerungen der Äthylenkörper. In: K. FREUDENBERG, Stereochemie, S. 913. Leipzig und Wien: F. Deuticke. 1933.

*152.* KUHN, R. und P. J. DRUMM: Umkehrbare Hydrierung und Dehydrierung bei Polyenen. Ber. dtsch. chem. Ges. **65**, 1458 (1932).

*153.* KUHN, R., P. J. DRUMM, M. HOFFER und E. F. MÖLLER: Farbreaktionen und Autoxydation von Hydropolyen-carbonsäure-estern. Ber. dtsch. chem. Ges. **65**, 1785 (1932).

*154.* KUHN, R., H. H. INHOFFEN, H. A. STAAB und W. OTTING: Vergleich des *cis*-Crocetin-dimethylesters aus Safran mit 8.8′-*cis*-Crocetin-dimethylester. Ber. dtsch. chem. Ges. **86**, 965 (1953).

*155.* KUHN, R. und E. LEDERER: Iso-carotin (Über das Vitamin des Wachstums, III. Mitt.). Ber. dtsch. chem. Ges. **65**, 637 (1932).

*156.* KUHN, R. und A. WINTERSTEIN: Über konjugierte Doppelbindungen. I. Synthese von Diphenyl-poly-enen. Helv. Chim. Acta **11**, 87 (1928).

*157.* — — Die Dihydroverbindung der isomeren Bixine und die Elektronen-Konfiguration der Polyene (Über konjugierte Doppelbindungen, XXIII. Mitt.). Ber. dtsch. chem. Ges. **65**, 646 (1932).

*158.* — — Über einen licht-empfindlichen Carotin-Farbstoff aus Safran. Ber. dtsch. chem. Ges. **66**, 209 (1933).

*159.* — — Über die Konstitution des Pikro-crocins und seine Beziehung zu den Carotin-Farbstoffen des Safrans. Ber. dtsch. chem. Ges. **67**, 344 (1934).

*160.* KUHN, R., A. WINTERSTEIN und E. LEDERER: Zur Kenntnis der Xanthophylle. Z. physiol. Chem. (Hoppe-Seyler) **197**, 141 (1931).

*161.* KUHN, W. und R. LANDOLT: Über den Photodichroismus fester Carotinoid-Lösungen. II. Helv. Chim. Acta **34**, 1929 (1951).

*162.* LANDOLT, R. und W. KUHN: Über den Photodichroismus fester Carotinoid-Lösungen. I. Helv. Chim. Acta **34**, 1900 (1951).

*163.* LEDERER, E. and M. LEDERER: Chromatography. A Review of Principles and Applications. 2nd ed. New York: Elsevier Publ. Co. 1957.

*164.* LEROSEN, A. L.: Continuous Washing Apparatus for Solutions in Organic Solvents. Ind. Eng. Chem., Analyt. Ed. **14**, 165 (1942).

*165.* — A Method for Standardization of Chromatographic Analysis. J. Amer. Chem. Soc. **64**, 1905 (1942).

*166.* LEROSEN, A. L. and L. ZECHMEISTER: Prolycopene. J. Amer. Chem. Soc. **64**, 1075 (1942).

*167.* LeRosen, A. L. and L. Zechmeister: The Carotenoid Pigments of the Fruit of *Celastrus scandens* L. Arch. Biochemistry **1**, 17 (1942).

*168.* Lewis, G. N. and M. Calvin: The Color of Organic Substances. Chem. Revs. **25**, 273 (1939).

*169.* Lindlar, H.: Ein neuer Katalysator für selective Hydrierungen. Helv. Chim. Acta **35**, 446 (1952).

*170.* Linner, E., C. H. Eugster und P. Karrer: Carotinoidsynthesen. XVIII. Synthese des 1,18-Di-$\beta$-naphtyl-3,7,12,16-tetramethyl-octadeca-nonaens. Helv. Chim. Acta **38**, 1869 (1955).

*170a.* Loeb, J. N., P. K. Brown and G. Wald: *Cis trans* Isomerism and Steric Hindrance. Nature (London) **184**, 617 (1959). With a Theoretical Discussion by G. Wald on p. 620.

*171.* Lunde, K.: Note on the Infrared Spectra of Pro-$\gamma$-carotene and Neo-$\gamma$-carotene P. Acta Chem. Scand. **13**, 2154 (1959).

*172.* Lunde, K. and L. Zechmeister: A Study of the Infrared Spectra of Some Stereoisomeric Diphenylpolyenes. Acta Chem. Scand. **8**, 1421 (1954).

*173.* — — *cis-trans* Isomeric 1,6-Diphenylhexatrienes. J. Amer. Chem. Soc. **76**, 2308 (1954).

*174.* — — Infrared Spectra and *cis-trans* Configurations of Some Carotenoid Pigments. J. Amer. Chem. Soc. **77**, 1647 (1955).

*175.* Mackinney, G.: Carotenoids. Annu. Rev. Biochem. **21**, 473 (1952).

*176.* Mackinney, G. and J. A. Jenkins: Inheritance of Carotenoid Differences in *Lycopersicon esculentum* Strains. Proc. Nat. Acad. Sci. (USA) **35**, 284 (1949).

*177.* — — Carotenoid Differences in Tomatoes. Proc. Nat. Acad. Sci. (USA) **38**, 48 (1952).

*178.* Magoon, E. F. and L. Zechmeister: On the *cis* Forms of Some Biphenylene Derivatives of Butadiene and Hexatriene. J. Amer. Chem. Soc. **77**, 5642 (1955).

*179.* — — On a *cis*-Neurosporene *ex Pyracantha* and the *in vitro* Stereoisomerization of Neurosporene. Arch. Biochem. Biophys. **68**, 263 (1957).

*180.* — — Stepwise Stereoisomerization of Prolycopene, a Poly*cis* Carotenoid, to all-*trans*-Lycopene. Arch. Biochem. Biophys. **69**, 535 (1957).

*181.* Manunta, C.: The Manner of Action of the Genes Responsible for the Color of the Fruit in Cultivated Strains of Tomato *(Lycopersicon esculentum)* and in Wild Species of the Genus. Genet. agrar. (Pavia) **3**, 38 pp. (1951) [Chem. Abstr. **48**, 9476 (1954)].

*182.* Martínez García, F.: The Carotenoids of Red Pepper. Rev. fac. ciênc., Univ. Coimbra **20**, 21 (1951) [Chem. Abstr. **46**, 8287 (1952)].

*183.* — The Presence of Stereoisomerism of Capsorubin in the Pimiento. Farma-cognosia (Madrid) **12**, 169 (1952) [Chem. Abstr. **48**, 5443 (1954)].

*184.* McConnell, H.: Catalysis of *Cis-Trans* Isomerization by Paramagnetic Substances. J. Chem. Physics **20**, 1043 (1952).

*185.* McNicholas, H. J.: The Visible and Ultraviolet Absorption Spectra of Carotene and Xanthophyll and the Changes Accompanying Oxydation. Bureau Standards J. Res. **7**, 171 (1931).

*186.* Meunier, P., J. Jouanneteau et G. Zwingelstein: Sur la coupure oxydante du $\beta$-carotène en rétinène (axérophtal) par MnO$_2$. C. R. hebd. Séances Acad. Sci. **231**, 1170 (1950).

*187.* — — — Sur l'isomérisation *cis trans* des caroténoides en C$_{40}$ provoquée par adsorption sur des oxydes métalliques. Bull. soc. chim. biol. (Paris) **33**, 1228 (1951).

*188.* MILAS, N. A., P. DAVIS, I. BELIČ and D. A. FLEŠ: Synthesis of β-Carotene. J. Amer. Chem. Soc. **72**, 4844 (1950).

*189.* MILLER, E. S.: A Precise Method, with Detailed Calibration for the Determination of Absorption Coefficients; the Quantitative Measurement of the Visible and Ultraviolet Absorption Spectra of α-Carotene, β-Carotene, and Lycopene. Plant Physiol. **12**, 667 (1937).

*190.* — Quantitative Biological Spectroscopy. Absorption Spectra, Vol. I. Minneapolis: Burgess Publ. Co. 1940.

*191.* MOORE, T.: Vitamin A and Carotene. I. The Association of Vitamin A Activity with Carotene in the Carrot Root. Biochemic. J. **23**, 803 (1929).

*192.* MULLIKEN, R. S.: Intensities of Electronic Transitions in Molecular Spectra. VII. Conjugated Polyenes and Carotenoids. J. Chem. Physics **7**, 364 (1939).

*193.* — Structure and Ultraviolet Spectra of Ethylene, Butadiene and their Alkyl Derivatives. Rev. Modern Physics **14**, 265 (1942).

*194.* — Quantum-mechanical Methods and the Electronic Spectra and Structure of Molecules. Chem. Revs. **41**, 201 (1947).

*195.* NASH, H. A. and F. P. ZSCHEILE: The *cis-trans* Isomerization of α-Carotene Isomers. Arch. Biochemistry **5**, 77 (1944).

*196.* NAYLER, P. and M. C. WHITING: Researches on Polyenes. II. The Synthesis of Cosmene. J. Chem. Soc. (London) **1954**, 4006.

*197.* OROSHNIK, W., G. KARMAS and A. D. MEBANE: Synthesis of Polyenes. I. *Retro*vitamin A Methyl Ether. Spectral Relationships between the β-Ionylidene and *Retro*ionylidene Series. J. Amer. Chem. Soc. **74**, 295 (1952).

*197a.* OROSHNIK, W. and A. D. MEBANE: Synthesis of Polyenes. VI. Isoprenoid Polyenes Containing Sterically Hindered *cis* Configurations. J. Amer. Chem. Soc. **76**, 5719 (1954).

*198.* PAAL, C. und W. HARTMANN: Über katalytische Wirkungen kolloidaler Metalle der Platingruppe. VIII. Die stufenweise Reduktion der Phenyl-propiolsäure. Ber. dtsch. chem. Ges. **42**, 3930 (1909).

*199.* PASCHKE, R. F., W. TOLBERG and D. H. WHEELER: *Cis,Trans* Isomerism of the Eleostearate Isomers. J. Amer. Oil Chem. Soc. **30**, 97 (1953).

*200.* PAUL, M. F., V. R. ELLS and H. E. PAUL: The Effect of Mineral Oil on Food Utilization. II. Changes of β-Carotene in Mineral Oil. Amer. J. Digestive Diseases **18**, 278 (1951).

*201.* PAULING, L.: Recent Work on the Configuration and Electronic Structure of Molecules; with some Applications to Natural Products. Fortschr. Chem. organ. Naturstoffe **3**, 203 (1939).

*202.* — A Theory of the Color of Dyes. Proc. Nat. Acad. Sci. (USA) **25**, 577 (1939).

*203.* — Zur *cis-trans*-Isomerisierung von Carotinoiden. Helv. Chim. Acta **32**, 2241 (1949).

*204.* PETRACEK, F. J. and L. ZECHMEISTER: Stereoisomeric Phytofluenes. J. Amer. Chem. Soc. **74**, 184 (1952).

*205.* — — Reaction of β-Carotene with N-Bromosuccinimide: The Formation and Conversions of Some Polyene Ketones. J. Amer. Chem. Soc. **78**, 1427 (1956).

*205a.* PINCKARD, J. H.: Thesis, California Institute of Technology, Pasadena, 1949.

*206.* PINCKARD, J. H., J. S. KITTREDGE, D. L. FOX, F. T. HAXO and L. ZECHMEISTER: Pigments from a Marine "Red Water" Population of the Dinoflagellate *Prorocentrum micans*. Arch. Biochem. Biophys. **44**, 189 (1953).

*207.* PLATT, J. R.: Electronic Structure and Excitation of Polyenes and Porphyrins. In: A. HOLLAENDER, Radiation Biology, Vol. III, p. 71. New York: McGraw-Hill. 1956.

208. Polgár, A., C. B. van Niel and L. Zechmeister: Studies on the Pigments of the Purple Bacteria. II. A Spectroscopic and Stereochemical Investigation of Spirilloxanthin. Arch. Biochemistry 5, 243 (1944).

209. Polgár, A. and L. Zechmeister: Isomerization of β-Carotene. Isolation of a Stereoisomer with Increased Adsorption Affinity. J. Amer. Chem. Soc. 64, 1856 (1942).

210. — — Action of Cold Concentrated Hydriodic Acid on Carotenes: Structure and cis-trans Isomerization of Some Reaction Products. J. Amer. Chem. Soc. 65, 1528 (1943).

211. — — A Spectroscopic Study in the Stereoisomeric Capsanthin Set. Cis-peak Effect and Configuration. J. Amer. Chem. Soc. 66, 186 (1944).

212. Porter, J. W.: Relationships Between Physical Properties and Structure of Carotenes and Colorless Polyenes. Arch. Biochem. Biophys. 45, 291 (1953).

213. Porter, J. W. and R. E. Lincoln: I. Lycopersicon Selections Containing a High Content of Carotenes and Colorless Polyenes. II. The Mechanism of Carotene Biosynthesis. Arch. Biochemistry 27, 390 (1950).

214. Porter, J. W. and F. P. Zscheile: Carotenes of Lycopersicon Species and Strains. Arch. Biochemistry 10, 537 (1946).

215. Poutet, J. J.: Procédé pour reconnaître la falsification de l'huile d'olive par celle de graines. Ann. chim. phys. [2] 12, 58 (1819).

216. Price, C. C. and G. Berti: The Polymerization of Stilbene in Boron Fluoride Etherate. J. Amer. Chem. Soc. 76, 1219 (1954).

217. Price, C. C. and M. Meister: cis-trans-Isomerization with Boron Fluoride. J. Amer. Chem. Soc. 61, 1595 (1939).

218. Quackenbush, F. W., H. Steenbock and W. H. Peterson: The Effect of Acids on Carotenoids. J. Amer. Chem. Soc. 60, 2937 (1938).

219. Rabourn, W. J. and F. W. Quackenbush: The Occurrence of Phytoene in Various Plant Materials. Arch. Biochem. Biophys. 44, 159 (1953).

220. — — The Structure of Phytoene. Arch. Biochem. Biophys. 61, 111 (1956).

221. Rabourn, W. J., F. W. Quackenbush and J. W. Porter: Isolation and Properties of Phytoene. Arch. Biochem. Biophys. 48, 267 (1954).

222. Rasmussen, R. S.: Infrared Spectroscopy in Structure Determination and its Application to Penicillin. Fortschr. Chem. organ. Naturstoffe 5, 331 (1948).

223. Rasmussen, R. S. and R. R. Brattain: Infra-red Absorption Spectra of the $C_2$ to $C_4$ Mono-olefins and of 2-Methyl-2-butene. J. Chem. Physics 15, 120 (1947).

224. — — Infra-red Absorption Spectra of Some $C_4$ and $C_5$ Dienes. J. Chem. Physics 15, 131 (1947).

225. — — Infrared Spectra of Some Carboxylic Acid Derivatives. J. Amer. Chem. Soc. 71, 1073 (1949).

226. Rosenberg, B.: Photoconduction and Cis-Trans Isomerism in β-Carotene. J. Chem. Physics 31, 238 (1959).

227. Rüegg, R., H. Lindlar, M. Montavon, G. Saucy, S. F. Schaeren, U. Schwieter und O. Isler: Synthesen in der Carotinoid-Reihe. 14. Mitt. Synthese von β-Apo-12'-carotinal ($C_{25}$). Helv. Chim. Acta 42, 847 (1959).

228. Rüegg, R., M. Montavon, G. Ryser, G. Saucy, U. Schwieter und O. Isler: Synthesen in der Carotinoid-Reihe. 15. Mitt. Synthesen in der β-Carotinal- und β-Carotinol-Reihe. Helv. Chim. Acta 42, 854 (1959).

229. Saperstein, S. and M. P. Starr: Association of Carotenoid Pigments with Protein Components in non-photosynthetic Bacteria. Biochim. Biophys. Acta 16, 482 (1955).

*230.* SAVINOV, B. G. and A. A. MIKHAÏLOVNINA: Neo-$\beta$-carotene as the Product of Primary Stereoisomeric Transformation of $\beta$-Carotene on Heating. Dokl. Akad. Nauk USSR **88**, 887 (1953) [Chem. Abstr. **48**, 3311 (1954)].

*231.* SCHROEDER, W. A.: Formation of Pro-carotenoids in "Monkey Flowers" under Some Conditions. J. Amer. Chem. Soc. **64**, 2510 (1942).

*232.* SEITZ, K., Hs. H. GÜNTHARD und O. JEGER: Veilchenriechstoffe. 37. Mitt. Über die Trennung von $\alpha$- und $\beta$-Jonon durch fraktionierte Destillation. Helv. Chim. Acta **33**, 2196 (1950).

*233.* SHEPPARD, N. and D. M. SIMPSON: The Infra-red and Raman Spectra of Hydrocarbons. I. Acetylenes and Olefins. Quart. Revs. Chem. Soc. (London) **6**, 1 (1952).

*234.* SHEPPARD, N. and G. B. B. M. SUTHERLAND: Vibration Spectra of Hydrocarbon Molecules. I. Frequencies due to Deformation Vibrations of Hydrogen Atoms Attached to a Double Bond. Proc. Roy. Soc. (London), Ser. A **196**, 195 (1949).

*235.* SIMPSON, W. T.: Resonance Force Theory of Carotenoid Pigments. J. Amer. Chem. Soc. **77**, 6164 (1955).

*236.* SLY, W. G.: A Preliminary Report on the Crystal-structure Determination of 15,15'-Dehydro-$\beta$-carotene. Acta Crystallogr. **8**, 115 (1955).

*237.* STARR, M. P. and S. SAPERSTEIN: Thiamine and the Carotenoid Pigments of *Corynebacterium poinsettiae*. Arch. Biochem. Biophys. **43**, 157 (1953).

*238.* STEENBOCK, H., M. T. SELL, E. M. NELSON and M. V. BUELL: The Fat-soluble Vitamine. J. Biol. Chem. **46**, XXXII (1921).

*239.* STITT, F., E. M. BICKOFF, G. F. BAILEY, C. R. THOMPSON and S. FRIEDLANDER: Spectrophotometric Determination of $\beta$-Carotene Stereoisomers in Alfalfa. J. Assoc. Official Agric. Chem. **34**, 460 (1951).

*240.* STRAIN, H. H.: Leaf Xanthophylls. Carnegie Inst. Washington Publ. No. 490. Washington. 1938.

*241.* — Carotene. XI. Isolation and Detection of $\alpha$-Carotene, and the Carotenes of Carrot Roots and of Butter. J. Biol. Chem. **127**, 191 (1939).

*242.* — Isomerizations of Polyene Acids and Carotenoids. Preparation of $\beta$-Eleostearic and $\beta$-Licanic Acids. J. Amer. Chem. Soc. **63**, 3448 (1941).

*243.* — Chromatographic Adsorption Analysis. New York: Interscience Publ. 1942.

*244.* — Problems in Chromatography and in Colloid Chemistry Illustrated by Leaf Pigments. J. Physic. Chem. **46**, 1151 (1942).

*245.* — Chloroplast Pigments. Annu. Rev. Biochem. **13**, 591 (1944).

*246.* — Molecular Structure and Adsorption Sequences of Carotenoid Pigments. J. Amer. Chem. Soc. **70**, 588 (1948).

*247.* — Leaf Xanthophylls: The Action of Acids on Violaxanthin, Violeoxanthin, Taraxanthin and Tareoxanthin. Arch. Biochem. Biophys. **48**, 458 (1954).

*248.* STRAIN, H. H. and W. M. MANNING: The Occurrence and Interconversion of Various Fucoxanthins. J. Amer. Chem. Soc. **64**, 1235 (1942).

*249.* — — A Unique Polyene Pigment of the Marine Diatom *Navicula Torquatum*. J. Amer. Chem. Soc. **65**, 2258 (1943).

*250.* STRAIN, H. H., W. M. MANNING and G. HARDIN: Xanthophylls and Carotenes of Diatoms, Brown Algae, Dinoflagellates, and Sea-anemones. Biol. Bull. **86**, 169 (1944).

*251.* SURMATIS, J. D., J. MARICQ and A. OFNER: Synthesis of Carotene Homologs. J. Organ. Chem. (USA) **23**, 157 (1958).

*252.* SUZUKI, N. and K. TSUKIDA: Carotenoids of the Flowers of *Osmanthus fragrans*. Chem. pharm. Bull. (Japan) **7**, 133 (1959).

*253.* SZASZ, G. J. and N. SHEPPARD: An Infra-red Spectroscopic Study of the Configuration of Some Chlorinated Butadienes. Trans. Faraday Soc. **49**, 358 (1953).

*254.* TAPPI, G. und P. KARRER: Über die Carotinoide aus den Staubbeuteln von *Lilium candidum. Cis*-Antheraxanthin. Helv. Chim. Acta **32**, 50 (1949).

*255.* TAPPI, G. e E. MENZIANI: Sui pigmenti del polline di *Lilium mantchiuricum*. Atti Soc. Nat. Mat. Modena **85**, 28 (1954).

*256.* THOMPSON, C. R., E. M. BICKOFF and W. D. MACLAY: Formation of Stereoisomers of $\beta$-Carotene in Alfalfa. Ind. Eng. Chem. **43**, 126 (1951).

*257.* TOMES, M. L., F. W. QUACKENBUSH, O. E. NELSON, Jr. and B. NORTH: The Inheritance of Carotenoid Pigment Systems in the Tomato. Genetics **38**, 117 (1953).

*258.* TROMBLY, H. H. and J. W. PORTER: Additional Carotenes and a Colorless Polyene of *Lycopersicon* Species and Strains. Arch. Biochem. Biophys. **43**, 443 (1953).

*259.* TSUKIDA, K. and L. ZECHMEISTER: The Stereoisomerization of $\beta$-Carotene Epoxides and the Simultaneous Formation of Furanoid Oxides. Arch. Biochem. Biophys. **74**, 408 (1958).

*260.* URUSHIBARA, Y. and O. SIMAMURA: The Effect of Oxygen and Reduced Nickel on the Catalytic Action of Hydrogen Bromide in the Isomerization of Isostilbene into Stilbene. Bull. Chem. Soc. Japan **12**, 507 (1937).

*261.* VALLENTYNE, J. R.: Carotenoids in a 20,000-Year-Old Sediment from Searles Lake, California. Arch. Biochem. Biophys. **70**, 29 (1957).

*261 a.* WAGNER, R., J. FINE, J. W. SIMMONS and J. H. GOLDSTEIN: Microwave Spectrum, Structure, and Dipole Moment of *s-trans* Acrolein. J. Chem. Physics **26**, 634 (1957).

*262.* WAIGHT, E. S. and R. L. ERSKINE: Absorption Spectra of Conjugated Carbonyl Compounds. In: G. W. GRAY, Steric Effects in Conjugated Systems (Chemical Society Symposium), p. 73. London: Butterworths. 1958.

*263.* WALLACE, V. and J. W. PORTER: Phytofluene. Arch. Biochem. Biophys. **36**, 468 (1952).

*264.* WALLCAVE, L.: Thesis, California Institute of Technology, Pasadena, 1953.

*265.* WALLCAVE, L., J. LEEMANN and L. ZECHMEISTER: Action of Boron Trifluoride Etherate on $\beta$-Carotene. Proc. Nat. Acad. Sci. (USA) **39**, 604 (1953).

*266.* WALLCAVE, L. and L. ZECHMEISTER: Conversion of Dehydro-$\beta$-carotene, via its Boron Trifluoride Complex, into an Isomer of Cryptoxanthin. J. Amer. Chem. Soc. **75**, 4495 (1953).

*267.* WARREN, C. K. and B. C. L. WEEDON: Carotenoids and Related Compounds. VI. Some Conjugated Polyene Diketones, and their Comparison with Capsorubin. J. Chem. Soc. (London) **1958**, 3972.

*268.* WILLSTÄTTER, R. und A. STOLL: Untersuchungen über Chlorophyll. Methoden und Ergebnisse. Berlin: Julius Springer. 1913.

*268 a.* WINTERSTEIN, A. und Mitarb. Helv. Chim. Acta (1960) (im Druck).

*269.* WISEMAN, H. G., S. S. STONE, H. L. SAVAGE and L. A. MOORE: Action of Celites on Carotene and Lutein. Analyt. Chemistry **24**, 681 (1952).

*270.* WITTIG, G. und W. WIEMER: Zur Valenztautomerie ungesättigter Systeme. Liebigs Ann. Chem. **483**, 144 (1930).

*271.* WÜRSCH, J. und U. SCHWIETER: Synthese von $\beta$-Carotin-[6,6'-[14]C]. Helv. Chim. Acta **39**, 1067 (1956).

*272.* WYMAN, G. M.: The *cis-trans* Isomerization of Conjugated Compounds. Chem. Revs. **55**, 625 (1955).

273. YAMAGUCHI, M.: Chemical Constitution of Renieratene. Bull. Chem. Soc. Japan **30**, 979 (1957).

274. — Chemical Constitution of Isorenieratene. Bull. Chem. Soc. Japan **31**, 51 (1958).

275. — Renieratene, a New Carotenoid Containing Benzene Rings, Isolated from a Sea Sponge. Bull. Chem. Soc. Japan **31**, 739 (1958).

276. ZALOKAR, M.: Isolation of an Acidic Pigment in *Neurospora*. Arch. Biochem. Biophys. **70**, 568 (1957).

277. ZECHMEISTER, L.: Carotinoide. Ein biochemischer Bericht über pflanzliche und tierische Polyenfarbstoffe. Berlin: Julius Springer. 1934.

278. — *cis-trans* Isomerization and Stereochemistry of Carotenoids and Diphenyl-polyenes. Chem. Revs. **34**, 267 (1944).

279. — Stereochemistry and Chromatography. Ann. New York Acad. Sci. **49**, 220 (1948).

280. — Adsorpt on and Some Constitutional and Steric Properties. Discuss. Faraday Soc. **7**, 54 (1949).

281. — Stereoisomeric Provitamins A. Vitamins and Horm. **7**, 57 (1949).

282. — Les provitamines A stéréoïsomériques. Bull. Soc. chim. biol. (Paris) **31**, 956 (1949).

283. — Progress in Chromatography 1938–1947. London: Chapman & Hall; New York: J. Wiley. 1950.

284. — Some Stereochemical Aspects of Polyenes. Experientia **10**, 1 (1954).

285. — Some in vitro Conversions of Naturally Occurring Carotenoids. Fortschr. Chem. organ. Naturstoffe **15**, 31 (1958).

286. — Bibliography of Papers Published by L. ZECHMEISTER and Co-authors in the Fields of Chemistry and Biochemistry 1913–1958. Wien: Springer-Verlag. 1958.

287. ZECHMEISTER, L. und L. v. CHOLNOKY: Die chromatographische Adsorptions-methode. Wien: Julius Springer. 1937.

288. — — Untersuchungen über den Paprika-Farbstoff. X. Citraurin aus Capsanthin. Liebigs Ann. Chem. **530**, 291 (1937).

289. — — Untersuchungen über den Paprika-Farbstoff. XI. Isomerisierungs-Erscheinungen. Liebigs Ann. Chem. **543**, 248 (1940).

290. ZECHMEISTER, L., L. v. CHOLNOKY und A. POLGÁR: Über die Isomerisierung des Zeaxanthins und Physaliens. Ber. dtsch. chem. Ges. **72**, 1678 (1939).

291. — — — Zur Isomerisierung von Xanthophyllen (Nachtrag). Ber. dtsch. chem. Ges. **72**, 2039 (1939).

292. ZECHMEISTER, L., H. J. DEUEL, Jr., H. H. INHOFFEN, J. LEEMANN, S. M. GREENBERG and J. GANGULY: Stereochemical Configuration and Provitamin A Activity. X. A Comparison of Synthetic 15,15'-Mono*cis*-β-carotene (Central Mono*cis*-β-carotene) with All-*trans*-β-carotene in the Rat and Chick. Arch. Biochem. Biophys. **36**, 80 (1952).

293. ZECHMEISTER, L. and R. B. ESCUE: New Stereoisomers of Methylbixin. Science (Washington) **96**, 229 (1942).

294. — — Isolation of Prolycopene and Pro-γ-carotene from *Evonymus fortunei*. J. Biol. Chem. **144**, 321 (1942).

295. — — A Stereochemical Study of Methylbixin. J. Amer. Chem. Soc. **66**, 322 (1944).

296. ZECHMEISTER, L. and G. KARMAKAR: The Occurrence of Phytofluene in Green Plant Organs. Arch. Biochem. Biophys. **47**, 160 (1953).

297. ZECHMEISTER, L. and B. K. KOE: Stepwise Dehydrogenation of the Colorless Polyenes Phytoene and Phytofluene with N-Bromosuccinimide to Carotenoid Pigments. J. Amer. Chem. Soc. **76**, 2923 (1954).

*298.* ZECHMEISTER, L. and R. M. LEMMON: Contribution to the Stereochemistry of Cryptoxanthin and Zeaxanthin. J. Amer. Chem. Soc. **66**, 317 (1944).

*299.* ZECHMEISTER, L. and A. L. LeROSEN: Stereoisomeric Diphenyloctatetraenes. J. Amer. Chem. Soc. **64**, 2755 (1942).

*300.* ZECHMEISTER, L., A. L. LeROSEN, W. A. SCHROEDER, A. POLGÁR and L. PAULING: Spectral Characteristics and Configuration of Some Stereoisomeric Carotenoids Including Prolycopene and Pro-γ-carotene. J. Amer. Chem. Soc. **65**, 1940 (1943).

*301.* ZECHMEISTER, L., A. L. LeROSEN, F. W. WENT and L. PAULING: Prolycopene, a Naturally Occurring Stereoisomer of Lycopene. Proc. Nat. Acad. Sci. (USA) **27**, 468 (1941).

*302.* ZECHMEISTER, L. and E. F. MAGOON: Spectral Maxima of Stereoisomeric Polyenes. Chem. and Ind. **1957**, 431.

*303.* ZECHMEISTER, L. and F. J. PETRACEK: Absence of Detectable Poly-*cis* Forms from Heat-isomerized Lycopene Solutions. J. Amer. Chem. Soc. **74**, 282 (1952).

*304.* — — On the Structure of the Deoxyluteins. Arch. Biochem. Biophys. **61**, 243 (1956).

*305.* ZECHMEISTER, L. and J. H. PINCKARD: Some Poly*cis*-lycopenes Occurring in the Fruit of *Pyracantha*. J. Amer. Chem. Soc. **69**, 1930 (1947).

*306.* — — Stereoisomeric Diphenyloctatetraenes. II. J. ·Amer. Chem. Soc. **76**, 4144 (1954).

*307.* ZECHMEISTER, L., J. H. PINCKARD, S. M. GREENBERG, E. STRAUB, T. FUKUI and H. J. DEUEL, Jr.: Stereochemical Configuration and Provitamin A Activity. VIII. Pro-γ-carotene (a Poly-*cis* Compound) and its All-*trans* Isomer in the Rat. Arch. Biochemistry **23**, 242 (1949).

*308.* ZECHMEISTER, L. and A. POLGÁR: *cis-trans* Isomerization and Spectral Characteristics of Carotenoids and Some Related Compounds. J. Amer. Chem. Soc. **65**, 1522 (1943).

*309.* — — *cis-trans* Isomerization and *cis*-Peak Effect in the α-Carotene Set and in Some Other Stereoisomeric Sets. J. Amer. Chem. Soc. **66**, 137 (1944).

*310.* — — Contributions to the Stereochemistry of γ-Carotene. J. Amer. Chem. Soc. **67**, 108 (1945).

*311.* ZECHMEISTER, L. and A. SANDOVAL: The Occurrence and Estimation of Phytofluene in Plants. Arch. Biochemistry **8**, 425 (1945).

*312.* ZECHMEISTER, L. and W. A. SCHROEDER: On the Occurrence of Stereoisomeric Carotenoids in Nature. Science (Washington) **94**, 609 (1941).

*313.* — — The Pigment of *Mimulus longiflorus* and the Isolation of its γ-Carotene Component. Arch. Biochemistry **1**, 231 (1942).

*314.* — — The Fruit of *Pyracantha angustifolia*: a Practical Source of Pro-γ-carotene and Prolycopene. J. Biol. Chem. **144**, 315 (1942).

*315.* — — Pro-γ-carotene. J. Amer. Chem. Soc. **64**, 1173 (1942).

*316.* — — *cis-trans* Isomerization and Spectral Characteristics of Gazaniaxanthin. Further Evidence of its Structure. J. Amer. Chem. Soc. **65**, 1535 (1943).

*317.* ZECHMEISTER, L. und P. TUZSON: Über das Polyen-Pigment der Orange, II. Mitt. Citraurin. Ber. dtsch. chem. Ges. **70**, 1966 (1937).

*318.* — — Spontaneous Isomerization of Lycopene. Nature (London) **141**, 249 (1938).

*319.* — — Isomerization of Carotenoids. Biochemic. J. **32**, 1305 (1938).

*320.* — — Umkehrbare Isomerisierung von Carotinoiden durch Jod-Katalyse. Ber. dtsch. chem. Ges. **72**, 1340 (1939).

*321.* ZECHMEISTER, L. and L. WALLCAVE: A Study of Some *cis-trans* Isomeric Dehydro-β-carotenes. J. Amer. Chem. Soc. **75**, 5341 (1953).

*322.* ZECHMEISTER, L. and F. W. WENT: Some Stereochemical Aspects in Genetics. Nature (London) **162**, 847 (1948).

*323.* ZELLER, P., F. BADER, H. LINDLAR, M. MONTAVON, P. MÜLLER, R. RÜEGG, G. RYSER, G. SAUCY, S. F. SCHAEREN, U. SCHWIETER, K. STRICKER, R. TAMM, P. ZÜRCHER und O. ISLER: Synthesen in der Carotinoid-Reihe. 13. Mitt. Synthese von Canthaxanthin. Helv. Chim. Acta **42**, 841 (1959).

*324.* ZIEGLER, H. H. v., C. H. EUGSTER und P. KARRER: Carotinoidsynthesen. XVI. Stereoisomere 1,3,7,12,16,18-Hexaphenyl-octadeca-nonaene. Helv. Chim. Acta **38**, 613 (1955).

*325.* ZSCHEILE, F. P. and B. W. BEADLE: Determination of $\beta$-Carotene and Neo-$\beta$-carotene with the Visual Spectrophotometer. Ind. Eng. Chem., Analyt. Ed. **14**, 633 (1942).

*326.* ZSCHEILE, F. P., R. H. HARPER and H. A. NASH: Photochemical Reaction of Iodine with Carotenoids. Arch. Biochemistry **5**, 211 (1944).

*327.* ZSCHEILE, F. P. and J. W. PORTER: Analytical Methods for Carotenes of *Lycopersicon* Species and Strains. Ind. Eng. Chem., Analyt. Ed. **19**, 47 (1947).

*(Received, October 2, 1959.)*

# The Gibberellins.

By P. W. Brian, John Frederick Grove and J. MacMillan,
Welwyn, Hertfordshire, England.

With 2 Figures.

## Contents.

*Acknowledgement.* The authors are indebted to numerous colleagues for helpful discussion in the compilation of this review.

# I. Introduction.

The discovery of the gibberellins stems from studies of a soil-borne disease of rice, the so-called bakanae disease, caused by infection by the fungus *Gibberella fujikuroi*. The classical publications, describing the critical steps in this discovery of a new series of plant growth-regulating compounds, are those of Kurosawa (*170*) who in 1926 first showed that cell-free filtrates from cultures of the fungus contained a growth-promoting principle, and Yabuta and Hayashi (*340*) who in 1939 first obtained crystalline active material for which they coined the name gibberellin. Until 1954 virtually all publications on these substances were Japanese; access to the numerous early Japanese publications has been greatly simplified for many readers by the publication by Stodola (*292*) of his invaluable *Source Book on Gibberellin, 1828–1957*.

It is striking that despite intensive searches (*93*), no other fungus besides *Gibberella fujikuroi*, with the possible exception of the yeast *Candida* (= *Torula*) *pulcherrima* (*162*), has been found to produce a gibberellin in culture. But the study of these interesting compounds received a further impetus with the discovery that compounds similar in their chemical and physiological properties were widely distributed in plants.

The purpose of this review is to summarise the present state of our knowledge concerning the chemistry of the gibberellins, their occurrence in plants, the developmental and growth responses which they induce in plants and their possible significance as plant hormones. We have set certain limits to the topics we discuss. For instance, we do not deal with cultural conditions for production of the gibberellins by *G. fujikuroi* (for which see *22, 95, 292, 294, 298, 340*). Neither do we discuss practical aspects of their use in agriculture or horticulture: this aspect has been reviewed recently by Wittwer and Bukovac (*337*).

## II. The Chemistry of the Gibberellins.

### 1. General Remarks.

The isolation of two crystalline biologically active compounds, gibberellin A, $C_{22}H_{26}O_7$, m. p. 242–244° (dec.), $[\alpha]_D + 36°$, and gibberellin B, $C_{19}H_{22}O_3$, m. p. 194–196°, $[\alpha]_D - 82°$, was first reported in 1938 by Yabuta and Sumiki (*342*) and Yabuta and Hayashi (*340*)*. The degradation of these compounds was described in a series of papers (*343–346, 348, 282*) from the University of Tokyo in the course of which it was established that gibberellin A was converted into gibberellin B by treatment with dilute mineral acid at 50–70° (*345*) and that gibberene, a hydrocarbon obtained by selenium dehydrogenation of gibberellins A and B, was a fluorene derivative (*346*). Because of the world war 1939–1945 this work did not become known outside Japan until several years after its publication.

More recently, in an attempt to repeat the isolation of gibberellin A, Curtis and Cross (*92, 82*) at the Akers Research Laboratories of Imperial Chemical Industries Ltd., Welwyn, Great Britain, obtained a new metabolite, gibberellic acid, $C_{19}H_{22}O_6$, m. p. 233–235° (dec.), $[\alpha]_D + 92°$, which differed from gibberellin A in its physical characteristics and in the properties of its derivatives. The same compound was also obtained (*294, 293*) by Stodola and his collaborators at the North Regional Research Laboratories, Peoria, U. S. A. who named it gibberellin X until the identity with gibberellic acid was established (*82*). alloGibberic acid, $C_{18}H_{20}O_3$, obtained from gibberellic acid by acid treatment, was shown (*82*) to be identical with gibberellin B and, after careful purification, to be devoid of growth promoting activity (*30*).

Gibberellic acid was produced by several strains of *G. fujikuroi* (and in particularly high yield by Akers Laboratories 917 strain) on the medium described by Borrow et al. (*22*). Using the high-nitrogen medium of

---

* The names gibberellin A and gibberellin B are assigned as in (*344*) and not as in (*342, 340*) where they were used in the reverse sense.

*References, pp. 417—433.*

the Japanese workers but a different strain of *G. fujikuroi* (N. R. R. L. 2284) the American workers (*294, 293*) obtained a mixture of gibberellins separated by chromatography on buffered Celite into gibberellic acid and gibberellin $A_1^*$, $C_{19}H_{24}O_6$, m. p. 255–258° (dec.), $[\alpha]_D + 36°$, which was similar in physical properties to gibberellin A. These results led the Japanese workers to a reexamination (*305*) of the crude gibberellin produced by the Tokyo University strains of *G. fujikuroi*, and gibberellin $A_1$, gibberellin $A_2$, $C_{19}H_{26}O_6$, and gibberellic acid (gibberellin $A_3$), were separated (as their methyl esters after alumina chromatography) from various batches of crude gibberellin obtained from several different strains. Another component of the mixture, gibberellin $A_4$ was described later (*308*): gibberellin $A_4$ is also produced by N. R. R. L. 2284 strain (*140*). The exact composition of the gibberellin A obtained in the period 1938 to 1941 remains obscure (*153*).

The physical properties of the gibberellins and their methyl esters are given in *Table 1*.

Four fungal gibberellins are known, gibberellic acid and gibberellins $A_1$, $A_2$ and $A_4$. All are metabolites of *G. fujikuroi*. Gibberellin $A_1$ has also been isolated from seed of runner bean (*208*) where it occurs together with gibberellin $A_5$ (*207*) (Chapter III, p. 390). Gibberellin $A_1$ (II; $R = H$) was shown (*118, 307*) to be the dihydro-derivative of gibberellic acid in which the ring A double bond of (I; $R = H$) had been reduced.

Table 1. Some Physical Constants of the Gibberellins.

| Compound | | Structure[e] | Acid, $R = H$ | | Methyl ester, $R = Me$ | |
|---|---|---|---|---|---|---|
| | | | M. p.[f] | $[\alpha]_D$[g] | M. p. | $[\alpha]_D$[g] |
| Gibberellic acid[a] | $C_{19}H_{22}O_6$ | I | 233–235° 285° (*140*) | $+ 92°$ | 209–210° | $+ 75°$ |
| Gibberellin $A_1$[b] . | $C_{19}H_{24}O_6$ | II | 255–258° 256° (*140*) | $+ 36°$ | 234–235° | $+ 46°$ |
| Gibberellin $A_2$ .. | $C_{19}H_{26}O_6$[c] | III | 235–237° 255° | $+ 12°$ | 190–192° | $+ 28°$ |
| Gibberellin $A_4$ .. | $C_{19}H_{24}O_5$ (*140*),[d] | IV | 214–215° | $- 3°$ | 176° | $0°$ |
| Gibberellin $A_5$ .. | $C_{19}H_{22}O_5$ | V | 260–261° | $- 77°$ | 190–191° | $- 75°$ |

a Alternative names now discarded: gibberellin X, gibberellin $A_3$.
b Alternative name now discarded: gibberellin A.
c TAKAHASHI et al. (*305*) gave $C_{20}H_{28}O_6$ subsequently revised (*155*) to $C_{19}H_{26}O_6$.
d TAKAHASHI et al. (*308*) gave $C_{18}H_{22}O_5$ and $C_{19}H_{24}O_5$ as alternative formulae and m. p. 222° (dec.), $[\alpha]_D - 21°$.
e Absolute configurations are depicted by the stereo-formulae (pp. 369, 386).
f Decomposition point: often variable and no guide to purity. The alternative values given are for polymorphic forms.
g In ethanol or methanol.

* Originally called gibberellin A but later renamed gibberellin $A_1$ (*118*).

The relationships between gibberellin $A_2$, gibberellin $A_4$ and gibberellin $A_1$ were elucidated by a complex series of transformations (*157*). Gibberellin $A_4$ can be converted directly into gibberellin $A_2$ (*140*) by treatment with acid which causes the addition of the elements of water to the terminal

(I.)         (II.)         (III.)

(IV.)         (V.)

methylene group of gibberellin $A_4$. The relationship between gibberellin $A_5$ and gibberellin $A_1$ was established by MacMillan et al. (*207*). At the present time the naturally occurring gibberellins consist of a group of tetracyclic lactonic carboxylic acids. All have one or more hydroxyl substituents and the majority have one or more ethylenic bonds present in or attached to the tetracyclic system; but there is no evidence that either of these structural features is essential to biological activity.

## 2. Nomenclature.

The nomenclature is based on the trivial name gibbane for the fully saturated tetracyclic system (VI), numbered as shown. The configuration of substituents is indicated by dotted ($\alpha$) and thick bonds ($\beta$) which are respectively below and above the plane of the $ABC$ ring system. $\xi$ is used in the conventional way when the configuration of a substituent is not known. In the naturally occurring gibberellins the $7 \rightarrow 9a$ (ring $C$) two carbon bridge is $\beta$, as in (VI) and the $1 \rightarrow 4a$ lactone bridge is $\alpha$: the hydrogen atoms at 4b and 10a are $\alpha$ and $\beta$ respectively. The ring system derived from gibbane by inversion at 7 and 9a is called $7\alpha$-gibbane.

Gibberellic acid (I; $R = H$) is thus $2\beta$:$4a\alpha$:$7\alpha$-trihydroxy-$1\beta$-methyl-8-methylene-$4b\alpha$,$10a\beta$-gibb-3-ene-$1\alpha$:$10\beta$-dicarboxylic acid $1 \rightarrow 4a$ lactone, or more conveniently, $2\beta$ : $4a$ : $7$-trihydroxy-$1$-methyl-8-methylenegibb-

3-ene-1 : $10\beta$-dicarboxylic acid $1 \to 4a$ lactone. Catalytic reduction (steric control) of an 8-methylene group in gibberellic acid and its relatives gives rise to pairs of $C_{(8)}$ epimers the absolute configuration of which is not known at present. Only one of these epimers is obtained by chemical

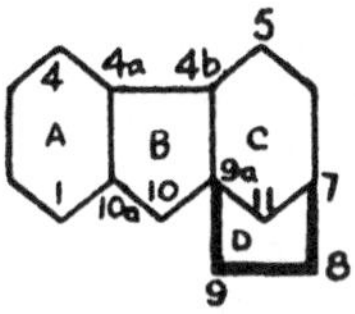

(VI.)

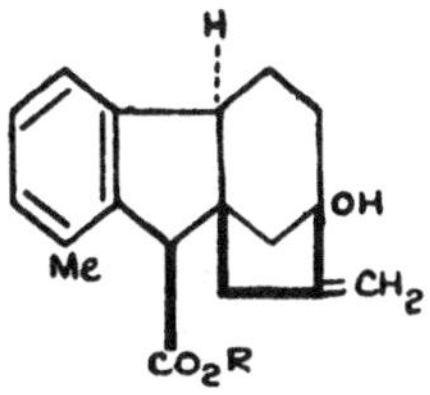

(VII.)

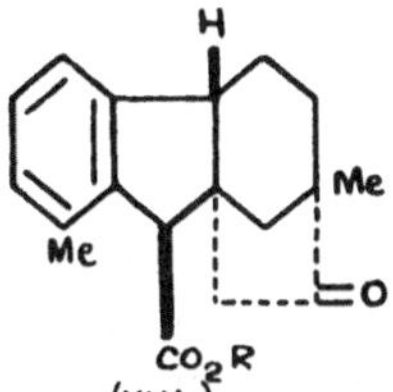

(VIII.)

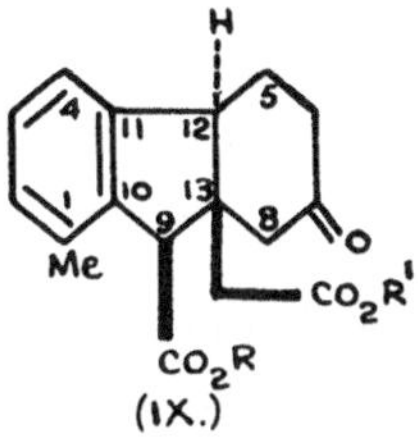

(IX.)

reduction (thermodynamic control) and this compound is arbitrarily called an 8-methylgibbane: the epimer is called an 8-*epi*methylgibbane. It is likely that compounds in the 8-methyl series will be shown to have the less hindered configuration and, on the projection formula (VI), will be the $8\alpha$ compounds.

Trivial names, e. g. allogibberic acid (VII; $R = H$), are retained for degradation products in which ring $A$ is aromatic and where the gibbane nomenclature cannot be used. For convenience within this class of compound the prefix *epi* is reserved for those compounds with $4b\beta$ e. g. *epi*gibberic acid (VIII; $R = H$).

Degradation products in which ring $D$ of gibbane has been opened are named as derivatives of fluorene e. g. $9\beta$-carboxy-1-methyl-7-oxo-5,6,7,8:$12\alpha$,13-hexahydrofluorene-$13\beta$-acetic acid (IX; $R = R' = H$).

This comprehensive system supercedes earlier conventions such as the use of "pseudo" (*305*) to differentiate $2\alpha$-hydroxygibbanes from the $2\beta$-hydroxy epimers and the use of $\alpha$ and $\beta$ (*118*) to differentiate isomeric reduction products of the gibberellins. In particular the name "$\alpha$-dihydrogibberellic acid" (*118*) for the gibbane (II; $R = H$) is withdrawn: the trivial name gibberellin $A_1$ may be used for this compound.

## 3. The Chemistry of Gibberellic Acid.

The chemistry of the gibberellins has been reviewed previously by BRIAN and GROVE (*29*), STOWE and YAMAKI (*298, 299*) and JOHNSON (*143*). For the relevant literature prior to 1958 cf. STODOLA (*292*).

Much of the Japanese work before 1955 is of doubtful value in view of the uncertainty about the composition of gibberellin A, and progress in the chemistry of the gibberellins really dates from the isolation by the British workers (*92*) of gibberellic acid, the first gibberellin to be obtained in pure form. Degradation of gibberellic acid put the chemistry of the gibberellins on a firm foundation; but the elucidation of the structure of ring *A* of gibberellic acid has been handicapped by lack of the more stable dihydro-derivative gibberellin $A_1$.

This work from the I. C. I. Akers Laboratories has been published in full mainly in the Journal of the Chemical Society (*82, 226, 227, 86, 118, 30, 87, 223, 225, 224, 88, 85, 84, 218, 121, 83, 90, 285, 217, 120, 89, 119*). Since 1955 the publications (*305, 153, 155, 281, 306, 307, 308, 284, 283, 309, 157*) of the University of Tokyo workers have been fragmentary and without much experimental detail.

In the following account priority is given to reactions having a direct bearing on the determination of the structure and stereochemistry of gibberellic acid. Gibberellic acid has the absolute configuration (I; $R = H$) and in general in considering degradation products stereo formulae with thick and dotted bonds (see p. 354) are only drawn in order to illustrate some stereochemical change that has occurred.

The structures of the gibberellic acid degradation products gibberene, gibberic acid and allogibberic acid are discussed below in the order in which they were established.

## A. Structure of Gibberene
[1:7-dimethylfluorene (X; $R = R' = $ Me)].

Dehydrogenation of gibberellins A and B (allogibberic acid) and gibberic acid with selenium was shown by Yabuta et al. (*346*) to give a ketone gibberone (p. 359) and a hydrocarbon gibberene, m. p. 107–108°. From its ultraviolet spectrum the latter was regarded, correctly, as a substituted fluorene but was incorrectly given the formula $C_{16}H_{16}$ and 4-ethyl-5-methylfluorene was suggested as a possible structure.

Mulholland and Ward (*226*) confirmed the formation of gibberene by selenium dehydrogenation of both gibberellic acid and gibberic acid and showed that it was 1:7-dimethylfluorene (X; $R = R' = $ Me) $C_{15}H_{14}$, as follows.

Stepwise oxidation of gibberene with potassium permanganate in acetone, sodium dichromate in boiling glacial acetic acid or potassium permanganate in aqueous pyridine at 100° and finally with alkaline potassium permanganate at 100° gave, sequentially, 1:7-dimethyl-fluorenone (gibberenone) (XI; $R = R' = $ Me) $C_{15}H_{12}O$, m. p. 77°; two acids, the yellow 1-methylfluorenone-7-carboxylic acid (XI; $R = $ Me, $R' = CO_2H$), m. p. 330–331°, and the red 7-methylfluorenone-1-carboxylic acid (XI; $R = CO_2H$, $R' = $ Me), m. p. 206–208°; and finally the known fluorenone-1:7-dicarboxylic acid (XI; $R = R' = CO_2H$) $C_{15}H_{18}O_5$, m. p. 350–354°, previously prepared from retene by BAMBERGER and HOOKER's method (10). Decarboxylation of 1-methylfluorenone-7-carboxylic acid with copper chromite in quinoline gave the known 1-methylfluorenone (XI; $R =$ Me, $R' =$ H) (203).

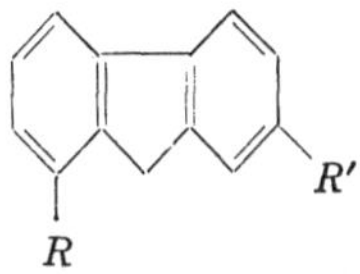

(X.) 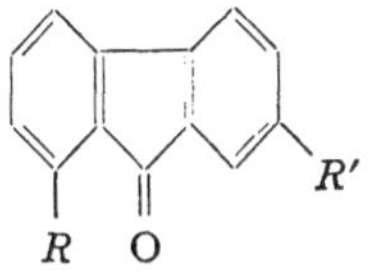　　　　(XI.) 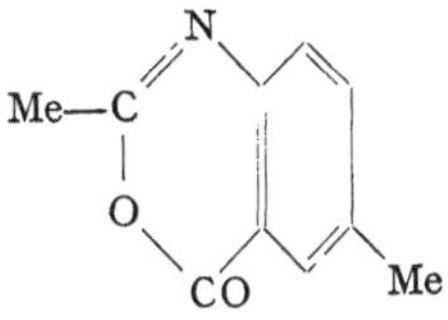　　　　(XII.)

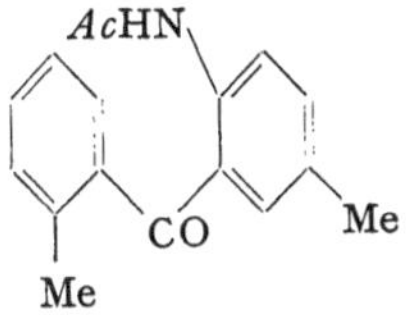

(XIII.) 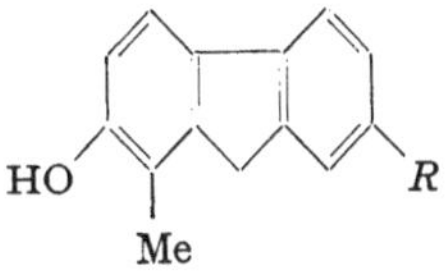　　　　(XIV.)

Finally, 1:7-dimethylfluorene and 1:7-dimethylfluorenone were synthesised (226) by an unambiguous route and shown to be identical with gibberene and gibberenone obtained by degradation of gibberic acid. 2-Amino-5-methylbenzoic acid was converted by acetic anhydride into 2:6-dimethyl-4-oxo-3:1-benzoxazine (XII) which with o-tolyl magnesium bromide gave 2-acetamido-5:2'-dimethylbenzophenone (XIII). Hydrolysis of the latter followed by diazotisation and ring closure gave 1:7-dimethylfluorenone (XI; $R = R' =$ Me) from which 1:7-dimethylfluorene (X; $R = R' =$ Me) was obtained by Wolff-Kishner reduction.

Earlier YABUTA et al. (346) had suggested that an acid fraction, m. p. 285°, which they had obtained by degradation of gibberene was fluorenone-4:5-dicarboxylic acid previously isolated by KRUBER (168) by degradation of 4:5-methylenephenanthrene. MULHOLLAND and WARD (227) therefore synthesised 4:5-dimethylfluorenone by pyrolysis of 6:6'-dimethyldiphenic acid: fluorenone-4:5-dicarboxylic acid, m. p. 284–286° (dec.), obtained by oxidation of 4:5-dimethylfluorenone was compared with the (1:7)-dicarboxylic acid obtained by oxidation of gibberene and shown to be a different compound. The general method (203) used by MULHOLLAND and WARD (226) for the synthesis of 1:7-dimethylfluorene was later adapted to the synthesis of other substituted fluorenes obtained in degradative studies of the gibberellins. Thus, 7-hydroxy-1-methylfluorene $C_{14}H_{12}O$ (X; $R =$ Me, $R' =$ OH), m. p. 166–168°, was obtained (224) from o-bromotoluene and 2-amino-5-methoxybenzoic acid; and 2-hydroxy-1:7-dimethylfluorene (90) $C_{15}H_{14}O$ (XIV; $R =$ Me), m. p. 202–204° (dec.), and 2-hydroxy-1-methylfluorene (284) $C_{14}H_{12}O$ (XIV; $R = H$), m. p. 185–187°, were obtained from the reaction of the magnesium derivative of 6-bromo-2-methoxy-toluene with 2-amino-5-methylbenzoic acid and anthranilic acid, respectively.

## B. Structure of Gibberic Acid.

### a) Gibberic Acid (XVI; $R = H$: Chart 1, p. 359).

Gibberic acid, m. p. 153–154°, together with isogibberic acid, m. p. 173–174°, and gibberellin C (p. 379) was first obtained (345) by the action of hot dilute mineral acid on gibberellin A. Subsequently, hydrolysis of gibberellic acid with boiling dilute mineral acid was shown (82) to give 1 mol. $CO_2$ and gibberic acid, m. p. 153–154° or 174–175°, $[\alpha]_D^{17.5} - 7°$, as the main product together with an isomer, *epi*gibberic acid (VIII; $R = H$), m. p. 227–230° or 252–255°, $[\alpha]_D^{15} + 131°$. Under the same conditions allogibberic acid (p. 365) was isomerised to gibberic acid. Gibberic acid is dimorphic, the high m. p. form being identical with the isogibberic acid of the Japanese workers. The latter proposed the formula $C_{19}H_{22}O_3$ for gibberic acid but this was later corrected by the British workers (82) to $C_{18}H_{20}O_3$.

The formation (82) of a neutral monomethyl ester (XVI; $R = Me$), m. p. 113–115°, $[\alpha] - 4°$, an oxime and an ester oxime showed that gibberic acid was a keto-acid: a band at 1741 cm.$^{-1}$ in the infrared spectrum indicated that the keto group was present in a five-membered ring. Gibberic acid contained no ethylenic double bonds detectable by microhydrogenation but the ultraviolet spectrum (see *Table 2*) was

Table 2. Ultraviolet Absorption Maxima in Ethanol (m$\mu$).

| Compound | $\lambda_{max.}$ | | | | log $\epsilon$ |
|---|---|---|---|---|---|
| Gibberic acid... | ~259, 265, | 274, | 300 | | 2.50, 2.56, 2.47, 1.49 |
| alloGibberic acid | ~259, 266, | 274.5, ~287, ~297 | | | 2.47, 2.50, 2.35, 1.28, 1.25 |
| Gibberdionic ⎰ a | 265, | 274 | | | 2.78, 2.71 |
| acid ⎱ | ~261, 265, | 274, ~286 | | | 2.68, 2.71, 2.63, 1.99 |
| *epi*Gibberic acid. | 265, | 274 | | | 2.46, 2.31 |
| *epi*alloGibberic | | | | | |
| acid ........ | ~261, 266, | ~274, | | | 2.40, 2.50, 2.37 |
| *epi*Gibber- ⎰ a | 260, 265, | ~275, | 292, | 307 | 2.73, 2.73, 2.58, 2.05, 2.07 |
| dionic acid ⎱ | 258, 263, | ~275, | 290, | 301 | 2.68, 2.66, 2.47, 2.07, 2.08 |
| Dehydrogibberic | | | | | |
| acid ........ | 260, | 269, | 290, 300, | | 4.14, 4.09, 3.50, 3.44 |
| Gibberone...... | ~249, 259.5, | 269, | 290, 301.5 | | 4.00, 4.13, 4.07, 3.62, 3.62 |
| Gibberdione . ⎰ b | ~250, 257.5, ~267, | | 290, 301 | | 4.12, 4.18, 4.07, 3.66, 3.62 |
| ⎱ | ~250, 257, | ~267, | 290, 301 | | 4.14, 4.21, 4.10, 3.65, 3.65 |
| XX (p. 359) .... | | 254, | 293, ~302 | | 4.18, 3.55, 3.54 |
| XXV (p. 359)... | | 233, | 284, ~293 | | 3.89, 3.21, 3.08 |

ᵃ In sodium hydroxide.    ᵇ In EtOH—0.1 N sodium hydroxide.

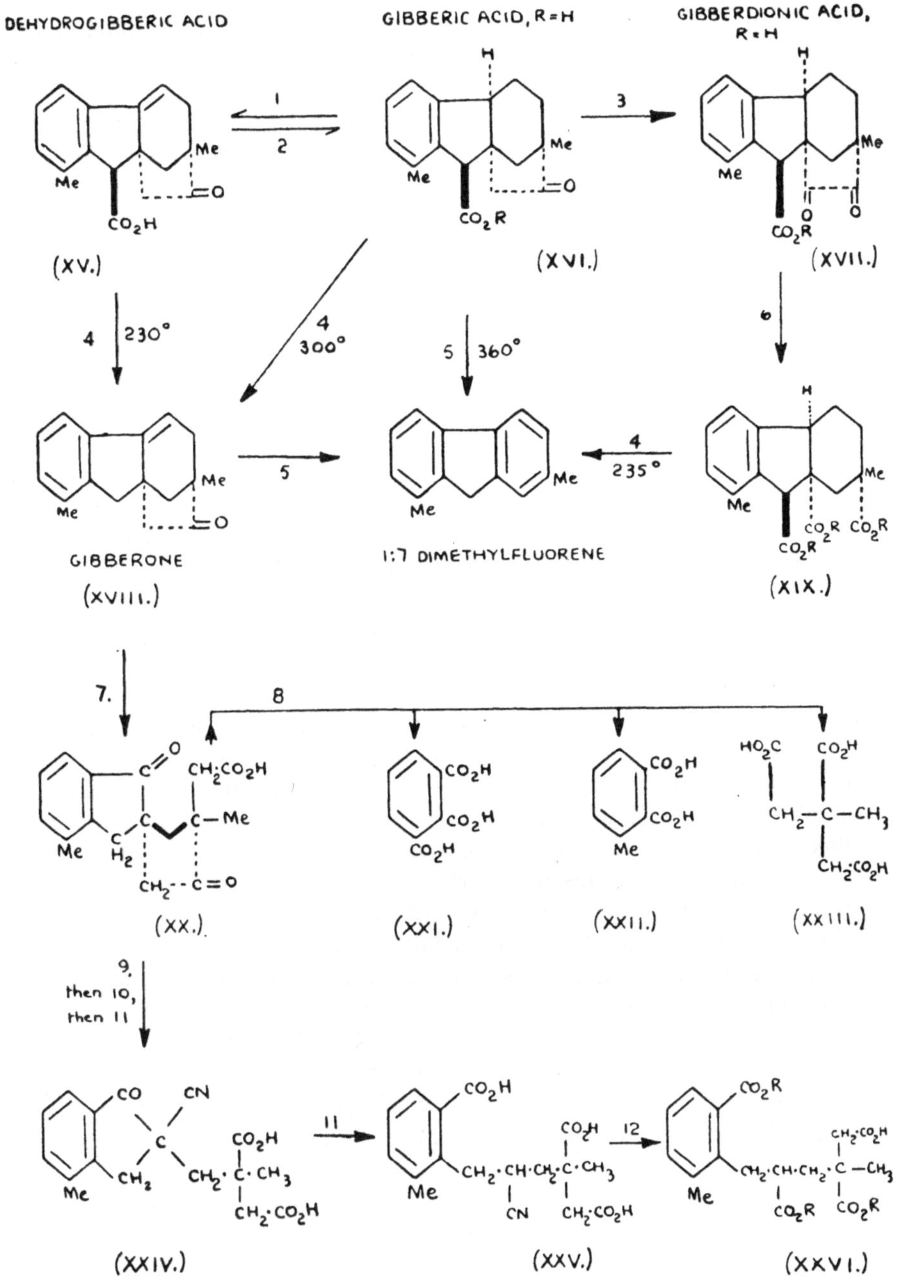

Chart 1. Degradation of Gibberic Acid.

(Reagents: 1. $KMnO_4$. 2. $H_2/Pd$. 3. $SeO_2$. 4. $Pd/C$. 5. Se. 6. $H_2O_2/NaOH$. 7. $CrO_3$. 8. $KMnO_4/Mg(NO_3)_2$. 9. $BuNO_2/NaOMe$. 10. $p\text{-}MeC_6H_4SO_2Cl$. 11. NaOH. 12. $H_2SO_4$.)

consistent with the presence of a benzenoid ring. These facts established that gibberic acid was a tetracyclic keto-acid containing an aromatic ring.

Yatazawa and Sumiki (348), on the basis of the $C_{19}H_{22}O_3$ formula had advanced the partial structure, $C_{16}H_{18}(CO_2H) \cdot COMe$, for gibberic acid and in support of this structure Seta and Sumiki (282) claimed that the crude product obtained by Beckmann rearrangement of gibberic acid oxime gave methylamine in good yield on acid hydrolysis. Cross, Grove, MacMillan, and Mulholland (87) repeated this experiment but failed to detect methylamine in the hydrolysis products of the crude mixture from the Beckmann rearrangement, only a portion of which was acidic. Instead, ammonia, the production of which is not inconsistent with a ring ketone structure, was obtained and an amide carboxylic acid, $C_{18}H_{21}O_3N$, isomeric with gibberic acid oxime was also isolated.

Two important degradations of gibberic acid had been effected by the Japanese workers but their significance in the elucidation of its structure was missed by the allocation of incorrect molecular formulae to the degradation products. In the first degradation gibberene, 1:7-dimethylfluorene (p. 356), was obtained by selenium dehydrogenation and gibberic acid was correctly recognised as containing a hexahydrofluorene nucleus although this was inconsistent with the partial structure, $C_{16}H_{18}(CO_2H) \cdot COMe$. In the second degradation gibberic acid was oxidised by selenium dioxide to gibberdionic acid, believed to have the partial structure, $C_{16}H_{18}(CO_2H) \cdot CO \cdot CHO$; further oxidation of gibberdionic acid with hydrogen peroxide was said to give carbon dioxide and a dicarboxylic acid, $C_{17-18}H_{22-24}O_6$.

Repetition (82) of this work, which is set out in *Chart 1*, gave the yellow gibberdionic acid, $C_{18}H_{18}O_4$ (XVII; $R = H$), m. p. 190–191°, which formed a colourless methanolate, $C_{18}H_{18}O_4 \cdot 2\,CH_3OH$, m. p. 93 to 96°, and was recognised as a cyclic α-diketone since the infrared spectrum of the orange methyl ester, $C_{19}H_{20}O_4$ (XVII; $R = Me$), m. p. 193–195°, showed $\nu_{max.}$ at 1736 (ester C=O), 1750 and 1764 cm.$^{-1}$ (five-membered ring α-diketone: cf. camphorquinone 1748 and 1769 cm.$^{-1}$). The ultraviolet spectrum of gibberdionic acid (Table 2, p. 358) was almost identical in ethanol and in 0.1N sodium hydroxide and closely resembled

(XXVII.)          (XXVIII.)          (XXIX.)          (XXX.) Gibberdione.

that of gibberic acid indicating that no enolisable hydrogen atom was present. As expected on structure (XVII; $R = H$) oxidation of gibberdionic acid with alkaline hydrogen peroxide under the conditions described by YATAZAWA and SUMIKI (*348*) gave only a trace of carbon dioxide and a tricarboxylic acid, $C_{18}H_{20}O_6 \cdot {}^1/_2 H_2O$, m. p. 158° (dec.), whose structure was deduced later as 1:7-dimethyl-5,6,7,8:12α,13-hexahydro-fluorene-7α:9β:13α-tricarboxylic acid (XIX; $R = H$, p. 359).

Gibberic acid therefore contains the partial structure (XXVII). The presence of the hexahydrofluorene nucleus (XXVIII) was established by dehydrogenation to 1:7-dimethylfluorene (*226*) and by oxidation to benzene-1 : 2 : 3-tricarboxylic acid (XXI) (*343*) in significant yield (*87*). Mild dehydrogenation (*82*) with palladised charcoal of the tricarboxylic acid (XIX; $R = H$) also gave 1:7-dimethylfluorene. Skeletal re-arrangement was therefore unlikely during the selenium dehydrogenation which involved the elimination from gibberic acid of a carboxyl group and the —$CH_2 \cdot CO$— group from the partial structure (XXVII).

The position of the carboxyl group in gibberic acid was established (*87*) by dehydrogenation of methyl gibberdionate (XVII; $R = Me$, p. 359) with palladium-charcoal at 230° to methyl 1:7-dimethylfluorene-9-carboxylate (XXIX), m. p. 121–122°, identical with a synthetic specimen prepared by carboxylation of the 9-lithium derivative of 1:7-dimethylfluorene followed by methylation.

The position of the methylene carbonyl bridge was determined by the series of degradation (*87*) depicted in Chart 1, p. 359. Oxidation of gibberic acid with alkaline potassium permanganate at 0° gave dehydrogibberic acid, $C_{18}H_{18}O_3$ (XV), m. p. 222–224° (dec.), $[\alpha]_D^{18} + 99°$, in which an ethylenic bond has been introduced in conjugation with the aromatic ring as shown by the ultraviolet spectrum (Table 2, p. 358). Catalytic reduction regenerated gibberic acid. Dehydrogibberic acid was decarboxylated at 230° by palladium-charcoal or charcoal alone to a ketone, gibberone, $C_{17}H_{18}O$ (XVIII, p. 359), m. p. 126–127°, $[\alpha]_D^{21} + 27°$, first obtained by YABUTA et al. (*346*) by selenium dehydrogenation of gibberic acid at 300–330° but wrongly formulated by them as $C_{18}H_{18}O$. Gibberone was also obtained (*87*) by dehydrogenation of gibberic acid with 30% palladium-charcoal at 210–300°: at the lower temperature some *epi*gibberic acid (VIII; $R = H$, p. 355) was formed. Selenium dehydrogenation of gibberone gave 1:7-dimethylfluorene. Gibberone was oxidised (*87*) by selenium dioxide to a 5-membered ring 1 : 2-diketone, gibberdione, $C_{17}H_{16}O_2$ (XXX), m. p. 181–182°, and by chromic oxide in high yield and without loss of carbon to a monobasic keto-acid, $C_{17}H_{18}O_4$ (XX), m. p. 155–156°, $[\alpha]_D^{18} + 27°$, presumably identical with an uncharacterised acid, m. p. 154–155°, obtained earlier by YABUTA et al. (*343*).

The keto-acid (XX) was recognised (87) as an indan-1-one by the ultraviolet spectrum (Table 2, p. 358), and the infrared spectrum of a dioxan solution showed $\nu_{max}$. 1715 cm.$^{-1}$ (indanone C=O), 1738 (monomeric carboxyl C=O) and 1744 (5-membered ring ketone C=O). The keto-acid was not hydrolysed by dilute mineral acid or by alkali and it was not therefore a $\beta$-keto acid or $\beta$-diketone: stability to permanganate in acetone at 20° and to chromic oxide in acetic acid at 75° indicated that it was a 2:2-disubstituted indanone; and in agreement with this formulation, bromination to completion yielded only a dibromo derivative, $C_{17}H_{16}O_4Br_2$, m. p. 205–207° (dec.), which did not eliminate hydrogen bromide on refluxing with collidine. Oxidation with potassium permanganate in the presence of magnesium nitrate at 85° gave benzene-1 : 2 : 3-tricarboxylic acid (XXI), 3-methylphthalic acid (XXII), isolated as the anhydride, and $\beta$-methyltricarballylic acid (XXIII). All these facts were consistent with the keto-acid being 4-methyl-1-oxoindan-2-*spiro*-1'-(4'-methyl-3'-oxo-*cyclo*pentane-4'-acetic acid) (XX): final proof was obtained by opening both non-benzenoid rings and by an unambiguous synthesis of the ultimate product.

With butyl nitrite in the presence of sodium methoxide the keto-acid formed an $\alpha$-oximino derivative, m. p. 238–240° (dec.), which on treatment with $p$-toluenesulphonyl chloride and alkali underwent Beck-mann rearrangement to a cyanotricarboxylic acid, $C_{17}H_{19}O_6N$, m. p. 188 to 189° (dec.), $\nu_{max}$. 2250 (C=N), 1696 cm.$^{-1}$ (carboxyl C=O), which must be 3-carboxy-6-(2-carboxy-6-methylphenyl)-5-cyano-3-methyl-hexanoic acid (XXV), formed by hydrolysis of the intermediate $\beta$-keto-nitrile (XXIV), since the ultraviolet spectrum (Table 2, p. 358) showed the absence of the indanone chromophore and was typical of a substituted benzoic acid. The amorphous tetracarboxylic acid (XXVI; $R = H$) obtained by hydrolysis of the cyanotricarboxylic acid with 50% sulphuric acid formed two isomeric tetramethyl esters, $C_{21}H_{28}O_8$ (XXVI; $R = Me$); $\alpha$, m. p. 83–84°, $[\alpha]_D^{23}$ — 6° and $\beta$, m. p. 47–48°, $[\alpha]_D^{23}$ + 12°, respectively, which are diastereoisomers arising from the racemisation of the spiran centre during hydrolysis of the keto-nitrile (XXIV).

The structures of the two esters were established by the synthesis of their racemates as outlined in *Chart 2*. Starting from 4-methylindan-1-one (XXXI), ethyl $\beta$-(2-ethoxycarbonyl-6-methylphenyl)-propionate (61) (XXXIV) was prepared via the hydroxymethylene ketone (XXXII) and the *bis*-(4-methyl-1-oxo-indanylidenemethyl)-hydroxylamine (XXXIII). The ester (XXXIV) on treatment with sodium followed by bromoacetone was converted into ethyl 2-acetonyl-4-methyl-1-oxoindan-2-carb-oxylate (XXXV). Hydrolysis of the latter followed by methylation gave methyl 2-(2-methoxycarbonyl-6-methylbenzyl)-4-oxopentanoate (XXXVI) which, unlike the ester (XXXV), condensed smoothly with ethyl cyanoacetate. Addition of potassium cyanide across the ethylenic bond of the product (XXXVII) followed by hydrolysis of the resulting dinitrile (XXXVIII) and decarboxylation gave crude, gummy 3:5-di-carboxy-6-(2-carboxy-6-methylphenyl)-3-methylhexanoic acid (XXVI; $R = H$). Methylation gave a mixture of tetramethyl esters separated by chromatography and fractional crystallisation into the $(\pm)$-$\alpha$-,

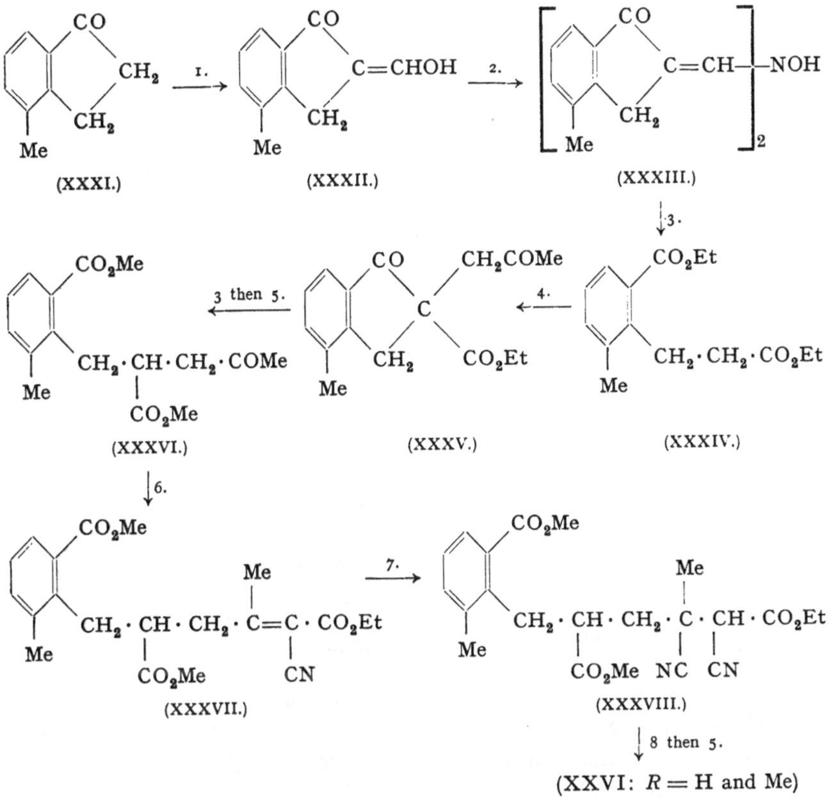

Chart 2. Synthesis of Methyl $(\pm)$-$\alpha$- and -$\beta$-3 : 5-dimethoxycarbonyl-6-(2-methoxycarbonyl-6-methylphenyl)-3-methylhexanoate (XXVI; $R =$ Me).

(Reagents: 1. H·CO₂Et/NaOMe. 2. NH₂OH. 3. KOH. 4. BrCH₂COCH₃. 5. CH₂N₂. 6. CN·CH₂CO₂Et. 7. KCN. 8. H₂SO₄.)

m. p. 78–80°, and $\beta$-, m. p. 36–38°, isomers. The synthetic esters did not depress the m. p.'s of the corresponding $(-)$-$\alpha$-, m. p. 83–84° and $(+)$-$\beta$-, m. p. 47–48° esters obtained by degradation of the keto-acid (XX) and were shown to be racemates of the natural compounds by comparison of the infrared spectra in solution. This synthesis completed the elucidation of the structure of the keto-acid (XX, p. 359) and consequently of gibberone (XVIII, p. 359). It followed that the methylene carbonyl bridge in gibberic acid must be attached as in (XVI, p. 359).

In earlier attempts (87) to locate the position of the carboxyl group in gibberic acid by degradation to a substituted 1:7-dimethylfluorene wherein the original carboxyl group would be retained as a methyl or hydroxymethyl group the reduction of gibberic acid was investigated; but this approach was abandoned (87) when it was found that the dehydrogenation conditions necessary for the removal of the 7 → 9a two carbon bridge were sufficiently drastic to eliminate a reduced carboxyl substituent at the 9-position of a fluorene nucleus. In the course of this work deoxogibberic acid, $C_{18}H_{22}O_2$ (XXXIX; $R = R' = R'' = H$), m. p. 108–110°, $[\alpha]_D^{23} + 19°$, was obtained by Clemmensen reduction of gibberic acid. Gibberic acid, methyl gibberate and methyl gibberate ethylene ketal yielded intractable products on lithium aluminium hydride reduction but methyl gibberdionate gave a cis 1:2-glycol, $C_{19}H_{24}O_4$ (XXXIX; $R = Me$, $R' = R'' = OH$), m. p. 156–158°, and a triol, $C_{18}H_{24}O_3$, m. p. 234–236°, presumed to be (XL) since the infrared spectrum showed no C=O absorption.

The evidence for the absolute configuration of gibberic acid (XVI; $R = H$) is more conveniently discussed under allogibberic acid (p. 369) where it is shown that rings B/C in gibberic acid are cis fused and the hydrogen atoms at 4b and 10 are $\alpha$. Gibberic acid was stable to hot aqueous acid or alkali, but under the latter conditions or with cold methanolic sodium hydroxide methyl gibberate (121) underwent epimerisation at $C_{(10)}$ with formation of the liquid 10$\alpha$ ester (XLI; $R = Me$). The oily ester, $C_{21}H_{26}O_6$, methyl 1:7-dimethyl-5,6,7,8:12$\alpha$,13-hexahydrofluorene-7$\alpha$:9$\beta$:13$\alpha$-tricarboxylate (XIX; $R = Me$: Chart 1, p. 359), $[\alpha]_D^{23} + 5°$, similarly gave (120) the 9$\alpha$ isomer (XLII), m. p. 119 to 120°, $[\alpha]_D^{25} - 22°$.

### b) epiGibberic Acid.

The chemistry of the 4b$\beta$ epimer, epigibberic acid (VIII above; $R = H$) is similar to that of gibberic acid but there are some interesting points of difference (*120*). Like gibberic acid, dehydrogenation of *epi*-gibberic acid with palladised charcoal or with selenium gave gibberone (XVIII) and 1:7-dimethylfluorene. Oxidation of *epi*gibberic acid with selenium dioxide followed by alkaline hydrogen peroxide gave sequentially, *epi*gibberdionic acid, $C_{18}H_{18}O_4$ (XLIII), m. p. 288–290°, $[\alpha]_D^{24} + 337°$, and 1:7-dimethyl-5,6,7,8:12$\beta$,13-hexahydrofluorene-7$\alpha$:9$\beta$:13$\alpha$-tricarb-

(XLIII.)　　　(XLIV.)　　　(XLV.)

oxylic acid, $C_{18}H_{20}O_6$ (XLIV; $R = H$), m. p. 283–284°, $[\alpha]_D^{23} + 87°$; both Clemmensen and Wolff-Kishner reduction afforded deoxo*epi*-gibberic acid, $C_{18}H_{22}O_2$ (XLV), m. p. 220–221°, $[\alpha]_D^{20} + 144°$. Unlike the corresponding 4b$\alpha$ compounds, *epi*gibberic acid was not oxidised by alkaline permanganate to a $\Delta$4b(5) dehydro-compound; and the $C_{(10)}$ centre in methyl *epi*gibberate (*121*) and the $C_{(9)}$ centre in the ester (XLIV; $R = Me$) were not inverted by hot aqueous or cold methanolic alkali.

### C. Structure and Stereochemistry of alloGibberic Acid.

#### a) alloGibberic Acid.

alloGibberic acid, $C_{18}H_{20}O_3$ (VII; $R = H$), m. p. 201–203°, $[\alpha]_D^{20} - 84°$, frequently isolated as the monohydrate, m. p. 125–130°, resetting 140° and remelting 195–197°, was obtained, together with 1 mol. carbon dioxide, by mild mineral acid degradation of gibberellic acid (*82, 30*). alloGibberic acid was also obtained by heating gibberellic acid with water and from gibberellenic acid (LVI, $R = H$; Chart 4, p. 371) (see also p. 375) with boiling water or cold mineral acid (*121*). Anhydrous allogibberic acid was prepared (*30*) from the hydrate by crystallisation from toluene and was shown to be a tetra-cyclic hydroxy-acid containing an aromatic ring by the following evidence (*82, 225*).

alloGibberic acid contains a carboxyl group which must be attached at the same position as in gibberic acid since the methyl ester (VII; $R = Me$), m. p. 98–99°, $[\alpha]_D^{22} - 82°$, was isomerised with mineral acid to methyl gibberate. alloGibberic acid contains a hydroxyl group ($\nu_{max.}$ 3460 cm.$^{-1}$) considered to be tertiary because of the difficulty of acylation and failure to oxidise dihydroallogibberic acid, $C_{18}H_{22}O_3$ (XLVI),

m. p. 204–207°, $[\alpha]_D^{21}$ — 71°, to a ketone: it contains a terminal methylene group since the methyl ester (VII, below; $R = Me$) showed a strong band near 890 cm.$^{-1}$ in the infrared spectrum and because formaldehyde was obtained in significant yield on ozonolysis. The ultraviolet spectrum (Table 2, p. 358) indicated the presence of an aromatic ring not conjugated with the exocyclic methylene group.

(XLVI.)          (VII.)          (XLVII.)

When dihydroallogibberic acid was oxidised with alkaline potassium permanganate (225) the product was the $\Delta^{4b(5)}$ dehydro-derivative, $C_{18}H_{20}O_3$ (XLVII), m. p. 170–172°, analogous to dehydrogibberic acid (XV, p. 359) obtained under similar conditions from gibberic acid. Dihydroallogibberic acid was regenerated as sole product on catalytic hydrogenation of (XLVII) under acidic, neutral and basic conditions.

The ozonolysis of allogibberic acid (225) is outlined in Chart 3 and gave in addition to formaldehyde, the $\alpha$-ketol (L; $R = H$), $C_{17}H_{18}O_4$,

Table 3. Constants for the $C_{17}H_{18}O_5$ Keto-acids

Degradation Products of allo- and epialloGibberic Acids.

| Compound | Configuration of substituent at | | | Acid $R = R' = H$ | | Ester $R = R' = Me$ | |
|---|---|---|---|---|---|---|---|
| | 9 | 12 H | 13 | M. p. | $[\alpha]_D^a$ | M. p. | $[\alpha]_D^a$ |
| IX (p. 367) . | $\beta$ | $\alpha$ | $\beta$ | 217–219° and 280° (dec.) | —112° | 205–207° | +17°* |
| LI (p. 367) . | $\alpha$ | $\alpha$ | $\beta$ | 252–255° (dec.) | —76° | oil | —53° |
| LII (p. 367) . | $\alpha$ | $\alpha$ | $\alpha$ | 205–207° | —77° | 167–169° | —78°* |
| LIII (p. 367) . | $\beta$ | $\alpha$ | $\alpha$ | 218–220° | +25° | oil | +55° |
| LXIV (p. 373) . | $\beta$ | $\beta$ | $\beta$ | 206–207° | +77° | 168–169° | +79°* |
| LXV (p. 373) . | $\alpha$ | $\beta$ | $\beta$ | 218–220° | —25° | oil | —57° |

$^a$ In ethanol except when * in acetone.

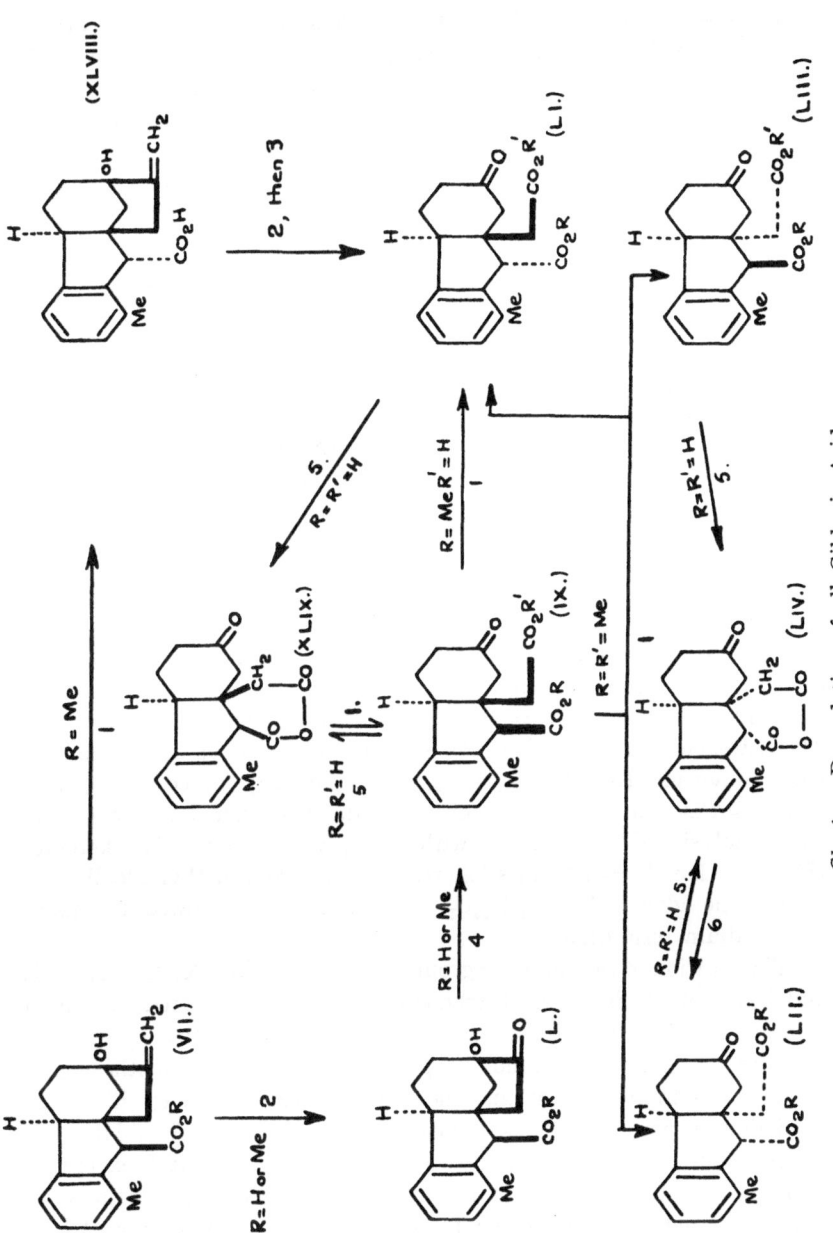

*Chart 3.* Degradation of alloGibberic Acid.

(Reagents: 1. OH⁻. 2. O₃. 3. H₂O₂. 4. NaBiO₃. 5. Ac₂O. 6. H₂O.)

m. p. 258–260° (dec.), $[\alpha]_D^{19}$ — 69°, which still contained the carboxyl group of allogibberic acid since ozonolysis of methyl allogibberate yielded the methyl ester (L; $R = Me$), m. p. 130–132°, $[\alpha]_D^{19}$ — 57°. The infrared spectrum of the latter showed $\nu_{max}$. 1752 cm.$^{-1}$ indicating that the ketone group was present in a saturated 5-membered ring. The ketols (L; $R = H$ or Me) were not oxidised by bismuth oxide in acetic acid but were split by sodium bismuthate to the keto-acids $9\beta$-carboxy-1-methyl-7-oxo-5,6,7,8:12$\alpha$,13-hexahydrofluorene-13$\beta$-acetic  acid   (IX, p. 367; $R = R' = H$; Chart 3), $C_{17}H_{18}O_5$ (Table 3, p. 366) and $9\beta$-methoxycarbonyl-1-methyl-7-oxo-5,6,7,8:12$\alpha$,13-hexahydrofluorene-13$\beta$-acetic

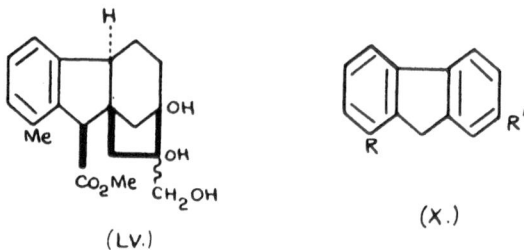

(LV.)                                                                    (X.)

acid   (IX;  $R = Me$,  $R' = H$),  $C_{18}H_{20}O_5$,   m.  p.  239–241°   (dec.), $[\alpha]_D^{21}$ — 66°, respectively, which contained a carbonyl group in a saturated 6-membered ring as shown by the infrared spectrum of the dimethyl ester (IX, p. 367; $R = R' = Me$), $\nu_{max}$. 1712 and 1739 cm.$^{-1}$ (ester $C=O$). The keto-acid (IX; $R = R' = H$), sometimes isolated as the 6-membered ring anhydride (XLIX) $C_{17}H_{16}O_4$, m. p. 285–287° (dec.), $\nu_{max}$. 1813, 1766, ~ 1726 cm.$^{-1}$, was also obtained directly from allogibberic acid by ozonolysis or by oxidation with zinc permanganate. The keto-acid (IX; $R = Me$, $R' = H$) was similarly obtained from methyl allogibberate or via the glycol (LV) $C_{19}H_{24}O_5$, m. p. 192–195°, followed by fission with sodium bismuthate.

The position of the carbonyl group in the keto-acid (IX; $R = R' = H$) and hence of one point of attachment of the five-membered ring in the $\alpha$-ketol (L; $R = H$) and in allogibberic acid was revealed by selenium dehydrogenation to a fluorenol, m. p. 166–168°, shown by synthesis (224) to be 7-hydroxy-1-methylfluorene (X, above; $R = Me$, $R' = OH$). In dehydrogenation of the keto-acid (IX, p. 367; $R = R' = H$) to the fluorenol all the non-skeletal carbon atoms were eliminated except the aromatic C—Me group and thus the second point of attachment of the five-membered ring was shown to be angular as in (VII, p. 367). The alternative angular position does not accommodate a five-membered ring $D$ or the formation of the dehydro-derivative (XLVII, p. 366) or allow the formation of a 6-membered ring anhydride (XLIX). These facts established the structure of allogibberic acid (225).

## b) Absolute Configuration.

The absolute configuration (VII, p. 367; $R = H$) in which rings $B/C$ are *trans* fused and the hydrogen atoms at 4b and 10 are $\alpha$, was deduced (*121, 84, 295*) from the following evidence. alloGibberic acid was stable to dilute aqueous alkali but methyl allogibberate underwent epimerisation (followed by hydrolysis) to give the oily 10$\alpha$ isomer (XLVIII; Chart 3, p. 367), $[\alpha]_D^{22} - 149°$, of allogibberic acid (*121*). The acid (XLVIII) with hot dilute mineral acid underwent rearrangement of rings $C/D$ (see

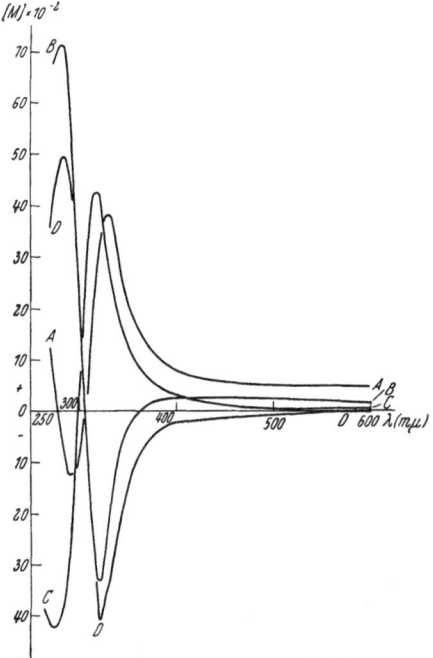

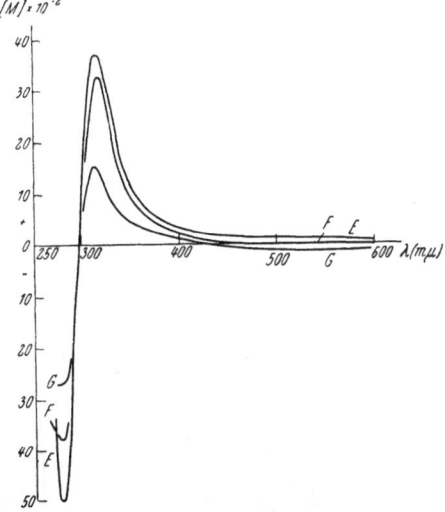

Fig. 1. Optical rotatory dispersion curves: $A$, ketol-lactone (LXVII) (p. 366); $B$, keto-ester (LXXXVIII, $R = H$) (p. 383); $C$, ketone (LIX) (p. 370); and $D$, gibberic acid (XVI, $R = H$) (p. 371).

Fig. 2. Optical rotatory dispersion curves (corrected for background dispersion): $E$, keto-ester (IX, $R = R' = $ Me) (p. 367); $F$, keto-ester (LXXII, $R = $ Me) (p. 377); and $G$, keto-ester (LXIV, $R = R' = $ Me) (p. 373).

below) to the 10$\alpha$ isomer (XLI; $R = H$; *Chart 4*, p. 371), $[\alpha]_D^{25} - 62°$, of gibberic acid and on ozonolysis furnished the 9$\alpha$ keto-acid (LI; $R = R' = H$) (Table 3, p. 366), also obtained by the action of alkali on the keto-acid (IX; $R = $ Me, $R' = H$).

The acid (LI; $R = R' = H$) with acetic anhydride furnished the anhydride (XLIX) but less readily than the 9$\beta$ epimer (IX; $R = R' = H$) did. Alkaline hydrolysis of the anhydride (XLIX) gave only the acid (IX; $R = R' = H$) and showed that the $C_{(9)}$ and $C_{(13)}$ substituents were *cis* related in that compound and *trans* in the acid (LI; $R = R' = H$). It follows that the $C_{(10)}$ carboxyl substituent and ring $C$ 2-carbon bridge are *cis* in allogibberic acid and hence in dehydrodihydroallo-gibberic acid (XLVII, p. 366). Since catalytic reduction of the latter will occur at the less hindered face of the molecule opposite to the $C_{(10)}$ substituent and ring $C$ bridge, and since this process regenerates the original stereochemistry at $C_{(4b)}$, it follows that rings $B$ and $C$ must be *trans* fused in allogibberic acid. The optical rotatory dispersion curves (*98, 96*) for the esters (IX; $R = R' = Me$) (see *Fig.* 2) and (LI; $R = R' = Me$) showed positive Cotton effects indicating (*121*) that these compounds have the 12$\alpha$, 13$\beta$ absolute configuration shown. It follows that allogibberic acid has the 4bS, 7S, 9aS, 10R absolute configuration (VII, p. 371; $R = H$).

The Wagner-Meerwein mechanism, via the intermediate cation (LVIII), proposed (*86, 225*) for the acid induced isomerisation (*82*) of allogibberic acid to gibberic acid indicated in Chart 4, p. 371 was disputed by Birch et al. (*17*) who believed that a mechanism involving hydration of the terminal methylene group

(LVIII.)

followed by pinacol-pinacolone rearrangement of the resulting glycol was more in accord with the results of their degradation of 8-$^{14}$C-gibberellic acid obtained by growing *G. fujikuroi* on a substrate of $CH_3 \cdot {}^{14}CO_2H$. No stereochemical change at $C_{(7)}$ and $C_{(9a)}$ is implied in the pinacol-pinacolone mechanism: but the Wagner-

(LIX.)    (LX.)    (LXI.)

Meerwein mechanism requires that the $7 \rightarrow 9a$ two carbon bridge in gibberic acid has the opposite configuration ($\alpha$) to that which it occupied in allogibberic acid ($\beta$). The rotatory dispersion curve for gibberic acid (*295, 120*) showed a negative Cotton effect (see *Fig. 1*), and was the mirror image of the curves obtained for the ketol (LXVII; Chart 6, p. 377) and for the ketone (LIX) (*140*) obtained by ozonolysis of gibberellin $A_4$ methyl ester. This evidence is conclusively in favour of the Wagner-Meerwein mechanism for the allogibberic acid $\rightarrow$ gibberic acid rearrangement and for the absolute configuration (XVI) for gibberic acid: and, indeed, some recent results (*120*) with 8-$^{14}$C-gibberellic acid indicate that the earlier work of Birch et al. was incorrect.

Catalytic reduction of dehydrogibberic acid (XV) regenerated the original (*B/C cis*) stereochemistry at $C_{(4b)}$ (cf. dehydrodihydrogibberic

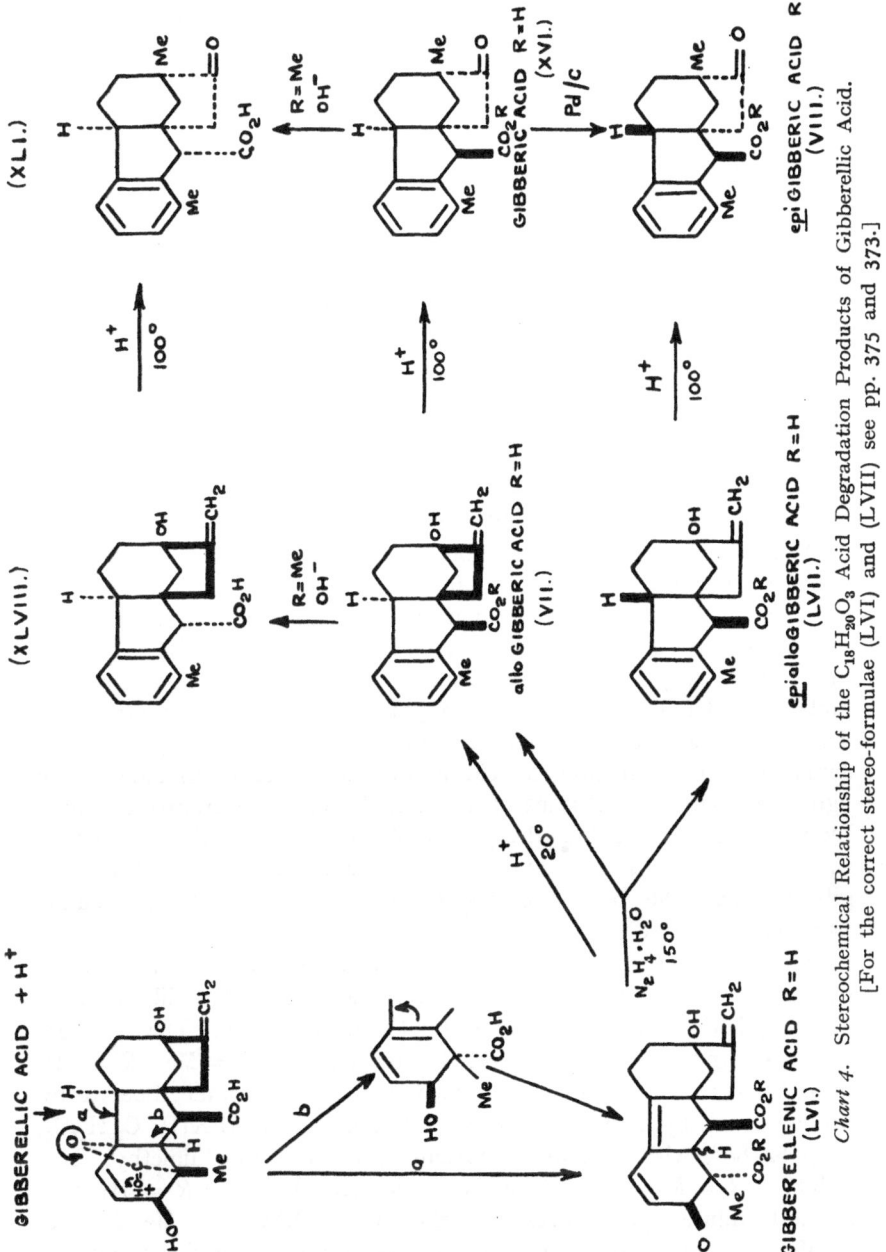

*Chart 4.* Stereochemical Relationship of the $C_{18}H_{20}O_3$ Acid Degradation Products of Gibberellic Acid.
[For the correct stereo-formulae (LVI) and (LVII) see pp. 375 and 373.]

acid, XLVII) and it is clear from these results that the configuration of the carboxyl substituent rather than the ring $C$ bridge determines the steric course of the reduction. In agreement with this observation, catalytic reduction of gibberone ($120$) gave a mixture of dihydro-gibberones (LX), $C_{17}H_{20}O$, m. p. 98° and 84°, epimeric at $C_{(4b)}$. The action of hot dilute aqueous alkali or cold methanolic alkali on the ester (IX; $R = R' = $ Me; p. 367) provided ($121$) another example of a reaction involving a change in the $B/C$ ring fusion in this series. In addition to the acid (LI; $R = R' = $ H) two new $C_{17}H_{18}O_5$ acids, the $9\alpha$, $12\alpha$, $13\alpha$ (LII; $R = R' = $ H) and $9\beta$, $12\alpha$, $13\alpha$ (LIII; $R = R' = $ H) acids (Table 3) were obtained. It is considered ($121$) that these compounds arise by intramolecular Claisen type condensation at $C_{(6)}$ giving the intermediate $1:3$-diketone (LXI) followed by fission of the $C_{(6)}$–$C_{(7)}$ bond. The acids (LII and LIII; $R = R' = $ H) both yielded the same anhydride (LIV), $C_{17}H_{16}O_4$, m. p. 154–155°, $[\alpha]_D^{22} + 190°$ which gave only the $9\alpha$, $13\alpha$ acid (LII; $R = R' = $ H) on hydrolysis. The $C_{(9)}$ and $C_{(13)}$ substituents must therefore be *cis* related in this compound.

### c) *epi*alloGibberic Acid.

When gibberellic acid or gibberellenic acid (LVI; $R = $ H) were boiled with hydrazine hydrate as indicated in Chart 4 both allogibberic acid and an isomer, which, from its mode of formation and properties was shown to be the 4b epimer, *epi*allogibberic acid (LVII; $R = $ H), m. p. 244°, $[\alpha]_D^{24} + 87°$ were formed. *epi*alloGibberic acid gave *epi*-gibberic acid (VIII; $R = $ H) (see p. 371) with hot dilute mineral acid. The ultraviolet spectrum of *epi*allogibberic acid (Table 2, p. 358) was similar to that of allogibberic acid, and ozonolysis gave formaldehyde in agreement with the presence of a terminal methylene group. Catalytic reduction gave two dihydro*epi*allogibberic acids (LXIII; Chart 5), $C_{18}H_{22}O_3$, m. p. 234–236° and 178–180°, epimeric at $C_{(8)}$. Like *epi*-gibberic acid the former compound was not oxidised by alkaline permanganate at 0°.

The degradation of *epi*allogibberic acid is outlined in *Chart 5*. Ozonolysis ($121$) gave in addition to formaldehyde a new dibasic keto-acid, $C_{17}H_{18}O_5$ (LXIV; $R = R' = $ H) (Table 3, p. 366). This $9\beta$, $12\beta$, $13\beta$ acid and its $C_{(9)}$ monomethyl ester (LXIV; $R = $ Me, $R' = $ H), m. p. 130–132°, $[\alpha]_D^{21} + 74°$, obtained from methyl *epi*allogibberate by ozonolysis or by fission of the corresponding glycol (LXII), $C_{19}H_{24}O_5$, m. p. 84–86°, with sodium bismuthate gave the same dimethyl ester (LXIV; $R = R' = $ Me). The keto-acid (LXIV; $R = R' = $ H) was stable to dilute aqueous alkali but the esters (LXIV; $R = $ Me, $R' = $ H or Me) were racemised at $C_{(9)}$ under these conditions giving a mixture of the acids (LXIV and LXV; $R = R' = $ H). Inversion at $C_{(13)}$ did not take place with the ester (LXIV; $R = R' = $ Me). Both acids readily gave the same anhydride, $C_{17}H_{16}O_4$ (LXVI), m. p. 154–155°, $[\alpha]_D^{24} - 191°$, with acetic anhydride. Since hydrolysis of the anhydride (LXVI) with water or alkali gave only the acid (LXIV; $R = R' = $ H) the $C_{(9)}$ and $C_{(13)}$ substituents are *cis* related in this acid and *trans* in the epimeric acid (LXV; $R = R' = $ H).

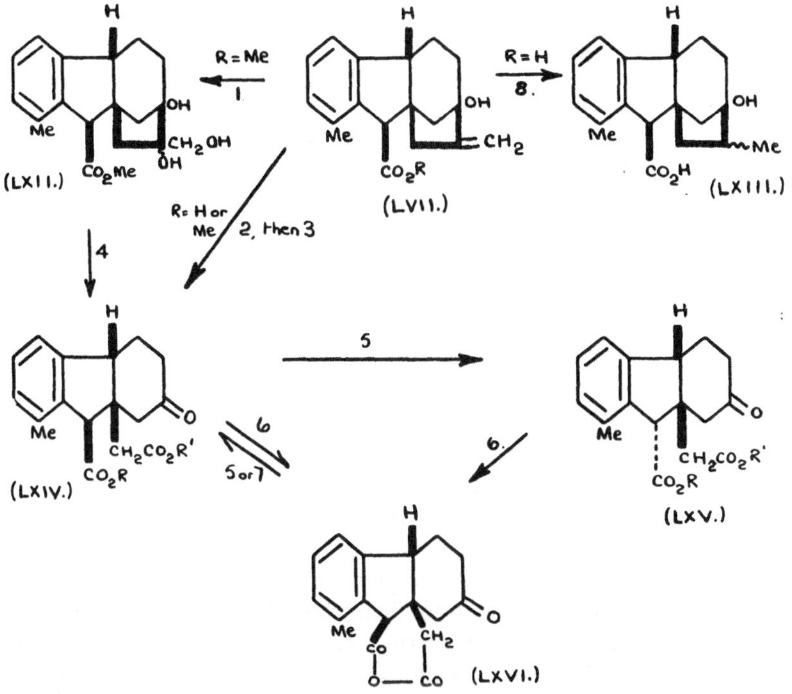

Chart 5. Degradation of *epi*alloGibberic Acid.

(Reagents: 1. $OsO_4$. 2. $O_3$. 3. $H_2O_2$. 4. $NaBiO_3$. 5. $OH^-$. 6. $Ac_2O$. 7. $H_2O$. 8. $H_2$.)

The acids (LXIV and LXV; $R = R' = $ H) were respectively the antipodes of the acids (LII and LIII; $R = R' = $ H) obtained by degradation of allogibberic acid: their m. p.'s and infrared spectra were identical but their rotatory dispersion curves were mirror images. The anhydrides (LXVI) and (LIV) were also antipodes. These relationships are to be expected if the acids (IX, p. 367; $R = R' = $ H) and (LXIV; $R = R' = $ H) differ only in configuration at $C_{(12)}$. Inversion of the remaining two asymmetric centres at $C_{(9)}$ and $C_{(13)}$ in (IX; $R = R' = $ H) then gives (LII; $R = R' = $ H), the mirror image of (LXIV; $R = R' = $ H).

This evidence provided proof that allo- and *epi*allogibberic acids differed only in configuration at 4b and that the ring $C$ bridge and $C_{(10)}$ carboxyl substituent were *cis* in both compounds. *epi*alloGibberic acid therefore has the absolute configuration (LVII; $R = H$), and the stereochemical relationships of the $C_{18}H_{20}O_3$ hydrolysis products of gibberellic acid are summarised in Chart 4. The positive Cotton effects, shown by the rotatory dispersion curves of the dimethyl esters (LXIV and LXV; $R = R' = Me$) indicated (*121*) that these compounds adopt preferentially the conformation in which the 12$\beta$ hydrogen atom is equatorial and the 13$\beta$ substituent axial.

Molecular models of the keto acids, $C_{17}H_{18}O_5$, show that when rings $B/C$ are *trans* fused the stable configuration of the $C_{(9)}$ carboxyl is $\alpha$ but when rings $B/C$ are *cis* fused there is little to choose on steric grounds between the two possible configurations for the $C_{(9)}$ substituent. The ready inversion at $C_{(9)}$ in the 9$\beta$, 12$\alpha$, 13$\beta$ esters (IX, p. 367; $R = Me$, $R' = H$ or Me) and the facile racemisation at $C_{(9)}$ in the esters (LXIV; $R = Me$, $R' = H$ or Me) are consistent with this observation as are the stability and lability of $C_{(9)}$ in the trimethyl esters (XLIV; $R = Me$) and (XIX; $R = Me$) respectively (see p. 364). The stability of the $C_{(13)}$ centre in the ester (LXIV; $R = R' = Me$) compared with the ester (IX; $R = R' = Me$) may be due to the difficulty of formation of an intermediate 1:3-diketone; the distance between $C_{(6)}$ and the appropriate methoxycarbonyl substituent is greater in those compounds with rings $B/C$ *trans* fused and ring $C$ in the chair conformation.

Models of the tetracyclic structures show that when rings $B/C$ are *trans* fused the least hindered configuration of the $C_{(10)}$ substituent is *trans* to the ring $C$ bridge; but when rings $B/C$ are *cis* fused the less hindered configuration is *cis* to the ring $C$ bridge although there is little difference between the two possibilities. The stability to alkali of the $C_{(10)}$ centre in methyl *epi*gibberate (see p. 364) and the lability under the same conditions of methyl gibberate and methyl allogibberate are consistent with these observations but the stability of methyl *epi*allogibberate is not wholly explained.

### D. Structure and Stereochemistry of Gibberellic Acid.

#### a) Gibberellic Acid.

Gibberellic acid, $C_{19}H_{22}O_6$ (I; $R = H$), first isolated by Curtis and Cross (*92*) is a colourless crystalline, optically active, tetracyclic di-hydroxy lactonic carboxylic acid, m. p. 233—235° (dec.). Kuhn-Roth oxidation showed the presence of (at least) one C—Me group.

Gibberellic acid formed (*82, 83*) a methyl ester (I; $R = Me$), m. p. 209 to 210°, and acetyl, m. p. 233—234° (dec.), and diacetyl, m. p. 188°,

$[\alpha]_D$ + 176°, derivatives; the latter formed a methyl ester, m. p. 167 to 169° (*218*). A large number of salts, esters and acyl derivatives have been described (*218*). One of the hydroxyl groups was shown to be secondary since, in the reduction products of gibberellic acid, it was readily oxidised to a ketone (*83*); the other was considered to be tertiary from the difficulty of acylation and the failure to prepare a ditosylate (*86*). When gibberellic acid was allowed to stand with excess of 0.1N alkali at room temperature a second equivalent of alkali was consumed: this,

(I.)          (LVI.)

together with the fact that the infrared spectra of the acid and its derivatives in dioxan solution showed a strong band near 1780 cm.$^{-1}$, indicated the presence of a saturated $\gamma$-lactone ring (*82*).

Microhydrogenation (*82*) showed the presence of two ethylenic bonds. The above evidence accounted for all the oxygen atoms and indicated that gibberellic acid was tetracyclic. The ultraviolet spectrum of pure gibberellic acid showed only end absorption and ruled out the presence of a conjugated system. Gibberellic acid is unstable in aqueous solution at room temperature and the 10a$\xi$-gibba-3,4a(4b)-diene-1,10-dicarboxylic acid, gibberellenic acid (LVI; $R = H$), $\lambda_{max.}$ 253 m$\mu$, $E_{1\,cm.}^{1\%}$ 648 (see p. 380), is one of the decomposition products (*113, 217*). The acid (LVI; $R = H$) is present in small amounts in gibberellic acid obtained by solvent extraction of *G. fujikuroi* broths (*113, 89*) followed by recrystallisation from ethyl acetate—light petroleum. It is readily detected and estimated by the characteristic ultraviolet absorption and can be removed from the gibberellic acid by chromatography on buffered Celite or by crystallisation from aqueous methanol (*89*). In the latter method the product is sometimes gibberellic acid hydrate, $C_{19}H_{22}O_6 \cdot H_2O$, m. p. 233–235° (dec.), after softening at 185°, conveniently recognised by its characteristic infrared spectrum. Anhydrous gibberellic acid was obtained on drying the hydrate at 100° in vacuo. With boiling water or with cold dilute mineral acid both gibberellic acid and the acid (LVI; $R = H$) gave allogibberic acid (VII, p. 371; $R = H$) together with 1 mol. of carbon dioxide (*82, 121*). Boiling dilute mineral acid gave gibberic acid (XVI, p. 371; $R = H$) (*82*).

Reduction of gibberellic acid with hydrazine in boiling ethanol (*83*) was stereospecific giving the tetrahydro compound (LXXIII; $R = H$;

Table 4. Reduction Products of Gibberellic Acid and their $C_{(2)}$ Epimers.

| Compound | Structure | Configuration of $C_{(2)}$ hydroxyl | Acid, $R = H$ | | Ester, $R = Me$ | |
|---|---|---|---|---|---|---|
| | | | M. p.[a] | $[\alpha]_D{}^f$ | M. p. | $[\alpha]_D{}^f$ |
| Gibberellin $A_1$.. | II (p. 377) | $\beta$ | 255–258° | +36° | 234–235° | +35° |
| | XC (p. 383) | $\alpha^b$ | 227–230°e | +43° | 193° | +42° |
| Tetrahydrogib-berellic acid . | LXXIII (p. 377) | $\beta$ | 300–301° | +64° | 271–272° | +54° |
| | LXXXIX (p. 383) | $\alpha$ | | | { 214° / 235–239° } | +59° |
| 8-*epi*Tetrahydro-gibberellic acid | LXXIII (p. 377) | $\beta$ | 289–291° | +42° | { 238° / 243–244° } | +43° |
| | LXXXIX (p. 383) | $\alpha$ | | | 166–168° | +46° |
| 2:4a-Dihydroxy-1:7-dimethyl-8-oxo-7α-gib-bane-1:10β-dicarboxylic acid 1 → 4a lactone...... | LXXXVIII (p. 383) | $\beta^c$ | 265–267° | +50° | 226–228° | +54° |
| | XCII (p. 383) | $\alpha^d$ | 262–264° | +48° | 238° | +42° |

a With decomposition.

Alternative names now discarded:
b pseudogibberellin $A_1$ (*305*), c gibberellin C (*305, 153*), d isogibberellin $A_1$ (*306*).
e Mono-hydrate, m. p. 138–145° decomp.
f In methanol or ethanol.

Chart 6), $C_{19}H_{26}O_6$, 2β:4a:7-trihydroxy-1:8-dimethylgibbane-1:10β-di-carboxylic acid 1 → 4a lactone as the only isolable product. Catalytic reduction in the presence of palladised charcoal was not reproducible but the resulting complex mixture of reduction and hydrogenolysis products could be separated by chromatography on buffered Celite (*119*) or on silica gel (*307*) into gibberellin $A_1$, $C_{19}H_{24}O_6$ (II, p. 377; $R = H$), and the above tetrahydro-compound (LXXIII; $R = H$) and its $C_{(8)}$ epimer (*119*) (Table 4).

The Japanese workers (*307*) have described two dibasic hydrogenolysis products considered to be hexahydro-compounds $C_{19}H_{28}O_6$. The stereochemistry of these compounds and their relationship to (LXXIV) a hydrogenolysis product of methyl gibberellate is not known. Chromic acid oxidation of one of them is said (*283*) to give a neutral ketone $C_{18}H_{26}O_4$, m. p. 198–200°.

Catalytic reduction of methyl gibberellate outlined in Chart 6 took a similar course and gave a mixture of acidic and neutral products; the latter were more readily purified and characterised. A mixture, separable by chromatography, of gibberellin $A_1$ methyl ester (II; $R = Me$)

and methyl 1-carboxy-2β:4a:7-trihydroxy-1:8ξ-dimethylgibb-3-ene-10β-carboxylate 1 → 4a lactone (LXXV), $C_{20}H_{26}O_6$, m. p. 227–230° (dec.), $[\alpha]_D^{17}$ —74°, was obtained (*118*) with 1 mol. hydrogen and palladised charcoal. With 2 mols. hydrogen and either Adams catalyst or palladised charcoal the neutral fraction was a mixture of tetrahydro compounds (LXXIII; *R* = Me) epimeric at $C_{(8)}$ (*83*). Hydrogenation in acetic acid with Adams catalyst until uptake ceased (*119*) gave the hydrogenolysis product 2β:7-dihydroxy-10β-methoxycarbonyl-1:8-*epi*dimethyl-4aξ-gibbane-1α-carboxylic acid (LXXIV) (*84*), $C_{20}H_{30}O_6$, m. p. 269–271° (dec.), $[\alpha]_D^{26}$ + 50° as a major product.

The presence of an allylic alcohol grouping in gibberellic acid was established (*85, 83*) (*Chart 6*) by the oxidation of methyl gibberellate

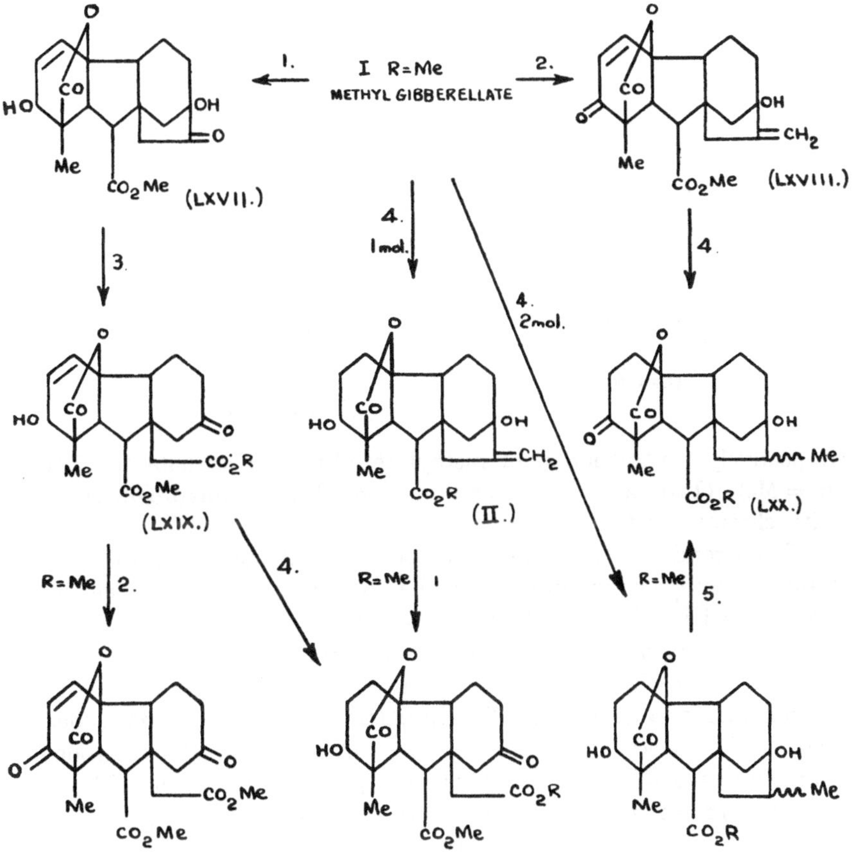


(LXXI.) (LXXII.) (LXXIII.)

*Chart 6.* Degradation of Methyl Gibberellate.

(Reagents: 1. $O_3$. 2. $MnO_2$. 3. $NaIO_4$. 4. $H_2$/Pd. 5. $CrO_3$.)

with manganese dioxide in chloroform* to the $\Delta^3$-2-one (LXVIII), $C_{20}H_{22}O_6$, m. p. 186–187°, $[\alpha]_D^{25} + 49°$, $\lambda_{max}$. 228 m$\mu$, $\varepsilon$ 9700. Catalytic hydrogenation of the latter afforded, after chromatographic separation, the $C_{(8)}$ epimeric 2-ketones, $C_{20}H_{26}O_6$ (LXX; $R = Me$), m. p. 161–163°, $[\alpha]_D^{16} + 144$, and m. p. 131–133°, $[\alpha]_D^{23} + 126°$, also obtained,

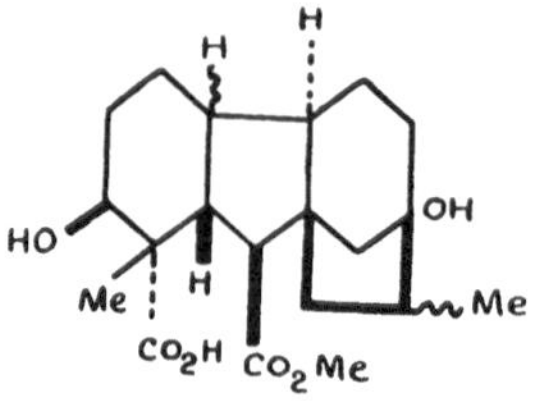

(LXXIV.)

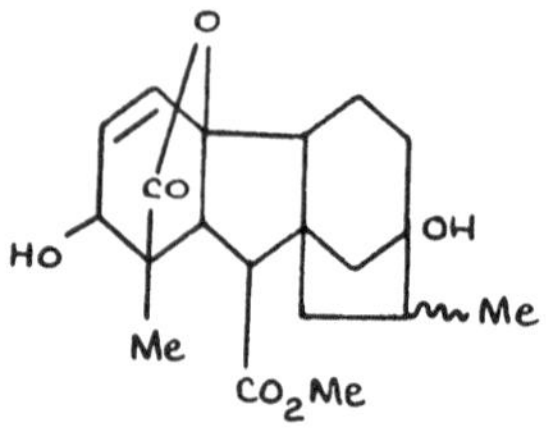

(LXXV.)

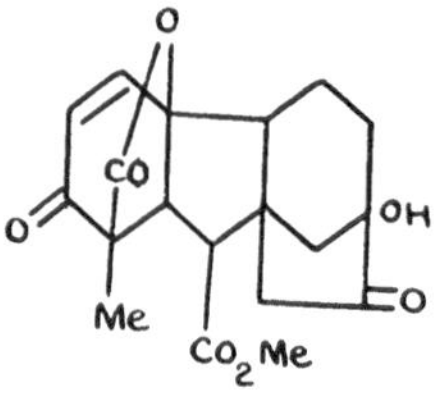

(LXXVI.)

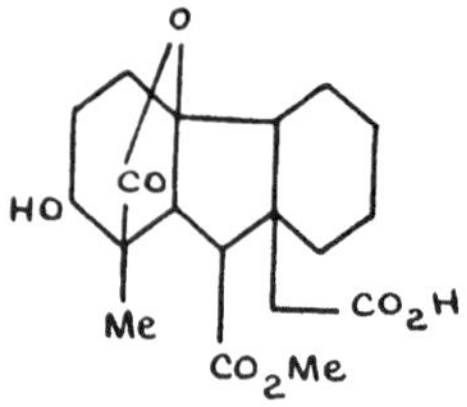

(LXXVII.)

respectively, by oxidation of methyl tetrahydrogibberellate (LXXIII; $R = Me$) (Table 4, p. 376) and methyl 8-*epi*-tetrahydrogibberellate with chromic oxide.

The second ethylenic bond in gibberellic acid was shown to be in a terminal methylene group (*118*) by the ozonolysis of methyl gibberellate (represented in Chart 6) which took a similar course to the ozonolysis of methyl allogibberate (p. 366) giving formaldehyde (*118*), a ketol, methyl 1-carboxy-2$\beta$:4a:7-trihydroxy-1-methyl-8-oxogibb-3-ene-10$\beta$-carboxylate 1 → 4a lactone, $C_{19}H_{22}O_7$ (LXVII), m. p. 230–232° (dec.), and a keto-acid, 1$\alpha$-carboxy-2$\beta$:11$\alpha$-dihydroxy-9$\beta$-methoxycarbonyl-1$\beta$-methyl-7-oxo-1,2:5,6,7,8:10$\beta$,11,12$\alpha$,13-decahydrofluorene-13$\beta$-acetic acid 1 → 11 lactone $C_{19}H_{22}O_8 \cdot H_2O$ (LXIX; $R = H$), m. p. 128–130° (dec.) (*85, 83*). The latter was also obtained by oxidation of the ketol (LXVII, p. 377) with periodate. Catalytic hydrogenation (*83*) of the keto-ester (LXIX; $R = Me$), m. p. 172–174°, $[\alpha]_D^{25} + 100°$, gave

---

* (*85*) corrects an earlier statement (*88*) to the contrary.

the corresponding perhydrofluorene (LXXII; $R = Me$) $C_{20}H_{26}O_8$, m. p. 169° (281) also obtained by the ozonolysis of gibberellin $A_1$ methyl ester to the keto-acid (LXXII; $R = H$), $C_{19}H_{24}O_8 \cdot H_2O$, m. p. 98°, followed by methylation. Oxidation of the keto-ester (LXIX, $R = Me$) with manganese dioxide in chloroform and of the ketol (LXVII) with chromic oxide, gave, respectively, the $\Delta^3$-2-ones $C_{20}H_{22}O_8$ (LXXI), m. p. 129–131°, $\lambda_{max.}$ 229 m$\mu$, $\varepsilon$ 7500, and $C_{19}H_{20}O_7$ (LXXVI), m. p. 215 to 229° (dec.) $\lambda_{max.}$ 229 m$\mu$, $\varepsilon$ 7050 (83).

In contrast to the facile oxidation of the secondary alcoholic group in methyl tetrahydrogibberellate (LXXIII; $R = Me$) and in the above ozonolysis products, the oxidation of methyl gibberellate with chromic oxide and a wide variety of other oxidising agents afforded intractable products (83). This is attributed to the ease with which both the terminal methylene group of gibberellic acid and the derived $C_{(8)}$ ketol are attacked by reagents which do not specifically oxidise secondary allylic alcohols.

The position of the carboxyl substituent on the gibbane nucleus of gibberellic acid is the same as in gibberic and therefore allogibberic acid since methyl gibberellate gave methyl gibberate with hot dilute mineral acid (153, 83). Both gibberellic acid and allogibberic acid therefore have tertiary hydroxyl, terminal methylene and carboxyl substituents and these facts suggested (86) that gibberellic acid and allogibberic acid had the same structure for rings $B/C/D$ and that the conversion of gibberellic acid to allogibberic acid involved only aromatisation of ring $A$.

This suggestion was confirmed (83) by the acid induced aromatisation of the keto-ester (LXIX; $R = Me$) to the keto ester (IX, p. 367; $R = R' = Me$; Chart 3, p. 367) obtained from methyl allogibberate; and it was supported by the selenium dehydrogenation (281) of the keto-acid (LXXII; $R = H$) to 7-hydroxy-1-methylfluorene (X, p. 368; $R = Me$, $R' = OH$) and by the observation that gibberellin $A_1$ with hot dilute mineral acid was isomerised (305, 153) to a keto-acid, $C_{19}H_{24}O_6 \cdot H_2O$ (LXXXVIII; $R = H$) (Chart 7, p. 383 and Table 4, p. 376), 2$\beta$:4a-dihydroxy-1:7-dimethyl-8-oxo-7$\alpha$-gibbane-1:10$\beta$-dicarboxylic acid 1 → 4a lactone, analogous to the isomerisation of allogibberic to gibberic acid. The keto-acid (LXXXVIII; $R = H$) was identical with the gibberellin C obtained in the early Japanese work on the degradation of gibberellin A. The rotatory dispersion curve of the keto-acid (120) (see Fig. 1, p. 369) showed a negative Cotton effect consistent with the absolute configuration (LXXXVIII; $R = H$).

Ring $A$ of gibberellic acid must accommodate the methyl group which appears at position 1 in the fluorene degradation products and also the saturated $\gamma$-lactone ring and the allylic secondary alcohol grouping. The position of this secondary hydroxyl group was established independently by the Japanese and British workers by the following degradations:

In the first method (*284*), Clemmensen reduction of the keto-acid (LXXII; $R = H$) gave the deoxo compound, $1\alpha$-carboxy-$2\beta$:$11\alpha$-dihydroxy-$9\beta$-methoxycarbonyl-$1$-methyl-$10\beta$:$12\alpha$-perhydrofluorene-$13\beta$-acetic acid $1 \rightarrow 11$ lactone (LXXVII), $C_{19}H_{26}O_7$, m. p. $250°$, which, after chromic acid oxidation and selenium dehydrogenation, afforded 2-hydroxy-1-methylfluorene (XIV, p. 357; $R = H$).

In the second method (*83*), 4a-hydroxy-$1$:$7$-dimethyl-$2$:$8$-dioxo-$7\alpha$-gibbane-$1$:$10$-dicarboxylic acid $1 \rightarrow 4a$ lactone $C_{19}H_{22}O_6$ (XCI; $R = H$, Chart 7, p. 383), m. p. $281$–$283°$, $[\alpha]_D^{16} + 154°$, obtained by oxidation of the keto-acid (LXXXVIII; $R = H$) with chromic oxide was dehydrogenated to 2-hydroxy-$1$:$7$-dimethylfluorene, $C_{15}H_{14}O$ (XIV; $R = Me$) (*90*).

The methyl and hydroxyl substituents in ring $A$ are therefore located at positions $1$ and $2$ respectively. This, since it was known from nuclear magnetic resonance studies (*88, 285*) that the ring $A$ methyl group was attached to a saturated carbon atom and was tertiary, left only two possible structures for ring $A$ of gibberellic acid namely (I, p. 375; $R = H$) and (LXXVIII) in which the carbon atom of the $\gamma$-lactone ring is attached at $C_{(4a)}$. The formation in high yield of acidic hydrogenolysis products from the gibb-3-ene methyl gibberellate (but not from the corresponding gibbane) suggested the presence of an allylic lactone system in the former compound favouring (I; $R = H$). Structure (LXXVIII) for ring $A$ of gibberellic acid was also excluded by two reactions in which, after opening the lactone ring but without removal of the carboxyl group, the products were shown to contain a conjugated system of double bonds. Thus, firstly, gibberellenic acid (*113*) $C_{19}H_{22}O_6$ (LVI, p. 375; $R = H$), m. p. $190$–$192°$ (dec.), cannot, from its ultraviolet absorption ($\lambda_{max}$ 253 m$\mu$, $\varepsilon$ 22400) have the homoannular diene structure (LXXIX) favoured by GERZON et al. (*113*) but must be the heteroannular 3,4a(4b)-diene (LVI, p. 375; $R = H$) (*85, 217*). Structure (LXXIX) may be an intermediate, by *trans* elimination, in its formation from gibberellic acid and because of this (see Chart 4, p. 371, route b)

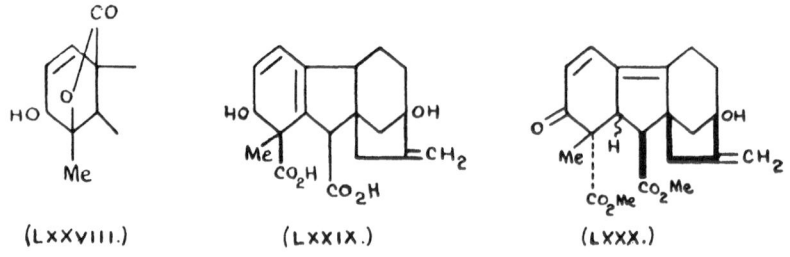

(LXXVIII.)          (LXXIX.)          (LXXX.)

the stereochemistry at $C_{(10a)}$ of gibberellenic acid is uncertain: the most probable configuration of the $C_{(10a)}$ hydrogen atom is $\alpha$ and the ring $A$ hydroxyl and carboxyl substituents are then equatorial. Manganese dioxide oxidation ($217$) of the ester (LVI, p. 375; $R = Me$) gave a dienone, $C_{21}H_{24}O_6$, m. p. 133–136°, $\lambda_{max}$. 309 m$\mu$, $\varepsilon$ 16,500, which was shown to be methyl 7-hydroxy-1-methyl-8-methylene-2-oxo-10a$\xi$-gibba-3,4a(4b)-diene-1:10$\beta$-dicarboxylate (LXXX) by comparison of the ultraviolet absorption intensity with the model homoannular dienone ethyl 1-methyl-2-oxo-cyclohexa-3:5-diene-1-carboxylate, (LXXXI), $\lambda_{max}$. 301 m$\mu$, $\varepsilon$ 3900 ($217$).

Secondly, in addition to methyl gibberate and methyl *epi*gibberate methanolic hydrogen chloride ($89$) on methyl gibberellate gave a conjugated triene, methyl 1:7-dimethyl-8-oxo-7$\alpha$:10a$\xi$-gibba-3,4a(4b),5-triene-1:10$\beta$-dicarboxylate (LXXXII), $C_{21}H_{24}O_5$, m. p. 197–198°, $[\alpha]_D^{19} + 762°$ ($\nu_{max}$. 1736 cm.$^{-1}$, no OH absorption), $\lambda_{max}$. 275, 287, 310 m$\mu$, log $\varepsilon$ 4.32, 4.42, 4.27. Catalytic reduction of (LXXXII) gave the 4a(4b)-ene, $C_{21}H_{28}O_5$, m. p. 127–128°, $[\alpha]_D^{17} + 24°$, $\lambda_{max}$. 204 m$\mu$, log $\varepsilon$ 4.06, while reduction with sodium borohydride gave the alcohol (LXXXIII), $C_{21}H_{26}O_5$, m. p. 149–153°, which retained the characteristic ultraviolet absorption of (LXXXII).

The British workers therefore concluded that gibberellic acid had structure (I, p. 375; $R = H$) ($85$) from which it followed that gibberellin $A_1$ had structure (II, p. 383; $R = H$). The structures of the other fungal and plant gibberellins have been related to gibberellin $A_1$ (see p. 354).

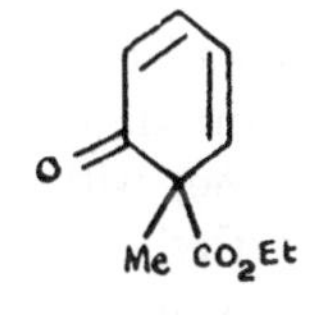

(LXXXI.)

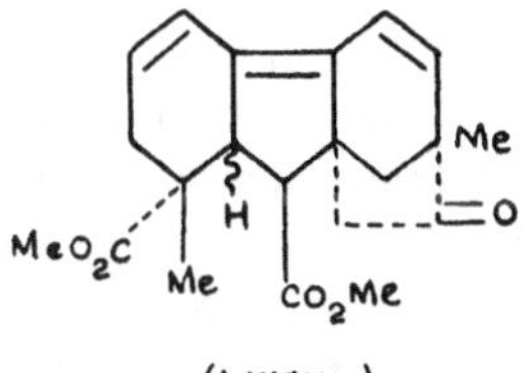

(LXXXII.)

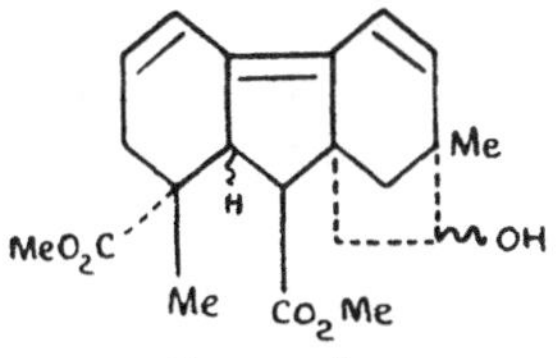

(LXXXIII.)

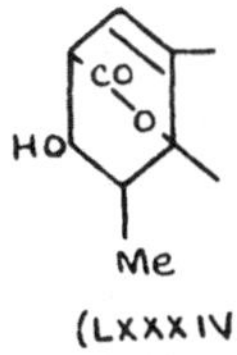

(LXXXIV.)

Several structures for ring $A$ of gibberellic acid were proposed prior to the proof ($85$) by the British workers of structure (I; $R = H$). The $\delta$-lactone structure (LXXXIV) favoured by the Japanese workers ($283$, $309$) is wholly inconsistent with the foregoing evidence and is based largely on the isolation, by selenium dehydrogenation of a degradation product of gibberellin A₁, of a fluorene, m. p. 94–97°, reported (283) to be identical with synthetic 1:3-dimethyl-fluorene, m. p. 99–100°. This report must be treated with some reserve since the m. p. of 1:3-dimethylfluorene has been recorded as 87° (65, 77, 140).

(Lxxxv.)        (Lxxxvi.)        (Lxxxvii.)

In the following paragraphs reactions reported by the University of Tokyo workers since 1957 are interpreted on the basis of structure (I, p. 375; $R = H$) for gibberellic acid. The structural formulae actually used in the Japanese papers are based on (LXXXIV) for ring $A$.

The British workers had earlier (88) proposed the structure (LXXXV; $R = H$) for gibberellic acid but before the reasons for this proposal are discussed two rearrangements of ring $A$ of the gibberellins must be considered.

### b) Epimerisation at Position 2 in Gibbanes.

$2\beta$-(axial)-Hydroxy compounds in which ring $A$ is saturated and (in gibbane analogues) the $1 \rightarrow 4a$ lactone bridge is intact (LXXXVI) are epimerised (88, 89) in very dilute aqueous alkali to the more stable $2\alpha$-(equatorial)-epimers and a retroaldol mechanism via the intermediate (LXXXVII) has been suggested (81). Thus, (88, 89) with 0.01N sodium hydroxide at room temperature, gibberellin A₁ methyl ester (II; $R = Me$) gave the $2\alpha$-hydroxy ester (XC; $R = Me$) (Table 4, p. 376) as illustrated in Chart 7: with acid the latter afforded the isomeric keto-ester (XCII; $R = Me$), the $2\alpha$-hydroxy epimer of the keto-ester (LXXXVIII; $R = Me$) previously obtained by acid rearrangement of gibberellin A₁ methyl ester. The Japanese workers (306) have carried out an identical series of transformations with the corresponding acids starting from gibberellin A₁ (II; $R = H$). Both the keto-esters (LXXXVIII; $R = Me$) and (XCII; $R = Me$) furnished the same 2:8-diketo-ester (XCI; $R = Me$), m. p. 219°, $[\alpha]_D + 149°$, on chromic oxide oxidation (88, 89).

Catalytic reduction of the $2\alpha$-hydroxy acid (XC; $R = H$), (Table 4, p. 376), obtained by alkaline hydrolysis of the corresponding ester (XC; $R = Me$) gave "dihydro-pseudogibberellin A₁" (305), $C_{19}H_{26}O_6$, m. p. 290 to 291° (dec.), which was shown (89) to be a mixture of $C_{(8)}$ epimers

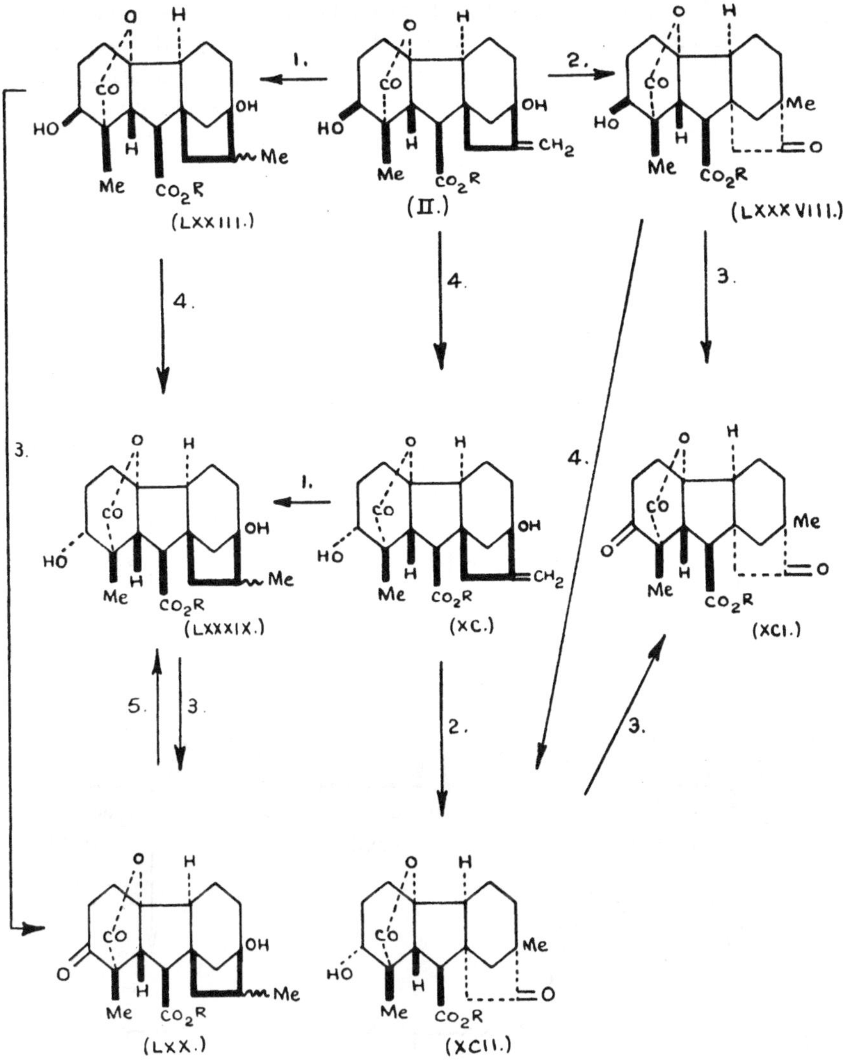

*Chart 7.* Epimerisation at $C_{(2)}$ in Gibbane $1 \rightarrow 4a$ Lactones.

(Reagents: 1. $H_2$/Pd. 2. $H^+$. 3. $CrO_3$. 4. $OH^-$. 5. $NaBH_4$.)

since methylation afforded the $2\alpha$-hydroxy ester (LXXXIX; $R = Me$) and its $C_{(8)}$ epimer.

With dilute alkali the $C_{(8)}$ epimeric methyl tetrahydrogibberellates (LXXIII; $R = Me$) likewise gave (*89*) the corresponding $2\alpha$-hydroxy compounds (LXXXIX; $R = Me$ (Table 4, p. 376) also obtained by catalytic reduction of the $2\alpha$-hydroxy ester (XC; $R = Me$). Chromic oxide oxidation of methyl tetrahydrogibberellate and its 2α-epimer gave the same 2-ketone (LXX; $R = Me$), m. p. 161–163°: the 8-epimeric ketone (LXX; $R = Me$), m. p. 131–133°, was obtained by oxidation of the corresponding two compounds in the 8-*epi* series. Sodium borohydride reduction of the ketone, m. p. 161–163°, gave predominantly the 2α-(equatorial) hydroxy ester (LXXXIX; $R = Me$).

METHYL GIBBERELLATE

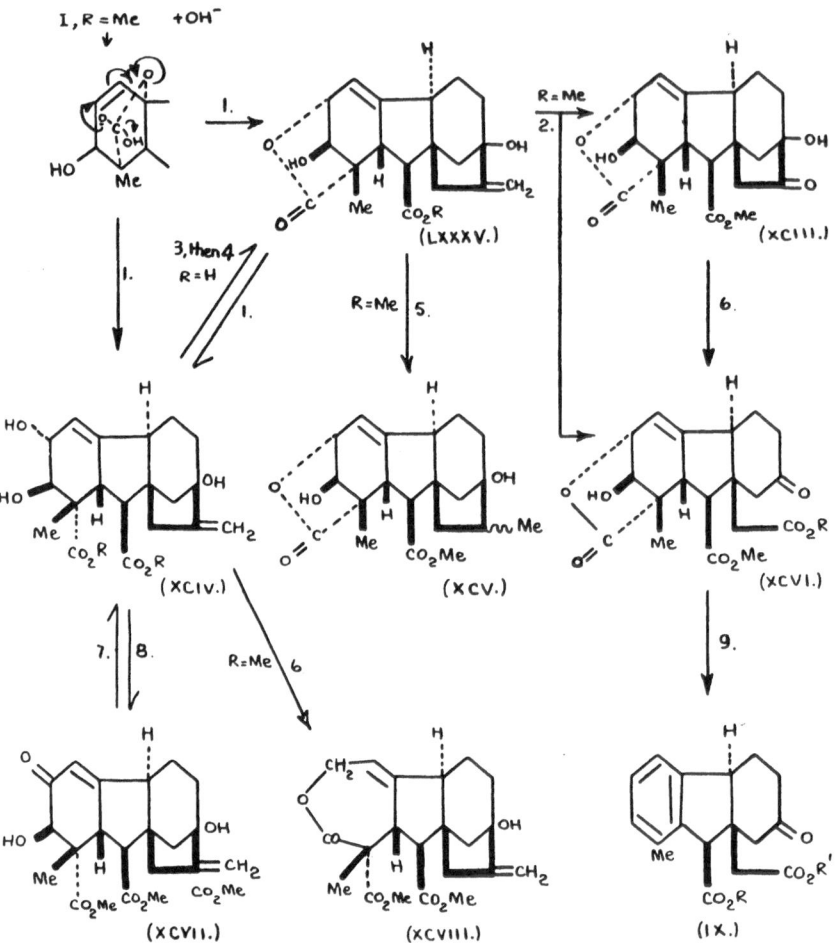

Chart 8. Degradation of Methyl 1-carboxy-2β,3α,7-trihydroxy-1-methyl-8-methylene-gibb-4-ene-10β-carboxylate 1 → 3 lactone.

(Reagents: 1. OH⁻. 2. O₃. 3. Heat 90°. 4. CH₂N₂. 5. H₂/Pd. 6. NaIO₄. 7. NaBH₄. 8. MnO₂. 9. H⁺.)

## c) Rearrangement of Gibb-3-ene $1 \rightarrow 4a$ Lactones.

The gibb-3-ene, methyl gibberellate, with 0.01N alkali furnished (*88, 89*), as shown in *Chart 8*, an isomer methyl 1-carboxy-2$\beta$,3$\alpha$:7-trihydroxy-1-methyl-8-methylenegibb-4-ene-10$\beta$-carboxylate $1 \rightarrow 3$ lactone (LXXXV; $R = Me$), m. p. 174°, $[\alpha]_D + 122°$, also obtained from the amorphous

$$
\begin{array}{ccc}
\text{OH} & \text{OH} & \\
| & | & \text{H} \quad | \\
-\text{CH}-\text{CH}-\overset{|}{\text{C}}=\text{C}- \\
\end{array}
\qquad
\begin{array}{c}
\text{CO}\!-\!\!-\!\!-\!\text{O} \\
| \quad\quad \text{OH} \quad | \\
-\text{C}-\text{C}-\text{CH}-\overset{|}{\text{C}}=\text{C}-
\end{array}
$$

(XCIX.)                     (C.)

2$\beta$,3$\alpha$:7-trihydroxy-1-methyl-8-methylenegibb-4-ene-1:10$\beta$-dicarboxylic acid (XCIV; $R = H$) $C_{19}H_{24}O_7 \cdot H_2O$, $[\alpha]_D + 41°$, by relactonisation followed by methylation (*83*).

The structure of the acid (XCIV; $R = H$), obtained by opening the lactone ring of gibberellic acid with alkali, was known from the evidence outlined below and the British workers therefore suggested (*88*) that the ester, m. p. 174°, was the 2$\alpha$-epimer of structure (LXXXV; $R = Me$) then assigned to methyl gibberellate. Some objections by the Japanese workers (*307*) were answered (*88*) but serious doubt was thrown on the suggestion when efforts to oxidise methyl gibberellate and the ester, m. p. 174° (or suitable derivatives), to the same ketone failed; and the proposal was finally abandoned when it was discovered that methyl gibberellate could be oxidised by manganese dioxide to a $\Delta^3$-2-one (see p. 377) and when catalytic hydrogenolysis of the two esters (*119*) was found to give the same acid (LXXIV, p. 378).

The ester, m. p. 174°, therefore has structure (LXXXV; $R = Me$) originally proposed (*88*) for methyl gibberellate. The facile alkali-induced epimerisation of a 2-hydroxy substituent in gibbane $1 \rightarrow 4a$ lactones is replaced in the gibb-3-ene $1 \rightarrow 4a$ lactones by an allylic type rearrangement to gibb-4-ene $1 \rightarrow 3$ lactones without concomitant epimerisation of the 2-hydroxy substituent.

The increase in absorption intensity at 210 m$\mu$ in the ester (LXXXV; $R = Me$), $\varepsilon$ 9,200, compared with methyl gibberellate, $\varepsilon$ 6,200, was consistent with the presence of a trisubstituted ethylenic bond in the former compound. Methyl gibberellate was not attacked by periodate but the dimethyl ester (XCIV; $R = Me$), m. p. 138–139°, methanolate, m. p. 95–100°, $[\alpha]_D^{22} + 16°$, rapidly consumed 1 mol. of the reagent (*83*) giving a compound, $C_{21}H_{26}O_7$, m. p. 138–139°, $[\alpha]_D^{23} + 127°$, believed to be the lactone (XCVIII) since three equivalents of alkali were consumed on hydrolysis. Oxidation of the ester (XCIV; $R = Me$) with manganese dioxide (*83*) in chloroform gave a ketol, methyl 2$\beta$:7-dihydroxy-1-methyl-8-methylene-3-oxogibb-4-ene-1:10$\beta$-dicarboxylate, $C_{21}H_{26}O_7 \cdot H_2O$ (XCVII), m. p. 105–120° (dec.), $[\alpha]_D^{15} + 121°$, which from its ultraviolet absorption at $\lambda_{max.}$ 240 m$\mu$, $\varepsilon$ 17,000, must also be an $\alpha,\beta$-unsaturated ketone of the type $R'R''C=CH \cdot CO—$. Sodium borohydride reduction of the ketol (XCVII) regenerated the ester (XCIV; $R = Me$) which must therefore contain the partial structure (XCIX). It followed that partial structure (C) must be present in the ester (LXXXV; $R = Me$) ($\nu_{max}$. 1774 cm.$^{-1}$, $\gamma$-lactone $C=O$) which since the nuclear magnetic resonance spectrum (285) showed the presence of a tertiary methyl group, must have structure (LXXXV, p. 384; $R = Me$).

In agreement with this structure the ester (LXXXV; $R = Me$) was not oxidised by manganese dioxide and catalytic reduction (89) gave mainly hydrogenolysis products. The ring $A$ trisubstituted double bond was difficult to reduce and with 1 mol. hydrogen the neutral fraction consisted essentially of a mixture of $C_{(8)}$ epimers, $C_{20}H_{26}O_6$ (XCV), m. p. 232–235°, $[\alpha]_D^{24} + 84°$, and m. p. 171–172°, $[\alpha]_D^{24} + 112°$. Ozonolysis (89) (see Chart 8) took the same course as the ozonolysis of methyl gibberellate and gave formaldehyde, a ketol, $C_{19}H_{22}O_7$ (XCIII), m. p. 175–177°, hydrate, $C_{19}H_{22}O_7 \cdot H_2O$, m. p. 104–106°, and a keto-acid, $C_{19}H_{22}O_8$ (XCVI; $R = H$), m. p. 176–178°, $[\alpha]_D^{22} + 52°$, also obtained by periodate oxidation of the ketol (XCIII). The keto-acid afforded the ester (IX; $R = R' = Me$; Chart 3, p. 367) on acid treatment followed by methylation.

As stated earlier aqueous solutions of gibberellic acid are unstable at room temperature and gibberellenic acid and the amorphous gibb-4-ene 1 → 3 lactonic acid (LXXXV; $R = H$) are among the decomposition products. Neither gibberellenic acid nor allogibberic acid were formed when the acid (LXXXV; $R = H$) was treated with dilute mineral acid at 20°; but the ester (LXXXV; $R = Me$) gave methyl gibberate with hot mineral acid and the triene (LXXXII, p. 381), in addition to methyl gibberate and methyl epigibberate with methanolic hydrogen chloride. Alkaline hydrolysis of the ester (LXXXV; p. 384; $R = Me$) regenerated the amorphous acid (XCIV; $R = H$).

### d) Absolute Configuration.

The absolute configuration of gibberellic acid was deduced by the British workers from the following evidence (84, 119). The 2-hydroxy substituent was known to be quasi-axial from the reactions described above and from the elimination reactions outlined on p. 388. The rotatory dispersion curve for the keto-ester (LXXII, $R = Me$: Chart 6, p. 377) was almost identical (see Fig. 2, p. 369) with the curve for the keto-ester (IX; $R = R' = Me$; Chart 3, p. 367) derived from allogibberic acid and significantly different from that for the keto-ester (LXIV; $R = R' = Me$; Chart 5, p. 373) obtained from epiallogibberic acid. Gibberellic acid therefore has the same trans $B/C$ ring fusion as allogibberic acid. Oxidation and decarboxylation of the acid (LXXIV) yielded methyl 7-hydroxy-1:8-*epi*dimethyl-2-oxo-4a$\xi$-gibbane-10$\beta$-carb-oxylate, $C_{19}H_{28}O_4$ (CI), m. p. 156–158°, the rotatory dispersion curve of which showed a negative Cotton effect in agreement with a $\beta$-hydrogen at $C_{(10a)}$; and in confirmation of this the nuclear magnetic resonance spectrum of methyl acetylgibberellate (*84*) showed that the protons at $C_{(10)}$ and $C_{(10a)}$ were *trans*.

The 1S, 2S, 4aR, 4bS, 7S, 9aS, 10S, 10aR absolute configuration of gibberellic acid (I, p. 384; $R = H$) was then deduced from the facile

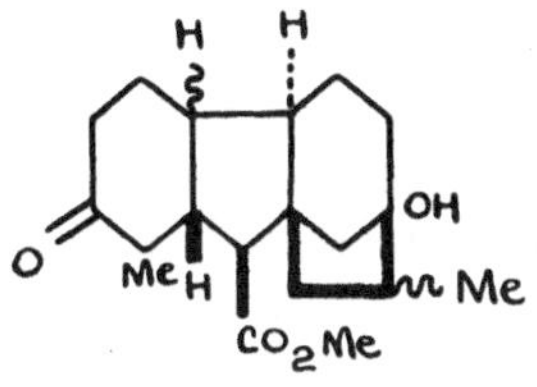

(CI.)

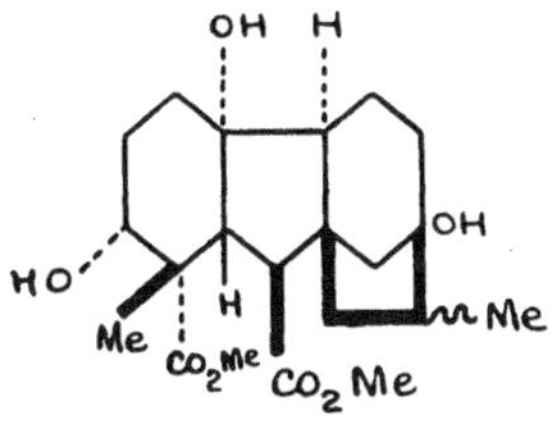

(CII.)

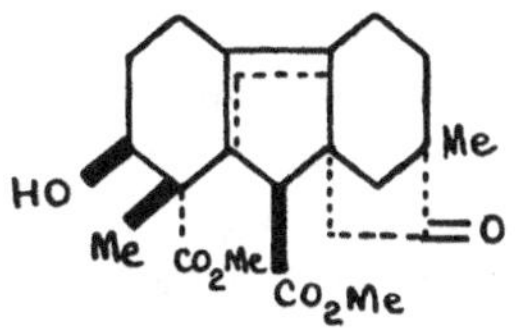

(CIII.)

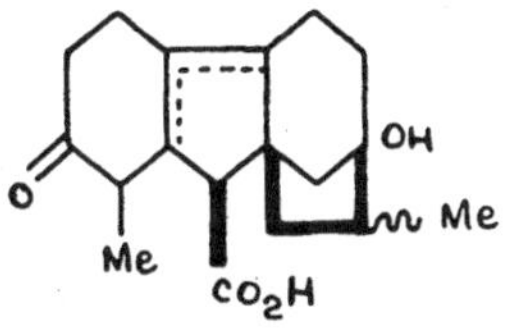

(CIV.)

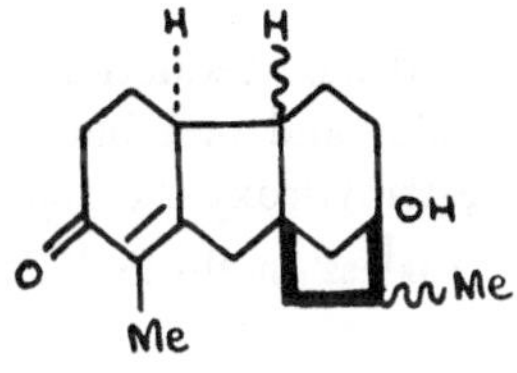

(CV.)

closure and stability of the lactone ring in gibberellin $A_1$ methyl ester and from the rotatory dispersion curves of the ketones (LXX, p. 383; $R = Me$) which showed strong positive Cotton effects as required both by the Octant rule (*96*) and by comparison with (+)-homo*epi*camphor whose absolute configuration is known.

The $\alpha$-orientation of the lactone ring has also been deduced (*296*), from the large positive rotation difference between the gibbane 1 → 4a lactone (LXXXIX; $R = Me$: Chart 7, p. 383: mixture of $C_{(8)}$ epimers) and the corresponding (?) $C_{(8)}$ epimeric mixture of hydroxyesters (CII).

## e) Miscellaneous Reactions of Ring A.

Several reactions associated with ring A of the gibberellins are readily explicable in terms of structure (I, p. 384; R = H) for gibberellic acid. Gibb-2-enes were formed when sulphonic acid esters of 2β-hydroxy-gibbanes were refluxed with collidine and this reaction was used (Chapter III) to prepare the gibberellin A₅ (V, p. 354; R = H) from gibberellin A₁ (II; R = H). The gibb-2-enes were formed directly, by *trans* elimination of the hydroxyl group from 2β-hydroxygibbanes, by the action of phosphorus oxychloride. Three examples of this reaction have been recorded by the Japanese workers (*284, 157*) but the structural formulae allotted by them to their products are incorrect.

Gibbane I → 4a lactones gave the corresponding gibb-4a(4b or 10a)-enes with methanolic hydrogen chloride. Thus the keto ester (LXXXVIII; R = Me; Chart 7, p. 383) afforded an unsaturated ester, $C_{21}H_{28}O_6$, m. p. 157–158°, $[\alpha]_D^{21} - 54°$, $\varepsilon$ 5,300 at 215 mμ, which must be methyl 2β-hydroxy-1:7-dimethyl-8-oxo-7α-gibb-4a(4b or 10a)-ene-1:10β-dicarboxylate (CIII) since it was not oxidised by manganese dioxide and the corresponding 2-ketone $C_{21}H_{26}O_6$, m. p. 119–120°, obtained by chromic oxide oxidation, was not an α,β-unsaturated ketone. The ketone (LXX, p. 383; R = Me), m. p. 161–163°, was only slowly attacked by hot dilute mineral acid (*83*) but more drastic treatment (*283*) of (LXX; R = H) gave 7-hydroxy-1:8-dimethyl-2-oxogibb-4a(4b or 10a)-ene-10β-carboxylic acid, $C_{18}H_{24}O_4$ (CIV), m. p. 215–217°, which readily lost $CO_2$ with movement of the double bond giving 7-hydroxy-1:8-dimethyl-2-oxo-4aα,4bξ-gibb-1(10a)-ene, $C_{17}H_{24}O_2$ (CV), m. p. 193°, $\lambda_{max}$. 247 mμ, $\varepsilon$ 18,000 (*83*).

Lithium aluminium hydride reduction (*83*) of the keto-ester (LXXXVIII, p. 383; R = Me) and of gibberellin A₁ methyl ester (*307, 283*) afforded 2β:4aα:8ξ-trihydroxy-1α:10β-*bis*hydroxymethyl-1:7-di-methyl-7α-gibbane (CVI), $C_{19}H_{32}O_5$, m. p. 177–180° and 2β:4aα:7-tri-

(CVI.)                    (CVII.)

hydroxy-1α:10β-*bis*hydroxymethyl-1-methyl-8-methylenegibbane (CVII), $C_{19}H_{30}O_5$, m. p. 208–210° respectively: both pentols were stable to periodate.

Compounds $C_{19}H_{30}O_5$, m. p. 208–210° *(307)*, and $C_{19}H_{26}O_4$ *(88)*, m. p. 217–219°, $[\alpha]_D^{21} + 37°$, have been obtained by lithium aluminium hydride reduction of methyl gibberellate but their structures are uncertain.

## 4. Biogenesis of the Gibberellins.

Inspection of structure (I, p. 384; $R = H$) of gibberellic acid showed that it could arise from a tricyclic diterpene skeleton (CVIII) in which the $C_{(7)}$ vinyl substituent is quasi-equatorial by (i) loss of the 17-methyl group (diterpene numbering), (ii) contraction of ring *B* to a five-membered ring with extrusion of the carboxyl group and

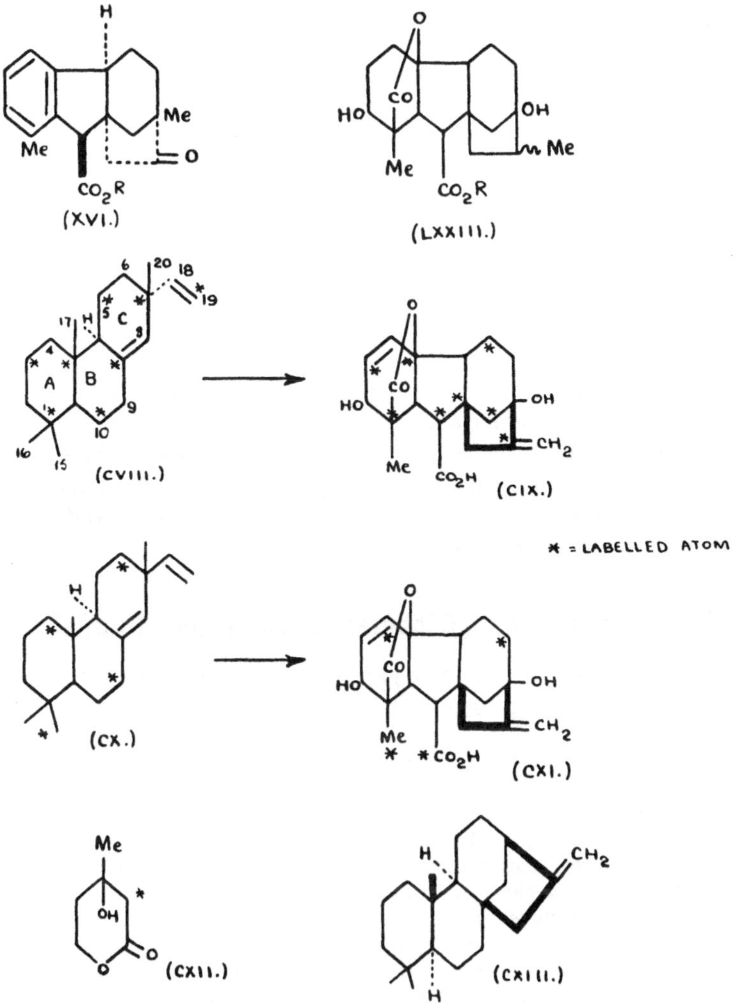

(XVI.)

(LXXIII.)

(CVIII.)

(CIX.)

✳ = LABELLED ATOM

(CX.)

(CXI.)

(CXII.)

(CXIII.)

(iii) rearrangement of ring $C$ according to the scheme proposed by Wenkert (*326*)* to give a phyllocladene-type bridged ring structure.

The correctness of these speculations was established (*17, 18*) by the degradation of gibberellic acid obtained from *G. fujikuroi* grown on $CH_3 \cdot {}^{14}CO_2H$ and $2\text{-}{}^{14}C$-mevalonic lactone (CXII) as substrates. The labelling patterns expected in a diterpene precursor from the usual mode of incorporation of these two compounds are shown in (CVIII) and (CX) and the corresponding patterns expected in the gibberellic acid molecule if the above speculations are fulfilled are shown in (CIX) and (CXI). The carbon dioxide obtained by decarboxylation (see Chart 1, p. 359) of gibberic acid (XVI; $R = H$) was labelled when derived from (CXI) but unlabelled when derived from (CIX) indicating that the carboxylic acid carbon of gibberellic acid is derived from position 9 of the diterpenoid precursor. The acetic acid obtained by Kuhn-Roth oxidation of tetrahydrogibberellic acid (LXXIII; $R = H$) derived from (CIX) was labelled indicating that formation of the phyllocladene-type $C/D$ ring system had occurred through movement of the $C_{(6)} \rightarrow C_{(7)}$ bond to $C_{(18)}$ (diterpene numbering) rather than by migration of the 20-methyl group.

In addition, this work showed that the methyl group attached to ring $A$ of gibberellic acid was derived specifically from $C_{(2)}$ of mevalonic acid and that there was no randomisation of label between the carbon atoms attached to $C_{(1)}$ of mevalonate-labelled gibberellic acid.

The absolute configuration of rings $B/C/D$ of gibberellic acid is the same as that found in phyllocladene (CXIII) but the $A/B$ ring fusion is antipodal to the phyllocladene *trans* $A/B$ ring junction. Gibberellic acid therefore joins cafestol (*97*), darutigenol (*260*), and andrographolide (*60*) which form a group of diterpenes, hydroxylated at $C_{(2)}$ (gibbane numbering) and possessing the "wrong" absolute configuration at the $A/B$ ring junction (*97*).

## III. Occurrence of Gibberellins in Higher Plants**.

The gibberellins can now be regarded as a new class of natural plant growth hormones. The presence of such hormones in higher plants was

---

* The $C_{(7)}$ vinyl substituent becomes quasi-axial in the half-boat conformation of ring $C$.

** **Glossary.** To assist readers unfamiliar with botanical nomenclature, the following definitions are provided: *Caulescent plants:* plants which at all stages of development after germination of the seed have leaf-bearing stems (see rosette-plants below). — *Deciduous:* those perennial plants which lose their leaves in the winter, developing new ones in the following spring. Cf. evergreens. — *Internode:* the positions at which leaves are borne on a stem are the *nodes*; that portion of the stem between two nodes is an internode. — *Flower-primordia:* at an early stage of development microscopic areas at a shoot growing-point may be distinguished first suggested by the fact that the gibberellins promote many normal processes of plant growth and development. Moreover, these fungal metabolites induce specific responses involving processes which are naturally under genetic or well-defined environmental control; such responses cannot be explained in terms of the auxins or other known growth substances. These considerations led LONA (*196*), RADLEY (*263*) and BRIAN (*24*) to suggest that the fungal gibberellins, or physiologically similar hormones, occur in higher plants and function in the growth regulating system.

This postulate was supported by the isolation of crude plant extracts with the requisite biological properties. Indeed, a plant extract with properties now recognisable as gibberellin-like was first described by MITCHELL, SKAGGS and ANDERSON (*215*) in 1951 before the first pure gibberellin was known. More recently, similar evidence indicating the widespread occurrence of gibberellin-like substances has been obtained by many workers, primarily by WEST and PHINNEY (*329*), PHINNEY, WEST, RITZEL and NEELY (*250*), RADLEY (*263, 264*), and LONA (*197*). Thus, active extracts have been isolated from the seeds (*329, 250, 264, 54, 139, 229, 230, 241, 269*), roots (*264, 288*) and shoot tissues (*263, 264, 205, 257, 287*) of many plants; from young inflorescences of *Brassica* (*197*); from some plant tissue cultures (*236*); and from coconut (*265*). The gibberellin-like nature of these extracts has been well established particularly in those cases where they have been shown to induce normal growth in certain genetic dwarfs (*139, 205, 236, 241, 250, 264, 265, 269, 329*) (see p. 395) or to induce flowering in certain types of plants (*179, 54*) (see p. 406). These are two specific properties of the gibberellins; indeed, the reversal of genetic dwarfing has been suggested (*26, 298*) as a definition

---

as destined to become flowers or leaves *(leaf-primordia)*; in many plants differentiation of flower-primordia is influenced by vernalisation or photoperiod (see below). — *Genetic-dwarfs:* plants of much reduced stem length, arising from normal plants by mutation; the dwarf-habit is inherited. — *Parthenocarpy:* formation of fruits (seedless) without the normal process of fertilisation of the ovary by pollen. — *Photoperiodism:* many stages of plant development (flowering, leaf-fall, etc.) are controlled by the length of daily periods of light and dark; plants sensitive to this environmental influence are sensitive to photoperiod, those not sensitive are *day-neutral.* Plants wich flower only when the daily light period exceeds a certain critical length are *long-day plants,* those which flower only when the daily dark-period exceeds a critical length are *short-day plants.* — *Protonemata:* the spores of mosses germinate to produce a web of simple filaments, or protonemata; on these upright leafy-buds develop eventually, forming shoots on which sexual organs are borne. — *Rosette plant:* plants in which stem-development is minimal in the first season of growth, leaves being borne in a close rosette round a central axis close to the ground. — *Vernalisation:* induction of some developmental change (e. g. flowering, breaking of dormancy) by exposure to low-temperature for a critical period.

of gibberellin-like activity. In particular, Phinney et al. (250) showed that extracts from the seed of six plant species induced the same differential response as the fungal gibberellins in a series of single-gene mutants of maize.

Attempts to isolate the active substances from crude plant extracts were first described by West and Phinney (330) who obtained a partially purified and highly active extract from the seed of both *Phaseolus vulgaris* and *Echinocystis macrocarpa*. Radley (264) had found that immature seed of *Phaseolus multiflorus* provided a highly active crude extract and from this source MacMillan and Suter (208) succeeded in isolating the first pure gibberellin from a higher plant. This gibberellin was identified as gibberellin $A_1$ (II, p. 393; $R = H$); it was isolated in a yield of two milligrams per kilogram of fresh weight of seed. Subsequently West and Murashige (328) briefly reported the isolation of gibberellin $A_1$ from the seed of *P. vulgaris*, although in a later paper from the same laboratory West and Phinney (331) referred to the isolation of a gibberellin-like substance, bean factor I, which was stated to be very similar to, but not positively identified as, gibberellin $A_1$. West (327) has since established conclusively that bean factor I and gibberellin $A_1$ are identical. The isolation of gibberellin $A_1$ from water sprouts of *Citrus unshui* has been reported recently by Kawarada and Sumiki (154).

In addition to gibberellin $A_1$, West and Phinney (331) isolated a second gibberellin-like substance, bean factor II, which was incompletely characterised but was shown to differ from the known gibberellins in its activity towards the single-gene dwarf mutants, *d*-1 and *d*-5, of maize. Bean factor II was at least ten times more active for *d*-5 than for *d*-1, whereas both mutants responded equally to gibberellic acid and gibberellin $A_1$.

MacMillan, Seaton and Suter (207) also isolated a second active acid from the immature seed of *P. multiflorus* using the same isolation technique as for gibberellin $A_1$. Although this new acid was distinct from the known fungal gibberellins these authors proposed the name, gibberellin $A_5$, for it because its chemical structure (V; $R = H$) and biological properties (140) closely resembled those of the known gibberellins. Gibberellin $A_5$ was isolated in a yield of about 1 mg. per kg. of seed and was shown to have structure (V; $R = H$) by the evidence summarised below.

The infrared spectra of gibberellin $A_5$, $C_{19}H_{22}O_5$, m. p. 260–261°, and of the methyl ester, $C_{20}H_{24}O_5$, m. p. 190–191°, showed absorption bands attributable to the groupings: carboxyl (or ester), saturated five-ring lactone, alcoholic hydroxyl, 1:1-disubstituted and 1:2-*cis*-disubstituted olefinic double bonds. Hydrogenation of the methyl ester (V; $R = Me$) gave a tetrahydro-derivative (CXIV), presumably a mixture of $C_{(8)}$ epimers (see p. 376). Gibberellin $A_5$ underwent acid-catalysed rearrangement to a keto-acid (CXV; $R = $ H), analogous to the keto-acid (LXXXVIII; $R = $ H) derived from gibberellin $A_1$ (II; $R = $ H); this fact indicated that the gibberellins $A_1$ and $A_5$ had the same $C/D$ ring structure. The structure of the keto-acid (CXV; $R = $ H), and hence of gibberellin $A_5$ (V; $R = $ H), was proved by the formation of the methyl ester (CXV; $R = $ Me) from the methyl ester (LXXXVIII; $R = $ Me) via the $p$-toluene sulphonyl derivative. Dehydration of (LXXXVIII; $R = $ H) to (CXV; $R = $ Me) has also been effected directly (*140*) by means of phosphorus oxychloride (p. 388).

Gibberellin $A_5$ (V; $R = $ H) has been directly related (*140*) to gibberellin $A_1$; treatment of the $C_{(2)}$-$p$-toluene sulphonyl derivative of gibberellin $A_1$ methyl ester with boiling collidine gave gibberellin $A_5$ methyl ester from which gibberellin $A_5$ was obtained on alkaline hydrolysis.

The absolute configuration (V) of gibberellin $A_5$ follows from the interrelation of gibberellin $A_5$ and gibberellic acid via gibberellin $A_1$.

Gibberellin $A_5$ and bean factor II have been shown to be identical despite the discrepancies in the published carbonyl frequencies in the infrared spectra of the two acids. The infrared absorption spectra of the two acids are identical when directly compared (*327*). Moreover gibberellin $A_5$ produces the same differential response as bean factor II in the dwarf mutants, $d$-1 and $d$-5 of maize (*327*).

Thus two gibberellins of established structure have been isolated from higher plants. There is evidence of others yet to be isolated.

First, LAZER and DAHLSTROM (*181*) claim to have demonstrated the presence of gibberellic acid in green malt by isotope dilution analysis and MANZELLI and ADLER (*210*) claim the tentative identification of gibberellic acid in an extract of Kudzu vine by chromatographic techniques and fluorescence detection. Secondly, although conclusions based on paper chromatographic studies of crude plant extracts should be treated with reserve, there is evidence for the occurrence of gibberellin-like substances with $R_f$ values which differ from those of gibberellic acid and the gibberellins $A_1$ and $A_5$. Active zones with much lower $R_f$ values than these gibberellins have been obtained from an extract of lupin seed (*250*) in solvent system 1 (see Table 5, p. 416) and from extracts of several other leguminous plants (*230*) in solvent system 7. On the other hand an extract of pea shoot tissue in solvent system 1 showed an active zone with a higher $R_f$ value then those of gibberellic acid and gibberellins $A_1$ and $A_5$.

With the isolation of gibberellin $A_1$ and gibberellin $A_5$ from plant tissue and the evidence for the widespread occurrence of gibberellin-like substances in higher plants, *there can be little doubt that gibberellins function in normal processes of plant growth and development.* The possible rôle of these newly-discovered hormones in the regulation of growth and flowering is discussed in a later section of this review (p. 399).

# IV. Effects on Plant Growth and Development.

The rate of publication of papers on effects of gibberellins on plants is so great that it is not possible here to refer to all of them. Readers wishing to obtain references to the earlier publications should consult the compilation made by Stodola (*292*). There are also other recent review articles which can be consulted with advantage (*26, 27, 29, 48, 67, 68, 69, 75, 159, 164, 298, 299, 349*).

The investigation of responses of plants to exogenous gibberellins has greater significance than was at first realised, since the discovery that they are to be found in many plant tissues carries with it the implication that they are natural hormones, with a regulating function in many phases of plant development. From this point-of-view the modification by gibberellins of plant responses to vernalisation and photoperiod are of special interest.

The evidence available at present indicates that the known gibberellins all have qualitatively similar biological activity. In most circumstances gibberellic acid is the most active, followed by gibberellin $A_1$ (*45, 47*). It has therefore been thought justifiable, and convenient, in the review of experimental work now presented, to use the more general term 'gibberellin', unless for some strong reason it has been thought desirable to specify the actual gibberellin used. In fact, in most cases either gibberellic acid, potassium gibberellate, or a mixture of gibberellic acid and gibberellin $A_1$ has been used.

## 1. Stem Growth.

It is convenient first to consider the effects of gibberellin on plants whose growth is not rigorously determined by photoperiod or temperature, including in this category day-neutral annuals, and facultative long-day plants growing under long-day conditions, and then to consider the more complex effects of gibberellin on plants whose development is closely regulated by photoperiod or vernalisation.

## A. Day-neutral Annuals, etc.

In *normally unbranched plants*, such as the garden pea *(Pisum sativum)*, the most characteristic effect of gibberellin is to increase the length of internodes without any increase in internode number. Only those internodes extending at the time of application respond by increased growth rate and the time for which the increased growth rate persists is related to the dose applied *(21, 24, 31, 147, 338)*. Internodes in plants treated with gibberellin actually become mature and cease growth sooner as a result of treatment, but the greater rate of extension is more than sufficient to counterbalance the more rapid maturation, and longer internodes are consequently produced *(35, 71)*. In *Sesame* plants treated with small doses of gibberellin newly formed internodes are longer until the effect of the gibberellin ceases, and then subsequently formed internodes are shorter than those of untreated plants *(63)*.

Gibberellins have the same effects on *bushy plants* such as 'Cupid' sweet pea *(Lathyrus odorata)* *(29, 36)* and dwarf bean *(Phaseolus vulgaris)* *(28, 43, 71, 212)*, but, in addition, branching is inhibited or suppressed. Consequently the whole form of the plant is changed, and in extreme cases, as in dwarf *Phaseolus*, the main axis develops the typical nutational movements of a climbing bean. This enhancement of apical dominance is a characteristic effect, also recorded in clover *(291)*, cereals *(28, 134, 238, 341)* and sugar cane *(14)*. It may sometimes appear that gibberellin has an opposite effect, encouraging branching. This is because gibberellin does not directly inhibit lateral buds; it reinforces the inhibiting activity of a functioning apical bud, but, if the apical bud has declined in activity either naturally or as a result of treatment with such substances as maleic hydrazide, or if the apical bud is excised, then gibberellin actually stimulates the development of lateral buds, causing a bushy growth *(33, 37, 114, 146, 334)*.

*Genetic dwarfs* usually show a much greater growth response than corresponding wild-type talls; this has been observed in peas *(11, 21, 24, 31)*, maize *(248, 249)*, rye-grass *(Lolium perenne)* *(79)*, cotton *(102)*, broad bean *(Vicia)* *(31)*, French bean *(Phaseolus vulgaris)* *(28, 31)*, coffee *(219)* and wheat *(300)* among others. The effect generally results in the production of a plant indistinguishable from, or very similar to the wild-type tall. However, not all mutant dwarfs respond in this way *(248, 249)*.

## B. Herbaceous Plants Sensitive to Vernalisation or Photoperiod.

The discovery by LANG *(173, 174)* in 1956 that biennial *Hyoscyamus* could be induced to bolt and flower by treatment with gibberellic acid, without the normally essential period of vernalisation, opened a new chapter in the study of the physiology of flowering. Though in most long-day plants vegetative bolting is associated with flowering, we shall deal with these two responses separately here, for reasons that will become apparent later.

Following Lang's discovery, many other vernalisation-requiring biennials and perennials have been found to bolt after gibberellin treatment even though kept at non-inductive temperatures; these include species of *Apium, Arabidopsis, Brassica, Campanula, Centaurium, Cheiranthus, Cichorium, Daucus, Digitalis, Geum, Oenothera, Petroselinum, Reseda, Scabiosa, Solidago* and *Scrophularia* (*44, 58, 67–69, 73–76, 129, 175, 176, 185, 199, 251, 277, 325, 336*).

In all these plants there are two blocks to bolting, one normally surmounted by vernalisation, the other by exposure to long-days. Most experiments with gibberellin have been carried out in long-day photoperiods and in such cases it is consequently known only that the gibberellin overcomes the first block; in some plants, however, as in *Hyoscyamus niger, Centaurium minus, Daucus carota* and *Oenothera biennis* (*44, 58, 174, 251*) gibberellin will cause bolting in short-days. It appears likely that in most cases gibberellin will overcome both blocks to bolting in biennials. The dormancy of buds on rhizomes of *Anemone nemorosa*, normally broken by exposure to low temperatures, is also terminated by gibberellin (*122*). It should be noted, however, that some cold-requiring biennials and perennials have not been induced to bolt by gibberellins, notably *Crepis virens* (*199*), *Lunaria biennis* (*325*), *Campanula* spp., *Erysimum* spp., *Eryngium variifolium, Isatis tinctoria* and *Saxifraga rotundifolia* (*76*).

All the plants mentioned above are rosette plants, with virtually no stalk before bolting begins. Other rosette plants have only one block to bolting, normally overcome by exposure to long-day photoperiods, no vernalisation being necessary. In 1956, Bünsow and Harder (*52*) using *Lapsana communis*, Lang (*175*) using an annual race of *Hyoscyamus niger* and several other long-day species, and Lona (*196*) using *Crepis leontodontoides* and *Lactuca scariola*, showed that rosette plants of this type also will bolt and flower in short-day photoperiods if treated with gibberellin.

Long-day rosette plants which can be induced to bolt in short days by gibberellin include species of *Adonis, Aethusa, Anethum, Arabidopsis, Aster, Brassica, Centaurea, Crepis, Hyoscyamus, Lactuca, Lapsana, Mimulus, Myosurus, Nicotiana, Papaver, Petunia, Polemonium, Raphanus, Rheum, Rudbeckia, Samolus, Sempervivum, Silene, Scabiosa, Spinacia, Trifolium* and *Scrophularia* (*52, 53, 56, 66–69, 75, 76, 176, 180, 196, 199, 201, 335*).

All caulescent long-day plants, which develop stalks even under short-days, but form longer internodes in long-day photoperiods, as far as is known also respond vegetatively to gibberellin by increased

internode extension in short-days; among those known to respond in
·this way are species of *Anagallis, Calamintha, Campanula, Circaea,
Euphorbia, Iberis, Isnardia, Pisum, Phaseolus, Sedum* and *Urtica* (*11, 21,
67–69, 75, 76, 100, 196, 201, 338*).

Short-day plants (i. e. those which *flower* only in inductive short-
day photocycles) also respond to long-day conditions by increased
vegetative vigour, and all tested also respond vegetatively to gibberellin
by stem elongation; these include species of *Begonia, Fragaria, Glycine,
Kalanchoë, Nicotiana, Perilla, Tinantia* and *Xanthium* (*56, 67–69, 93,
122, 126–128, 175, 176, 184, 198–200, 257, 311, 325*). The same applies
to the long-short-day plants *Bryophyllum crenatum* and *B. daigremontianum*
growing under short-day conditions (*51, 100, 325*).

*The responses of these plants to gibberellin show a well-defined
pattern,* viz. that gibberellin will in most cases induce responses similar
to those normally induced by long-day photoperiods or by vernalisation.
The exceptions are discussed further in a later section.

## C. *Photoperiod-controlled Growth of Woody Plants.*

The seasonal character of growth in deciduous shrubs and trees is
essentially a response to photoperiod. Development of autumn foliage
colours, leaf-fall, cessation of growth, and bud-dormancy are responses
to the decreasing length of day in autumn, and the termination of this
dormant period and the onset of renewed growth are responses to
vernalisation or the return of long-day photoperiods, or both. In a
number of such plants it has been found possible to terminate dormancy
in short-day conditions by treatment with gibberellin, including species
of *Acer, Camellia, Paeonia, Pinus, Prunus, Rubus, Vitis* and *Weigela*
(*3, 13, 23, 42, 99, 136, 193, 202, 240, 319*). On the other hand some woody
plants failed to respond in such experiments (*193, 240*). The onset of
dormancy in autumn can be prevented by gibberellin treatment. This
is the case with some species of *Acer, Betula, Fraxinus, Liriodendron,
Parthenocissus, Prunus* and *Rhododendron*; on the other hand little or
no effect was observed with *Acer rubrum, Castanea sativa, Fagus sylvatica,
Quercus robur* and *Ulmus procera*; and dormancy in *Taxodium distichum*
was actually accelerated (*38*).

An analogous dormancy, the physiological dwarfing of *Malus arnoldiana*,
peach and cherry seedlings, resulting from use of unvernalised seed, is
also broken by gibberellin (*12, 108, 161*). *Citrus* grows in a series of
flushes, so that there are periods of summer dormancy as well as winter
dormancy; both types are broken by gibberellin (*80*). An interesting
observation in this species is that new shoots formed on gibberellin
treated plants are frequently thorny; thorniness is a juvenile character,
and in this connection it is noteworthy that treatment of mature shoots

of Ivy *(Hedera helix)* leads to a resumption of juvenile shoot formation, with characteristic development of long-internodes, abundant adventitious roots and characteristically shaped leaves *(272)*.

Thus in all these examples of effects of gibberellin on woody plants *there is again a marked tendency for gibberellin to simulate the effects of long-days*. The exceptions to this generalisation, mentioned above, need further investigation, as also does the very unexpected observation that gibberellin treatment of some woody plants (spp. of *Acer, Betula, Fagus, Fraxinus, Sorbus* and *Vitis*) before the onset of dormancy, actually leads to a prolongation of dormancy in the following spring *(39, 271, 322)*.

### D. Cell-extension or Cell-multiplication?

Growth of a plant tissue or organ may involve cell-division and cell-extension, and it is important that we should know whether gibberellin influences only one or both of these processes. The extension of leaves and stems of rice seedlings infected by *Gibberella fujikuroi*, and showing "bakanae" symptoms, was attributed by Shimada *(286)* mainly, but not entirely, to cell-extension. In their early work on the effect of crude gibberellin on a range of plants, Yabuta and Hayashi *(341)*, while emphasising the importance of cell-extension, came to the conclusion that cell-multiplication was also involved. In the extension of pea internodes cell-extension is certainly involved *(28, 117)*, but Griffith has shown that cell-multiplication is also an important factor, particularly in some tissues, the growth response in epidermal tissue being almost entirely due to cell-multiplication, whereas in pith growth cell-extension is possibly the more important process *(117)*. In stem extension of *Phaseolus* and *Begonia* both processes are involved *(105, 115, 122)*; in addition there is an increase in stem thickness due to greatly increased cambial activity with consequent formation of new secondary xylem cells *(71, 122)*. Similarly gibberellic acid increases the length of petioles in strawberry, and this is reflected in increase in number and length of epidermal cells *(123)*. Skjegstad (quoted in *275*) has shown that gibberellin induces increased mitotic activity in the intercalary meristem of the leaf-sheath of dwarf maize, and increased cell-length in those parts of the blade above the meristem. *Thus we have a general picture that gibberellin-induced stem extension in caulescent plants is a consequence both of increased cell-multiplication and of increase in cell-size.*

It never seemed likely that the great increases in stem-length involved in the bolting response of a rosette plant could be explained purely in terms of cell extension; since the stem in the rosette stage is almost non-existent, cell multiplication must be involved. This has been demonstrated to be the case by the detailed investigations of Sachs, Bretz and Lang *(275)*. They studied the gibberellin-induced bolting of biennial *Hyoscyamus niger* and of the long-day plant *Samolus parviflorus*. Within 24 hr. of treatment with gibberellin they were able to detect increased mitotic activity in the pith, cortical and vascular tissues of apical and subapical regions. There was no evidence of any effect on cell elongation for 72 hr.; thus the considerable initial extension of stem taking place during that period was due entirely to cell-multiplication. The histological picture was indistinguishable from that seen in plants induced to bolt by the natural stimuli of vernalisation or photoperiod, the orientation of divisions induced by gibberellin being in every way orderly and natural. The response of dormant woody stems is essentially a re-awakening of cambial activity, viz. of cell multiplication (*319*).

Because of the pronounced mitotic activity frequently induced by gibberellin, it may not be out of place here briefly to mention some responses of tissue cultures. The picture is very unclear; growth of some, perhaps most, cultures is inhibited, that of others stimulated, increases in cell-multiplication and cell expansion being recorded (*135, 233—235, 237*). This is a field of investigation needing much more attention. Of organ cultures studied, accelerated growth (involving cell-multiplication) of unpollinated *Coopersia* ovules has been recorded (*273*) and increased root-production on adventive embryos from *Citrus* nucellus cultures (*266*).

### E. Mode of Action Studies: Interactions with Auxins.

Since gibberellins do induce cell-extension and since this, rightly or wrongly, is considered to be the most characteristic of the various responses of plant tissues to auxins, several detailed comparisons of the spectrum of responses of gibberellins and auxins have been made. It is clear from these comparisons that though in some systems they act similarly, e. g. in inducing cell extension, in others they are quite different. For example, auxins inhibit root growth, gibberellins have little effect; auxins suppress growth of lateral buds in shoots from which the apical bud has been removed, gibberellins encourage the growth of such laterals; auxins stimulate production of adventitious roots on stem cuttings, gibberellins inhibit rooting (*26, 37, 146*).

Nevertheless, since both do induce cell-extension we need to know more of the nature of their functional relationship. There are significant differences in the circumstances in which the two kinds of hormone affect cell-extension. Whereas auxins induce great increases in extension of tissue fragments, e. g. coleoptile sections, they have little growth-promoting effect on intact plants (*1, 31*); gibberellins on the other hand, though they have striking effects on intact plants, induce only relatively small growth increments in coleoptile or shoot sections (*32, 34, 131—133, 148, 150, 171, 242, 261, 262*). The fact that gibberellins do induce responses in some section growth tests is sometimes overlooked; broadly speaking it is in sections with a high endogenous growth rate that gibberellin induces the greatest response (*26, 262*). An indication of the nature of the relationship between auxins and gibberellins was first clearly demonstrated by BRIAN and HEMMING (*32*, *34*), studying the extension in high-intensity light of pea internode sections cut from light-grown plants (green sections). Such sections have a low endogenous growth rate, show little or no response to gibberellic acid alone, a large response to auxins, and a consistent extra response to gibberellic acid in the presence of an auxin. *It was concluded that growth responses to gibberellic acid were dependent on the presence of an auxin.* The synergistic relationship has since been demonstrated in other extension growth systems and, as will appear below, in systems involving quite different physiological responses.

It is an interesting fact that this synergism is exhibited in green sections far more than in etiolated sections; for example in sections from pea plants grown in complete darkness or in darkness broken by occasional red light, no synergism is demonstrable, the effects of gibberellin and auxin being additive (*150*, *261*). However, GALSTON and WARBURG (*112*) have shown that if 100 mm. lengths of etiolated pea epicotyl are treated basally with gibberellin before the sections are cut from the apical region, then there is a synergistic relation between the effects of this gibberellin pre-treatment and subsequent exposure of sections to indolylacetic acid. The magnitude of the synergism is related to the length of epicotyl through which the gibberellin has to pass before reaching the apical region. GALSTON and WARBURG have interpreted the synergism in terms of a 'third-factor' required for the interaction, a factor which may be presumed to be limiting in etiolated tissues but abundant in green tissues. Considerable interest is therefore aroused by the discovery by STOWE (*297*) that fatty acid esters promote growth of etiolated pea epicotyl sections and, indeed, seem to promote the gibberellin-auxin synergism.

Other interpretations of the 'third-factor' have been proposed. Indeed, before the work of GALSTON and WARBURG mentioned above, BRIAN and HEMMING (*34*) attributed the synergism to interaction with a third, inhibitory factor. They based their hypothesis on the fact that sections in an auxin : gibberellin : sucrose medium grow faster than comparable tissues in an untreated intact plant growing in light, but at much the same rate as comparable tissues in a gibberellin treated plant. Assuming that auxin was not limiting in the untreated intact plant, it was concluded that the action of gibberellin was to block the action of some inhibitory influence in the intact plant, releasing the full growth-promoting potential of the endogenous auxins. No direct evidence concerning the nature of this inhibitory influence was presented.

Another interpretation has been that either directly or indirectly gibberellin reduces enzymatic destruction of the auxin indolylacetic acid

(IAA), and that the growth responses are due to a consequent increase in endogenous IAA levels.

This view has encountered considerable criticism (26, 262, 299). The interpretation has been put forward in two forms, the simplest being that of PILET. He has claimed that gibberellin inhibits the in vitro activity of an IAA-oxidase system from carrot tissue culture (252, 253), and that treatment of intact *Trifolium* plants decreases IAA-oxidase activity in sections cut from petioles of such plants (254), and he considers that such an effect in the intact plant can explain the growth responses to gibberellin. However, other workers have found no in vitro inhibition of IAA-oxidase preparations (34, 151) or decrease in IAA-oxidase content in tissues of treated plants (151). A variant of this theory, which avoids the objections mentioned, is that of GALSTON (110); he has produced evidence for the existence of an inhibitor of IAA-oxidase in plant tissues, which increases in concentration after treatment of plants with gibberellin.

However, both variants of the IAA-oxidase theory of gibberellin action can be criticised on five grounds: (i) the auxin:gibberellin interaction described by BRIAN and HEMMING (34) is effective with such synthetic auxins as naphthalene-acetic acid or 2:4-dichlorophenoxyacetic acid which are unaffected by IAA-oxidase; (ii) gibberellin induces a growth response in green (34) or etiolated (262) pea sections even in the presence of supra-optimal IAA, whereas if it were preventing IAA destruction it should increase the inhibitory effect of excess IAA; (iii) it cannot possibly account for circumstances where gibberellin and auxin have opposite effects, e. g. in root initiation on stem cuttings; (iv) exogenously applied auxin has little or no effect on stem extension; (v) the evidence that IAA-oxidase has a regulating effect on plant growth (109) is entirely circumstantial, and there is weighty evidence against that view (40).

However, McCUNE and GALSTON (206) have recently provided sound evidence that gibberellin treatment causes a marked qualitative change in peroxidase patterns in intact peas and dwarf maize and PHILLIPS, VLITOS and CUTLER (247) and NITSCH (quoted in 299) have demonstrated a big build-up of endogenous growth substances after application of gibberellin to peas. There is, therefore, a kernel of hard fact in these observations whose true significance is not yet apparent.

All the work on auxin : gibberellin interactions so far discussed has been carried out with parts of plants, above all with stem sections. Under such conditions cell-extension is of far greater importance than cell-multiplication, indeed there is probably no cell-multiplication in such systems, yet we have seen that cell-multiplication is, in intact plants, probably more important in responses to gibberellin than cell-extension. It is worthy of note therefore, that removal of the apical bud of light-grown pea plants reduces the response to exogenous gibberellin (315, 316) and that treatment of the apical stump with IAA in lanolin partially restores the response (34).

This is apparently not true of etiolated plants, where growth of plants with the apical bud removed can be restored to normal by apical application of gibberellin (189).

More important still is the demonstration by Lang (*177*) that the bolting response in the long-day plant *Silene armeria* shows a marked auxin : gibberellin synergism, and by Wareing (*319*) that the re-awakening of cambial activity in dormant *Acer* twigs is much more effectively induced by a mixture of IAA and gibberellic acid than by either alone. The latter case is particularly illuminating, in so far as there are qualitative differences in the tissue formed when both hormones are supplied. A different form of IAA : gibberellin interaction in intact plants, which does not seem to be typical, has been observed by Weijer (*324*). Here the stem-elongation response of plants of *Impatiens balsamina* to gibberellin was completely *prevented*, if the plants were also sprayed with 0.1 μg./ml. IAA.

Thus we may *summarise* this aspect of investigations of the mode of action of gibberellin as follows: (i) A synergistic relationship exists between auxins and gibberellin in stem extension; this has been shown most clearly in intact stems. (ii) It appears likely that an auxin is essential for stem extension and that gibberellins only further promote stem extension if auxins are non-limiting. (iii) There is no evidence yet that gibberellins are *essential* for stem extension, though the extreme response of rosette plants suggests that this may be the case. (iv) The nature of the synergism remains to be elucidated; almost certainly other hormonal factors will be found to be involved, and it may well be, as Purves and Hillman (*262*) have concluded, that the primary actions of gibberellin and auxin are not closely connected.

### F. Mode of Action Studies: Interactions with Light.

Several of the earlier workers on the gibberellins commented on the similarity of the effects produced to etiolated growth (*28, 146, 303, 304*). More recently there have been reported a number of more quantitative investigations of gibberellin effects in relation to red-light inhibitions of growth of plants previously grown in darkness; these have added interest because the action spectrum of these inhibitions is usually the same as that of photoperiodic phenomena [but see (*194*)]. Thus it has been shown that red light reduces the extension rate of untreated pea seedlings, but does not do so if the plants are treated with gibberellin (*188, 190, 192, 194, 316*). Similar results have been obtained with *Perilla* (*200*) and *Helianthus* (*190*).

Two exceptions to this general relationship are interesting. In *Cucumis sativus* and *Cucurbita pepo* gibberellic acid fails to overcome red-light inhibition completely. Lockhart (*190*), who first noticed this, suggested that the result might be explicable in terms of gibberellin specificity, the natural gibberellin of cucurbits possibly having properties somewhat different from those of gibberellic acid. The strength of this suggestion is greatly reinforced by the subsequent discovery that in cucurbits, though in no other plants yet examined, gibberellin $A_4$ is much more active in inducing stem-extension than gibberellic acid (47). The second exception is in *Phaseolus*, where gibberellin induces no response in the absence of light, though in untreated plants light is inhibitory; it appears that in this plant, if gibberellin is non-limiting, light actually stimulates growth (191).

Other investigations (91, 101, 200) have shown that in its effects on stem-extension of this kind, gibberellin acts in the same sense as far-red light, but whereas the effects of far-red irradiation can be reversed by subsequent exposure to red, gibberellin effects cannot be so reversed. In more recent studies LOCKHART (192) and LOCKHART and GOTTSCHALL (194) have produced indirect evidence that red-light growth inhibitions of Alaska pea are due to a reduction of endogenous gibberellin level.

However, inhibition of stem extension by light is a phenomenon only demonstrable clearly when dark-grown plant material is used. In light-grown, photoperiodically sensitive plants, increasing the daily irradiation by lengthening the day causes increased stem extension. In such material gibberellin acts in the same sense as light and their effects are frequently less than additive, *suggesting that the initial response to increased day-length is a building up of endogenous gibberellins to saturating levels*, that is the opposite of the situation envisaged by LOCKHART in dark-grown peas. Results of this type have been obtained with Alderman (late) pea, rhubarb, *Weigela* and *Pinus* (23, 42, 338). With rosette plants there is sometimes a less easily interpretable relation between day-length and response to gibberellin; in lettuce day length and gibberellin effects are additive (46); in *Centaurium*, on the other hand, though during continued treatment gibberellin had a greater effect on stem extension in short-days than in long-days, growth stopped almost immediately after gibberellin dosage ceased in short-days, but carried on in long-days (58). Indeed, shoots of differing age may show a similar difference in response on the same plant, as in raspberry (136).

Such observations suggest that under certain circumstances the response to gibberellin is related to the size of the dose, but that under other circumstances the gibberellin acts by a 'trigger mechanism'. The picture is complicated further by observations that in some day-neutral plants increased day-length affects neither normal growth nor response to gibberellin, as in Dwarf Progress pea (338), whereas in others it seems to increase the response to gibberellin, as in *Zinnia* (4).

### G. *Other Interactions.*

DANCER (94) has reported an interesting synergism between gibberellic acid and zinc salts in *Phaseolus* stem extension. Zinc deficient plants show symptoms suggestive of auxin deficiency, and the symptoms can be relieved by application of indolylacetic acid or its precursor tryptophan (*313*); thus zinc is believed to be necessary for tryptophan biosynthesis. Dancer's observations can plausibly be explained by supposing that in *Phaseolus*, if gibberellin is non-limiting, auxin becomes limiting as a consequence of zinc-limited tryptophan production. Marré and Arrigoni (*211*) have claimed that reduced glutathione potentiates the activity of gibberellic acid, using growth of etiolated pea epicotyl sections.

## 2. Leaf Growth.

The effects of gibberellin on leaf growth are less well documented than their effects on stem growth.

Cereal leaves become longer, slightly narrower, but of increased surface area (*28*). Leaves of dicotyledons may show no noticeable response, as in *Pisum* (*28, 169*) or may show increased surface area, as in *Cucumis*, Cupid sweet pea, *Galinsoga*, *Phaseolus*, *Raphanus*, *Trifolium* and *Vitis* (*3, 36, 114, 137, 158, 165, 167, 169, 291*). A change in leaf-shape is usually involved, leaves typically becoming relatively elongated, whether leaf-area is affected or not (*3, 36, 59, 63, 80, 114, 122, 137, 167, 172, 243, 291*). Transition from a juvenile to a mature leaf-form may be involved, as in *Ipomoea* and *Eucalyptus* (*243, 280*) or from mature to juvenile leaf-form, as in *Hedera* (*272*). Chakravarti (*63*) has recorded a change from opposite to alternate insertion of leaves on stems of *Sesamum indicum*, the alternate condition normally being seen only on flowering shoots.

Such recorded increases in leaf-size may only reflect a more rapid growth, final size being unaffected (*137, 138*). Indeed, the initial acceleration of leaf expansion induced by gibberellin may be followed by a secondary depression of the rate of expansion (*137, 243*). Such a compensatory secondary depression of growth rate has also been observed in stem extension (Chapter IV, p. 395). Total leaf area per plant may also be increased temporarily of permanently (*138, 183*). Dry weight of plants is frequently found to be increased by gibberellin; there is little evidence for any increase or decrease in net assimilation rate and, in consequence, any such increases in dry weight must be attributed to increased carbon-fixation consequent upon increased leaf-area (*28, 103, 124, 138, 183, 221, 337*).

Gibberellic acid induces extension of wheat-leaf sections in the dark (*133, 262*) but the response is not entirely specific as auxins sometimes have an effect and there may be synergistic relations between gibberellin and auxin. On the other hand, though auxins have little or no effect on expansion of etiolated *Phaseolus* leaf discs, gibberellins do. The effect is greater in light than in the dark, and is apparently reversed by exposure to far-red light. A number of other substances, including adenine, kinetin and 1:3-diphenylurea promote leaf-disc expansion and the inter-relations of their effects need clarification (*279, 321*).

### 3. Root Growth.

#### A. Root Extension.

Gibberellin has no striking effect on the growth of roots of intact plants. In the early Japanese work, in which crude gibberellins were used, inhibitory effects were reported (341, 347), but these were much smaller than the well-known inhibitory effects of auxins. More recent investigations with purer materials have given the impression that, in concentrations below 100 $\mu$g./ml., gibberellin has little effect on root growth, and that concentrations higher than this are mildly inhibitory (37, 122, 303, 304). A fairly consistent effect is to increase the shoot : root dry weight ratio (28, 71, 142). Early radicle extension of two gymnosperms is stimulated, viz. in *Pseudotsuga menziesii* and *Pinus lambertiana* (41, 267). Root sections from two strains of maize respond to gibberellin by increased extension, though other strains show no response (333).

Two studies of the effect of gibberellin on growth of excised tomato root cultures have been reported (182, 301). The effects of gibberellin are complex and interesting interactions with auxins can be demonstrated. STREET (301) concludes that if endogenous auxins in the root are supra-optimal, the gibberellin accelerates the normal loss of meristematic activity (ageing) and that it is similarly inhibitory if endogenous auxin is markedly sub-optimal. At intermediate auxin levels gibberellin may be stimulatory to growth.

The lack of effect of exogenous gibberellin on roots of intact plants may have been due to endogenous levels being sufficient to saturate growth systems. *The work on excised root culture shows that gibberellins can, after all, intervene in root growth regulation* and that there, as in shoot growth, gibberellin : auxin interactions are of great importance.

#### B. Root Initiation on Stem Cuttings.

Unlike auxins, which induce formation of adventitious roots on stem cuttings, gibberellins inhibit rooting and antagonise the promoting effects of auxins: this has been demonstrated in *Citrus* (70), *Pisum* (24, 37) and *Salix* (122). Similarly it inhibits formation of adventitious roots on discs cut from *Begonia* leaves (278) and on stems of intact *Bryophyllum* plants (314). An apparent exception to this general trend is a stimulation of production of base roots in maize (238).

#### C. Development of Root-nodules on Leguminous Plants.

It has recently been shown (111, 312) that root-nodule formation is reduced in *Phaseolus* by treating the foliage with gibberellin. Similar experiments with *Trifolium repens* showed no effect on nodule number (107).

It has long been suspected that nodule formation is subject to hormonal influences emanating from root meristems (*244*) and the relationship of gibberellin to nodule formation obviously merits further attention.

## 4. Flowering.

### A. *Rosette Plants Requiring Vernalisation or Long-days.*

Most of these rosette plants which have been shown to bolt vegetatively in non-inductive conditions in response to gibberellin (see Chapter IV, p. 395) also produce flowers. Nevertheless, a few species which bolt completely fail to initiate flower primordia. These include such vernalisation-requiring species as *Scrophularia vernalis* and *Geum urbanum*, and such long-day plants as *Lactuca scariola*, *Mimulus luteus*, *Sempervivum funkii* and *Trifolium pratense* (*76*, *196*, *199*, *325*).

Though a few species, such as these, failed to flower in response to gibberellin, in many more the gibberellin treatment failed to reproduce *exactly* the effects of the natural stimuli to flowering. Rosette plants which bolt in response to vernalisation, or long-days, or both, initiate flower buds more or less at the same time as stem-extension commences. Frequently, gibberellin treatment results in considerable stem-extension before any initiation of flower primordia can be detected, the bolting and flowering responses thus becoming partially dissociated. This was first observed by Lang (*174*) in *Hyoscyamus*, and has since been noticed in other plants (*44*, *76*, *129*, *199*, *277*, *325*). The complete failure to flower mentioned in the preceding paragraph can thus be regarded as an extreme case of a fairly general tendency, though in some cases, it must be stressed, as in *Brassica napus* (*199*) and *Arabidopsis thaliana* (*277*), gibberellin treatment does result in simultaneous differentiation of flower primordia and stem elongation.

### B. *Caulescent Plants Induced to Flower by Vernalisation or Long-day Photoperiods.*

Chouard (*76*) has pointed out the remarkable fact that no caulescent plants flowering in response to vernalisation or exposure to long days have been induced to flower by gibberellin treatment, though they do show the typical vegetative long-day response of longer internodes (see Chapter IV, p. 396). This is true also of such cereals as Petkus rye (*176*, *196*) and winter wheat (*163*, *196*). In late varieties of *Pisum*, which are facultative long-day plants, gibberellin actually delays flowering if applied under inductive conditions, in the sense that the first flower appears at a later node (*11*, *21*, *220*). It would be interesting to know whether gibberellin has such an effect on other caulescent plants.

## C. Short-day Plants.

We have seen (Chapter IV, p. 395) that gibberellin induces vegetative responses in short-day plants, growing under short-day conditions, characteristic of plants grown in long-day photoperiods. It is therefore not surprising to find that if gibberellin is applied in long-day conditions (i. e. non-inductive for flowering) flowering is not induced. This has been shown to be the case in *Glycine max, Kalanchoë blossfeldiana, Nicotiana* spp., *Panicum miliaceum, Perilla nankinensis* and *Xanthium* spp. (*68, 69, 126, 127, 175, 176, 199*).

Less work has been done on the influence of gibberellin on flowering of short-day plants under short-day conditions, but, insofar as it is justifiable to generalise from so few observations, it appears that *two kinds of response are encountered.*

In some species gibberellin inhibits flowering, i. e. it simulates the effect of long-day conditions, as it does in the vegetative responses invoked; this type of response is clearly shown in *Fragaria* (*257, 311*) and in *Kalanchoë* (*126—128*). In other species gibberellin has little effect, perhaps causing a slight delay or a slight advancement in the development of flower primordia; this type of response is obtained in *Pharbitis* (*244a*), *Xanthium* (*116, 184*), *Perilla* (*67–69, 199*) and *Nicotiana* spp. (*56, 67–69*). In none of these cases does gibberellin appear to act as a stimulus causing initiation of flower primordia.

Gibberellin treatment of *Bryophyllum* spp. (so-called longshort-day plants) causes flowering in short-day photoperiods, but not in long-days, viz. it will replace the necessary preliminary exposure to long days (*51, 325*).

## D. Discussion.

The main impression given by these data on flowering is that *gibberellin treatment again shows a tendency to substitute for vernalisation, or simulates exposure to long-day photoperiods.* This is generally true of rosette plants, where with few exceptions it induces flowering, and of one group of short-day plants, exemplified by *Fragaria* and *Kalanchoë*, where flowering is inhibited. But there are two groups of plants which do not fall into this pattern, viz. (a) those short-day plants typified by *Xanthium* and *Nicotiana*, where flowering is not inhibited, and (b) the caulescent long-day plants, where flowering is not stimulated and is even delayed. Observations of this kind have led BARBER (*11*) and CHOUARD (*76*) to question whether there is any justification for continuing to suppose that there is one single mechanism for control of flowering, and BARBER has pointed out that it would be most unlikely that evolutionary changes in adaptive control of flowering should be based on one chemical evocator. Nevertheless it is striking that in the rosette plants, where control of flowering appears to be mainly based on a stimulator of flowering, and in the late pea where it is probably based on a flowering inhibitor (*11*), gibberellins should in fact have a differential effect on flowering, though not on stem extension.

As we have seen, most rosette plants bolt vegetatively in response to gibberellin, and a high proportion of these then flower. But in many cases initiation of flower primordia is delayed until appreciable shoot extension has resulted. This partial dissociation of vegetative and flowering responses has led Lang (*175, 176*) and others (*62, 76, 199, 277, 325*) to suggest that the primary action of gibberellin is on stem elongation, and that this elongation itself sets in train other processes which lead to flower initiation, flowering thus being a secondary response to gibberellin. This conclusion is scarcely justifiable, for three reasons: (a) in some plants bolting induced by gibberellin is *not* followed by flowering; (b) the processes of stem elongation and flower initiation are experimentally separable in other ways, and flowering can be induced in some long-day plants without stem elongation (*231*); (c) there is an essential difference between flowering and vegetative responses to long-day photoperiods, the former needing a specific induction period, exposure to long-day photoperiods over a longer period than needed for optimum induction being without further effect, whereas the size of the vegetative response is related to the length of exposure to long-day photoperiods (*245*). Against this background, it appears that the frequent differential effect of gibberellin on vegetative development and on flowering is yet another indication of the potential independence of the two processes and, indeed, in gibberellin we have a useful tool for study of their interaction.

It has been commonly supposed that control of flowering is based on a single hormone (the hypothetical 'florigen'), and that this induces flowering both in long-day plants and short-day plants. Quite clearly, from the results presented above, the gibberellins are not 'florigens'. A stronger case could be made for a close relationship between the gibberellins and 'vernalin', the hypothetical hormone produced in response to vernalisation, since in broad outline the actual properties of gibberellin are similar to the postulated properties of 'vernalin'; nevertheless, there are detailed differences (*277*).

### 5. Fertilisation and Fruit Growth.

#### A. Pollen Growth.

Gibberellin has been shown to stimulate tube growth of pollen of species of *Lilium, Linaria, Lobelia, Lonicera, Petunia, Pseudotsuga, Tradescantia* and *Verbena* (*64, 72, 147, 304, 317*). On the other hand inhibition of extension and malformation of pollen-tubes of species of *Delphinium* and *Saintpaulia* have also been recorded (*64*). Chandler (*64*) has suggested that the stimulating effect on pollen growth may possibly be of use in circumventing certain kinds of incompatibility.

### B. Parthenocarpy.

Parthenocarpic fruit-set induced by gibberellin was first recorded in tomato by WITTWER et al. (*339*); in their experience it was far more effective than any indole compound. The effect is systemic and treatment of foliage results in fruit-set (*246*). LUCKWILL (*204*) has demonstrated a marked synergism between auxins (IAA and 2-naphthoxyacetic acid) and gibberellic acid in the induction of fruit-set in tomatoes. Since then, parthenocarpic fruit-set has also been induced in *Zephyranthes* and in some non-apomictic species of *Rosa*. In *Zephyranthes*, as in tomato, gibberellin is more effective than auxins (*274*). There are interesting differences in the response of different species of *Rosa*. Fruit-set can be induced readily in *R. rugosa* with both auxins and gibberellin. *R. spinosissima* responds weakly both to auxins and gibberellins if applied singly, but a combination of naphthalene-acetamide and gibberellic acid induces almost complete fruit set, the effects of the two substances being more than additive. *R. arvensis* does not set parthenocarpic fruit if treated with auxin but does so after gibberellin treatment (*141*, *258*).

### C. Fruit Swelling.

Under conditions where there is an excess of far-red wave-lengths in the light supplied, swelling of tomato fruits may cease; this 'summer dormancy' can be overcome by supplementary irradiation with red light, or by gibberellin treatment (*187*). In an extensive experimental programme with many varieties WEAVER (*323*) has shown that enlargement of berries is one of the most significant responses of seedless grapes to gibberellin. Seeded varieties were usually less responsive in his experiments, and he suggests that the poor swelling of the untreated seedless varieties may be due to low endogenous levels of the growth-hormones, which may include gibberellins, normally formed in the developing seeds, in seeded varieties. However, RIVES and POUGET (*270*) have reported greatly improved swelling of some seeded grapes after gibberellin treatment.

### 6. Seed Germination.

### A. Breaking of Dormancy.

The most interesting effect of gibberellin in connection with germination is its capacity to break dormancy. Dormancy in seeds may be due to a requirement for low temperature treatment (after-ripening) or for light, or may be of a less specific nature; gibberellin has been found to induce seeds of each of these classes to germinate without further treatment.

The cold requirements of *Arabidopsis* (*166*), peach (*99*), Douglas fir (*268*) and sweet cherry (*108*) seeds can be wholly or partially replaced by gibberellin. In the latter case gibberellin treatment of the seed only partially substitutes for optimum after-ripening (six months at $3°$); seeds soaked in gibberellin germinate normally but after a time growth ceases and the seedling forms a rosette, a condition characteristic of insufficient after-ripening. The rosette may be induced to form a new stem by further treatment with gibberellin.

Gibberellins will replace the light requirement of seeds of *Arabidopsis (166)*, the cactus *Carnegiea gigantea (2)*, *Kalanchoë (49, 50, 55)*, lettuce *(104, 125, 144, 195, 255, 289, 290)* and carpet grass *(290)*. Marked temperature-controlled synergisms and antagonisms between gibberellin and 6-substituted purines (kinins) have been demonstrated in lettuce *(125, 290)*; these two kinds of hormone appear to act independently, as gibberellins and auxins probably do in stem elongation. The effect on these light-requiring seeds is of particular interest, since the action spectrum is again similar to that of photoperiodism; the promoting effect of red light can be reversed by far red, but gibberellin-induced germination cannot be reversed completely *(125, 144, 255)*. Light-insensitive lettuce seed can be made sensitive by treatment with coumarin; gibberellin will reverse the inhibition of germination induced by coumarin *(214)*.

Non-specific dormancy is partially or completely broken by gibberellin in seeds of several grass species *(57)* and in species of *Bartsia*, *Diapensia*, *Draba*, *Erysimum*, *Gentiana*, *Geranium*, *Luzula* and *Trollius (145)*. On the other hand, seed-dormancy in some plants has been found to be unaffected by gibberellin *(9, 145)*. If false-oats *(Avena fatua)* are treated with gibberellin while the seeds are developing, the seeds fail to develop natural dormancy *(20)*. Seasonal barley dormancy is broken by gibberellic acid, and in this case a marked synergism with sulphides has been demonstrated *(106, 222, 256)*.

## B. Acceleration of Germination.

The early Japanese workers were the first to report acceleration of seed germination and, for commercial reasons concerned with the production of malt, interest has since centred on barley seed, where not only is germination accelerated in the morphological sense but the biochemical changes associated with germination are also accelerated and even carried further than normal; respiration is enhanced and production of certain enzymes, e. g. $\alpha$-amylase, cellulases, proteinases and transaminases, is accelerated and increased *(16, 130, 213, 229, 239, 276)*.

### 7. Effects on Lower Plants.

Gibberellin has been found to influence growth and development of most other groups of green plant, in addition to the angiosperms which have mainly concerned us in the preceding pages. Gymnosperms are somewhat variable in their response, and there is a suspicion that exogenous gibberellin affects their development less frequently than is the case with angiosperms. Nevertheless, stem extension responses, breaking of bud dormancy and improved seed-germination or seedling growth have been recorded (*41, 212, 232, 267, 332*). Growth of prothallia of the ferns *Dryopteris filix-mas* and *Camptosorus rhizophyllus* is accelerated (*160*) and morphogenetic changes in apices of *Dryopteris austriaca* are induced by gibberellin (*318*). Protonemata of the moss *Splachnum ampullaceum* show more rapid growth in the presence of gibberellin (*209*). In studies of the moss *Pohlia nutans* no effect of gibberellin on growth of protonemata was observed, but there was a marked increase in the number and rate of production of leafy buds (*216*); this is a response normally induced by exposure to red light. Another light-simulating effect has been recorded in connection with elongation of the sporogonial setae of the liverwort *Pellia epiphylla*. The sporogonium is differentiated early in the winter, but elongation of the setae does not occur until the following spring, in response to increasing day-length. Dormant setae will extend under short day conditions if treated with gibberellin (*7*). Cell multiplication and cell-extension in green algae of the genera *Ulothrix* and *Ulva* are also stimulated (*78, 259*). Fungal growth is usually unaffected by gibberellin (*28*) but there is one unconfirmed report of inhibition of fungal growth (*310*).

These scattered observations suggest that *a gibberellin growth-regulating system arose early in the evolutionary history of green plants*; the fact that the gibberellins were first known as metabolic products of a fungus, though apparently they are functionless in fungi, seems all the more extraordinary.

## 8. Discussion.

As we said earlier, the discovery that gibberellins are widely distributed in plants, taken together with all the information now available concerning the responses they induce when applied exogenously to plants or parts of plants, carries with it the implication that they are *natural plant hormones*. If we adopt that standpoint, then two aspects stand out as meriting further consideration, (a) their interactions with other hormones, and (b) their function in responses to photoperiod and vernalisation. These are dealt with in turn below.

### A. Hormonal Interactions.

We have presented evidence of interactions between gibberellins and auxins in the following plant responses: (a) in stem extension, using stem sections or intact plants, including rosette plants sensitive to photoperiod; (b) in initiation of cambial activity in woody shoots; (c) in extension of cereal leaves; (d) in root growth, at least in excised root cultures; (e) in initiation of roots on stem cuttings; and (f) in induction of parthenocarpic fruit development. Thus such interactions are not uncommon, and this lends probability to the supposition that auxin : gibberellin interactions are in fact involved in responses where they have not yet been demonstrated, e. g. in flowering or seed-germination. It is, of course, difficult to demonstrate such interactions unless endogenous levels of both kinds of hormone are limiting in the experimental system.

It should be noted, also, that several investigators have had to invoke the intervention of other, unidentified hormones to explain the observed characters of auxin : gibberellin interactions.

The nature of the auxin : gibberellin interaction varies from system to system. In some cases, e. g. in stem extension, the relation is synergistic; in others, e. g. root initiation, it is antagonistic. Some of the synergisms involve evocation of a qualitatively different response by combinations of the two hormones. Thus whatever the merits of the various "auxin-sparing" hypotheses concerning the auxin : gibberellin interaction in stem extension, such a mechanism certainly cannot explain all auxin : gibberellin interactions. It seems much more likely that both act independently in chains of reactions, but this is a matter which can only be cleared up by further experimental work.

There is also evidence of gibberellin : kinin interactions in (a) correlative inhibition of lateral buds (*334*); (b) in leaf-expansion; (c) in breaking of dormancy of some light-sensitive seeds; and possibly (d) in formation of sporophyte buds on moss protonemata. The natural significance of the synthetic kinins used in the work referred to is in some doubt, but these results strongly suggest that gibberellins will be found to be associated with other hormones besides the auxins.

*It is a corollary of the existence of such interactions that the failure of one hormone to induce a response, in some given experimental system, may be due to another hormone being limiting.*

## B. The Rôle of Gibberellins as Plant Hormones.

Gibberellin simulates, to a greater or lesser extent, the effects of vernalisation in so far as it promotes: (a) vegetative bolting in most biennials and perennials requiring vernalisation; (b) breaking of dormancy of woody shoots and other kinds of shoot dormancy, including that of tubers, rhizomes, etc., normally terminated by exposure to low temperature; (c) differentiation of flower primordia in many of the biennials and perennials which form a stem in response to gibberellin; and (d) germination of cold-requiring seeds. *It simulates exposure to long-day photoperiods* in so far as: (a) like long-day photoperiods it stimulates stem extension in photoperiodically-sensitive plants, the biggest effect being seen in rosette plants; (b) it prevents senescence and autumnal fall of leaves of many deciduous trees; and (c) it promotes flowering in many long-day plants and strongly inhibits flowering in some short-day plants. Analogous light-simulating effects are: (a) promotion of germination of light-sensitive seeds; (b) promotion of expansion of leaf tissues; and (c) promotion of formation of sporophyte buds on moss protonemata.

LANG (*176*) pointed out in 1957 that the gibberellins are the first substances found capable of inducing flowering of biennials and long-day plants under strictly non-inductive conditions and that, since gibberellin-like substances are widely distributed in plants, it seems reasonable to assume that they play an important part in the regulation of flowering in such plants. Since then the observed range of vernalisation- and long-day-simulating effects of exogenous gibberellin has been greatly extended, as indicated in the preceding paragraph, and it is now quite certain that gibberellins occur widely in plants. BRIAN (*25, 26*) carried LANG's argument further to include all processes of development normally controlled by vernalisation or by photoperiod. He showed that responses to these environmental influences could in most cases be explained in terms of cold- or light-induced biosynthesis of gibberellins, basing his scheme on one proposed earlier by LIVERMAN and BONNER (*186*) in terms of auxin synthesis. It was claimed that the 'gibberellin theory' was stronger than LIVERMAN and BONNER's 'auxin theory' in so far as exogenous gibberellins promote developmental changes far more effectively and regularly than auxins. There is now some direct evidence that gibberellins are involved in responses to photoperiod; LANG (*178*) has demonstrated greater concentrations of gibberellin-like materials in photo-induced *Hyoscyamus* than in non-induced plants, and PORLINGIS and BOYNTON (*257*) have demonstrated more gibberellin-like substances in the runners of strawberries, formed in response to long-day photoperiods, than in the short apical axes.

However, *account must be taken of circumstances where gibberellins fail to simulate vernalisation or exposure to long-days*. There are five such circumstances which have been mentioned in preceding pages: (a) a minority of rosette plants fail to bolt when exogenous gibberellic acid is applied; (b) vegetative bolting and initiation of flower primordia in some long-day plants fail to keep in step after application of gibberellin, and in extreme cases no initiation of flower primordia occurs; (c) in caulescent long-day plants, gibberellin fails to induce flowering, and may inhibit, though the normal vegetative responses to long-days are induced; (d) gibberellin does not inhibit flowering of some short-day plants, though long-day photoperiods do; and (e) gibberellin failed to prevent autumnal leaf senescence in some deciduous trees.

These anomalies could be explained in terms of gibberellin specificity, gibberellic acid not being able to replace the native gibberellin. This is possible, but unlikely to be the case where a whole group of taxonomically unrelated plants, e. g. the caulescent long-day plants, fail to respond. In view of the accumulating evidence that interactions between two or more hormones are commonly involved in responses to gibberellin, it seems more likely that failures to respond to gibberellin may be a consequence of other hormones being limiting in non-inductive conditions. It will be recalled that Lang (*177*) found that *Silene armeria* bolted more readily in response to an auxin : gibberellin mixture than to gibberellin alone, and that Wareing (*319*) made similar observations in connection with the termination of cambial dormancy in *Acer campestris*. Indeed, Lang (*176*), Brian (*26*) and Chaylakhian (*67—69*) have envisaged the existence of hormonal interactions in responses of plants to vernalisation and photoperiod. Nevertheless, it is striking that endogenous gibberellin seems to be limiting more frequently than other hormones.

In addition to the light-effects just mentioned, we have to consider the fact that light-inhibition of seedlings is apparently reversed by exogenous gibberellin, and the possibility that such inhibitions may be due to lowering of the level of endogenous gibberellin. It may seem that the postulate of light-induced decrease of gibberellin levels in such circumstances is incompatible with the postulate of increased endogenous gibberellin levels resulting from exposure to long-day photoperiods. There is probably no real incompatibility, since light-inhibitions seem to be restricted to early phases of growth and noticeable only in plants previously grown in darkness.

*In conclusion*, it can be said with some confidence, despite our present uncertainties concerning the natural rôle of the gibberellins, that the use of the gibberellins as an experimental tool will greatly assist investigations aimed at increasing our knowledge of the regulation of plant growth and development.

## V. Detection and Estimation of Gibberellins.

### 1. Bioassay.

The early Japanese gibberellin research was based on growth measurements of rice seedlings, usually grown in culture solutions (*292, 341*). This method has recently been standardised (*8*) and is claimed to be sensitive to gibberellic acid at a concentration of 0.25 µg./ml. and to give a linear response to log dose in the range 0.25–8.0 µg./ml. The method should be applicable to other cereals, but the results of

SUDIA (302) suggest that complex dose : response relationships may sometimes be encountered, limiting the usefulness of the method. The use of genetic dwarfs provides a more sensitive bioassay, again with a linear log dose : response relationship; with the dwarf pea, variety Meteor, 0.01 $\mu$g. gibberellic acid can be detected (35), and with some of the dwarf mutants of maize as little as 0,001 $\mu$g. (248, 249).

Bioassays based on measurement of growth of excised parts of plants give more rapid results, but in general are not truly quantitative, and serve only for detection of gibberellins. Gibberellins induce, as we have seen, extension in cereal coleoptile and mesocotyl sections, and in pea stem sections; such responses cannot be used to detect gibberellins as they are not specific, since auxins, thiols and a variety of other substances also elicit a response. Sections from the first leaf of barley, oats and wheat have been shown by HAYASHI and MURAKAMI (133) and RADLEY (264) to be a good deal more specific and this test can be used with some confidence, especially if sections near the meristem are used; occasionally, however, auxins will induce a response. Discs from dark-grown *Phaseolus* leaves respond to gibberellin (279), but again are not specific, responding also to kinins. Expansion of discs from light-grown *Phaseolus* leaves, carrying out the test in light, affords the basis for a quite highly specific method of detecting gibberellins, but it is not quantitative (140).

It has been suggested that metabolic bioassays might be developed, based on the respiration rate (239) or development of enzyme activity (16) in barley seeds, but at the moment the quantitative assays based on dwarf maize and peas are undoubtedly the most valuable.

## 2. Chemical and Physical Methods.

$R_f$ values in a number of solvent systems for paper chromatograms are given in Table 5, p. 416 (140) and are in reasonable agreement with those obtained elsewhere (250).

Gibberellic acid is easily seen as a spot fluorescent in ultraviolet light after the paper has been sprayed with 5% ethanolic sulphuric acid (263) and heated at 80° for a few minutes. The other gibberellins give only faint, barely detectable, spots under these conditions. Gibberellins $A_1$, $A_4$ and $A_5$ are detected as yellow spots on a pink background by means of an 0.5% aqueous potassium permanganate spray but gibberellin $A_2$ can only be detected by a bromophenol blue spray after chromatography using systems 1 or 4. Gibberellic acid and gibberellin $A_1$ were not separated by any of the solvent systems in Table 5 or by others (305). $R_f$ values for gibberellin $A_2$ were not significantly different from those for gibberellic acid and gibberellin $A_1$ in solvent systems 1 to 5. Gibberellins $A_4$ and $A_5$ were readily distinguished from the other three gibberellins in solvent system 1, and were separated from each other by solvent system 6. Gibberellic acid and gibberellin $A_1$ can be separated by the method of BIRD and PUGH (19) using solvent system 6

Table 5. Some $R_f$ Values for the Gibberellins.

$R_f$ values[e] for the gibberellins (50 $\mu$g. spots) run on Whatman's No. 1 paper (20″ × 20″) at 20 ± 2°.

| Compound | Solvent System | | | | | | |
|---|---|---|---|---|---|---|---|
| | 1[a] | 2[a] | 3[a] | 4[a] | 5[a] | 6[a] | 7[b] |
| Gibberellic acid ...... | 0.31[c] | 0.64 | 0.90 | 0.75 | 0.89 | 0.0 | 0.64 |
| Gibberellin $A_1$........ | 0.31 | 0.64 | 0.89 | 0.74 | 0.90 | 0.0 | 0.64 |
| Gibberellin $A_2$........ | 0.32 | 0.61[d] | 0.87[d] | 0.75 | 0.90 | | |
| Gibberellin $A_4$........ | 0.58 | | | 0.81 | 0.95 | 0.74 | |
| Gibberellin $A_5$........ | 0.54 | | | | | 0.0 | 0.74 |

Key to Solvent systems (running time in parenthesis):

1. n-Butanol:1.5N-ammonium hydroxide (3:1) (11 hr.).
2. n-Amylalcohol:pyridine:water (7:7:6) (10 hr.).
3. n-Butanol:acetic acid:water (19:1:6) (10.5 hr.)..
4. Ethanol:3N-ammonium hydroxide (4:1) (8 hr.).
5. Chloroform:ethanol:water:formic acid (20:4:2:1) (6 hr.).
6. Benzene:acetic acid:water (4:2:1) (2 hr.).
7. Isopropanol:7N-ammonium hydroxide (5:1) (24 hr.).

[a] Descending solvent flow.    [b] Ascending solvent flow.
[c] This solvent system gave variable results and $R_f$ values for gibberellic acid were obtained in the range 0.22–0.40. The *relative* $R_f$ values for the gibberellins were constant nevertheless.
[d] Detected by bioassay (250).    [e] For sprays see text.

but eluting the paper for about 40 hr. and allowing the solvent front to run off the bottom of the paper.

With concentrated sulphuric acid gibberellic acid gives a characteristic intense wine-red solution which develops a blue fluorescence and this provides the basis both for the spray used (above) for detecting gibberellic acid on paper and for methods of estimating gibberellic acid by colourimetry (5) and by fluorimetry (140, 152). Gibberellenic acid (LVI, $R = H$) and other 2-hydroxy-gibb-3(or 4)-enes such as (LXXXV, $R = H$) give similar colours with concentrated sulphuric acid and interfere in the estimation but gibberellins $A_1$, $A_2$, $A_4$ and $A_5$ give only straw yellow solutions (with a weak green fluorescence) with the reagent and do not interfere (140). Gibberellenic acid can be estimated separately (140) by its ultraviolet absorption (see p. 381) at 253–254 m$\mu$. Kato (149) has incorrectly attributed this absorption band to gibberellic acid.

(LVI.)                    (LXXXV.)

The gibberellic acid content of fermentation broths containing no other gibberellins (which form solid solutions with gibberellic acid) has been estimated by an isotope dilution method (6) and by the intensity of the infrared absorption near 1784 cm.$^{-1}$ in dioxan solution due to the $\gamma$-lactone $C=O$ group (140). The last method can be used to estimate the total gibberellins in a mixture. Infrared (320) and radioisotopic methods (15) have been used for the determination of gibberellic acid and gibberellin $A_1$ in mixtures and in fermentation broths and a polarographic method of estimating gibberellins has been described (156).

### References.

1. ABRAMS, G. J. v.: Auxin Relations of a Dwarf Pea. Plant Physiol. **28**, 443 (1953).
2. ALCORN, S. M. and E. B. KURTZ: Some Factors Affecting the Germination of Seed of the Saguaro Cactus *(Carnegiea gigantea)*. Amer. J. Bot. **46**, 526 (1959).
3. ALLEWELDT, G.: Die Wirkung der Gibberellinsäure auf einjährige Reben bei verschiedener Photoperiode. Vitis **2**, 23 (1959).
4. APPLEGATE, H. G.: Photoperiod and Temperature Effects on Gibberellin-sprayed Plants. Bot. Gaz. **120**, 39 (1958).
5. ARISON, B., G. V. DOWNING, R. A. GRAY, M. A. MANZELLI, J. D. NEUSS, O. SPETH, N. R. TRENNER and F. J. WOLF: Gibberellic Acid Assay Methods. 132nd meeting Amer. Chem. Soc., New York, 1957, Abstracts p. 35 C.
6. ARISON, B. H., O. C. SPETH and N. R. TRENNER: Mass Isotope Dilution Assay for Gibberellic Acid. Analyt. Chemistry **30**, 1083 (1958).
7. ASPREY, G. F., K. BENSON-EVANS and A. G. LYON: Effect of Gibberellin and Indoleacetic Acid on Seta Elongation in *Pellia epiphylla*. Nature (London) **181**, 1351 (1958).
8. AYTOUN, R. S. C., A. T. DUNN and D. A. L. SEILER: A Biological Assay for Gibberellic Acid with Rice Seedlings. Analyst **84**, 216 (1959).
9. BAKER, J. N.: Effect of Gibberellic Acid, 2,4-D, and Indoleacetic Acid on Seed Germination and Epicotyl and Radicle Growth of Intermediate and Pubescent Wheatgrass. J. Range Management **11**, 227 (1958).
10. BAMBERGER, E. und S. C. HOOKER: Über das Reten. Liebigs Ann. Chem. **229**, 102 (1885).
11. BARBER, H. N.: Physiological Genetics of *Pisum*. II. The Genetics of Photoperiodism and Vernalisation. Heredity **13**, 33 (1958).
12. BARTON, L. V.: Growth Response of Physiologic Dwarfs of *Malus arnoldiana Sarg.* to Gibberellic Acid. Contr. Boyce Thompson Inst. **18**, 311 (1956).
13. BARTON, L. V. and C. CHANDLER: Physiological and Morphological Effects of Gibberellic Acid on Epicotyl Dormancy of Tree Peony. Contr. Boyce Thompson Inst. **19**, 201 (1957).
14. BATES, J. F.: Preliminary Experiments on the Effects of Gibberellic Acid on Germination, Growth and Flowering in Sugar Cane. Proc. 1957 meeting B. W. I. Sugar Technologists 165 (1958).
15. BAUMGARTNER, W. E., L. S. LAZER, A. M. DALZIEL, E. V. CARDINAL and E. L. VARNER: Determination of Gibberellins by Derivative Labelling with Diazomethane-$C^{14}$ and by Isotopic Dilution Analysis with Tritium Labelled Gibberellins. J. Agric. Food Chem. **7**, 422 (1959).

16. Bergquist, G., A.-M. Stensgard and N. Nielsen: The Influence of Gibberellic Acid on the Transaminase Content of Germinating Barley Seeds. Physiol. Plantarum 12, 386 (1959).

17. Birch, A. J., R. W. Rickards and H. Smith: The Biosynthesis of Gibberellic Acid. Proc. Chem. Soc. (London) 1958, 192.

18. Birch, A. J. and H. Smith: Ciba Found. Sympos. Biosynthesis of Terpenes and Steroids, p. 253. London: Churchill. 1959.

19. Bird, H. L. and C. T. Pugh: A Paper Chromatographic Separation of Gibberellic Acid and Gibberellin A₁. Plant Physiol. 33, 45 (1958).

20. Black, M. and J. M. Naylor: Prevention of the Onset of Seed Dormancy by Gibberellic Acid. Nature (London) 184, 468 (1959).

21. Bonde, E. K. and T. C. Moore: Effect of Gibberellic Acid on the Growth and Flowering of Telephone Peas. Physiol. Plantarum 11, 451 (1958).

22. Borrow, A., P. W. Brian, V. E. Chester, P. J. Curtis, H. G. Hemming, C. Henehan, E. G. Jefferys, P. B. Lloyd, I. S. Nixon, G. L. F. Norris and M. Radley: Gibberellic Acid, a Metabolic Product of the Fungus Gibberella fujikuroi: Some Observations on its Production and Isolation. J. Sci. Food Agric. 6, 340 (1955).

23. Bourdeau, P. F.: Interaction of Gibberellic Acid and Photoperiod in the Vegetative Growth of Pinus elliotti. Nature (London) 182, 118 (1958).

24. Brian, P. W.: The Effects of some Microbial Metabolic Products on Plant Growth. Sympos. Soc. exp. Biol. 11, 166 (1957).

25. — Role of Gibberellin-like Hormones in Regulation of Plant Growth and Flowering. Nature (London) 181, 1122 (1958).

26. — Effects of Gibberellins on Plant Growth and Development. Biol. Revs. 34, 37 (1959).

27. — Morphogenetic Effects of the Gibberellins. J. Linnean Soc. London (Bot.) 56, 237 (1959).

28. Brian, P. W., G. W. Elson, H. G. Hemming and M. Radley: The Plant-growth-promoting Properties of Gibberellic Acid, a Metabolic Product of the Fungus Gibberella fujikuroi. J. Sci. Food Agric. 5, 602 (1954).

29. Brian, P. W. and J. F. Grove: Gibberellic Acid. Endeavour 16, 161 (1957).

30. Brian, P. W., J. F. Grove, H. G. Hemming, T. P. C. Mulholland and M. Radley: Gibberellic Acid. Part VI. The Biological Activity of alloGibberic Acid and its Identity with Gibberellin B. Plant Physiol. 33, 329 (1958).

31. Brian, P. W. and H. G. Hemming: The Effect of Gibberellic Acid on Shoot-growth of Pea Seedlings. Physiol. Plantarum 8, 669 (1955).

32. — — A Relation between the Effects of Gibberellic Acid and Indolylacetic Acid on Plant Cell Extension. Nature (London) 179, 417 (1957).

33. — — The Effect of Maleic Hydrazide on the Growth Response of Plants to Gibberellic Acid. Ann. appl. Biol. 45, 489 (1957).

34. — — Complementary Action of Gibberellic Acid and Auxins in Pea Internode Extension. Ann. Botany 22, 1 (1958).

35. Brian, P. W., H. G. Hemming and D. Lowe: Effect of Gibberellic Acid on Rate of Extension and Maturation of Pea Internodes. Ann. Botany 22, 539 (1958).

36. — — — The Effect of Gibberellic Acid on Shoot Growth of Cupid Sweet Peas. Physiol. Plantarum 12, 15 (1959).

37. Brian, P. W., H. G. Hemming and M. Radley: A Physiological Comparison of Gibberellic Acid with some Auxins. Physiol. Plantarum 8, 899 (1955).

38. Brian, P. W., J. H. P. Petty and P. T. Richmond: Effects of Gibberellic Acid on Development of Autumn Colour and Leaf-fall of Deciduous Plants. Nature (London) 183, 58 (1959).

39. BRIAN, P. W., J. H. P. PETTY and P. T. RICHMOND: Extended Dormancy of Deciduous Woody Plants Treated in Autumn with Gibberellic Acid. Nature (London) 184, 69 (1959).

40. BRIGGS, W. R., T. A. STEEVES, I. M. SUSSEX and R. H. WETMORE: A Comparison of Auxin Destruction by Tissue Extracts and Intact Tissues of the Fern *Osmunda cinnamomea*. Physiol. Plantarum 8, 899 (1955).

41. BROWN, C. L. and E. M. GIFFORD: The Relation of the Cotyledons to Root Development of Pine Embryos in vitro. Plant Physiol. 33, 57 (1958).

42. BUKOVAC, M. J. and H. DAVIDSON: Gibberellin Effects on Photoperiod-controlled Growth of *Weigela*. Nature (London) 183, 59 (1959).

43. BUKOVAC, M. J. and S. H. WITTWER: Gibberellic Acid and Higher Plants: I. General Growth Responses. Quart. Bull. Michigan agric. Exp. Station 39, 307 (1956).

44. — — Gibberellin and Higher Plants: II. Induction of Flowering in Biennials. Quart. Bull. Michigan agric. Exp. Station 39, 650 (1957).

45. — — Comparative Biological Effectiveness of the Gibberellins. Nature (London) 181, 1484 (1958).

46. — — Reproductive Responses of Lettuce (*Lactuca sativa*, Variety Great Lakes) to Gibberellin as Influenced by Seed Vernalisation, Photoperiod and Temperature. Proc. Amer. Soc. hort. Sci. 71, 407 (1958).

47. — — Comparative Biological Activities of the Gibberellins. Paper read at 4th Internat. Conf. Plant Growth Regulation, Yonkers, Aug. 1959.

48. BÜNSOW, R.: Anwendungsmöglichkeiten der Gibberelline. Angew. Bot. 32, 186 (1958).

49. BÜNSOW, R. und K. v. BREDOW: Einfluß der Gibberelline auf die Tages-längenabhängigkeit der Samenkeimung von Kalanchoë. Naturwiss. 45, 95 (1958).

50. — — Wirkung von Licht und Gibberellin auf die Samenkeimung der Kurztag-pflanze *Kalanchoë blossfeldiana*. Biol. Zentralbl. 77, 132 (1958).

51. BÜNSOW, R. und R. HARDER: Blütenbildung von *Bryophyllum* durch Gibberellin. Naturwiss. 43, 479 (1956).

52. — — Blütenbildung von *Lapsana* durch Gibberellin. Naturwiss. 43, 527 (1956).

53. — — Blütenbildung von *Adonis* und *Rudbeckia* durch Gibberellin. Naturwiss. 44, 453 (1957).

54. BÜNSOW, R., J. PENNER und R. HARDER: Blütenbildung bei *Bryophyllum* durch Extrakt aus Bohnensamen. Naturwiss. 45, 46 (1958).

55. BÜNSOW, R. und C. SEIFERTH: Förderung der Samenkeimung durch Gibberellin in Abhängigkeit vom Reifezustand. Naturwiss. 46, 153 (1959).

56. BURK, L. G. and T. C. TSO: Effects of Gibberellic Acid on *Nicotiana* Plants. Nature (London) 181, 1672 (1958).

57. BUTTON, E. F.: Effect of Gibberellic Acid on Laboratory Germination of Creeping Red Fescue *(Festuca rubra)*. Agron. J. 51, 60 (1959).

58. CARR, D. J., A. J. McCOMB and L. D. OSBORNE: Replacement of the Requirement for Vernalisation in *Centaurium minus* MOENCH. by Gibberellic Acid. Naturwiss. 44, 428 (1957).

59. CARVAJAL, J. F.: Estudio preliminar sobre la respuesta del cafeto el acido gibberellico. Rev. biol. trop. Univ. Costa Rica 6, 273 (1958).

60. CAVA, M. P. and B. WEINSTEIN: Structure of Andrographolide. Chem. and Ind. 1959, 851.

61. CHAKRAVARTI, R. N.: A new Method of Formation of Eudalene. V. J. Indian Chem. Soc. 20, 393 (1943).

62. Chakravarti, S. C.: Gibberellic Acid and Vernalisation. Nature (London) **182**, 1612 (1958).
63. — Some Effects of Gibberellic Acid on *Sesamum indicum* L. Phyton (Buenos Aires) **11**, 75 (1958).
64. Chandler, C.: The Effect of Gibberellic Acid on Germination and Pollentube Growth. Contr. Boyce Thompson Inst. **19**, 215 (1957).
65. Chardonnens, L. et A. Würmli: Sur les dérivés de la fluorénone. IV. La diméthyl-1,3-fluorénone. Helv. Chim. Acta **33**, 1338 (1950).
66. Chaylakhian, M. K.: Growth and Flowering of Plants as Affected by Gibberellins. Doklady Akad. Nauk SSSR **117**, 1077 (1957).
67. — The Effect of Gibberellin on the Growth and Development of Plants. Bot. Zhurn. **43**, 927 (1958).
68. — Hormone Factors Involved in Flowering of Plants. Fiziol. Rast. Akad. Nauk SSSR **5**, 541 (1958).
69. — Hormonale Faktoren des Pflanzenblühens. Biol. Zentralbl. **77**, 641 (1958).
70. Chaylakhian, M. K. and T. V. Nekrassova: Physiologically Active Substances as a Means to Surmount Polarity in Lemon Cuttings. Doklady Akad. Nauk SSSR **119**, 826 (1958).
71. Chiang, Y. L. and S. H. Chiang: Effects of Gibberellic Acid on Growth and Xylem Development in Adzuki Bean Plants (*Phaseolus radiatus* L. var. *aurea* Prain). Formosan Science **13**, 49 (1959).
72. Ching, K. K. and T. M. Ching: Extracting Douglas-fir Pollen and Effects of Gibberellic Acid on its Germination. Forest Sci. **5**, 74 (1959).
73. Chouard, P.: La journée courte ou l'acide gibbérellique comme succédanés du froid pour la vernalisation d'une plante vivace en rosette, le *Scabiosa Succisa* L. C. R. hebd. Séances Acad. Sci. **245**, 2520 (1957).
74. — Présentation de quelques plantes en cours d'expérimentation sur les facteurs de la floraison. Bull. soc. bot. France **105**, 135 (1958).
75. — Les gibbérellines. Rev. horticole **1958**, 1793.
76. — Diversité des mécanismes des dormances, de la vernalisation et du photopériodisme, révélée notamment par l'action de l'acide gibbérellique. Mem. soc. bot. France **1956—1957**, 51 (1958).
77. Cologne, J. et J. Sibeud: Synthèse d'hydrocarbones fluoréniques. C. R. hebd. Séances Acad. Sci. **232**, 845 (1951).
78. Conrad, H., P. Saltman and R. Eppley: Effects of Auxin and Gibberellic Acid on Growth of *Ulothrix*. Nature (London) **184**, 556 (1959).
79. Cooper, J. P.: The Effect of Gibberellic Acid on a Genetic Dwarf in *Lolium perenne*. New Phytologist **57**, 235 (1958).
80. Cooper, W. C. and A. Peynado: Effect of Gibberellic Acid on Growth and Dormancy in Citrus. Proc. Amer. Soc. hort. Sci. **72**, 284 (1958).
81. Cornforth, J. W.: Personal communication (cf. Chem. and Ind. **1959**, 184).
82. Cross, B. E.: Gibberellic Acid. Part I. J. Chem. Soc. (London) **1954**, 4670.
83. — Gibberellic Acid. Part XIII. The Structure of Ring A. J. Chem. Soc. (London) (to be published).
84. Cross, B. E., J. F. Grove, P. McCloskey, T. P. C. Mulholland and W. Klyne: Stereochemistry of Gibberellic Acid. Chem. and Ind. **1959**, 1345.
85. Cross, B. E., J. F. Grove, J. MacMillan, J. S. Moffatt, T. P. C. Mulholland, J. C. Seaton and N. Sheppard: A Revised Structure for Gibberellic Acid. Proc. Chem. Soc. (London) **1959**, 302.
86. Cross, B. E., J. F. Grove, J. MacMillan and T. P. C. Mulholland: Gibberellic Acid. Part IV. The Structure of Gibberic and Allogibberic Acids and Possible Structures for Gibberellic Acid. Chem. and Ind. **1956**, 954.

87. CROSS, B. E., J. F. GROVE, J. MACMILLAN and T. P. C. MULHOLLAND: Gibberellic Acid. Part VII. The Structure of Gibberic Acid. J. Chem. Soc. (London) 1958, 2520.

88. CROSS, B. E., J. F. GROVE, J. MACMILLAN, T. P. C. MULHOLLAND and N. SHEPPARD: The Structure of Gibberellic Acid. Proc. Chem. Soc. (London) 1958, 221.

89. CROSS, B. E., J. F. GROVE and A. MORRISON: Gibberellic Acid. Part XVIII. Some Rearrangements of Ring A. J. Chem. Soc. (London) (to be published).

90. CROSS, B. E. and P. H. MELVIN: Gibberellic Acid. Part XIV. 2-Hydroxy-1:7-dimethylfluorene. J. Chem. Soc. (London) (in press).

91. CURRY, G. M. and E. C. WASSINK: Photoperiodic and Formative Effects of Various Wavelength Regions in Hyoscyamus niger as Influenced by Gibberellic Acid. Mededel. Landbouwhogesch. Wageningen 56, 1 (1956).

92. CURTIS, P. J. and B. E. CROSS: Gibberellic Acid. A new Metabolite from the Culture Filtrates of Gibberella fujikuroi. Chem. and Ind. 1954, 1066.

93. CURTIS, R. W.: Survey of Fungi and Actinomycetes for Compounds Possessing Gibberellin-like Activity. Science (Washington) 125, 646 (1957).

94. DANCER, J.: Synergistic Effect of Zinc and Gibberellin. Nature (London) 183, 901 (1959).

95. DARKEN, M. A., A. L. JENSEN and P. SHU: Production of Gibberellic Acid by Fermentation. Applied Microbiol. 7, 301 (1959).

96. DJERASSI, C.: Some Recent Applications of Optical Rotatory Dispersion Studies to Organic Chemical Problems. Record Chem. Progr. 20, 101 (1959).

97. DJERASSI, C., M. CAIS and L. A. MITSCHER: Terpenoids. XXXVII. The Structure of the Pentacyclic Diterpene Cafestol. On the Absolute Configuration of Diterpenes and Alkaloids of the Phyllocladene Group. J. Amer. Chem. Soc. 81, 2386 (1959).

98. DJERASSI, C. and W. KLYNE: Recording and Nomenclature of Optical Rotatory Dispersion. Proc. Chem. Soc. (London) 1957, 302.

99. DONOHO, C. W. and D. R. WALKER: Effect of Gibberellic Acid on Breaking of Rest Period in Elberta Peach. Science (Washington) 126, 1178 (1957).

100. DOSTAL, R.: Gibberellic Acid and Growth Correlations. Nature (London) 183, 1338 (1959).

101. DOWNS, R. J., S. B. HENDRICKS and H. A. BORTHWICK: Photoreversible Control of Elongation of Pinto Beans and other Plants under Normal Conditions of Growth. Bot. Gaz. 118, 199 (1957).

102. ERGLE, D. R.: Some Responses of Normal and Mutant Cottons to Gibberellic Acid. 54th Annu. Proc. Assoc. Southern Agric. Workers 1957, 227.

103. — Compositional Factors Associated with the Growth Responses of Young Cotton Plants to Gibberellic Acid. Plant Physiol. 33, 344 (1958).

104. EVENARI, M., G. NEUMANN, S. BLUMENTHAL-GOLDSCHMIDT, A. M. MAYER and A. POLJAKOFF-MAYBER: The Influence of Gibberellic Acid and Kinetin on Germination and Seedling Growth of Lettuce. Bull. Res. Council Israel 6 D, 65 (1958).

105. FEUCHT, J. R.: The Effect of Gibberellins on Internodal Tissues of Phaseolus vulgaris. Amer. J. Bot. 45, 520 (1958).

106. FISCHICH, O., M. THIELEBEIN und A. GRAHL: Brechung der Keimruhe bei Gerste durch Gibberellinsäure und Rindite. Naturwiss. 44, 642 (1957).

107. FLETCHER, W. W., J. W. S. ALCORN and J. C. RAYMOND: Effect of Gibberellic Acid on the Nodulation of White Clover (Trifolium repens L.). Nature (London) 182, 1319 (1958).

*108.* Fogle, H. W.: Effects of Duration of After-ripening, Gibberellin and other Pretreatments on Sweet Cherry Germination and Seedling Growth. Proc. Amer. Soc. hort. Sci. **72**, 129 (1958).

*109.* Galston, A. W.: Some Metabolic Consequences of the Administration of Indoleacetic Acid to Plant Cells. In: R. L. Wain and F. Wightman, The Chemistry and Mode of Action of Plant Growth Substances, p. 219. London: Butterworth. 1956.

*110.* — Studies on Indoleacetic Acid Oxidase and its Inhibitor in Light-grown Peas. Plant Physiol. (Suppl.) **32**, xxi (1957).

*111.* — Gibberellins and Nodulation. Nature (London) **183**, 545 (1959).

*112.* Galston, A. W. and H. Warburg: An Analysis of Auxin-Gibberellin Interaction in Pea Stem Tissue. Plant Physiol. **34**, 16 (1959).

*113.* Gerzon, K., H. L. Bird and D. O. Woolf: Gibberellenic Acid, a By-product of Gibberellic Acid Fermentation. Experientia **13**, 487 (1957).

*114.* Gray, R. A.: Alteration of Leaf Size and Shape and other Changes Caused by Gibberellins in Plants. Amer. J. Bot. **44**, 674 (1957).

*115.* Greulach, V. A. and J. G. Haesloop: The Influence of Gibberellic Acid on Cell Division and Cell Elongation in *Phaseolus vulgaris.* Amer. J. Bot. **45**, 566 (1958).

*116.* — — Influence of Gibberellin on *Xanthium* Flowering as Related to Number of Photoinductive Cycles. Science (Washington) **127**, 646 (1958).

*117.* Griffith, M. M.: Some Anatomical Effects of Gibberellic Acid on Dwarf Peas. Quart. J. Florida Acad. Sci. **20**, 238 (1957).

*118.* Grove, J. F., P. W. Jeffs and T. P. C. Mulholland: Gibberellic Acid. Part V. The Relation between Gibberellin $A_1$ and Gibberellic Acid. J. Chem. Soc. (London) **1958**, 1236.

*119.* Grove, J. F. and P. McCloskey: Gibberellic Acid. Part XIX. Stereochemistry. J. Chem. Soc. (London) (to be published).

*120.* Grove, J. F., J. MacMillan, T. P. C. Mulholland and W. B. Turner: Gibberellic Acid. Part XVII. The Stereochemistry of Gibberic and *epi*Gibberic Acids. J. Chem. Soc. (London) (in press).

*121.* Grove, J. F. and T. P. C. Mulholland: Gibberellic Acid. Part XII. The Stereochemistry of alloGibberic Acid. J. Chem. Soc. (London) (in press).

*122.* Gundersen, K.: Some Experiments with Gibberellic Acid. Acta Horti Gotoburgensis **22**, 87 (1958).

*123.* Guttridge, C. G. and P. A. Thompson: Effect of Gibberellic Acid on Length and Number of Epidermal Cells in Petioles of Strawberry. Nature (London) **183**, 197 (1959).

*124.* Haber, A. H. and N. E. Tolbert: Photosynthesis in Gibberellin-treated Leaves. Plant Physiol. **32**, 152 (1957).

*125.* — — Effects of Gibberellic Acid, Kinetin and Light on the Germination of Lettuce Seed. In: Photoperiodism and Related Phenomena in Plants and Animals. Washington: Amer. Assoc. Adv. Sci. 1959.

*126.* Harder, R. und R. Bünsow: Einfluß des Gibberellins auf die Blütenbildung bei *Kalanchoë blossfeldiana.* Naturwiss. **43**, 544 (1956).

*127.* — — Zusammenwirken von Gibberellin mit photoperiodisch bedingten blühfördernden und blühhemmenden Vorgängen bei *Kalanchoë blossfeldiana.* Naturwiss. **44**, 454 (1957).

*128.* — — Über die Wirkung von Gibberellin auf Entwicklung und Blütenbildung der Kurztagpflanze *Kalanchoë blossfeldiana.* Planta **51**, 201 (1958).

*129.* Harrington, J. F., L. Rappaport and K. J. Hood: Influence of Gibberellins on Stem Elongation and Flowering of Endive. Science (Washington) **125**, 601 (1957).

*130.* HAYASHI, T.: Action of Gibberellin on the Production of Amylase during Germination. J. Agric. Chem. Soc. Japan **16**, 531 (1940).

*131.* HAYASHI, T. and Y. MURAKAMI: Biochemistry of the "Bakanae" Fungus. XXIX. The Physiological Action of Gibberellin. (5) The Effect of Gibberellin on Straight Growth of Etiolated Pea Epicotyl Sections. J. Agric. Chem. Soc. Japan **27**, 675 (1953).

*132.* — — Biochemistry of the "Bakanae" Fungus. XXX. The Physiological Action of Gibberellin. (6) The Effect of Gibberellin on Straight Growth of Isolated Sections of Cereal Coleoptiles. J. Agric. Chem. Soc. Japan **27**, 797 (1953).

*133.* — — Biochemistry of the "Bakanae" Fungus. XXXII. The Physiological Action of Gibberellin. (7) The Response of Different Parts of Cereal Leaves to Gibberellin. J. Agric. Chem. Soc. Japan **28**, 543 (1954).

*134.* HAYASHI, T., Y. TAKIJIMA and Y. MURAKAMI: Biochemistry of the "Bakanae" Fungus. XXVIII. The Physiological Action of Gibberellin. (4). J. Agric. Chem. Soc. Japan **27**, 672 (1953).

*135.* HENDERSON, J. H. M.: Effect of Gibberellin on Sunflower Tissue Culture. Nature (London) **182**, 880 (1958).

*136.* HUDSON, J. P.: Effect of Weather on Plant Behaviour. Nature (London) **182**, 1337 (1958).

*137.* HUMPHRIES, E. C.: Effect of Gibberellic Acid and Kinetin on Growth of the Primary Leaf of Dwarf Bean *(Phaseolus vulgaris)*. Nature (London) **181**, 1081 (1958).

*138.* — The Effect of Gibberellic Acid and Kinetin on the Growth of Majestic Potato. Ann. appl. Biol. **46**, 346 (1958).

*139.* IMAMURA, S., Y. OGAWA and Y. HIRONO: Effects of Gibberellin and Gibberellin-like Substances on a Dwarf Strain of *Pharbitis nil* CHOIS. Proc. IXth Internat. Bot. Congr., Montreal, Abstr. **2**, 176 (1959).

*140.* Imperial Chemical Industries Ltd., Akers Research Laboratories: unpublished results.

*141.* JACKSON, G. A. D. and M. V. PROSSER: The Induction of Parthenocarpic Development in *Rosa* by Auxins and Gibberellic Acid. Naturwiss. **46**, 407 (1959).

*142.* JASKA, F. V.: The Effect of Gibberellic Acid on Kentucky Bluegrass Root Production. Agron. J. **51**, 184 (1959).

*143.* JOHNSON, A. W.: Gibberellic Acid. Science Progress **46**, 501 (1958).

*144.* KAHN, A., J. A. Goss and D. E. SMITH: Effect of Gibberellin on Germination of Lettuce Seed. Science (Washington) **125**, 645 (1957).

*145.* KALLIO, P. and P. PIIROINEN: Effect of Gibberellin on the Termination of Dormancy in some Seeds. Nature (London) **183**, 1830 (1959).

*146.* KATO, J.: Studies on the Physiological Effect of Gibberellin. I. On the Differential Activity between Gibberellin and Auxin. Mem. Coll. Sci. Kyoto B **20**, 189 (1953).

*147.* — Responses of Plant Cells to Gibberellin. Bot. Gaz. **117**, 16 (1955).

*148.* — Effect of Gibberellin on Elongation, Water Uptake and Respiration of Pea-stem Sections. Science (Washington) **123**, 1132 (1956).

*149.* — Nonpolar Transport of Gibberellin through Pea Stem and a Method for its Determination. Science (Washington) **128**, 1008 (1958).

*150.* — Studies on the Physiological Effect of Gibberellin. II. On the Interaction of Gibberellin with Auxins and Growth Inhibitors. Physiol. Plantarum **11**, 10 (1958).

*151.* KATO, J. and M. KATSUMI: Effect of Gibberellins on IAA-oxidase. Naturwiss. **45**, 344 (1958).

*152.* KAVANAGH, F. and N. R. KUZEL: Fluorometric Determination of Gibberellic and Gibberellenic Acids in Fermentation Products, Commercial Formulations and Purified Materials. J. Agric. Food Chem. **6**, 459 (1958).

*153.* KAWARADA, A., H. KITAMURA, Y. SETA, N. TAKAHASHI, M. TAKAI, S. TAMURA and Y. SUMIKI: Biochemical Studies on "Bakanae" Fungus. Part XXXV. Relation between Gibberellins $A_1$, $A_2$ and Gibberellic Acid. Bull. Agric. Chem. Soc. Japan **19**, 278 (1955).

*154.* KAWARADA, A. and Y. SUMIKI: Occurrence of Gibberellin $A_1$ in Water Sprouts of Citrus. Bull. Agric. Chem. Soc. Japan **23**, 343 (1959).

*155.* KITAMURA, H., Y. SETA, N. TAKAHASHI, A. KAWARADA and Y. SUMIKI: Biochemistry of "Bakanae" Fungus. Part 37. Chemical Structure of Gibberellins. Part VII. Bull. Agric. Chem. Soc. Japan **21**, 71 (1957).

*156.* KITAMURA, H. and Y. SUMIKI: The Biochemistry of "Bakanae" Fungus. Part 31. The Physico-chemical Determination of Gibberellin. J. Agric. Chem. Soc. Japan **28**, 449 (1954).

*157.* KITAMURA, H., N. TAKAHASHI, Y. SETA and Y. SUMIKI: Biochemical Studies on "Bakanae" Fungus. Part 47. Chemical Structure of Gibberellins. Part XIV. Bull. Agric. Chem. Soc. Japan **22**, 434 (1958).

*158.* KNAPP, R.: Über die Wirkung von Gibberellin auf Wachstum und Blüten-bildung bei verschiedenen Temperatur- und Licht-Verhältnissen. Z. Naturforsch. **11** b, 698 (1956).

*159.* — Die Gibberelline und ihre Bedeutung für die Pflanzenphysiologie. Naturwiss. **45**, 408 (1958).

*160.* KNOBLOCH, I. W.: Gibberellic Acid and Ferns. Amer. Fern J. **47**, 134 (1957).

*161.* KOVACOVA-FERJANCIKOVA, V.: The Effect of Gibberellic Acid on the Stem Development in a Culture of Peach Seedlings. Naturwiss. **46**, 454 (1959).

*162.* KRASILNIKOV, N. A., M. K. CHAYLAKHIAN, I. V. ASEEVA and L. P. KHLO-PENKSVA: On a Gibberellin-like Substance Produced by Soil Yeasts. Doklady Akad. Nauk SSSR **123**, 1124 (1959).

*163.* KREKULE, J. and A. MARTINOVSKA: The Effect of Gibberellic Acid on the Development of *Triticum* and *Panicum*. Bot. Zhurn. **43**, 953 (1958).

*164.* KREKULE, J. and J. ULLMAN: Nové poznatky z chemie a biologie gibberellinu a jejich vyuzutí v zemedelitví. Prehled 8, 1297 (1958).

*165.* KRIBBEN, F. J.: Gibberellinsäure und Blattwachstum. Naturwiss. **44**, 429 (1957).

*166.* — Die Abkürzung der Samenruhe bei *Arabidopsis* durch Gibberellinsäure. Naturwiss. **44**, 313 (1957).

*167.* KRIBBEN, F. J. und H. J. REISENER: Über das Wachstum des Gurkenkotyledos unter dem Einfluß von Gibberellinsäure. Beitr. Biol. Pflanzen **34**, 379 (1958).

*168.* KRUBER, O.: Über neue Kohlenwasserstoffe aus dem Anthracenöl des Stein-kohlenteers. Ber. dtsch chem. Ges. **67**, 1000 (1934).

*169.* KURAISHI, S. and T. HASHIMOTO: Promotion of Leaf Growth and Acceleration of Stem Elongation by Gibberellin. Bot. Mag. (Tokyo) **70**, 86 (1957).

*170.* KUROSAWA, E.: Experimental Studies on the Nature of the Substance Excreted by the "Bakanae" Fungus. Trans. Nat. Hist. Soc. Formosa **16**, 213 (1926).

*171.* KUSE, G.: Necessity of Auxin for the Growth Effect of Gibberellin. Bot. Mag. (Tokyo) **71**, 151 (1958).

*172.* LAIBACH, F.: Gibberellinsäurewirkungen bei Platyopuntien. Ber. dtsch. bot. Ges. **70**, 199 (1957).

*173.* LANG, A.: Stem Elongation in a Rosette Plant, Induced by Gibberellic Acid. Naturwiss. **43**, 257 (1956).

*174.* LANG, A: Induction of Flower Formation in Biennial *Hyoscyamus* by Treatment with Gibberellin. Naturwiss. **43**, 284 (1956).

*175.* — Gibberellin and Flower Formation. Naturwiss. **43**, 544 (1956).

*176.* — The Effect of Gibberellin upon Flower Formation. Proc. Nat. Acad. Sci. (USA) **43**, 709 (1957).

*177.* — The Gibberellins and their Role in Plant Growth and Development. Internat. Sympos. Photo-thermoperiodism, Parma, 1957. Internat. Union Biol. Sci., Ser. B (Colloques) **34**, 55 (1959).

*178.* — Induction of Reproductive Growth in Plants. Proc. Internat. Congr. Biochem., 1958: Symposium VI, Biochemistry of Morphogenesis. London: Pergamon Press. 1959, p. 126.

*179.* LANG, A., J. A. SANDOVAL and A. BEDRI: Induction of Bolting and Flowering in *Hyoscyamus* and *Samolus* by a Gibberellin-like Material from a Seed Plant. Proc. Nat. Acad. Sci. (USA) **43**, 960 (1957).

*180.* LANGRIDGE, J.: Effect of Day-length and Gibberellic Acid on the Flowering of *Arabidopsis*. Nature (London) **180**, 36 (1957).

*181.* LAZER, L. and R. DAHLSTROM: personal communication.

*182.* LEE, A. E.: The Effects of Various Substances on the Comparative Growth of Excised Tomato Roots of Clones Carrying Dwarf and Normal Alleles. Amer. J. Bot. **46**, 16 (1959).

*183.* LEH, H.-O.: Versuche über die Wirkung von Gibberellinsäure auf einige Kulturpflanzen. Z. Pflanzenernährung, Düngung, Bodenkunde **83**, 234 (1958).

*184.* LINCOLN, R. G. and K. C. HAMNER: An Effect of Gibberellic Acid on the Flowering of *Xanthium*, a Short Day Plant. Plant Physiol. **33**, 101 (1958).

*185.* LINDSTROM, R. S., S. H. WITTWER and M. J. BUKOVAC: Gibberellin and Higher Plants: IV. Flowering Responses of some Flower Crops. Quart. Bull. Michigan agric. Exp. Station **39**, 673 (1957).

*186.* LIVERMAN, J. L. and J. BONNER: The Interaction of Auxin and Light in the Growth Responses of Plants. Proc. Nat. Acad. Sci. (USA) **39**, 905 (1953).

*187.* LIVERMAN, J. L. and S. P. JOHNSON: Control of Arrested Fruit Growth in Tomato by Gibberellins. Science (Washington) **125**, 1086 (1957).

*188.* LOCKHART, J. A.: Reversal of the Light Inhibition of Pea Stem Growth by the Gibberellins. Proc. Nat. Acad. Sci. (USA) **42**, 841 (1956).

*189.* — Studies on the Organ of Production of the Natural Gibberellin Factor in Higher Plants. Plant Physiol. **32**, 204 (1957).

*190.* — The Response of Various Species of Higher Plants to Light and Gibberellic Acid. Physiol. Plantarum **11**, 478 (1958).

*191.* — The Influence of Red and Far-red Radiation on the Response of *Phaseolus vulgaris* to Gibberellic Acid. Physiol. Plantarum **11**, 487 (1958).

*192.* — Studies on the Mechanism of Stem Growth Inhibition by Visible Radiation. Plant Physiol. **34**, 457 (1959).

*193.* LOCKHART, J. A. and J. BONNER: Effects of Gibberellic Acid on the Photoperiod-controlled Growth of Woody Plants. Plant. Physiol. **32**, 492 (1957).

*194.* LOCKHART, J. A. and V. GOTTSCHALL: Growth Responses of Alaska Pea Seedlings to Visible Radiation and Gibberellic Acid. Plant Physiol. **34**, 460 (1959).

*195.* LONA, F.: L'acido gibberellico determina la germinazione dei semi di *Lactuca Scariola* in fase di scoto-inhibizione. Ateneo Parmense **27**, 641 (1956).

*196.* — L'azione dell'acido gibberellico sull'accrescimento caulinare di talune piante erbacea in condizioni esterne controllate. Nuovo giorn. bot. ital. **63**, 61 (1956).

*197.* Lona, F.: Azione gibberellicosimile di estrati ottenuti da giovani strutture fiorali di *Brassica napus* L. var. *oleifera*. Ateneo Parmense **28**, 111 (1957).

*198.* — Results of Twelve Years of Work on the Photoperiodic Responses of *Perilla ocymoides*. Proc. Koninkl. Akad. Wetensch. Amsterdam, Ser. C **62**, 1 (1959).

*199.* — Brief Accounts on the Physiological Activities of Gibberellic Acid and other Substances in Relation to Photothermal Conditions. Internat. Sympos. Photo-thermoperiodism, Parma, 1957. Internat. Union Biol. Sci., Ser. B (Colloques) **34**, 143 (1959).

*200.* Lona, F. ed A. Bocchi: Interferenza dell'acido gibberellico nell'effetto della luce rossa e rosso-estrema sull'allungamento del fusto di *Perilla ocymoides* L. Ateneo Parmense **27**, 645 (1956).

*201.* — — Sviluppo vegetativo e riproduttivo di alcune longidiurne in rapporto all'azione dell'acido gibberellico. Nuovo giorn. bot. ital. **68**, 469 (1956).

*202.* Lona, F. e R. Borghi: Germogliazione di gemme di *Fagus sylvatica* L. in periodo di quiescenza invernale, a fotoperiodo breve, per azione dell'acido gibberellico. Ateneo Parmense **28**, 116 (1957).

*203.* Lothrop, W. C. and P. A. Goodwin: A New Modification of the Ullmann Synthesis of Fluorene Derivatives. J. Amer. Chem. Soc. **65**, 363 (1943).

*204.* Luckwill, L. C.: Fruit Growth in Relation to Internal and External Chemical Stimuli. Sympos. Soc. Study Develop. Growth **17**, 223 (1958).

*205.* McComb, A. J. and D. J. Carr: Evidence from a Dwarf Pea Bioassay for Naturally Occurring Gibberellins in the Growing Plant. Nature (London) **181**, 1548 (1958).

*206.* McCune, D. C. and A. W. Galston: Inverse Effects of Gibberellin on Peroxidase Activity and Growth in Dwarf Strains of Peas and Corn. Plant Physiol. **34**, 416 (1959).

*207.* MacMillan, J., J. C. Seaton and P. J. Suter: A new Plant Growth Promoting Acid—Gibberellin $A_5$ from the seed of *Phaseolus multiflorus*. Proc. Chem. Soc. (London) **1959**, 325.

*208.* MacMillan, J. and P. J. Suter: The Occurrence of Gibberellin $A_1$ in Higher Plants: Isolation from the Seed of Runner Bean *(Phaseolus multiflorus)*. Naturwiss. **45**, 46 (1958).

*209.* Maltzahn, K. E. v. and I. G. MacQuarrie: Effect of Gibberellic Acid on the Growth of Protonemata in *Splachnum ampullaceum* (L.) Hedw. Nature (London) **181**, 1139 (1958).

*210.* Manzelli, M. A. and N. Adler: personal communication.

*211.* Marré, E. ed O. Arrigoni: Aumento del glutatione ridotto in segmenti di internodio di pisello trattati con acido gibberellico. Atti accad. naz. Lincei, Rendic. **25**, 309 (1958).

*212.* Marth, P. C., W. V. Audia and J. W. Mitchell: Effects of Gibberellic Acid on Growth and Development of Plants of Various Genera and Species. Bot. Gaz. **118**, 106 (1956).

*213.* Mastovsky, J.: Application of Gibberellins in Breweries and Malt Plants. Kvasny Prumysl **5**, 81 (1959).

*214.* Mayer, A. M.: Joint Action of Gibberellic Acid and Coumarin in Germination. Nature (London) **184**, 826 (1959).

*215.* Mitchell, J. W., D. P. Skaggs and W. P. Anderson: Plant Growth Stimulating Hormones in Immature Bean Seeds. Science (Washington) **114**, 159 (1951).

*216.* Mitra, G. C. and A. Allsopp: Effects of Kinetin, Gibberellic Acid and Certain Auxins on the Development of Shoot Buds on the Protonema of *Pohlia nutans*. Nature (London) **183**, 974 (1959).

*217.* MOFFATT, J. S.: Gibberellic Acid. Part XVI. The Chromophore of Gibberellenic Acid. J. Chem. Soc. (London) (in press).

*218.* MOFFATT, J. S. and M. RADLEY: Gibberellic Acid. Part XI. The Growth-promoting Activities of some Functional Derivatives of Gibberellic Acid. J. Sci. Food Agric. (in press).

*219.* MONACO, L. C. y A. CARVALHO: Efeito da giberelina em mutantes de cafe. Biol. Suptda. Serr. Cafe, S. Paulo 33, 17 (1958).

*220.* MOORE, T. C. and E. K. BONDE: Interaction of Gibberellic Acid and Vernalisation in the Dwarf Telephone Pea. Physiol. Plantarum 11, 752 (1958).

*221.* MORGAN, D. G. and G. C. MEES: Gibberellic Acid and the Growth of Crop Plants. J. Agric. Sci. 50, 49 (1959).

*222.* MORRIS, E. O.: Effect of Gibberellic Acid upon the Germination of Barley. Chem. and Ind. 1958, 97.

*223.* MORRISON, A. and T. P. C. MULHOLLAND: Gibberellic Acid. Part VIII. Synthesis of Methyl $(\pm)$-$\alpha$- and $\beta$-3:5-Dimethoxycarbonyl-6-(2-methoxycarbonyl-6-methylphenyl)-3-methylhexanoate. J. Chem. Soc. (London) 1958, 2536.

*224.* — — Gibberellic Acid. Part X. 7-Hydroxy-1-methylfluorene. J. Chem. Soc. (London) 1958, 2702.

*225.* MULHOLLAND, T. P. C.: Gibberellic Acid. Part IX. The Structure of allo-Gibberic Acid. J. Chem. Soc. (London) 1958, 2693.

*226.* MULHOLLAND, T. P. C. and G. WARD: Gibberellic Acid. Part II. The Structure and Synthesis of Gibberene. J. Chem. Soc. (London) 1954, 4676.

*227.* — — Gibberellic Acid. Part III. Synthesis of Fluorenone-4:5-dicarboxylic Acid. J. Chem. Soc. (London) 1956, 2425.

*228.* MUNEKATA, H. and S. KATO: Biochemical Studies of the "Bakanae" Fungus. XL. Application of Gibberellin to the Malting Industry. Bull. Brewing Sci. (Tokyo) 3, 1 (1957).

*229.* MURAKAMI, Y.: The Effect of the Extract of Immature Bean Seeds on the Growth of Coleoptile and Leaf of Rice Plant. Bot. Mag. (Tokyo) 70, 376 (1957).

*230.* — A Paper Chromatographic Survey of Gibberellins and Auxins in Immature Seeds of Leguminous Plants. Bot. Mag. (Tokyo) 72, 36 (1959).

*231.* MURNEEK, A. E.: Length of Day and Temperature Effects in *Rudbeckia*. Bot. Gaz. 102, 269 (1940).

*232.* NELSON, T. C.: Early Responses of some Southern Tree Species to Gibberellic Acid. J. Forestry 55, 518 (1957).

*233.* NÉTIEN, G.: Action des gibberellines sur la culture des tissus végétaux cultivés *in vitro*. C. R. hebd. Séances Acad. Sci. 244, 2732 (1957).

*234.* — Action de la gibberelline sur différents types de tissus végétaux cultivés *in vitro*. C. R. hebd. Séances Acad. Sci. 247, 1645 (1958).

*235.* NICKELL, L. G.: Gibberellin and the Growth of Plant Tissue Cultures. Nature (London) 181, 499 (1958).

*236.* — Production of Gibberellin-like Substances by Plant Tissue Cultures. Science (Washington) 128, 88 (1958).

*237.* NICKELL, L. G. and W. TULECKE: Responses of Plant Tissue Cultures to Gibberellin. Bot. Gaz. 120, 245 (1959).

*238.* NICKERSON, N. H.: Sustained Treatment with Gibberellic Acid of five Different Kinds of Maize. Ann. Missouri Bot. Gard. 46, 19 (1959).

*239.* NIELSEN, N. and G. BERGQUIST: The Stimulation of the Respiration of Seeds with Gibberellic Acid and its Analytical Application. Physiol. Plantarum 11, 329 (1958).

*240.* Nitsch, J. P.: Photoperiodism in Woody Plants. Proc. Amer. Soc. hort. Sci. **70**, 526 (1957).

*241.* — Présence de gibberellines dans l'albumen immature du Pommier. Bull. soc. bot. France **105**, 479 (1958).

*242.* Nitsch, J. P. and C. Nitsch: Studies on the Growth of Coleoptile and First Internode Sections. A New, Sensitive, Straight-growth Test for Auxins. Plant Physiol. **31**, 94 (1956).

*243.* Njoku, E.: Effect of Gibberellic Acid on Leaf Form. Nature (London) **182**, 1097 (1958).

*244.* Nutman, P. S.: Sources of Incompatibility Affecting Nitrogen Fixation in Legume Symbiosis. Sympos. Soc. Exp. Biol. **13**, 42 (1959).

*244a.* Ogawa, Y. und S. Imamura: Über die fördernde Wirkung von Gibberellin auf die Blütenbildung einer Kurztagpflanze, *Pharbitis Nil* Chois. Proc. Japan Acad. **34**, 629 (1958).

*245.* Parker, M. W., S. B. Hendricks and H. A. Borthwick: Action Spectrum for the Photoperiodic Control of Floral Initiation of the Long-day Plant *Hyoscyamus niger*. Bot. Gaz. **111**, 242 (1950).

*246.* Persson, A. and L. Rappaport: Gibberellin-induced Systemic Fruit-set in a Male Sterile Tomato. Science (Washington) **127**, 816 (1958).

*247.* Phillips, I. D. J., A. J. Vlitos and H. Cutler: The Influence of Gibberellic Acid upon the Endogenous Growth Substances of the Alaska Pea. Contr. Boyce Thompson Inst. **20**, 111 (1959).

*248.* Phinney, B. O.: Growth Response of Single Gene Dwarf Mutants in Maize to Gibberellic Acid. Proc. Nat. Acad. Sci. (USA) **42**, 185 (1956).

*249.* Phinney, B. O. and C. A. West: The Growth Response of Single Gene Dwarf Mutants of *Zea mays* to Gibberellins and to Gibberellin-like Substances. Cytologia (Suppl. Vol., Proc. Internat. Genetics Symp., 1956) **1957**, 384.

*250.* Phinney, B. O., C. A. West, M. Ritzel and P. M. Neely: Evidence for "Gibberellin-like" Substances from Flowering Plants. Proc. Nat. Acad. Sci. (USA) **43**, 398 (1957).

*251.* Picard, C.: Remarques sur l'action de l'acide gibbérellique sur *Oenothera biennis*. C. R. hebd. Séances Acad. Sci. **247**, 2184 (1958).

*252.* Pilet, P.-E.: Action des gibberellines sur l'activité auxines-oxydasique de tissus cultivés *in vitro*. C. R. hebd. Séances Acad. Sci. **245**, 1327 (1957).

*253.* — Acide gibbérellique et destruction auxinique *in vitro*. C. R. hebd. Séances Acad. Sci. **249**, 298 (1959).

*254.* Pilet, P.-E. et W. Wurgler: Action des gibbérellines sur la croissance et l'activité auxines-oxydasique du *Trifolium ochroleucum* Hudson. Ber. schweiz. bot. Ges. **68**, 54 (1958).

*255.* Poljakoff-Mayber, A., M. Evenari and G. Neumann: Effect of Red Light and Gibberellic Acid on the Temperature-inhibited Germination of Lettuce Seeds. Bull. Res. Council Israel **6** D, 99 (1958).

*256.* Pollock, J. R. A.: Growth Substances in Relation to Dormancy in Barley. Chem. and Ind. **1958**, 387.

*257.* Porlingis, I. C. and D. Boynton: Growth Responses of the Strawberry Plant to Gibberellic Acid and to Environmental Conditions. Plant Physiol. (Suppl.) **34**, xvi (1959).

*258.* Prosser, M. V. and G. A. D. Jackson: Induction of Parthenocarpy in *Rosa arvensis* Huds. with Gibberellic Acid. Nature (London) **184**, 108 (1959).

*259.* Provasoli, L.: Effect of Plant Hormones on Seaweeds. Biol. Bull. **113**, 321 (1957).

260. PUDLES, J., A. DIARA et E. LEDERER: Sur la constitution chimique du darutigénol, diterpène triol tricyclique. Bull. soc. chim. France **1959**, 693.

261. PURVES, W. K. and W. S. HILLMAN: Response of Pea Stem Sections to Indoleacetic Acid, Gibberellic Acid, and Sucrose as Affected by Length and Distance from Apex. Physiol. Plantarum **11**, 29 (1958).

262. — — Experimental Separation of Gibberellin and Auxin Actions in Etiolated Pea Epicotyl Sections. Physiol. Plantarum **12**, 786 (1959).

263. RADLEY, M.: Occurrence of Substances Similar to Gibberellic Acid in Higher Plants. Nature (London) **178**, 1070 (1956).

264. — The Distribution of Substances Similar to Gibberellic Acid in Higher Plants. Ann. Botany **22**, 297 (1958).

265. RADLEY, M. and E. DEAR: Occurrence of Gibberellin-like Substances in the Coconut. Nature (London) **182**, 1098 (1958).

266. RANGASWAMY, N. S.: *In vitro* Culture of Nucellus and Embryos of *Citrus*. Proc. Delhi Univ. Seminar "Modern Developments in Plant Physiology", **1957**, 104 (1958).

267. RICHARDSON, S. D.: Radicle Elongation of *Pseudotsuga menziesii* in Relation to Light and Gibberellic Acid. Nature (London) **181**, 429 (1958).

268. — Germination of Douglas-fir Seed as Affected by Light, Temperature and Gibberellic Acid. Forest Sci. **5**, 174 (1959).

269. RITZEL, M.: The Distribution and Time of Occurrence of Gibberellin-like Substances from Plants. Plant Physiol. (Suppl.) **32**, xxxi (1957).

270. RIVES, M. et R. POUGET: Action de la gibberelline sur la dormance de la vigne (*Vitis vinifera* L.). C. R. hebd. Séances Acad. Sci. **248**, 3600 (1959).

271. — — Action de la gibberelline sur la compacité des grappes de deux variétés de vigne. C. R. hebd. Séances Acad. agric. France 15 Avril 1959.

272. ROBBINS, W. J.: Gibberellic Acid and the Reversal of Adult *Hedera* to a Juvenile State. Amer. J. Bot. **44**, 743 (1957).

273. SACHAR, R. C. and M. KAPOOR: Influence of Kinetin and Gibberellic Acid on the Test Tube Seeds of *Coopersia pedunculata* HERB. Naturwiss. **45**, 552 (1958).

274. — — Gibberellin in the Induction of Parthenocarpy in *Zephyranthes*. Plant Physiol. **34**, 168 (1959).

275. SACHS, R. M., C. F. BRETZ and A. LANG: Shoot Histogenesis: the Early Effects of Gibberellin upon Stem Elongation in two Rosette Plants. Amer. J. Bot. **46**, 376 (1959).

276. SANDEGREN, E. und H. BELING: Versuche mit Gibberellinsäure bei der Malzherstellung. Brauerei, wiss. Beilage **11**, 231 (1958).

277. SARKAR, S.: Versuche zur Physiologie der Vernalisation. Biol. Zentralbl. **77**, 1 (1958).

278. SCHRAUDOLF, H. and J. REINERT: Interaction of Plant Growth Regulators in Regeneration Processes. Nature (London) **184**, 465 (1959).

279. SCOTT, R. A. and J. L. LIVERMAN: Control of Etiolated Bean Leaf-disk Expansion by Gibberellins and Adenine. Science (Washington) **126**, 122 (1957).

280. SCURFIELD, G. and E. F. BIDDISCOMBE: Effects of Gibberellic Acid on Winter Pasture Production. Nature (London) **183**, 1196 (1959).

281. SETA, Y., H. KITAMURA, N. TAKAHASHI and Y. SUMIKI: Biochemistry of "Bakanae" Fungus. Part 38. Chemical Structure of Gibberellins. VIII. Bull. Agric. Chem. Soc. Japan **21**, 73 (1957).

282. SETA, Y. and Y. SUMIKI: The Biochemistry of "Bakanae" Fungus. Part 26. The Chemical Constitution of Gibberellin. Part 6. J. Agric. Chem. Soc. Japan **26**, 508 (1953).

283. Seta, Y., N. Takahashi, H. Kitamura and Y. Sumiki: Biochemical Studies on "Bakanae" Fungus. Part 45. Chemical Structure of Gibberellins. Part XI. Bull. Agric. Chem. Soc. Japan 22, 429 (1958).

284. Seta, Y., N. Takahashi, H. Kitamura, M. Takai, S. Tamura and Y. Sumiki: Biochemistry of "Bakanae" Fungus. Part 44. Chemical Structure of Gibberellins. Part XI. Bull. Agric. Chem. Soc. Japan 22, 61 (1958).

285. Sheppard, N.: Gibberellic Acid. Part XV. The Nuclear Magnetic Resonance Spectra and Structures of Gibberellic Acid Derivatives. J. Chem. Soc. (London) (in press).

286. Shimada, S.: Internal Structure of Rice Plants Affected by Bakanae Fungus. Ann. Phytopath. Soc. Japan 2, 471 (1932).

287. Simpson, G. M.: A Colorimetric Test for Gibberellic Acid and Evidence from a Dwarf Pea Assay for the Occurrence of a Gibberellin-like Substance in Wheat Leaves. Nature (London) 182, 528 (1958).

288. Sircar, S. M. and M. Kundu: Effect of Root Extract of Water Hyacinth (Eichornia speciosa Kunth.) on the Growth and Flowering of Rice. Sci. and Culture (India) 24, 332 (1959).

289. Skinner, C. G. and W. Shive: Synergistic Effect of Gibberellin and 6-(Substituted)-purines on Germination of Lettuce Seed. Arch. Biochem. Biophys. 74, 283 (1958).

290. Skinner, C. G., F. D. Talbert and W. Shive: Effect of 6-(Substituted)-purines and Gibberellin on the Rate of Seed Germination. Plant Physiol. 33, 190 (1958).

291. Stoddart, J. L.: The Effects of Gibberellic Acid upon Growth Habit and Heading in Late-flowering Red Clover (Trifolium pratense L.). J. agric. Sci. 52, 161 (1959).

292. Stodola, F. H.: Source Book on Gibberellin, 1828—1957. Agricult. Res. Serv. U. S. Depart. Agricult., 1958.

293. Stodola, F. H., G. E. Nelson and D. J. Spence: The Separation of Gibberellin A and Gibberellic Acid on Buffered Partition Columns. Arch. Biochem. Biophys. 66, 438 (1957).

294. Stodola, F. H., K. B. Raper, D. I. Fennell, H. F. Conway, V. E. Sohns, C. T. Langford and R. W. Jackson: The Microbiological Production of Gibberellins A and X. Arch. Biochem. Biophys. 54, 240 (1955).

295. Stork, G. and H. Newman: The Stereochemistry of Allogibberic Acid and of Gibberic Acid. J. Amer. Chem. Soc. 81, 3168 (1959).

296. — — The Stereochemistry of Gibberellic Acid. J. Amer. Chem. Soc. 81, 5518 (1959).

297. Stowe, B. B.: Growth Promotion in Pea Epicotyl Sections by Fatty Acid Esters. Science (Washington) 128, 421 (1958).

298. Stowe, B. B. and T. Yamaki: The History and Physiological Action of the Gibberellins. Annu. Rev. Plant Physiol. 8, 181 (1957).

299. — — Gibberellins: Stimulants of Plant Growth. Science (Washington) 129, 807 (1959).

300. Stoy, V. and A. Hagberg: Effects of Gibberellic Acid on Erectoides Mutations in Barley. Hereditas 44, 516 (1958).

301. Street, H. E.: Regulation of Growth and Differentiation in Roots. Proc. Internat. Congr. Biochem., 1958: Symposium VI, Biochemistry of Morphogenesis. London: Pergamon Press. 1959, p. 98.

302. Sudia, T. W.: Influence of Temperature on the Response of Germinating Barley Grains to Potassium Gibberellate. Plant Physiol. 34, 473 (1959).

*303.* SUMIKI, Y.: Biochemistry of the "Bakanae" Fungus. XXV. The Physiological Action of Gibberellin (3). J. Agric. Chem. Soc. Japan **26**, 393 (1952).

*304.* — Gibberellin and the Abnormal Growth of Plants. Studies on the Hormone Produced by *Gibberella fujikuroi* WR. Kagaku (Science) **25**, 563 (1955).

*305.* TAKAHASHI, N., H. KITAMURA, A. KAWARADA, Y. SETA, M. TAKAI, S. TAMURA and Y. SUMIKI: Biochemical Studies on "Bakanae" Fungus. Part XXXIV. Isolation of Gibberellins and their Properties. Bull. Agric. Chem. Soc. Japan **19**, 267 (1955).

*306.* TAKAHASHI, N., Y. SETA, H. KITAMURA, A. KAWARADA and Y. SUMIKI: Biochemistry of "Bakanae" Fungus. Part XXXIX. Chemical Structure of Gibberellins. IX. Bull. Agric. Chem. Soc. Japan **21**, 75 (1957).

*307.* TAKAHASHI, N., Y. SETA, H. KITAMURA and Y. SUMIKI: Biochemical Studies of "Bakanae" Fungus. Part XLI. Chemical Structure of Gibberellins. X. Bull. Agric. Chem. Soc. Japan **21**, 327 (1957).

*308.* — — — — Biochemical Studies on "Bakanae" Fungus. Part XLII. A New Gibberellin, Gibberellin $A_4$. Bull. Agric. Chem. Soc. Japan **21**, 396 (1957).

*309.* — — — — Biochemical Studies on "Bakanae" Fungus. Part XLVI. Chemical Structure of Gibberellins. Part XIII. Bull. Agric. Chem. Soc. Japan **22**, 432 (1958).

*310.* TERUI, M. and H. KAGAWA: The Influence of Gibberellin on the Growth of some Plant-parasitic Fungi. Bull. Fac. Agric. Hirosaki Univ. **4**, 88 (1958).

*311.* THOMPSON, P. A. and C. G. GUTTRIDGE: Effect of Gibberellic Acid on the Initiation of Flowers and Runners in the Strawberry. Nature (London) **184**, 72 (1959).

*312.* THURBER, G. A., J. R. DOUGLAS and A. W. GALSTON: Inhibitory Effect of Gibberellins on Nodulization in Dwarf Beans, *Phaseolus vulgaris*. Nature (London) **181**, 1082 (1958).

*313.* TSUI, C.: The Role of Zinc in Auxin Synthesis in the Tomato. Amer. J. Bot. **37**, 257 (1948).

*314.* VIANA, M. J.: Differential Action of Indoleacetic Acid and Gibberellin on Decapitated Plants of *Bryophyllum daigremontianum*. Substitution of Apical Dominance for a Basal Dominance. Portugaliae Acta Biol., Ser. A **5**, 282 (1958).

*315.* VLITOS, A. J. and W. MEUDT: Relationship between Shoot Apex and Effect of Gibberellic Acid on Elongation of Pea Stems. Nature (London) **180**, 284 (1957).

*316.* — — The Effect of Light and of the Shoot Apex on the Action of Gibberellic Acid. Contrib. Boyce Thompson Inst. **19**, 55 (1957).

*317.* WADA, B.: Cytological Studies on the Effect of Gibberellin upon Mitotic Cells. Jap. J. Genet. (Suppl.) **2**, 24 (1949).

*318.* WARDLAW, C. W. and G. C. MITRA: Responses of a Fern Apex to Gibberellic Acid, Kinetin and α-Naphthaleneacetic Acid. Nature (London) **181**, 400 (1958).

*319.* WAREING, P. F.: Interaction between Indoleacetic Acid and Gibberellic Acid in Cambial Activity. Nature (London) **181**, 1744 (1958).

*320.* WASHBURN, W. H., F. A. SCHESKE and J. R. SCHENCK: Infrared Determination of Gibberellins. J. Agric. Food Chem. **7**, 420 (1959).

*321.* WATSON, D. J.: Botany Department. Rothamsted Exp. Station, Annu. Rept. **1958**, 75 (1959).

*322.* WEAVER, R. J.: Prolonging Dormancy in *Vitis vinifera* with Gibberellin. Nature (London) **183**, 1198 (1959).

*323.* Weaver, R. J. and S. B. McCune: Response of Certain Varieties of *Vitis vinifera* to Gibberellin. Hilgardia **28**, 297 (1959).

*324.* Weijer, J.: Interaction of Gibberellic Acid and Indoleacetic Acid in Impatiens. Science (Washington) **129**, 896 (1959).

*325.* Wellensiek, S. J.: Gibberellazuur, stengelstrekking en bloei. Verslag Gew. Vergad. Afdel. Natuurkunde, Koninkl. Nederl. Akad. Wetenschap. **67**, 44 (1958).

*326.* Wenkert, E.: Structural and Biogenetic Relationships in the Diterpene Series. Chem. and Ind. **1955**, 282.

*327.* West, C. A.: personal communication.

*328.* West, C. A. and K. H. Murashige: The Isolation of Gibberellin A$_1$ and Bean Factor II and the Chemical Properties of other Gibberellin-like Factors in Beans and Peas. Plant Physiol. (Suppl.) **33**, xxxviii (1958).

*329.* West, C. A. and B. O. Phinney: Properties of Gibberellin-like Factors from Extracts of Higher Plants. Plant Physiol. (Suppl.) **31**, xx (1956).

*330.* — — Purification and Properties of Gibberellin-like Substances from Flowering Plants. Plant Physiol. (Suppl.) **32**, xxxii (1957).

*331.* — — Gibberellins from Flowering Plants. I. Isolation and Properties of a Gibberellin from *Phaseolus vulgaris* L. J. Amer. Chem. Soc. **81**, 2424 (1959).

*332.* Westing, A. H.: Effect of Gibberellin on Conifers: Generally Negative. J. Forestry **57**, 120 (1959).

*333.* Whaley, W. G. and J. Kephart: Effect of Gibberellic Acid on Growth of Maize Roots. Science (Washington) **125**, 234 (1957).

*334.* Wickson, M. and K. V. Thimann: The Antagonism of Auxin and Kinetin in Apical Dominance. Physiol. Plantarum **11**, 62 (1958).

*335.* Wittwer, S. H. and M. J. Bukovac: Gibberellin and Higher Plants: III. Induction of Flowering in Long-day Annuals Grown under Short Days. Quart. Bull. Michigan Agric. Exp. Station **39**, 661 (1957).

*336.* — — Gibberellin Effects on Temperature and Photoperiodic Requirements for Flowering of some Plants. Science (Washington) **126**, 30 (1957).

*337.* — — The Effects of Gibberellin on Economic Crops. Econ. Bot. **12**, 213 (1958).

*338.* Wittwer, S. H., M. J. Bukovac, G. R. McVey and J. C. Ballard: Gibberellin Modifications of Photoperiod Controlled Growth in Herbaceous Plants. Naturwiss. **46**, 117 (1959).

*339.* Wittwer, S. H., M. J. Bukovac, H. M. Sell and L. E. Weller: Some Effects of Gibberellin on Flowering and Fruit Setting. Plant Physiol. **32**, 39 (1957).

*340.* Yabuta, T. and T. Hayashi: Biochemical Studies on "Bakanae" Fungus. Part II. Isolation of Gibberellin, the Active Principle which Makes the Rice Seedlings Grow Slenderly. J. Agric. Chem. Soc. Japan **15**, 257 (1939).

*341.* — — Biochemical Studies on "Bakanae" Fungus. Part III. On the Action of Gibberellin, a Growth-promoting Substance, on the Physiology of Plants. J. Agric. Chem. Soc. Japan **15**, 403 (1939).

*342.* Yabuta, T. and Y. Sumiki: (Communication to the Editor.) J. Agric. Chem. Soc. Japan **14**, 1526 (1938).

*343.* Yabuta, T., Y. Sumiki and K. Aso: Biochemical Studies of "Bakanae" Fungus. Part XXIII. The Chemical Constitution of Gibberellin. (4). J. Agric. Chem. Soc. Japan **25**, 159 (1951).

*344.* Yabuta, T., Y. Sumiki, K. Aso, T. Tamura, H. Igarashi and K. Tamari: Biochemistry of the "Bakanae" Fungus of Rice. Part X. The Chemical Constitution of Gibberellin (1). J. Agric. Chem. Soc. Japan **17**, 721 (1941).

*345.* YABUTA, T., Y. SUMIKI, K. ASO, T. TAMURA, H. IGARASHI and K. TAMARI: Biochemistry of the "Bakanae" Fungus of Rice. Part XI. The Chemical Constitution of Gibberellin (2). J. Agric. Chem. Soc. Japan **17**, 894 (1941).

*346.* — — — — — — Biochemistry of the "Bakanae" Fungus. Part XII. The Chemical Constitution of Gibberellin (3). J. Agric. Chem. Soc. Japan **17**, 975 (1941).

*347.* YABUTA, T., Y. SUMIKI, K. FUKUNAGA and M. HORIUCHI: Biochemistry of the "Bakanae" Fungus. XXII. The Effect of Gibberellin on the Chemical Composition of Rice Seedlings. J. Agric. Chem. Soc. Japan **24**, 396 (1951).

*348.* YATAZAWA, M. and Y. SUMIKI: Biochemistry of the "Bakanae" Fungus. Part XXIV. The Chemical Constitution of Gibberellin (5). J. Agric. Chem. Soc. Japan **25**, 503 (1952).

*349.* ZEEVART, J. A. D.: Produktie, fysiologische werking en mogelijkheden voor toepassing van een nieuwe groep stoffen de gibberellienen. Landbouwk. Tijdschr. **70**, 123 (1958).

*(Received, January 25, 1960.)*

# Selected Subjects in Sedimentation Analysis, with Some Applications to Biochemistry.

By J. W. WILLIAMS, Madison, Wisconsin.

With 7 Figures.

## Contents.

*Acknowledgments.* The writer is indebted to the National Institutes of Health (RG-4912 and RG-4196), the Rockefeller Foundation, the Research Committee of the Graduate School and the Wisconsin Alumni Research Foundation. The original manuscript was written during the tenure of a John Simon Guggenheim Memorial Foundation Fellowship in 1956–57; it has now been thoroughly revised. — I should like to thank friends both in and out of Madison who have helped me with advice and by discussion, and in the preparation of the manuscript. Notable among them are Drs. R. L. BALDWIN, J. M. CREETH, L. J. GOSTING and L. PELLER. Dr. BALDWIN was kind enough to permit the use of several figures original, in whole or in part, with him. In the preparation of the typewritten manuscript I am indebted to PATRICIA M. CREETH.

## Glossary of Terms.

### (The Romans designate Chapters.)

$A$   area of a cylindrical surface in the ultracentrifuge cell parallel to the axis of rotation and cut by the sector angle.

$A$   total number of antibody molecules in the system (III).

$\text{Å}$   Ångström unit.

$a$   length of molecule (II).

$a, b, c$   molar concentration of constituents A, B, C at position $x$ (III).

$a_0, b_0, c_0$   initial concentrations of constituents A, B, C (III).

$A_i G_j$   antibody-antigen complex (III).

$b$   cross-sectional dimension of molecule (II).

$c$   concentration, g./cc.

$c^0$   initial solute concentration in a sedimentation velocity experiment.

$c_{01}$   free antigen concentration (g./l.) (III).

$c_G$   total antigen concentration (g./l.) (III).

$c_s$   concentration of those solute molecules with sedimentation coefficient $s$ at position $x$.

$c_t$   constant solute concentration in the region ahead of the boundary and at time $t$.

$\bar{c}$   average concentration at some intermediate position in the cell, $= c_0^0 \, (x_0/x_b)^2$, when the value of $x_b$ is taken as the midpoint of the displacement.

$D$   diffusion coefficient.

$D_A$   reduced height-area diffusion coefficient.

$D_i$   diffusion coefficient of $i$th component.

$D_{ij}$   cross-term diffusion coefficient.

$D_1$   diffusion coefficient of a main solute constituent.

$D_{12}$   cross-term diffusion coefficient — macromolecule 1 and solvent 2.

$D_{11}$   main macromolecule diffusion coefficient ($= D_1 = D$ in a two-component system).

$D^*$   apparent diffusion coefficient.

$\bar{D}$   mean diffusion coefficient.

$e$   natural log base $= \exp$.

$f_i$   friction coefficient of the $i$th component, per molecule.

$f_0$   friction coefficient of a spherical, unsolvated molecule.

$f/f_0$   asymmetry number.

$G$   total number of antigen molecules in the system (III).

$G(s)$   integral distribution function of sedimentation coefficients.

28*

$g(s)$    differential distribution function of sedimentation coefficients after removal of diffusion and of $s$ upon $c$ effects.

$g^*(\mathfrak{S})$    apparent sedimentation coefficient distribution.

$g(\mathfrak{S})$    apparent sedimentation coefficient distribution after removal of diffusion effects.

$H$    maximum height of boundary gradient curve.

$h$    ordinate of boundary gradient curve.

$J_i$    flow of the $i^{\text{th}}$ component; amount of solute $i$ crossing unit area per unit time.

$K$    equilibrium constant.

$L_{ik}$    phenomenological coefficient — related to friction coefficients in sedimentation and diffusion.

$M$    monomer concentration (III).

$M_i$    molecular weight of the $i^{\text{th}}$ component.

$M_1$    molecular weight of a main solute constituent.

$M_{10}$    molecular weight of a complex, components 0 and 1 (solvated solute).

$M_{12}$    molecular weight of a complex, components 1 and 2.

$m$    concentration in molalities.

$m_{ijk}$    number of antigen-antibody aggregates composed of $i$ bivalent antibody molecules, $j$ univalent antibody molecules and $k$ antigen molecules (III).

$m_i$    amount of solute $i$ in a given volume.

$N$    Avogadro's number.

$n$    number of monomer molecules which associate to form a polymer (III).

$n$    refractive index.

$\Delta n_0$    difference in refractive index between solvent and original solution before sedimentation commenced.

$P$    pressure.

$P$    polymer concentration (III).

$p$    extent of reaction (fraction of antigen sites in system which have reacted) (III).

$p$    standard deviation of distribution curve of sedimentation coefficient (III).

$q$    number of free antibody sites on an aggregate (III).

$R$    molar gas constant.

$R$    differential refractive increment (I).

$r$    radius of a spherical, unsolvated molecule.

$S$    Svedberg unit of sedimentation, $10^{-13}$ second.

$\mathfrak{S}$    reduced coordinate with the units of sedimentation coefficient (III).

$s$    sedimentation coefficient; unit, second.

$s_f$    unit equivalent to $-s$.

$s_i$    sedimentation coefficient of the $i^{\text{th}}$ component.

$s^0$    sedimentation coefficient for initial concentration $c^0$.

$s_{12}$    sedimentation coefficient of a smaller, slower-moving macromolecule in the presence of a larger, second macromolecule.

$\bar{s}$    mean sedimentation coefficient.

$T$    absolute temperature.

$t$    time.

$u$    electrophoretic mobility.

$V$    volume.

$V_e$    volume of an equivalent ellipsoid of revolution of protein (II).

$v$    velocity.

$\bar{v}_1$    partial specific volume of the solute (solvent = component 0).

$w$    grams of solvent bound per gram solute (II).

$X_i$    "forces" causing the irreversible processes, i. e. the flows (I).

$x$    radial distance in the ultracentrifuge.

$x_0$    radial distance to the meniscus.

$x_1$    radial distance to a position $(x_1)$ in the cell behind the boundary.

$x_2$    radial distance to a position $(x_2)$ in the region of constant concentration ahead of the boundary in the cell.

$x_3$    radial distance to a position $(x_3)$ in the region of constant concentration ahead of a second boundary in the cell.

$x_b$ — displacement of boundary from the axis of rotation measured by the square root of the second moment of the boundary gradient curve.

$x_H$ — distance from the axis of rotation to the maximum ordinate of the boundary gradient curve.

$y$ — activity coefficient on the $c$ concentration scale.

$z$ — valence of an ion, plus or minus.

$\alpha$ — "number of molecules of solvent bound per mole of solute" (IV).

$\beta$ — parameter involving sedimentation coefficient and intrinsic viscosity which is related to the axial ratio of an "effective" hydrodynamic ellipsoid of revolution (II).

$\Gamma$ — measure of relative solvation concentration, on a mole per mole basis (IV).

$\Gamma'$ — $\Gamma (M_2/M_1)$, measure of relative solvation concentration, on a gram per gram basis (IV).

$\delta$ — coordinate transformation in the Gilbert theory (III).

$\varepsilon$ — charge per mole of protons.

$\eta$ — coefficient of viscosity of the solvent.

$[\eta]$ — intrinsic viscosity.

$\mu$ — chemical potential per gram.

$\mu^0$ — reference $\mu$, temperature and pressure dependent, on the $c$ concentration scale.

$\bar{\mu}$ — total potential per gram.

$\nu$ — number of ions.

$\varrho$ — density of a solution.

$\varrho^0$ — density of a mixed solvent for which no sedimentation flow takes place.

$\sigma$ — rate of production of entropy per unit volume by the irreversible processes (I).

$\sigma^2$ — standard deviation of experimental Gaussian error equilibrium distribution of macromolecules (IV).

$\sigma^2$ — second moment about the mean position $x$ of the boundary gradient curve (III).

$\Phi^{1/3} P^{-1}$ — parameter involving sedimentation coefficient and intrinsic viscosity which is related to an "effective" hydrodynamic sphere (II).

$\psi$ — electrical potential.

$\omega$ — angular velocity ($2\pi$ times frequency).

## Introduction.

There are in general two kinds of articles which have to do with sedimentation behavior. In the one the subject matter is descriptive of the theory and practice of the use of the ultracentrifuge in analysis; in the other consideration is given to the application of the methods so provided in the solution of problems in biology and medicine. This time the author has felt that an attempt to review some selected topics, both as to the main outlines of the theory and the applications of the technique, might be of interest. Because of limitations of space it could not be a balanced, objective account of the subject, so that of choice we have written principally about those aspects of the subject in which we are currently interested and supposedly better informed.

The several Chapters will begin with statements about the related theory, but except in the case of a few footnotes detailed mathematical analyses will be avoided.

At this particular time the theory which is fundamental to the experiment is in a significant period of development and extension, and the newer, more general equations suggest that some of the classical methods for the description, measurement and interpretation of sedimentation velocity behaviors are only approximately correct. It is the newer derivation of the general flow equations for transport processes that has had such an impact on the theory of velocity sedimentation. Now, more exact and definitive equations are made available for the correlation of sedimentation and diffusion measurements, for the evaluation of solute molecular weight, for the description of solute heterogeneity and for the study of interacting systems. Even in the case of the traditional two-component, incompressible system with no volume change on mixing, some questions which remained unresolved in the kinetic theory, force-friction concept derivations are now answered, but more importantly, the new equations mark the directions in which the expressions may be extended for application to multicomponent systems.

Unfortunately the report is being prepared at a time when the few available expressions for macromolecular or macro-ion solute molecular weight in a three-component system have not been subjected to experimental study and verification. A typical example of such a system is that of the protein chemist: a macro-ion for which a neutral salt, "supporting electrolyte", must be added to the aqueous system to repress charge effects. Indeed, the flow equations derived by thermodynamics of irreversible processes are believed to be far more accurate than the experiments of today. All of the equations used in sedimentation analysis depend upon a proper formulation of the flows, along with statements of the conservation of mass.

So, a significant task of the next period of years is the further development and the application of the more complete and more sophisticated theory of sedimentation analysis; it is just beginning to be possible to see the course ahead. Pending the extension to the multi-component systems, it is essential for our purpose that we have at hand a knowledge of the system of equations on which the experiments of today can be based and interpreted. We shall largely restrict our considerations to theory and practice with the classical two-component, incompressible system, but in doing so the attempt will be made to arrive at the essential, but restricted, working equations in such a way that an inquisitive and friendly attitude toward the more general basic principles and their extensions will be produced. These analyses, Chapter I, though brief and perforce oversimplified, will provide the basic reasoning upon which the practical applications are based.

In the biophysics or physical biochemistry laboratory the ultra-centrifuge has many uses, indeed the instrument may be said still to enjoy spectacular success as a means to study the behavior and interactions of protein molecules in solution. Soon after the construction of the first successful machines observations of convection-free and vibration-free sedimentation used in connection with diffusion measurements had established the fact that the native soluble proteins form systems of essentially monodisperse large molecules. It was not long before some of them could be shown to dissociate reversibly with change in pH and ionic strength. These early studies showed the respiratory proteins to be exceedingly well defined compounds, with the normal serum proteins representing a more complex group. In pathological sera the presence of new constituents could be demonstrated. Now, in addition to characterization of proteins and other macromolecules the instrument is invaluable as a means to follow the purification of viruses, enzymes, polypeptides, nucleic acids; indeed plant and animal proteins and polysaccharides in general. It provides opportunities to study small changes in these macromolecules and, within limits, to draw conclusions about the shape of such solutes. It is with the more traditional determination of the so-called "molecular characteristic constants" of proteins and other macromolecules that Chapter II is concerned.

It was the original intent of that great pioneer scientist, THE SVEDBERG, to use his ultracentrifuge to make size distribution analyses; the Uppsala program of the characterization of the proteins as macromolecules began at a later date. One of the most important attributes of the sedimentation velocity experiment is its sensitivity to heterogeneity. This is a property shared in common with other moving boundary methods; indeed it was of great help in the development of the equations required for sedimentation analysis to be able on occasion to refer to the (sometimes) less involved mathematical descriptions of reversible boundary spreading in electrophoresis. It turns out that a very simple transformation of the common boundary or boundary gradient curve gives a description of the heterogeneity of the solute, as measured by the sedimentation coefficient, for the limiting case in which the boundary spreading due to diffusion is negligible and in which the boundary sharpening due to sedimentation coefficient variation with concentration can be neglected. Either integral or differential distributions of sedimentation coefficient are thus provided, with the latter being used only for continuous distributions, i. e. polydisperse systems. It will also appear that there are a number of situations in which the effects of the two complicating factors, spreading of the boundaries by diffusion and their sharpening by the dependence of sedimentation coefficient on concentration, can be corrected out of apparent distribution curves. It is the purpose of Chapter III to indicate how the mathematical analyses are made and to demonstrate the use of the equations in the study of some biological systems.

The interpretation of experiments with systems which contain a third component presents complications. By a completely arbitrary

decision the ternary solution may be described as being either a one-
or a two-solvent system, retaining the notion that a substance which
is present in large quantity is usually called a solvent. It is with the
solutions of the second kind, with solvent combinations such as water
to which urea, sucrose, or glycine, etc., has been added in appreciable
amounts that the more serious difficulties of interpretation arise; good
approximations are available when two of the components are present
as "solutes" in relatively very small amounts. It is certain that the binary
solvent may not be treated as one component, but it is another problem
to provide adequate explanations. To date, progress in this direction
has been more satisfactory in connection with the sedimentation
equilibrium experiment; then there has been the introduction of the
density gradient or zone sedimentation experiment which involves a
sedimentation equilibrium in a binary solvent. We have included as
Chapter IV a brief consideration of such ternary systems which is for
the most part restricted to the equilibrium as contrasted with the transport
sedimentation problem. In this way at least one may be made aware
of some of the pitfalls ahead if one would attempt to extend and use
the simple force-friction concept to describe the transport problem in
the binary solvent.

An effort has been made to write the review in such a way that
Chapters II and III might form an entity, but there are contained in
them some references to equations which appear in Chapter I. Except
where noted, use has been made of the same symbols as were adopted
for our recent review of some of the theoretical aspects of sedimentation (*162*).

# I. Basic Theory.

The modern laboratory for the practice of physical biochemistry
contains apparatus for the study of diffusion and electrophoresis in
addition to the ultracentrifuge. Even in the absence of an externally
applied field, directed flows of matter will result whenever gradients
of chemical potential are present in a system; in other words, diffusion
takes place. And, superimposed upon the diffusional flow, one may have
additional flows caused by the application of external force fields. These
fields may be gravitational, centrifugal or electrical in character, giving
rise to the processes known as velocity sedimentation and electrophoresis.
The mathematical description of each of these three experiments requires
two fundamental equations: a continuity equation from the law of the
conservation of mass, and a flow equation. For each of the three processes
in a binary system the combination of the two equations leads to a
basic differential equation. The solutions of the three basic equations
then provide methods for measuring the three coefficients (or mobilities):

those of diffusion, $D$, velocity sedimentation, $s$, and electrophoresis, $u$. Furthermore, they may give detailed information about the form of the boundary gradient curves that are provided by optical systems which serve to record the time rate of progress of the experiment, and in so doing, furnish the means for the solution of practical problems in biology and medicine.

Diffusion is the process by which concentration differences in a solution spontaneously disappear. As a consequence of their thermal energy the molecules in a fluid are in constant motion, but after the system has reached macroscopic homogeneity there is no longer a directed (diffusional) flow of components due to gradients of chemical potential, the forces causing the flow. Any directed flow is opposed by a frictional force.

In the more common sedimentation velocity experiment redistributions of solute concentration are produced along a column in a sector-shaped cell due to the combined effects of the externally applied mechanical force field and of diffusion. Thus, as the molecules move along radial lines toward the bottom of the ultracentrifuge cell the boundary of concentration spreads out with time. The driving force is now the negative of the gradient of total potential, i. e. the algebraic sum of chemical, centrifugal and electrical potentials (field strength $= -$ potential gradient). There is again a frictional force resisting the action; the observed velocity of sedimentation represents an almost steady-state condition.

Two boundaries are formed in electrophoresis, one in each limb of a rectangular "U"-shaped cell; the upper solution in each case is made up of buffer and supporting electrolyte, with protein solution in Donnan equilibrium with it beneath. Diffusion causes a blurring of these two boundaries while electrical migration, resulting from the applied electrical field, tends to separate them into a series of boundaries, one for each protein constituent and a stationary dilution boundary. The driving force is again the negative gradient of total potential, in this instance the so-called electrochemical potential, the sum of chemical and electrostatic potentials, and there is a friction force to oppose the driving force.

Electrophoresis will be removed from further consideration by making note of a significant difference between this process and the other two phenomena. In electrophoresis the ions, macro-ions and gegen-ions, are caused to move in *opposite* directions, but in diffusion and in sedimentation they are constrained to move *together*. In terms of the simpler kinetic theory concepts, applicable to *ideal* two-component systems, the quantities called friction coefficients in diffusion and sedimentation are then the same, but the corresponding quantity in electrophoresis is different.

Since this review pertains to the basic equations of sedimentation and some application to the study of problems in biology and medicine, we shall make note of the fact that diffusion as such is often encountered in biological systems, but discuss the subject only as it is involved in the sedimentation processes.

## 1. Charge Effects.

In the hands of the protein physical chemist the ultracentrifuge has been a very successful instrument. In certain respects this individual has been fortunate. In the first place, the protein solutes often are either nearly monodisperse or mixtures of two or three discrete components and "globular" in form. Most important of all, concentration-dependence effects are small; in the language of the polymer chemist the usual aqueous salt medium constitutes a "poor" solvent and solute-solvent interaction, or non-ideality effects, are relatively unimportant. In practice sedimentation measurements are usually made at high pressures, but in aqueous systems at least any consequential effects are small enough so that the results may be compared directly with those from diffusion measurements, carried out at atmospheric pressure. Perhaps the chief complication is a result of charge effects. Basically, there are two of them, the primary and the secondary effects of SVEDBERG and PEDERSEN (*144*, *112*). In sedimentation, the more important primary effect acts to produce a reduction in the rate of movement of the macro-ion, and means had to be devised whereby the large ion could be "released" from its gegen-ions. We give qualitative descriptions of the means by which the protein ions are freed to move with a rate which is characteristic of their own friction coefficients, both in diffusion and in sedimentation.

### a) Diffusion.

Those ions which have a concentration gradient will move spontaneously to reduce this gradient. Because of the condition of electrical neutrality the two kinds of ions, macro-ions and gegen-ions, of the protein must move together rather than diffuse at rates which are controlled by individual friction coefficients. This results in an increase in the diffusional velocity of the macro-ions and a decrease in the mobility of the gegen-ions, an effect which is dependent on the macro-ion concentration.

Diffusion measurements on proteins are ordinarily performed in the presence of a buffer salt, "supporting electrolyte", in sufficient concentration to render "negligible" the effects of any net charge possessed by the protein ion. But in removing such effects in this manner the system now becomes one of more than two components and the basic theory becomes more complicated, because of a coupling of solute flows. It is not strictly valid to treat the salt and water solvent as a single component.

## b) Sedimentation.

The ultracentrifugal field exerts a force on the macro-ions and on the gegen-ions; this force is different for the several ions. If this were the only force the resulting mobilities of these ions would be different, and characteristic, leading to a separation of charge and accompanied by the performance of electrical work in the solution. However, the macro-ions and gegen-ions must move together, thus the excess electrical force retards the heavier macro-ions and accelerates the slower small ones. The potential of the centrifugal and electrical forces is included with the chemical potential to form the total potential per gram, $\bar{\mu}$, and

$$\bar{\mu}_i = \mu_i - \frac{\omega^2 x^2}{2} + \frac{z_i \varepsilon}{M_i} \psi \tag{1}$$

where $\mu_i$ is the chemical potential per gram of the $i^{\text{th}}$ solute, $\omega$ is the angular velocity ($2\pi$ times frequency), $x$ is the radial distance*, $z$ is the valence of the ion, plus or minus, $\varepsilon$ is the charge per mole of protons, $M_i$ is the molecular weight of the solute component, and $\psi$ is the electrical potential.

In a region of constant concentration the corresponding force is

$$-\left(\frac{\partial \bar{\mu}_i}{\partial x}\right)_t = (1 - \bar{v}_i \varrho)\, \omega^2\, x - \frac{z_i \varepsilon}{M_i}\left(\frac{d\psi}{dx}\right) \tag{2}$$

where $\bar{v}_i$ is the partial specific volume of the solute and $\varrho$ is the density of the solution.

Now as has been noted, proteins may be present in ionic form; they are weak electrolytes. The most obvious way to make the electrostatic term negligible is to work with near isoelectric protein ($z_i \simeq 0$), but this is not generally practical. The common procedure by which the protein ions are released so as to permit them to move with a rate which is substantially characteristic of their own mobility is again to add supporting electrolyte to the system. This is yet another manifestation of the Gibbs-Donnan effect, already well-known to investigators of osmotic pressure in protein systems. The early quantitative descriptions of charge effects are due to SVEDBERG (*142*), TISELIUS (*150, 152*), LAMM (*70*) and PEDERSEN (*110*); they represent treatments at an advanced level, but they are at times difficult to understand and apply mainly to the sedimentation equilibrium experiment. A more recent discussion of the

---

* Since the path of the sedimentation is in a radial direction, some authors use the symbol *r* for this quantity. The equations descriptive of the several sedimentation processes are actually written for a cylindrical coordinate system and there is much justification for the change from $x$ to $r$. However, in the early days of the ultracentrifuge the symbol $r$ was used to represent the radius of a particle, with $x$ as radial distance, and we conform to the usage of the Uppsala Laboratory.

problem, restricted entirely to sedimentation equilibrium, is that of JOHNSON, KRAUS and SCATCHARD (64). In the presence of the excess of slightly sedimenting supporting electrolyte the protein solute is considered as a neutral molecule, an approximation which becomes less exact as the charge to weight ratio of the solute increases. In general, and over the years, too little attention has been given to the problem of determining just how much of this added electrolyte is required to suppress the primary charge effect under any given set of conditions, and to the effects of ion-binding. Nonetheless, the addition of the electrolyte has served a most important purpose.

In addition to the primary effects which are caused by the separation of macro-ion from the gegen-ions there are secondary charge effects which result from the separation of the ions of the supporting electrolyte. They have been subjected to a comprehensive study by PEDERSEN (112).

Unfortunately, and as in the case of the diffusion experiment, a new complication is introduced when the salt is added in that a multicomponent system is now to be considered when the working equations are derived. With the presence of each actual component a concentration gradient to correspond is produced in the rotating cell. There is a multiplicity of diffusional flows, usually with coupling between them, and the experimental flow equations must be adequate for the description of the combined sedimentation-diffusion processes in the general case of a system with any arbitrary number of components. These complications are not avoided by an extrapolation of diffusion effects of *one* of the solutes to infinite dilution; the solute flows uncouple completely only in the limit of infinite dilution of *all* the solutes.

## 2. The Theory of Sedimentation Analysis.

As has been already indicated, there are two fundamental types of equations for the description of velocity sedimentation, the law of the conservation of mass and the expression for the flow of a solute. From the combination of these two equations there can be obtained the relationships by which sedimentation coefficients are defined and evaluated, molecular weights are found, heterogeneity is measured and interacting systems are studied. The combination of the two basic equations gives an equation (or set of equations in multicomponent systems) which is known as the Lamm differential equation of the ultracentrifuge. It is the analog of Fick's second law which is descriptive of the diffusion process in two-component systems.

Most, but not all, of the methods for measuring the sedimentation coefficient are based on solutions of this Lamm differential equation, just as are the methods for the evaluation of a diffusion coefficient related to solutions of Fick's second law. Furthermore, the mathematical expressions for the form and other properties of the boundary or boundary gradient versus distance curves as a function of time and distance are derived from these differential equations. Accordingly, the sources of the Lamm equation are indicated.

### a) Conservation of Mass. The Continuity Equation.

Sedimentation in the ultracentrifuge may be considered to take place in a section of volume such as is suggested by *Figure 1*. The cylindrical surfaces of the cell at the several radial distances, $x$, are parallel to the axis of rotation. In a given element of volume, $V$, bounded by the surfaces denoted as $x_1$ and $x_2$, and in the absence of chemical reactions which involve the solute we write the equation,

$$(AJ_i)_{x_1} - (AJ_i)_{x_2} = \frac{dm_i}{dt} \quad (3)$$

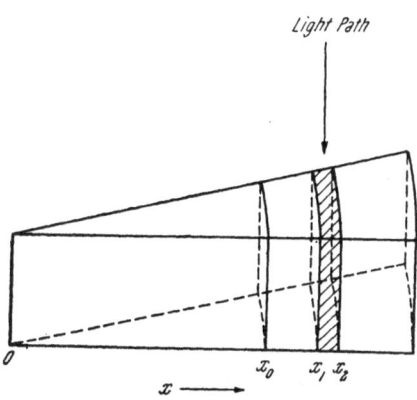

It states that per unit time the amount of sedimenting solute, component $i$, entering the volume element through the surface at distance $x_1$, minus the amount flowing out at $x_2$, is equal to the time rate of change of the amount, $m_i$, remaining in this volume. The amount of this solute crossing unit area per unit time is the quantity designated as its flow, $J_i$, and the

Fig. 1. Schematic representation of the sector-shaped cell used in the ultracentrifuge, with a volume element located between radial distances $x_1$ and $x_2$. The angle made by the walls of the cell is such that sedimentation takes place along radial lines from the center of rotation $O$. The light path in the optical system is directed through the transparent windows of the cell. With this system, sensitive to differences in refractive index or light absorption, the effects of the mechanical force field in displacing the boundary between solvent and solution and of other causes in modifying its form are recorded.

cross-sectional area of the cell is denoted by $A$. In the ultracentrifuge, the area $A$ is proportional to the radial distance from the center of rotation, or $A = k x$. The volume contained in the sector is $V = 1/2 k x^2$, or $dV/dx = k x$. The amount of this component in the section of volume between $x_1$ and $x_2$ is

$$m_i = \int_{x_1}^{x_2} c_i (dV/dx) \, dx = k \int_{x_1}^{x_2} x c_i \, dx \quad (4)$$

The concentration, $c_i$, must be expressed on a volume based scale, usually grams per cubic centimeter. Thus, from the definition of $m_i$, we have from equations (3) and (4),

$$(x J_i)_{x_1} - (x J_i)_{x_2} = \frac{\partial}{\partial t} \left[ \int_{x_1}^{x_2} x c_i \, dx \right] \quad (5)$$

The continuity equation follows when this equation is differentiated with respect to $x$ at constant $t$, i. e.

$$\left(\frac{\partial c_i}{\partial t}\right)_x = -\frac{1}{x}\left(\frac{\partial(x\,J_i)}{\partial x}\right)_t \tag{6}$$

When suitable flow equations in the form of expressions for $J_i$ are provided, the problem which remains is to solve the resulting differential equation, first derived by LAMM, to provide the working equations for the interpretation of the experimental data.

### b) The Flow Equations.

Before looking into the actual source of the general flow equations some remarks about them are required. There are both experimental and theoretical flow equations to describe the macroscopic movement of components at the different points throughout the liquid in a sedimentation experiment. The experimental flow, $J_i$, may be also described as the product of a velocity in cm./sec. and the concentration, thus $J_i = c_i\,v_i$. This description of a flow, equivalent to the earlier one, is important because the experimental definition of the sedimentation coefficient depends upon it.

In theoretical treatments of the experiment, flows of components are expressed as sums of forces, gradients of *total* potential, each multiplied by an appropriate coefficient (or diffusional mobility). For the linear displacements at constant temperature and pressure the number of terms contained in a general flow equation will depend upon the number of components present in the system.

An important consideration with flow equations is the frame of reference. In the experimental flow equations the flow is referred to a mark on the cell. However, if a bulk flow of liquid is superimposed on the flow due to sedimentation, as occurs when there is a volume change on mixing, it must be taken into account. The theoretical flow equations ordinarily are written for flows relative to the local center of mass or relative to the mean volume velocity. For practical laboratory operations equations descriptive of flows relative to the cell must be provided. They are obtained by conversions from the theoretical equations. In this report only the flows in incompressible systems and thus relative to the cell will be involved.

The recent derivation of general flow equations from the thermodynamics of irreversible processes by DE GROOT, OVERBEEK, HOOYMAN and others (35, 62, 63, 113, 162) has had a marked influence on the theory of the sedimentation velocity experiment. Now, this experiment is being placed on the same sound theoretical basis as is the one which involves equilibrium, and means are provided by which transport in multicomponent systems may be described.

Obviously, it is impossible to give here more than a very brief description of the basic thought and to take from the more general formulation some of the simpler results which are of aid in the interpretation and practical use of the conventional procedures with the ultracentrifuge. The basic concept is that in a system not in equilibrium irreversible changes take place, leading to a production of entropy. The deviations from the equilibrium conditions may be considered as forces, $X_i$, causing the irreversible processes, i. e. the flows, $J_i$. These forces and flows are contained in the expression for the entropy production per unit volume, $\sigma$, as the sum of products of the two terms. Thus, for the linear flows of solutes $i$ (solvent = o), with the flows referred to the solvent-fixed frame,

$$T\sigma = \sum_{i=1}^{q} (J_i)\, X_i \quad (i = 1 \ldots q) \tag{7}$$

where $T$ represents the absolute temperature.

When the displacements from equilibrium are not large, the flows can be taken to be linear functions of the forces, while the forces causing the flows are given by the negative gradients of the total potential. Provided the system is incompressible and there is no volume change on mixing, the flow of solute $i$ in a system of $q + 1$ components, relative to the cell, may be written

$$J_i = \sum_{k=1}^{q} L_{ik} \cdot X_k = - \sum_{k=1}^{q} L_{ik} \left( \frac{\partial \bar{\mu}_k}{\partial x} \right)_t \quad (i = 1 \ldots q) \tag{8}$$

The coefficients $L_{ik}$ are called phenomenological coefficients and are related to the practical sedimentation and diffusion coefficients which are obtained experimentally.

The chemical potential per gram, $\mu_i$, for a neutral molecule is related to the concentration by the usual expression

$$\mu_i = (\mu_i)_c^0 + \frac{RT}{M_i} \ln y_i c_i \tag{9}$$

where $R$ is the molar gas constant, $(\mu)_c^0$ is a reference potential whose value is temperature and pressure dependent and $y$ is the activity coefficient on the $c$ concentration (g./cc.) scale. Since the experiments are carried out at constant temperature, $\mu_i$ is a function of pressure and of the $q$ solute concentrations. Also, $\bar{\mu}_i = \mu_i - \omega^2 x^2/2$.

With these descriptions of the potentials which are involved, the fundamental and practical coefficients can be now related, and a useful general flow equation is soon obtained

$$J_i = c_i s_i \omega^2 x - \sum_{j=1}^{q} D_{ij} \left( \frac{\partial c_j}{\partial x} \right)_t \quad (i = 1, \ldots, q) \tag{10}$$

in which
$$s_i = \frac{1}{c_i} \sum_{k=1}^{q} L_{ik} (1 - \bar{v}_k \varrho) \quad (i = 1, \ldots, q) \tag{11}$$

and
$$D_{ij} = \sum_{k=1}^{q} L_{ik} \left( \frac{\partial \mu_k}{\partial c_j} \right)_{T, P, c_m} \quad (i, j = 1, \ldots, q) \tag{12}$$

It is to be noted that in this flow equation the flows due to diffusion and to sedimentation are directly additive.

The sedimentation coefficient, $s_i$, and diffusion coefficients, $D_{ij}$, are related to the same set of coefficients $L_{ik}$. In the expression for $s_i$ the density of the solution is represented by $\varrho$, and the partial specific volume of the $k^{th}$ constituent is given by $\bar{v}_k$. The description of the diffusion coefficients $D_{ij}$ and their experimental determination have been given in detail by GOSTING (56).

### 3. Velocity Equation for Two-Component, Incompressible Systems and Its Uses.

For two-component systems, with no volume change on mixing, $q = 1$ and the expressions (10), (11) and (12) reduce to the classical ones of SVEDBERG; they are the ones which must be used because no experimental data are as yet available where multiplicity of components has been properly considered. The errors involved in using the simple equations for the system protein ion, buffer and supporting electrolyte, water, have not been evaluated, but they are believed to be small provided the protein is near its isoelectric point.

The flow equation becomes

$$J_1 = L_1 X_1 = -\frac{c_1 M_1}{N f_1} \left[\frac{\partial \bar{\mu}_1}{\partial x}\right]_t \tag{13}$$

or, since $J_1 = c_1 (dx/dt)_1$, a *velocity equation* may be written as

$$v_1 = \left(\frac{dx}{dt}\right)_1 = -\frac{M_1}{N f_1} \left[\frac{\partial \bar{\mu}_1}{\partial x}\right]_t \tag{13a}$$

where $f_1$ is the frictional coefficient per molecule and $N$ is Avogadro's number. With the appearance of this equation we are reminded of the earlier use of the force-friction concept in which for a solute undergoing frictional flow, a driving force, now $M_1 (-\partial \bar{\mu}_1/\partial x)_t$, is set equal to the force of frictional resistance, $N f_1 v_1$. With the definitions of chemical and total potential and the fact that

$$\frac{\partial \mu_1}{\partial c_1} = \frac{RT}{c_1 M_1} \left(1 + c_1 \frac{\partial \ln y_1}{\partial c_1}\right) \tag{9a}$$

there are now found to be contained in the velocity equation the two-component system analogs of equations (10), (11) and (12)*. They are the ones in common use, even when the experiments have been performed in multicomponent systems.

$$J_1 = c_1 s_1 \omega^2 x - D (\partial c_1/\partial x)_t \tag{14}$$

$$s_1 = \frac{L_1 (1 - \bar{v}_1 \varrho)}{c_1} = \frac{M_1 (1 - \bar{v}_1 \varrho)}{N f_1} \tag{15}$$

$$D = L_1 (\partial \mu_1/\partial c_1) = \frac{RT}{N f_1} \left(1 + c_1 \frac{\partial \ln y_1}{\partial c_1}\right) \tag{16}$$

To obtain the expression for the sedimentation coefficient it is necessary to insert the definition for $\bar{\mu}$ into equation (13a), and, remembering that it is a function of pressure and composition at constant temperature, perform the indicated simple calculus operations*.

Equations (10) and (14) agree when the system contains two components, as only one diffusion coefficient is then involved. However, in a multicomponent system, $q^2$ diffusion coefficients are required [equation (10)], and the classical equations derived on the basis of the simple force-friction concept are approximate relations. Thus a process which appears to be simple from the point of view of a mechanistic picture is subject to complications which are sometimes serious.

The use of mixed solvents should be avoided wherever possible, at least for the present time. LANSING and KRAEMER (71) showed that the treatment of multicomponent systems as binary ones might lead to errors in the estimate of solute molecular weight, even though infinite dilution values of sedimentation and diffusion coefficients for the solute were combined. The use of a "binding-coefficient" was introduced by WALES and WILLIAMS (157) to deal with the sedimentation equilibrium experiment in a ternary system of neutral molecules, but it is only very

---

* The derivation of the flow equation is here indicated:

$$\frac{dx}{dt} = - \frac{M_1}{N f_1} \left( \frac{\partial \bar{\mu}_1}{\partial x} \right)$$

$$\frac{d\bar{\mu}_1}{dx} = \left( \frac{\partial \mu_1}{\partial x} \right)_P + \left( \frac{\partial \mu_1}{\partial P} \right)_x \frac{dP}{dx} - \omega^2 x$$

The pressure variation with distance $dP/dx = \omega^2 x \varrho$, and for the incompressible system, $(\partial \mu_i / \partial P)_c = \bar{v}_i$.

Therefore, and with the use of equation (9),

$$v_1 = \left( \frac{dx}{dt} \right)_1 = - \frac{RT}{N f_1 c_1} \cdot \frac{dc_1}{dx} + \frac{M_1 \omega^2 x (1 - \bar{v}_1 \varrho)}{N f_1}$$

$$J_i = c_1 v_1 = - D \frac{dc_1}{dx} + \frac{M_1 \omega^2 x (1 - \bar{v}_1 \varrho) c_1}{N f_1}.$$

Again with the use of the velocity equation, and in the region of the cell in which $dc_1/dx = 0$,

$$\frac{dx}{dt} = - \frac{M_1}{N f_1} \frac{\partial}{\partial x} \left( \mu - \frac{\omega^2 x^2}{2} \right) = - \frac{M_1}{N f_1} \left( \frac{\partial \mu}{\partial P} \cdot \frac{dP}{dx} - \omega^2 x \right) =$$

$$= \frac{M_1 (1 - \bar{v}_1 \varrho) \omega^2 x}{N f_1}$$

So, $J_1 = - D (dc_1/dx) + c_1 s_1 \omega^2 x$, where $s = (dx/dt)/\omega^2 x = \dfrac{M_1 (1 - \bar{v}_1 \varrho)}{N f_1}$.

recently that BALDWIN (v. *162*) has worked out more exact equations to treat the corresponding sedimentation velocity problem and to extend it to cases where charged macromolecules are present. The new flow equation, equation (10), contains in it the classical one, equation (14), for the two-component system but one may not use the latter [or equation (15)] as a source to derive expressions which will be valid in multicomponent systems (*56*).

## 4. The Differential Equation of the Ultracentrifuge and Its Solutions. Two Components.

By combining the continuity equation (6) and the flow-equation (14) one obtains directly the basic equation for the description of the sedimentation behavior of a single homogeneous solute in the sector-shaped cell when it is subjected to a centrifugal force. Called the Lamm differential equation of the ultracentrifuge, it may be written as follows:

$$\left(\frac{\partial c}{\partial t}\right)_x = \frac{1}{x}\frac{\partial}{\partial x}\left[\left(D\frac{\partial c}{\partial x} - c_1 s_1 \omega^2 x\right)x\right]_t \tag{17}$$

ARCHIBALD (*4, 5*) has obtained a rigorous solution of the Lamm equation, but the functions in which the solution is expressed are so complicated that they have hardly been useful as yet. A much simpler series solution applicable to rapid sedimentation has been provided by FAXÉN (*42*). The use of this solution is restricted to situations where the sedimentation and diffusion processes are not disturbed by the meniscus and when the boundary between solvent and solution is still far from the bottom of the cell; furthermore in obtaining the solution it was assumed that $s$ and $D$ are constants, independent of solute concentration. Under these conditions, and writing only the leading term,

$$\frac{\partial c}{\partial x} = \frac{c^0 \exp(-2s\omega^2 t)\exp[-(x_0\exp(s\omega^2 t)-x)^2/2\sqrt{\pi D t}]}{2\sqrt{\pi D t}}[1+\ldots] \tag{18}$$

The quantity $s\omega^2 t$ is usually small compared to unity. The meaning of the symbols is as follows: $x$ and $x_0$ are the distances from the center of rotation to the boundary position and meniscus, $c^0$ is the original solute concentration, and $t$ is the time of sedimentation at constant angular velocity, $\omega$. The factor $c^0 \exp(-2s\omega^2 t)$, obtained from the radial dilution law for the process in the sector-shaped cell, represents the change in concentration with distance in the cell, and the term $x_0\exp(x\omega^2 t)$ in the exponent shifts the curve, by the sedimentation process, along the $x$-axis. The quantity $\partial c/\partial x$ defines the $y$-axis of a plot, with distances in the cell, $x$, as abscissae.

Thus, this equation is descriptive of the form of the familiar boundary gradient curve which is observed in the experiment. By the experimental boundary gradient curve is meant the graph obtained, after the subtraction of a reference base line, when concentration gradient (or refractive index gradient) is plotted as a function of distance and in the region of the boundary. In the experiment refractive index is generally used as a measure of concentration, it being assumed that the refractive index is a linear function of the solute concentration, or

$$n = n\,(\bar{c}_1) + R\,(c_1 - \bar{c}_1) \tag{19}$$

Here, $n\,(\bar{c})$ is the refractive index at concentration $\bar{c}$. The constant $R$ is the differential refractive increment, i. e. the change in refractive index which corresponds to unit change of solute concentration. If the relationship is non-linear and equation (19) is written for $\partial n/\partial x$, further complications and inaccuracies are introduced.

If the initial boundary conditions could be satisfied the Faxén equation, even in this simplest form, would provide the means for the estimation of both $s$ and $D$ in a single experiment. Indeed, in the early days of ultracentrifugal analysis such estimates were made and molecular weights were computed from these data. Also, by using the Faxén solution, again assuming the initial and boundary conditions to be satisfied, it would be possible to calculate the theoretical sedimentation curve for the homogeneous solute at any time during an experiment. Such curves, and also theoretical curves based on a value of $D$ obtained in an independent experiment, had been used to judge of the uniformity of a solute.

It is now a well-recognized experimental fact that sedimentation coefficients of macromolecules usually, if not always, vary with solute concentration. Also there are theoretical reasons and experimental evidence that $D$ varies with $c$. It is now understood that the effects of even relatively small variations of $s$ with $c$ are such that the earlier estimates of $D$ from the spreading of boundary gradient curves in sedimentation experiments (and consequently $M$) were considerably in error and that the tests for homogeneity based on the Faxén equation are not at all reliable. It is much more difficult to solve the Lamm equation when $s$ is allowed to vary linearly with concentration, but Fujita (45) has been successful in this endeavor. The general solution, which is of the Faxén type, is so complicated that it will not be reproduced, but, for the case where $s$ decreases with increasing $c$, it shows two things: 1) The shape of the boundary gradient curve is markedly sharpened. 2) The velocity of the position of the maximum gradient becomes noticeably less as the initial concentration in the cell is increased.

The symmetry of the boundary gradient curves will be well maintained except at the edges. The sharpening of boundary gradient curves was already a well-known experimental phenomenon in solutions of organic high polymers where $s$ upon $c$ effects are abnormally large (flexible molecules as contrasted with the "globular" proteins). Also, it is at once evident that diffusion coefficients, computed either by using the maximum height-area or reduced second moment method from the sedimentation patterns must have been appreciably in error.

At the present time, it is unlikely that physical biochemists will use the Fujita equation to construct the $dc/dx$ versus $x$ curves for a protein solute. However, his solution of the Lamm equation does provide a relatively simple expression for the ratio of the maximum height of the boundary gradient curve to its area, $H/A$, an equation analogous to the relationship which is used to compute the reduced height-area diffusion coefficient, $D_A$. BALDWIN (10) has used this expression with telling effect, both to study the homogeneity of a bovine plasma albumin sample and to evaluate the diffusion coefficient. He used a method of successive approximations to determine the values of $D$ from the measurements of $H/A$. This could be done because $D$ is quite well known for the albumin. Quite recently, FUJITA (46) has described a new method whereby his equation is applied directly to observed data in the absence even of an approximate value of the result to obtain the diffusion coefficient. One of the interesting observations about the experiments with BSA was a quite large boundary sharpening effect even though for this protein the dependence of $s$ upon $c$ is not at all pronounced.

So it can be expected that the Fujita solution gradually will replace that of Faxén for computations of $s$ and $D$ and for tests of heterogeneity. It is to be expected that diffusion coefficient data obtained in this way will be less precise than those of independent diffusion measurements, but the error is now greatly reduced and it is an obvious advantage to be able to measure both $s$ and $D$ for a biological material which can be obtained only in extremely small, e. g. 10 mg., quantities. As a by-product, this measurement of diffusion coefficient may constitute a sensitive test for the homogeneity of the solute.

Lacking homogeneity, mathematical descriptions obtained from the differential equation of the ultracentrifuge permit one to take into account the effects of heterogeneity, diffusion and dependence of sedimentation coefficient on concentration in modifying the form of the boundary or boundary gradient curve. To this study we have assigned the descriptive term boundary spreading in sedimentation velocity experiments. Such analyses have found many uses in investigations of biological systems, and some of them will be mentioned in Chapter III, p. 465.

## II. The Sedimentation Velocity Experiment
## and the Determination of Molecular Weight.
## (Two-Component, Idealized Systems.)

### 1. Definition and Evaluation of the Sedimentation Coefficient.

Perhaps the most elementary of all experimental flow equations is the expression known as Fick's first law, $J_1 = - D (\partial c_1/\partial x)_t$. This relation defines the diffusion coefficient in terms of $J_1$, a one-dimensional transport or flow of the solute, component 1, and the solute concentration gradient, $(\partial c_1/\partial x)_t$, at a time $t$ and position $x$ relative to the cell. It is strictly applicable for diffusion at constant temperature and pressure in two-component systems showing no volume change on mixing, but it has also been used to describe the results for a number of systems which contain three or more components, and often within the error of measurement until the advent of the Gouy and Rayleigh diffusiometers.

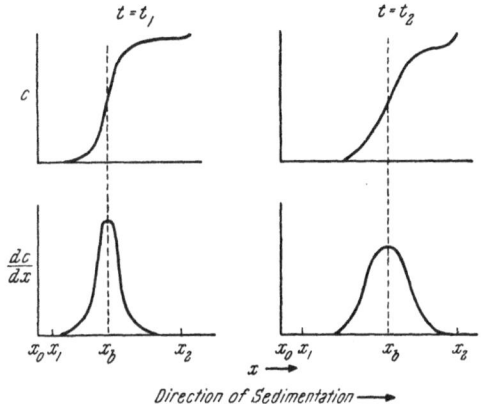

Fig. 2. Plan diagrams of concentration versus distance in the cell and concentration gradient versus distance in the cell for the simple two-component system in the ultracentrifuge at two times, $t_1$ and $t_2$. At the shorter time $t_1$ the $c$ vs. $x$ boundary is the steeper, and the $dc/dx$ vs. $x$ boundary is the narrower, one. At time $t = 0$ the $c$ vs. $x$ curve would have been infinitely sharp. The effect of continued diffusion is to spread out the boundary between solvent and solution. In the region between the boundary and near the bottom of the cell the concentration is constant and $dc/dx = 0$, but this concentration is higher at $t_1$ than at $t_2$. The law describing the change of concentration with time is called the radial dilution law.

In a similar way the expression for the flow by sedimentation, equation (20), provides a definition of the sedimentation coefficient, $s$, in a region of constant concentration, i. e. where $(\partial c_1/\partial x) = 0$.

$$s_1 \equiv J_1/c_1 \, \omega^2 \, x \qquad (20)$$

This is an operational, experimental definition, provided the curves of concentration versus distance in the cell at several times are made available in the experiment, and these data are used in combination with a statement of the conservation of mass, in the form

$$(x \, J_1)_{x_2} = - \frac{\partial}{\partial t} \int_{x_0}^{x_2} x \, c_1 \, dx \qquad (21)$$

[cf. equation (5)]. There is no flow of solute through the meniscus, so in this case $(x\,J_1)_{x_0} = 0$. The c. g. s. unit of $s$ is the second; the Svedberg unit is $10^{-13}$ sec. and is denoted by $S$.

Typical curves representative of concentration and concentration gradient distributions in the cell are shown in *Figure 2*, p. 453. Effects of translational displacement by sedimentation and of boundary blurring by diffusion are indicated.

This definition is written in a form to be applicable to the two-component, incompressible system with no volume change on mixing. Unlike Fick's first law it may be applied to a multicomponent system. If the subscripts one were replaced by $i$, it would describe the sedimentation coefficient of the $i^{th}$ component, solvent or arbitrary solute, in a multicomponent system and subject to the same restrictions.

In this Chapter we are largely concerned with the behavior of what we term an *idealized* system, one of two neutral components at constant temperature and pressure with no volume change on mixing. The descriptive word is used not at all in connotation with activity coefficients. Our simpler equations, including (20), are almost always utilized by investigators of protein behavior in application to what are strictly multicomponent systems. The sedimentation coefficient of the protein solute is measured from the velocity of a boundary formed by this solute. Since the mass transport is independent of the shape of the boundary (*162*)

$$\int_{x_0}^{x_2} x\,c\,dx = \frac{(x_2{}^2 - x_b{}^2)}{2} c_{x_2} \tag{22}$$

The equations (20), (21) and (22) provide the essential working equation,

$$s = \frac{1}{\omega^2\,x_b}\frac{dx_b}{dt} \tag{23}$$

where $x_b$ represents the position the boundary would have at any time $t$, provided no blurring due to diffusion had taken place during the course of the experiment. Thus $s$, in this way related to the rate of movement of the boundary, is the sedimentation coefficient of the molecules in the region of constant concentration beyond the boundary.

Neither the boundary position nor the concentration ahead of the boundary is altered by diffusion. So, the infinitely sharp boundary moves with the same velocity per unit acceleration as does a solute molecule ahead of the boundary. If the position of this hypothetical boundary can be located the problem is solved. It is achieved by giving consideration to a region of the boundary enclosed between planes $x_1$ and $x_2$, Figure 2, and writing expressions for the amount of solute between the planes: a) for an infinitely sharp boundary and b) for a blurred boundary, equation (22). By the law of the conservation of mass the two amounts are equal, and it is readily shown that (53)

$$x_b^2 = \frac{\int_{x_1}^{x_2} x^2 \left(\frac{dc}{dx}\right)_x dx}{\int_{x_1}^{x_2} \left(\frac{dc}{dx}\right)_x dx} = \frac{\int_{x_1}^{x_2} x^2 \left(\frac{dc}{dx}\right)_x dx}{c_t} \qquad (23\,a)$$

where $c_t$ is the constant solute concentration in the region ahead of the boundary. Thus, when $c = 0$ at $x_1$, this equation gives the boundary position in terms of the square root of the *second* moment of the boundary gradient curve, $dc/dx$ versus $x$.

When the boundary gradient curve is symmetrical, the position of the maximum gradient, $x_H$, coincides with the *first* moment of the boundary gradient curve. The position $x_b$ does not quite correspond to $x_H$, but is displaced somewhat towards the solution side of the boundary. For a protein of molecular weight 60,000 and at the speeds customarily used for sedimentation velocity determinations the error in $s$ will be of the order of 0.1% if $x_b$ and $x_H$ are equated. The actual non-coincidence of $x_b$ and $x_H$ is a consequence of the fact that the sedimentation takes place in a sector-shaped cell.

Equation (23) forms the common basis for the computation of $s$. When this coefficient is assumed to be a constant, independent of solute concentration, it may be integrated to give

$$\ln (x_b/x_0) = s\,\omega^2 t \qquad (23\,b)$$

and if $\ln (x_b/x_0)$ is plotted against $\omega^2 t$, the slope of the line gives the sedimentation coefficient.

With the recognition that $s$ is not a constant but a coefficient which is a function of solute concentration there arises the problem of assignment of proper concentration to each $s$ value. Up to within relatively recent years the procedure had been to refer $s$ as measured in this way to the initial solute concentration, $c^0$, without concern for the changing concentration with distance in the sector-shaped cell. For the region ahead of the boundary where concentration does not vary with distance, equation (17) becomes

$$dc_t/dt = -2\,c_t\,\omega^2 s \qquad \text{and}$$

$$c_t = c^0\,e^{-2s\,\omega^2 t} = c^0\,(x_0/x_b)^2 \qquad (24)$$

This is the radial dilution law. Since there is such a dilution effect, a more exact procedure is to consider the value of $s$ measured in this way as referring to the average concentration, $\bar{c}$, defined as $\bar{c} = c^0\,(x_0/x_b)^2$, when the value of $x_b$ represents the mid-point of the actually observed distance of displacement of the boundary (66, 67).

The plot of $\ln (x_b/x_0)$ versus $\omega^2 t$ deviates somewhat from a straight line because $s$ is not a constant, as assumed for the integration of equation (23). Since $s$ varies with concentration and $c$ varies with time, it is an improvement to write $s$ as a simple function of $t$ for the integration and obtain a more exact relationship of the general form

$$\ln (x_b/x_0) = s_0 (1 - k c^0) \omega^2 t + \ldots \qquad (23c)$$

Now $s$ is replaced by the quantity $s_0 (1 - k c^0)$ and is the value of the sedimentation coefficient for the given initial concentration, $c^0$. The dependence of $s$ on concentration in dilute protein solutions obeys quite well the linear equation, $s = s_0 (1 - k c)$, but it is better to consider this to be an empirical relationship.

The use of the more exact equation (23c) requires a knowledge of the slope $k$ of the $s$ vs. $c$ curve. When account is taken of the concentration dependence of $s$ in this way, the computed sedimentation coefficients are somewhat smaller, about $1\%$ for bovine serum albumin, as compared to the result obtained when the calculations are carried out in the more conventional way.

According to equation (15), $s$ is inversely proportional to a friction coefficient, which is in turn dependent upon the coefficient of viscosity. Since the viscosity of an aqueous solution changes by a factor of about $2\%$ per degree change of temperature, it becomes obvious that to obtain precision sedimentation coefficient data the temperature of the solution in the cell must be accurately known.

A decade ago there was much discussion of the fact that such data showed large discrepancies, with those figures obtained with the Svedberg oil-turbine machine being consistently higher than the corresponding data from experiments with the Spinco ultracentrifuge. The difficulty was removed when it could be shown that the observed thermocouple readings in the oil-turbine machine indicated cell temperatures which were too low. Insufficient attention had been given to this problem by the users of the oil-turbine apparatus. In recent years much progress has been made in increasing both the precision and the accuracy with which a sedimentation coefficient can be established.

BALDWIN (*10*) has made a detailed study and analysis of the methods of computing $s$ with a view to determining whether full use is being made of the accuracy now obtainable in the ultracentrifuges of the present day. He has included in his article a sample calculation of $s$ which is illustrative of what now can be achieved.

### 2. Molecular Weights from Sedimentation and Diffusion.

For the two-component, idealized system in which the solute does not dissociate into ions, solute molecular weight may be obtained by elimination of the quantity $N f_1$ between equations (15) and (16) to give

$$M_1 = \frac{R T s_1}{D (1 - \bar{v}_1 \varrho)} \left[ 1 + c_1 \frac{\partial \ln y_1}{\partial c_1} \right] \qquad (25)$$

In order to avoid the necessity of a determination of the thermodynamic term, a difficult problem in itself, it is customary to extrapolate sedimentation and diffusion coefficient data, along with the partial specific volume, to zero solute concentration. The thermodynamic term then becomes unity and the familiar Svedberg equation is obtained

$$\underset{(c \to 0)}{M_1} = \frac{R T s_1}{D (1 - \bar{v}_1 \varrho)} \tag{26}$$

If, however, each solute molecule dissociates as a strong electrolyte into $\nu_+$ cations of mass $M_+$ and $\nu_-$ anions of mass $M_-$, each per mole of ions, the relationship would have to be written

$$\frac{M_1}{\nu_+ + \nu_-} = \frac{R T s_1}{D (1 - \bar{v}_1 \varrho)} \tag{26a}$$

where $\bar{v}_1$ becomes the partial specific volume of the neutral salt. An apparent molecular weight, computed by using equation (26), would be the number average molecular weight of all the ions.

Proteins are weak electrolytes, and the established procedure is to dissolve them in a combination buffer and supporting electrolyte aqueous medium to repress the charge effects. Then the system is treated as being one in which a neutral solute, the protein component, is dissolved in the aqueous electrolyte system, considered as the other component. Actually there are at least three components (protein, salt and water), and the study of coupled flows in diffusion (56) demonstrates that the Svedberg equation in its simplest form, equation (26), cannot give precise molecular weight data. Using the new, more complete flow equations BALDWIN has derived expressions analogous to SVEDBERG's equation for use with three-component non-electrolyte systems (11) and with three-component systems which contain ionized solutes (162). As might have been predicted the resulting expressions bear a strong resemblance to the Svedberg equation; in addition they contain terms which are reminiscent of the expression for $M_1$ from sedimentation equilibrium studies of three-component systems (157). There has been no experimental verification of such more elaborate equations and the exact magnitude of the corrective terms is unknown. However, it seems unlikely that in the conventional experiments with relatively dilute buffer these terms could significantly affect the magnitude of the sedimentation coefficient.

Such being the case our interest must be somewhat restricted in outlook. We shall consider, as have all others, that we are dealing with two-component, idealized systems and make a few remarks about the quantities $s_1$, $D$ and $\bar{v}_1$, their evaluation, and the degree of precision by which protein molecular weight data can be obtained by using the experiments which involve transport.

First of all, the measurements of $s_1$ and $D$ for use in equation (26) should be made at the same temperature and in the same buffer for the extrapolation to zero concentration. When these coefficients have been measured at somewhat different temperatures or in different buffer systems they may be converted to a common basis by using the relations

$$s\,\eta/(1 - \bar{v}\,\varrho) = \text{constant} \qquad (27)$$

$$D\,\eta/T = \text{constant} \qquad (28)$$

where $\eta$ is the coefficient of viscosity of the solvent. The sources of these equations are found in equations (15) and (16), with the additional assumption that the ratio $f/\eta$ is constant, one not quite justified by experiment (81). Thus, the measurements of $s$ and $D$ should be carried out in the same buffer-supporting electrolyte system and at temperatures as close together as possible and not too far removed from the reference temperature, $20°$ or $25°$, as the case may require.

One of the great difficulties in the assignment of a precision value to the molecular weight of a protein stems from the fact that truly homogeneous solute materials have not been made available for the experiment. It is obviously the molecular weight of the main component which is desired, but that which is provided is usually an average value for a particular sample. Both sedimentation and diffusion patterns must be examined for the presence of impurities, and methods of evaluation of the data must be applied to give quantities which are adjusted to be characteristic of the main component.

It has been demonstrated by GOSTING (56) that when refractometric data for diffusion are obtained with an interferometric optical system, sufficient accuracy is available to provide a fairly sensitive test for heterogeneity and an analysis for the relative amount of the main component. But as he has indicated, those procedures by which an apparent diffusion coefficient is computed to give the datum for the main constituent have to be used with caution, because multicomponent solutions have been used and in them some coupling or interaction of the flows between the salt and protein transport processes may have taken place. Indeed, a buffer and salt aqueous system with completely homogeneous protein solute might show a Gouy fringe deviation graph, the earlier interpretation of which would have been solute heterogeneity. The effects of such interactions are eliminated only in the limit of infinite dilution of *all* the solutes, and the use of a too dilute buffer-electrolyte medium may cause difficulties of another nature, those due to electrostatic charges on the protein. The study of diffusion in the typical protein systems and the precise evaluation of $D$ is a difficult task and much remains to be achieved before reproducibility of the measurements can be placed upon a satisfactory basis and data of high precision attained.

It is one of the important sources of trouble that even were the protein solute homogeneous the effect of the so-called cross-term diffusion coefficients in the actual experimental system as they modify $D$ is at present unknown.

When the sedimentation coefficients are not too greatly dependent upon concentration, and when the sedimentation boundary gradient curve shows minima, with full or partial resolution, to correspond to the presence of several different solutes the sedimentation coefficient, $s_1$, of the main constituent is made available by the experiment. SVEDBERG and PEDERSEN (*144*) have discussed the theory and methods for the analysis in this case. Also, if the sedimentation coefficients are known functions of concentration, allowances for them can be made for the non-ideality thus introduced to provide a precise determination of the sedimentation coefficients. It is of interest to note that the relative amount of an impurity or impurities determined in this way can be often used with great advantage in the computation of the diffusion coefficient of the main component from the measured average value (*10*).

A third quantity for which experimental data are required for use in the Svedberg equation is the partial specific volume, $\bar{v}_1$. For protein solutions the product $\bar{v}_1 \varrho$ is approximately 0.75. It is subtracted from unity to give the bouyancy correction, so if the quantity $(1 - \bar{v}_1 \varrho)$ is to have the same precision as $s_1$ and $D$, the product should be known with a precision which is three times as great (*80*). The determination of the density of the solution, $\varrho$, presents no serious problems, but the situation is quite different with respect to $\bar{v}_1$. Here, it is required to know precisely what is the protein concentration, and such factors as again uncertainty about the homogeneity and reproducibility of the protein and also lack of knowledge as to moisture content of the sample limit the precision with which a result can be obtained. Literature values for $\bar{v}_1$ are often greatly at variance; for instance, for a protein as well defined as ribonuclease one finds values of $\bar{v}_1$ which range from 0.695 (*58*) to 0.728 (*26*). Methods are available for the determination and it would be a "labor of love" on the part of a competent person to brighten an otherwise dreary situation. It is to be regretted that the experiments begun by DAYHOFF, PERLMANN and MacINNES (*34*) and by CHARLWOOD (*30*) have not been continued. All too often the practice has been either to assume a value of $\bar{v}_1$, or to compute one which is based upon an observation of McMEEKIN (*92*) that the volumes of many proteins are the sums of the volumes of their constituent amino-acids. These practices are not sufficient when precision molecular weight data are required.

There have been written a number of excellent reviews on the subject of sedimentation in the ultracentrifuge (*38, 39, 53, 68, 84, 98, 144*), some of which contain extensive tables of what are sometimes called the molecular characteristic constants of proteins. Along with the values of $s_1$, $D$, $\bar{v}_1$ and $M_1$, one may find values for an "asymmetry number", defined by the relation $f/f_0$, where $f_0$ $(= 6\,\pi\,\eta\,r)$ is the friction coefficient a solute molecule of the same molecular weight would have if it were both spherical in form and unsolvated. Also in the general literature elaborate formulae have been provided whereby such asymmetry numbers can be converted to molecular dimensions of assumed ellipsoids of revolution, but no satisfactory method has evolved whereby effects of solvation and of actual asymmetry of the model can be differentiated. When the value for the ratio is unity it can be concluded that the molecule is substantially spherical in form and unsolvated. Furthermore, it is believed that in protein systems solvation alone can be hardly expected to give values of $f/f_0$ which are higher than 1.4, so that higher values would be indicative of molecular elongation.

In view of our remarks about evaluations of $D$ and $\bar{v}_1$, it does not seem advisable at this time to make any attempt to extend or amend these tables. Without meaning to cast doubt as to the essential accuracy of such data, we have indicated that in spite of elaborate studies there remains considerable uncertainty in the so-called precision values of $s_1$, $D$, $\bar{v}_1$ and $M_1$. This is true even for such proteins as $\beta$-lactoglobulin, ribonuclease and bovine serum albumin, about which we make a few remarks.

The *serum albumin*, in particular, has been the subject of a recent thorough critique (*40*). Earlier considered to be a simple homogeneous protein, the physical chemistry of its solutions became the subject of much investigation, especially in the United States. However, it was soon found both by the form of the boundary gradient curves in sedimentation and by the fringe deviation graphs of AKELEY and GOSTING (*1*) for the Gouy diffusiometer experiments that the solute is not pure; it contains a relatively small amount of heavier material, presumably dimer molecules. In the better articles descriptive of attempts to evaluate the diffusion coefficient of the monomer, one finds two diffusion coefficients: $D_A$, the observed average value, and $D_1$, the computed or corrected value for the main monomer constituent. For systems at pH 4.60 and $\bar{c} \simeq 0.5$ and using the Gouy method, GOSTING (v. *14*) found $(D_A)_{25,\,w} = 6.66\text{--}6.70 \times 10^{-7}$. This value corresponds to $(D_1)_{25,\,w} = 6.79\text{--}6.84 \times 10^{-7}$, figures which are dependent upon the analyses for the amount of the slower diffusing "dimer" which is present, and the assumption that no coupling of flows has taken place. CREETH (*31*), using the Rayleigh diffusiometer, found the comparable $(D_1)_{25,\,w}$ to be in the neighborhood of $6.87 \times 10^{-7}$, a fact which seems to justify the two-component analysis of the fringe deviation graphs.

*References, pp. 495—502.*

The sedimentation coefficient as a function of concentration is given by BALDWIN (*10*) to be $s_{25, w} = 5.01$ $(1 - 0.054\,c - 0.004\,c^2)$, with a standard deviation of $0.1\%$. Since $\bar{v}_1 = 0.734\,\text{ml/g}$ and $\varrho_{H_2O} = 0.9971$ at $25°$, the value of $M_1 = 65,600$ is computed. Inasmuch as the diffusion coefficient value corresponds to an average concentration $\bar{c} = 0.5$, and the $D$ upon $c$ relationship is unknown, it is necessary to use the sedimentation coefficient which corresponds to this concentration, $s_{25, w}$ $(\bar{c} = 0.5) = 4.87\,S$ for insertion in the Svedberg equation, reasoning that the use of the ratio $s_1/D_1$ at this concentration will give a better approximation than could be obtained by using limiting values of $s_1$ and $D_1$, with the latter still subject to much uncertainty. If one wishes to compare these bovine serum albumin data for $s_1$, $D_1$ and $M_1$ with other literature values, and there are many, articles of HARRINGTON, JOHNSON and OTTEWILL (*57*) and of LOEB and SCHERAGA (*79*) may be consulted.

Even when the application of the two-component analysis is justified, the calculation of $D_1$ values from the observed $D_A$ can only be made when the deviation graphs indicate that the "impurity" has a markedly different diffusion coefficient. Thus in the case of ribonuclease, CREETH (*31*) studied the diffusion of a 5 × crystallized sample, finding $(D_A)_{20, w} = 1.068 \times 10^{-6}\,\text{cm}^2\,\text{sec}^{-1}$ at $\bar{c} = 0.44$, and a deviation plot indicative of significant heterogeneity. However, it was not possible to compute any reliable value of $D_1$ from these data because either faster or slower diffusing impurities (or both) could have produced the observed deviations. It may be remarked that even in favorable cases, a significant loss of precision accompanies the correction of $D_A$ to give $D_1$.

Thus the calculation of the molecular weight of ribonuclease from sedimentation and diffusion coefficient data (*3*, *22*, *122*) for comparison with the values based on amino-acid analysis (*61*) and sedimentation equilibrium experiments (*156*) must await a clarification of the diffusion problem.

Both OGSTON (*29*, *103*) and TIMASHEFF (*149*, *153*, *154*) and their collaborators have devoted much thought and care to physico-chemical studies of *β-lactoglobulin* solutions. This protein, again once thought to be homogeneous, has turned out to be a mixture of two genetically different modifications, β-lactoglobulins A and B (*6*). As the studies with the protein became more and more detailed there came the realization that again one must recognize tendencies to association, limited to be sure to the pH region between 3.5 and 5.25. In the neutral pH region the difference between the A and B forms is slight when measured in terms of titration curves, optical rotatory dispersion and ultraviolet light absorption. None the less, there are differences in immuno-chemical response, solubility behavior and electrophoretic patterns (*138*).

Sedimentation and diffusion data again indicate some heterogeneity, but there may be something inherent in the method of preparation of the material which is a contributing factor. Both CECIL and OGSTON ($29$) and CREETH ($31$) have made attempts to compute $D_1$ data from observed $D_A$ values to give $(D_1)_{20, w} = 7.73 \times 10^{-7}$ and $7.59 \times 10^{-7}$, respectively, for $\bar{c} = 0.47$ g./100 ml.

The sedimentation coefficient data seem to be in need of further revision. Earlier values of $s_{20, w}$, extrapolated to infinite dilution, from OGSTON's laboratory ($28$, $65$) are almost certainly too high and lacking in reproducibility, no doubt due to some uncertainties about the actual temperature of the solution in the ultracentrifuge cell. The more recent articles by TOWNEND and TIMASHEFF ($149$) give sedimentation coefficients at several concentrations, but it is not clear whether these correspond to average concentrations or initial concentrations. For $s_{20, w}$ in near neutral solution, one finds the datum $2.85$ S, "measured at a protein concentration of 10 g./l.". It seems certain from measurements of light scattering that the molecular weight must be in the neighborhood of $35,000$ ($154$), although the values given in the tables to which reference has been made from the combined sedimentation and diffusion data are substantially higher. The partial specific volume is $0.751$ ($111$).

### 3. Molecular Weight Determination by Combination of Sedimentation Coefficient and Intrinsic Viscosity.

It has been indicated that sedimentation is opposed by the viscous resistance which the macromolecule encounters as it moves through the medium. The friction coefficient $f$, which characterizes the resistance, is proportional to the viscosity of the medium and to a factor which is a function of the size and shape of the solute. In the simplest case, that of an unsolvated spherical molecule, we have seen that $f = f_0 = 6 \pi \eta r$.

It is an experimental fact that the frictional ratio $f/f_0$ for proteins and polymers in general is always greater than unity, but its definitive meaning is both ambiguous and complex. If solvation, or hydration in the case of proteins, were the sole cause of the deviation from unity, the molecular volume of the solute would be expressed as $(M \bar{v}_1 + M w/\varrho) N$ where $\varrho$ is the density of the bound solvent (itself unknown) and $w$ is the number of grams of solvent bound per gram of solute. Further, protein chemists are well aware of the use of the HERZOG, ILLIG and KUDAR ($60$) and the PERRIN ($114$) equations which express the frictional ratio as a function of the ellipticity of molecules whose shape is assumed to be that either of prolate or oblate ellipsoids of revolution, used sometimes but not always after an allowance has been made for the effect of solvation. There grew up an extensive literature in which actual dimensions were assigned to protein molecules in solution and models of them became popular.

However, there is no sound basis for the belief that the real shape of a protein or other macromolecular solute conforms to the ellipsoid, a model taken in order to simplify a mathematical problem. And, as already suggested, no satisfactory method has been developed whereby the contributions of solvation and of actual asymmetry could be resolved, and by 1953 there came almost simultaneous revolt in the form of three independent articles (*102*, *123*, *127*). Of these we make a few remarks about the one of SCHERAGA and MANDELKERN (*127*) because it had an unexpected consequence; it appears to make available a means to combine sedimentation coefficient and viscosity data to give a solute molecular weight. In this analysis the assumption that the molecular volume of a solvated protein can be expressed as in the preceding paragraph is discarded, and along with it the notion that this volume is really the hydrodynamically effective volume of the solute in motion. The authors interpret the observed physical behavior of the protein as regards sedimentation, diffusion and viscosity in terms of a rigid *equivalent* ellipsoid of revolution and its axial ratio; the actual shape of the molecule is probably quite different. Means are provided whereby both $V_e$, the volume of this supposed equivalent ellipsoid, and its axial ratio may be computed.

The interesting parameter of SCHERAGA and MANDELKERN is a quantity called $\beta$, which is evaluated from the sedimentation coefficient, the molecular weight, the intrinsic viscosity, $[\eta]$, the partial specific volume, and the viscosity of the solvent. The function $\beta$ is not a constant; it changes in value with axial ratio. Tabular values of this dependence of $\beta$ on axial ratio for prolate and oblate ellipsoids are given in the original article. A quantity to correspond is the universal *constant* for the flexible chains of the polymer chemist, $\Phi^{1/3} P^{-1}$, of MANDELKERN and FLORY (*87*). Here, as far as friction effects are concerned the polymer molecule is replaced by an effective hydrodynamic sphere.

Expressions for $[\eta]$, $f/f_0$ (from PERRIN), and the extrapolated value of sedimentation coefficient, $s^0$, are combined to give the function $\beta$,

$$\beta \equiv \frac{N s^0 [\eta]^{1/3} \cdot \eta}{M^{2/3} (1 - \bar{v}_1 \varrho)} \tag{29}$$

where $N$ is the Avogadro number. The value of $\beta$ for a sphere would be $2.12 \times 10^6$. For oblate ellipsoids of axial ratio 300, $\beta$ has increased only to $2.15 \times 10^6$, but for prolate ellipsoids it changes significantly with axial ratio. SCHERAGA and MANDELKERN themselves emphasize the fact that data of high precision are required for the computation of this parameter when it is to be used to determine the axial ratio and the dimensions of the effective hydrodynamic ellipsoid.

This disadvantageous situation actually has been put to some gain by FLORY (88) and DOTY (20, 37) and their associates, among others, for the computation of solute molecular weight. The procedure amounts to using observed values of sedimentation coefficient in conjunction with intrinsic viscosity data. If the molecules are known not to be in the form of random coils an assignment of a value for $\beta$ is involved. For an oblate ellipsoid $\beta$ is substantially a constant, but if other data indicate the prolate form to be the more probable there must be more arbitrariness in this process. In the case of a substance such as the soluble denatured *collagen* (as contrasted with the "procollagen" of OREKHOVICH) known to be in the form of random coils (161) the constant $\Phi^{1/3} P^{-1}$ is involved, and not $\beta$.

For greatly elongated molecules this may be a more satisfactory procedure than to attempt to base the molecular weight determination on the classical Svedberg equation, requiring as it does measurements of sedimentation and diffusion coefficients which can be reliably extrapolated to infinite dilution, but there still must be doubts as to the real precision of the result. The Scheraga-Mandelkern equation may be looked upon as being of the same general type as the classical Svedberg equation. To obtain the latter, friction coefficients of sedimentation and diffusion were equated and eliminated to produce the resulting expression; now the friction coefficients of expressions for limiting sedimentation coefficient and an intrinsic viscosity that had been derived for ellipsoids of revolution are involved in like fashion. Viscosity involves an interchange of momentum and sedimentation means transport of mass and it seems as if more theoretical work is required to establish a real validity as compared to an empirical use of such a relationship as the Scheraga-Mandelkern equation.

In the case of such elongated molecules the dependence of sedimentation coefficient on molecular weight is quite small. If $a$ measures the length and $b$ the cross-sectional dimension of the molecule and if $a \gg b$ it can be shown with the use of Perrin's equation for prolate ellipsoids that

$$b = k \sqrt{\ln 2 \, (a/b)} \cdot \sqrt{s} \tag{30}$$

$$(a = k' \ln 2 \, (a/b) \cdot 1/D) \tag{31}$$

Thus the sedimentation coefficient measures the thickness of the elongated molecule; and incidentally, the diffusion coefficient would measure its length. Certainly, it is an experimental fact that the sedimentation coefficient changes but little with length of this kind of molecule, but whether the postulates and boundary conditions upon which the Perrin equation is based remain good at such high elongation is another question.

At any rate, the *deoxyribose nucleic acids* (DNA) are now popular subjects of research in biochemistry, and sedimentation and viscosity methods, among others, are being regularly employed to characterize their solution behavior. We mention two articles of DOTY and associates in this connection. In the first one (*116*) sodium deoxypentose nucleate samples from calf thymus glands have been shown to have a weight average molecular weight of about six million by light scattering and a molecular shape resembling that of a stiff coil whose root mean square end to end separation is 5000 Å and whose contour length is 20,000 Å. These data are corroborated by the molecular weight computed from equation (29). However, BUTLER et al. (*24*) indicate there is a discrepancy between data from light scattering and from sedimentation-viscosity sources which can amount to a factor of three with DNA systems.

In the more recent DOTY report (*37*) DNA solutions have been exposed to nine kilocycle sonic waves for varying times in the absence of oxygen and the effects of the irradiation examined. It was found that the molecules were degraded through double-chain scission to produce fragments which have the base-paired helical structures intact. Equation (30) was used to demonstrate that the original DNA and its fragments in solution assume the form of somewhat extended random coils but with constant molecular cross-sectional distance.

The researches of NISHIHARA and DOTY (*99*) on the sonic fragmentation of *collagen* macromolecules are also here of interest. These macro-molecules in solution are believed to consist of three polypeptide chains held together by hydrogen bonds (*117*); they have been shown to be 3100 Å in length, 13.5 Å in diameter and to have a molecular weight of 360,000 (*19*, *20*). On being broken by sonic irradiation the shorter pieces retain the three-stranded, helical structure. Again, by using equations (29) and (30) their molecular weights and cross-sectional diameters are evaluated, with the diameter remaining unchanged. It is to be noted that when the molecular weight drops from 370,000 to 149,000, the limiting sedimentation coefficient $s_{20, w}$ changes only from 3.02 S to 2.54 S, but the intrinsic viscosity shows a five-fold decrease.

# III. Heterogeneity
## and the Form of the Boundary Gradient Curve.

Of all the subjects in ultracentrifugal analysis, the study of poly-dispersity is one of the original items (*143*, *145*, *146*). With non-uniform solute material and as a result of the applied force field, a partial separation is affected to provide opportunities to determine the distribution of sedimentation coefficients, or, in the case of the sedimentation equilibrium experiment, to learn about molecular weight distribution. It is indeed a fractionation, but one of a physical nature. For the distribution of sedimentation coefficients it is necessary to carry out mathematical analyses of the form either of the concentration-distance or of the concentration gradient-distance curves obtained from the optical systems employed in the transport experiment.

In the long record of the researches on heterogeneity by using the sedimentation velocity experiment two general avenues of approach are to be recognized: a) Direct computation of the distribution of sedimentation coefficients without making any assumptions as to the form of the particular distribution. b) Use of the experimental parameters in connection with a continuous distribution function, the mathematical form of which has been invariably assumed. Of the two approaches we prefer at present the first one; it will be our hope here to be able to define the basic distribution functions which go with it, and to show how they have been used in the solution of selected biological problems.

## 1. Boundary Spreading: Negligible Diffusion, Constant Sedimentation Coefficient.

The fundamental relationship, neglecting effects of diffusion, between $dc/ds$, the ordinate of the distribution of sedimentation coefficients, and $dc/dx$, the concentration gradient which is obtained by experiment, appears to have been derived for the first time by BRIDGMAN $(21)^*$ in the form,

$$\frac{dc}{ds} = \frac{dc}{dx} x \omega^2 t \left(\frac{x}{x_0}\right)^2 \tag{32}$$

To complete the distribution curve an additional transformation is required; the distances, $x$, which form the abscissae of the experimental boundary gradient curve must be converted to the sedimentation coefficients, $s$, which correspond. This is accomplished by the use of the formula

$$\ln (x/x_0) = s \omega^2 t \tag{33}$$

The prototype for the analysis appears in the very early papers descriptive of sedimentation analysis. Here, for the sedimentation velocity experiment, either in the earth's gravitational field or with an optical centrifuge of

---

* This research marked the beginning in this Laboratory of an extensive and continuing program devoted to the study of solute heterogeneity as revealed by boundary spreading in the sedimentation velocity and electrophoresis experiments. After this start it was found that the immediate problems in electrophoresis were more readily treated because of the use of the rectangular cell. With the additional experience it then became possible to include the effects of radial dilution while extending the mathematical analysis to the sedimentation transport case.

*References, pp. 495—502.*

low power, the particles were assumed to be spherical in form and of radius $r$. The distribution function used was

$$\frac{dc}{dr} = \frac{dc}{dx}\frac{dx}{dr}$$

This formula was written down but not used by Svedberg and Nichols in their original description of an optical centrifuge (143). The transformation of abscissae, $x$ to $r$, was made by using Stokes' law.

In the literature Signer and Gross (130) are usually accredited as being the originators of this transformation of coordinates. It appears, however, that they wrote only about displacements of curves without any change of the coordinate axes to provide a distribution curve, a quite different matter. Equations sometimes described as being theirs (98) and on other occasions attributed to Kraemer (69) provide means to reduce a sedimentation pattern obtained at any time $t_i$ to a standard reference time $t_1$; they are

$$x_1 = x_0 \left(\frac{x_i}{x_0}\right)^{t_1/t_i} \quad \text{and} \tag{33 a}$$

$$h_1 = h_i \left(\frac{x_i}{x_1}\right)^2 \frac{(x_i - x_0)}{(x_1 - x_0)} \tag{34}$$

where the parameters $x_1$, $h_1$ and $t_1$ correspond to the convenient reference time, $t_1$, with $x_i$, $h_i$ and $t_i$ being the parameters at the time of measurement, $t_i$; and $h$ is the ordinate of the diagram. Equation (33a) is readily obtainable from equation (33), but equation (34) is not a distribution function in the sense we require, nor do we believe it is to be derived from equation (32).

Equation (32) of Bridgman may be normalized to give the expression now familiarly known as $g(s)$, thus

$$g(s) = \frac{1}{c^0}\left(\frac{dc_s}{dx}\right)_t x\,\omega^2\,t\left(\frac{x}{x_0}\right)^2 \tag{32 a}$$

The total original solute concentration is again denoted by $c^0$ while $c_s$ gives the concentration of those solute molecules with sedimentation coefficient $s$, at a position $x$.

The value of the distribution function, $g(s)$, represents the relative frequency of material with sedimentation coefficient $s$ $(= \ln (x/x_0)/\omega^2 t)$ within the solute. It is often expressed on a refractive index basis, i. e.

$$g(s) = \frac{(dn/dx)_t}{\Delta n_0} x\,\omega^2\,t\left(\frac{x}{x_0}\right)^2 \tag{32 b}$$

Here $\Delta n_0$ is the difference in refractive index between solvent and the original solution before the sedimentation commenced and $dn/dx$ is the refractive index gradient at the position $x$, a quantity closely related by constants to the concentration gradient [see equation (19)].

The integral distribution curve may be also used. It is designated by $G(s)$, where

$$G(s) = \int_0^s g(s)\,ds \tag{35}$$

It may represent either concentration or refractive index fraction of material with sedimentation coefficients $\leqq s$, depending upon whether equation (32a) or (32b) has been used. The derivation of these equations is to be found in a number of places (*55, 158, 162*).

One of the most important considerations when computations are made is that a reference base line correction for solvent redistribution in the cell be carried out prior to the transformation of the coordinates. In this way one is assured that $dc/dx$ or $dn/dx$ represents solute contributions to the particular gradient. Illustrative curves are included as *Figure 3*. The distribution curves of $g(s)$ or of $G(s)$ are of course descriptive of the solute, so that transformation of each one of the three boundary gradient curves at the three times should produce a single distribution, whether in the differential or integral form.

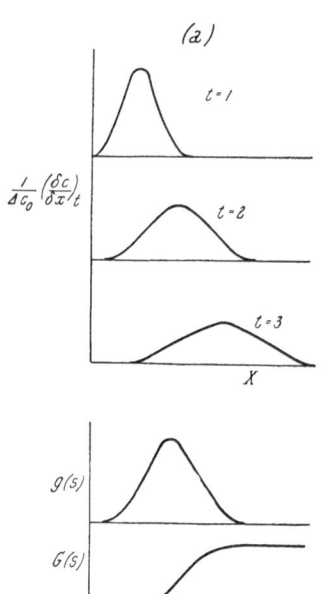

Fig. 3. Schematic curves for the case when the influence of diffusion is negligible and with no dependence of sedimentation coefficients on concentration. The boundary gradient curves at three different times for a sample with a continuous distribution of sedimentation coefficients are converted to the distribution curves; the lower distribution curve is the integral of the upper one.

This general type of analysis was applied by BRIDGMAN in his study of the sedimentation behavior of some rabbit liver *glycogens*. In order to use it one must demonstrate the absence of any broadening of the boundary gradient curve due to diffusion effects. One may perform the sedimentation experiments at different ultracentrifuge speeds to see if the boundary gradient curves have the same outline at a specified distance from the meniscus. Also, if the experimental boundary gradient curves, taken at the several intervals of time, can be substantially superimposed when transformed by the relationships (32) and (33), we have proof that diffusion has been negligible during the course of the experiment. In general, the experiments with glycogen met these tests satisfactorily, but any effects of concentration dependence of $s$ were not considered. (Treatments for the complications of diffusion and $s$ upon $c$ effects will be the subjects of subsequent considerations.)

The bulk of the glycogen could be shown to have sedimentation coefficients in the range from 20 to 120 S, with the most probable value, $s_{20, w} = 70$ S. It could not be proven whether the glycogen in the tissue is of corresponding particle size, or if this particle represents the chemical molecule rather than an aggregate. Observations of BELL et al. (*18*) are consistent with these data, but $g(s)$ vs. $s$ curves were not given.

However, a remark was made to the effect that all observed boundary gradient curves were much broader than could be accounted for by diffusion alone.

In connection with their study of the binding of corn glycogen and phosphorylase MADSEN and CORI (85) required a measure of the molecular weight of the glycogen. As expected the material was found to be polydisperse, with the bulk of it having sedimentation coefficients in the range of $s_{20, w} = 150$–$350$ S. POLGLASE, BROWN and SMITH (115) found normal human liver glycogen to contain two polydisperse components with sedimentation coefficients $s_{20, w}$ between 60 to 100 S for one, and for the other from 150–300 S, but it was not considered to be worthwhile to obtain the definitive distributions.

The β-lipoproteins from human plasma are heterogeneous in chemical composition and therefore density, and presumably in size as well. They appear to consist of a continuum of closely similar compounds of widely varying lipid and protein content rather than being discrete entities with constant composition. They have been the subject of much investigation because GOFMAN (50) has claimed to see the outlines of an integrated understanding of the vital problem of coronary artery hardening and heart attacks in his analyses of them. His "atherogenic index" value is based upon estimates of the relative amounts of certain lipoprotein classes contained in the whole β-lipoprotein material.

Under the conditions of study the lipoproteins of densities not far from that of water (average density $\simeq 1.026$ g./ml.) are dissolved in a salt solution of somewhat higher but controlled density, often 1.063. Now, in the ultracentrifuge the motion of the solute macromolecules is toward the center of rotation, and by means of the standard computational methods they are assigned $s_f$ values, in Svedberg units, instead of using a negative sign with the conventional sedimentation coefficient. The molecules in four classes, $s_f = 0$–$12$ S; $s_f = 12$–$2)$ S; $s_f = 20$–$100$ S; and $s_f = 100$–$400$ S are now indicated to be related to excessive hardening of the arteries. This is a broader claim by far as compared to some of the earlier ones. For instance, LINDGREN, ELLIOTT and GOFMAN (78) had concluded that the molecules of the $s_f = 10$–$20$ S class are "intimately related to the common and highly important disease atherosclerosis". There has been considerable controversy on the general subject (51); we believe much of it might have been avoided if some of the arbitrariness of analysis could have been removed to put the results upon a more mathematical basis. The GOFMAN procedures have undergone a more or less continuous revision and improvement with the years (36).

The sedimentation coefficient is a function of molecular weight, molecular shape and density. Assuming for the moment that the

$\beta$-lipoprotein molecules really are spherical, then heterogeneities both with respect to molecular weight and density remain, and they must be resolved. One should compute by means of the distribution function, equation (32), or its equivalent, values for $g(s)$ vs. $s$ for $\beta$-lipoprotein fractions of substantially constant density, in this way to learn about quantities much more closely related to distribution of molecular size. ONCLEY and associates (*106*, *108*) in doing this have made substantial progress with $\beta$-lipoprotein analysis.

## 2. Boundary Spreading:
### Appreciable Diffusion, Constant Sedimentation Coefficient.

To this point in Chapter III the assumptions have been that we are dealing with systems characterized by negligible diffusion and by constant sedimentation coefficients. In this Section the effects of sedimentation coefficient dependence on concentration will be further deferred but those of diffusion will be involved. Such macromolecules as the *glycogens* and *β-lipoproteins* are extremely large, with molecular weights of the order $10^6$–$10^7$, so the effects of diffusion are small, but, as we have already seen in the case of the more common substantially homogeneous protein solutes (Chapter I), appreciable diffusion takes place during the course of a sedimentation experiment, and the Faxén, Archibald and Fujita solutions of the differential equation of the ultracentrifuge take it into account.

Here again, with the heterogeneous solutes, it is usually necessary to give consideration to the effects of diffusion in broadening the boundary gradient curve when it is desired to obtain a distribution curve which is a true measure of the heterogeneity. It is evident that if equations (32) and (33) were used directly for the computation, the resulting distribution curve would be apparent rather than real in character; it is required to correct the individual apparent distribution curves, corresponding to several arbitrary times, for the effect of diffusion. The correction is based upon analyses by BALDWIN, GOSTING and WILLIAMS (*9*, *16*, *55*, *159*); it is made possible because the spreading of the boundary by heterogeneity is closely proportional to the time, $t$, while that caused by diffusion goes to good approximation with $t^{1/2}$. So, extrapolation to infinite time of the several apparent distributions should separate out boundary broadening due to diffusion and produce the true distribution, the quantity $g(s)$. GOSTING made a careful mathematical study of the extrapolation to infinite time by solving the boundary spreading equation

$$\left(\frac{dc}{dx}\right)_t = c^0 \int_0^\infty \left(\frac{d(c/c^0)}{dx}\right)_s g(s) \qquad (36)$$

for the case when FAXÉN's equation may be used to give the form of boundary gradient curve produced by a single sedimenting component, thus putting the procedure upon a sound mathematical basis. From this solution was found the proper function of time for use in the extrapolation, that is $1/x\,t$, and also the range of time in an experiment in which the extrapolation is substantially linear. For details the original article should be consulted; suffice it to note here the conclusion that

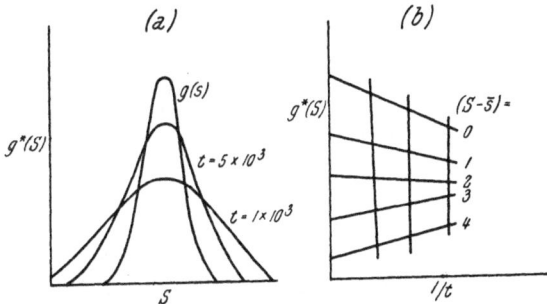

Fig. 4. Theoretical Gaussian curves showing the elimination of the effects of diffusion by extrapolation to infinite time to provide the distribution of sedimentation coefficients. Curve a shows the "apparent distribution" of $s$, $g^*(\mathfrak{S})$, at two different finite times and also the true extrapolated distribution, $g(s)$. Curve b shows how $g(s)$ is obtained by extrapolating $g^*(\mathfrak{S})$, at fixed values of $\mathfrak{S}$, vs. $1/t$. The letter $\mathfrak{S}$ denotes the variable defined by equation (37), while $\bar{s}$ is the mean sedimentation coefficient of the distribution, $g(s)$.
[BALDWIN (9)].

a satisfactory representation of $g(s)$ is obtained in this way only when the boundary spreading due to heterogeneity is fairly large relative to that caused by diffusion.

There are instances in which special types of computation are required. When the effects of diffusion are somewhat larger ERIKSSON (41) and BALDWIN (13) have devised and implemented an improved method of procedure. It was tested by BALDWIN with data for *thyroglobulin*.

When the effects of diffusion are present it is convenient to introduce a reduced coordinate $\mathfrak{S}$ which has the unit second, just as does the sedimentation coefficient, such that

$$\mathfrak{S} \equiv \ln\,(x/x_0)/\omega^2\,t \qquad (37)$$

The apparent distribution is then defined as

$$g^*(\mathfrak{S}) = \frac{dc_t}{dx} \cdot \frac{x\,\omega^2\,t}{c_0} \left(\frac{x}{x_0}\right)^2 \qquad (38)$$

in which $dc_t/dx$ is the experimentally observed solute concentration gradient. The use of the asterisk will indicate that diffusion effects are involved; when the diffusion effects have been removed the designation will be $g(\mathfrak{S})$ until the concentration dependence of sedimentation coefficient effects have been eliminated, then the function becomes the corrected $g(s)$.

The procedure by which the extrapolation to infinite time is carried out to eliminate the effects of diffusion to give a true distribution of sedimentation coefficients is illustrated by *Figure 4*, p. 471. The apparent distributions at finite times are defined by the Bridgman relationship, equation (32).

There are some systems in which the resolving power of the ultra-centrifuge is insufficient to allow of the analysis involving the extrapolation to infinite time, and another type of investigation is required. The

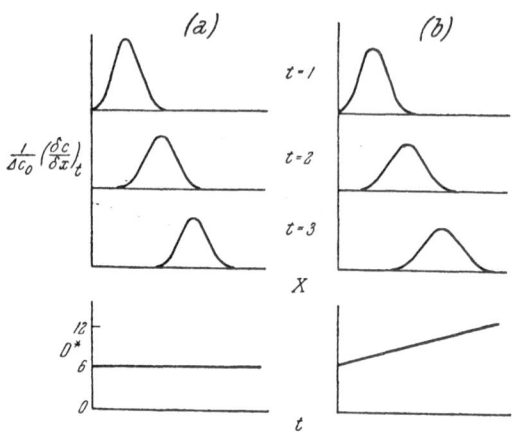

Fig. 5. Theoretical curves based upon FAXEN's solution showing the influence of diffusion when the dependence of sedimentation coefficients on concentration is negligible. Figure 5 a shows the pattern for a single sedimenting solute of $s = 4 \times 10^{-13}$ sec., $D = 6 \times 10^{-7}$ cm.² sec.⁻¹; Figure 5 b shows the corresponding pattern for two solutes, A and B, initially present in equal amounts, with $s_A = 3.5 \times 10^{-13}$ sec. and $s_B = 4.5 \times 10^{-13}$ sec., but with uniform $D = 6 \times 10^{-7}$ cm.² sec.⁻¹ for each. The upper curves show the boundary gradient curves at three successive times; the lower ones show the change with time of the apparent diffusion coefficient $D^*$, as defined by equation (39a) (BALDWIN).

standard deviation, $p$, of the distribution of $s$ can be found from $\sigma^2$, the second moment about the mean position, $\bar{x}$, of the boundary gradient curve. Again, treating the case for constant $s$, $\sigma^2$ is given by the formula (*159*)

$$\sigma^2 = (p\,\omega^2\,\bar{x}\,t)^2\,\{1 + \ldots\} + 2\,\bar{D}\,t\,\{1 + \bar{s}\,\omega^2\,t + \ldots\} \tag{39}$$

where $\bar{s}$ and $\bar{D}$ are the mean sedimentation and diffusion coefficients. The use of this equation is also limited because $s$ was assumed to be constant instead of a function of solute concentration in order to avoid a complex mathematical problem.

From equation (39) we may write an expression for an apparent diffusion coefficient, $D^*$, computed from the standard deviations of the boundary gradient curves as observed at several times, $t$, in an experiment. It is,

$$\frac{\sigma^2}{2\,t} = D^* = \bar{D} + \bar{D}\,\omega^2\,\bar{s}\,t + \frac{p^2\,\omega^4}{2}\,\bar{x}\,t \tag{39 a}$$

For the homogeneous solute $D^* = D$, but with heterogeneity there will be an abnormally large boundary spreading, and $D^*$ will be always larger than $\bar{D}$ at finite times.

In *Figure 5* are presented theoretical curves showing the influence of diffusion when the sedimentation coefficients are assumed not to vary with concentration. As in Figure 3, p. 468 the three boundary gradient curves correspond to three different times.

In spite of the ever present problem of concentration dependence equations (32), (33), (38) and (39) can have real application. It was originally intended that these types of analyses in which the effects of diffusion are removed by extrapolation to infinite time might be very useful in connection with researches having to do with enzymatic actions on *proteins* and *polysaccharides*. For example, it was so used by us (*159*) in a program of research which had to do with the action of such enzymes as pepsin and papain on gamma-globulin substrates, an object being to ascertain to what extent the antibody molecules could be degraded and still retain biological activity. Although our earlier thinking about the nature of the cleavage was thereby modified, the results were not entirely satisfactory because the effects of concentration dependence remained and methods were not at the time worked out by which they could be removed. Even so, it is probably true that with the globular proteins the correction for diffusion is the more important one.

Another case in point is the work of BALDWIN (7) with Shiga *neurotoxin*. Since but very small amounts of the toxin were available it was not possible to make concentration dependence corrections (see this Chapter below), but it could be demonstrated that the material as prepared was distinctly heterogeneous in spite of the experimental record of a single, fairly symmetrical boundary gradient curve in the ultracentrifuge. For the characterization an apparent diffusion coefficient was plotted against time and extrapolated to zero time to give an estimate of the true diffusion coefficient. The distribution of sedimentation coefficients was characterized by the two general procedures we have outlined.

### 3. Boundary Spreading: Sharpening Effects of Concentration Dependence of Sedimentation Coefficient.

In a sedimentation experiment the concentration changes across a boundary. So, if $s$ is a function of concentration, the coefficient will vary continuously throughout the boundary region. This is just as true for each of the different molecular species in a mixture as it is for the case in which the solute molecules are all of one kind. (It will be recalled that in the homogeneous system the true boundary position was defined by requiring that it move with the same mobility as the solute molecules ahead of the boundary.) Since the macromolecules on the solvent side of the boundary have a higher sedimentation coefficient than do those on the solution side the result will be an abnormal sharpening of the boundary. It is the magnitude of this boundary sharpening effect that was studied by FUJITA (45) for the homogeneous solute with a linear dependence of $s$ upon $c$ (Chapter I). Here we have an excellent solution of a fundamental problem in sedimentation, but unfortunately it cannot be simply extended to the heterogeneous solute system (162).

For the polydisperse systems empirical means have been provided whereby the $s$ upon $c$ dependence effects may be taken into account.

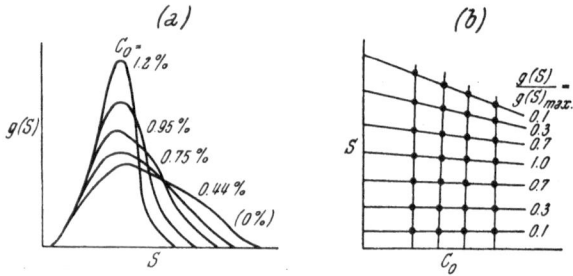

Fig. 6. Some experimental data for a dextran fraction showing how, after the effects of diffusion have been removed by extrapolation to infinite time, the effects of concentration dependence are taken into account. Diagram $a$ shows curves of $g(\mathfrak{S})$ vs. $\mathfrak{S}$, obtained for several initial concentrations $c_1^0$ of the sample. The method of extrapolation of these $g(\mathfrak{S})$ vs. $\mathfrak{S}$ curves to infinite dilution is suggested in diagram $b$, and the resulting $g(s)$ vs. $s$ curve is identified by the designation o%.

So, in the complete analysis there is now involved a double procedure; first of all an apparent distribution of $s$, $g^*(\mathfrak{S})$ is extrapolated to infinite time to remove the effects of diffusion, and there are obtained several $g(\mathfrak{S})$ vs. $\mathfrak{S}$ curves, each of which corresponds to a definite concentration. Then, these individual curves can be extrapolated to infinite dilution (160) or a calculated correction can be made (8, 66) to account for the effects of concentration dependence. In one extrapolation, illustrated by *Figure 6*, values of $\mathfrak{S}$ corresponding to a fixed ratio of $g(\mathfrak{S})/g(\mathfrak{S})_{max}$ are plotted against concentration to the limiting value. In another graphical procedure the extrapolated curve is obtained by plotting $g(\mathfrak{S})$ versus concentration at fixed values of $\mathfrak{S}$.

Fortunately, all these procedures to remove concentration dependent boundary sharpening effects are consistent in that they give substantially the same limiting $g(s)$ vs. $s$ curve provided they are used with systems that are sufficiently dilute to produce only mild effects of concentration dependence. They are not yet guided by mathematical treatments which would be of aid in the performance of the extrapolations.

It should be mentioned that multiple JOHNSTON-OGSTON effects (65) are also present in these boundaries of heterogeneous solutes; the distortions in the form of the boundary gradient curves so produced are presumably removed in referring the behavior to infinite dilution. It appears that the modification of the form of the boundary gradient curve due to the Johnston-Ogston effects is not as significant as that due to the so-called "self-sharpening". The nature of the Johnston-Ogston effect will be more closely examined in the next section.

Objective tests of the applicability and efficacy of our methods of analysis have been carried out in several laboratories. ROSENKRANZ and BENDICH (120) have made observations of the sedimentation behavior of two artificial mixtures of degraded and undegraded DNA samples at high total dilution and have shown quite satisfactory agreement between observed and computed integral distributions of s for the mixtures. No attempt was made to make adjustments for s upon c effects but their magnitude was indicated to be small in the extremely dilute solutions. Each component of the mixture had been the subject of prior sedimentation analysis. ERIKSSON (41) had earlier conducted exactly the same type of investigation with two polymethyl methacrylate fractions and with comparable suitable verification. His data were extrapolated to infinite dilution.

These descriptions of heterogeneity in sedimentation velocity experiments have found useful application in the study of certain materials which have been proposed for use as plasma extenders, notably gelatin (161), dextran (105, 160) and polyvinylpyrrolidone (PVP) (95). In solution these materials produce polydisperse systems of molecules of the random coil rather than the globular type. Both diffusion and concentration dependence effects are pronounced in these solutions and must be removed by the procedures which have been indicated before significant descriptions of the true heterogeneity are obtained. The distribution of sedimentation coefficient curves for a dextran fraction have been already presented in Figure 6. Similar data for some gelatin plasma extender solutions are available. The analyses of polyvinyl-pyrrolidone have been less complete because difficulties were encountered in the extrapolations by which diffusion boundary broadening effects were removed and because there remain concentration dependence effects which must have been pronounced.

Of particular interest to the writer have been the descriptions of the pattern of action of crystalline muscle phosphorylase on glycogen as revealed by sedimentation coefficient analysis (73). The procedures in discussion have yielded the distribution function of the sedimentation coefficients; these could be subsequently recomputed to molecular weight distributions through the combination of sedimentation and diffusion coefficients obtained for several discrete fractions. The calculations for the $g^*(\mathfrak{S})$ vs. $\mathfrak{S}$ and the $g(\mathfrak{S})$ vs. $\mathfrak{S}$ curves were encoded for solution on a digital computer resulting in a very substantial reduction in labor with both corrections, diffusion and concentration dependence, being carefully considered.

The *enzymatic* degradations, carried out to various extents and under different pH conditions, showed that always a less heterogeneous sample was produced in the reaction, one which with increase in extent of reaction had an increased proportion of sedimentation coefficients centered about approximately the same maximum as before degradation. This is an observation which is consistent with the notion of greater probability of enzymatic action on the outer glucose chains of the largest molecules. From the point of view of one interested in polymer chemistry this result might be thought to conflict with one of the fundamental postulates of reactivity between large molecules, to wit: any unreacted site is as reactive as any other site regardless of the size or shape of the aggregate to which it is attached (*44*). LARNER et al. (*73*) have pointed out that there does exist the possibility of a splitting of a weak linkage in the polysaccharide by either an enzymatic or non-enzymatic reaction. We would also suggest there may have been problems in connection with the complete elimination of concentration dependence effects.

Another research, one which has opened up new vistas, is that of SHOOTER and BUTLER (*25, 129*) on the sedimentation behavior of *deoxyribonucleic acid* at extremely low concentrations. In all of the prior record of sedimentation experiments with DNA preparations it will be found that scale-line displacement or Philpot-Svensson optical systems were used to record the progress of the redistributions in the ultracentrifuge cell; these optical systems in use with the conventional cells of 6 or 12 mm. depth require DNA solutions at such concentrations that self-sharpening of the boundary is all too evident. At concentrations in the region 0.02 to 0.04% the area under the boundary gradient curve is too small for any accurate analysis.

By the expedient of using the ultraviolet absorption optical system, the practice of which is already described in detail in the SVEDBERG-PEDERSEN monograph (*144*), it was shown to be possible to conduct convection-free and stable boundary sedimentation experiments with DNA solutions at concentrations as low as 0.001%. In all, the behavior of five DNA samples was observed and integral distribution curves, $G(s)$ vs. $s$, were obtained. For the five samples, the individual limiting distribution curves differed considerably, but they covered the range 12 to 40 S, or greater. It was demonstrated that the heterogeneity was not caused by aggregation; presumably it is due in part to variations in molecular shape as well as mass. Effects of diffusion were shown to be negligible, but proper corrections were made for concentration dependence of sedimentation coefficient. Because the light absorption method was applied it was more convenient to use the integral form of distribution curve, otherwise the procedures were identical with those which had been used for gelatin and dextran, for example. It was noted that the curve at infinite dilution differs but little from that obtained at a concentration of 0.001%, another fact which gives one confidence in the suitability of the descriptions of the heterogeneity, and indeed that DNA is polydisperse with respect to sedimentation coefficient.

Many investigators were now quick to recognize new opportunities, and as a consequence there have been printed in the intervening years the results of a number of researches of this general kind with DNA preparations from various sources. A few selected references are given (33, 37, 119, 121). SCHUMAKER and SCHACHMAN (128) have provided directions whereby the concentration-distance plots of the light absorption method are differentiated, so as to make the necessary conversions to produce $g(s)$ vs. $s$ curves, as contrasted with those of $G(s)$ vs. $s$. Some may prefer the differential distribution functions and curves; they give the same information. It often occurs in articles descriptive of DNA solutions that $g(\mathfrak{S})$ vs. $\mathfrak{S}$ curves are provided for finite concentrations. This practice is to be compared with that of reporting a sedimentation coefficient for a simple homogeneous solute at an arbitrary concentration, in both cases at a fixed temperature.

Experiments with the Rayleigh and Jamin interference optical systems have hardly reached the point where they may be considered to be routine because some mechanical problems remain to be solved. They do offer substantial possibilities for making observations and measurements at relatively low concentrations.

## 4. The Johnston-Ogston Effect.

Nowhere in the literature of ultracentrifugal analysis is there a better demonstration of the necessity to make allowances for concentration dependence of sedimentation coefficient than in the system of two solute components of different sedimentation coefficients. By using the Faxén solution ($s$ and $D$ constant) one might construct the boundary gradient curve diagrams for the mixture in a known proportion at different total solute concentrations and compare them with the corresponding experimental curves. In the one case, with sufficient resolving power, the ratio of the amounts of the two components as measured by the areas under the individual curves and compared to the total area would be constrained to remain constant, but in the other, the apparent ratio of the amount of the slower component to the faster one would depend upon the total concentration of the sedimenting material, with the anomaly always becoming more and more prominent with increase in total concentration. In the case of pre-war investigations with normal serum protein solutions made up of *albumins* and *globulins* to give essentially two maxima in the boundary gradient curve, the observations were understood in terms of chemical reactivities of the proteins themselves and not, as was shown by JOHNSTON and OGSTON (65) as a

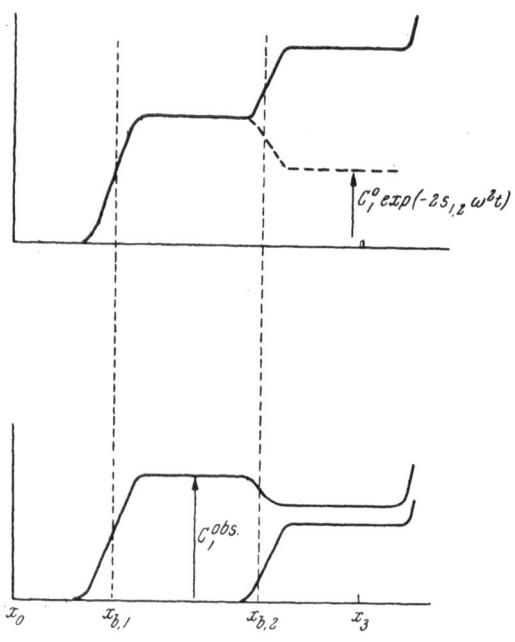

$$C_1^0 exp(-2s_{1,2}\,\omega^2 t)$$

$$C_1^{obs.}$$

$x_0 \qquad x_{b,1} \qquad x_{b,2} \qquad x_3$

Fig. 7. Representative sketch of concentration vs. distance in the cell, showing sedimentation patterns for two solute components, with the slower sedimenting one being designated as 1. The direction of sedimentation is from left to right in both cases. The upper curve is suggestive of the actual experimental diagram while the lower one has been drawn to describe the concentrations of the individual solutes as they change with distance in the cell.

necessary consequence of a physical phenomenon. Such changes in apparent concentration of the slower component are a direct result of differences in its rate of sedimentation in the absence and in the presence of the faster component. The slower component will sediment more rapidly in the region behind the boundary of the faster component, but as it moves into the other boundary its sedimentation rate will decrease. The slower component will then accumulate or "build-up" behind the fast boundary and be present at a lower concentration beyond this boundary than before it. This situation is indicated by *Figure 7*. The result is that in the mixture, and as recorded by the optical methods of measurement, there will be indicated too high a value for the concentration of the slower component and too low a value for the faster component.

*References, pp. 495—502.*

Let us consider the two solute system, components 1 and 2, with complete or nearly complete resolution. If there is no concentration dependence of sedimentation coefficient simple expressions for the individual original concentrations may be written down. They are

$$c_1^0 = (c_1)_t \left(\frac{x_1}{x_0}\right)^2 = (c_1)_t \cdot e^{2 s_1 \omega^2 t} \quad \text{and} \tag{40}$$

$$c_2^0 = (c_2)_t \left(\frac{x_2}{x_0}\right)^2 = (c_2)_t \cdot e^{2 s_2 \omega^2 t} \tag{41}$$

The concentrations at the several times $t$ are measured by the areas under the curves
$$c_1 = \int_{\text{across 1}} (dc_1/dx)\, dx \quad \text{and} \tag{40a}$$

$$c_2 = \int_{\text{across 2}} (dc_2/dx)\, dx \tag{41a}$$

In the more complicated real, as contrasted with the ideal, system, and with an analysis which makes provision for the sector-shaped cell, it is found that*

---

* The derivation of equation (42) for the most general case of diffuse boundaries involves tedious algebra. To simplify the presentation and yet convey an idea of the method of approach we may assume the boundaries to be infinitely sharp, corresponding to positions $x_{b,2}$ and $x_{b,1}$ in Figure 7. Then, one may write the following expression for $m_1^0$, the total amount of the slow component 1 originally present in the cell, as
$$m_1^0 = \frac{k}{2} c_1^0 [x_{b,2}{}^2 - x_0{}^2]$$

Making use of the formula, $dm_1/dt = (A\, J_1)\, x_{b,2} = k\, x_{b,2}\, c_1^t\, \omega^2\, x_{b,2}\, s_{1,2}$, there is obtained by integration an expression for $m_1$, the amount of component 1 which passes the position $x_{b,2}$ in time $t$.

$$m_1 = \frac{k}{2} c_1^0\, x_{b,2}{}^2 [1 - \exp(- 2 s_{1,2}\, \omega^2 t)]$$

Then $m_1^t$, the difference between $m_1^0$ and $m_1$, is equal to the amount of the slower component which is left behind the position $x_{b,2}$ at time $t$,

$$m_1^t = \frac{k}{2} c_1^0 [x_{b,2}{}^2 \exp(- 2 s_{1,2}\, \omega^2 t) - x_0{}^2]$$

This amount of component 1 is confined to the volume between the positions $x_{b,2}$ and $x_{b,1}$, so
$$V_1^t = \frac{k}{2} (x_{b,2}{}^2 - x_{b,1}{}^2)$$

Remembering that $x_t = x_0 \exp(s\, \omega^2 t)$, we then have

$$\frac{c_1^{\text{obs}}}{c_1^0} = \frac{\exp[2\, \omega^2 t\, (s_2 - s_{1,2})] - 1}{\exp(2 s_2\, \omega^2 t) - \exp(2 s_1\, \omega^2 t)}$$

This equation is equivalent to equation (42) when the substitutions $\exp(x) = = 1 + x + \dfrac{x^2}{2!} + \dots$ are made. The result obtained by the simpler analysis is identical with that of the more detailed derivation by using the position $x_3$ provided that the sedimentation coefficient $s_{1,2}$ can be assumed to be independent of time.

$$\frac{c_1^{obs}}{c_1^0} = \exp\left(-2\,s_{1,2}\,\omega^2 t\right)\left[\frac{s_2 - s_{1,2}}{s_2 - s_1}\right] =$$

$$= \exp\left(-2\,s_1\,\omega^2 t\right)\left[\frac{s_2 - s_{1,2}}{s_2 - s_1}\right]\left\{1 + 2\,\omega^2 t\,(s_1 - s_{1,2}) + \ldots\right\} \quad (42)$$

Since the sedimentation coefficient of component 1 in the mixture, $s_{1,2}$, is always less than its value in the absence of the faster component, $s_1$, the apparent concentration, $c_1^{obs}$, is always greater than the true concentration, $c_1^0$. The term in the square brackets is the original Johnston-Ogston expression; it had not been corrected for radial dilution.

In electrophoresis there is a parallel anomaly which arises from changes in the concentrations of buffer and protein ions across the moving boundary to produce differences of electrical conductivity. It had been treated in a similar way, in a rectangular channel to be sure and as required, prior to the Johnston-Ogston analysis.

There are several procedures by which the analysis of unknown mixtures may be conducted. The most common one of them, already suggested, involves conventional analyses of the areas ·under the two boundary gradient portions of the sedimentation patterns, with progressive dilution of the total solute. The values for the apparent concentrations of the slower component, plotted against total solute concentration, may be then extrapolated to infinite dilution of total solute to eliminate the effect of the concentration dependence of $s_1$ as it moves in the $s_2$ boundary. Trautman et al. (155) have presented a more sophisticated analytical procedure. The measurement of the sedimentation coefficient $s_{1,2}$ presents a number of special experimental problems. For instance, if the sedimentation coefficient of the faster component is lowered in the presence of the slower component there will be an accumulation in the boundary which gives rise to density inversion and convective disturbance.

Obviously, it is often of value to be able to make an analysis of the composition of a mixture of inert discrete solutes and the sedimentation pattern can give much valuable information about the presence of impurities as *viruses, enzymes, plant proteins*, etc., are passed through various stages of preparation. The patterns also provide the means to obtain the amounts of macromolecular reactants in an equilibrated system, as we shall see.

## 5. Protein-Protein Interactions.

To this point in the review much of the discussion of the several problems in sedimentation analysis has been based on solutions to the simple continuity equation (6), written for the two-component, idealized system. Often, however, chemical reactions involving macromolecules occur and it is necessary to give consideration to the effect of their re-equilibration in the boundary layers on the spreading and form of the boundary gradient curve. When studying systems in which a) monomer-polymer equilibria, b) bimolecular associations, and c) interconversions of isomers take place, ways must be found to extend the fundamental theory for measuring concentrations and sedimentation coefficients. First of all, one may seek to determine the composition of each complex and the equilibrium constants for the reactions. Then, too, the theory for the chemically reacting systems is important in the analysis of mixtures because if the existence of reactions is ignored, false constructions and conclusions will result.

We note that the corresponding problems in diffusion (56) and in electrophoresis (23, 82) with simultaneous chemical reactions in the boundary have been discussed by a number of workers. The parallelism between the extended theory for electrophoresis and velocity sedimentation is quite close, and the experiments are often complementary.

In writing the theoretical equations the guiding principle is the fact that the law of the conservation of mass must continue to apply to the sum of all the forms in which any given component exists. Thus, the simpler concentrations and sedimentation coefficients of before are now replaced by the corresponding constituent quantities (151). If a constituent $A$ exists in several forms, $n$ in number, and using a bar over the symbol to designate a constituent quantity, the more complex equations take the following forms (162):

$$\bar{J}_A = \sum_1^n (J_A)_i = \bar{c}_A \, \bar{s}_A \, \omega^2 \, x - \bar{D}_A \, (\partial \bar{c}_A / \partial x)_t \tag{43}$$

$$\bar{c}_A = \sum_1^n (c_A)_i \tag{44}$$

$$\bar{s}_A = \frac{\sum_1^n (c_A)_i \, (s_A)_i}{\bar{c}_A} \tag{45}$$

$$\bar{D}_A = \frac{\sum_1^n (\partial c_A / \partial x)_i \, (D_A)_i}{\partial \bar{c}_A / \partial x} \tag{46}$$

$$\left(\frac{\partial \bar{c}_A}{\partial t}\right)_x = -\frac{1}{x}\left[\frac{\partial(x \, \bar{J}_A)}{\partial x}\right]_t \tag{47}$$

$$(x_b{}^2)_A = \frac{\int_{x_1}^{x_2} x^2 \, (\partial \bar{c}_A / \partial x) \, dx}{(\bar{c}_A)_{x_2} - (\bar{c}_A)_{x_1}} \tag{48}$$

If the constituent $A$ disappears in moving across the boundary the constituent sedimentation coefficient, $\bar{s}_A$, is given by the rate of movement of the boundary position $(x_b)_A$, as defined by this last equation. It is the values of $\bar{s}_A$ which provide one method to study complex formation. Some of the relationships derive from those of SMITH and BRIGGS (137) and of ALBERTY and MARVIN (2) for their electrophoretic investigations of the interaction between methyl orange and bovine serum albumin; LONGSWORTH and JACOBSEN (83) had made earlier observations of this interaction. The treatments of SCHACHMAN (124) and WILLIAMS et al. (162) are quite like those of ALBERTY and MARVIN, except that the problem is discussed in relation to the sedimentation velocity experiment.

### a) Monomer-Polymer Equilibria.

For the time being it would appear to be most practical to contemplate what happens to the shape of a reaction boundary when a rapid re-equilibration reaction takes place within it. Let us assume a *protein* in solution as polymer $P$, in rapidly reversible equilibrium with its monomer $M$, allowing the letters to represent concentrations. Since the polymer has the higher sedimentation coefficient the proportion of monomer in the dilute trailing portion of the boundary is greater than it is in the more concentrated leading layers, and the boundary broadens abnormally as it moves away from the center of rotation. To develop anything like a complete mathematical description of the several situations which may arise is out of the question at this time, but progress in this direction is being made.

For this case of sufficiently rapid reactions, there have been several theoretical analyses (43, 47-49, 139); the papers of GILBERT have attracted the most attention. His treatment is based on an analogy between the sedimentation or electrophoresis of associating molecules and chromatography. The equilibrium constant is defined by the equation

$$KP = M^n$$

where $n =$ number of molecules of monomer in the polymer. Also, let $c = P + M$ in grams per unit volume. The quantities $M$ and $P$ are thus the concentrations by weight of monomer and polymer. GILBERT derived an equation to describe the variation of concentration $(P + M)$ of the substance through the boundary, finding,

$$c = (P + M) = \left[\frac{K}{n} \cdot \frac{\delta}{1 - \delta}\right]^{\frac{1}{n-1}} \left\{1 + \frac{1}{n} \cdot \frac{\delta}{1 - \delta}\right\} \qquad (49)$$

for the form of the boundary curve when $c$ is plotted vs. $x$. In this expression

$$\frac{\delta}{1 - \delta} = \frac{n M^{n-1}}{K} \qquad (49a)$$

with $\delta$, a coordinate transformation, being defined as

$$\delta = \frac{(x/t) - v_M}{v_P - v_M} \tag{49b}$$

The velocities $v_P$ and $v_M$ are related to the corresponding flows by the equations

$$J_P = P\,v_P \tag{50}$$

$$J_M = M\,v_M \tag{51}$$

With the usual schlieren optical systems the quantity obtained is $\partial(P + M)/\partial x$; it is thus necessary to differentiate equation (49) to obtain the expression which describes the outline of the boundary gradient curves*. In the derivation, the effects of diffusion have been neglected and the sedimentation coefficients for each species are assumed constant in order to simplify the mathematical apparatus.

GILBERT has drawn several patterns, $d(P + M)/dx$ vs. $x$, for different values of $n$ to compare them with some actual cases. He was especially interested to explain the observations of MASSEY, HARRINGTON and HARTLEY (90) for the polymerization of $\alpha$-chymotrypsin, and found that with his equation and using the value $n = 6$ he could reproduce quite well the form of the experimental boundary gradient curves and also the variation of sedimentation coefficient with protein concentration.

---

* By analogy with equations used in chromatography theory and with rapid re-equilibration Gilbert begins his derivation with the equation

$$dM + dP + v\left(\frac{\partial P}{\partial x}\right)_t = 0$$

In it the polymer of concentration $P$ sediments with a velocity $v$ relative to stationary monomer, it having been imagined that the monomer molecules are halted by giving the system an equal and opposite velocity to them. We prefer to write

$$\left(\frac{\partial M}{\partial t}\right)_x + v_M\left(\frac{\partial M}{\partial x}\right)_t + \left(\frac{\partial P}{\partial t}\right)_x + v_P\left(\frac{\partial P}{\partial x}\right)_t = 0$$

Now, making use of the definition of the equilibrium constant, it can be shown that

$$\frac{dx}{dt}\left[1 + \frac{n\,M^{n-1}}{K}\right] = v_M + v_P\left(\frac{n\,M^{n-1}}{K}\right)$$

The velocity is really $(\partial x/\partial t)_M = -(\partial M/\partial t)_x/(\partial M/\partial x)_t$, or the rate of movement of a layer of *constant* monomer concentration in the boundary.

In another similar analysis, one might cause to be involved a velocity $(\partial x/\partial t)_c$, which would refer to the rate of movement of a layer of *constant* $(P + M)$ in the boundary, a more graphic quantity. The expression (49) purports to provide the mathematical form of the boundary curve, but the result has not been obtained through the customary route of arriving at a particular solution of the Lamm differential equation (17) of the ultracentrifuge.

He also considered the dimerization of chymotrypsin; his idealized schlieren pattern for a solution containing equal weights of monomer and dimer was asymmetrical in form, with a trailing edge towards the meniscus, much as the experiments of Neurath and Dreyer (97) have indicated.

It is of interest that when $n \geq 3$, the concentration gradient curve shows two maxima in the boundary region. The outline does not touch the base-line, so the reaction case can be distinguished, in principle at least, from that of two independent and non-reacting solutes.

There are many proteins which are known to undergo association-dissociation reactions. Among them are *arachin, insulin, casein, hemoglobin* and *β-lactoglobulin*. Of these, the β-lactoglobulin has received especially detailed study in the hands both of Ogston and Tilley (103) and of Townend and Timasheff (149). By comparing the results of sedimentation and electrophoresis experiments the former demonstrated the presence of a polymerization reaction at pH 4.65. Townend and Timasheff later established the pH interval 3.5 to 5.25 to be the region of association. After the observation of Aschaffenburg and Drewry (6) that the β-lactoglobulin as usually prepared is made up of two genetically different proteins A and B, it could be shown that the β-lactoglobulin A is more susceptible to association than is the B variety (104, 149, 154).

Using the Gilbert equation (49) with $n = 4$, a datum obtained by light-scattering and the observation that the re-equilibration is rapid, Townend and Timasheff computed an equilibrium constant and the standard free energy of association for the reaction

$$4 \, \beta^A \rightleftarrows \beta_4{}^A$$

Light-scattering data, observations of the concentration dependence of sedimentation coefficient and electrophoretic patterns are now all brought into reasonable coherence with these equilibrium constant data.

### b) Bimolecular Associations.

As before, the continuity equation forms the point of origin for the analysis, this time by Gilbert and Jenkins (49), for the situation when reactions of the type $A + B \rightleftharpoons C$ occur and are re-equilibrated in a sedimentation boundary. Their starting equation is based upon the conservation of mass:

$$\partial a/\partial t + v_A \, (\partial a/\partial x) = \partial b/\partial t + v_B \, (\partial b/\partial x) = - \, \partial c/\partial t - v_C \, (\partial c/\partial x) \quad (52)$$

where $v_A$, $v_B$ and $v_C$ are the velocities of the constituents A, B and C, and $a$, $b$ and $c$ are the corresponding molar concentrations at the position $x$ in the cell as modified from the values $a_0$, $b_0$ and $c_0$ in the unchanged body of the solution.

With the assumption of ideal sedimentation and in the rectangular cell these partial differential equations are solved, but only the results are set down. In order to give an indication of their use the shape of a sedimenting reaction boundary has been worked out for an arbitrary example, the purpose being to show how profoundly the form of the $c$ vs. $x$ and $dc/dx$ vs. $x$ curves are modified by the re-equilibration. It is made clear that any interpretation of the form of these curves which neglects the effect of the reaction in the boundary will provide an improper estimate of the nature of the complex and of the concentrations of the unreacted molecules.

It is highly desirable that theoretical researches of this kind be extended because there are a number of problems of interest to the biologist which conform to this reaction type and for which the sedimentation velocity and electrophoresis methods had seemed to offer great possibilities of investigation. Reactions involving macro-molecules such as the combination of *proteolytic enzymes* with inhibitors and the interaction of *antigens* with their *antibodies* have been observed in these ways, but the resolution of the components at the boundaries between solution and solvent is disarranged by the re-equilibrations as the reactants and products tend to be separated by the force of the applied external field.

Fortunately, there are some cases in which it can be demonstrated that the effects of the re-equilibration are not too significant and progress can be made with them. One such area of research is the determination of the "valence" of precipitating antibodies and the determination of the equilibrium constants of the several reactions which are involved. Following the early studies of HEIDELBERGER and PEDERSEN (59) and PAPPENHEIMER, LUNDGREN and WILLIAMS (109) a number of investigators have applied the techniques of sedimentation and electrophoretic analysis to the problem of antibody valence (17, 89, 107, 131–136). Then, with the aid of the theory of GOLDBERG (52, 54) for reactions of multivalent antigen with bivalent and univalent antibody, possibilities to compute the several equilibrium constants are opened up. The investigations of PAPPENHEIMER et al. with *diphtheria toxin*, G, and its equine antibody, A, were sufficient in extent and detail to demonstrate that the empirical composition of the dissolved reaction complex in the region of toxin excess varied from AG to $AG_2$, approaching the latter value for high antigen excess. In other words, the "valence" of the precipitating antibody is two. This result was based entirely upon sedimentation analysis.

It is one of the difficulties of the ultracentrifugal analysis that the total protein concentration of the solutions studied, as calculated from the area under the sedimentation diagrams, is generally less than the usual analytical value. This has been assumed to be due to the fact that large complexes of very high sedimentation coefficients are soon removed from solution. However, the sedimentation diagrams do demonstrate the existence of several distinct antigen-antibody complexes which are not resolved on electrophoresis. It is significant that the relative amounts of such higher satellites decrease with increase in percentage of free antigen until eventually a single complex corresponding to the composition $AG_2$ remains. It seems to be established that the composition of the several complexes, perhaps $AG_2$, $A_2G_3$, $A_3G_4$, depends upon the amounts of antigen and antibody in the system and that they are in equilibrium with each other and with the free antigen in the system. With the GOLDBERG theory one may make estimates of the concentrations of each of these complexes, thus providing data for comparison with direct ultracentrifugal analysis.

The more rapidly moving components must be relatively richer in antibody. As they move through the cell molecules of the lighter complexes are left behind, and there must be a continual readjustment of the several equilibria. Assuming the interactions to be reversible, there is as yet no information about the relative magnitudes of the velocity constants for the forward and backward reactions. If both are small, finite concentrations of the several components will exist at equilibrium since the adjustment of the equilibrium is slow in comparison with the rate of resolution. On the other hand, if both reaction rates are high and of the same general magnitude, the equilibrium would be adjusted as rapidly as required by the resolution. In the evaluation of the sedimentation experiments (and ascending boundary electrophoresis experiments, which correspond) resolution has been quite good, and it is customary to take the relative area under the free antigen portion of the diagram to measure the percentage of this component. Perhaps it can be said that the experiments have been performed in a region of sufficiently high antigen excess so that the effects of the re-equilibration have become negligible.

Having determined the amount of the free antigen in the system, after correction for the Johnston-Ogston boundary anomaly, the extent of the reaction may be computed by using the GOLDBERG theory (52). In the derivation of the general expression, $m_{ijk}$, for the number of aggregates each of which is composed of $i$ bivalent antibody molecules, $j$ univalent antibody molecules and $k$ antigen molecules, it is assumed among other things that any unreacted site is as reactive as any other site regardless of the size or shape of the aggregate to which it is attached. There is some evidence that *rabbit serum antibody* molecules have two like and equivalent reactive groups, while the antibody molecules formed in *horse serum* may have two unlike valences. Another assumption which may require further consideration is that of the complete solubility of protein complexes which are less than infinite in size, especially as one may elect to vary the ionic strength of the reaction medium.

The equation (18) of GOLDBERG for $m_{ijk}$ when *written for bivalent antibodies only*, becomes

$$m_{ijk} = m_{i0k} = m_{ik} =$$
$$= \frac{fG\,(f\,k - k)!}{(f\,k - 2\,k + 2 - q)!\,k!\,q!} \left(\frac{fG}{2\,A}\right)^{k-1} p^{k+i-1}\,(1-p)^{fk-k-i+1}\left(1 - p\,\frac{fG}{2\,A}\right)^{i-k+1}$$
(53)

where $f$ is the valence of the antigen $\simeq 4$ to $8$, $G$ is the total number of antigen molecules in the system, $A$ is the total number of antibody molecules in the system, $p$ is the extent of the reaction (fraction of antigen sites in the system which have reacted), $q$ is the number of free antibody sites on an aggregate, and $m_{ik}$ is the number of aggregates in the system, each of which is composed of $i$ bivalent antibody molecules and $k$ antigen molecules. ($j = 0$, because monovalent antibody is assumed to be absent.)

By making the substitutions $k = 1$, $i = 0$ and $q = 0$ there is obtained the expression for the concentration of free antigen

$$c_{01} = c_G\,(1-p)^f$$
(53a)

where $c_G$ is the total antigen concentration and $c_{01}$ is the free antigen concentration, both measured in grams per liter. The sedimentation analysis gives $c_{01}$, and $c_G$ is known, so the extent of the reaction, $p$, is made directly available. Again, subject to the limitations of the theory, one may now use the general formula to compute the other concentrations, grams or moles (if A and G molecular weights are known) per liter of AG complex, $AG_2$ complex, etc. Having obtained the individual concentrations it is standard procedure to calculate the equilibrium constants and the standard free energies of the several reactions.

For the reaction $G + AG \rightleftharpoons AG_2$ and using equation (53) the equilibrium constant in terms of concentrations in moles per liter can be shown to be

$$K = \frac{f \cdot M_A \cdot p}{4\,c_A\,(1-p)\left(1 - \frac{p\,f}{2} \cdot \frac{c_G}{c_A} \cdot \frac{M_A}{M_G}\right)}$$
(54)

For the system *bovine serum albumin* and its rabbit antibody SINGER and CAMPBELL (*133*) have determined the equilibrium constant, $K$, to be $2.5 \pm 0.5 \times 10^4$ for this reaction at $0°$ C in veronal buffer at pH $= 8.5$. Electrophoresis rather than sedimentation analysis was used in this experiment. It has become more common to make use of the electrophoresis experiment wherever possible, but this is not necessary. Indeed there are situations when the electrophoretic mobilities of antigen and antibody are so nearly alike that one must resort to sedimentation experiments.

While it has been amply demonstrated that the precipitating rabbit antibodies to bovine serum albumin (BSA) and ovalbumin antigens are largely, if not all, bivalent the situation with respect to the chicken serum antibodies is as yet uncertain. BANOVITZ et al. (17) carried out physical chemical studies, sedimentation and electrophoresis, on complexes of BSA and its chicken antibodies with the apparent result that antigen and antibody combined in the mole ratio of unity, actually about 0.9. More recently MAKINODAN and associates (86) have demonstrated that there is a normal serum macroglobulin which co-precipitates with the BSA-chicken anti-BSA aggregates. The presumption is that when correction is made for the presence of this inert material in the aggregates the antibody "valence" will be more in line with the figure two.

Often complexing agents are used to modify the solubility of a protein (148). A good example is the use of protamine binding to help to keep *insulin* in solution. During the war years attempts were made by us to observe this binding by ultracentrifugal analysis, but the pressure of other tasks prevented any definitive study. As a by-product of this research it was found (96) that insulin in acid solution exists largely in units of molecular weight 12,000 in mobile equilibrium with at least an association product of molecular weight 36,000. This equilibrium reaction has since become a popular subject of investigation, but there are still many points about it which require clarification. As we found, the equilibrium is very sensitive to both pH and ionic strength (cf. 107). Any discussion of these several equilibria would properly belong in the previous sub-section of this Chapter III, were it to be presented.

### c) Interconversion of Isomers.

CANN, KIRKWOOD and BROWN (27) have made a theoretical study of the electrophoretic behavior of systems, rectangular cell and constant applied field, containing a protein which can exist in two isomeric states; the subject is reviewed both by LONGSWORTH (82) and by BROWN and TIMASHEFF (23). In these places it is shown how the half-time of the reaction determines the resolution of the boundaries, thus to provide criteria for the recognition of boundary outline when a re-equilibration of isomers is taking place. The equations could be rewritten for the sedimentation-velocity experiment without too great difficulty.

If the reaction is slow, and if the sedimentation coefficients for the two isomeric molecules are sufficiently different, two moving boundaries will be observed experimentally. For the special case in which the rates of forward and backward reactions are equal, the computations of CANN et al. show that two maxima will be resolved in the boundary gradient curve when the duration of the sedimentation experiment is less than the half-time of the reaction. Although the apparent amounts of the two forms do not depend upon the degree of resolution, the apparent sedimentation coefficients will depend upon the half-time of the reaction and on the time of sedimentation. If the rates of reaction are negligible the usual procedures for the analysis of two solutes apply. When the rates are instantaneous the system behaves as if it contained but one solute.

Although such interconversions of isomers are presumed to be of significance in connection with the mechanisms of the enzymatic degradations of protein substrates, little is really known about their nature (76, 77, 100).

# IV. Three-Component Systems.

## 1. Solvation.

It has been already indicated in Chapter I that we believe the use of a mixed solvent in sedimentation experiments should be avoided wherever possible. The flow equation in general use (14) is really restricted to use with the two-component, idealized system, but it is a good approximation for the macro-ion, buffer and salt, water system of the protein chemist when the electrical potential gradient, $\partial \psi / \partial x$, is made negligible. In this instance the relative amounts both of the electrolyte and the protein are small.

The system* solvent (0)—macromolecular solute (1)—solvent (2) does require consideration because there are occasions when a mixed solvent is required to bring a *protein* into solution and it is also the system adopted by Meselson, Stahl and Vinograd (94) for their zone-sedimentation experiment. These situations differ from that of the traditional protein solution in that both solvents are present in substantial amounts. Now the use of the mixed solvent system is unsatisfactory if it is the aim of the experiment to determine the solute molecular weight, owing to the probability of selective binding of one of the solvent components by the macromolecule. The literature of this subject may be said to have begun with articles published by McBain (91) and Lansing and Kraemer (72) in 1936.

In our sketch we resort, first of all, to equations which have to do with sedimentation equilibrium. For the two-component, neutral molecule system the differential equation descriptive of the equilibrium may be put in the form

$$\frac{M_1 \left(1 - \bar{v}_1 \varrho\right) \omega^2 x \, c_1}{RT \left(1 + c_1 \left[\partial \ln y_1 / \partial c_1\right]\right)} = \frac{dc}{dx} \tag{55}$$

The solute component may be considered as being either unsolvated or solvated. In the second instance, the component 1,0 is defined as having molecular weight $M_{1,0} = M_1 + \alpha M_0$, where $\alpha$ represents the number of moles of solvent bound per mole of solute. However, $c_{1,0} = c_1$, and $dc_{1,0}/dx = dc_1/dx$. Therefore,

$$\frac{M_1 \left(1 - \bar{v}_1 \varrho\right)}{1 + c_1 \dfrac{\partial \ln y_1}{\partial c_1}} = \frac{M_{1,0} \left(1 - \bar{v}_{1,0} \varrho\right)}{1 + c_{1,0} \dfrac{\partial \ln y_{1,0}}{\partial c_{1,0}}} \tag{55a}$$

---

* In Chapter III, Section 4, the discussion was devoted to what were called "two-solute" systems. The term "three-component systems" is all inclusive; in making the separation convenience is the only criterion.

Thus, if $\bar{v}$ is measured for the solvated solute,

$$\lim_{c_1 \to 0} M_1 (1 - \bar{v}_1 \varrho) = M_{1,0} (1 - \bar{v}_{1,0} \varrho) \tag{55b}$$

and the true molecular weight of the unsolvated material is to be obtained with proper evaluation of the experiment.

The equations descriptive of sedimentation equilibrium for an $n$-component system, again of neutral molecules, have been written in compact form by GOLDBERG (53). It is assumed that the density of the solution and the partial specific volumes of all components are independent of composition and pressure. Now, instead of the single equation (55), a set of $n - 1$ independent equations is obtained. In terms of the concentration and activity coefficient of the $i^{\text{th}}$ component,

$$\frac{M_i (1 - \bar{v}_i \varrho) \omega^2 x c_i}{RT} = \frac{dc_i}{dx} + c_i \sum_{k=1}^{n-1} \left( \frac{\partial \ln y_i}{\partial c_k} \right) \frac{dc_k}{dx} \tag{56}$$

So, for the three-component system, we write for the solute

$$\frac{M_1 (1 - \bar{v}_1 \varrho) \omega^2 x c_1}{RT} = \frac{dc_1}{dx} + c_1 \left( \frac{\partial \ln y_1}{\partial c_1} \right) \frac{dc_1}{dx} + c_1 \left( \frac{\partial \ln y_1}{\partial c_2} \right) \frac{dc_2}{dx} \tag{56a}$$

or

$$\frac{M_1 (1 - \bar{v}_1 \varrho) \omega^2 x}{RT} \cdot \frac{c_1}{dc_1/dx} = 1 + c_1 \frac{\partial \ln y_1}{\partial c_1} + \frac{c_1}{dc_1/dx} \cdot \frac{dc_2}{dx} \left( \frac{\partial \ln y_1}{\partial c_2} \right) \tag{56b}$$

Now, as $c_1 \to 0$, the quantity $c_1/(dc_1/dx)$ is finite. The right hand side of the equation then becomes

$$1 + \left( \frac{c_1}{dc_1/dx} \right)_0 \frac{dc_2}{dx} \left( \frac{\partial \ln y_1}{\partial c_2} \right)$$

For the two-component system it would be simply unity, with all quantities required being readily accessible from the experiment. It is now the more apparent why the use of the mixed solvent system is to be avoided if possible.

This discussion is not irrelevant to the sedimentation velocity problem. Reference has been already made to the fact that the Baldwin expression, to appear as equation (60), for the sedimentation coefficient, $s_1$, in the three-component system, bears a strong resemblance to equation (56a), as it should; it has been obtained by using the thermodynamics of irreversible processes. Also, those who are familiar with the Archibald equations which are descriptive of the transient state during the approach to sedimentation equilibrium will recognize equation (26) in the modification

$$\frac{s_1 \omega^2 x}{D} = \frac{M (1 - \bar{v}_1 \varrho) \omega^2 x}{RT} \tag{57}$$

The McBAIN article (91) purported to demonstrate that solvation affects in a different manner molecular weight data determined by the sedimentation methods as compared to those obtained by the more classical methods of physical chemistry, such as osmotic pressure. The very title of the communication, "The Determination of Bound Water by Means of the Ultracentrifuge", was undoubtedly intended to attract the attention of students to the subject; in this success was achieved. The quick reply by LANSING and KRAEMER (72) has stood the test of time to a far greater extent, even though it is not without fault. In it there was demonstrated the fact that in binary solutions the change in partial specific volume resulting from solvation does *not* affect the molecular weight determined by sedimentation equilibrium in any way which is different from that for conventional methods [cf. also (15)]. There was also clear indication and delineation of the misconceptions in the McBAIN remarks concerning "bound water" and its determination in mixed solvent systems, but they were not always heeded.

*References, pp. 495—502.*

The now extensive literature descriptive of later attempts to measure the amount of water bound per gram of protein has been reviewed in another place (74). Suffice it to remark that the "hydrated densities" of a number of *proteins* have been evaluated by determining the density of a solution, water plus a second solvent component such as glycerol, sucrose or urea, in which the protein particles neither sediment nor float. The density of the mixed solvent corresponding to zero sedimentation coefficient was then determined by extrapolation procedures, but it became eventually apparent that this density for a given macromolecule is a function of the nature of the second solvent component. Thus for *tobacco mosaic virus protein* in sucrose solutions SCHACHMAN and LAUFFER (125, 126) found the solvent density necessary to prevent sedimentation to be 1.27 g./ml., but when the second solvent was changed to serum albumin this density decreased to the value 1.13 g./ml.

The earlier theoretical attempts to treat the subject were based upon a mechanistic picture or kinetic theory approach in which attempts were made to define the components of the system in terms of the species as they existed in solution, for consideration as hydrodynamic units. Thus, investigators started with a model and relied on the assumption that all interactions between component molecules in the system could be accounted for by this degree of solvation. In the more successful thermodynamic treatments of today, all such interactions, of which solvation and aggregation may be a part, manifest themselves as activity coefficient terms in expressions for the chemical potentials, $(RT/M_i) \ln y_i$ [cf. equation (9)]. To provide a thermodynamic measure of the solvation in a sedimentation equilibrium experiment, use is made of a number defined by the expression (157, 162)

$$\Gamma = - \frac{(\partial \mu_2/\partial m_1)_{m_2, P}}{(\partial \mu_2/\partial m_2)_{m_1, P}} = \left( \frac{\partial m_2}{\partial m_1} \right)_{\mu_2, P} \tag{58}$$

where the concentrations are now expressed in molalities. This number can be considered as a measure of the *relative* solvation of the macromolecules by the solvent components 2 and 0; it is a precise and definitive number and it is a thermodynamic quantity.

For the sedimentation velocity case more theoretical work appears to be required before an answer can be given to the question as to whether the quantity discussed by SCHACHMAN (124),

$$\alpha = - \frac{M_1 (1 - \bar{v}_1 \varrho^0)}{M_2 (1 - \bar{v}_2 \varrho^0)} \tag{59}$$

is equivalent to our $\Gamma$. In the expression, $\varrho^0$ is the density of mixed solvent for which no sedimentation flow of macromolecules takes place. In the SCHACHMAN monograph the identical symbol $\alpha$ is used for the two quantities, and it is presumed from the text they are meant to be identified. To look a bit further into the situation we at last write down the Baldwin equation for the sedimentation coefficient of a neutral macromolecular solute in the mixed solvent, neutral components 0 and 2.

$$\lim_{c_1 \to 0} s_1 = \frac{M_1 (1 - \bar{v}_1 \varrho) D_{11}}{RT} \cdot$$

$$\cdot \left\{ 1 + c_2 \left[ \frac{M_2 (1 - \bar{v}_2 \varrho)}{M_1 (1 - \bar{v}_1 \varrho)} \right] \cdot \left[ \frac{\frac{1}{D_{11}} \left( \frac{\partial D_{12}}{\partial c_1} \right)_{T, P, c_2} - \left( \frac{\partial \ln y_1}{\partial c_2} \right)_{T, P, c_1}}{1 + c_2 \left( \frac{\partial \ln y_2}{\partial c_2} \right)_{T, P, c_1}} \right] \right\} \tag{60}$$

The symbol $D_{11}$ represents the main macromolecular diffusion coefficient and the coefficient $D_{12}$ is a cross-term diffusion coefficient, macromolecule and solvent 2. Perhaps one might call $D_{12}$ a kinetic coupling term. Only in the event that the slope represented by the ratio $\partial D_{12}/\partial c_1$ goes to zero as the concentration of macromolecule approaches infinite dilution could the two quantities $\Gamma$ and $\alpha$ become equal. Although $D_{12}$ must go to zero as $c_1 \to 0$ this does not mean that the partial derivative, $\partial D_{12}/\partial c_1$, must vanish under this condition. Little if anything is known experimentally about the magnitude of this differential coefficient, but the data of O'DONNELL and GOSTING (101) about the interaction of flows in diffusion suggest that it will not vanish. There might be experimental conditions under which $(1/D_{11})(\partial D_{12}/\partial c_1)$ would be very small compared to $(\partial \ln y_1/\partial c_2)_{\Gamma, P, c_1}$ but we do not know as yet what they are. Even in this event and when $s_1$ is set equal to zero in equation (60), the transformation of the terms involving activity coefficients from the "$c$" concentration scale to the "$m$" molality scale of equation (59) is a formidable task.

The problems in transport are more involved, and while arguments such as those of SCHACHMAN (124) about them may be persuasive they are sometimes incomplete in logic. As a concluding sentence and as PELLER (113) has written, the evaluation of "binding" data from sedimentation velocity experiments in ternary systems without reference to diffusion measurements may be tenuous.

## 2. Sedimentation Equilibrium in a Density Gradient.

Although the use of the ternary system necessitates cautious interpretation, the MESELSON et al. (94) experiment which has to do with density gradient sedimentation is full of fascination and worthy of discussion. It was originally intended to be a new method for the study of the molecular weight and partial specific volume of macromolecules. Involved are observations of the equilibrium distribution of macromolecular material in a density gradient, itself at equilibrium. The density gradient is established by the sedimentation of a low-molecular weight second solvent (as we use the term) in a solution subject to a constant centrifugal field. The experiment answers the description of zone sedimentation, with the mixed solvent being selected which has a density comparable to $1/\bar{v}_1$ for the solute, so that the latter will form a band near the center of the cell.

In their derivation of the basic equations, MESELSON et al. have assumed that a) deviations due to thermodynamic non-ideality may be neglected, b) the solution is incompressible and partial specific volumes are constant, independent of $x$, c) concentration of the second solvent, in this case cesium chloride, and solution density are linear functions of distance over short distances in the cell.

With these assumptions, and using a simplified notation, the concentration distribution of the macromolecules at equilibrium in the constant density gradient is*

$$c_1 = c_1^0 \, e^{-(x-x_0)^2/2\,\sigma^2} \tag{61}$$

* A simplified derivation of the MESELSON type is here given. Since terms involving thermodynamic non-ideality are neglected, equation (55) may be put in the form

$$\frac{dc_1}{c_1} = \frac{M_1(1 - \bar{v}_1\,\varrho)\,\omega^2\,x\,dx}{RT}$$

It is further assumed that solution density is linear in distance, so

$$\varrho = \varrho_{x_0} + \left(\frac{\partial\varrho}{\partial x}\right)_{x_0}(x - x_0)$$

From the observed value of the standard deviation $\sigma^2$ of the experimental Gaussian error equilibrium distribution of macromolecules in the band, the solute molecular weight may be computed from the equation

$$M_1 = \frac{RT}{\omega^2 \, x_0 \, \bar{v}_1 \, (d\varrho/dx) \, \sigma^2} \tag{62}$$

In these equations, $x_0$ represents the position of rest for the macromolecular component such that $(1 - \bar{v}_1 \varrho_{x_0}) = 0$. The density of the solution at $x$ is $\varrho$, and its rate of change with distance in the region of the band is $d\varrho/dx$, actually by assumption. The partial specific volume of a neutral macromolecule then becomes equal to the reciprocal of the solution density at the position $x_0$.

$$\frac{1}{\varrho_{x_0}} = \bar{v}_1 \tag{63}$$

The band width, as measured by the standard deviation of the "Gaussian" band, is inversely proportional to the square root of the molecular weight, $M_1$. The corresponding linear plot, $\log c$ vs. $(x - x_0)^2$, has been used as a test for homogeneity, and $M_1$ has been computed from the slope of this line. In addition to MESELSON et al., YEANDLE (163) has considered the question of charge effects caused by the dissociation of a macro-electrolyte solute into ions; both parties have concluded that the shape of the band continues to be Gaussian.

When the effects of preferential interaction are present, solute for one of the solvents, a different significance must be assigned to $\sigma^2$ and to $\varrho_{x_0}$, even though equation (61) retains its mathematical form. Now, using a binding coefficient, $\Gamma'$, (157, 162)

$$\frac{1}{\varrho_{x_0}} = \frac{\bar{v}_1 + \Gamma' \, \bar{v}_2}{1 + \Gamma'} \tag{64}$$

where $\Gamma' = \Gamma(M_2/M_1)$. The coefficient $\Gamma$ is given by equation (58). It is multiplied by $(M_2/M_1)$ in order to use a gram-per-gram rather than a mole-per-mole basis. If $\Gamma'$ grams of component 2 and none of component 0 are bound to each gram of macromolecule 1 to produce a complex component, the partial specific volume of this complex becomes

$$\bar{v}_{1,2} = \frac{\bar{v}_1 + \Gamma' \, \bar{v}_2}{1 + \Gamma'} = \frac{1}{\varrho_{x_0}} \tag{65}$$

So, the partial specific volume obtained in the experiment must refer to the complex which we call $M_{1,2}$. Since $\bar{v}_{1,2}$ may be quite different from $\bar{v}_1$, the molecular weight computed with equation (62) may be appreciably affected. The band for a single solute is still Gaussian in outline, but its position in the cell and its width are altered.

Another difficulty is not unrelated. In addition to the spreading of the band due to diffusion (gradient of chemical potential in the boundary) there may be

and, at the point in the cell where $(1 - \bar{v}_1 \varrho_{x_0}) = 0$, $\omega^2 \, x = \omega^2 \, x_0$. Then,

$$RT \frac{dc_1}{c_1} = M_1 \left[ 1 - \bar{v}_1 \varrho_{x_0} - \bar{v}_1 \left( \frac{\partial \varrho}{\partial x} \right)_{x_0} (x - x_0) \right] \omega^2 \, x_0 \, dx$$

On integration,

$$\ln \frac{c_1}{c_1^0} = -\frac{M_1 \, \omega^2 \, x_0 \, \bar{v}_1 \, (\partial \varrho/\partial x) \, \dfrac{(x - x_0)^2}{2}}{RT}$$

to provide equations (61) and (62).

another spreading due to a distribution of effective densities in the macromolecula,' solute. If this distribution of densities happens to be Gaussian, or for practical purposes nearly so, the macromolecular substance will again form a Gaussian band in the zone and thus appear to be homogeneous. The standard deviation of the resulting distribution becomes the sum of the two standard deviations, one for diffusion and one for the distribution of densities (12). Consequently, any molecular weight computed by using the enhanced band width will be less than the true value, often by a large amount.

The experiment has found spectacular success not as a means for moleculaɪ weight determination but as an instrument of high sensitivity in the separation of solutes of different effective densities. In this application Meselson and Stahl (93) were able to indicate the pathways by which DNA replicates in E. coli. Until this time radioisotopic labels had been employed in experiments bearing on the distribution of parental atoms among the progeny molecules (75, 147). Now, Meselson and Stahl have provided a further proof that nitrogen in a DNA molecule is divided equally between two sub-units within it and that each daughter molecule receives one sub-unit when DNA multiplies.

As starting material they used a DNA labeled with $N^{15}$ to impart to the molecule an increased density. By using the density gradient sedimentation it was possible to observe the distribution of the $N^{15}$ among molecules of DNA following the transfer of a uniformly $N^{15}$-labeled, exponentially growing bacterial population to a growth medium which contained only the ordinary $N^{14}$. The resolving power of density available in the experiment is such that bands of DNA fully-labeled, half-labeled and unlabeled are separated. For $N^{14}$-DNA and $N^{15}$-DNA the difference in buoyant density is 0.014 g./ml.; the separation of bands in this case is essentially complete. The sedimentation patterns were obtained by using the ultraviolet light absorption method.

The results may be summarized as follows: One generation time after the addition of $N^{14}$ only half-labeled molecules are present. When two generation times have elapsed after the addition of the $N^{14}$, half-labeled and unlabeled DNA are present in equal amounts. Thus, the observations are in exact accord with the expectation of the Watson-Crick (32) model for DNA duplication.

In Chapter III it is recorded that DNA samples are heterogeneous with respect to sedimentation coefficient and molecular weight. Very recently it has been shown (118, 141) that the molecular density of DNA is related to its base composition, and another heterogeneity is involved. The Baldwin analysis was based upon two assumptions, homogeneous molecular weight and Gaussian distribution of effective density. More recently, Sueoka (140) has attempted a statistical treatment of DNA distribution in the density gradient sedimentation experiment which will be general for both density and molecular weight heterogeneities. The theory is applied in the evaluation of experiments with sonicated molecules of calf thymus and pneumococcus DNA, with the results being given in the form of distributions both of sedimentation coefficient and of molecular weight.

The situation may be compared to the one already encountered, Chapter III, in the traditional sedimentation velocity experiment with the $\beta$-lipoproteins as solutes. Here were molecular weight and density heterogeneities to be resolved; effects of diffusion were considered to be negligible in order to simplify the problem. We believe the resolving power as regards density and molecular weight of the two experiments, the older and the newer, to be substantially the same. Important difficulties are met in the theory and practice of each and it remains to be seen which will become the more useful for the descriptions of the heterogeneities so often characteristic of biochemical materials.

*References, pp. 495—502.*

## References.

*1.* AKELEY, D. F. and L. J. GOSTING: Studies of the Diffusion of Mixed Solutes with the Gouy Diffusiometer. J. Amer. Chem. Soc. **75,** 5685 (1953).

*2.* ALBERTY, R. A. and H. H. MARVIN, Jr.: Protein-Ion Interaction by the Moving Boundary Method. Theory of the Method. J. Physic. Coll. Chem. **54,** 47 (1950).

*3.* ANFINSEN, C. B., R. R. REDFIELD, W. L. CHOATE, J. PAGE and W. R. CARROLL: Studies on the Gross Structure, Cross-Linkages, and Terminal Sequences in Ribonuclease. J. Biol. Chem. **207,** 201 (1954).

*4.* ARCHIBALD, W. J.: The Process of Diffusion in a Centrifugal Field of Force. Physic. Rev. **53,** 746 (1938).

*5.* — The Process of Diffusion in a Centrifugal Field of Force. II. Physic. Rev. **54,** 371 (1938).

*6.* ASCHAFFENBURG, R. and J. DREWRY: Occurrence of Different $\beta$-Lactoglobulins in Cow Milk. Nature (London) **176,** 218 (1955).

*7.* BALDWIN, R. L.: The Neurotoxin of Shigella Shigae. II. Examination of the Toxin in the Oil-Turbine Ultracentrifuge. Brit. J. exp. Pathol. **34,** 217 (1953).

*8.* — Boundary Spreading in Sedimentation Velocity Experiments. II. The Correction of Sedimentation Coefficient Distributions for the Dependence of Sedimentation Coefficient on Concentration. J. Amer. Chem. Soc. **76,** 402 (1954).

*9.* — Boundary Spreading in Sedimentation Velocity Experiments. III. Effects of Diffusion on the Measurement of Heterogeneity when Concentration Dependence is Absent. J. Physic. Chem. **58,** 1081 (1954).

*10.* — Boundary Spreading in Sedimentation Velocity Experiments. 5. Measurement of the Diffusion Coefficient of Bovine Albumin by Fujita's Equation. Biochemic. J. **65,** 503 (1957).

*11.* — Molecular Weights from Studies of Sedimentation and Diffusion in Three-Component Systems. J. Amer. Chem. Soc. **80,** 496 (1958).

*12.* — Equilibrium Sedimentation in a Density Gradient of Materials Having a Continuous Distribution of Effective Densities. Proc. Nat. Acad. Sci. (USA) **45,** 939 (1959).

*13.* — Boundary Spreading in Sedimentation Velocity Experiments. VI. A Better Method for Finding Distributions of Sedimentation Coefficient when the Effects of Diffusion are Large. J. Physic. Chem. **63,** 1570 (1959).

*14.* BALDWIN, R. L., L. J. GOSTING, J. W. WILLIAMS and R. A. ALBERTY: Transport Processes and the Heterogeneity of Proteins. Discuss. Faraday Soc. **20,** 13 (1955).

*15.* BALDWIN, R. L. and A. G. OGSTON: The Diffusion and Sedimentation Coefficients of a Liquid Two-Component System in Terms of Macroscopic Properties of the System. Trans. Faraday Soc. **50,** 749 (1954).

*16.* BALDWIN, R. L. and J. W. WILLIAMS: Boundary Spreading in Sedimentation Velocity Experiments. J. Amer. Chem. Soc. **72,** 4325 (1950).

*17.* BANOVITZ, J., S. J. SINGER and H. R. WOLFE: Precipitin Production in Chickens. XVIII. Physical Chemical Studies on Complexes of Bovine Serum Albumin and its Chicken Antibodies. J. Immunology **82,** 481 (1959).

*18.* BELL, D. J., H. GUTFREUND, R. CECIL and A. G. OGSTON: Physicochemical Observations of Some Glycogens. Biochemic. J. **42,** 405 (1948).

*19.* BOEDTKER, H. and P. DOTY: On the Nature of the Structural Element of Collagen. J. Amer. Chem. Soc. **77,** 248 (1955).

*20.* — — The Native and Denatured States of Soluble Collagen. J. Amer. Chem. Soc. **78,** 4267 (1956).

21. BRIDGMAN, W. B.: Some Physical Chemical Characteristics of Glycogen. J. Amer. Chem. Soc. 64, 2349 (1942).
22. BRIDGMAN, W. B. and J. W. WILLIAMS: Optical Problems of the Ultracentrifuge. Ann. New York Acad. Sci. 43, 195 (1942).
23. BROWN, R. A. and S. N. TIMASHEFF: Applications of Moving Boundary Electrophoresis to Protein Systems. In: M. BIER, Electrophoresis. Theory, Methods, and Applications, p. 317. New York and London: Academic Press. 1959.
24. BUTLER, J. A. V., D. J. R. LAURENCE, A. B. ROBINS and K. V. SHOOTER: Molecular Weights and Physical Properties of Deoxyribonucleic Acid. Nature (London) 180, 1340 (1957).
25. BUTLER, J. A. V., D. M. PHILLIPS and K. V. SHOOTER: Influence of Protein Heterogeneity of Deoxyribonucleic Acid (DNA). Arch. Biochem. Biophys. 71, 423 (1957).
26. BUZZELL, J. G. and C. TANFORD: The Effect of Charge and Ionic Strength on the Viscosity of Ribonuclease. J. Physic. Chem. 60, 1204 (1956).
27. CANN, J. R., J. G. KIRKWOOD and R. A. BROWN: Theory of Isomerization Equilibrium in Electrophoresis. I. Arch. Biochem. Biophys. 72, 37 (1957).
28. CECIL, R. and A. G. OGSTON: The Accuracy of the Svedberg Oil-Turbine Ultracentrifuge. Biochemic. J. 43, 592 (1948).
29. — — The Sedimentation Constant, Diffusion Constant and Molecular Weight of Lactoglobulin. Biochemic. J. 44, 33 (1949).
30. CHARLWOOD, P. A.: Partial Specific Volumes of Proteins in Relation to Composition and Environment. J. Amer. Chem. Soc. 79, 776 (1957).
31. CREETH, J. M.: Studies of Free Diffusion in Liquids with the Rayleigh Method. III. The Analysis of Known Mixtures and Some Preliminary Investigations with Proteins. J. Physic. Chem. 62, 66 (1958).
32. CRICK, F. H. C. and J. D. WATSON: The Complementary Structure of Deoxyribonucleic Acid (DNA). Proc. Roy. Soc. (London) A 223, 80 (1954).
33. DAVISON, P. F.: The Effect of Hydrodynamic Shear on the Deoxyribonucleic Acid from $T_2$ and $T_4$ Bacteriophages. Proc. Nat. Acad. Sci. (USA) 45, 1560 (1959).
34. DAYHOFF, M. O., G. E. PERLMANN and D. A. MacINNES: The Partial Specific Volumes, in Aqueous Solution, of Three Proteins. J. Amer. Chem. Soc. 74, 2515 (1952).
35. DE GROOT, S. R., P. MAZUR and J. T. G. OVERBEEK: Nonequilibrium Thermodynamics of the Sedimentation Potential and Electrophoresis. J. Chem. Physics 20, 1825 (1952).
36. DE LALLA, O. F. and J. W. GOFMAN: Ultracentrifugal Analysis of Serum Lipoproteins. In: D. GLICK, Methods of Biochemical Analysis, Vol. 1, p. 459. New York: Interscience Publ., Inc. 1954.
37. DOTY, P., B. B. McGILL and S. A. RICE: The Properties of Sonic Fragments of Deoxyribose Nucleic Acid. Proc. Nat. Acad. Sci. (USA) 44, 432 (1958).
38. DUCLAUX, J.: Centrifuges et ultracentrifuges. Traité de Chimie Physique, No. 1228. Paris: Hermann et Cie. 1955.
39. EDSALL, J. T.: The Size, Shape and Hydration of Protein Molecules. In: H. NEURATH and K. BAILEY, The Proteins, Vol. I, Part B, p. 549. New York: Academic Press. 1953.
40. — Aspects actuels de la biochimie des acides aminés et des protéines. Actualités Biochimiques, No. 20. Paris: Masson et Cie. 1958.
41. ERIKSSON, A. F. V.: Mass Distribution of Unfractionated and Fractionated Polymethyl Methacrylates Determined by Ultracentrifugation and Fractional Precipitation. Acta Chem. Scand. 10, 360 (1956).

*42.* FAXÉN, H.: Über eine Differentialgleichung aus der physikalischen Chemie. Ark. Mat. Astron. Fysik **21** B, Nr. 3 (1929).

*43.* FIELD, E. O. and A. G. OGSTON: Boundary Spreading in the Migration of a Solute in Rapid Dissociation Equilibrium. Theory and its Application to the Case of Human Hemoglobin. Biochemic. J. **60**, 661 (1955).

*44.* FLORY, P. J.: Principles of Polymer Chemistry. Ithaca: Cornell Univ. Press. 1953.

*45.* FUJITA, H.: Effects of a Concentration Dependence of the Sedimentation Coefficient in Velocity Ultracentrifugation. J. Chem. Physics **24**, 1084 (1956).

*46.* — Evaluation of Diffusion Coefficients from Sedimentation Velocity Measurements. J. Physic. Chem. **63**, 1092 (1959).

*47.* GILBERT, G. A.: General Discussion. Discuss. Faraday Soc. **20**, 68 (1955).

*48.* — Sedimentation and Electrophoresis of Interacting Substances. I. Idealized Boundary Shape for a Single Substance Aggregating Reversibly. Proc. Roy. Soc. (London) A **250**, 377 (1959).

*49.* GILBERT, G. A. and R. C. L. JENKINS: Boundary Problems in the Sedimentation and Electrophoresis of Complex Systems in Rapid Reversible Equilibrium. Nature (London) **177**, 853 (1956). — Sedimentation and Electrophoresis Interacting Systems. II. Proc. Roy. Soc. (London) A **253**, 420 (1959).

*50.* GOFMAN, J. W.: What We *Do* Know about Heart Attacks. New York: G. P. Putnam's Sons. 1958.

*51.* GOFMAN, J. W., M. A. LAUFFER, I. H. PAGE, F. J. STARE, et al.: Evaluation of Serum Lipoprotein and Cholesterol Measurements as Predictors of Clinical Complications of Atherosclerosis. Circulation **14**, 691 (1956).

*52.* GOLDBERG, R. J.: A Theory of Antibody-Antigen Reactions. I. Theory for Reactions of Multivalent Antigen with Bivalent and Univalent Antibody. J. Amer. Chem. Soc. **74**, 5715 (1952).

*53.* — Sedimentation in the Ultracentrifuge. J. Physic. Chem. **57**, 194 (1953).

*54.* GOLDBERG, R. J. and J. W. WILLIAMS: Antigen-Antibody Reactions in Theory and Practice. Discuss. Faraday Soc. **13**, 224 (1953).

*55.* GOSTING, L. J.: Solution of Boundary Spreading Equations for Electrophoresis and the Velocity Ultracentrifuge. J. Amer. Chem. Soc. **74**, 1548 (1952).

*56.* — Measurement and Interpretation of Diffusion Coefficients of Proteins. Adv. Protein Chem. **11**, 429 (1956).

*57.* HARRINGTON, W. F., P. JOHNSON and R. H. OTTEWILL: Bovine Serum Albumin and its Behavior in Acid Solution. Biochemic. J. **62**, 569 (1956).

*58.* HARRINGTON, W. F. and J. A. SCHELLMAN: Evidence for the Instability of Hydrogen-Bonded Peptide Structures in Water, Based on Studies of Ribonuclease and Oxidized Ribonuclease. C. R. Trav. Lab. Carlsberg, Sér. chim. **30**, 21 (1956).

*59.* HEIDELBERGER, M. and K. O. PEDERSEN: Molecular Weight of Antibodies. J. exp. Medicine **65**, 393 (1937).

*60.* HERZOG, R. O., R. ILLIG und H. KUDAR: Über die Diffusion in molekulardispersen Lösungen. Z. physik. Chem. A **167**, 329 (1934).

*61.* HIRS, C. H. W., W. H. STEIN and S. MOORE: Peptides Obtained by Chymotryptic Hydrolysis of Performic Acid-Oxidized Ribonuclease. A Partial Structural Formula for the Oxidized Protein. J. Biol. Chem. **221**, 151 (1956).

*62.* HOOYMAN, G. J.: Thermodynamics in Sedimentation of Paucidisperse Systems. Physica **22**, 761 (1956).

*63.* HOOYMAN, G. J., H. HOLTAN, Jr., P. MAZUR and S. R. DE GROOT: Thermodynamics of Irreversible Processes in Rotating Systems. Physica **19**, 1095 (1953).

64. JOHNSON, J. S., K. A. KRAUS and G. SCATCHARD: Distribution of Charged Polymers at Equilibrium in a Centrifugal Field. J. Physic. Chem. 58, 1034 (1954).

65. JOHNSTON, J. P. and A. G. OGSTON: A Boundary Anomaly Found in the Ultracentrifugal Sedimentation of Mixtures. Trans. Faraday Soc. 42, 789 (1946).

66. JULLANDER, I.: Studies on Nitrocellulose Including the Construction of an Osmotic Balance. Ark. Kemi, Mineral. Geol. 21 A, No. 8 (1945).

67. KEGELES, G. and F. J. GUTTER: The Determination of Sedimentation Constants from Fresnel Diffraction Patterns. J. Amer. Chem. Soc. 73, 3770 (1951).

68. KINELL, P. O. and B. G. RÅNBY: Ultracentrifugal Sedimentation of Poly-molecular Substances. Adv. Colloid Sci. 3, 161 (1950).

69. KRAEMER, E. O.: In: T. SVEDBERG and K. O. PEDERSEN, The Ultracentrifuge, p. 327. Oxford: Clarendon Press. 1940.

70. LAMM, O.: Messung und Berechnung von Sedimentations-gleichgewichten an hochmolekularen Metaphosphaten. Ark. Kemi, Mineral. Geol. 17 A, No. 25 (1944).

71. LANSING, W. D. and E. O. KRAEMER: Molecular Weight Analysis of Mixtures by Sedimentation. J. Amer. Chem. Soc. 57, 1369 (1935).

72. — — Solvation and the Determination of Molecular Weights by Means of the Svedberg Ultracentrifuge. J. Amer. Chem. Soc. 58, 1471 (1936).

73. LARNER, J., B. R. RAY and H. F. CRANDALL: Pattern of Action of Crystalline Muscle Phosphorylase on Glycogen as Determined from Molecular Size Distribution Studies. J. Amer. Chem. Soc. 78, 5890 (1956).

74. LAUFFER, M. A. and I. J. BENDET: The Hydration of Viruses. Adv. Virus Research 2, 241 (1954).

75. LEVINTHAL, C.: The Mechanism of DNA Replication and Genetic Recombination in Phage. Proc. Nat. Acad. Sci. (USA) 42, 394 (1956).

76. LINDERSTRÖM-LANG, K. U.: Structure and Enzymatic Breakdown of Proteins. Cold Spring Harbor Sympos. Quant. Biol. 14, 117 (1950).

77. — Proteins and Enzymes. Stanford Univ. Publ., Univ. Ser., Med. Sci., Lane Medical Lectures, Vol. VI, 1952.

78. LINDGREN, F. T., H. A. ELLIOTT and J. W. GOFMAN: The Ultracentrifugal Characterization and Isolation of Human Blood Lipids and Lipoproteins, with Applications to the Study of Atherosclerosis. J. Physic. Coll. Chem. 55, 80 (1951).

79. LOEB, G. I. and H. A. SCHERAGA: Hydrodynamic and Thermodynamic Properties of Bovine Serum Albumin at Low pH. J. Physic. Chem. 60, 1633 (1956).

80. LONGSWORTH, L. G.: National Academy of Sciences Conference on the Ultra-centrifuge. Proc. Nat. Acad. Sci. (USA) 36, 502 (1950).

81. — Temperature Dependence of Diffusion in Aqueous Solutions. J. Physic. Chem. 58, 770 (1954).

82. — Moving Boundary Electrophoresis—Theory. In: M. BIER, Electrophoresis. Theory, Methods, and Applications, p. 91. New York and London: Academic Press. 1959.

83. LONGSWORTH, L. G. and C. F. JACOBSEN: An Electrophoretic Study of the Binding of Salt Ions by $\beta$-Lactoglobulin and Bovine Serum Albumin. J. Physic. Coll. Chem. 53, 126 (1949).

84. LUNDGREN, H. P. and W. H. WARD: Molecular Size of Proteins. In: D. M. GREENBERG, Amino Acids and Proteins, p. 312. Springfield: Charles C. Thomas. 1951.

85. MADSEN, N. B. and C. F. CORI: The Binding of Glycogen and Phosphorylase. J. Biol. Chem. **233**, 1251 (1958).
86. MAKINODAN, T., N. GENGOZIAN and R. E. CANNING: Demonstration of a Normal Serum Macroglobulin Coprecipitating with the Bovine Serum Albumin (BSA)-Chicken Anti-BSA Aggregate. Science (Washington) **130**, 1419 (1959).
87. MANDELKERN, L. and P. J. FLORY: The Frictional Coefficient for Flexible Chain Molecules in Dilute Solution. J. Chem. Physics **20**, 212 (1952).
88. MANDELKERN, L., W. R. KRIGBAUM, H. A. SCHERAGA and P. J. FLORY: Sedimentation Behavior of Flexible Chain Molecules: Polyisobutylene. J. Chem. Physics **20**, 1392 (1952).
89. MARRACK, J. R., H. HOCH and R. G. S. JOHNS: The Valency of Antibodies. Brit. J. exp. Pathol. **32**, 212 (1951).
90. MASSEY, V., W. F. HARRINGTON and B. S. HARTLEY: Physical Properties of Chymotrypsin and Chymotrypsinogen Using the Depolarization of Fluorescence Technique. Discuss. Faraday Soc. **20**, 24 (1955).
91. McBAIN, J. W.: The Determination of Bound Water by Means of the Ultracentrifuge. J. Amer. Chem. Soc. **58**, 315 (1936).
92. McMEEKIN, T. L. and K. MARSHALL: Specific Volumes of Proteins and the Relationship to their Amino Acid Contents. Science (Washington) **116**, 142 (1952).
93. MESELSON, M. and F. W. STAHL: The Replication of DNA in *Escherichia Coli*. Proc. Nat. Acad. Sci. (USA) **44**, 671 (1958).
94. MESELSON, M., F. W. STAHL and J. VINOGRAD: Equilibrium Sedimentation of Macromolecules in Density Gradients. Proc. Nat. Acad. Sci. (USA) **43**, 581 (1957).
95. MILLER, L. E. and F. A. HAMM: Macromolecular Properties of Polyvinylpyrrolidone: Molecular Weight Distribution. J. Physic. Chem. **57**, 110 (1953).
96. MOODY, L. S.: II. The Molecular Behavior of Insulin in Acid Solution. Dissert., University of Wisconsin, 1944.
97. NEURATH, H. and W. J. DREYER: Mechanism of Activation of Trypsinogen and Chymotrypsinogen. Discuss. Faraday Soc. **20**, 32 (1955).
98. NICHOLS, J. B. and E. D. BAILEY: Determinations with the Ultracentrifuge. In: A. WEISSBERGER, Physical Methods of Organic Chemistry, 2nd ed., p. 621. New York: Interscience Publ. Inc. 1949.
99. NISHIHARA, T. and P. DOTY: The Sonic Fragmentation of Collagen Macromolecules. Proc. Nat. Acad. Sci. (USA) **44**, 411 (1958).
100. O'DONNELL, I. J., R. L. BALDWIN and J. W. WILLIAMS: Correlation of the N ⇌ α Reaction of Thyroglobulin with the Type of Breakdown Produced by Papain. Biochim. Biophys. Acta **28**, 294 (1958).
101. O'DONNELL, I. J. and L. J. GOSTING: The Concentration Dependence of the Four Diffusion Coefficients of the System NaCl-KCl-H$_2$O at 25°C. In: W. J. HAMER, The Structure of Electrolytic Solutions, p. 160. New York: J. Wiley and Sons, Inc. 1959.
102. OGSTON, A. G.: Dimensions of Solute Particles from Dynamic Properties of their Solutions. Trans. Faraday Soc. **49**, 1481 (1953).
103. OGSTON, A. G. and J. M. A. TILLEY: Studies on the Heterogeneity of Crystallized β-Lactoglobulin. Biochemic. J. **59**, 644 (1955).
104. OGSTON, A. G. and M. P. TOMBS: Heterogeneity of Bovine β-Lactoglobulin. Biochemic. J. **66**, 399 (1957).
105. OGSTON, A. G. and E. F. WOODS: Sedimentation of Some Fractions of Degraded Dextran. Trans. Faraday Soc. **50**, 635 (1954).
106. ONCLEY, J. L.: private communication.

*107.* ONCLEY, J. L., E. ELLENBOGEN, D. GITLIN and F. R. N. GURT: Protein-Protein Interactions. J. Physic. Chem. **56**, 85 (1952).

*108.* ONCLEY, J. L., K. W. WALTON and D. G. CORNWELL: A Rapid Method for the Bulk Isolation of β-Lipoproteins from Human Plasma. J. Amer. Chem. Soc. **79**, 4666 (1957).

*109.* PAPPENHEIMER, A. M., Jr., H. P. LUNDGREN and J. W. WILLIAMS: Studies on the Molecular Weight of Diphtheria Toxin, Antitoxin, and their Reaction Products. J. exp. Medicine **71**, 247 (1940).

*110.* PEDERSEN, K. O.: Über das Sedimentationsgleichgewicht von anorganischen Salzen in der Ultrazentrifuge. Z. physik. Chem. A **170**, 41 (1934).

*111.* — Ultracentrifugal and Electrophoretic Studies on the Milk Proteins. II. The Lactoglobulin of Palmer. Biochemic. J. **30**, 961 (1936).

*112.* — On Charge and Specific Ion Effects on Sedimentation in the Ultracentrifuge. J. Physic. Chem. **62**, 1282 (1958).

*113.* PELLER, L.: Sedimentation in Multicomponent Systems. J. Chem. Physics **29**, 415 (1958).

*114.* PERRIN, F.: Mouvement brownien d'un ellipsoïde. II. Rotation libre et dépolarisation des fluorescences. Translation et diffusion de molécules ellipsoïdales. J. phys., Radium [7] **7**, 1 (1936).

*115.* POLGLASE, W. J., D. M. BROWN and E. L. SMITH: Studies on Human Glycogen. II. Sedimentation in the Ultracentrifuge. J. Biol. Chem. **199**, 105 (1952).

*116.* REICHMANN, M. E., S. A. RICE, C. A. THOMAS and P. DOTY: Further Examination of the Molecular Weight and Size of Desoxypentose Nucleic Acid. J. Amer. Chem. Soc. **76**, 3047 (1954).

*117.* RICH, A. and F. H. C. CRICK: Structure of Collagen. Nature (London) **176**, 915 (1955).

*118.* ROLFE, R. and M. MESELSON: The Relative Homogeneity of Microbial DNA. Proc. Nat. Acad. Sci. (USA) **45**, 1039 (1959).

*119.* ROSENKRANZ, H. S. and A. BENDICH: Sedimentation Studies of Fractions of Deoxyribonucleic Acid. J. Amer. Chem. Soc. **81**, 902 (1959).

*120.* — — Studies on the Sedimentation Behavior of Artificial Mixtures of Deoxyribonucleic Acid. J. Amer. Chem. Soc. **81**, 2842 (1959).

*121.* — — Studies on the Effect of Heat on Deoxyribonucleic Acid. J. Amer. Chem. Soc. **81**, 6255 (1959).

*122.* ROTHEN, A.: Molecular Weight and Electrophoresis of Crystalline Ribonuclease. J. Gen. Physiol. **24**, 203 (1940).

*123.* SADRON, C.: Methods of Determining the Form and Dimensions of Particles in Solution: a Critical Survey. Progr. Biophys. Biophys. Chem. **3**, 237 (1953).

*124.* SCHACHMAN, H. K.: Ultracentrifugation in Biochemistry. New York and London: Academic Press. 1959.

*125.* SCHACHMAN, H. K. and M. A. LAUFFER: The Hydration, Size and Shape of Tobacco Mosaic Virus. J. Amer. Chem. Soc. **71**, 536 (1949).

*126.* — — The Density Correction of Sedimentation Constants. J. Amer. Chem. Soc. **72**, 4266 (1950).

*127.* SCHERAGA, H. A. and L. MANDELKERN: Consideration of the Hydrodynamic Properties of Proteins. J. Amer. Chem. Soc. **75**, 179 (1953).

*128.* SCHUMAKER, V. N. and H. K. SCHACHMAN: Ultracentrifugal Analysis of Dilute Solutions. Biochim. Biophys. Acta **23**, 628 (1957).

*129.* SHOOTER, K. V. and J. A. V. BUTLER: Sedimentation of Deoxyribonucleic Acid at Low Concentrations. Trans. Faraday Soc. **52**, 734 (1956).

*130.* SIGNER, R. and H. GROSS: Ultrazentrifugale Polydispersitätsbestimmungen an hochpolymeren Stoffen. 95. Mitt. über hochpolymere Verbindungen. Helv. Chim. Acta **17**, 726 (1934).

*131.* SINGER, S. J. and D. H. CAMPBELL: Physical Chemical Studies of Soluble Antigen-Antibody Complexes. I. The Valence of Precipitating Rabbit Antibody. J. Amer. Chem. Soc. **74,** 1794 (1952).

*132.* — — Physical Chemical Studies of Soluble Antigen-Antibody Complexes. II. Equilibrium Properties. J. Amer. Chem. Soc. **75**, 5577 (1953).

*133.* — — Physical Chemical Studies of Soluble Antigen-Antibody Complexes. III. Thermodynamics of the Reaction between Bovine Serum Albumin and its Rabbit Antibodies. J. Amer. Chem. Soc. **77**, 3499 (1955).

*134.* — — Physical Chemical Studies of Soluble Antigen-Antibody Complexes. IV. The Effect of pH on the Reaction between Bovine Serum Albumin and its Rabbit Antibodies. J. Amer. Chem. Soc. **77**, 3504 (1955).

*135.* — — Physical Chemical Studies of Soluble Antigen-Antibody Complexes. V. Thermodynamics of the Reaction between Ovalbumin and its Rabbit Antibodies. J. Amer. Chem. Soc. **77**, 4851 (1955).

*136.* SINGER, S. J., L. EGGMAN and D. H. CAMPBELL: Physical Chemical Studies of Soluble Antigen-Antibody Complexes. VI. The Effect of pH on the Reaction between Ovalbumin and its Rabbit Antibodies. J. Amer. Chem. Soc. **77**, 4855 (1955).

*137.* SMITH, R. F. and D. R. BRIGGS: Electrophoretic Analysis of Protein Interaction. I. Interaction of Bovine Serum Albumin and Methyl Orange. J. Physic. Coll. Chem. **54**, 33 (1950).

*138.* SMITHIES, O.: The Application of Four Methods for Assessing Protein Homogeneity to Crystalline $\beta$-Lactoglobulin: an Anomaly in Phase Rule Solubility Tests. Biochemic. J. **58**, 31 (1954).

*139.* STEINER, R. F.: Reversible Association Processes of Globular Proteins. V. The Study of Associating Systems by the Methods of Macromolecular Physics. Arch. Biochem. Biophys. **49**, 400 (1954).

*140.* SUEOKA, N.: A Statistical Analysis of Deoxyribonucleic Acid Distribution in Density Gradient Centrifugation. Proc. Nat. Acad. Sci. (USA) **45**, 1480 (1959).

*141.* SUEOKA, N., J. MARMUR and P. DOTY: Heterogeneity of Deoxyribonucleic Acids. II. Dependence of the Density of Deoxyribonucleic Acids on Guanine-Cytosine Content. Nature (London) **183**, 1429 (1959).

*142.* SVEDBERG, T.: Zentrifugierung, Diffusion und Sedimentationsgleichgewicht von Kolloiden und hochmolekularen Stoffen. Kolloid-Z. **36**, Erg.-Bd., 53 (1925).

*143.* SVEDBERG, T. and J. B. NICHOLS: Determination of Size and Distribution of Size of Particle by Centrifugal Methods. J. Amer. Chem. Soc. **45**, 2910 (1923).

*144.* SVEDBERG, T. and K. O. PEDERSEN: The Ultracentrifuge. Oxford: Clarendon Press. 1940.

*145.* SVEDBERG, T. and H. RINDE: Determination of the Distribution of Size of Particles in Disperse Systems. J. Amer. Chem. Soc. **45**, 943 (1923).

*146.* — — The Ultra-centrifuge, a New Instrument for the Determination of Size and Distribution of Size of Particle in Amicroscopic Colloids. J. Amer. Chem. Soc. **46**, 2677 (1924).

*147.* TAYLOR, J. H., P. S. WOODS and W. L. HUGHES: The Organization and Duplication of Chromosomes as Revealed by Autoradiographic Studies Using Tritium-Labeled Thymidine. Proc. Nat. Acad. Sci. (USA) **43**, 122 (1957).

*148.* TIMASHEFF, S. N. and J. G. KIRKWOOD: Electrophoresis-Convection Applied to the Complexed Insulin-Protamine System. J. Amer. Chem. Soc. **75**, 3124 (1953).

*149.* TIMASHEFF, S. N. and R. TOWNEND: The Association Behavior of β-Lacto-globulins A and B. J. Amer. Chem. Soc. **80**, 4433 (1958).

*150.* TISELIUS, A.: Über die Berechnung thermodynamischer Eigenschaften von kolloiden Lösungen aus Messungen mit der Ultrazentrifuge. Z. physik. Chem. **124**, 449 (1926).

*151.* — Study of the Electrophoresis of Proteins by the Moving-Boundary Method. Nova Acta Regiae Soc. Sci. Upsaliensis **7**, No. 4 (1930).

*152.* — Über den Einfluß der Ladung auf die Sedimentationsgeschwindigkeit von Kolloiden, besonders in der Ultrazentrifuge. Kolloid-Z. **59**, 306 (1932).

*153.* TOWNEND, R. and S. N. TIMASHEFF: The pH Dependence of the Association of β-Lactoglobulin. Arch. Biochem. Biophys. **63**, 482 (1956).

*154.* — — The Molecular Weight of β-Lactoglobulin. J. Amer. Chem. Soc. **79**, 3613 (1957).

*155.* TRAUTMAN, R., V. N. SCHUMAKER, W. F. HARRINGTON and H. K. SCHACHMAN: The Determination of Concentrations in the Ultracentrifugation of Two-Component Systems. J. Chem. Physics **22**, 555 (1954).

*156.* VAN HOLDE, K. E. and R. L. BALDWIN: Rapid Attainment of Sedimentation Equilibrium. J. Physic. Chem. **62**, 734 (1958).

*157.* WALES, M. and J. W. WILLIAMS: Effect of Solvation on Sedimentation Experiments. J. Polymer Sci. **8**, 449 (1952).

*158.* WILLIAMS, J. W.: Sedimentation Analysis and Some Related Problems. J. Polymer Sci. **12**, 351 (1954).

*159.* WILLIAMS, J. W., R. L. BALDWIN, W. M. SAUNDERS and P. G. SQUIRE: Boundary Spreading in Sedimentation Velocity Experiments. I. The Enzymatic Degradation of Serum Globulins. J. Amer. Chem. Soc. **74**, 1542 (1952).

*160.* WILLIAMS, J. W. and W. M. SAUNDERS: Size Distribution Analysis in Plasma Extender Systems. II. Dextran. J. Physic. Chem. **58**, 854 (1954).

*161.* WILLIAMS, J. W., W. M. SAUNDERS and J. S. CICIRELLI: Size Distribution Analysis in Plasma Extender Systems. I. Gelatin. J. Physic. Chem. **58**, 774 (1954).

*162.* WILLIAMS, J. W., K. E. VAN HOLDE, R. L. BALDWIN and H. FUJITA: The Theory of Sedimentation Analysis. Chem. Rev. **58**, 715 (1958).

*163.* YEANDLE, S.: Effect of Electric Field on Equilibrium Sedimentation of Macromolecules in a Density Gradient of Cesium Chloride. Proc. Nat. Acad. Sci. (USA) **45**, 184 (1959).

*(Received, February 1, 1960.)*

# Structure and Immunological Specificity of Polysaccharides.

By MICHAEL HEIDELBERGER, New York.

### Contents.

This paper is an adaptation and extension, with new data, of "All Polysaccharides are Immunologically Specific," Proceedings of the Fourth International Congress of Biochemistry, Vienna, 1958, Vol. I, Symposium I, 52–66; I. U. B. Symposium Series, Vol. 3, Pergamon Press, Ltd., London, New York and Paris, 1959.

## I. Introduction.

Immunological specificity is a distinguishing characteristic of the combination of antigens and antibodies. Antigens are substances foreign to a living animal which, when injected into the animal, or, if microorganisms or viruses, having passed the normal barriers of defense, stimulate the production of new substances called antibodies. These combine with and assist in the elimination of the foreign substances, the antigens. In general, antibodies combine only with the substances which initiated their production, or with chemically closely related substances, and this behavior is termed specific. Antibodies which circulate in the blood are serum globulins, newly modified in their synthesis as a result of the presence of the antigen in the body. Many antigens are proteins, but other classes of substances, such as carbohydrates or polysaccharides, may also function as antigens, at least in mice, cattle, and men, to stimulate the production of antibodies.

Antigens and antibodies are recognized by their mutual combination: if the antigens are particulate, such as microbes, these are agglutinated, or clumped together, by homologous antibodies, and this is readily observed. If the antigen is in solution, it may precipitate with its homologous antibody, just as sulfate precipitates with barium ion, and just as chloride, bromide and iodide precipitate with silver ion. As with the halogens, immunological specificity is not absolute, for closely related antigens may precipitate with antibodies to one or the other, as duck egg albumin precipitates a portion of the antibodies resulting from the injection of hen's egg albumin into animals, and hen's egg albumin precipitates a portion of the antibodies to duck egg albumin.

For a review of immune precipitation and agglutination and quantitative theories of these reactions, see HEIDELBERGER (34).

The participation of carbohydrates in immune reactions such as bacterial agglutination and the mutual precipitation of antigen and antibody was securely established in 1923, when AVERY and the writer (42, 49) showed that the capsular slimes of virulent pneumococci, responsible for both virulence and type-specificity, were composed of polysaccharides which differed for each of the serological types of *Pneumococcus* studied. This led to the recognition of polysaccharides as determinants of the immunological specificities of many varieties of microorganisms. These polysaccharides were first considered to be a special class, set aside from the more widely distributed plant and animal carbohydrates by their haptenic properties; that is, by their capacity to react with antibodies stimulated in animals by the injection of intact, Gram-positive, homologous pneumococci, or by the complex, relatively undegraded, carbohydrate-containing antigens of other microorganisms. As we shall see, this distinction has turned out to be illusory. The original classification of the pneumococcal polysaccharides as haptens survived only briefly, for SCHIEMANN and CASPER (98), PERLZWEIG and KEEFER (89), FRANCIS and TILLETT (27), and FELTON (25) showed that these substances were fully antigenic in the mouse, and, indeed in man*. In both of these species, immunization against pneumococcal pneumonia

---

* D. HAMMER has now shown that cows may be stimulated to form antibodies by injection of pneumococcal polysaccharides (Habilitationsschrift, Tierhygienisches Institut der Universität, Freiburg i. Br., 1959).

*References*, pp. 529—536.

by vaccination with small amounts of the purified type-specific polysaccharides proved to be feasible and effective. [For a summary see (36).]

The present review, however, is concerned with the organic chemistry of these substances, their use in the elucidation of the mechanism of immune reactions, and the application of the knowledge so gained to carbohydrate chemistry in general.

In their initial studies on the mechanism of the precipitin reaction, that is, the mutual precipitation of antigen and antibody, KENDALL and the writer (53) made use of nitrogen-free pneumococcal polysaccharides as antigens and so simplified their analytical chemical problem, for none of the nitrogen precipitated could be derived from the antigen.

The quantitative theory of the precipitin reaction which resulted from these studies postulates the mutual precipitation of antigen and antibody because each is multivalent with respect to the other. After a molecule of antigen combines with a molecule of antibody, there are still available on the antigen other groupings capable of combining with other molecules of antibody, and there are still available on the antibody one or more groupings capable of combining with an additional molecule or molecules of antigen (56). Antigen-antibody combination is thus a dynamic process which goes on until enormous aggregates are built up. These either fall out of solution because of their weight, or, as MARRACK (78) has suggested, lose their affinity for water because of the juxtaposition and consequent discharge of innumerable positively and negatively charged groups. According to the theory, then, immunological specificity, as evidenced in the precipitin reaction or in bacterial agglutination, is a function of the interaction of multiple reactive groupings on both antigen and antibody.

Ideal material for tests of the validity of this theory is furnished by the sugars, oligo-, and polysaccharides. On the other hand, because of the enormous complexity of the proteins and the tedium and labor involved in sorting out the amino acids of which they are composed and determining their order and arrangement, only a mere beginning has been made on an urgently needed body of knowledge of the fine structure of these all-important substances. In the case of the carbohydrates and polysaccharides, however, such a body of knowledge already exists and is being rapidly extended, for the methods of sugar chemistry, while also involved and laborious, are much less so than those as yet proposed in the chemistry of proteins. The carbohydrates, therefore, present a wide choice of materials for the study of main interest to this report; namely, the relation between chemical constitution and immunological specificity.

Many polysaccharides are considered to be polymers of definite, repeating units composed of one or more sugars. That such units, or small multiples of them, might function immunologically as the multiple reactive groupings postulated by our theory, was shown by KENDALL and the writer (55)*. We found that partial hydrolytic products of the specific capsular polysaccharides of Types I and III pneumococcus having an average molecular size of 700 to 2000 could still precipitate a portion of the antibodies in Type I and Type III antipneumococcal horse sera. The multivalence of the intact specific polysaccharides was thereby clearly indicated.

## II. Chemistry of Capsular Polysaccharides of *Pneumococcus* and O-Polysaccharides of *Salmonella*.

Let us now survey briefly what is known of the chemical constitution of some of the specific capsular polysaccharides which characterize the 70 or 80 pneumococcal types**, and of the O-carbohydrates of members of the vast family of *Salmonella*. We shall refer later to precipitin reactions with the antibodies stimulated by the antigens of which these polysaccharides are the specific determinants. In *Table 1* (p. 519) are set forth the sugars and chemical components of those of the polysaccharides in these groups of which we have definite knowledge. That the list is continually expanding is due to the efforts of Drs. PAUL A. REBERS in our laboratory, M. STACEY and S. A. BARKER in Birmingham, England, J. K. N. JONES in Kingston, Ontario, who are working on various pneumococcal polysaccharides, and Drs. ANNE-MARIE STAUB in Paris, O. WESTPHAL and coworkers in Freiburg, and D. A. L. DAVIES of Porton, who are studying *Salmonella****. In *Chart 1* (p. 507) are given possible structural formulas of five of the pneumococcal polysaccharides—for only three of them are the structures uniquely determined.

One cannot peruse Table 1 without noting the frequency with which the same sugars, *D*-glucose, *D*-galactose, and *L*-rhamnose appear. *Pneumococcus* and *Salmonella* are able to build these sugars into polysaccharides of more than a hundred different immunological specificities, not only by shifting the positions of linkage of the reducing groups and their anomeric and occasionally enantiomorphic relationships, but also by abandoning one or another of these three sugars (or four, with mannose, in *Salmonella*) and substituting for it an amino sugar, a sugar acid, a deoxysugar, a polyol, or even phosphoric acid. Not only are these

---

* Similar, unpublished studies were also carried out with S I and S VI.

** The pneumococcal types will be designated in this review as Pn with the appropriate Roman numeral, and the corresponding specific capsular polysaccharides as S I, S II, etc.

*** A detailed review of the chemistry of these polysaccharides by Dr. DAVIES is now in press (Advances in Carbohydrate Chemistry, Vol. 15).

*References, pp. 529—536.*

Type II*   | [3)-L-rham(1 → 3)-L-rham(1 → 4)-D-glucur(1 → 3)-L-rham(1 → 4)-α-D-glu(1 ····]$x$
           D-glucur(1 ↑6)

Type III   | [3)-β-D-glucur(1 → 4)-β(?)-D-glu(1 →]$x$

Type VI    | [2)-α-D-gal(1 → 3)-D-glu(1 → 3)-L-rham(1 → 3)ribitol(1[5]-OPO─]$x$
           OH
           O

Type VIII  | [4)-β-D-glucur(1 → 4)-β-D-glu(1 → 4)-α-D-glu(1 → 4)-α-D-gal(1 →]$x$

Type XIV* † | [6)-D-NAc glucosm(1 → 3)-D-gal(1 → 6)-D-NAc glucosm(1 → 3)-D-gal(1 → 6)-D-NAc glucosm(1 → 3)-D-gal(1 →]$x$
            β-D-gal(1 → 4)-D-glu(1 ↑4)        α-D-gal(1 ↑4)
            D-glu(1 ↑4)

Chart 1. Structural Formulas of Pneumococcal Polysaccharides.

(All sugars are apparently in the pyranose form.)

* Not uniquely determined.

† Lactose has now been isolated among the products of partial hydrolysis (12). S XIV is susceptible to the action of both α- and β-galactosidases (KABAT). Most of the other linkages are probably β-.

serological specificities different, but many of them seem almost absolute
within the species, for the number of overlappings, or cross-reactions,
so far encountered among the pneumococci, at least, has been amazingly
small. If all of these diverse and complicated syntheses are made possible
by intermediates of the type of uridine-diphospho-sugars, as now appears
likely [cf. for example (100)], the flexibility of such a mechanism is
indeed astounding.

If the theory of multiple reactive groups has any validity, it is evident
from Table 1 and Chart 1 that there should be at least one strong cross-
reaction, that between pneumococcal Types III and VIII, for the specific
polysaccharides of both of these types, S III and S VIII, contain multiple
units of cellobiuronic acid. This cross-reaction was first recorded
in 1928 (107). Later, GOEBEL and his coworkers showed that S III
was a polycellobiuronic acid, with each unit linked to the next by a
probable β-union to position 3 of the glucuronic acid (64), and that
cellobiuronic acid made up about one-half of the molecule of S VIII (29).
Now that JONES and PERRY have worked out the fine structure of
S VIII (68), the chemical basis for this instance of cross-reactivity in
agglutination and precipitation has been fully elucidated and is in accord
with the theory of multiple reactive groupings; for the theory permits
the prediction of cross-reactivity when two polysaccharides contain
multiple reactive groupings in common. Moreover, cross-reactivity
does not depend, in such instances, on an "antigen in common" as the
bacteriologists so often and so loosely express the reason for cross-
reactivity: S III and S VIII are *different* antigens which do, however,
possess *multiple reactive groups* in common. The writer hopes that he
will be pardoned for the satisfaction he takes in the apparent linearity
of the structure of S VIII, since extensive branching seemed excluded
on the basis of a quantitative study of the cross-reaction made long ago
with KABAT, MAYER, and SHRIVASTAVA (52, 52a).

Additional aspects of the chemistry of the pneumococcal polysaccharides will be discussed in connection with each type for which
something, at least, of its chemistry is known and the cross-reactions
encountered in antisera to the type in question. Attention is called
here to the cross-reactivity of S II in antityphoid serum (47). The
quantity of nitrogen precipitated from the available serum was 34 μg.
per ml., or 8% of the antibody precipitable by the typhoid O-polysaccharide used. Since the same three sugars, D-galactose, D-glucose,
and L-rhamnose, which occur so frequently among the pneumococcal
type-specific substances are generally present in the O-antigens of
*Salmonella*, it is probable that many more cross-reactions between these
classes of microorganisms would be found if they were systematically
looked for. Type-specificity in the *Salmonella* group appears to be

mediated mainly by certain peculiar deoxysugars (*105*), a class which, except for rhamnose and amino sugars, has not yet turned up among the pneumococci.

## III. Cross-reactions of Antipneumococcal Sera.

(Tables 2–4, pp. 520–525.)

Evaluation of tests for cross-reactivity is complicated by the great variability of antisera. A negative test, for example, may simply mean that the serum used, if a single one, contains no cross-precipitating antibody, while that of another animal, even of the same species, immunized with the same antigen, might yield a precipitate. Examples in which this actually occurred are given in the pages which follow. In many instances only a single serum was available, or small samples of different sera, so that more comprehensive testing would be necessary before any polysaccharide could definitely be said *not* to react with antibody of a given serological type. On the other hand, positive tests are more meaningful, and it has been shown in many instances that precipitates are actually made up of cross-reacting polysaccharide and antibody [cf. for example HEIDELBERGER et al. (*38, 48, 51*)].

Antisera to the pneumococcal types will be considered in their numerical order, although our present knowledge of the chemistry of the appropriate antigenic determinants is unevenly distributed along the way. In fact, little is known of the components of the specific polysaccharide of the very first, NEUFELD's "typical" pneumococcus, later designated Type I, except that it is amphoteric and contains galacturonic acid (*57*) and possibly glucose. *Table 2* (p. 520) presents a summary of all polysaccharides which gave moderate to strong cross-reactions, ++ or better, on an arbitrary scale of — to ++++. The individual cross-reactions are given in *Tables 3* and *4* (pp. 523, 525). In a number of instances in which the ++ rating was followed by quantitative measurements of the amounts of nitrogen precipitated by the cross-reacting polysaccharide, values ranging from 8 to 19 $\mu$g. per ml. were found. All qualitative tests were carried out in a refrigerator at 0 to 4°, while the tubes for quantitative estimations were allowed to stand in a bath at 0° for 6 to 20 days (*51, 59*). Many of the reactions would not have been seen at room temperature or at 37°. In the case of cross-reactions of pneumococcal polysaccharides in antipneumococcal sera, any antibody to pneumococcal C-polysaccharide was first removed [cf. for example (*52, 52a*)].

Many of the sulfated polysaccharides, such as soluble agar, $\varkappa$- and $\lambda$-carrageenans, and that of *Irideae laminaroides*, precipitate antipneumococcal horse sera heavily, the last two practically regardless of pneumococcal type. Such precipitation is at least in part unspecific.

The gum of a green seaweed, *Acrosiphonia centralis* (*87*), on the other hand, contains but 8% sulfate and appears to precipitate certain sera specifically if formation of its insoluble calcium salt is avoided by prior addition of ethylenediamine tetraacetate. Data on this gum are therefore included in the Tables.

*1.* Type I. — [S I: glu?, galacturonic acid + unknown nitrogenous components.] Almost all tests were made with New York City Dept. of Health horse serum 884\*. An additional horse serum from the same source, no. 592\*\*, reacted less strongly. Relatively few polysaccharides have so far been found which cross-react with the antiserum; namely, oat glucan (*91*), hualtaco gum (source unknown), ketha gum (from *Feronia elephantum*) (*80*) and the gums of *Khaya grandifolia* and *K. senegalensis* (*4a*). Because of the strong cross-reaction, the presence of galacturonic acid was suspected in the *Khaya* gums, and Prof. E. L. Hirst, who had kindly supplied the samples, was notified of this. As it turned out, he had already identified the acid and found that it existed in double units, linked $1 \rightarrow 4$ (*4a*). Ketha gum is stated not to contain galacturonic acid (*80*), but the writer has requested Prof. S. Mukherjee to reinvestigate this question, as the gum precipitates more than one-third of the antibody in serum 884. Galacturonic acid has been identified in nori utsugi (*75*), which reacts weakly. Gums containing galacturonic acid which do not cross-react appreciably in Type I antiserum are, karaya (*61*), olibanum II (*23*), pectin and pectic acid, tororo aoi *(Hibiscus manihot)* (*76*), and okra (*111*).

Chromatograms of hydrolyzed S I show a weak spot corresponding to glucose. The cross-reactivity of oat glucan, which is made up of $\beta$-1,3- and $\beta$-1,4-linked glucose, in anti-Pn I might be considered as a confirmation of the presence of glucose in S I.

*2.* Type II. — [S II: *D*-glu, *L*-rham (50%), *D*-glucur, see Chart 1, p. 507.] Three antisera were used: 930\*\*\* and 1054†, New York City Dept. of Health, and 513††, Bureau of Laboratories, New York State Dept. of Health. These varied greatly in their cross-reactivities. Although the first contained 20% more antibody than the second, it gave weaker cross-reactions than the other two sera. While Pn types V and VI are partially agglutinated by Type II antisera (*8*) the three available sera failed to precipitate S V and S VI. Twenty-five polysaccharides have thus far been found reactive with this type of antibody. These will be discussed according to the sugar, multiple occurrences of which appear to be mainly responsible for the precipitation.

---

   \* Bleeding, Oct. 6, 1939.
  \*\* Bleeding, Dec. 29, 1939.
 \*\*\* Bleeding, June 25, 1937.
   † Bleeding, Aug. 14, 1939.
  †† Bleeding, Feb. 26, 1940.

*References, pp. 529—536.*

*a. D*-Glucuronic Acid.

The cross-reaction of gum arabic in Type II antiserum was discovered as a result of Dr. O. T. Avery's belief that there must exist, "free in nature" as he said, other gums related to the pneumococcal capsular polysaccharides. Tests were made with antisera to Types I, II, and III and a number of commercially available gums. Gum arabic was found to react most strongly, and with Type II antiserum (*43*). At the time, the chemical basis of this reactivity was not clearly understood, but it now appears certain that this reaction and that of related materials such as the gum of *Acacia pycnantha* (*63a*), ketha (*80*), and gum ghatti (*4*) are referable to their glucuronic acid end groups. This is shown by the greater avidity of the reaction of gum arabic after removal of obstructing arabofuranose residues on mild hydrolysis, and by the failure of most of the rhamnose, when present, to appear in the specific precipitate (*16, 39*). In fact, rhamnose-containing samples of gum arabic are fractionated far more efficiently by Type II antiserum than by several alternative methods (*39*).

The highly branched gum of *Brachychiton diversifolium*, with numerous end groups of glucuronic acid (*63*), precipitated 80% of the antibodies in Type II antipneumococcal serum 513 and 50% from no. 1054 (*37*). Another powerful precipitant of these sera was the gum of *Albizzia zygea*, which contains both glucuronic acid and its 4-O-methyl derivative*.

Corn fiber (hull) hemicellulose also contains glucuronic acid (*86, 110, 112*), while flax hemicellulose is made up of 4-O-methylglucuronic acid, *L*-rhamnose in 1,3-linkage, and xylose (*28*). The flax substance precipitates far less antibody than does corn fiber hemicellulose. This is perhaps due to partial interference by the 4-O-methyl group of the glucuronic acid, although this would be countered by a slight reinforcement due to the rhamnose (cf. Type VI). Mesquite gum, however, which also contains 4-O-methyl-*D*-glucuronic acid and apparently no unsubstituted glucuronic acid (*20, 101a, 114*) reacted more strongly, as was found by Marrack and Carpenter (*79*) with the degraded gum. *Khaya grandifolia* gum also appears to react because of its content of 4-O-methyl-*D*-glucuronic acid (*4a*). The polysaccharides of *Azotobacter chroococcum* (*74*) and *Cryptococcus neoformans* A (*24*) also appear to cause precipitation because they contain multiple groupings of glucuronic acid, although participation of the glucose of the former is not excluded. Precipitation of lung galactan is due to a glucuronic acid-containing impurity which was first identified in this fashion (*48*). Although the slime of *Aerobacter aerogenes* A 3 (*6, 116*) is said probably to contain end-groups of glucuronic acid (*6*), these must be very disadvantageously spaced, since there is no reaction with Type II antiserum (*Table 3*, p. 523). The polysaccharide of strain 418 (*13*), in

---

* Private communication from Prof. E. L. Hirst and Mrs. E. Percival.

which end-groups of glucuronic acid have been demonstrated, does precipitate Type II antiserum, as does that of A 4 (22) in which the uronic acid has not been otherwise identified.

### b. D-Glucose, in 1,4,6-Linkage.

According to BUTLER and STACEY (18) all of the glucose in S II is in the form of 1,4,6-branch-points. This led to the prediction, from the quantitative theory of the precipitin reaction, that all polysaccharides with such linkages of glucose would precipitate Type II antipneumococcal sera. This was immediately verified with glycogen from the most diverse sources and with the amylopectins (41). Since all of the linkages in these substances are α-, the result may also indicate that the glucose in S II is α-linked, for immunological specificity has long been known to be sensitive to the anomeric configuration of the glycoside linkage (9) at least when this is in a terminal position. Another polysaccharide in which two of every three glucose residues are 1,4,6-linked is that of tamarind seed (115), and since this also reacts, the configuration may be α- in this instance as well. Fractions of synthetic polyglucose also precipitate (40), as do the dextrans, even those containing only α-1,6-linked glucose. As all of the 1,6-linkages of glycogen and amylopectin are at the 1,4,6-branch-points, this either indicates that the branches of S II entering at position 4 are sterically unimportant for Type II specificity, or that some of the glucose in S II is exceedingly difficult to methylate completely, and so failed to appear as 2,3,4-tri-O-methyl-glucose in reference (18).

The first cross-reaction found in antipneumococcal sera with another class of microorganisms was that of the *Klebsiella pneumoniae*, Type B capsular substance (11). Although glucose is the only constituent in common with S II thus far identified, it is possible that the uronic acid in the B substance, as yet uncharacterized, may be glucuronic acid and so take part in the reaction [cf. GOODMAN and KABAT (31), also for an alternative explanation of the reactivity of polyglucoses].

### c. L-Rhamnose, Linked 1,3-.

Since the group-specific C-polysaccharide of Group A hemolytic streptococci reacted very weakly with Type II antiserum (38) it could be predicted that the V variant (81, 82), in which much of the N-acetyl-glucosamine of the C-polysaccharide is lacking, and the C-polysaccharide of Group A degraded by a streptococcal enzyme which removes N-acetyl-glucosamine would both precipitate more strongly in Type II antiserum than did the original C substance. This prediction has been verified for one of two antisera showing the greater extent of cross-reactivity, no. 513, while the differences in the other, 1054, though probably real, are very small (58). The result also points to the probability that some at least, of the rhamnose residues in the C and V substances of Group A streptococci are linked 1,3-, a point not touched upon by SCHMIDT (99).

It is not clear why olibanum I, said to contain D-galactose and L-arabinose (23), should precipitate even weakly in Type II antiserum.

3. Type III. — [S III : ($\beta$-(?)cellobiuronic acid 1,3-)$_x$, see Chart 1, p. 507.] Only a single horse serum was available, no. 792*, New York City Dept. of Health. Aside from the cross-reactions of S VIII, already mentioned, and the similar one with oxidized cotton (50), the rather weak precipitate of the azotobacter polysaccharide is the only appreciable one found, and is probably also due to the presence of cellobiuronic acid units in the polysaccharide (74). The reaction in Type VIII antiserum was still weaker.

4. Type IV. — [S IV: gal, unidentified $Ac$NH-sugar.] Several antisera were used but very few tests were positive. The rather limited cross-reaction of the arabogalactan of Jeffrey pine (109) in the weaker sera and possibly those of gum ghatti and okra in the most cross-reactive, New York State 6c9**, would seem referable to multiple units of similarly linked galactose.

5. Type V. — [S V: D-glu, D-glucur, and unidentified basic sugars (15).] Type V was originally called II A because pneumococci of this type were "weakly and incompletely" agglutinated by Type II antisera (8). Most of the tests were carried out with New York City Dept. of Health serum 606, which precipitated somewhat more antibody with S II than did New York State serum 555***. Knowledge of the chemistry of S V is still too incomplete to permit much comment on the cross-reactions. Five of the reactive gums contain D-glucuronic acid, so that multiple recurrences of this acid are possibly responsible.

6. Type VI. — [S VI: D-glu, D-gal, L-rham, ribitol, $PO_4$; see Chart 1, p. 507.] This type was designated II B by AVERY (8) and VI by COOPER and her associates (19). Three antisera of widely different cross-reactivities were used: New York State horse sera 614 and 681, and New York City horse serum 771†. The first gave only traces of precipitate with S II, while from the last, one-fifth of the antibody was precipitated (60). The relation of the pneumococcal Types II and VI has been shown to be due to the presence, in each of the otherwise widely different polysaccharide structures, of multiple groupings of 1,3-linked L-rhamnose (18, 60, 95). Accordingly, flax straw hemicellulose (28) and the rhamnogalactan of Bacillus polymyxa, also known to contain rhamnose linked in this fashion (84, 85), both precipitate serum 771, especially. Although

---

*  Bleeding, Feb. 1, 1937.
**  Bleeding. Sept. 5, 1941.
***  Bleedings, Jul. 16, 1934 and Aug. 20, 1941, respectively.
  †  Bleedings, Nov. 2, 1938, Jun. 29, 1940, and a pool of bleedings of May 10 and 31, 1938, respectively.

the polymyxa substance contains galactose, little or none of it is present as end-groups (*84*). As would be expected, the Group A and V-variant streptococcal C carbohydrates show the same relative behavior in this Type VI antiserum as in Type II antiserum 513 (*58, 60*).

Sera 614 and 681 precipitated oat and barley glucans presumably because of the content of $\beta$-1,3-linked glucose (*7*) in these carbohydrates. Dextran 1355-S-4, which contains about one-third of $\alpha$-1,3-like linkages of glucose (*66, 93*), also precipitated these sera, but dextran N 236, with only 4% of such bonds (*71a*), did not.

Precipitation by the galactomannans carob (*62, 101*), guar (*2, 92, 117*), and Kentucky coffee bean (*73*), by S XIV, and by *Khaya grandifolia* would seem to be due to the known presence in these substances of multiple non-reducing end-groups of *D*-galactose. Although S VI is linear and therefore does not appear to possess such end-groups, its galactose is linked 1,2-, and the free hydroxyls at positions 3, 4, and 6 might easily exert much the same effect on antibody production and reactivity as though the entire sugar were terminal. The heavy precipitation by okra is probably due, at least in part, to the same cause, but the fine structure of this substance has not yet been worked out (*111*). A sample of *Khaya senegalensis* gum, treated with alkali to remove acetyl groups, as had been done with the sample of purified *Khaya grandifolia* gum furnished by Prof. E. L. Hirst, precipitated nine times as much antibody from anti-Pn VI as did the intact gum.

7. Type VII. — [S VII: gal, glu?, rham, amino sugar.] Several horse sera were available but few quantitative estimations have been made since so little is known of the chemistry of S VII. Many cross-reactions have been noted in Type VII antisera. Those of synthetic polyglucose, glycogen and dextran would seem to afford a confirmation of glucose in S VII, but little comment can be made on the others; conceivably, 1,3-linked *D*-galactose is indicated in S VII. S XVIII precipitates Type VII antisera, in accordance with the long-known cross-agglutination between the two types (*19*), but not as strongly as in the reverse cross-reaction (Table 3, p. 523).

8. Type VIII. — [S VIII: (cellobiuronic acid, cellobiose, $\alpha$-gal-1,4-)$_x$, see Chart 1, p. 507.] Several horse sera of not greatly differing cross-reactivities were available, as well as an unusual pool of strongly cross-reactive rabbit antisera. Of the numerous carbohydrates which have been found to precipitate in Type VIII antiserum, the classical example, that of S III, is, as already noted, due to the presence in both S III and S VIII of multiple units of cellobiuronic acid. Oxidized cotton also reacts strongly for the same reason (*50*), while the slight reactivity of the azotobacter polysaccharide (*74*), which also contains cellobiuronic acid units, is surprising. Another group of cross-reactions in Type VIII

antiserum is referable to multiple units of cellobiose*: barley and oat glucans (*1, 7, 91*) and Iles glucomannan (*96, 102*) are the substances involved. Indeed, the positive serological test (*59*) preceded the chemical isolation of cellobiose from barley and oat glucans by Prof. I. A. PREECE and confirmed the deductions made from their degradation by enzymes (*3*). The rather limited precipitation of the antisera by the galactomannans of carob, guar, and Kentucky coffee bean is apparently due to their multiple end groups of galactose, which seem able to engage the antibody surfaces complementary to the -α-1,4-galactose groupings of S VIII. Possibly, also, some of the adjacent hydrogens and hydroxyls of the adjoining mannose residue of the galactomannans are so placed sterically as to favor combination with the antibody.

9. Type IX. — [S IX: glu, uronic acid, amino sugar(s) (*45*).] All tests were carried out with New York State serum 623**. The reaction of the α-linked polyglucoses in this antiserum provides an obvious clue to its chemical basis. Quantitative evidence also points to the presence of α-1,4-linked glucose in S IX rather than solely α-1,6- or α-1,4,6-linked units (*51*), although a different interpretation is given to inhibitory experiments in Type IX antiserum by GOODMAN and KABAT (*32*).

10. Types X, XI, and XIII. — Since there are no data on the sugars comprising the specific polysaccharides of these types, aside from qualitative tests for amino sugars in X and XIII (*17*), little comment can be made on the appreciable cross-reactivity of the corresponding antisera. Since, however, barley and oat glucans react fairly strongly in anti-X and anti-XIII, it is probable that both S X and S XIII will be found to contain β-1,3- and/or β-1,4-linked glucose. Surprisingly many polysaccharides cross-react in antisera to all three of the types IX, X, and XI. S XIV gave a precipitate in the one anti-Pn X serum on hand, New York State 627***.

11. Type XII. — [S XII: gal, glu, amino sugar(s) (*45*).] New York State antiserum 625† and New York City antiserum 296†† were used. In these antisera the reaction of the α-polyglucoses appears due to their multiple α-1,6-, or α-1,4,6-linkages (*51*). Precipitation of the galactomannans and anthrax polysaccharide would seem referable to their end groups of galactose.

12. Type XIV. — [S XIV: D-gal, D-glu, NAc-glucosm, see Chart 1, p. 507.] As in the case of Type II, some two dozen cross-reactions have

---

* Comprising the glucose of the cellobiuronic acid and the adjoining glucose (cf. Chart 1, p. 507).
** Bleeding date, Jun. 9, 1939.
*** Bleeding date, May 24, 1939.
† Bleeding date, Dec. 21, 1938.
†† New York City preparation 15.

been found in antisera to this type*. Both S II and S XIV are highly branched, and while this may have no connection with the great cross-reactivity of the antibodies engendered by these determinants of specificity, branching results in an increase in the number of end-groups. The importance of end-groups in specificity has frequently been stressed [LANDSTEINER and VAN DER SCHEER (72), GOEBEL, AVERY and BABERS (30), KABAT (69, 70)]. It is to be hoped that more will be learned in the next few years of this structural chemical aspect of immunological specificity. So many of the polysaccharides precipitated by Type XIV antiserum contained non-reducing end-groups of galactose (35, 44, 48) that the prediction could be made that S XIV would also be found to possess such end-groups (48), and this has since been verified (14). The presence of $\beta$-galactose 1,3-, 1,6-, or 1,3,6- in S XIV could also be deduced, and this linkage has now been established as $\beta$-1,3- (14). There is a difference of opinion as to whether the anthrax cross-reaction also depends upon such a grouping (83) or in part upon the non-reducing end-group (44).

*13.* Types XV, XVI, and XVII. — [S XVII: gal, glu, rham, unknown sugar, pentitol? (67).] There is as yet no basis for discussion of the reactions encountered in these antisera.

*14.* Type XVIII. — [S XVIII: $D$-glu, $L$-rham, sec. $PO_4$ (77).] Most of the tests were carried out with New York State horse serum 495. Multiple recurrences of similar glucose linkages are obviously responsible for the precipitation of glycogen, synthetic polyglucoses, and dextran by this antiserum, but one cannot say more than that they appear to be $\alpha$-1,4-, 1,6-, or 1,4,6-. S VII also precipitates the antiserum, in conformity with a long-known cross-agglutination (19).

*15.* Type XIX. — [S XIX: glu, rham, N$Ac$ sugar, $PO_4$ (67).] Seventeen cross-reactions have been noted with an antiserum, New York State 631**, to this type. Several would appear to be due to glucose linkages similar to those in S XIX, which is now under investigation.

*16.* Types XX and XXII. — New York State anti-Pn XX No. 616***, and anti-Pn XXII New York City No. 243 and New York State No. 566 were used†. Both S XX and S XXII will undoubtedly be found to contain $\alpha$-linked glucose, since antisera to both types react with the $\alpha$-polyglucoses so far tested. Anti-XXII also precipitates barley and oat glucans, so that S XXII probably contains $\beta$-1,3- and/or $\beta$-1,4-bound glucose in addition. It may possess a uronic acid component as well, since many gums with such constituents also react.

---

* New York State 635, bleeding of May 25, 1939, was used for most tests.
** Bleeding date, Feb. 2, 1938.
*** Bleeding, Apr. 17, 1939.
† Respective bleeding dates, Apr. 17, and May 22, 1939.

*References, pp. 529—536.*

*17.* Types XXI, XXIV, XXV, XXVII, XXXII, and XXXIII. — Relatively few substances have been tested in the rather limited amounts of these antisera which were available. *Rhizobium radicicolum (98a)* gum precipitated heavily in anti-XXVII, New York City No. 668*. Gum arabic and gum ghatti gave precipitates with a rabbit anti-Pn XXXII pool.

*18.* Type XXIII. — [S XXIII: *D*-gal, *D*-glu, *L*-rham, $PO_4$, 2:2:2:1 (67).] New York City horse serum No. 912** was used. Reactivity of the galacto-mannans, gum arabic and gum ghatti in this antiserum would appear due to some of the galactose residues in these gums.

*19.* Types XXVIII, XXIX and XXXI. — A few cross-reactions have turned up among the limited number of substances tested in antisera to these types.

## IV. Cross-reactions in Anti-*Salmonella* Sera.

*1.* *S. typhi* [O-polysaccharide: *D*-gal, *D*-glu, *D*-man, *L*-rham, tyvelose (*21*)]. Two antisera, Nos. 154*** and 834†, obtained from horses at the Institut Pasteur, Paris, were available through the courtesy of Dr. ANNE-MARIE STAUB. Of the polysaccharides tested in these antisera, those of types II and VI pneumococcus and of anthrax yielded precipitates, also yeast mannan (*33, 33a*) and the gums of *Aloe vera*, corn fiber and ghatti. The reaction of S II and S VI may be due either to glucose or rhamnose, or both. That of yeast mannan appears to locate the linkages of the mannose in the O-polysaccharide of *S. typhi* at position 2- and/or 3- (*47*). Oxidation of the polysaccharide with periodate indicates that this is correct (*106*), but an alternative explanation of the cross-reaction might involve precipitation with antibody reactive with multiple tyvelose groups (*105*), since tyvelose is 3,6-di-deoxy-*D*-mannose (*26*).

*2.* *S. paratyphi* B [O-polysaccharide: *D*-gal, *D*-glu, *D*-man, *L*-rham, abequose (*21*)]. Horse serum 1137†† was also supplied by Dr. STAUB. The strong reactivity of the galactomannans appears to indicate that the mannose linkages in the B polysaccharide occur at position 4- or 6-, in contrast to 2- or 3- in the typhoid substance. Although the evidence of oxidation with periodate favors this interpretation, it is not excluded that the reaction takes place through the portion of antibody reactive with multiple groupings of abequose (*105*). This sugar is 3,6-di-deoxy-*D*-galactose (*26*), and the B polysaccharide contains *D*-galactose. The precipitation with oat glucan may point to β-1,3- or β-1,4-linkages for the glucose of the B substance.

---

* Preparation 3.
** Bleeding, May 4, 1939.
*** Bleeding of Apr. 5, 1957.
† Bleedings of Apr. 11, 1950 and Mar. 15, 1951.
†† Bleeding of Nov. 4, 1952.

## V. Cross-reactions in Anti-*Mycoplasma mycoides* Serum.

*Mycoplasma mycoides* [Specific polysaccharide: D-gal, (*89a*)]. A single bovine antiserum from cow no. 578* was available through the courtesy of Dr. A. W. Rodwell, Animal Health Research Laboratories, Parkville, Victoria, Australia, who also furnished information as to the polysaccharide. Only a few substances were tested in this serum, but a remarkably high proportion of positive reactions was obtained. The precipitation with oat glucan would point to the presence of glucose in this "galactan" or in another antigen of the mycoplasma.

## VI. Conclusion.

The extensive cross-reactivities which have been demonstrated and discussed in the course of this study are the result, with a few duly noted classical exceptions, of only a limited number of tests made during the last few years. It is evident that their scope could be greatly widened by the inclusion of other antisera containing antibodies to antigenic determinants of known constitution, by further chemical study of the type-specific polysaccharides of pneumococcus, some of which are happily under investigation in laboratories in several countries, and by further tests, with polysaccharides of known constitution, of sera containing antibodies to unknown determinants. Enough has already emerged from these studies to justify the belief that a small, well-selected stock of antisera will eventually be a part of the standard equipment of chemical laboratories engaged in the study of carbohydrates and polysaccharides. Such immunochemical reagents have, in these preliminary essays, furnished: (a) rapid tests for the possible presence of certain positions or linkages of a given sugar, numerous examples of which have been given in the course of this review; (b) materials for testing for heterogeneity in hetero-polysaccharides, as exemplified by the removal of most of the rhamnose from gum arabic by precipitation of the gum with anti-Pn II (*16, 39*); and (c) tests for heterogeneity in homopolysaccharides such as glycogen, by precipitation of successive portions with purified antibody and determination of the antigen-antibody ratios of the precipitates, as explained in detail in (*51*). These new developments in the chemistry of polysaccharides have been recognized by a chapter in Smith and Montgomery's book (*101 b*).

Since several pneumococcal polysaccharides and at least one of *Hemophilus influenzae* (*118*) contain organic phosphate, and others number amino sugars and polyols among their constituents, an important task remains to determine the as yet unknown function of these constituents in the over-all immunological specificity of the substances involved.

* Autopsy of July 1, 1953.

*References, pp. 529—536.*

That the amino sugars play a definite part would seem evident from the studies already carried out on blood-group substances [reviewed in (71)].

# VII. Tables.

Table 1. Specific Capsular Polysaccharides of Pneumococcus and Polysaccharides of Cross-reacting Microorganisms.

| Pn Type | Neutral sugars | Sugar acids | NH$_2$- or AcNH-sugars | Other |
|---|---|---|---|---|
| I | glu (?) | D-galacturonic | | unknown N-containing compounds |
| II | L-rham (50%), D-glu | D-glucuronic | | |
| III | D-glu | D-glucuronic | | |
| IV | gal | | AcNH-unidentified | |
| V | D-glu | D-glucuronic | two, unidentified | |
| VI | D-gal, D-glu, L-rham | | | ribitol, sec.-PO$_4$ |
| VII | gal, glu?, rham | | unidentified | |
| VIII | D-gal, D-glu | D-glucuronic | | |
| IX | glu | unidentified | unidentified | |
| XII | gal, glu | | unidentified | |
| XIV | D-gal, D-glu | | N-Ac-glucosamine | |
| XVII | gal, glu, rham, unidentified | | | polyol |
| XVIII | D-glu, L-rham | | | sec.-PO$_4$ |
| XIX | glu, rham | | AcNH-unidentified | PO$_4$ |
| XXIII | D-gal, D-glu, L-rham | | | PO$_4$ |

| Microorganisms | Neutral sugars | Sugar acids | NH$_2$- or AcNH-sugars |
|---|---|---|---|
| Aerobacter-Klebsiella | | | |
| A. aerogenes A 3 (S) ... | D-glu, L-fucose | glucur | |
| A. aerogenes 8172...... | D-glu, L-rham, D-man | glucur | |
| A. aerogenes 418....... | D-glu, D-man | D-glucur, D-mannuronic | |
| A. aerogenes 5920 (cloacae) (33b)...... | D-glu, D-gal, L-fuc | uronic | |
| A. aerogenes (Klebsiella pneum. B).......... | D-glu | glucur? | |
| Anthrax ............... | D-gal | | N-Ac-glucosm |
| Azotobacter chroococcum .. | D-gal, D-glu | D-glucur | |
| Cryptococcus neoformans A | gal, man, xylose | glucur | |
| Mycoplasma mycoides .... | D-gal | | |
| Rhizobium radicicolum.... | D-glu | D-glucur | |
| Salmonella typhi ........ | D-gal, D-glu, L-rham, D-man, tyvelose | | |
| Salmonella paratyphi B .. | D-gal, D-glu, L-rham, D-man, abequose | | |

Table 2. Polysaccharides Which Precipitate Antipneumococcal and Other Sera Moderately to Strongly.

(++ or more on a scale of — to ++++. Pn = pneumococcus, S = specific polysaccharide of pneumococcus, followed by type number.)

### Anti-Pn I.

Hualtaco, ketha, khaya grandifolia, khaya senegalensis, oat glucan, S III.

### Anti-Pn II.

Aerobacter aerogenes NCTC 418, also A 4, acrosiphonia, albizzia, amylopectins, azotobacter chroococcum, brachychiton, corn (maize) cob heteroglycan (*113*), corn fiber hemicellulose, dextrans, flax straw hemicellulose, glycogens and limit dextrins, gum arabic, gum ghatti, house dust, karaya, ketha, khaya grandifolia, klebsiella pneumoniae B, lung galactan (impurity), olibanum I, streptococcus "V" carbohydrate, synthetic polyglucoses, tamarind seed, taxus cuspidata (*104*).

### Anti-Pn III.

Azotobacter chroococcum, oxidized cotton, S VIII.

### Anti-Pn IV.

Arabogalactan of Jeffrey pine, okra.

### Anti-Pn V.

Acrosiphonia, aerobacter aerogenes 418, also A 4, albizzia, brachychiton, corn fiber hemicellulose, cryptococcus neof. A, gum arabic, gum ghatti, olibanum I, S II, degraded tororo aoi.

### Anti-Pn VI.

Barley and oat glucans, carob, guar, flax straw hemicellulose, Kentucky coffee bean, both khayas after removal of acetyl, okra, rhamnogalactan of B. polymyxa, S II.

### Anti-Pn VII.

Corn fiber hemicellulose, glycogen, gum ghatti, house dust, ketha, both khayas, okra, pseudomonas fluorescens*, tamarind seed, taxus cusp., S XVIII.

### Anti-Pn VIII.

Aerobac. aerog. A3 Sl (klebsiella type 54), aloe vera (*97*), barley and oat glucans, carob, guar, gum ghatti, hualtaco, Iles glucomannan, Kentucky coffee bean, ketha, nori utsugi, olibanum I, oxidized cotton, S III, synthetic polyglucoses, taxus.

### Anti-Pn IX.

Aerobac. aerog. 418, also 8172, aloe vera, amylopectins, azotobac., dextrans, E. coli O 8**, glycogen and limit dextrins, gum ghatti, ketha, khaya grandifolia, pseudomonas fluoresc., synthetic polyglucoses, tamarind seed, taxus.

---

* Fraction 12 of Drs. Y. Takeda and Kawai of the National Institute for Infectious Diseases, Tokyo.

** O-polysaccharide furnished by Prof. G. F. Springer.

*References, pp. 529—536.*

*(Table 2, continued.)*

### Anti-Pn X.

Azotobac., barley and oat glucans, cedrela\*, coccidioides immitis\*\*, cryptococcus neof. A, E. coli O 8, gum ghatti, house dust, ketha, khaya grandif., pseudomonas fluor., rhizobium radicic., S XIV, tamarind seed, taxus.

### Anti-Pn XI.

Aloe vera, cryptococcus A, ketha, pseudomonas fl., synth. polyglucoses, taxus.

### Anti-Pn XII.

Amylopectins, anthrax (from guinea pig), carob, dextrans, glycogen and limit dextrins, guar, house dust, Kentucky coffee bean, okra, synthetic polyglucoses, tamarind seed.

### Anti-Pn XIII.

Barley and oat glucans, brachychiton.

### Anti-Pn XIV.

Acrosiphonia, albizzia, aloe vera, anthrax (g. p.), arabogalactans of Jeffrey pine and larch (5, *18 a*), azotobac., barley and oat glucans, blood group substs. (*71*), brachychiton, carob, cedrela, corn cob heteroglycan, corn fiber hemicell., cryptococcus neof. A, dextran N 236, E. coli O 8, fagara (*108*), guar, gum arabic, gum ghatti, hualtaco, karaya, Kentucky coffee bean, ketha, khaya seneg., lung galactan, okra, olibanum I, physarum\*\*\*, pseudomon. fluor., rhizobium radicic., synthetic polyglucose, tamarind seed, taxus.

### Anti-Pn XV, XVI, XVII.

Aerobac. aerog. 418, aloe vera, azotobacter, house dust in XV. Acrosiphonia and physarum in XVI. Okra, both khayas in XVII.

### Anti-Pn XVIII.

Aerobac. aerog. K 26, aloe vera, dextran N 236, horse liver glycogen, house dust, synth. polyglucoses, S VII.

### Anti-Pn XIX.

Acrosiphonia, aerobac. aerog. 418, albizzia, aloe vera, anthrax, brachychiton, gum ghatti, hualtaco, ketha, both khayas, olibanum I and II, pseudomon. fluor., rhizob. radic., synth. polyglucose, tamarind seed, taxus.

### Anti-Pn XX.

Acrosiphonia, aerobac. aerog. A 4, albizzia, amylopectins, brachychiton, cryptococcus neof. A, dextrans, glycogen and limit dextrins, house dust, pseudomon. fluor., S II, synth. polyglucoses, tamarind seed.

---

\* Supplied by Prof. R. L. WHISTLER.

\*\* Polysaccharide sent by Prof. W. Z. HASSID.

\*\*\* Polysaccharide furnished by Prof. H. P. RUSCH. The substance should also be listed as reactive in anti-Pn VII.

(Table 2, continued.)

### Anti-Pn XXI, XXIV, XXV, XXVII.

Only rhizobium radic. precipitated type XXVII anti-Pn horse serum.

### Anti-Pn XXII.

Aerobac. aerog. 418 and K 26, albizzia, aloe vera, amylopectins, azotobac., barley and oat glucans, brachychiton, cedrela, coccid. immitis, corn cob heteroglycan, corn fiber hemicell., dextrans, E. coli O 8, glycogen and limit dextrins, gum ghatti, house dust., hualtaco, ketha, both khayas, olibanum I, pseudomon. fluor., rhizob. radicic., S II, synthetic polyglucoses, tamarind seed, taxus.

### Anti-Pn XXIII.

Aerobac. aerog. 8172, albizzia, carob, flax straw hemicellulose, guar, gum arabic, gum ghatti, hualtaco, Kentucky coffee bean, ketha, oat glucan, okra, olibanum I, taxus.

### Anti-Pn XXVIII, XXIX, XXXI, XXXII, and XXXIII.

Carob, guar, Kentucky coffee bean in XXVIII; aloe vera, corn fiber hemicell., gum ghatti in XXIX; gum arabic, gum ghatti in XXXI; gum arabic, gum ghatti in XXXII.

### Anti-Salmonella typhi.

Aloe vera, anthrax (g. p.), arabogalactan of Jeffrey pine, corn cob heteroglycan, corn fiber hemicell., E. coli O 8, glycogen, gum ghatti, ketha, S II, S VI, synth. poly-glucose, tamarind seed, taxus, yeast mannan.

### Anti-Salmonella paratyphi B.

Acrosiphonia, albizzia, arabogalactan of Jeffrey pine, carob., fagara, guar, gum ghatti, Kentucky coffee bean, oat glucan, okra, S XVIII, synth. polyglucose, taxus.

### Anti-Mycoplasma mycoides.

Acrosiphonia, albizzia, azotobacter, fagara, flax straw hemicellulose, oat glucan.

References, pp. 529—536.

Table 3. Cross-reactions of Polysaccharides of Microorganisms in Antipneumococcal and Other Antisera.

| Polysaccharide | Anti-Pn | | | | | | | | | | | | | |
|---|---|---|---|---|---|---|---|---|---|---|---|---|---|---|
| | I | II | III | IV | V | VI | VII | VIII | IX | X | XI | XII | XIII* | XIV |
| Pn S II | ++± | | | | | +++± | ±±± | ±+++ | | ±±± | – – – | ±±± | | ± |
| S III | | + | ± | | | +++± | ±±± | ±++± | | ±±± | – – – | ±±± | | ± |
| S VI | | | ± | | ±±+ | ±± | ±±± | ±+++ | | ±±± | – – – | ±±± | | ±± |
| S VII | | ±± | ±±±± | ± | ±± | + | ±±± | – | | ±+± | – | ±±± | | ±±± – |
| S VIII | | ±±±± | ±±±± | ±±± – ±±± | ±± – ±±± | – – – – – – | – + – – – – | – ±±++ | ± – ± – – | ±+ – – ±± | – – – ± | – – – – | – – – ± | ±±± – |
| S XIV | | ±± | | ±±± | ±++ | + | ±++ | ±+++ | ± – ± – – | ±+ | – – – ± | ±± | ±± – | ±±± – |
| S XVIII | | ± – | | ±± – ±±± | ±± – ±±± | – – – – – – | – + – – – – | ± – ± – – | ± – ± – – | + | – – | ±± | ± | ±±±± – |
| Aerobacter aerogenes, Friedländer B | – | ±± – – – | – – – | ±± – | ±±±±±± ± | – ± – – – – | ± – – – – – | – + – + ±± – ±± | ± ±± – – – ±± | ±±±±±± ±±±±±± ±±± | – ±± – + ±±±±++ ±±±±±± | ±±±± ± ±± ±±±±++ | ±± – – ±±±± – + | ±±±±++ ±±±±++ ±±±±++ ±±±±++ ±±±±++ |
| A. aerogenes, A 3 S1 | | | | | | | | | | | | | | |
| A. aerogenes, A 4 | | | | | | | | | | | | | | |
| A. aerogenes, A 29 | | | | | | | | | | | | | | |
| A. aerogenes, K 26 | | | | | | | | | | | | | | |
| A. aerogenes, NCTC 418 | | | | | | | | | | | | | | |
| A. aerogenes, NCTC 8172 | | | | | | | | | | | | | | |
| A. aerogenes, N 5920 | | | | | | | | | | | | | | |
| Anthrax | | | | | | | | | | | | | | |
| Azotobacter chrooc. | | | | | | | | | | | | | | |
| Coccidioides immitis | | | | | | | | | | | | | | |
| Cryptococcus neof. A.† | | | | | | | | | | | | | | |
| E. coli O 8 | + | | | | | | | | | | | | | |
| Physarum polycephal. | + | ±±± | | | ±± | ±± | ±±± | ±± | ±± | ±± | | – ± | + | ±±±±++ |
| Pseudomonas fluoresc. | ±± | ±±± | – | | ±± | ±± | | | ±± | ±± | | | | |
| Rhizobium radicic. | – | | | | | | | | | | | | | |

\* Rabbit antisera.

† Polysaccharides of *Cryptococcus neoformans* B and C did not cross-react in anti-Pn I–XII, XIV.

(Table 3, continued.)

| Polysaccharide | Anti-Pn | | | | | | | | | | | | Other | | |
|---|---|---|---|---|---|---|---|---|---|---|---|---|---|---|---|
| | XV | XVI | XVII* | XVIII | XIX | XX | XXII | XXIII | XXVII | XXVIII | XXIX* | XXXI* | Ty | p-Ty B | Bov. Myco-plasma |
| Pn S II | | | | ± | ±+ | ++ | ++ | | | | | | +++ | + | |
| S III | | | | +++ | ± | ± | ± | | | | | | +++ | ± | |
| S VI | | | | ±++ | | | | | | | | | ±+ | ±+ | |
| S VII | | | | +++ | | | | | | | | | +± | ++ | |
| S VIII | | | | +++ | | | | | | | | | + | ++ | |
| S XIV | | | | | | | | | | | | | ±+ | | |
| S XVIII | | | | — | | | | | | | | | ±± | | |
| *Aerobacter aerogenes,* | | | | | | | | | | | | | | | |
| Friedländer B | | | | | ±+ | ±+ | ++ | | | | | | | | |
| *A. aerogenes,* A 3 S 1 | — | — | | | ±—+ | ±+ | ++ | ± | | | | | ± | | |
| *A. aerogenes,* A 4 | — | +±+± | | | ±—+ | —+± | ++± | | | | | | | | |
| *A. aerogenes,* A 29 | — | ±+ | ±—+ | | +± | +± | ±+± | +± | | | | | ±+ | ±+ | +++ |
| *A. aerogenes,* K 26 | — | | | +± | ±+±+ | +±++ | +±++ | ++±+ | | | | | ±+ | ±+± | + |
| *A. aerogenes,* NCTC 418 | ++++ | — | ±—+ | — | ±+±+ | ±+±+ | ±+±+ | +++± | | | | | ±+ | +++± | |
| *A. aerogenes,* NCTC 8172 | ±— | ±+ | ±—+ | | ±+±+ | ±+±+ | +++± | ± | +±+± | +±+ | | | ±+ | ±+± | |
| *A. aerogenes,* N 5920 | ++ | — | —+ | +±—+ | +±++ | +±++ | +++ | + | +++ | + | | | +++ | +++ | + |
| Anthrax | — | | ±—+ | +± | ±—+ | ±—+ | +± | | | | | | | | |
| Azotobacter chrooc. | — | | —+ | — | ±+±+ | ±+±+ | ±+++ | | | | | | | | |
| Coccidioides immitis | ±± | | ±—+ | | +±+± | +±+± | +±+± | + | | | | | | | |
| Cryptococcus neof. A.† | — | | +±—+ | | +±++ | +±++ | +++ | | | | | | | | |
| E. coli O 8 | +±± | | | | +±+± | +±+± | +++ | | | | | | | | |
| Physarum polycephal. | | | | | ±+±+ | ±+++ | +++ | | | | | | ±++ | ±+± | |
| Pseudomonas fluoresc. | — | +± | | | +±++ | +±++ | +++ | | +±+ | | | | +++ | +++± | |
| Rhizobium radicic. | — | — | — | — | +±+± | +±+± | +++ | | +±+ | + | | | ±+± | ±+± | |
| Yeast mannan | — | | | | | | | | | | | | +++± | ±++ | + |

* Rabbit antisera.
† Polysaccharides of *Cryptococcus neoformans* B and C did not cross-react in anti-Pn XV, XVIII–XX, XXII, XXVIII.

References, pp. 529—536.

Table 4. Cross-reactions of Animal and Plant Polysaccharides in Antipneumococcal and Other Antisera.

| Polysaccharide | Anti-Pn | | | | | | | | | |
|---|---|---|---|---|---|---|---|---|---|---|
| | I | II | III | IV | V | VI | VII | VIII | IX | X |
| Synthetic polyglucose A | −+ | +++ | −− | +±+ | ±+ | −± | +++ | +++ | +++ | −+ |
| Dextran N 236 | + | +++ | − | +± | ±+ | ± | ±±± | ±±+ | ±±± | + |
| Oyster glycogen A | ++ | +±+ | + | ±± | ±± | ++ | ±±++ | +++± | ±±+ | +++ |
| Amylopectin | ++ | +± | | | | | ++± | +++± | ±±+ | +±+ |
| Barley, oat glucans | +±± | ±± | | | | | +± | +++ | − | ±± |
| Lung galactan | ++ | ±+ | | | | | | | | +++ |
| Iles glucomannan | +±±± | −+ | +− | +±±+ | ++− | ++++ | ±±++ | +++ | ±±+ | ±±++ |
| Carob, guar | +±±± | +±±+ | −+ | +±++ | +±++ | ++± | +±±+ | ±±+ | ±±+ | +++± |
| Tamarind seed | +±± | +++± | ±± | ±± | ++± | −+ | +±±+ | −+ | − | +++ |
| Okra | ++ | ++ | −+ | ±± | +++ | − | +++ | +± | + | −+ |
| Karaya | +±+ | +±++ | −− | −− | ±+ | −±+ | +++ | ±±++ | ±±++ | +±++± |
| Gum arabic | ±±±± | ++±+ | −+ | +± | +++ | ±++ | +++ | ±±++ | +++ | +±++ |
| Gum ghatti | ++±± | +++± | ±± | − | +++± | ± | ±±++ | +±±+ | ++++ | +±+± |
| Khaya grandifolia | ±±± | ++± | −− | − | ++ | ±+ | +++± | ±±++ | +++ | +±±± |
| Khaya senegalensis | ±±++ | +++ | −+ | − | +++ | ±+ | ±±±+ | +++ | + | +±±++ |
| Beech hemicellulose | ±± | ++ | +±± | − | +++ | −± | ++±± | ±++ | + | ±±++ |
| Flax straw hemicellulose | | | ±±+ | − | +++ | ± | +±±± | +±++ | + | +±++ |
| Corn fiber (hull) hemicellulose | −− | | ±++± | +± | ±++ | ± | −+ | +±+ | −+ | ±±± |
| Corn cob heteroglycan | −− | ±±+ | +±± | − | +++ | −+ | +±±+ | +± | ±++ | +++ |
| Degraded nori utsugi | −− | | +±± | − | ±±++ | −± | +±± | − | − | ±++ |
| Degraded tororo aoi | −± | ++ | −+ | − | ++ | − | ±±+ | +± | +± | +±+± |
| Olibanum I | | | | | | | | | | |
| Olibanum II | ±±+ | +±+ | ±±+ | − | ±±± | −± | +±± | ±++ | −+ | +±+ |
| Aloe vera | +±++ | ±±+ | ±± | − | +±+ | − | +±±+ | +± | +±+ | +±++ |
| "Hualtaco" | +++± | +++± | ±± | − | ±+ | − | +++ | ±±+ | ±±+ | ±+++ |
| Ketha | +±± | ±+ | − | − | ±+ | − | +++ | − | − | +++ |
| Taxus cuspidata | | | | | | | | | | |
| Cedrela | | | | | | | | | | |

(Table 4, continued.)

| Polysaccharide | Anti-Pn | | | | | | | | | |
|---|---|---|---|---|---|---|---|---|---|---|
| | I | II | III | IV | V | VI | VII | VIII | IX | X |
| Acrosiphonia | + | +++ | + | — | ++ | — | — | +++ | — | ± |
| Albizzia | ± | +++ | — | ± | +++ | ± | ± | ++ | ± | ± |
| Brachychiton | ± | +++ | ± | — | ++ | — | + | ± | — | ± |
| Fagara | — | + | + | ± | + | — | + | — | — | + |
| Arabogalactan, Jeffrey pine | — | — | — | ± | — | — | + | — | — | — |
| House dust | + | ++± | + | — | ± | — | ++± | ± | ± | ++ |

| Polysaccharide | Anti-Pn | | | | | | | | | |
|---|---|---|---|---|---|---|---|---|---|---|
| | XI | XII | XIII* | XIV | XV | XVI | XVII* | XVIII | XIX | XX |
| Synthetic polyglucose A | ++ | ++++ | ±± | ++± | — | — | — | ++ | — | ++++ |
| Dextran N 236 | — | +++ | ++ | + | — | + | — | +++ | — | +++ |
| Oyster glycogen A | + | +++ | — | — | ± | — | — | ++ | — | +++ |
| Amylopectin | ±± | +± | ++ | +++ | ± | — | ± | +± | ± | ++ |
| Barley, oat glucans | — | ± | ±± | +++ | ±± | — | ++ | — | — | ++ |
| Lung galactan | + | +++ | — | +++ | — | — | +± | + | + | + |
| Carob, guar | ± | ++± | — | +++ | ± | ± | +±± | + | ++ | ++ |
| Tamarind seed | — | ++ | — | +++ | — | ± | ++ | + | — | ± |
| Okra | — | — | — | ++± | — | — | +±± | — | — | ++ |
| Karaya | — | — | — | +± | — | — | ++± | — | — | + |
| Gum arabic | — | — | — | ++ | ± | — | — | — | — | — |
| Gum ghatti | ± | — | — | + | — | — | — | — | — | ± |
| Khaya grandifolia | ± | — | — | ++± | — | — | ++± | — | ++± | ++ |
| Khaya senegalensis | ++± | — | — | +± | ±± | — | +++ | — | +++± | + |
| Beech hemicellulose | +++ | — | — | ++ | ± | — | +± | — | +± | — |
| Corn fiber (hull) hemicellulose | +± | — | — | +++ | — | — | +± | ± | ++ | ++± |

\* Rabbit antisera. Some of the tests in anti-Pn XVI were also carried out in such sera.

References, pp. 529—536.

(Table 4, continued.)

| Polysaccharide | Anti-Pn | | | | | | | | | | Other | | |
|---|---|---|---|---|---|---|---|---|---|---|---|---|---|
| | XI | XII | XIII* | XIV | XV | XVI | XVII* | XVIII | XIX | XX | Ty | p-Ty B | Bov. Myco-plasma |
| Corn cob heteroglycan | ⧺ | — | — | ⧺ | — | — | — | — | + | — | | | |
| Degraded nori utsugi | ⧺ | | — | +++ | ⧺ | | | + | +++ | ⧺ | | | |
| Degraded tororo aoi | — | — | — | +++ | ⧺ | — | — | ⧺ | +++ | + | | | |
| Oiibanum I | + | | — | ⧺ | — | | | — | +++ | + | | | |
| Oiibanum II | — | | — | ⧺ | — | | | — | +++ | ⧺ | | | |
| Aloe vera | +++ | | — | +++ | ⧺ | — | + | +++ | +++ | ⧺ | | | |
| "Hualtaco" | +++ | | — | +++ | + | | ⧺ | +++ | +++ | + | | | |
| Ketha | ⧺ | — | — | +++ | ⧺ | ⧺ | ⧺ | — | +++ | ⧺ | | | |
| Taxus cuspidata | ⧺ | — | ⧺ | +++ | — | + | ⧺ | — | +++ | + | | | |
| Cedrela | + | | — | +++ | — | ⧺ | ⧺ | — | ⧺ | — | | | |
| Acrosiphonia | ⧺ | ⧺ | — | +++ | — | ⧺ | ⧺ | ⧺ | +++ | +++ | | | |
| Albizzia | + | | ⧺ | +++ | — | ⧺ | — | — | +++ | +++ | | | |
| Brachychiton | — | ⧺ | ⧺ | +++ | — | ⧺ | + | ⧺ | +++ | +++ | | | |
| Fagara | — | | + | +++ | — | ⧺ | — | — | ⧺ | ⧺ | | | |
| Arabogalactan, Jeffrey pine | | ⧺ | ⧺ | ++++ | — | ⧺ | | — | ⧺ | | | | |
| House dust | ⧺ | +++ | ⧺ | ⧺ | ++ | ⧺ | ⧺ | +++ | ⧺ | ++ | | | ++ |

| Polysaccharide | Anti-Pn | | | | | | | Other | | |
|---|---|---|---|---|---|---|---|---|---|---|
| | XXII | XXIII | XXVII | XXVIII | XXIX* | XXXI* | XXXII* | Ty | p-Ty B | Bov. Myco-plasma |
| Synthetic polyglucose A | ++++ | | | | | | | +++ | ++ | |
| Dextran N 236 | +++ | | | | | | | +++ | | |
| Oyster glycogen A | +++ | | | | | | | | | |
| Amylopectin | +++ | | — | + | | | | ⧺ | ⧺ | +++ |
| Barley, oat glucans | ⧺ | ++ | ⧺ | | | | | + | | |
| Lung galactan | + | | | | | | | | | |
| Flax straw hemicellulose | | ++ | | | | | | | | ++ |

* Rabbit antisera. Some of the tests in anti-Pn XVI were also carried out in such sera.

(Table 4, continued.)

| Polysaccharide | Anti-Pn | | | | | | | Other | | |
| --- | --- | --- | --- | --- | --- | --- | --- | --- | --- | --- |
| | XXII | XXIII | XXVII | XXVIII | XXIX* | XXXI* | XXXII* | Ty | p-Ty B | Bov. Mycoplasma |
| Iles glucomannan | ± | ++± | | +++± | | +±|± | | ++± | ++++ | |
| Carob, guar | ++± | ++++ | | ++++ | ++ | ++++ | ++ | ±| | ±+++ | |
| Tamarind seed | + | ++++ | || | ± | | | ++ | | ++ | |
| Okra | | +± | | | +++ | | | + | + | |
| Gum arabic | +++± | +++± | | | | | | | | |
| Gum ghatti | ++± | ++± | | | | | | | | |
| Khaya grandifolia | +++ | + | | | | | | | | |
| Khaya senegalensis | + | | | | | | | | | |
| Beech hemicellulose | ++ | ++± | ||| | ++±|+±| | ++ | ++ | +± | ++±+++±++ | ++±+++± | |
| Corn fiber (hull) hemicellulose | | | | | | ++± | | | | |
| Corn cob heteroglycan | + | | | | ++ | | | | | |
| Degraded nori utsugi | +± | | | | | | | | | |
| Degraded tororo aoi | ± | | | | | | | | | |
| Olibanum I | ++± | +± | | +± | | | | | | |
| Olibanum II | ++ | +± | | | | | | | | |
| Aloe vera | +± | +± | | | | | | | | |
| "Hualtaco" | +++ | +++ | | | | | | | | |
| Ketha | +±++ | + | | | +± | ++ | ±± | ++ | ++ | ++ |
| Taxus cuspidata | +++ | +++ | | ++ | | | | +++±| | +++± | ++++ |
| Cedrela | ++ | ++± | | | | | | ± | +| | ++ |
| Acrosiphonia | +++ | + | | + | | | | ++++±++ | +++±| | ++ |
| Albizzia | +++ | | | +| | | | | +++± | +++± | |
| Brachychiton | | ± | | ++±+| | | | | ++± | ++++ | |
| Fagara | | | | + | | | | ±+± | ++± | |
| Arabogalactan, Jeffrey pine | ++ | +± | | | | |||| | | +++± | +++± | |
| House dust | | +± | + | ± | ± | | | | | |

* Rabbit antisera. Some of the tests in anti-Pn XVI were also carried out in such sera.

References, pp. 529—536.

### References.

*1.* ACKER, L., W. DIEMAIR und E. SAMHAMMER: Über das Lichenin des Hafers. II. Molekulargewichtsbestimmungen und weitere Untersuchungen zu den Konstitutionen. Z. Lebensm. Untersuch. Forschg. **102**, 225 (1955).

*2.* AHMED, Z. F. and R. L. WHISTLER: The Structure of Guaran. J. Amer. Chem. Soc. **72**, 2524 (1950).

*3.* AITKEN, R. A., B. P. EDDY, M. INGRAM and C. WEURMAN: The Action of Culture Filtrates of the Fungus *Myrothecium verrucaria* on β-Glucosans. Biochemic. J. **64**, 63 (1956).

*4.* ASPINALL, G. O., B. J. AURET and E. L. HIRST: Gum Ghatti (Indian Gum). III. Neutral Oligosaccharides formed on Partial Acid Hydrolysis of the Gum. J. Chem. Soc. (London) **1958**, 4408.

*4 a.* ASPINALL, G. O., E. L. HIRST and N. K. MATHESON: Plant Gums of the Genus *Khaya*. The Structure of *Khaya grandifolia* Gum. J. Chem. Soc. (London) **1956**, 989.

*5.* ASPINALL, G. O., E. L. HIRST and E. RAMSTAD: The Constitution of Larch ε-Galactan. J. Chem. Soc. (London) **1958**, 593.

*6.* ASPINALL, G. O., R. S. P. JAMIESON and J. F. WILKINSON: The Structure of the Extracellular Polysaccharide of *Aerobacter aerogenes* A 3 (S l) (*Klebsiella* type 54). J. Chem. Soc. (London) **1956**, 3483.

*7.* ASPINALL, G. O. and R. G. J. TELFER: Cereal Gums. I. The Methylation of Barley Glucosans. J. Chem. Soc. (London) **1954**, 3519.

*8.* AVERY, O. T.: A Further Study of the Biologic Classification of Pneumococci. J. exp. Medicine **22**, 804 (1915).

*9.* AVERY, O. T., W. F. GOEBEL and F. H. BABERS: Chemo-immunological Studies on Conjugated Carbohydrate-Proteins. VII. Immunological Specificity of Antigens Prepared by Combining α- and β-Glucosides of Glucose with Proteins. J. exp. Medicine **55**, 769 (1932).

*10.* AVERY, O. T. and M. HEIDELBERGER: Immunological Relationships of Cell Constituents of Pneumococcus. J. exp. Medicine **38**, 81 (1923).

*11.* AVERY, O. T., M. HEIDELBERGER and W. F. GOEBEL: The Soluble Specific Substance of Friedländer's Bacillus. II. Chemical and Immunological Relationships of Pneumococcus Type II and of a Strain of Friedländer's Bacillus. J. exp. Medicine **42**, 709 (1925).

*12.* BARKER, S. A.: unpublished data.

*13.* BARKER, S. A., A. B. FOSTER, I. R. SIDDIQUI and M. STACEY: Structure of the Capsular Polysaccharide of *Aerobacter aerogenes* (N. C. T. C. 418). J. Chem. Soc. (London) **1958**, 2358.

*13 a.* — — — — Structure of an Acidic Polysaccharide Elaborated by *Aerobacter aerogenes*. Nature (London) **181**, 999 (1958).

*14.* BARKER, S. A., M. HEIDELBERGER, M. STACEY and D. J. TIPPER: Immunopolysaccharides. X. The Structure of the Immunologically Specific Polysaccharide of Pneumococcus Type XIV. J. Chem. Soc. (London) **1958**, 3468.

*15.* BARKER, S. A. and J. M. WILLIAMS: unpublished data.

*16.* BEISER, S. M., E. A. KABAT and J. M. SCHOR: Immunochemical Studies on the Specific Polysaccharide of Type II Pneumococcus. J. Immunology **69**, 297 (1952).

*17.* BROWN, R.: Chemical and Immunological Studies of the Pneumococcus. V. The Soluble Specific Substances of Types I-XXXII. J. Immunology **37**, 445 (1939).

*18.* BUTLER, K. and M. STACEY: Immunopolysaccharides. IV. Structural Studies on the Type II Pneumococcus Specific Polysaccharide. J. Chem. Soc. (London) **1955**, 1537.

*18a.* CAMPBELL, W. G., E. L. HIRST and J. K. N. JONES: The ε-Galactan of Larch Wood *(Larix decidua).* J. Chem. Soc. (London) **1948**, 774.

*19.* COOPER, G., C. ROSENSTEIN, A. WALTER and L. PEIZER: The Further Separation of Types among the Pneumococci hitherto Included in Group IV and the Development of Therapeutic Antisera for these Types. J. exp. Medicine **55**, 531 (1932).

*20.* CUNNEEN, J. I. and F. SMITH: The Constitution of Mesquite Gum. I. Isolation of 6- and 4-Glucuronosidogalactose. — II. Methylated Mesquite Gum. J. Chem. Soc. (London) **1948**, 1141, 1146.

*21.* DAVIES, D. A. L., A. M. STAUB, I. FROMME, O. LÜDERITZ and O. WESTPHAL: Contribution of Deoxymethylpentoses to the Serological Specificity of some Bacterial Polysaccharides, and the Recognition of a new Sugar, Paratose. Nature (London) **181**, 822 (1958).

*22.* DUDMAN, W. F. and J. F. WILKINSON: The Composition of the Extracellular Polysaccharides ot *Aerobacter-Klebsiella* Strains. Biochemic. J. **62**, 289 (1956).

*23.* EL-KHADEM, H. and M. M. MEGAHED: The Gum Component of Olibanum. J. Chem. Soc. (London) **1956**, 3953.

*24.* EVANS, E. E. and R. J. THERIAULT: The Antigenic Composition of *Cryptococcus neoformans.* IV. The Use of Paper Chromatography for Following Purification of the Capsular Polysaccharide. J. Bacteriol. **65**, 571 (1953).

*25.* FELTON, L. D.: Studies on the Immunizing Substances in Pneumococci. II. Separation of the Organism into Acid Soluble and Acid Insoluble Fractions. J. Immunology **27**, 379 (1934).

*26.* FOUQUEY, C., E. LEDERER, O. LÜDERITZ, I. POLONSKY, A. M. STAUB, S. STIRM, R. TINELLI et O. WESTPHAL: Synthèses de 3,6-didésoxyhexoses; détermination de la structure des sucres naturels: abéquose, colitose, tyvelose et ascarylose. C. R. hebd. Séances Acad. Sci. **246**, 2417 (1958).

*27.* FRANCIS, T., Jr. and W. S. TILLETT: Cutaneous Reactions in Pneumonia. The Development of Antibodies Following the Intradermal Injection of Type-specific Polysaccharide. J. exp. Medicine **52**, 573 (1930).

*28.* GEERDES, J. D. and F. SMITH: The Constitution of the Hemicellulose of the Straw of Flax *(Linum usitatissimum* Sp.). I. Identification of 2-O-(4-O-Methyl-D-glucuronosyl)-D-xylose. — II. Hydrolysis of the Methylated Hemicellulose. J. Amer. Chem. Soc. **77**, 3569, 3572 (1955).

*29.* GOEBEL, W. F.: Chemoimmunological Studies on the Soluble Specific Substance of Pneumococcus. II. The Chemical Basis of the Immunological Relationship between the Capsular Polysaccharides of Types III and VIII Pneumococcus. J. Biol. Chem. **110**, 391 (1935).

*30.* GOEBEL, W. F., O. T. AVERY and F. H. BABERS: Chemo-immunological Studies on Conjugated Carbohydrate-Proteins. IX. The Specificity of Antigens Prepared by Combining the *p*-Aminophenol Glycosides of Disaccharides with Proteins. J. exp. Medicine **60**, 599 (1934).

*31.* GOODMAN, J. W. and E. A. KABAT: Immunochemical Studies on Cross-reactions of Antipneumococcal Sera. I. Cross-reactions of Type II and XX Antipneumococcal Sera with Dextrans and of Type II Antipneumococcal Sera with Glycogen and Friedländer Type B Polysaccharide. J. Immunology **84**, 333 (1960).

*32.* — — Immunochemical Studies on Cross-reactions of Antipneumococcal Sera. II. Cross-reactions of Types IX and XII Antipneumococcal Sera with Dextrans. J. Immunology **84**, 347 (1960).

*33.* HAWORTH, W. N., R. L. HEATH and S. PEAT: The Constitution of Yeast Mannan. J. Chem. Soc. (London) **1941**, 833.

*33a.* HAWORTH, W. N., E. L. HIRST and F. A. ISHERWOOD: Polysaccharides. XXIV. Yeast Mannan. J. Chem. Soc. (London) **1937**, 784.

*33b.* HEATH, E. C. and S. ROSEMAN: The Conversion of *D*-Glucose to *L*-Fucose by *Aerobacter cloacae.* J. Biol. Chem. **230**, 511 (1958).

*34.* HEIDELBERGER, M.: Quantitative Absolute Methods in the Study of Antigen-Antibody Reactions. Bacteriol. Rev. **3**, 49 (1939). — Chemical Aspects of the Precipitin and Agglutinin Reactions. Chem. Revs. **24**, 323 (1939).

*35.* — Immunological Specificities Involving Multiple Units of Galactose. II. J. Amer. Chem. Soc. **77**, 4308 (1955).

*36.* — The Formation of Antibodies in Man after Injection of Pneumococcal Polysaccharides. Proc. Nat. Acad. Sci. (USA) **43**, 883 (1957).

*37.* — The Immunological Specificity of Type II Pneumococcus and its Separation into Partial Specificities. II. J. exp. Medicine **111**, 33 (1960).

*38.* HEIDELBERGER, M. and J. ADAMS: The Immunological Specificity of Type II Pneumococcus and its Separation into Partial Specificities. J. exp. Medicine **103**, 189 (1956).

*39.* HEIDELBERGER, M., J. ADAMS and Z. DISCHE: Fractionation of Gum Arabic by Chemical and Immunological Procedures. J. Amer. Chem. Soc. **78**, 2853 (1956).

*40.* HEIDELBERGER, M. and A. C. AISENBERG: Serological Reactivity of Synthetic Polyglucoses. Proc. Nat. Acad. Sci. (USA) **39**, 453 (1953).

*41.* HEIDELBERGER, M., A. C. AISENBERG and W. Z. HASSID: Glycogen, an Immunologically Specific Polysaccharide. J. exp. Medicine **99**, 343 (1954).

*42.* HEIDELBERGER, M. and O. T. AVERY: The Soluble Specific Substance of Pneumococcus. I. J. exp. Medicine **38**, 73 (1923). — II. J. exp. Medicine **40**, 301 (1924).

*43.* HEIDELBERGER, M., O. T. AVERY and W. F. GOEBEL: A "Soluble Specific Substance" Derived from Gum Arabic. J. exp. Medicine **49**, 847 (1929).

*44.* HEIDELBERGER, M., S. A. BARKER and B. BJÖRKLUND: Immunological Specificities Involving Multiple Units of Galactose. III. J. Amer. Chem. Soc. **80**, 113 (1958).

*45.* HEIDELBERGER, M., S. A. BARKER and M. STACEY: Components of the Specific Polysaccharides of Types IX, XII, and XIV Pneumococcus. Science (Washington) **120**, 781 (1954).

*46.* HEIDELBERGER, M., B. BJÖRKLUND and J. LARNER: Cross Reactions of Polyglucoses in Antipneumococcal Sera. V. Precipitation by Glycogens and Limit Dextrins. J. Immunology **78**, 431 (1957).

*47.* HEIDELBERGER, M. and F. CORDOBA: Cross-reactions of Antityphoid and Antiparatyphoid B Horse Sera with Various Polysaccharides. J. exp. Medicine **104**, 375 (1956).

*48.* HEIDELBERGER, M., Z. DISCHE, W. B. NEELY and M. L. WOLFROM: Immunochemistry and the Structure of Lung Galactan. J. Amer. Chem. Soc. **77**, 3511 (1955).

*49.* HEIDELBERGER, M., W. F. GOEBEL and O. T. AVERY: The Soluble Specific Substance of Pneumococcus. III. J. exp. Medicine **42**, 727 (1925).

*50.* HEIDELBERGER, M. and G. L. HOBBY: Oxidized Cotton, an Immunologically Specific Polysaccharide. Proc. Nat. Acad. Sci. (USA) **28**, 516 (1942).

*51.* HEIDELBERGER, M., H. JAHRMÄRKER, B. BJÖRKLUND and J. ADAMS: Cross Reactions of Polyglucoses in Antipneumococcal Sera. III. Reactions in Horse Sera. J. Immunology **78**, 419 (1957).

*52.* HEIDELBERGER, M., E. A. KABAT and M. MAYER: A Further Study of the Cross Reaction between the Specific Polysaccharides of Types III and VIII Pneumococci in Horse Antisera. J. exp. Medicine **75**, 35 (1942).

*52 a.* Heidelberger, M., E. A. Kabat and D. L. Shrivastava: A Quantitative Study of the Cross Reaction of Types III and VIII Pneumococci in Horse and Rabbit Antisera. J. exp. Medicine **65**, 487 (1937).

*53.* Heidelberger, M. and F. E. Kendall: A Quantitative Study of the Precipitin Reaction between Type III Pneumococcus Polysaccharide and Purified Homologous Antibody. J. exp. Medicine **50**, 809 (1929), and later papers.

*54.* — — Specific and Non-specific Polysaccharides of Type IV Pneumococcus. J. exp. Medicine **53**, 625 (1931). [Also unpublished experiments by Dr. P. A. Rebers.]

*55.* — — Studies on the Precipitin Reaction. Precipitating Haptens; Species Differences in Antibodies. J. exp. Medicine **57**, 373 (1933).

*56.* — — The Precipitin Reaction between Type III Pneumococcus Polysaccharide and Homologous Antibody. III. A Quantitative Study and a Theory of the Reaction Mechanism. J. exp. Medicine **61**, 563 (1935).

*57.* Heidelberger, M., F. E. Kendall and H. W. Scherp: The Specific Polysaccharides of Types I, II, and III Pneumococcus. A Revision of Methods and Data. J. exp. Medicine **64**, 559 (1936).

*58.* Heidelberger, M. and M. McCarty: Cross-reactions of Streptococcal A and V Carbohydrates in Antipneumococcal Horse Sera. Proc. Nat. Acad. Sci. (USA) **45**, 235 (1959).

*59.* Heidelberger, M. and P. A. Rebers: Cross Reactions of Polyglucoses in Antipneumococcal Sera. VI. Precipitation of Type VIII and Type III Antisera by β-Glucans. J. Amer. Chem. Soc. **80**, 116 (1958).

*60.* — — Immunochemistry of the Pneumococcal Types II, V, and VI (II, II A, II B). I. The Relation of Type VI to Type II and other Correlations between Chemical Constitution and Precipitation in Antisera to Type VI. J. Bacteriol. 1960 (in press).

*61.* Hirst, E. L. and S. Dunstan: The Structure of Karaya Gum *(Cochlospermum gossypium)*. J. Chem. Soc. (London) **1953**, 2332.

*62.* Hirst, E. L. and J. K. N. Jones: The Galactomannan of Carob-seed Gum (Gum Gatto). J. Chem. Soc. (London) **1948**, 1278.

*63.* Hirst, E. L., E. Percival and R. S. Williams: The Structure of *Brachychiton diversifolium* Gum *(Sterculia caudata)*. J. Chem. Soc. (London) **1958**, 1942.

*63 a.* Hirst, E. L. and A. S. Perlin: The Gum of *Acacia pycnantha*. J. Chem. Soc. (London) **1954**, 2622.

*64.* Hotchkiss, R. D. and W. F. Goebel: Chemo-immunological Studies on the Soluble Specific Substance of Pneumococcus. III. The Structure of the Aldobionic Acid from the Type III Polysaccharide. J. Biol. Chem. **121**, 195 (1937), and later papers.

*65.* Ivánovics, G.: Untersuchungen über das Polysaccharid der Milzbrandbazillen. Z. Immunitätsforsch. **97**, 402 (1939).

*65 a.* — Die immunbiologische Verwandtschaft zwischen dem Anthrax-Polysaccharid, der Pneumokokkus-Typus-XIV-Kapselsubstanz und der spezifischen Substanz der menschlichen roten Blutkörper der Gruppe A. Z. Immunitätsforsch. **98**, 373 (1940).

*66.* Jeanes, A., W. C. Haynes, C. A. Wilham, J. C. Rankin, E. H. Melvin, M. J. Austin, J. E. Cluskey, B. E. Fisher, H. M. Tsuchiya and C. E. Rist: Characterization and Classification of Dextrans from Ninety-six Strains of Bacteria. J. Amer. Chem. Soc. **76**, 5041 (1954).

*67.* Jones, J. K. N.: unpublished data.

*68.* Jones, J. K. N. and M. B. Perry: The Structure of the Type VIII Pneumococcus Specific Polysaccharide. J. Amer. Chem. Soc. **79**, 2787 (1957).

69. KABAT, E. A.: Some Configurational Requirements and Dimensions of the Combining Site on an Antibody to a Naturally Occurring Antigen. J. Amer. Chem. Soc. 76, 3709 (1954).

70. — Heterogeneity in Extent of the Combining Regions of Human Antidextran. J. Immunology 77, 377 (1956).

71. — Blood Group Substances. Their Chemistry and Immunochemistry. New York: Academic Press. 1956.

71a. KABAT, E. A. and D. BERG: Dextran, an Antigen in Man. J. Immunology 70, 514 (1953).

72. LANDSTEINER, K. and J. VAN DER SCHEER: On the Serological Specificity of Peptides. J. exp. Medicine 55, 781 (1932).

73. LARSON, E. B. and F. SMITH: The Constitution of the Galactomannan of the Seeds of the Kentucky Coffee Bean (Gymnocladus dioica). J. Amer. Chem. Soc. 77, 429 (1955).

74. LAWSON, G. J. and M. STACEY: Immunopolysaccharides. I. Preliminary Studies of a Polysaccharide from Azotobacter chroococcum, Containing a Uronic Acid. J. Chem. Soc. (London) 1954, 1925.

75. MACHIDA, S. and M. INANO: Chemical Studies on Polyuronides. VI. On the Mucilage of Nori-utsugi Plant, Hydrangea paniculata, SIEB. Bull. Chem. Soc. Japan 28, 629 (1955).

76. MACHIDA, S. and N. UCHINO: A Summary of Chemical Studies on Polyuronides. I. Bull. Fac. Textile Fibers, Kyoto Univ. Industr. Arts Textile Fibers 1, 116 (1954).

77. MARKOWITZ, H. and M. HEIDELBERGER: Chemical Constitution of the Specific Polysaccharide of Type XVIII Pneumococcus. J. Amer. Chem. Soc. 76, 1317 (1954).

78. MARRACK, J. R.: The Chemistry of Antigens and Antibodies. Special Report Series, No. 194, 2nd ed. London: His Majesty's Stationery Office. 1938.

79. MARRACK, J. and B. R. CARPENTER: The Cross Reactions of Vegetable Gums with Type II Antipneumococcal Serum. Brit. J. exp. Pathol. 19, 53 (1938).

80. MATHUR, G. P. and S. MUKHERJEE: Investigations on the Structure of Ketha (Feronia elephantum: Family Rutaceae) Gum. J. Sci. Ind. Res. (India) 11 B, 544 (1952).

81. McCARTY, M.: Variation in the Group-specific Carbohydrate of Group A Streptococci. II. Studies on the Chemical Basis for Serological Specificity of the Carbohydrates. J. exp. Medicine 104, 629 (1956).

82. McCARTY, M. and R. C. LANCEFIELD: Variation in the Group-specific Carbohydrate of Group A Streptococci. I. Immunochemical Studies on the Carbohydrate of Variant Strains. J. exp. Medicine 102, 11 (1955).

83. MESTER, L. and G. IVÁNOVICS: The Structure of the Immunospecific Polysaccharide of Bacillus anthracis. Chem. and Ind. 1957, 493.

84. MISAKI, A., H. ISHIKAWA and S. TERAMOTO: Studies on the Bacterial Polysaccharide, Rhamnogalactan, Produced by B. polymyxa var. XIII. Preparation of Partially Depolymerized Materials and Determination of the Molecular Weights. J. Fermentn. Technol. (Japan) 36, 320 (1958).

85. MISAKI, A. and S. TERAMOTO: Studies on the Bacterial Polysaccharide, Rhamnogalactan, Produced by B. polymyxa var. XI. Methyl Sugar Components of the Methylated Rhamnogalactan. J. Fermentn. Technol. (Japan) 36, 266 (1958).

86. MONTGOMERY, R., F. SMITH and H. C. SRIVASTAVA: Structure of Corn Hull Hemicellulose. I. Partial Hydrolysis and Identification of 2-O-(α-D-Glucopyranosyluronic Acid)-D-xylopyranose. — II. Identification of the α- and β-Forms of Methyl 2-O-[Methyl-(2,3,4-tri-O-acetyl-α-D-glucopyranosyl)-

uronate]-3,4-di-O-acetyl-D-xylopyranoside. J. Amer. Chem. Soc. **78**, 2837, 6169 (1956). — Montgomery, R. and F. Smith: III. Identification of the Methylated Aldobiouronic Acid Obtained from Methyl Corn Hull Hemicellulose. J. Amer. Chem. Soc. **79**, 695 (1957). — Montgomery, R., F. Smith and H. C. Srivastava: IV. Partial Hydrolysis and Identification of 3-O-α-D-Xylopyranosyl-L-arabinose and 4-O-β-D-Galactopyranosyl-β-D-xylose. J. Amer. Chem. Soc. **79**, 698 (1957).

87. O'Donnell, J. J. and E. Percival: Structural Investigations on the Water-soluble Polysaccharides from the Green Seaweed *Acrosiphonia centralis (Spongomorpha arcta)*. J. Chem. Soc. (London) **1959**, 2168.

88. Oshibuchi, T. and H. Kusunose: Studies on the Mucilage of the Root of "Tororo-aoi" *(Abelmoschus manihot)*. IV. Formation of Rhamnodigalacturonic Acid Salt by Hydrolysis. J. Agric. Chem. Soc. Japan **31**, 481 (1957).

89. Perlzweig, W. A. and C. S. Keefer: The Immunizing Antigen of the Pneumococcus. III. The Purification of the Water-soluble Antigen. J. exp. Medicine **42**, 747 (1925), and earlier papers.

89a. Plackett, P. and S. H. Buttery: A Galactan from *Mycoplasma mycoides*. Nature (London) **182**, 1236 (1958).

90. Pon, G. et A. M. Staub: Étude chimique des polyosides somatiques des Salmonelles. II. Analyse comparée des polyosides extraits de *S. para* B, *S. typhi murium* et *S. typhi*. Bull. soc. chim. biol. (Paris) **37**, 1283 (1955).

91. Preece, I. A. and K. G. Mackenzie: Non-starchy Polysaccharides of Cereal Grains. II. Distribution of Water-soluble Gum-like Materials in Cereals. J. Inst. Brewing **58**, 457 (1952).

92. Rafique, C. M. and F. Smith: The Constitution of Guar Gum. J. Amer. Chem. Soc. **72**, 4634 (1950).

93. Rankin, J. C. and A. Jeanes: Evaluation of the Periodate Oxidation Method for the Structural Analysis of Dextrans. J. Amer. Chem. Soc. **76**, 4435 (1954).

94. Rebers, P. A., S. A. Barker, M. Heidelberger, Z. Dische and E. E. Evans: Precipitation of the Specific Polysaccharide of *Cryptococcus neoformans* A by Types II and XIV Antipneumococcal Sera. J. Amer. Chem. Soc. **80**, 1135 (1958).

95. Rebers, P. A. and M. Heidelberger: The Specific Polysaccharide of Type VI Pneumococcus. I. Preparation, Properties, and Reactions. J. Amer. Chem. Soc. **81**, 2415 (1959).

96. Rebers, P. A. and F. Smith: The Constitution of Iles Mannan. J. Amer. Chem. Soc. **76**, 6097 (1954).

97. Roboz, E. and A. J. Haagen-Smit: A Mucilage from *Aloe vera*. J. Amer. Chem. Soc. **70**, 3248 (1948).

98. Schiemann, O. und W. Casper: Sind die spezifisch präcipitablen Substanzen der drei Pneumokokkentypen Haptene? Z. Hyg. u. Infektionskrankh. **108**, 220 (1927), and earlier and later papers.

98a. Schlüchterer, E. and M. Stacey: The Capsular Polysaccharide of *Rhizobium radicicolum*. J. Chem. Soc. (London) **1945**, 776.

99. Schmidt, W. C.: Group A Streptococcus Polysaccharide: Studies on its Preparation, Chemical Composition and Cellular Localization after Intravenous Injection into Mice. J. exp. Medicine **95**, 105 (1952).

100. Smith, E. E. B., G. T. Mills and E. M. Harper: The Isolation of Uridine Pyrophosphogalacturonic Acid from a Type I Pneumococcus. Biochim. Biophys. Acta **23**, 662 (1957). — A Comparison of the Uridine Pyrophosphoglycosyl Metabolism of Capsulated and Non-capsulated Pneumococci. J. Gen. Microbiol. **16**, 426 (1957).

*101.* SMITH, F.: The Constitution of Carob Gum. J. Amer. Chem. Soc. **70**, 3249 (1948).

*101a.* — The Constitution of Mesquite Gum. III. The Structure of the Monomethyl Glucuronic Acid Component. J. Chem. Soc. (London) **1951**, 2646.

*101b.* SMITH, F. and R. MONTGOMERY: The Chemistry of Plant Gums and Mucilages, and some Related Polysaccharides. New York: Reinhold Publ. Corp. 1959.

*102.* SMITH, F. and H. C. SRIVASTAVA: Acetolysis of the Glucomannan of Iles Mannan. J. Amer. Chem. Soc. **78**, 1404 (1956).

*103.* SMITH, H. and H. T. ZWARTOUW: The Polysaccharide from *Bacillus anthracis* Grown *in vivo*. Biochemic. J. **63**, 447 (1956).

*104.* SPRINGER, G. F., N. ANSELL, W. BRANDES and R. F. NORRIS: Bibliotheca Haematol., Suppl. Acta Haematol. No. 7, 190 (1958).

*105.* STAUB, A. M.: Specificity and Saccharides. In: G. F. SPRINGER, Polysaccharides in Biology. Trans. 5th Conference. New York: Josiah Macy Jr. Foundn. 1960.

*106.* STAUB, A. M. et R. TINELLI: Étude immunochimique sur les Salmonelles. IV. Structure chimique de certains motifs antigéniques présents dans les « antigènes » O 9 et 12 du tableau de White-Kauffmann. Bull. soc. chim. biol. (Paris) **39**, Suppl. I, 65 (1957).

*107.* SUGG, J. Y., E. L. GASPARI, W. L. FLEMING and J. M. NEILL: Studies on Immunological Relationships among the Pneumococci. I. A Virulent Strain of Pneumococcus which is Immunologically Related to but not Identical with Typical Strains of Type III Pneumococci. J. exp. Medicine **47**, 917 (1928).

*108.* TORTO, F. G.: The Gum of *Fagara xanthoxyloides*. Nature (London) **180**, 864 (1957).

*109.* WADMAN, W. H., A. B. ANDERSON and W. Z. HASSID: The Structure of an Arabogalactan from Jeffrey Pine *(Pinus Jeffreyi)*. J. Amer. Chem. Soc. **76**, 4097 (1954).

*110.* WHISTLER, R. L. and J. N. BeMILLER: Hydrolysis Components from Methylated Corn Fiber Gum. J. Amer. Chem. Soc. **78**, 1163 (1956).

*111.* WHISTLER, R. L. and H. E. CONRAD: A Crystalline Galactobiose from Acid Hydrolysis of Okra Mucilage. J. Amer. Chem. Soc. **76**, 1673 (1954). — 2-*O*-(*D*-Galactopyranosyluronic Acid)-*L*-rhamnose from Okra Mucilage. J. Amer. Chem. Soc. **76**, 3544 (1954).

*112.* WHISTLER, R. L. and W. M. CORBETT: Oligosaccharides from Partial Acid Hydrolysis of Corn Fiber Hemicellulose. J. Amer. Chem. Soc. **77**, 6328 (1955).

*113.* WHISTLER, R. L. and G. E. LAUTERBACH: Hydrolysis Products from Methylated Arabinoxyloglycan and Arabinogalacto-mono-*O*-methylglucuronoxyloglycan of Corn Cobs. J. Amer. Chem. Soc. **80**, 1987 (1958). — Isolation of Two Further Polysaccharides from Corn Cobs. Arch. Biochem. Biophys. **77**, 62 (1958).

*114.* WHITE, E. V.: The Constitution of Mesquite Gum. I. The Methanolysis Products of Methylated Mesquite Gum. J. Amer. Chem. Soc. **68**, 272 (1946). — II. Partial Hydrolysis of Mesquite Gum. J. Amer. Chem. Soc. **69**, 622 (1947). — III. Hexamethyl-3-glucuronosido-methylgalactoside Methyl Ester. J. Amer. Chem. Soc. **69**, 2264 (1947). — IV. 4-Methoxy-*D*-glucuronic Acid. J. Amer. Chem. Soc. **70**, 367 (1948).

*115.* WHITE, E. V. and P. S. RAO: Constitution of the Polysaccharide from Tamarind Seed. J. Amer. Chem. Soc. **75**, 2617 (1953).

116. WILKINSON, J. F., W. F. DUDMAN and G. O. ASPINALL: The Extracellular Polysaccharide of *Aerobacter aerogenes* A 3 (S l) (Klebsiella Type 54). Biochemic. J. **59**, 446 (1955).
117. WISE, L. E. and J. W. APPLING: Quantitative Determination of *D*-Galactose by Selective Fermentation with Special Reference to Plant Mucilages. Ind. Eng. Chem., Analyt. Ed. **16**, 28 (1944).
118. ZAMENHOF, S., G. LEIDY, P. L. FITZGERALD, H. E. ALEXANDER and E. CHARGAFF: Polyribophosphate, the Type-specific Substance of *Hemophilus influenzae*, Type b. J. Biol. Chem. **203**, 695 (1953).

*(Received, January 22, 1960.)*

# Namenverzeichnis. Index of Names. Index des Auteurs.

# Sachverzeichnis. Index of Subjects. Index des Matières.